W9-BTL-119

LOGIC AND COMPUTER DESIGN FUNDAMENTALS

M. Morris Mano
California State University, Los Angeles
Charles R. Kime
University of Wisconsin, Madison

Prentice Hall
Upper Saddle River, New Jersey 07458

Library of Congress Cataloging-in-Publication Data

Mano, M. Morris,
Logic and computer design fundamentals / M. Morris Mano and Charles R. Kime.
p. cm.
Includes bibliographical references and index.
ISBN 0-13-182098-2
1. Electronic digital computers--Circuits. 2. Logic circuits. 3. Logic design. I. Kime, Charles R. II. Title.
TK7888.4.M36 1997
621.39'2--dc20 96-26502
CIP

Editor-in-Chief: **Marcia Horton**
Acquisitions editor: **Tom Robbins**
Managing editor: **Bayani Mendoza de Leon**
Production editor: **Irwin Zucker**
Art director: **Amy Rosen**
Assistant art director: **Rod Hernandez**
Creative director: **Paula Maylahn**
Cover design: **Bruce Kenselaar**
Cover illustration: **David Bishop**
Interior design: **Judy Matz-Coniglio**
Manufacturing buyer: **Donna Sullivan**
Editorial assistant: **Nancy Garcia**

Trademark Information

Actel and ACT are registered trademarks of Actel Corporation. Altera and MAX are registered trademarks of Altera, Inc. MAX 7000 is a trademark of Altera Corporation. PAL is a registered trademark of Advanced Micro Devices, Inc. Viewlogic, ViewDraw and ViewSim are registered trademarks of Viewlogic Systems, Inc. ViewTrace is a trademark of Viewlogic Systems, Inc. Xilinx is a registered trademark of Xilinx, Inc. XC4000 is a trademark of Xilinx, Inc.

The author and publisher of this book have used their best efforts in preparing this book. These efforts include the development, research, and testing of the theories and programs to determine their effectiveness. The author and publisher make no warranty of any kind, expressed or implied, with regard to these programs or the documentation contained in this book. The author and publisher shall not be liable in any event for incidental or consequential damages in connection with, or arising out of, the furnishing, performance, or use of these programs.

Printed in the United States of America

10 9 8 7 6 5 4 3 2 1

ISBN 0-13-182098-2

Prentice-Hall International (UK) Limited, London
Prentice-Hall of Australia Pty. Limited, Sydney
Prentice-Hall Canada Inc., Toronto
Prentice-Hall Hispanoamericana, S.A., Mexico
Prentice-Hall of India Private Limited, New Delhi
Prentice-Hall of Japan, Tokyo
Simon & Schuster Asia Pte. Ltd., Singapore
Editora Prentice-Hall do Brasil, Ltda., Rio de Janeiro

CONTENTS

□ CHAPTER **3**

COMBINATIONAL LOGIC DESIGN 99

CHAPTER 4

SEQUENTIAL CIRCUITS 172

□ CHAPTER 8

□ CHAPTER 9

INSTRUCTION SET ARCHITECTURE 426

CHAPTER 10

□ CHAPTER 11

□ CHAPTER 12

PREFACE

This book is based on the book *Computer Engineering: Hardware Design* (1988) by one of the authors (Mano). The general premise of both books is a combined treatment of logic design, digital system design and computer design basics. The advancing scale of integration of digital electronic circuits and accelerated use of logic synthesis tools have raised much of the digital system design process to increasingly higher levels. Register transfer level treatment of datapaths and controls, including the use of a hardware description language, is now central to industrial practice. Moreover, the movement of computer design toward the so-called RISC (Reduced Instruction Set Computer) has penetrated to the very core of the domain of the CISC (Complex Instruction Set Computer). The RISC architecture, with the accompanying pipelined implementation, has not only established itself as the dominating architectural form for processors, but now is embedded as a primary component in CISC designs as well. Additionally, the use of multiple levels of memory hierarchy has become a dominant factor in contemporary computers.

Based on the above trends, we have updated Chapters 7, 8, 10, and 12 of the previously mentioned book. Chapters 7 and 8 deal with datapaths and control, 10 with central processing units (CPUs), and 12 with memory systems. The first three chapters successively refine datapath and control design illustrations in order to limit the number of design details with which the student must deal. To this end, a single datapath is developed in Chapter 7 and modified into a pipelined datapath. The initial datapath from Chapter 7 is used in Chapter 8 to construct a simple computer with a hardwired control and a more complex computer with a microprogrammed control. The pipelined datapath from Chapter 7 serves as the core of a pipelined computer with a hardwired control presented in Chapter 8. Finally, more elaborate and complete designs derived from those in Chapter 8 appear in Chapter 10 for both the microprogrammed and pipelined computers.

At the beginning of each chapter, a drawing of a generic computer of the PC genre serves as a framework for illustrating the role of concepts in each chapter of the book. The drawing is accompanied by an overview of the chapter including a paragraph relating the material in the chapter to its role in the computer. The parts of the computer using some aspect of the chapter material are shaded in blue. Additionally, at the end of each chapter, a summary of the important items from the chapter appears.

Rather than using MSI parts, we use their generic counterparts which we refer to as *functional blocks*. MSI parts are becoming less important due to the increase in the scale of integration, the growth in the use of synthesis tools and more frequent use of programmable logic devices in course laboratories. Also, obsolete technology-based assumptions that shaped the 7400 series MSI parts are avoided by using these generic functional blocks. We have chosen IEEE standard symbols for gates and storage elements and more commonly used rectangular symbols for functional blocks.

Chapters 1 through 5 of the book treats logic design, and chapters 6 through 8 deal with digital system design using computer subsystem examples. Chapters 9 through 12 focus directly on computer design. This arrangement provides solid digital system design fundamentals while accomplishing a gradual, bottom-up development of fundamentals for use in top-down computer design in later chapters. Summaries of the topics covered in each chapter follow.

Chapter 1 introduces digital and computer systems and information representation. Binary number and character representations and basic binary arithmetic are the focus of most of the chapter. Included in the discussion of character representations is Unicode, a 16-bit international character code standard.

Chapter 2 introduces logic gates and deals with basic concepts and techniques for designing gate circuits. The chapter presents Boolean algebra and Karnaugh maps as logic simplification tools and covers design using NAND, NOR and XOR gates. A section on switch level design of CMOS circuits may be covered at the instructor's option, depending on the audience.

Chapter 3 covers combinational circuit analysis and design. It introduces design hierarchy, computer-aided design and top-down design and presents computer-based digital simulation as an analysis tool. Several types of functional blocks which correspond to medium scale integrated (MSI) circuits are introduced or designed. Simplification of arithmetic hardware serves as motivation for the presentation of complement arithmetic. Finally, standard IEEE symbols for combinational functional blocks are described.

Chapter 4 presents sequential circuit concepts and design. This chapter introduces latches and flip-flops as storage elements and demonstrates sequential circuit analysis procedures. Design procedures include a brief introduction to developing state diagrams and state tables from specifications. The design procedures focus on design using D flip-flops or J-K flip-flops.

Chapter 5 deals with structured sequential circuits, notably registers and counters. Shift registers are introduced and applied to serial operations. The discussion of counters focuses on the synchronous binary type. Standard IEEE symbols for sequential circuits are introduced.

Chapter 6 presents random access memory (RAM) and various forms of programmable logic. The RAM section deals with the structure of RAM integrated circuits and the interconnection of such circuits to form a memory. Hamming codes provide a focus for discussing error detection and correction. In addition to treating three basic forms of programmable logic, this chapter covers large scale programmable logic including field programmable gate arrays.

Chapter 7 introduces a simple hardware description language and deals in detail with the design of a basic computer datapath including a pipelined version. It treats register transfer operations and introduces methods for implementing transfers with and without buses. The datapath design done here serves as the foundation for all datapaths treated in the remainder of the book.

Chapter 8 deals with the sequencing of register transfer operations. It introduces the algorithmic state machine (ASM) chart as a representation for sequencing and controlling operations. Hardwired and microprogrammed versions of a binary multiplier illustrate ASM use and sequential control design. A simple computer is built upon the basic datapath from Chapter 7 by adding hardwired control. A more complex microprogrammed control is described and the chapter finishes with a simple pipelined computer.

Chapter 9 introduces many facets of instruction set architecture. It deals with address count, addressing modes, and the various types of instructions including data transfer, data manipulation, floating point, program control and program interrupt. Addressing modes and other aspects of instructions are illustrated with brief segments of instruction code.

Chapter 10 illustrates and compares two different CPU designs, one CISC and one RISC. Except for a bit of comparison, the designs are independent, so either one or both may be covered based on the audience and on the time available. The CISC design uses a conventional datapath and microprogrammed control. The RISC design uses a pipelined datapath and hardwired control. The chapter concludes with a brief overview of more advanced concepts in CPU design.

Chapter 11 deals with data transfer between the CPU, input-output interfaces and peripheral devices. Discussion of a keyboard, a CRT display and a hard disk as peripherals is included and a keyboard interface is illustrated. Other topics covered range from serial communication to I/O processors.

Chapter 12 covers memory systems with a particular focus on memory hierarchies. The concept of locality of reference is introduced and illustrated by consideration of the cache/main memory and main memory/hard disk relationships. An overview of cache design parameters is provided. The treatment of memory management focuses on paging and a translation lookaside buffer structure.

The book incorporates many homework problems, including challenging, more open-ended problems marked with an asterisk. An instructor's manual which includes suggestions for use of the book, information for obtaining CAD tools, and problem solutions is available to course instructors. A World Wide Web site for the book, accessible at `http://www.prenhall.com`, includes down-loadable overhead transparency originals for all complex figures and tables from the book. The use of computer-based logic simulation in solving a portion of the homework problems is recommended. The logic simulation examples in the text were produced using Prentice Hall's *Workview Office Student Edition – Schematic Entry and Digital Analysis* by R. James Duckworth. This CD-ROM-based logic simulation package for Windows incorporates commercial CAD products and provides an introductory tutorial and exercises in a convenient logic simulation package for students.

Because of its broad coverage of both logic and computer design, with the proper selection of material, this book can serve several different objectives in sophomore through junior level courses. Chapters 1 through 9 plus a part of Chapter 10 provide an overview of logic and computer hardware for computer science, computer engineering, electrical engineering or engineering students in general in a single semester course. Chapters 1 through 6 give a basic introduction to logic design which can be completed in a single quarter for electrical and computer engineering students. Coverage of Chapters 1 through 8 in a semester, with perhaps some supplementary material, provides a stronger more contemporary logic design treatment. The entire book, covered in two quarters, provides the basics of logic and computer design for computer engineering and science students. Coverage of the entire book with appropriate supplementary material including a laboratory can fill a two-semester sequence in logic design and computer architecture. Finally, due to its moderately paced treatment of a wide range of topics, the book is ideal for self-study by engineers and computer scientists.

Among the contributions of many people to this book, we particularly grateful for the reviews of initial drafts of the book chapters by UW colleague Don Dietmeyer and UW graduate student Mark Aurora. We wish to thank colleagues at Wisconsin including Yu Hen Hu, Jennifer Hou, Rajiv Jain, Suresh Chalasani, and Danial Neebel who used various drafts of selected chapters in the Digital Systems Fundamentals course. Feedback provided by reviewers R. Danadpani, University of Colorado–Colorado Springs; Niraj K. Jha, Princeton University; and Jim Smith, University of Wisconsin–Madison, was incredibly valuable in improving the final version of the book. Comments by Janak Patel, University of Illinois–Urbana-Champaign, also strongly influenced the development of a portion of the chapters. The understanding and support of UW ECE department chairs, Bahaa Saleh and Willis Tompkins, during this project is very much appreciated.

We thank those at Prentice Hall who have had a role in bringing about this book, notably, Don Fowley for his efforts in initiating the project, Tom Robbins for his sound guidance through the bulk of it, and Irwin Zucker for effectively dealing with the many details of the book production.

M. Morris Mano

Charles R. Kime

A very special thanks goes to my wife, Val, for understanding my dedication to this project, enduring being ignored, and listening patiently, even when I was in my grouchiest of moods. Her tolerance of the computer in the dining L, the overflowing bookcases flanking the fireplace, and the seemingly perpetual piles of books and papers on the floor in between has been truly amazing.

Charles R. Kime

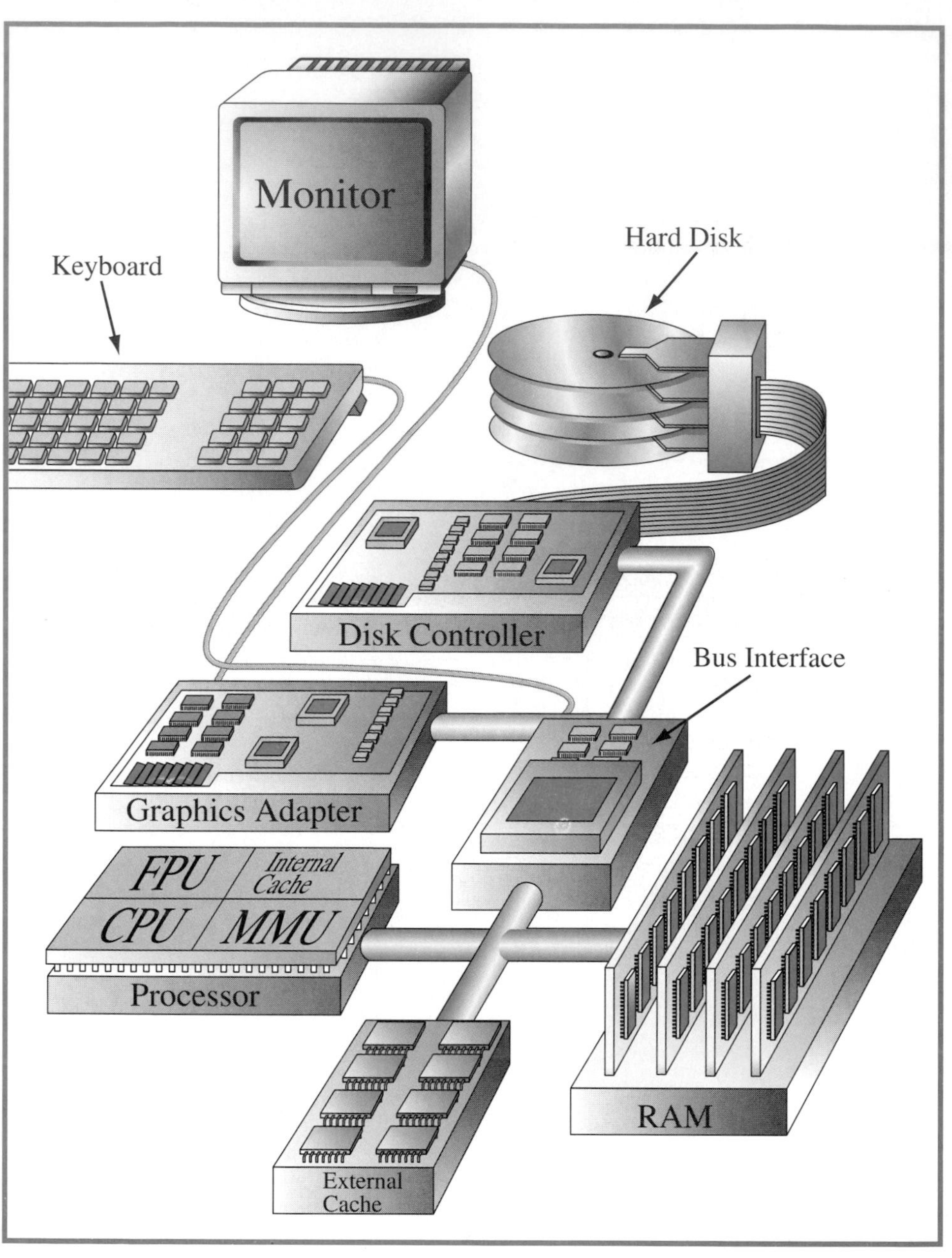

Monitor
Hard Disk
Keyboard
Disk Controller
Bus Interface
Graphics Adapter
FPU
Internal Cache
CPU
MMU
Processor
RAM
External Cache

CHAPTER 1

DIGITAL COMPUTERS AND INFORMATION

Logic design fundamentals and computer design fundamentals are the topics of this book. Logic design deals with the basic concepts and tools used to design digital hardware consisting of logic circuits. Computer design deals with the additional concepts and tools used to design computers and other complex digital hardware. Computers and digital hardware in general are referred to as digital systems. Thus, this book is about understanding and designing digital systems. Due to its generality and complexity, the computer provides an ideal vehicle for learning the concepts of and tools for digital system design. In addition, due to its widespread use, the computer itself is deserving of study. Hence, the focus in this book is on computers and their design.

The computer will be not only a vehicle, but also a motivator for study. To this end, we use the exploded pictorial diagram of a computer of the class commonly referred to as a PC (personal computer) given on the opposite page. We use this generic computer to highlight the significance of the material covered and its relationship to the overall system. A bit later in the chapter, we will discuss the various major components of the generic computer and see how they relate to a block diagram often used to describe a computer.

1-1 DIGITAL COMPUTERS

Today, digital computers have such a prominent and growing role in modern society, that we often say we are in the "information age." Computers are involved in our business transactions, communications, transportation, medical treatment, and entertainment. They monitor our weather and environment. In the industrial world, they are heavily employed in design, manufacturing, distribution, and sales. They have contributed to many scientific discoveries and engineering developments that would have been unattainable otherwise. Notably, the design of a new processor for a modern computer could not be done without the use of many computers!

The most striking property of the digital computer is its generality. It can follow a sequence of instructions, called a program, that operates on given data. The user can specify and change the program or the data according to specific needs. As a result of this flexibility, general-purpose digital computers can perform a variety of information-processing tasks that range over a very wide spectrum of applications. The general-purpose digital computer is the best known example of a *digital system*. Characteristic of a digital system is its manipulation of discrete elements of information. Any set that is restricted to a finite number of elements contains discrete information. Examples of discrete sets are the 10 decimal digits, the 26 letters of the alphabet, the 52 playing cards, and the 64 squares of a chessboard. Early digital computers were used mostly for numeric computations. In this case, the discrete elements used were the digits. From such an application, the term *digital computer* emerged.

Discrete elements of information are represented in a digital system by physical quantities called *signals*. Electrical signals such as voltages and currents are most common. Electronic devices called transistors predominate in the circuitry that implements these signals. The signals in all present-day electronic digital systems use just two discrete values and are therefore said to be *binary*.

We typically represent the two discrete values by ranges of voltage values called HIGH and LOW. Output voltage ranges and input voltage ranges are illustrated in Figure 1-1. The HIGH output voltage value ranges between 4.0 and 5.5 volts, and the LOW output voltage value ranges between −0.5 and 1.0 volt. The high input range allows 3.0 to 5.5 volts to be recognized as a HIGH, and the low input range allows −0.5 to 2.0 volts to be recognized as a LOW. The fact that the input ranges are longer than the output ranges allows the circuits to function correctly in spite of variations in their behavior and undesirable "noise" voltages that may be added to or subtracted from the outputs.

We give the output and input voltage ranges a number of different names. Among these are HIGH and LOW, TRUE and FALSE, and 1 and 0. It is clear that

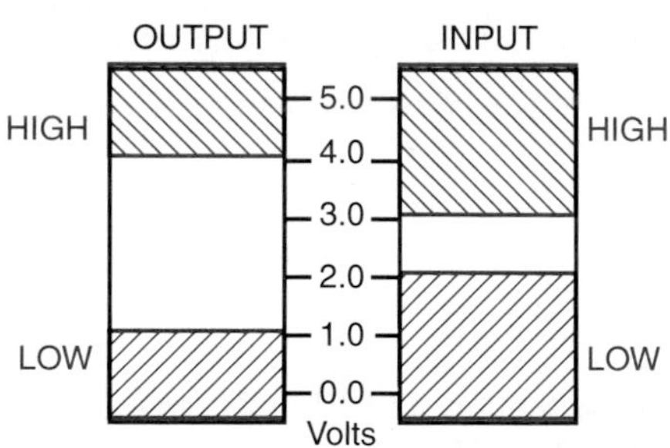

□ **FIGURE 1-1**
An Example of Voltage Ranges for Binary Signals

the higher voltage ranges are associated with HIGH, abbreviated H, and the lower voltage ranges with LOW, abbreviated L. We find, however, that for TRUE and 1 and FALSE and 0, there is a choice. TRUE and 1 can be associated with either the higher or lower voltage range and FALSE and 0 with the other range. Unless otherwise indicated, we assume that TRUE and 1 are associated with the higher of the voltage ranges, H, and that FALSE and 0 are associated with the lower of the voltage ranges, L.

Why is binary used? In contrast to the situation in Figure 1-1, consider a system with 10 values representing the decimal digits. In such a system, the voltages available—say, 0 to 5.0 volts—could be divided into 10 ranges, each of length 0.5 volt. A circuit would provide an output voltage within each of these 10 ranges. An input of a circuit would need to determine in which of the 10 ranges an applied voltage lies. If we wish to allow for noise on the voltages, then output voltage might be permitted to range over less than 0.25 volt, and boundaries between inputs could only vary by less than 0.25 volt. This would require complex and costly electronic circuits and still could be disturbed by small "noise" voltages or small variations in the circuits occurring during their manufacture or use. As a consequence, the use of such multivalued circuits is very limited. Instead, binary circuits are used in which correct circuit operation can be achieved with significant variations in both the two output voltages and the two input ranges. The resulting transistor circuit with an output that is either HIGH or LOW is simple, easy to design, and extremely reliable.

Information Representation

Since 0 and 1 are associated with the binary number system, they are the preferred names for the signal ranges. A binary digit is called a *bit*. Information is represented in digital computers by groups of bits. By using various coding techniques, groups of bits can be made to represent not only binary numbers, but also other groups of discrete symbols. Groups of bits, properly arranged, can even specify to the computer the instructions to be executed and the data to be processed.

Discrete quantities of information either emerge from the nature of the data being processed or may be purposely quantized from continuous values. For example, a payroll schedule is inherently discrete data containing employee names, social security numbers, weekly salaries, income taxes, and so on. An employee's paycheck is processed using discrete data values such as letters of the alphabet (for the employee's name), digits (for the salary), and special symbols like $. On the other hand, an engineer may measure the speed of rotation of an automobile wheel, which varies continuously with time, but may record only specific values at specific times in tabular form. The engineer is thus quantizing the continuous data, making each number in the table a discrete quantity of information. In a case such as this, if the measurement can be converted to an electronic signal, the quantization of the signal in both value and time can be performed automatically by an analog-to-digital conversion device.

Computer Structure

A block diagram of a digital computer is shown in Figure 1-2. The memory stores programs as well as input, output, and intermediate data. The datapath performs arithmetic and other data-processing operations as specified by the program. The control unit supervises the flow of information between the various units. A datapath, when combined with the control unit, forms a component referred to as a *central processing unit*, or CPU.

The program and data prepared by the user are transferred into memory by means of an input device such as a keyboard. An output device, such as a CRT (cathode-ray tube) monitor, displays the results of the computations and presents them to the user. A digital computer can accommodate many different input and output devices, such as hard disks, floppy disk drives, CD-ROM drives, and scanners. These devices use some digital logic, but often include analog electronic circuits, optical sensors, CRTs or LCDs (liquid crystal displays), and electromechanical components.

The control unit in the CPU retrieves the instructions, one by one, from the program stored in the memory. For each instruction, the control unit manipulates the datapath to execute the operation specified by the instruction. Both program and data are stored in memory. A digital computer is a powerful system. It can perform arithmetic computations, manipulate strings of alphabetic characters, and be programmed to make decisions based on internal and external conditions.

More on the Generic Computer

At this point, we will briefly discuss the generic computer and relate its various parts to the block diagram in Figure 1-2. At the lower left of the picture on page 2 is the heart of the computer, an integrated circuit called the *processor*. Modern processors such as this one are quite complex and consist of millions of transistors. The processor contains four functional modules: the CPU, the FPU, the MMU, and the internal cache.

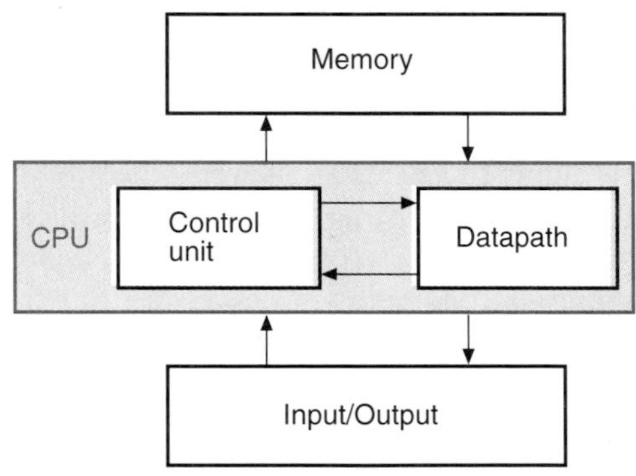

□ **FIGURE 1-2**
Block Diagram of a Digital Computer

We have already discussed the CPU. The FPU (floating-point unit) is somewhat like the CPU, except that its datapath and control unit are specifically designed to perform floating-point operations. In essence, these operations process information represented in the form of scientific notation, e.g., 1.234×10^7, permitting the generic computer to handle very large and very small numbers. The CPU and the FPU, in relation to Figure 1-2, each contain a datapath and a control unit.

The MMU is the memory management unit. The MMU plus the internal cache and the separate blocks near the bottom of the computer labeled "External Cache" and "RAM" (random access memory) are all part of the memory in Figure 1-2. The two caches are special kinds of memory that allow the CPU and FPU to get at the data to be processed much faster than with RAM alone. RAM is what is most commonly referred to as memory. As its main function, the MMU causes the memory that appears to be available to the CPU to be much, much larger than the actual size of the RAM. This is accomplished by data transfers between the RAM and the hard disk shown at the top of the picture of the generic computer. So the hard disk, which we discuss later as an input/output device, appears conceptually as a part of the memory and input/output.

The connection paths shown between the processor, memory, and external cache are the pathways between integrated circuits. These are typically implemented as fine copper conductors on a printed circuit board. The connection paths below the bus interface are referred to as the processor bus. The connections above the bus interface are referred to as the input/output (I/O) bus. The processor bus and the I/O bus attached to the bus interface carry different numbers of bits of data and have different ways of controlling the movement of data. They may also operate at different speeds. The bus interface hardware handles these differences so that data can be communicated between the two buses.

All of the remaining structures in the generic computer are considered part of I/O in Figure 1-2. In terms of sheer physical volume, these structures dominate. In order to enter information into the computer, a keyboard is provided. In order to view output in the form of text or graphics, a graphics adapter card and CRT monitor are provided. The hard disk discussed previously is an electromechanical magnetic storage device. It stores large quantities of information in the form of magnetic flux on spinning disks coated with magnetic materials. In order to control the hard disk and transfer information to and from it, a disk controller is used. The keyboard, graphics adapter card, and disk controller card are all attached to the I/O bus. This allows these devices to communicate through the bus interface with the CPU and other circuitry connected to the processor buses.

The generic computer consists mainly of an interconnection of digital modules. To understand the operation of each module, it is necessary to have a basic knowledge of digital systems and their general behavior. Chapters 1 through 5 of this book deal with logic design of digital circuits in general. Chapters 6 through 8 discuss the primary components of a digital system, their operation, and their design. The operational characteristics of RAM are explained in Chapter 6. The organization and design of datapaths are discussed in Chapter 7. Methods for designing control units are introduced in Chapter 8. Chapters 9 through 11 present the basics of computer design. Some typical instructions employed in digital computer CPUs are presented in Chapter 9. The organization and design of typical CPUs are examined in Chapter 10. Input and output devices and the various ways that a CPU can communicate with them are discussed in Chapter 11. Finally, memory hierarchy concepts related to the caches and MMU are introduced in Chapter 12.

To guide the reader through this material and to keep in mind the "forest" as we carefully examine many of the "trees," at the beginning of each chapter we provide a copy of the generic computer. Appropriate blue shading represents the coverage of the material in the chapter. In addition, accompanying discussion appears in a blue box to tie the topics in the chapter to the associated components of the computer. At the completion of our journey, we will have covered most of the various modules of the computer and will have an understanding of the fundamentals that underlie both its function and design.

Earlier we mentioned that a digital computer manipulates discrete elements of information and that all information in the computer is represented in binary form. Operands used for calculations may be expressed in the binary number system or in the decimal system by means of a binary code. The letters of the alphabet are also converted into a binary code. The purpose of the remainder of this chapter is to introduce the binary number system, binary arithmetic, and selected binary codes as a basis for further study in the succeeding chapters. In relation to the generic computer, this material is very important and spans all of the components except some in I/O that involve mechanical operations and analog (as contrasted with digital) electronics. Thus, almost all of the generic computer is shaded in blue.

1-2 NUMBER SYSTEMS

The decimal number system is employed in everyday arithmetic to represent numbers by strings of digits. Depending on its position in the string, each digit has an associated value of an integer raised to the power of 10. For example, the decimal number 724.5 is interpreted to represent 7 hundreds plus 2 tens plus 4 units plus 5 tenths. The hundreds, tens, units, and tenths are powers of 10 implied by the position of the digits. The value of the number is computed as follows:

$$724.5 = 7 \times 10^2 + 2 \times 10^1 + 4 \times 10^0 + 5 \times 10^{-1}$$

The convention is to write only the digits and infer the corresponding powers of 10 from their positions. In general, a decimal number with a decimal point is represented by a string of coefficients:

$$A_n A_{n-1} \ldots A_1 A_0 . A_{-1} A_{-2} \ldots A_{-m+1} A_{-m}$$

Each coefficient A_i is one of 10 digits (0, 1, 2, 3, 4, 5, 6, 7, 8, 9). The subscript value i gives the position of the coefficient and, hence, the weight 10^i by which the coefficient must be multiplied.

The decimal number system is said to be of *base* or *radix* 10, because the coefficients are multiplied by powers of 10 and the system uses 10 distinct digits. In general, a number in base r contains r digits, 0, 1, 2, ..., $r - 1$, and is expressed as a power series in r with the general form

$$A_n r^n + A_{n-1} r^{n-1} + \ldots + A_1 r^1 + A_0 r^0 + A_{-1} r^{-1} + A_{-2} r^{-2} + \ldots + A_{-m+1} r^{-m+1} + A_{-m} r^{-m}$$

When the number is expressed in positional notation, only the coefficients and the radix point are written down:

$$A_n A_{n-1} \ldots A_1 A_0 . A_{-1} A_{-2} \ldots A_{-m+1} A_{-m}$$

In general, the . is called the *radix point*. A_n is referred to as the *most significant digit* (msd), and A_{-m} is referred to as the *least significant digit* (lsd), of the number. Note that if $m = 0$, the lsd is $A_{-0} = A_0$. To distinguish between numbers of different bases, it is customary to enclose the coefficients in parentheses and place a subscript after the right parenthesis to indicate the base of the number. However, when the context makes the base obvious, it is not necessary to use parentheses. The following illustrates a base-5 number with $n = 3$ and $m = 1$ and its conversion to decimal:

$$\begin{aligned}(312.4)_5 &= 3 \times 5^2 + 1 \times 5^1 + 2 \times 5^0 + 4 \times 5^{-1} \\ &= 75 + 5 + 2 + 0.8 = (82.8)_{10}\end{aligned}$$

Note that for all the numbers not enclosed in parentheses, the arithmetic is performed with decimal numbers. Note also that the base-5 system uses only five digits, and, therefore, the values of the coefficients in a number can be only 0, 1, 2, 3, and 4 when expressed in that system.

In addition to decimal, three number systems are used in computer work: binary, octal, and hexadecimal. These are base-2, base-8, and base-16 number systems, respectively.

Binary Numbers

The binary number system is a base 2 system with two digits: 0 and 1. A binary number such as 11010.11 is expressed with a string of l's and 0's and, possibly, a binary point. The decimal equivalent of a binary number can be found by expanding the numbers as a power series with a base of 2. For example,

$$(11010)_2 = 1 \times 2^4 + 1 \times 2^3 + 0 \times 2^2 + 1 \times 2^1 + 0 \times 2^0 = (26)_{10}$$

As noted earlier, the digits in a binary number are called bits. When a bit is equal to 0, it does not contribute to the sum during the conversion. Therefore, the conversion to decimal can be obtained by adding the numbers with powers of two corresponding to the bits that are equal to 1. For example,

□ **TABLE 1-1**
Powers of Two

n	2^n	n	2^n	n	2^n
0	1	8	256	16	65,536
1	2	9	512	17	131,072
2	4	10	1,024	18	262,144
3	8	11	2,048	19	524,288
4	16	12	4,096	20	1,048,576
5	32	13	8,192	21	2,097,152
6	64	14	16,384	22	4,194,304
7	128	15	32,768	23	8,388,608

$$(110101.11)_2 = 32 + 16 + 4 + 1 + 0.5 + 0.25 = (53.75)_{10}$$

The first 24 numbers obtained from 2 to the power of n are listed in Table 1-1. In computer work, 2^{10} is referred to as K (kilo), 2^{20} as M (mega), and 2^{30} as G (giga). Thus,

$$4\text{K} = 2^2 \times 2^{10} = 2^{12} = 4{,}096 \text{ and } 16\text{M} = 2^4 \times 2^{20} = 2^{24} = 16{,}777{,}216$$

The conversion of a decimal number to binary can be easily achieved by a method that successively subtracts powers of two from the decimal number. To convert the decimal number N to binary, first find the greatest number that is a power of two (see Table 1-1) and that, subtracted from N, produces a positive difference. Let the difference be designated N_1. Now find the greatest number that is a power of two and that, subtracted from N_1, produces a positive difference N_2. Continue this procedure until the difference is zero. In this way, the decimal number is converted to its powers-of-two components. The equivalent binary number is obtained from the coefficients of a power series that forms the sum of the components. 1's appear in the binary number in the positions for which terms appear in the power series, and 0's appear in all other positions. This method is demonstrated by the conversion of decimal 625 to binary as follows:

$$625 - 512 = 113 = N_1 \qquad 512 = 2^9$$

$$113 - 64 = 49 = N_2 \qquad 64 = 2^6$$

$$49 - 32 = 17 = N_3 \qquad 32 = 2^5$$

$$17 - 16 = 1 = N_4 \qquad 16 = 2^4$$

$$1 - 1 = 0 = N_5 \qquad 1 = 2^0$$

$$(625)_{10} = 2^9 + 2^6 + 2^5 + 2^4 + 2^0 = (1001110001)_2$$

Octal and Hexadecimal Numbers

As previously mentioned, all computers and digital systems use the binary representation. The octal (base-8) and hexadecimal (base-16) systems are useful for representing binary quantities indirectly because they possess the property that their bases are powers of two. Since $2^3 = 8$ and $2^4 = 16$, each octal digit corresponds to three binary digits and each hexadecimal digit corresponds to four binary digits.

The more compact representation of binary numbers in either octal or hexadecimal is much more convenient for people than using bit strings in binary that are three to four times as long. Thus, most computer manuals use either octal or hexadecimal numbers to specify binary quantities. A group of 15 bits, for example, can be represented in the octal system with only five digits. A group of 16 bits can be represented in hexadecimal with four digits. The choice between an octal and a hexadecimal representation of binary numbers is arbitrary, although hexadecimal tends to win out, since bits often appear in a group of size divisible by four.

The octal number system is the base-8 system with digits 0, 1, 2, 3, 4, 5, 6, 7. An example of an octal number is 127.4. To determine its equivalent decimal value, we expand the number in a power series with a base of 8:

$$(127.4)_8 = 1 \times 8^2 + 2 \times 8^1 + 7 \times 8^0 + 4 \times 8^{-1} = (87.5)_{10}$$

Note that the digits 8 and 9 cannot appear in an octal number.

It is customary to use the first r digits from the decimal system, starting with 0, to represent the coefficients in a base-r system when r is less than 10. The letters of the alphabet are used to supplement the digits when r is greater than 10. The hexadecimal number system is a base-16 system with the first 10 digits borrowed from the decimal system and the letters A, B, C, D, E, and F used for the values 10, 11, 12, 13, 14, and 15, respectively. An example of a hexadecimal number is

$$(\text{B65F})_{16} = 11 \times 16^3 + 6 \times 16^2 + 5 \times 16^1 + 15 \times 16^0 = (46687)_{10}$$

The first 16 numbers in the decimal, binary, octal, and hexadecimal number systems are listed in Table 1-2. Note that the sequence of binary numbers follows a prescribed pattern. The least significant bit alternates between 0 and 1, the second significant bit alternates between two 0's and two 1's, the third significant bit alternates between four 0's and four l's, and the most significant bit alternates between eight 0's and eight 1's.

The conversion from binary to octal is easily accomplished by partitioning the binary number into groups of three bits each, starting from the binary point and proceeding to the left and to the right. The corresponding octal digit is then assigned to each group. The following example illustrates the procedure:

$$(010\ 110\ 001\ 101\ 011.\ 111\ 100\ 000\ 110)_2 = (26153.7406)_8$$

The corresponding octal digit for each group of three bits is obtained from the first eight entries in Table 1-2. Note that 0's can be freely added to the left and right ends of the string of bits to make the total count of bits into multiples of three, both to the left and to the right of the binary point.

□ **TABLE 1-2**
Numbers with Different Bases

Decimal (base 10)	Binary (base 2)	Octal (base 8)	Hexadecimal (base 16)
00	0000	00	0
01	0001	01	1
02	0010	02	2
03	0011	03	3
04	0100	04	4
05	0101	05	5
06	0110	06	6
07	0111	07	7
08	1000	10	8
09	1001	11	9
10	1010	12	A
11	1011	13	B
12	1100	14	C
13	1101	15	D
14	1110	16	E
15	1111	17	F

Conversion from binary to hexadecimal is similar, except that the binary number is divided into groups of four digits. The previous binary number is converted to hexadecimal as follows:

$$(0010\ 1100\ 0110\ 1011.\ 1111\ 0000\ 0110)_2 = (2C6B.F06)_{16}$$

The corresponding hexadecimal digit for each group of four bits is obtained by reference to Table 1-2.

Conversion from octal or hexadecimal to binary is done by a procedure which is the reverse of that just performed. Each octal digit is converted to a three-digit binary equivalent. Similarly, each hexadecimal digit is converted to its four-digit binary equivalent. This is illustrated in the following examples:

$$(673.12)_8 = 110\ 111\ 011.\ 001\ 010 = (110111011.00101)_2$$

$$(3A6.C)_{16} = 0011\ 1010\ 0110.\ 1100 = (1110100110.11)_2$$

Number Ranges

In digital computers, the range of numbers that can be represented is based on the number of bits available in the hardware structures that store and process information. The number of bits in these structures is most frequently a power of two, such as 8, 16, 32, and 64. Since the numbers of bits is fixed by the structures, the addition of leading or trailing zeros to represent numbers is necessary, and the range of numbers that can be represented is also fixed.

For example, for a computer processing 16-bit unsigned integers, the number 537 is represented as 0000001000011001. The range of integers that can be handled by this representation is from 0 to $2^{16} - 1$, that is, from 0 to 65,535. If the same computer is processing 16-bit unsigned fractions with the binary point to the left of the most significant digit, then the number 0.375 is represented by .0110000000000000. The range of fractions that can be represented is from 0 to $(2^{16} - 1)/2^{16}$, or from 0.0 to 0.9999847412.

In later chapters, we will deal with fixed-bit representations and ranges for binary signed numbers and floating-point numbers. In both of these cases, some of the bits are used to represent information other than simple integer or fraction values.

1-3 Arithmetic Operations

Arithmetic operations with numbers in base r follow the same rules as for decimal numbers. However, when a base other than the familiar base 10 is used, one must be careful to use only r allowable digits and perform all computations with base-r digits. Examples of the addition of two binary numbers are as follows (note the names of the operands for addition):

Carries:	00000	101100
Augend:	01100	10110
Addend:	+10001	+10111
Sum:	11101	101101

The sum of two binary numbers is calculated following the same rules as for decimal numbers, except that the sum digit in any position can be only 1 or 0. Also, a carry in binary occurs if the sum in any bit position is greater than 1. (A carry in decimal occurs if the sum in any digit position is greater than 9.) Any carry obtained in a given position is added to the bits in the column one significant position higher. In the first example, since all of the carries are 0, the sum bits are simply the sum of the augend and addend bits. In the second example, the sum of the bits in the second column from the right is 2, giving a sum bit of 0 and a carry bit of 1 ($2 = 2 + 0$). The carry bit is added with the 1's in the third position, giving a sum of 3, which produces a sum bit of 1 and a carry of 1 ($3 = 2 + 1$).

The following are examples of the subtraction of two binary numbers; as with addition, note the names of the operands:

Borrows:	00000	00110		00110
Minuend:	10110	10110	10011	11110
Subtrahend:	−10010	−10011	−11110	−10011
Difference:	00100	00011		−01011

The rules for subtraction are the same as in decimal, except that a borrow into a given column adds 2 to the minuend bit. (A borrow in the decimal system adds 10 to the minuend digit.) In the first example shown, no borrows occur, so the difference bits are simply the minuend bits minus the subtrahend bits. In the second example, in the right position, the subtrahend bit is 1 with the minuend bit 0, so it is necessary to borrow from the second position as shown. This gives a difference bit in the first position of 1 $(2 + 0 - 1 = 1)$. In the second position, the borrow is subtracted, so a borrow is again necessary. Recall that, in the event that the subtrahend is larger than the minuend, we subtract the minuend from the subtrahend and give the result a minus sign. This is the case in the third example, in which this interchange of the two operands is shown.

The final operation to be illustrated is binary multiplication, which is quite simple. The multiplier digits are always 1 or 0. Therefore, the partial products are equal either to the multiplicand or to 0. Multiplication is illustrated by the following example:

Multiplicand:	1011
Multiplier:	× 101
	1011
	0000
	1011
Product:	110111

Arithmetic operations with octal, hexadecimal, or any other base-r system will normally require the formulation of tables from which one obtains sums and products of two digits in that base. An easier alternative for adding two numbers in base r is to convert each pair of digits in a column to decimal, add the digits in decimal, and then convert the result to the corresponding sum and carry in the base-r system. Since addition is done in decimal, we can rely on our memories for obtaining the entries from the familiar decimal addition table. The sequence of steps for adding the two hexadecimal numbers 59F and E46 is shown in Example 1-1.

■ EXAMPLE 1-1

Perform the addition $(59F)_{16} + (E46)_{16}$.

Hexadecimal	Equivalent Decimal Calculation			
	1 ← (carry)		1 ← (carry)	
5 9 F	5	Carry	9	15 Carry
E 4 6	14		4	6
1 3 E 5	1 19 = 16 + 3		14 = E	21 = 16 + 5

The equivalent decimal calculation columns on the right show the mental reasoning that must be carried out to produce each digit of the hexadecimal sum. Instead of adding F + 6 in hexadecimal, we add the equivalent decimals, 15 + 6 = 21. We then convert back to hexadecimal by noting that 21 = 16 + 5. This gives a sum digit of 5 and a carry of 1 to the next higher order column of digits. The other two columns are added in a similar fashion. ■

The multiplication of two base-r numbers can be accomplished by doing all the arithmetic operations in decimal and converting intermediate results one at a time. This is illustrated in the multiplication of two octal numbers shown in Example 1-2.

■ EXAMPLE 1-2

Perform the multiplication $(762)_8 \times (45)_8$.

Octal	Octal		Decimal		Octal
762	5×2	=	$10 = 8 + 2$	=	12
45	$5 \times 6 + 1$	=	$31 = 24 + 7$	=	37
4672	$5 \times 7 + 3$	=	$38 = 32 + 6$	=	46
3710	4×2	=	$8 = 8 + 0$	=	10
43772	$4 \times 6 + 1$	=	$25 = 24 + 1$	=	31
	$4 \times 7 + 3$	=	$31 = 24 + 7$	=	37

The computations on the right show the mental calculations for each pair of octal digits. The octal digits 0 through 7 have the same value as their corresponding decimal digits. The multiplication of two octal digits plus a carry, derived from the calculation on the previous line, is done in decimal, and the result is then converted back to octal. The left digit of the two-digit octal result gives the carry that must be added to the digit product on the next line. The blue digits from the octal results of the decimal calculations are copied to the octal partial products on the left. For example, $(5 \times 2)_8 = (12)_8$. The left digit, 1, is the carry to be added to the product $(5 \times 6)_8$, and the blue least significant digit, 2, is the corresponding digit of the octal partial product. When there is no digit product to which the carry can be added, the carry is written directly into the octal partial product, as in the case of the 4 in 46. ■

Conversion from Decimal to Other Bases

The conversion of a number in base r to decimal is done by expanding the number in a power series and adding all the terms, as shown previously. We now present a general procedure for the reverse operation of converting a decimal number to a number in base r. If the number includes a radix point, it is necessary to separate the number into an integer part and a fraction part, since each part must be converted differently. The conversion of a decimal integer to a number in base r is done by dividing the number and all successive quotients by r and accumulating the remainders. This procedure is best explained by example.

■ EXAMPLE 1-3

Convert decimal 153 to octal.

The conversion is to base 8. First, 153 is divided by 8 to give a quotient of 19 and a remainder of 1, as shown in blue. Then 19 is divided by 8 to give a quotient of 2 and a remainder of 3. Finally, 2 is divided by 8 to give a quotient of 0 and a remainder of 2. The coefficients of the desired octal number are obtained from the remainders:

$153/8 = 19 + 1/8$	Remainder $= 1$	↑	Least significant digit
$19/8 = 2 \ + 3/8$	$= 3$		
$2/8 = 0 \ + 2/8$	$= 2$		Most significant digit

$(153)_{10} = (231)_8$ ■

Note in the preceding example that the remainders are read from last to first, as indicated by the arrow, to obtain the converted number. The quotients are divided by r until the result is 0. We also can use this procedure to convert decimal numbers to binary. In this case, the base of the converted number is 2, and therefore, all the divisions must be done by 2.

■ EXAMPLE 1-4

Convert decimal 41 to binary.

$41/2 = 20 + 1/2$	Remainder $= 1$	↑	Least significant digit
$20/2 = 10$	$= 0$		
$10/2 = 5$	$= 0$		
$5/2 = 2 + 1/2$	$= 1$		
$2/2 = 1$	$= 0$		
$1/2 = 0 + 1/2$	$= 1$		Most significant digit

$(41)_{10} = (101001)_2$

Of course, the decimal number could be converted by the sum of powers of two:

$$(41)_{10} = 32 + 8 + 1 = (101001)_2$$ ■

The conversion of a decimal fraction to base r is accomplished by a method similar to that used for integers, except that multiplication by r is used instead of division, and integers are accumulated instead of remainders. Again, the method is best explained by example.

■ EXAMPLE 1-5

Convert decimal 0.6875 to binary.

First, 0.6875 is multiplied by 2 to give an integer and a fraction. The new fraction is multiplied by 2 to give a new integer and a new fraction. This process is continued until the fractional part equals 0 or until there are enough digits to give sufficient accuracy. The coefficients of the binary number are obtained from the integers as follows:

$0.6875 \times 2 = 1.3750$	Integer $= 1$	Most significant digit
$0.3750 \times 2 = 0.7500$	$= 0$	
$0.7500 \times 2 = 1.5000$	$= 1$	
$0.5000 \times 2 = 1.0000$	$= 1$	Least significant digit

$(0.6875)_{10} = (0.1011)_2$ ■

Note in the foregoing example that the integers are read from first to last, as indicated by the arrow, to obtain the converted number. In that example, a finite number of digits appears in the converted number. The process of multiplying fractions by r does not necessarily end with zero, so we must decide how many digits of the fraction to use from the conversion. Also, remember that the multiplying number is equal to r. Therefore, to convert a decimal fraction to octal, we must multiply the fractions by 8.

■ EXAMPLE 1-6

Convert decimal 0.513 to a three digit octal fraction.

$0.513 \times 8 = 4.104$	Integer $= 4$	Most significant digit
$0.104 \times 8 = 0.832$	$= 0$	
$0.832 \times 8 = 6.656$	$= 6$	
$0.656 \times 8 = 5.248$	$= 5$	Least significant digit

The answer, to three significant figures, is obtained from the integer digits; note that the last integer digit, 5, is used for rounding the second-to-the-last digit, 6, to obtain

$$(0.513)_{10} = (0.407)_8$$ ■

The conversion of decimal numbers with both integer and fractional parts is done by converting each part separately and then combining the two answers. Using the results of Example 1-3 and Example 1-6, we obtain

$$(153.513)_{10} = (231.407)_8$$

1-4 Decimal Codes

The binary number system is the most natural system for a computer, but people are accustomed to the decimal system. One way to resolve this difference is to convert decimal numbers to binary, perform all arithmetic calculations in binary, and then convert the binary results back to decimal. This method requires that we store the decimal numbers in the computer in a way that they can be converted to binary. Since the computer can accept only binary values, we must represent the decimal digits by a code that contains l's and 0's. It is also possible to perform the arithmetic operations directly with decimal numbers when they are stored in the computer in coded form.

An n-bit *binary code* is a group of n bits that assume up to 2^n distinct combinations of 1's and 0's, with each combination representing one element of the set that is being coded. A set of four elements can be coded with a 2-bit binary code, with each element assigned one of the following bit combinations: 00, 01, 10, 11. A set of 8 elements requires a 3-bit code, and a set of 16 elements requires a 4-bit code. The bit combination of an n-bit code is determined from the count in binary from 0 to $2^n - 1$. Each element must be assigned a unique binary bit combination, and no two elements can have the same value; otherwise the code assignment will be ambiguous.

A binary code will have some unassigned bit combinations if the number of elements in the set is not a power of 2. The 10 decimal digits form such a set. A binary code that distinguishes among 10 elements must contain at least four bits, but six out of the 16 possible combinations will remain unassigned. Numerous different binary codes can be obtained by arranging four bits into 10 distinct combinations. The code most commonly used for the decimal digits is the straightforward binary assignment listed in Table 1-1. This is called *binary-coded decimal* and is commonly referred to as BCD. Other decimal codes are possible, and a few of them are presented in Chapter 3.

□ **TABLE 1-3**
Binary-Coded Decimal (BCD)

Decimal Symbol	BCD Digit
0	0000
1	0001
2	0010
3	0011
4	0100
5	0101
6	0110
7	0111
8	1000
9	1001

Table 1-3 gives a 4-bit code for each decimal digit. A number with n decimal digits will require $4n$ bits in BCD. Thus, decimal 396 is represented in BCD with 12 bits as

$$0011 \quad 1001 \quad 0110$$

with each group of four bits representing one decimal digit. A decimal number in BCD is the same as its equivalent binary number only when the number is between 0 and 9, inclusive. A BCD number greater than 10 has a representation different from its equivalent binary number, even though both contain 1's and 0's. Moreover, the binary combinations 1010 through 1111 are not used and have no meaning in the BCD code. Consider decimal 185 and its corresponding value in BCD and binary:

$$(185)_{10} = (0001\ \ 1000\ \ 0101)_{BCD} = (10111001)_2$$

The BCD value has 12 bits, but the equivalent binary number needs only 8 bits. It is obvious that a BCD number needs more bits than its equivalent binary value. However, there is an advantage in the use of decimal numbers because computer input and output data are handled by people who use the decimal system. BCD numbers are decimal numbers and not binary numbers, even though they are represented in bits. The only difference between a decimal and a BCD number is that decimals are written with the symbols 0, 1, 2, ..., 9, and BCD numbers use the binary codes 0000, 0001, 0010, ..., 1001.

BCD Addition

Consider the addition of two decimal digits in BCD, together with a possible carry of 1 from a previous less significant pair of digits. Since each digit does not exceed 9, the sum cannot be greater than $9 + 9 + 1 = 19$, the 1 being a carry. Suppose we add the BCD digits as if they were binary numbers. Then the binary sum will produce a result in the range from 0 to 19. In binary, this will be from 0000 to 10011, but in BCD, it should be from 0000 to 1 1001, the first 1 being a carry and the next four bits being the BCD digit sum. When the binary sum is less than 1010 (without a carry), the corresponding BCD digit is correct. But when the binary sum is greater than or equal to 1010, the result is an invalid BCD digit. The addition of binary 6, $(0110)_2$, to the sum converts it to the correct digit and also produces a decimal carry as required. This is because the difference between a carry from the most significant bit position of the binary sum and a decimal carry is $16 - 10 = 6$. Thus, the decimal carry and the correct BCD sum digit are forced by adding 6 in binary. Consider the following three-digit BCD addition:

11	BCD carry	1	1	
448		0100	0100	1000
+489		+0100	+1000	+1001
937	Binary sum	1001	1101	0001
	Add 6		+0110	+0110
	BCD sum		1 0011	1 0111
	BCD result	1001	0011	0111

In each position, the two BCD digits are added as if they were two binary numbers. If the binary sum is greater than 1001, we add 0110 to obtain the correct BCD digit sum and a carry. In the right column, the binary sum is equal to 17. The presence of the carry indicates that the sum is greater than 16 (certainly greater than 9), so a correction is needed. The addition of 0110 produces the correct BCD digit sum, 0111 (7), and a carry of 1. In the next column, the binary sum is 1101 (13), again an invalid BCD digit. Addition of 0110 produces the correct BCD digit sum, 0011 (3), and a carry of 1. In the final column, the binary sum is equal to 1001 (9) and is the correct BCD digit.

1-5 Alphanumeric Codes

Many applications of digital computers require the handling of data consisting not only of numbers, but also of letters. For instance, an insurance company with thousands of policyholders will use a computer to process its files. To represent the names and other pertinent information, it is necessary to formulate a binary code for the letters of the alphabet. In addition, the same binary code must represent numerals and special characters such as $. Any alphanumeric character set for English is a set of elements that includes the 10 decimal digits, the 26 letters of the alphabet, and several (more than three) special characters. Such a set requires at least 32 codes if only capital letters are included and at least 64 codes if both uppercase and lowercase letters are included. In the first case, we need a binary code of at least six bits, and in the second, we need a binary code of at least seven bits. Binary codes play an important role in digital computers. The codes must be in binary because computers can handle only 1's and 0's. Note that binary encoding merely changes the symbols, not the meaning of the elements of information being encoded.

ASCII Character Code

The standard binary code for the alphanumeric characters is called ASCII (American Standard Code for Information Interchange). It uses seven bits to code 128 characters, as shown in Table 1-4. The seven bits of the code are designated by B_1 through B_7, with B_7 being the most significant bit. Note that the most significant three bits of the code determine the column of the table and the least significant four bits the row of the table. The letter A, for example, is represented in ASCII as 1000001 (column 100, row 0001). The ASCII code contains 94 characters that can be printed and 34 nonprinting characters used for various control functions. The printing characters consist of the 26 uppercase letters, the 26 lowercase letters, the 10 numerals, and 32 special printable characters such as %, @, and $.

The 34 control characters are designated in the ASCII table with abbreviated names. They are listed again below the table with their full functional names. The control characters are used for routing data and arranging the printed text into a prescribed format. There are three types of control characters: format effectors, information separators, and communication control characters. Format effectors are characters that control the layout of printing. They include the familiar typewriter controls such as backspace (BS), horizontal tabulation (HT), and carriage return (CR). Information separators are used to separate the data into divisions—for example, paragraphs and pages. They include characters such as record separator (RS) and file separator (FS). The communication control characters are used during the transmission of text from one location to the other. Examples of communication control characters are STX (start of text) and ETX (end of text), which are used to frame a text message transmitted via communication wires.

ASCII is a 7-bit code, but most computers manipulate an 8-bit quantity as a single unit called a *byte*. Therefore, ASCII characters most often are stored one per byte, with the most significant bit set to 0. The extra bit is sometimes used for specific purposes, depending on the application. For example, some printers recognize an additional 128 8-bit characters, with the most significant bit set to 1. These char-

□ TABLE 1-4
American Standard Code for Information Interchange (ASCII)

<table>
<tr><th></th><th colspan="8">$B_7B_6B_5$</th></tr>
<tr><th>$B_4B_3B_2B_1$</th><th>000</th><th>001</th><th>010</th><th>011</th><th>100</th><th>101</th><th>110</th><th>111</th></tr>
<tr><td>0000</td><td>NULL</td><td>DLE</td><td>SP</td><td>0</td><td>@</td><td>P</td><td>`</td><td>p</td></tr>
<tr><td>0001</td><td>SOH</td><td>DC1</td><td>!</td><td>1</td><td>A</td><td>Q</td><td>a</td><td>q</td></tr>
<tr><td>0010</td><td>STX</td><td>DC2</td><td>"</td><td>2</td><td>B</td><td>R</td><td>b</td><td>r</td></tr>
<tr><td>0011</td><td>ETX</td><td>DC3</td><td>#</td><td>3</td><td>C</td><td>S</td><td>c</td><td>s</td></tr>
<tr><td>0100</td><td>EOT</td><td>DC4</td><td>$</td><td>4</td><td>D</td><td>T</td><td>d</td><td>t</td></tr>
<tr><td>0101</td><td>ENQ</td><td>NAK</td><td>%</td><td>5</td><td>E</td><td>U</td><td>e</td><td>u</td></tr>
<tr><td>0110</td><td>ACK</td><td>SYN</td><td>&</td><td>6</td><td>F</td><td>V</td><td>f</td><td>v</td></tr>
<tr><td>0111</td><td>BEL</td><td>ETB</td><td>'</td><td>7</td><td>G</td><td>W</td><td>g</td><td>w</td></tr>
<tr><td>1000</td><td>BS</td><td>CAN</td><td>(</td><td>8</td><td>H</td><td>X</td><td>h</td><td>x</td></tr>
<tr><td>1001</td><td>HT</td><td>EM</td><td>)</td><td>9</td><td>I</td><td>Y</td><td>i</td><td>y</td></tr>
<tr><td>1010</td><td>LF</td><td>SUB</td><td>*</td><td>:</td><td>J</td><td>Z</td><td>j</td><td>z</td></tr>
<tr><td>1011</td><td>VT</td><td>ESC</td><td>+</td><td>;</td><td>K</td><td>[</td><td>k</td><td>{</td></tr>
<tr><td>1100</td><td>FF</td><td>FS</td><td>,</td><td><</td><td>L</td><td>\</td><td>l</td><td>|</td></tr>
<tr><td>1101</td><td>CR</td><td>GS</td><td>-</td><td>=</td><td>M</td><td>]</td><td>m</td><td>}</td></tr>
<tr><td>1110</td><td>SO</td><td>RS</td><td>.</td><td>></td><td>N</td><td>^</td><td>n</td><td>~</td></tr>
<tr><td>1111</td><td>SI</td><td>US</td><td>/</td><td>?</td><td>O</td><td>_</td><td>o</td><td>DEL</td></tr>
</table>

Control Characters:

NULL	NULL	DLE	Data link escape
SOH	Start of heading	DC1	Device control 1
STX	Start of text	DC2	Device control 2
ETX	End of text	DC3	Device control 3
EOT	End of transmission	DC4	Device control 4
ENQ	Enquiry	NAK	Negative acknowledge
ACK	Acknowledge	SYN	Synchronous idle
BEL	Bell	ETB	End of transmission block
BS	Backspace	CAN	Cancel
HT	Horizontal tab	EM	End of medium
LF	Line feed	SUB	Substitute
VT	Vertical tab	ESC	Escape
FF	Form feed	FS	File separator
CR	Carriage return	GS	Group separator
SO	Shift out	RS	Record separator
SI	Shift in	US	Unit separator
SP	Space	DEL	Delete

acters enable the printer to produce additional symbols, such as those from the Greek alphabet or characters with accent marks as used in languages other than English.

Parity Bit

To detect errors in data communication and processing, an eighth bit is sometimes added to the ASCII character to indicate its parity. A *parity bit* is an extra bit included to make the total number of 1's either even or odd. Consider the following two characters and their even and odd parity:

	With Even Parity	With Odd Parity
ASCII A = 1000001	01000001	11000001
ASCII T = 1010100	11010100	01010100

In each case, we use the extra bit in the leftmost position of the code to produce an even number of 1's in the character for even parity or an odd number of 1's in the character for odd parity. In general, one parity or the other is adopted, with even parity being more common.

The parity bit is helpful in detecting errors during the transmission of information from one location to another. Assuming that even parity is used, the simplest case is handled as follows: An even-parity bit is generated at the sending end for each character; the 8-bit characters that include parity bits are transmitted to their destination; the parity of each character is then checked at the receiving end; if the parity of the received character is not even, it means that at least one bit has changed its value during the transmission. This method detects one, three, or any odd number of errors in each character transmitted. An even number of errors is undetected. Other error-detection codes, some of which are based on additional parity bits, may be needed to take care of an even number of errors. What is done after an error is detected depends on the particular application. One possibility is to request retransmission of the message on the assumption that the error was random and will not occur again. Thus, if the receiver detects a parity error, it sends back a NAK (negative acknowledge) control character consisting of the even-parity eight bits, 10010101, from Table 1-4. If no error is detected, the receiver sends back an ACK (acknowledge) control character, 00000110. The sending end will respond to a NAK by transmitting the message again, until the correct parity is received. If, after a number of attempts, the transmission is still in error, a message can be sent to the operator to check for malfunctions in the transmission path.

Parity may be used with binary numbers representing items other than ASCII characters. In Chapter 6, we will see how parity bits can be used to perform error correction for binary data stored in computer memories.

Unicode

Unicode is a new standard for 16-bit alphanumeric codes. It is incorporated in the standard ISO/IEC (International Organization for Standardization/International Electrotechnical Commission) 10646 and is sometimes referred to as Unicode/10646. Since 16 bits provide 65,536 code words, Unicode has the capacity to represent the symbols and ideographs of the world's languages. A character code in Unicode is represented by four hexadecimal digits. Standard ASCII codes with $(00)_{16}$ appended to the left (excluding the control codes) constitute the first 95 characters in Unicode. The codes used are from $(0000)_{16}$ through $(007F)_{16}$ (excluding the control codes). Table 1-5 gives the first 191 character codes in Unicode. The codes are in hexadecimal, with the three most significant digits determining the table column and the least significant digit the table row. Note that different graphic characters may be associated with a given character code, such as for the dollar sign ($) and tilde (~). The codes from $(00A0)_{16}$ through $(00FF)_{16}$ are referred to as Latin 1. These codes provide additional letters used in the major languages of Europe, based on the Latin alphabet. Carrying on the tradition of ASCII, a miscellaneous set of mathematical signs and punctuation is included. Latin 1 codes are based on standard ISO 8859-1. In Unicode, the mathematical symbols, signs, and punctuation not included in ASCII and Latin 1 are included at a higher range.

Figure 1-3 gives the four major zones for the assignment of Unicode code words. The hexadecimal digit pairs shown are the leftmost two digits of the code words. Each block in the figure represents 4,096 codes. The A-Zone contains codes for alphabets, syllables, and symbols. The I-Zone contains codes for ideographs. An ideographic character stands for a word or inseparable grammatical unit, rather than a sound. Chinese script used in various languages is an example of ideographic characters. Since the ideographs represent words rather than characters, there are large numbers of them. The O-Zone is currently vacant, but is likely to be used for ideographs in the future.

The R-Zone is for restricted use. It is further broken down into the Private Use Area, the Compatibility Zone, and Special Codes. FFFE and FFFF are not character codes and are specifically excluded from Unicode. The Private Use Area is available to those needing special characters for their application pro-

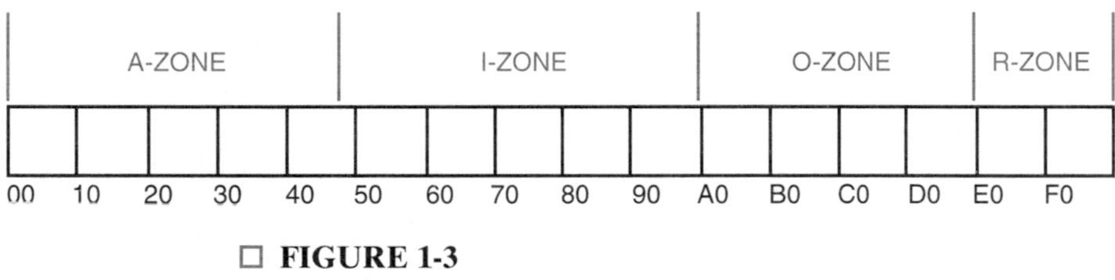

□ **FIGURE 1-3**
Major Zones for Unicode Code Assignment

TABLE 1-5
First 256 Codes for Unicode[a]

	Control		ASCII						Control		Latin 1					
	000	001	002	003	004	005	006	007	008	009	00A	00B	00C	00D	00E	00F
0	CTRL	CTRL	SPACE	0	@	P	`	p	CTRL	CTRL	NB SP	°	À	Đ	à	D
1	CTRL	CTRL	!	1	A	Q	a	q	CTRL	CTRL	¡	±	Á	Ñ	á	ñ
2	CTRL	CTRL	"	2	B	R	b	r	CTRL	CTRL	¢	^{2}	Â	Ò	â	ò
3	CTRL	CTRL	#	3	C	S	c	s	CTRL	CTRL	£	^{3}	Ã	Ó	ã	ó
4	CTRL	CTRL	$\|$	4	D	T	d	t	CTRL	CTRL	¤	´	Ä	Ô	ä	ô
5	CTRL	CTRL	%	5	E	U	e	u	CTRL	CTRL	¥|¥	µ	Å	Õ	å	õ
6	CTRL	CTRL	&	6	F	V	f	v	CTRL	CTRL	¦	¶	Æ	Ö	æ	ö
7	CTRL	CTRL	'	7	G	W	g	w	CTRL	CTRL	§	·	Ç	×	ç	÷
8	CTRL	CTRL	(	8	H	X	h	x	CTRL	CTRL	¨	¸	È	Ø	è	ø
9	CTRL	CTRL	)	9	I	Y	i	y	CTRL	CTRL	©	^{1}	É	Ù	é	ù
A	CTRL	CTRL	*	:	J	Z	j	z	CTRL	CTRL	ª	º	Ê	Ú	ê	ú
B	CTRL	CTRL	+	;	K	[	k	{	CTRL	CTRL	«	»	Ë	Û	ë	û
C	CTRL	CTRL	,	<	L	\\	l	|	CTRL	CTRL	¬	¼|1/4	Ì	Ü	ì	ü
D	CTRL	CTRL	-	=	M	]	m	}	CTRL	CTRL	-	½|1/2	Í	Ý	í	ý
E	CTRL	CTRL	.	>	N	^	n	~|¯	CTRL	CTRL	®	¾|3/4	Î	þ	î	þ
F	CTRL	CTRL	/	?	O	_	o	CTRL	CTRL	CTRL	¯	¿	Ï	ß	ï	ÿ

[a] Unicode, Inc., The Unicode Standard: Worldwide Character Encoding, Version 1.0, Volume 1, © 1990, 1991 by Unicode, Inc. Reprinted by permission of Addison-Wesley Publishing Company, Inc.

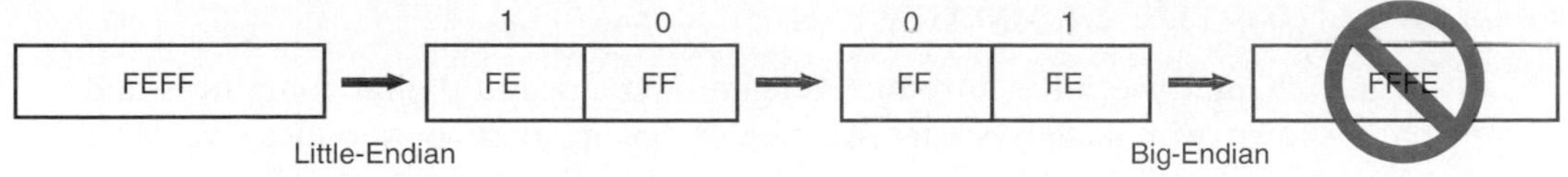

□ **FIGURE 1-4**
Byte-Ordering Problem for Unicode Codes

grams. For example, icons used in menus could be specified by character codes in this range. The Compatibility Zone contains characters that are mapped to other areas in the overall code space. The characters available in this special area are in widespread use, but are not directly compatible with the way in which Unicode encoding deals with character representation, so they could not be included directly in other areas.

Although Unicode characters are defined as 16 bits, they can be implemented in computers by two bytes. A computer *word* is composed of multiple bytes. Suppose that we consider a 16-bit word consisting of two bytes. If byte 0 is on the right of the word (the little or least significant end) and byte 1 on the left, the computer is said to be *little-endian*. If byte 0 is on the left of the word (the big or most significant end) and byte 1 on the right, the computer is said to be *big-endian*. Now suppose that a string of $2n$ bytes representing n Unicode characters is transferred from a little-endian computer to a big-endian computer. Then the byte order in a 16-bit word is interchanged, thereby garbling the characters. This byte-ordering problem for Unicode is illustrated in Figure 1-4. If the code $(\text{FEFF})_{16}$, the nonprinting Unicode value for BYTE ORDER MARK, is included at the beginning of the original string, then it will appear as the invalid code $(\text{FFFE})_{16}$, due to the pairs of bytes being reversed. This signals that all pairs are in the wrong order and must be swapped before the 16-bit words are interpreted as characters. Thus, by placing Unicode $(\text{FEFF})_{16}$ at the beginning of a character string, an application can determine whether the bytes need to be swapped before interpretation. A similar situation occurs in going from a big-endian to a little-endian computer.

■ **EXAMPLE 1-7**

Find the word represented by the Unicode string FFFE, 4300, 6F00, 6400, 6500.

Since the string begins with FFFE, the bytes are in the wrong order and need to be swapped before being interpreted as Unicode characters. Swapping the bytes gives the Unicode string

FEFF, 0043, 006F, 0064, 0065

which can be decoded as the word "Code" by using Table 1-5. ■

1-6 Chapter Summary

In this chapter, we have introduced digital systems and digital computers and have shown why such systems use signals having only two values. We have referred to a generic computer in order to provide a frame of reference and motivation for the topics to be studied. Number system concepts, including base (radix) and radix point, were presented. Because of their correspondence to two-valued signals, binary numbers were discussed in detail. Octal (base 8) and hexadecimal (base16) were also emphasized, since they are useful as shorthand notation for binary. Arithmetic operations in bases other than base 10 and the conversion of numbers from one base to another were covered. Because of the predominance of decimal in normal use, Binary Coded Decimal (BCD) was treated.

The representation of information in the form of characters instead of numbers by means of the ASCII code for the English alphabet was presented, and Unicode for representing the world's languages was discussed. In addition, the parity bit was presented as a technique for error detection.

In subsequent chapters, we will treat the representation of signed numbers and floating-point numbers. We will also introduce additional codes for the decimal digits. Although these topics fit well with the topics in this chapter, they are difficult to motivate without associating them with the hardware used to implement the operations they denote. Thus, we delay their presentation until we examine the associated hardware.

References

1. Hwang, K. *Computer Arithmetic.* New York: Wiley, 1979.
2. Cavanagh, J. *Digital Computer Arithmetic.* New York: McGraw-Hill, 1984.
3. Mano, M. M. *Computer Engineering: Hardware Design.* Englewood Cliffs, NJ: Prentice Hall, 1988.
4. Mano, M. M. *Digital Design*, 2nd ed. Englewood Cliffs, NJ: Prentice Hall, 1991.
5. Mano, M. M. *Computer System Architecture,* 3rd ed. Englewood Cliffs, NJ: Prentice Hall, 1993.
6. Patterson, D. A., and Hennessy, J. L. *Computer Organization and Design: The Hardware/Software Interface.* San Mateo, CA: Morgan Kaufmann, 1994.
7. White, R. *How Computers Work.* Emeryville, CA: Ziff-Davis Press, 1993.
8. The Unicode Consortium. *The Unicode Standard: Worldwide Character Encoding*, version 1.0, vol. 1. Reading, MA: Addison-Wesley, 1991.
9. Bettels, J., and Bishop, E. A. "Unicode: A Universal Character Code," *Digital Technical Journal*, vol. 5, no. 3 (Summer 1993), pp. 21-31.
10. Williams, M. R. *A History of Computing Technology.* Englewood Cliffs, NJ: Prentice Hall, 1985.

PROBLEMS

The asterisk (*) indicates a more advanced problem.

1–1. List the binary, octal, and hexadecimal numbers from 16 to 31.

1–2. What is the exact number of bits in a memory that contains **(a)** 12K bits; **(b)** 128M bits; **(c)** 2G bits?

1–3. What is the decimal equivalent of the largest binary number that can be obtained with **(a)** 8 bits and **(b)** 32 bits?

1–4. Convert the following binary numbers to decimal: 111010, 10101111.101, and 110110110.

1–5. Convert the following decimal numbers to binary: 1940, 1056, 138, and 1995.

1–6. Convert the following numbers with the indicated bases to decimal: $(12021)_3$, $(4321)_5$, and $(A98)_{12}$.

1–7. Convert the following numbers from the given base to the other three bases listed in the table:

Decimal	Binary	Octal	Hexadecimal
369.3125	?	?	?
?	10111101.101	?	?
?	?	326.5	?
?	?	?	F3C7.A

1–8. Add, subtract, and multiply the following numbers without converting to decimal:
(a) $(715)_8$ and $(367)_8$ **(b)** $(15F)_{16}$ and $(A7)_{16}$ **(c)** $(110101)_2$ and $(110110)_2$

1–9. Convert the following decimal numbers to the indicated bases using the methods of Examples 1-3 and 1-6:
(a) 7562.45 to octal **(b)** 1938.257 to hexadecimal **(c)** 175.175 to binary.

1–10. *An alternative method of converting from a number in base r to a decimal number is as follows:

$$(N)_{10} = ((((\ldots(A_{n-1}r + A_{n-2})r + A_{n-3})\ldots)r + A_2)r + A_1)r + A_0$$

Assuming a base-2 conversion, is this a more efficient method in terms of the number of additions and multiplications involved than the method given using powers of 2? Assume that each power of 2 must be calculated. Give a quantitative argument to support your answer.

1–11. Perform the following conversion by using base 2 instead of base 10 as the intermediate base for the conversion:
(a) $(764.7)_8$ to hexadecimal **(b)** $(F6D.C)_{16}$ to octal **(c)** $(147.5)_8$ to base 4

1–12. *Division is composed of multiplications and subtractions. Perform the binary division 1111110 ÷ 110 to obtain a quotient and remainder.

1–13. There is considerable evidence to suggest that base 20 has historically been used for number systems in a number of cultures.
(a) Write the digits for a base-20 system, using an extension of the same digit representation scheme employed for hexadecimal.
(b) Convert $(1995)_{10}$ to base 20. **(c)** Convert $(AGH.F)_{20}$ to decimal.

1–14. *Devise a specialized algorithm for converting from base 20 to decimal that is based on the fact that 20 is 2 times 10.

1–15. *In each of the following cases, determine the radix *r*:
(a) $(B1)_r = (144)_{10}$ **(b)** $(436)_r = (357)_{10}$

1–16. Represent the decimal numbers 831 and 793 in BCD, and then show the steps necessary to form their sum.

1–17. *Find an algorithm for BCD subtraction similar to the one used for BCD addition. For simplicity, assume that the minuend is always at least as large as the subtrahend.

1–18. Find the binary representations for each of the following BCD numbers:
(a) 0011 1001 0111 1000 **(b)** 0100 0110 1001.0111 0101

1–19. Write your full name in ASCII, using an 8-bit code **(a)** with the leftmost bit always 0 and **(b)** with the leftmost bit selected to produce even parity. Include a space between names and a period after the middle initial.

1–20. Decode the following ASCII code: 1001010 1101111 1101000 1101110 0100000 1000100 1101111 1100101.

1–21. Show the bit configuration that represents the decimal number 295 in **(a)** binary, **(b)** BCD, **(c)** ASCII, **(d)** Unicode.

1–22. A computer represents information in groups of 32 bits. How many different integers can be represented in **(a)** binary, **(b)** BCD, and **(c)** 8-bit ASCII, all using 32 bits?

1–23. List the 10 BCD digits with a parity bit giving odd parity in the leftmost position (a total of five bits per digit). Repeat with a parity bit for even parity.

1–24. What bit must be complemented to change an ASCII letter from uppercase to lowercase and vice versa?

1–25. Write your full name in Unicode. Include a space between names and a period after the middle initial.

1–26. Decode the following Unicode given in hexadecimal:

FEFF 0031 00F7 0034 003D 00BC 0020 0069 0073 0020 0061 0020 0046 0052 0041 0043 0054 0049 004F 004E 002E

1–27. *Decode the following Unicode given in binary. The result is a question in Spanish.

1111111111111110	1011111100000000	0100001100000000
1111001100000000	0110110100000000	0110111100000000
0011111100000000		

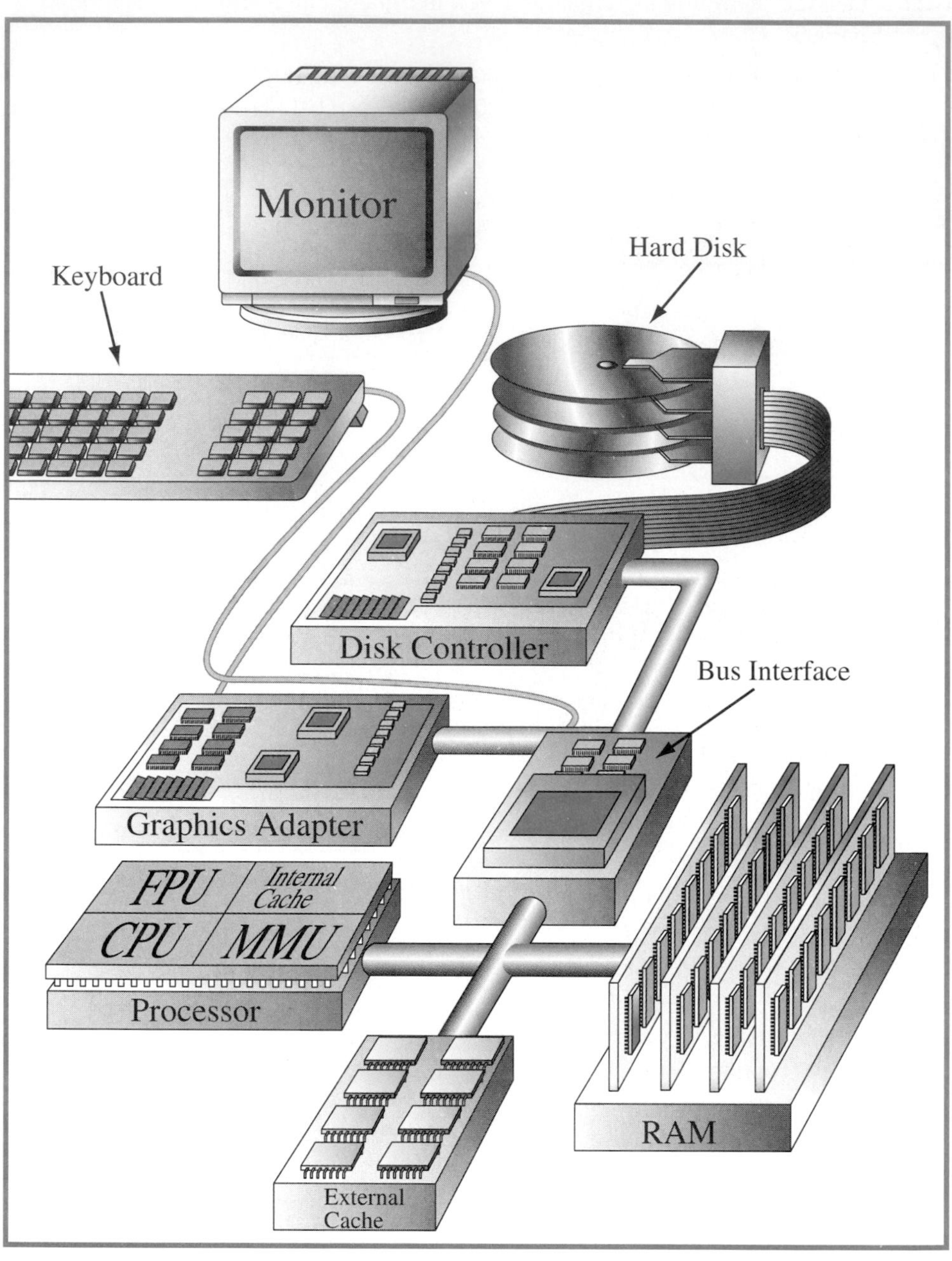
Monitor
Hard Disk
Keyboard
Disk Controller
Bus Interface
Graphics Adapter
FPU
Internal Cache
CPU
MMU
Processor
RAM
External Cache

CHAPTER

2

COMBINATIONAL LOGIC CIRCUITS

In this chapter, we will learn about gates, the most primitive logic elements used in digital systems. In addition, we will learn the mathematical techniques used in designing circuits from these gates and learn how to design cost-effective circuits. These techniques are fundamental to the design of almost all digital circuits, so we find the generic computer shaded in blue over most of the electronic portions of the hardware. The circuits involving memory are only lightly shaded, however, since large portions of memory are designed as electronic circuits, which do not use the logic circuit concept employed here. Because of its use throughout the design of almost all of the computer, what we study in this chapter is widely applied. The chapter contains the most fundamental material for an in-depth understanding of computers and digital systems and how they are designed.

2-1 BINARY LOGIC AND GATES

Digital circuits are hardware components that manipulate binary information. The circuits are implemented using transistors and interconnections in complex semiconductor devices called integrated circuits. Each basic circuit is referred to as a *logic gate*. For simplicity in design, we model the transistor-based electronic circuits as logic gates. Thus, the designer need not be concerned with the internal electronics of the individual gates, but only with their external logic properties. Each gate performs a specific logical operation. The outputs of gates are applied to the inputs of other gates to form a digital circuit.

In order to describe the operational properties of digital circuits, it is necessary to introduce a mathematical notation that specifies the operation of each gate and that can be used to analyze and design circuits. This binary logic system is one of a class of mathematical systems referred to generally as *Boolean algebras*. The name is in honor of the English mathematician George Boole, who in 1854 published a book introducing the mathematical theory of logic. The specific Boolean algebra we will study is used to describe the interconnection of digital gates and to design logic circuits through the manipulation of Boolean expressions. We first introduce the concept of binary logic and show its relationship to digital gates and binary signals. We then present the properties of the Boolean algebra, together with other concepts and methods useful in designing logic circuits.

Binary Logic

Binary logic deals with binary variables that take on two discrete values and with the operations of mathematical logic applied to these variables. The two values the variables take may be called by different names, as mentioned in Section 1-1, but for our purpose, it is convenient to think in terms of binary values and assign 1 or 0 to each variable. In the first part of this book, variables are designated by letters of the alphabet, such as A, B, C, X, Y, and Z. Later this notation will be expanded to include strings of letters, numbers, and special characters. Associated with the binary variables are three basic logical operations called AND, OR, and NOT:

1. AND. This operation is represented by a dot or by the absence of an operator. For example, $Z = X \cdot Y$ or $Z = XY$ is read "Z is equal to X AND Y." The logical operation AND is interpreted to mean that $Z = 1$ if and only if $X = 1$ and $Y = 1$; otherwise $Z = 0$. (Remember that X, Y, and Z are binary variables and can be equal to just 1 or 0.)
2. OR. This operation is represented by a plus symbol. For example, $Z = X + Y$ is read "Z is equal to X OR Y," meaning that $Z = 1$ if $X = 1$ or if $Y = 1$, or if both $X = 1$ and $Y = 1$. $Z = 0$ if and only if $X = 0$ and $Y = 0$.
3. NOT. This operation is represented by a bar over the variable. For example, $Z = \overline{X}$ is read "Z is equal to NOT X," meaning that Z is what X is not. In other words, if $X = 1$, then $Z = 0$; but if $X = 0$, then $Z = 1$. The NOT operation is also referred to as the *complement* operation, since it changes a 1 to 0 and a 0 to 1.

Binary logic resembles binary arithmetic, and the operations AND and OR have similarities to multiplication and addition, respectively. This is why the symbols used for AND and OR are the same as those used for multiplication and addition. However, binary logic should not be confused with binary arithmetic. One should realize that an arithmetic variable designates a number that may consist of many digits, whereas a logic variable is always either a 1 or a 0. The following equations define the logical OR operation:

$$0 + 0 = 0$$

$$0 + 1 = 1$$

$$1 + 0 = 1$$

$$1 + 1 = 1$$

These resemble binary addition, except for the last operation. In binary logic, we have 1 + 1 = 1 (read "one OR one is equal to one"), but in binary arithmetic, we have 1 + 1 = 10 (read "one plus one is equal to two"). To avoid ambiguity, the symbol $\vee$ is sometimes used for the OR operation instead of the + symbol. But as long as arithmetic and logic operations are not mixed, each can use the + symbol with its own independent meaning.

The next equations define the logical AND operation:

$$0 \cdot 0 = 0$$

$$0 \cdot 1 = 0$$

$$1 \cdot 0 = 0$$

$$1 \cdot 1 = 1$$

This operation is identical to binary multiplication, provided that we use only a single bit. An alternative symbol to the $\cdot$ for AND is $\wedge$, which is often used in conjunction with $\vee$ for OR.

For each combination of the values of binary variables such as X and Y, there is a value of Z specified by the definition of the logical operation. The definitions may be listed in compact form in a truth table. A *truth table* for an operation is a table of combinations of the binary variables showing the relationship between the values that the variables take on and the values of the result of the operation. The truth tables for the operations AND, OR, and NOT are shown in Table 2-1. The tables list all possible combinations of values for two variables and the results of the operation. They clearly demonstrate the definition of the three operations.

□ **TABLE 2-1**
Truth Tables for the Three Basic Logical Operations

AND

X	Y	$Z = X \cdot Y$
0	0	0
0	1	0
1	0	0
1	1	1

OR

X	Y	$Z = X + Y$
0	0	0
0	1	1
1	0	1
1	1	1

NOT

X	$Z = \overline{X}$
0	1
1	0

Logic Gates

Logic gates are electronic circuits that operate on one or more input signals to produce an output signal. Electrical signals such as voltages or currents exist throughout a digital system in either of two recognizable values. Voltage-operated circuits respond to two separate voltage ranges that represent a binary variable equal to logic 1 or logic 0, as illustrated in Figure 1-1. The input terminals of logic gates accept binary signals within the allowable range and respond at the output terminals with binary signals that fall within a specified range. The intermediate regions

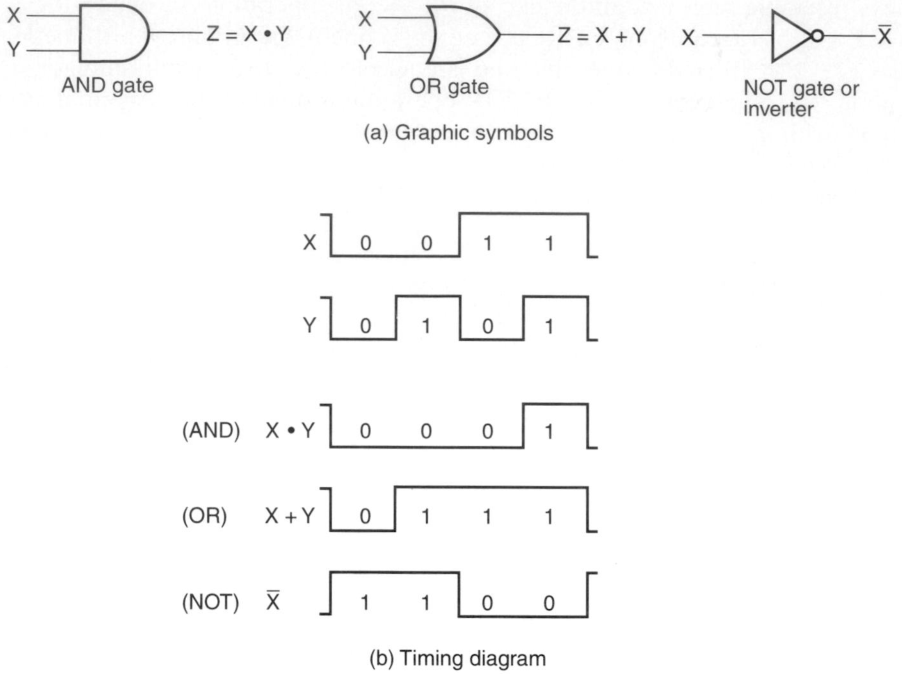

(a) Graphic symbols

(b) Timing diagram

FIGURE 2-1
Digital Logic Gates

between the allowed ranges in the figure are crossed only during changes from 1 to 0 or from 0 to 1. These changes are called *transitions*, and the intermediate regions are called the *transition regions.*

The graphics symbols used to designate the three types of gates—AND, OR, and NOT—are shown in Figure 2-1(a). The gates are electronic circuits that produce the equivalents of logic-1 and logic-0 output signals in accordance with their respective truth tables if the equivalents of logic-1 and logic-0 input signals are applied. The two input signals X and Y to the AND and OR gates take on one of four possible combinations: 00, 01, 10, or 11. These input signals are shown as timing diagrams in Figure 2-1(b), together with the timing diagrams for the corresponding output signal for each type of gate. The horizontal axis of a *timing diagram* represents time, and the vertical axis shows a signal as it changes between the two possible voltage levels. The low level represents logic 0 and the high level represents logic 1. The AND gate responds with a logic-1 output signal when both input signals are logic 1. The OR gate responds with a logic-1 output signal if either input signal is logic 1. The NOT gate is more commonly referred to as an *inverter*. The reason for this name is apparent from the response in the timing diagram. The output logic signal is an inverted version of input logic signal X.

AND and OR gates may have more than two inputs. An AND gate with three inputs and an OR gate with six inputs are shown in Figure 2-2. The three-

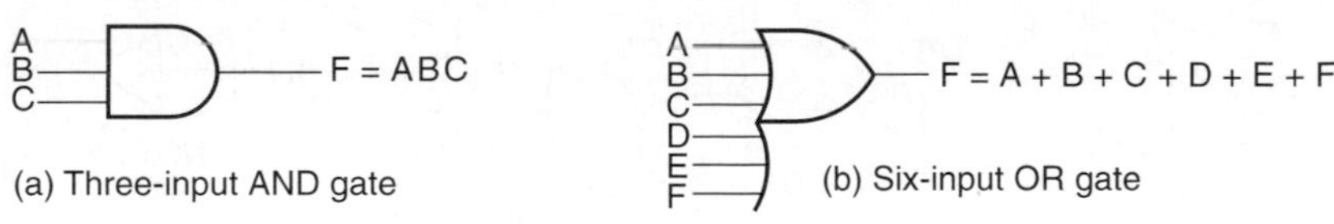

□ **FIGURE 2-2**
Gates with More than Two Inputs

input AND gate responds with a logic-l output if all three inputs are logic l. The output is logic 0 if any input is logic 0. The six-input OR gate responds with a logic 1 if any input is logic 1; its output becomes a logic 0 only when all inputs are logic 0.

2-2 Boolean Algebra

The Boolean algebra we present is an algebra dealing with binary variables and logic operations. The variables are designated by letters of the alphabet, and the three basic logic operations are AND, OR, and NOT (complementation). A *Boolean function* consists of a binary variable denoting the function, an equals sign, and an algebraic expression formed by using binary variables, the constants 0 and 1, the logic operation symbols, and parentheses. For a given value of the binary variables, a Boolean function can be equal to either 1 or 0. Consider as an example the Boolean function

$$F = X + \overline{Y}Z$$

The two parts of the expression, X, and $\overline{Y}Z$, are called *terms* of the function F. The function F is equal to 1 if term X is equal to 1 or if term $\overline{Y}Z$—i.e., both $\overline{Y}$ and Z—are equal to 1. Otherwise, F is equal to 0. The complement operation dictates that if $\overline{Y} = 1$, then Y must equal 0. Therefore, we can say that $F = 1$ if $X = 1$, or if $Y = 0$ and $Z = 1$. A Boolean function expresses the logical relationship between binary variables. It is evaluated by determining the binary value of the expression for all possible combinations of values for the variables.

A Boolean function can be represented in a truth table. A *truth table* for a function is a list of all combinations of l's and 0's that can be assigned to the binary variables and a list that shows the value of the function for each binary combination. The truth tables for the logic operations given in Table 2-1 are special cases of truth tables for functions. The number of rows in a truth table is 2^n, where n is the number of variables in the function. The binary combinations for the truth table are the n-bit binary numbers that correspond to counting in decimal from 0 through $2^n - 1$. Table 2-2 shows the truth table for the function $F = X + \overline{Y}Z$. There are eight possible binary combinations that assign bits to the three variables X, Y, and Z. The column labeled F contains either 0 or 1 for each of these combinations. The table shows that the function is equal to 1 if $X = 1$ and if $Y = 0$ and $Z = 1$. Otherwise, the function is equal to 0.

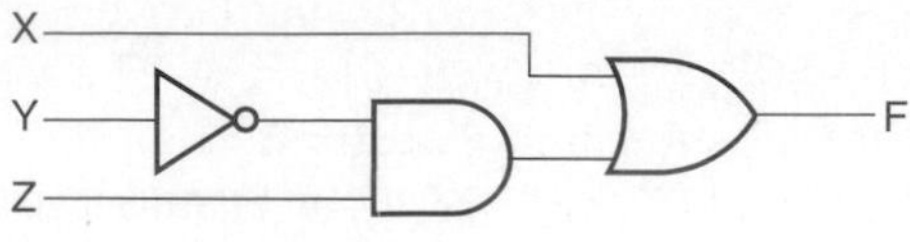

□ **FIGURE 2-3**
Logic Circuit Diagram for $F = X + \overline{Y}Z$

□ **TABLE 2-2**
Truth Table
for the Function $F = X + \overline{Y}Z$

X	Y	Z	F
0	0	0	0
0	0	1	1
0	1	0	0
0	1	1	0
1	0	0	1
1	0	1	1
1	1	0	1
1	1	1	1

A Boolean function can be transformed from an algebraic expression into a circuit diagram composed of logic gates. The logic circuit diagram for F is shown in Figure 2-3. An inverter on input Y generates the complement, $\overline{Y}$. An AND gate operates on $\overline{Y}$ and Z, and an OR gate combines X and $\overline{Y}Z$. In logic circuit diagrams, the variables of the function are taken as the inputs of the circuit, and the binary variable F is taken as the output of the circuit. The gates are interconnected by wires that carry logic signals. Logic circuits of this type are called *combinational* logic circuits, since the variables are "combined" by the logical operations. This is in contrast to the sequential logic to be treated in Chapter 4, in which variables are stored over time as well as being combined.

There is only one way that a Boolean function can be represented in a truth table. However, when the function is in algebraic form, it can be expressed in a variety of ways. The particular expression used to designate the function dictates the interconnection of gates in the logic circuit diagram. By manipulating a Boolean expression according to Boolean algebraic rules, it is often possible to obtain a simpler expression for the same function. This simpler expression reduces both the number of gates in the circuit and the numbers of inputs to the gates. To see how this is done, it is necessary first to study the basic rules of Boolean algebra.

Basic Identities of Boolean Algebra

Table 2-3 lists the most basic identities of Boolean algebra. The notation is simplified by omitting the symbol for AND whenever doing so does not lead to confusion. The first nine identities show the relationship between a single variable X, its complement $\overline{X}$, and the binary constants 0 and 1. The next five identities, 10 through 14, have counterparts in ordinary algebra. The last three, 15 through 17, do not apply in ordinary algebra, but are very useful in manipulating Boolean expressions.

The basic rules listed in the table have been arranged into two columns that demonstrate the property of duality of Boolean algebra. The *dual* of an algebraic expression is obtained by interchanging OR and AND operations and replacing 1's by 0's and 0's by 1's. An equation in one column of the table can be obtained from the corresponding equation in the other column by taking the dual of the expressions on both sides of the equals sign. For example, relation 2 is the dual of relation 1 because the OR has been replaced by an AND and the 0 by 1. It is important to note that most of the time the dual of an expression is not equal to the original expression, so that an expression usually cannot be replaced by its dual.

The nine identities involving a single variable can be easily verified by substituting each of the two possible values for X. For example, to show that $X + 0 = X$, let $X = 0$ to obtain $0 + 0 = 0$, and then let $X = 1$ to obtain $1 + 0 = 1$. Both equations are true according to the definition of the OR logic operation. Any expression can be substituted for the variable X in all the Boolean equations listed in the table. Thus, by identity 3 and with $X = AB + C$, we obtain

$$AB + C + 1 = 1$$

Note that identity 9 states that double complementation restores the variable to its original value. Thus, if $X = 0$, then $\overline{X} = 1$ and $\overline{\overline{X}} = 0 = X$.

☐ TABLE 2-3
Basic Identities of Boolean Algebra

1.	$X + 0 = X$	2.	$X \cdot 1 = X$	
3.	$X + 1 = 1$	4.	$X \cdot 0 = 0$	
5.	$X + X = X$	6.	$X \cdot X = X$	
7.	$X + \overline{X} = 1$	8.	$X \cdot \overline{X} = 0$	
9.	$\overline{\overline{X}} = X$			
10.	$X + Y = Y + X$	11.	$XY = YX$	Commutative
12.	$X + (Y + Z) = (X + Y) + Z$	13.	$X(YZ) = (XY)Z$	Associative
14.	$X(Y + Z) = XY + XZ$	15.	$X + YZ = (X + Y)(X + Z)$	Distributive
16.	$\overline{X + Y} = \overline{X} \cdot \overline{Y}$	17.	$\overline{X \cdot Y} = \overline{X} + \overline{Y}$	DeMorgan's

Identities 10 and 11, the commutative laws, state that the order in which the variables are written will not affect the result when using the OR and AND operations. Identities 12 and 13, the associative laws, state that the result of forming an operation among three variables is independent of the order that is taken, and therefore, the parentheses can be removed altogether as follows:

$$X + (Y + Z) = (X + Y) + Z = X + Y + Z$$

$$X(YZ) = (XY)Z = XYZ$$

These two laws and the first distributive law, identity 14, are well known from ordinary algebra, so they should not impose any difficulty. The second distributive law, given by identity 15, is the dual of the ordinary distributive law and does not hold in ordinary algebra. As illustrated previously, each variable in an identity can be replaced by a Boolean expression, and the identity still holds. Thus, consider the expression $(A + B)(A + CD)$. Letting $X = A$, $Y = B$, and $Z = CD$, and applying the second distributive law, we obtain

$$(A + B)(A + CD) = A + BCD$$

□ **TABLE 2-4**
Truth Tables to Verify DeMorgan's Theorem

A) X	Y	X + Y	$\overline{X+Y}$	B) X	Y	$\overline{X}$	$\overline{Y}$	$\overline{X}\cdot\overline{Y}$
0	0	0	1	0	0	1	1	1
0	1	1	0	0	1	1	0	0
1	0	1	0	1	0	0	1	0
1	1	1	0	1	1	0	0	0

The last two identities in Table 2-3,

$$\overline{X + Y} = \overline{X} \cdot \overline{Y} \text{ and } \overline{X \cdot Y} = \overline{X} + \overline{Y}$$

are referred to as DeMorgan's theorem. This is a very important theorem and is used to obtain the complement of an expression. DeMorgan's theorem can be illustrated by means of truth tables that assign all the possible binary values to X and Y. Table 2-4 shows two truth tables that verify the first part of DeMorgan's theorem. In A, we evaluate $\overline{X + Y}$ for all possible values of X and Y. This is done by first evaluating $X + Y$ and then taking its complement. In B, we evaluate $\overline{X}$ and $\overline{Y}$ and then AND them together. The result is the same for the four binary combinations of X and Y, which verifies the identity of the equation.

Note the order in which the operations are performed when evaluating an expression. In part B of the table, the complement over a single variable is evaluated and then the AND operation, just as we do in ordinary algebra with multiplication and addition. In part A, the OR operation is evaluated first. Then, noting that the complement over an expression such as $X + Y$ is considered as specifying NOT $(X + Y)$, we evaluate the expression within the parentheses and take the complement of the result. It is customary to exclude the parentheses when complementing an expression, since a bar is drawn over the entire expression. Thus, $\overline{(X + Y)}$ is expressed as $\overline{X + Y}$ when designating the complement of $X + Y$.

DeMorgan's theorem can be extended to three or more variables. The general DeMorgan's theorem can be expressed as

$$\overline{X_1 + X_2 + \ldots + X_n} = \overline{X}_1\overline{X}_2\ldots\overline{X}_n$$
$$\overline{X_1 X_2 \ldots X_n} = \overline{X}_1 + \overline{X}_2 + \ldots + \overline{X}_n$$

Observe that the logic operation changes from OR to AND or from AND to OR. In addition, the complement is removed from the entire expression and placed instead over each variable. For example,

$$\overline{A + B + C + D} = \overline{A}\,\overline{B}\,\overline{C}\,\overline{D}\,.$$

Algebraic Manipulation

Boolean algebra is a useful tool for simplifying digital circuits. Consider, for example, the Boolean function

$$F = \overline{X}YZ + \overline{X}Y\overline{Z} + XZ$$

The implementation of this function with logic gates is shown in Figure 2-4(a). Input variables X and Z are complemented with inverters to obtain $\overline{X}$ and $\overline{Z}$. The three terms in the expression are implemented with three AND gates. The OR gate forms the logical OR of the terms. Now consider a simplification of the function by applying some of the identities listed in Table 2-3:

$$\begin{aligned} F &= \overline{X}YZ + \overline{X}Y\overline{Z} + XZ \\ &= \overline{X}Y(Z + \overline{Z}) + XZ && \text{by identity 14} \\ &= \overline{X}Y \cdot 1 + XZ && \text{by identity 7} \\ &= \overline{X}Y + XZ && \text{by identity 2} \end{aligned}$$

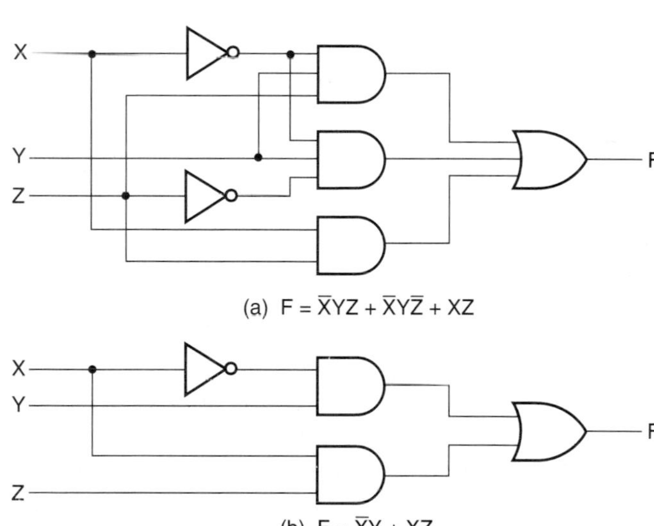

□ **FIGURE 2-4**
Implementation of Boolean Function with Gates

The function is reduced to only two terms and can be implemented with gates as shown in Figure 2-4(b). It is obvious that the circuit in (b) is simpler than the one in (a); yet both implement the same function. It is possible to use a truth table to verify that the two implementations are equivalent. This is shown in Table 2-5. As expressed in Figure 2-4(a), the function is equal to 1 if $X = 0$, $Y = 1$, and $Z = 1$; if $X = 0$, $Y = 1$, and $Z = 0$; or if X and Z are both 1. This produces the four 1's for F in the table. As expressed in Figure 2-4(b), the function is equal to 1 if $X = 0$ and $Y = 1$ or if $X = 1$ and $Z = 1$. This produces the same four 1's in the table. Since both expressions produce the same truth table, they are equivalent. Therefore, the two circuits have the same output for all possible binary combinations of the three input variables. Each circuit implements the same function, but the one with fewer gates is preferable because it requires fewer components.

□ **TABLE 2-5**
Truth Table for Boolean Function

X	Y	Z	F
0	0	0	0
0	0	1	0
0	1	0	1
0	1	1	1
1	0	0	0
1	0	1	1
1	1	0	0
1	1	1	1

When a Boolean expression is implemented with logic gates, each term requires a gate, and each variable within the term designates an input to the gate. We define a *literal* as a single variable within a term that may or may not be complemented. The function of Figure 2-4(a) has three terms and eight literals; the one in Figure 2-4(b) has two terms and four literals. By reducing the number of terms, the number of literals, or both in a Boolean expression, it is often possible to obtain a simpler circuit. Boolean algebra is applied to reduce an expression for the purpose of obtaining a simpler circuit. For highly complex functions, finding the best expression based on counts of terms and literals is very difficult, even by the use of computer programs. Certain methods, however, for reducing expressions are often included in computer tools for synthesizing logic circuits. These methods can obtain good, if not the best, solutions. The only manual method for the general case is a cut-and-try procedure employing the basic relations and other manipulations that become familiar with use. The following examples use identities from Table 2-3 to illustrate a few of the possibilities:

1. $X + XY = X(1 + Y) = X$
2. $XY + X\overline{Y} = X(Y + \overline{Y}) = X$
3. $X + \overline{X}Y = (X + \overline{X})(X + Y) = X + Y$

Note that the intermediate step $X = X \cdot 1$ has been omitted when X is factored out in equation 1. The relationship $1 + Y = 1$ is useful for eliminating redundant terms, as is done with the term XY in this same equation. The relation $Y + \overline{Y} = 1$ is useful for combining two terms, as is done in equation 2. The two terms being combined must be identical except for one variable, and that variable must be complemented in one term and not complemented in the other. Equation 3 is simplified by means of the second distributive law (identity 15 in Table 2-3). The following are three more examples of simplifying Boolean expressions:

4. $X(X + Y) = X + XY = X$
5. $(X + Y)(X + \overline{Y}) = X + Y\overline{Y} = X$
6. $X(\overline{X} + Y) = X\overline{X} + XY = XY$

Note that the intermediate steps $XX = X = X \cdot 1$ have been omitted during the manipulation of equation 4. The expression in equation 5 is simplified by means of the second distributive law. Here again, we omit the intermediate steps $Y\overline{Y} = 0$ and $X + 0 = X$.

Equations 4 through 6 are the duals of equations 1 through 3. Remember that the dual of an expression is obtained by changing AND to OR and OR to AND throughout (and l's to 0's and 0's to 1's if they appear in the expression). The *duality principle* of Boolean algebra states that a Boolean equation remains valid if we take the dual of the expressions on both sides of the equals sign. Therefore, equations 4, 5, and 6 can be verified by taking the dual of equations 1, 2, and 3, respectively.

The following *consensus theorem* is sometimes useful when simplifying Boolean expressions:

$$XY + \overline{X}Z + YZ = XY + \overline{X}Z$$

The theorem shows that the third term, YZ, is redundant and can be eliminated. Note that Y and Z are associated with X and $\overline{X}$ in the first two terms and appear together in the term that is eliminated. The proof of the consensus theorem is obtained by first ANDing YZ with $(X + \overline{X}) = 1$ and proceeds as follows:

$$\begin{aligned} XY + \overline{X}Z + YZ &= XY + \overline{X}Z + YZ(X + \overline{X}) \\ &= XY + \overline{X}Z + XYZ + \overline{X}YZ \\ &= XY + XYZ + \overline{X}Z + \overline{X}YZ \\ &= XY(1 + Z) + \overline{X}Z(1 + Y) \\ &= XY + \overline{X}Z \end{aligned}$$

The dual of the consensus theorem is

$$(X + Y)(\overline{X} + Z)(Y + Z) = (X + Y)(\overline{X} + Z)$$

The following example shows how the consensus theorem can be applied in manipulating a Boolean expression:

$$\begin{aligned}(A+B)(\overline{A}+C) &= A\overline{A}+AC+\overline{A}B+BC \\ &= AC+\overline{A}B+BC \\ &= AC+\overline{A}B\end{aligned}$$

Note that $A\overline{A} = 0$ and $0+AC = AC$. The redundant term eliminated in the last step by the consensus theorem is BC.

Complement of a Function

The complement of a function, $\overline{F}$, is obtained from an interchange of 1's to 0's and 0's to 1's for the values of F in the truth table. The complement of a function can be derived algebraically by applying DeMorgan's theorem. The generalized form of this theorem states that the complement of an expression is obtained by interchanging AND and OR operations and complementing each variable and constant.

EXAMPLE 2-1

Find the complement of each of the following two functions: $F_1 = \overline{X}Y\overline{Z}+\overline{X}\,\overline{Y}Z$ and $F_2 = X(\overline{Y}\,\overline{Z}+YZ)$.

Applying DeMorgan's theorem as many times as necessary, we obtain the complements as follows:

$$\begin{aligned}\overline{F_1} &= \overline{\overline{X}Y\overline{Z}+\overline{X}\,\overline{Y}Z} = (\overline{\overline{X}Y\overline{Z}})\cdot(\overline{\overline{X}\,\overline{Y}Z}) \\ &= (X+\overline{Y}+Z)(X+Y+\overline{Z}) \\ \overline{F_2} &= \overline{X(\overline{Y}\,\overline{Z}+YZ)} = \overline{X}+(\overline{\overline{Y}\,\overline{Z}+YZ}) \\ &= \overline{X}+(\overline{\overline{Y}\,\overline{Z}}\cdot\overline{YZ}) \\ &= \overline{X}+(Y+Z)(\overline{Y}+\overline{Z})\end{aligned}$$

A simpler method for deriving the complement of a function is to take the dual of the function and complement each literal. This method follows from the generalization of DeMorgan's theorem. Remember that the dual of an expression is obtained by interchanging AND and OR operations and 1's and 0's. To avoid confusion in handling complex functions, adding parentheses around terms before taking the dual is helpful.

EXAMPLE 2-2

Find the complements of the functions in Example 2-1 by taking their duals and complementing each literal.

$$F_1 = \overline{X}Y\overline{Z} + \overline{X}\,\overline{Y}Z = (\overline{X}Y\overline{Z}) + (\overline{X}\,\overline{Y}Z)$$

$$\text{The dual of } F_1\text{:} \quad (\overline{X} + Y + \overline{Z})(\overline{X} + \overline{Y} + Z)$$

$$\text{Complement each literal:} \quad (X + \overline{Y} + Z)(X + Y + \overline{Z}) = \overline{F}_1$$

$$F_2 = X(\overline{Y}\,\overline{Z} + YZ) = X((\overline{Y}\,\overline{Z}) + (YZ))$$

$$\text{The dual of } F_2\text{:} \quad X + (\overline{Y} + \overline{Z})(Y + Z)$$

$$\text{Complement each literal:} \quad \overline{X} + (Y + Z)(\overline{Y} + \overline{Z}) = \overline{F}_2$$

■

2-3 Standard Forms

A Boolean function can be written in a variety of ways when expressed algebraically. There are, however, a few ways of writing algebraic expressions that are considered to be standard forms. The standard forms facilitate the simplification procedures for Boolean expressions and frequently result in more desirable logic circuits.

The standard forms contain *product* terms and *sum* terms. An example of a product term is $X\overline{Y}Z$. This is a logical product consisting of an AND operation among three literals. An example of a sum term is $X + Y + \overline{Z}$. This is a logical sum consisting of an OR operation among the literals. It must be realized that the words "product" and "sum" do not imply arithmetic operations in Boolean algebra; instead, they specify the logical operations AND and OR, respectively.

Minterms and Maxterms

It has been shown that a truth table defines a Boolean function. An algebraic expression representing the function is derived from the table by finding the logical sum of all product terms for which the function assumes the binary value 1. A product term in which all the variables appear exactly once, either complemented or uncomplemented, is called a *minterm*. Its characteristic property is that it represents exactly one combination of the binary variables in a truth table. It has the value 1 for that combination and 0 for all others. There are 2^n distinct minterms for n variables. The four minterms for the two variables X and Y are $\overline{X}\,\overline{Y}$, $\overline{X}Y$, $X\overline{Y}$, and XY. The eight minterms for the three variables X, Y, and Z are listed in Table 2-6. The binary numbers from 000 to 111 are listed under the variables. For each binary combination, there is a related minterm. Each minterm is a product term of exactly three literals. A literal is a complemented variable if the corresponding bit of the related binary combination is 0 and is an uncomplemented variable if it is 1. A symbol m_j for each minterm is also shown in the table, where the subscript j denotes the decimal equivalent of the binary combination for which the minterm has the value 1. This list of minterms for any given n variables can be formed in a similar manner from a list of the binary numbers from 0 through $2^n - 1$. In addition, the truth table for each minterm is given in the right half of the table. These truth tables clearly show that each minterm is 1 for the corresponding binary com-

□ **TABLE 2-6**
Minterms for Three Variables

X	Y	Z	Product Term	Symbol	m_0	m_1	m_2	m_3	m_4	m_5	m_6	m_7
0	0	0	$\overline{X}\overline{Y}\overline{Z}$	m_0	1	0	0	0	0	0	0	0
0	0	1	$\overline{X}\overline{Y}Z$	m_1	0	1	0	0	0	0	0	0
0	1	0	$\overline{X}Y\overline{Z}$	m_2	0	0	1	0	0	0	0	0
0	1	1	$\overline{X}YZ$	m_3	0	0	0	1	0	0	0	0
1	0	0	$X\overline{Y}\overline{Z}$	m_4	0	0	0	0	1	0	0	0
1	0	1	$X\overline{Y}Z$	m_5	0	0	0	0	0	1	0	0
1	1	0	$XY\overline{Z}$	m_6	0	0	0	0	0	0	1	0
1	1	1	XYZ	m_7	0	0	0	0	0	0	0	1

□ **TABLE 2-7**
Maxterms for Three Variables

X	Y	Z	Sum Term	Symbol	M_0	M_1	M_2	M_3	M_4	M_5	M_6	M_7
0	0	0	$X+Y+Z$	M_0	0	1	1	1	1	1	1	1
0	0	1	$X+Y+\overline{Z}$	M_1	1	0	1	1	1	1	1	1
0	1	0	$X+\overline{Y}+Z$	M_2	1	1	0	1	1	1	1	1
0	1	1	$X+\overline{Y}+\overline{Z}$	M_3	1	1	1	0	1	1	1	1
1	0	0	$\overline{X}+Y+Z$	M_4	1	1	1	1	0	1	1	1
1	0	1	$\overline{X}+Y+\overline{Z}$	M_5	1	1	1	1	1	0	1	1
1	1	0	$\overline{X}+\overline{Y}+Z$	M_6	1	1	1	1	1	1	0	1
1	1	1	$\overline{X}+\overline{Y}+\overline{Z}$	M_7	1	1	1	1	1	1	1	0

bination and 0 for all other combinations. Such truth tables will be helpful later in using minterms to form Boolean expressions.

A sum term that contains all the variables in complemented or uncomplemented form is called a *maxterm*. Again, it is possible to formulate 2^n maxterms with n variables. The eight maxterms for three variables are listed in Table 2-7. Each maxterm is a logical sum of the three variables, with each variable being complemented if the corresponding bit of the binary number is 1 and uncomplemented if it is 0. The symbol for a maxterm is M_j, where j denotes the decimal equivalent of the binary combination for which the maxterm has the value 0. In the right half of the table, the truth table for each maxterm is given. Note that the value of the maxterm is 0 for the corresponding combination and 1 for all other combinations. It is now clear where the terms "minterm" and "maxterm" come from: a minterm is a function, not equal to 0, having the minimum number of 1's in its truth table; a maxterm is a function, not equal to 1, having the maximum of 1's in its truth table. Note from Table 2-6 and Table 2-7 that a minterm and maxterm with the same subscript are the complements of each other; that is, $M_j = \overline{m_j}$. For example, for $j = 3$, we have

□ **TABLE 2-8**
Boolean Functions of Three Variables

(a) X	Y	Z	F	$\overline{F}$	(b) X	Y	Z	E
0	0	0	1	0	0	0	0	1
0	0	1	0	1	0	0	1	1
0	1	0	1	0	0	1	0	1
0	1	1	0	1	0	1	1	0
1	0	0	0	1	1	0	0	1
1	0	1	1	0	1	0	1	1
1	1	0	0	1	1	1	0	0
1	1	1	1	0	1	1	1	0

$$\overline{m_3} = \overline{\overline{X}YZ} = X + \overline{Y} + \overline{Z} = M_3$$

A Boolean function can be expressed algebraically from a given truth table by forming the logical sum of all the minterms that produce a 1 in the function. This expression is called a *sum of minterms*. Consider the Boolean function F in Table 2-8(a). The function is equal to 1 for each of the following binary combinations of the variables X, Y, and Z: 000, 010, 101 and 111. These combinations correspond to minterms 0, 2, 5, and 7. By examining Table 2-8 and the truth tables for these minterms in Table 2-6, it is evident that the function F can be expressed algebraically as the logical sum of the stated minterms:

$$F = \overline{X}\,\overline{Y}\,\overline{Z} + \overline{X}Y\overline{Z} + X\overline{Y}Z + XYZ = m_0 + m_2 + m_5 + m_7$$

This can be further abbreviated by listing only the decimal subscripts of the minterms:

$$F(X, Y, Z) = \Sigma m(0, 2, 5, 7)$$

The symbol Σ stands for the logical sum (Boolean OR) of the minterms. The numbers following it represent the minterms of the function. The letters in parentheses following F form a list of the variables in the order taken when the minterms are converted to product terms.

Now consider the complement of a Boolean function. The binary values of $\overline{F}$ in Table 2-8(a) are obtained by changing 1's to 0's and 0's to 1's in the values of F. Taking the logical sum of minterms of $\overline{F}$, we obtain

$$\overline{F} = \overline{X}\,\overline{Y}Z + \overline{X}YZ + X\overline{Y}\,\overline{Z} + XY\overline{Z} = m_1 + m_3 + m_4 + m_6$$

or, in abbreviated form,

$$\overline{F}(X, Y, Z) = \Sigma m(1, 3, 4, 6)$$

Note that the minterm numbers for $\overline{F}$ are the ones missing from the list of the minterm numbers of F. We now take the complement of $\overline{F}$ to obtain F:

$$F = \overline{m_1 + m_3 + m_4 + m_6} = \overline{m_1} \cdot \overline{m_3} \cdot \overline{m_4} \cdot \overline{m_6}$$

$$= M_1 \cdot M_3 \cdot M_4 \cdot M_6 \text{ (Since } \overline{m_j} = M_j)$$

$$= (X + Y + \overline{Z})(X + \overline{Y} + \overline{Z})(\overline{X} + Y + Z)(\overline{X} + \overline{Y} + Z)$$

This shows the procedure for expressing a Boolean function as a *product of maxterms.* The abbreviated form for this product is

$$F(X, Y, Z) = \Pi M(1, 3, 4, 6)$$

where symbol Π denotes the logical product (Boolean AND) of the maxterms whose numbers are listed in parentheses. Note that the decimal numbers included in the product of maxterms will always be the same as the minterm list of the complemented function, such as (1, 3, 4, 6) in the foregoing example. Maxterms are seldom used directly when dealing with Boolean functions, as we can always replace them with the minterm list of $\overline{F}$.

The following is a summary of the most important properties of minterms:

1. There are 2^n minterms for n Boolean variables. These minterms can be evaluated from the binary numbers from 0 to $2^n - 1$.
2. Any Boolean function can be expressed as a logical sum of minterms.
3. The complement of a function contains those minterms not included in the original function.
4. A function that includes all the 2^n minterms is equal to logic 1.

A function that is not in the sum-of-minterms form can be converted to that form by means of a truth table, since the truth table always specifies the minterms of the function. Consider, for example, the Boolean function

$$E = \overline{Y} + \overline{X}\,\overline{Z}$$

The expression is not in sum-of-minterms form, because each term does not contain all three variables X, Y, and Z. The truth table for this function is listed in Table 2-8(b). From the table, we obtain the minterms of the function:

$$E(X, Y, Z) = \Sigma m(0, 1, 2, 4, 5)$$

The minterms for the complement of E are given by

$$\overline{E}(X, Y, Z) = \Sigma m(3, 6, 7)$$

Note that the total number of minterms in E and $\overline{E}$ is equal to eight, since the function has three variables, and three variables produce a total of 8 minterms. With 4 variables, there will be a total of 16 minterms, and for two variables, there will be 4 minterms. An example of a function that includes all the minterms is

$$G(X, Y) = \Sigma m(0, 1, 2, 3) = 1$$

Since G is a function of two variables and contains all four minterms, it is always equal to logic 1.

Sum of Products

The sum-of-minterms form is a standard algebraic expression that is obtained directly from a truth table. The expression so obtained contains the maximum number of literals in each term and usually has more product terms than necessary. This is because, by definition, each minterm must include all the variables of the function, complemented or uncomplemented. Once the sum of minterms is obtained from the truth table, the next step is to try to simplify the expression to see whether it is possible to reduce the number of product terms and the number of literals in the terms. The result is a simplified expression in *sum-of-products* form. This is an alternative standard form of expression that contains product terms with one, two, or any number of literals. An example of a Boolean function expressed as a sum of products is

$$F = \overline{Y} + \overline{X}Y\overline{Z} + XY$$

The expression has three product terms, the first with one literal, the second with three literals, and the third with two literals.

The logic diagram for a sum of products form consists of a group of AND gates followed by a single OR gate, as shown in Figure 2-5. Each product term requires an AND gate, except for a term with a single literal. The logical sum is formed with an OR gate that has single literals and the outputs of the AND gates as inputs. It is assumed that the input variables are directly available in their complemented and uncomplemented forms, so inverters are not included in the diagram. The AND gates followed by the OR gate form a circuit configuration referred to as a *two-level implementation*.

If an expression is not in sum-of-products-form, it can be converted to the standard form by means of the distributive laws. Consider, for example, the expression

$$F = AB + C(D + E)$$

This is not in sum-of-products form, because the term D + E is part of a product, but is not a single literal. The expression can be converted to a sum of products by applying the appropriate distributive law as follows:

$$F = AB + C(D + E) = AB + CD + CE$$

The function F is implemented in a nonstandard form in Figure 2-6(a). This requires two AND gates and two OR gates. There are three levels of gating in the

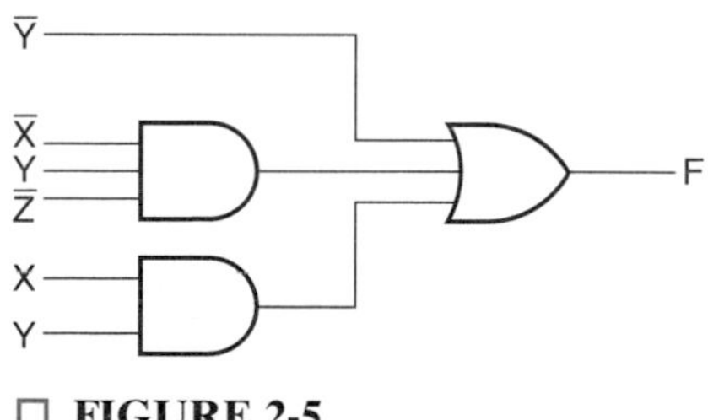

□ **FIGURE 2-5**
Sum-of-Products Implementation

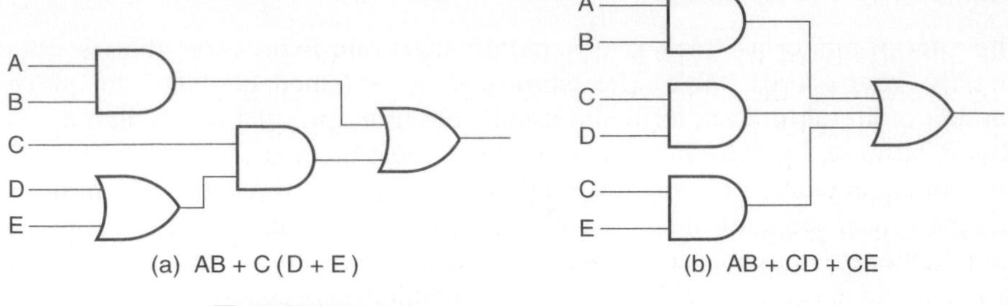

□ **FIGURE 2-6**
Three-Level and Two-Level Implementation

circuit. F is implemented in sum-of-products form in Figure 2-6(b). This circuit requires three AND gates and an OR gate and uses two levels of gating. The decision as to whether to use a two-level or multiple-level (three levels or more) implementation is complex. Among the issues involved are the numbers of gates and gate inputs and the amount of delay between the time the input values are set and the time the resulting output values appear. Also, as we will see in Chapter 6, two-level implementations are the natural form for certain implementation technologies.

Product of Sums

Another standard form of expressing Boolean functions algebraically is the *product of sums*. This form is obtained by forming a logical product of sum terms. Each logical sum term may have any number of distinct literals. An example of a function expressed in product-of-sums form is

$$F = X(\overline{Y} + Z)(X + Y + \overline{Z})$$

This expression has sum terms of one, two, and three literals. The sum terms perform an OR operation, and the product is an AND operation.

The gate structure of the product-of-sums expression consists of a group of OR gates for the sum terms (except for a single literal terms), followed by an AND gate. This is shown in Figure 2-7 for the preceding function F. As with the sum of products, this standard type of expression results in a two-level gating structure.

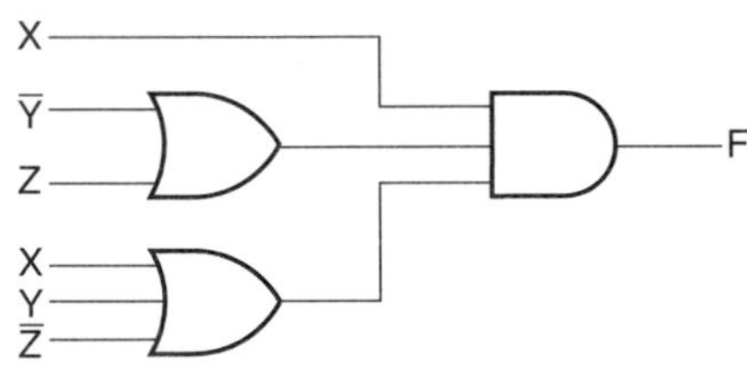

□ **FIGURE 2-7**
Product-of-Sums Implementation

2-4 Map Simplification

The complexity of the digital logic gates that implement a Boolean function is directly related to the algebraic expression from which the function is implemented. Although the truth table representation of a function is unique, when expressed algebraically, the function appears in many different forms. Boolean expressions may be simplified by algebraic manipulation as discussed in Section 2-2. However, this procedure of simplification is awkward because it lacks specific rules to predict each succeeding step in the manipulative process and it is difficult to determine whether the simplest expression has been achieved. By contrast, the map method provides a straightforward procedure for simplifying Boolean functions of up to four variables. Maps for five and six variables can be drawn as well, but are more cumbersome to use. The map is also known as the *Karnaugh map*, or *K-map*. The map is a diagram made up of squares, with each square representing one minterm of the function. Since any Boolean function can be expressed as a sum of minterms, it follows that a Boolean function is recognized graphically in the map by those squares whose minterms are included in the function. In fact, the map presents a visual diagram of all possible ways a function may be expressed in a standard form. By recognizing various patterns, the user can derive alternative algebraic expressions for the same function, from which the simplest can be selected.

The simplified expressions produced by the map are always in sum-of-products or product-of-sums form. Thus, maps handle simplification for two-level implementations, but do not apply directly to possible simpler implementations for the general case with three or more levels. It will be assumed that the simplest algebraic expression is one with a minimum number of terms and with the fewest possible number of literals in each term. This produces a two-level implementation having a logic circuit diagram with a minimum number of gates and the minimum number of inputs to the gates. We will see subsequently that the simplest expression is not necessarily unique. It is sometimes possible to find two or more expressions that satisfy the simplification criterion. In that case, either solution is satisfactory. This section covers only sum-of-products simplification. In the next section, we will show how to perform product-of-sums simplification.

Two-Variable Map

There are four minterms for a Boolean function with two variables. Hence, the two-variable map consists of four squares, one for each minterm, as shown in Figure 2-8(a). The map is redrawn in Figure 2-8(b) to show the relationship between the squares and the two variables X and Y. The 0 and 1 marked on the left side and the top of the map designate the values of the variables. The variable X appears complemented in row 0 and uncomplemented in row 1. Similarly, Y appears complemented in column 0 and uncomplemented in column 1. Note that the four combinations of these binary values correspond to the truth table rows associated with the four minterms.

A function of two variables can be represented in a map by marking the squares that correspond to the minterms of the function. As an example, the func-

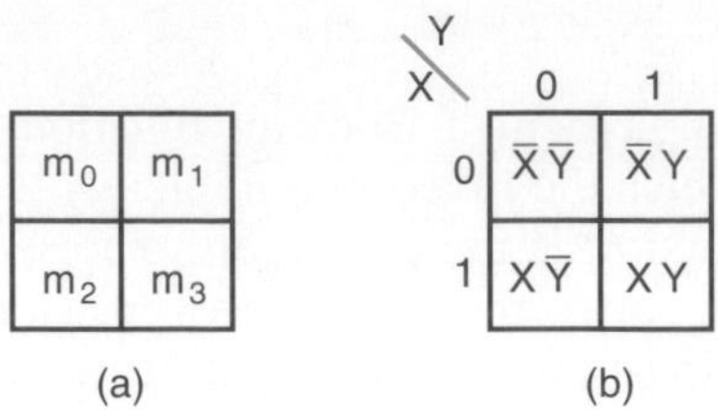

□ **FIGURE 2-8**
Two-Variable Map

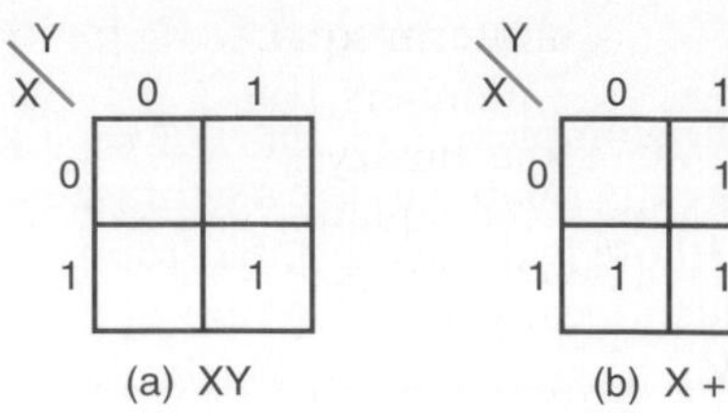

□ **FIGURE 2-9**
Representation of Functions in the Map

tion XY is shown in Figure 2-9(a). Since XY is equal to minterm m_3, a 1 is placed inside the square that belongs to m_3. Figure 2-9(b) shows the map for the logical sum of three minterms

$$m_1 + m_2 + m_3 = \overline{X}Y + X\overline{Y} + XY = X + Y$$

The simplified expression $X + Y$ is determined from the two-square area for the variable X in the second row and the two-square area for Y in the second column. Together, these two areas enclose the three squares belonging to X or Y. This simplification can be justified by algebraic manipulation:

$$\overline{X}Y + X\overline{Y} + XY = \overline{X}Y + X(\overline{Y} + Y) = (\overline{X} + X)(Y + X) = X + Y$$

The exact procedure for combining squares in the map will be clarified in the examples that follow.

Three-Variable Map

There are eight minterms for three binary variables. Therefore, a three-variable map consists of eight squares, as shown in Figure 2-10. The map drawn in part (b) is marked with binary numbers in each row and each column to show the binary values of the minterms. Note that the numbers along the columns do not follow the binary count sequence. The characteristic of the listed sequence is that only one bit changes in value from one adjacent column to the next.

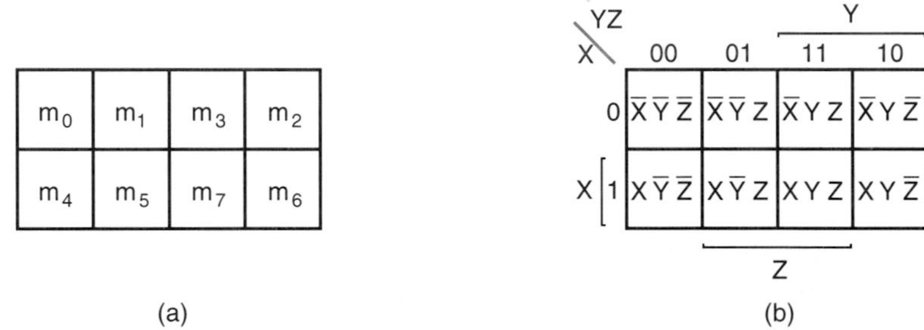

□ **FIGURE 2-10**
Three-Variable Map

A minterm square can be located in the map in two ways. First, we can memorize the numbers listed in Figure 2-10(a) for each minterm location, or we can refer to the binary numbers along the rows and columns in Figure 2-10(b). For example, the square assigned to m_5 corresponds to row 1 and column 01. When these two numbers are combined, they give the binary number 101, whose decimal equivalent is 5.

Another way of looking at square $m_5 = X\overline{Y}Z$ is to consider it to be in the row marked X and the column belonging to $\overline{Y}Z$ (column 01). Note that there are four squares where each variable is equal to 1 and four where each is equal to 0. The variable appears uncomplemented in the four squares where it is equal to 1 and complemented in the four squares where it is equal to 0. For convenience, we write the variable name along the four squares where it is uncomplemented. After one becomes familiar with maps, the use of the variable names alone is sufficient to label the map regions. To this end, it is important to note the location of these labels to obtain all minterms on the map.

In the two-variable map, the function XY demonstrated that a function or a term for a function can consist of a single square of the map. But to achieve simplification, we need to consider multiple squares corresponding to product terms. To understand how combining squares simplifies Boolean functions, we must recognize the basic property possessed by adjacent squares: Any two adjacent squares placed horizontally or vertically (but not diagonally) to form a rectangle correspond to minterms that differ in only a single variable. The single variable appears uncomplemented in one square and complemented in the other. For example, m_5 and m_7 lie in two adjacent squares. Variable Y is complemented in m_5 and uncomplemented in m_7, while the other two variables match in both squares. The logical sum of two such adjacent minterms can be simplified into a single product term of two variables:

$$m_5 + m_7 = X\overline{Y}Z + XYZ = XZ(\overline{Y} + Y) - XZ$$

Here the two squares differ in the variable Y, which can be removed when the logical sum (OR) of the two minterms is formed. Thus, on a 3-variable map, any two minterms in adjacent squares that are ORed together produce a product term of two variables.

EXAMPLE 2-3

Simplify the Boolean function

$$F(X, Y, Z) = \Sigma m(2, 3, 4, 5)$$

First, a 1 is marked in each minterm that represents the function. This is shown in Figure 2-11, where the squares for minterms 010, 011, 100, and 101 are marked with 1's. The next step is to explore collections of squares on the map representing product terms to be considered for the simplified expression. We call such objects *rectangles*, since their shape is that of a rectangle (including, of course, a square). Rectangles that correspond to product terms, however, are restricted to contain numbers of squares that are powers of 2, such as 1, 2, 4, 8, ... So our goal is to find

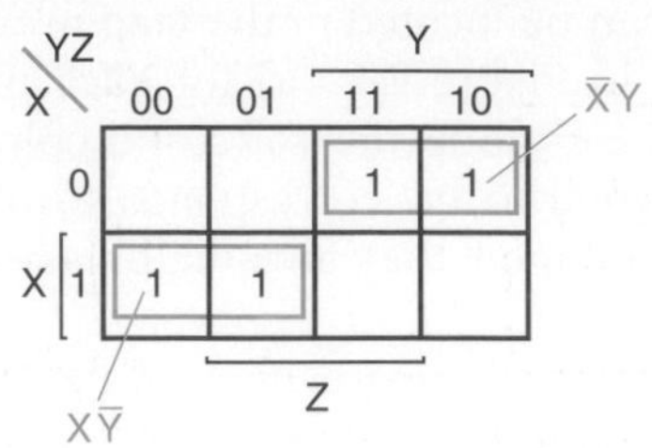

□ **FIGURE 2-11**
Map for Example 2-3: $F(X, Y, Z) = \Sigma m(2,3,4,5) = \overline{X}Y + X\overline{Y}$

the fewest such rectangles that include all of the minterms marked with 1's. This will give the fewest product terms. In the map in the figure, two rectangles enclose all four squares containing 1's. The upper right rectangle represents the product term $\overline{X}Y$. This is determined by observing that the rectangle is in row 0, corresponding to $\overline{X}$, and the last two columns, corresponding to Y. Similarly, the lower left rectangle represents the product term $X\overline{Y}$. (The second row represents X and the two left columns represent $\overline{Y}$.) Since these two rectangles include all of the 1's in the map, the logical sum of the corresponding two product terms gives the simplified expression for F:

$$F = \overline{X}Y + X\overline{Y}$$

In some cases, two squares in the map are adjacent and form a rectangle of size two, even though they do not touch each other. For example, in Figure 2-10, m_0 is adjacent to m_2 and m_4 is adjacent to m_6 because the minterms differ by one variable. This can be readily verified algebraically:

$$m_0 + m_2 = \overline{X}\,\overline{Y}\,\overline{Z} + \overline{X}Y\overline{Z} = \overline{X}\,\overline{Z}(\overline{Y} + Y) = \overline{X}\,\overline{Z}$$

$$m_4 + m_6 = X\overline{Y}\,\overline{Z} + XY\overline{Z} = X\overline{Z}(\overline{Y} + Y) = X\overline{Z}$$

The rectangles corresponding to these two product terms, $\overline{X}\,\overline{Z}$ and $X\overline{Z}$, are shown on the map in Figure 2-12(a). Based on the location of these rectangles, we must modify the definition of adjacent squares to include this and other, similar cases.

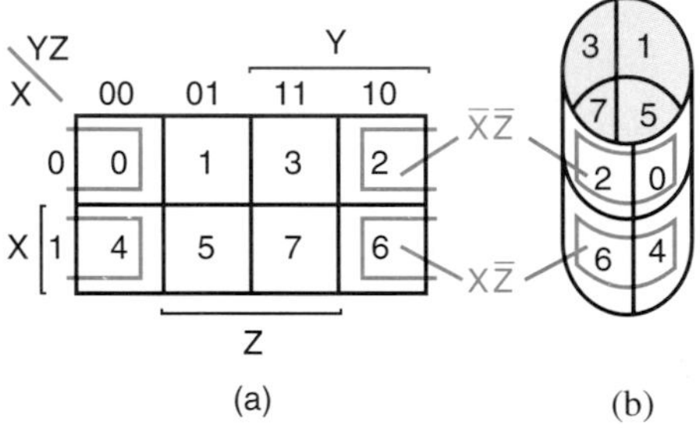

□ **FIGURE 2-12**
Three-Variable Map: Flat and on a Cylinder to Show Adjacent Squares

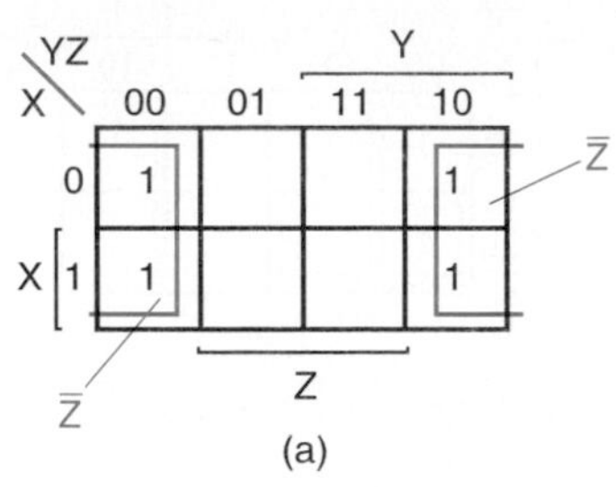

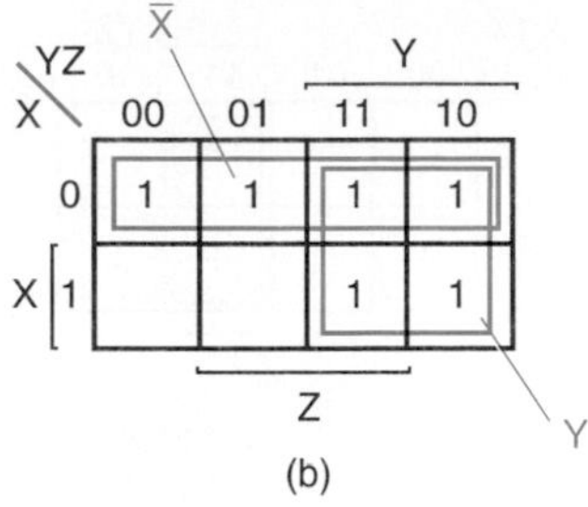

□ **FIGURE 2-13**
Product Terms Using Four Minterms

We do so by considering the map as being drawn on a cylinder, as shown in Figure 2-12(b), where the right and left edges touch each other to correctly establish minterm adjacencies and form the rectangles. In the maps in Figure 2-12, we have simply used numbers rather than *m*'s to represent the minterms. Both of these notations will be used freely.

A four-square rectangle represents a product term that is the logical sum of four minterms. For the three-variable case, such a product term is only one literal. As an example, the logical sum of the four adjacent minterms 0, 2, 4, and 6 reduces to a single literal term $\overline{Z}$:

$$m_0 + m_2 + m_4 + m_6 = \overline{X}\,\overline{Y}\,\overline{Z} + \overline{X}Y\overline{Z} + X\overline{Y}\,\overline{Z} + XY\overline{Z}$$

$$= \overline{X}\,\overline{Z}(\overline{Y} + Y) + X\overline{Z}(\overline{Y} + Y)$$

$$= \overline{X}\,\overline{Z} + X\overline{Z} = \overline{Z}(\overline{X} + X) = \overline{Z}$$

The rectangle for this product term is shown in Figure 2-13(a). Note that the product term $\overline{Z}$ uses the fact that the left and right edges of the map are adjacent in order to form the rectangle. Two other examples of rectangles corresponding to product terms derived from four minterms are shown in Figure 2-13(b).

In general, as more squares are combined, we obtain a product term with fewer literals. For three-variable maps:

One square represents a minterm of three literals.

A rectangle of two squares represents a product term of two literals.

A rectangle of four squares represents a product term of one literal.

A rectangle of eight squares encompasses the entire map and produces a function that is always equal to logic 1.

EXAMPLE 2-4

Simplify the following two Boolean functions:

$$F_1(X, Y, Z) = \Sigma m(3, 4, 6, 7)$$

$$F_2(X, Y, Z) = \Sigma m(0, 2, 4, 5, 6)$$

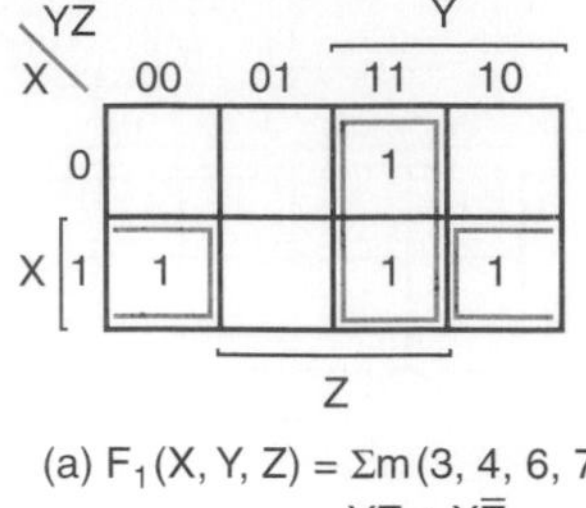

(a) $F_1(X, Y, Z) = \Sigma m(3, 4, 6, 7)$
$= YZ + X\overline{Z}$

(b) $F_2(X, Y, Z) = \Sigma m(0, 2, 4, 5, 6)$
$= \overline{Z} + X\overline{Y}$

□ **FIGURE 2-14**
Maps for Example 2-4

The map for F_1 is shown in Figure 2-14(a). There are four squares marked with 1's, one for each minterm of the function. Two adjacent squares are combined in the third column to give a two-literal term YZ. The remaining two squares with 1's are also adjacent by the cylinder-based definition and are shown in the diagram with their values enclosed in half rectangles. When combined, these two squares give the two-literal term $X\overline{Z}$. The simplified function thus becomes

$$F_1 = YZ + X\overline{Z}$$

The map for F_2 is shown in Figure 2-14(b). First, we immediately combine the four adjacent squares in the first and last columns based on what we learned from Figure 2-13, to give the single literal term $\overline{Z}$. The remaining single square representing minterm 5 is combined with an adjacent square that already is being used once. This is not only permissible, but desirable, since the two adjacent squares give the two-literal term $X\overline{Y}$, while the single square represents the three-literal minterm $X\overline{Y}Z$. The simplified function is

$$F_2 = \overline{Z} + X\overline{Y}$$ ■

On occasion there are alternative ways of combining squares to produce equally simplified expressions. An example of this is demonstrated in the map of Figure 2-15. Minterms 1 and 3 are combined to give the term $\overline{X}Z$, and minterms 4 and 6 produce the term $X\overline{Z}$. However, there are two ways that the square of minterm 5 can be combined with another adjacent square to produce a third two-literal term. Combining it with minterm 4 gives the term $X\overline{Y}$; combining it instead with minterm 1 gives the term $\overline{Y}Z$. Each of the two possible simplified expressions listed in Figure 2-15 has three terms of two literals each, so there are two possible simplified solutions for this function.

If a function is not expressed as a sum of minterms, we can use the map to obtain the minterms of the function and then simplify the function. It is necessary, however, to have the algebraic expression in sum-of-products form, from which each product term is plotted in the map. The minterms of the function are then read directly from the map. As an example, consider the Boolean function

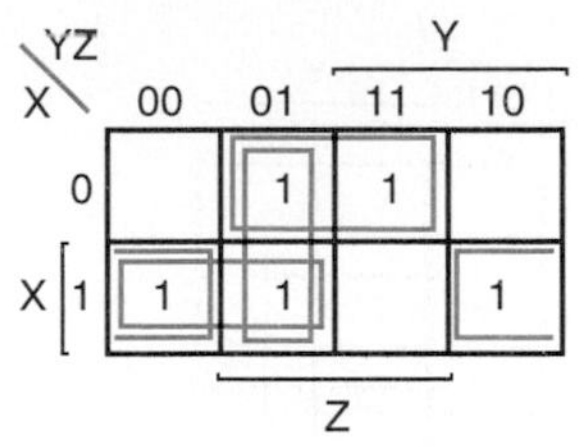

☐ **FIGURE 2-15**

$$
\begin{aligned}
F(X, Y, Z) &= \Sigma m(1,3,4,5,6) \\
&= \overline{X}Z + X\overline{Z} + X\overline{Y} \\
&= \overline{X}Z + X\overline{Z} + \overline{Y}Z
\end{aligned}
$$

☐ **FIGURE 2-16**

$F(X, Y, Z) = \Sigma m(1,2,3,5,7) = Z + \overline{X}Y$

$$F = \overline{X}Z + \overline{X}Y + X\overline{Y}Z + YZ$$

Three product terms in the expression have two literals and are represented in a three-variable map by two squares each. The two squares corresponding to the first term, $\overline{X}Z$, are found in Figure 2-16 from the coincidence of $\overline{X}$ (first row) and Z (two middle columns), to give squares 001 and 011. Note that when marking l's in the squares, it is possible to find a 1 already placed there from a preceding term. This happens with the second term, $\overline{X}Y$, which has 1's in squares 011 and 010; but square 011 is common with the first term, $\overline{X}Z$, so only one 1 is marked in it. Continuing in this fashion, we find that the function has five minterms, as indicated by the five l's in the figure. The minterms are read directly from the map to be 1, 2, 3, 5, and 7. The function as originally given has four product terms. It can be simplified on the map to only two simple terms, $F = Z + \overline{X}Y$, giving a significant reduction in the cost of implementation.

Four-Variable Map

There are 16 minterms for four binary variables, and therefore, a four-variable map consists of 16 squares, as shown in Figure 2-17. The minterm assignment in each square is indicated in part (a) of the diagram. The map is redrawn in (b) to show the relationship of the four variables. The rows and columns are numbered so that only one bit of the binary number changes in value between any two adjacent columns or rows, guaranteeing the same property for adjacent squares. The minterms corresponding to each square can be obtained by combining the row number with the column number. For example, when combined, the numbers in the third row (11) and the second column (01) give the binary number 1101, the binary equivalent of 13. Thus, the square in the third row and second column represents minterm m_{13}. In addition, each variable is marked on the map to show the eight squares in which it appears uncomplemented. The other eight squares, in which no label is indicated, correspond to the variable being complemented. Thus, W appears complemented in the first two rows and uncomplemented in the second two rows.

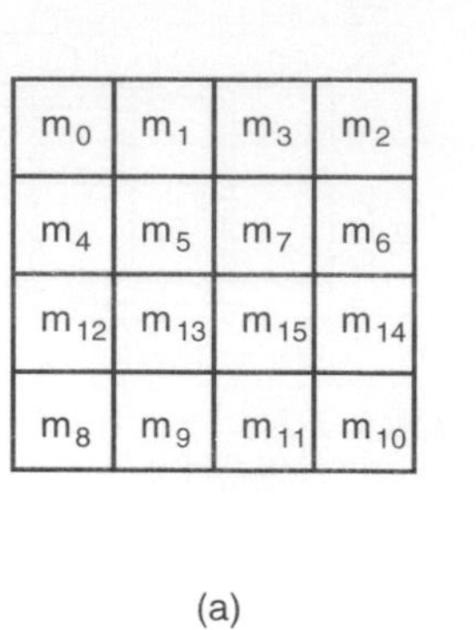

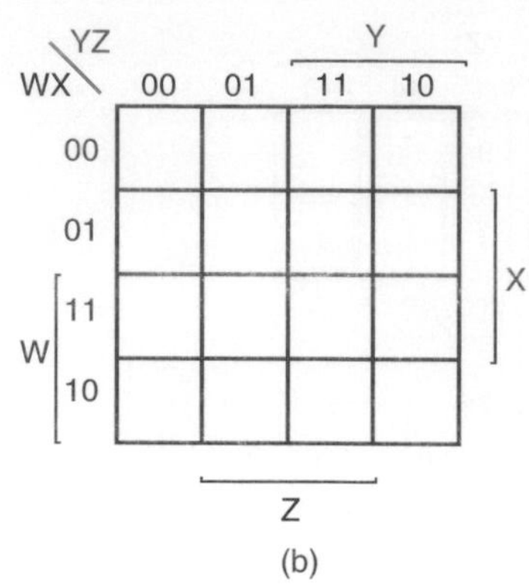

FIGURE 2-17
Four-Variable Map

The method used to simplify four-variable functions is similar to that used to simplify three-variable functions. Adjacent squares are defined to be squares next to each other, as for two- and three-variable maps. To show adjacencies between squares, the map of Figure 2-18(a) is drawn as a torus in Figure 2-18(b), with the top and bottom edges, as well as the right and left edges, touching each other to show adjacent squares. For example, m_0 and m_2 are two adjacent squares, as are m_0 and m_8. The combinations of squares that can be chosen during the simplification process in the four-variable map are as follows:

One square represents a minterm of four literals.

A rectangle of 2 squares represents a product term of three literals.

A rectangle of 4 squares represents a product term of two literals.

A rectangle of 8 squares represents a product term of one literal.

A rectangle of 16 squares produces a function that is always equal to logic 1.

No other combination of squares can be used. An interesting product term of two literals $\overline{X}\,\overline{Z}$, is shown in Figure 2-18. In (b), when the map is viewed as a torus, the adjacencies of the squares that represent this product term are clear, but in (a) these squares are on the four corners of the map and appear quite removed from each other. This product term is important to recall, since it is often missed. It also serves as a reminder that the left edge and the right edge of the map are adjacent, as are the top edge and the bottom edge. Thus, in general, rectangles on a map cross the left and right edges, top and bottom edges, or both.

The following examples show the procedure for simplifying four-variable Boolean functions.

EXAMPLE 2-5

Simplify the Boolean function

$$F(W, X, Y, Z) = \Sigma m(0, 1, 2, 4, 5, 6, 8, 9, 12, 13, 14)$$

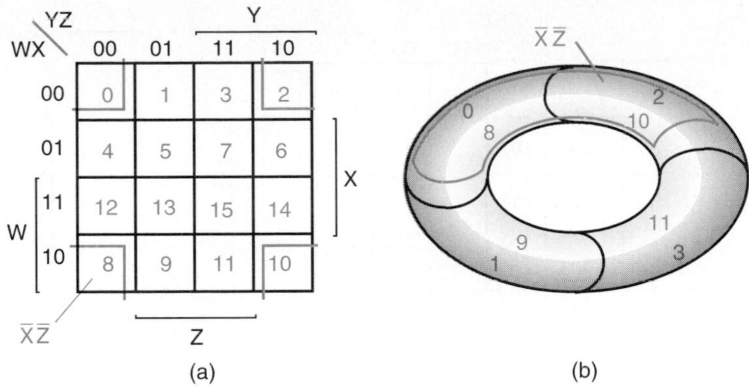

☐ **FIGURE 2-18**
Four-Variable Map: Flat and on a Torus to Show Adjacencies

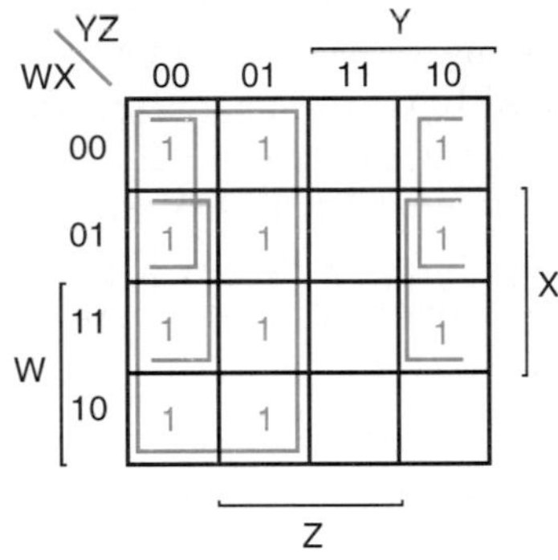

☐ **FIGURE 2-19**
Map for Example 2-5: $F = \overline{Y} + \overline{W}\,\overline{Z} + X\overline{Z}$

The minterms of the function are marked with 1's in the map of Figure 2-19. Eight squares in the two left columns are combined to form a rectangle for the one literal term, $\overline{Y}$. The remaining three 1's cannot be combined to give a simplified term; rather, they must be combined as two- or four square rectangles. The top two 1's on the right are combined with the top two 1's on the left to give the term $\overline{W}\,\overline{Z}$. Note again that it is permissible to use the same square more than once. We are now left with a square marked with a 1 in the third row and fourth column (minterm 1110). Instead of taking this square alone, which will give a term of four literals, we combine it with squares already used to form a rectangle of four squares in the two middle rows and the two end columns, giving the term $X\overline{Z}$. The simplified expression is the logical sum of the three terms:

$$F = \overline{Y} + \overline{W}\,\overline{Z} + X\overline{Z}$$ ■

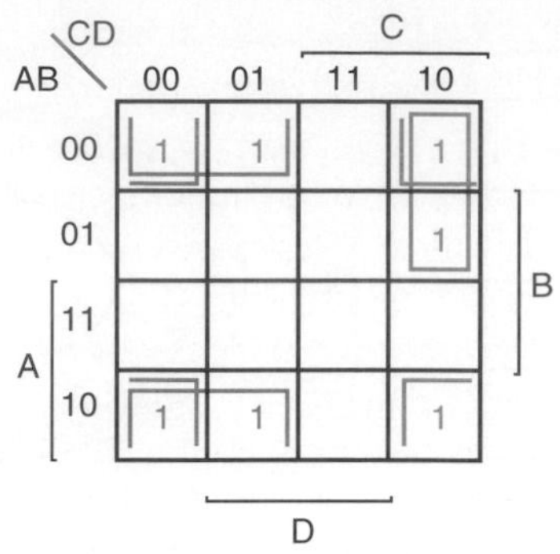

□ **FIGURE 2-20**
Map for Example 2-6: $F = \overline{B}\,\overline{D} + \overline{B}\,\overline{C} + \overline{A}C\overline{D}$

■ **EXAMPLE 2-6**

Simplify the Boolean function

$$F = \overline{A}\,\overline{B}\,\overline{C} + \overline{B}C\overline{D} + A\overline{B}\,\overline{C} + \overline{A}BC\overline{D}$$

This function has four variables: *A*, *B*, *C*, and *D*. It is expressed in sum-of-products form with three terms of three literals each and one term of four literals. The area in the map covered by the function is shown in Figure 2-20. Each term of three literals is represented in the map by two squares. $\overline{A}\,\overline{B}\,\overline{C}$ is represented by squares 0000 and 0001, $\overline{B}C\overline{D}$ by squares 0010 and 1010, and $A\overline{B}\,\overline{C}$ by squares 1000 and 1001. The term with four literals is minterm 0110. The function is simplified on the map by taking the 1's in the four corners, to give the term $\overline{B}\,\overline{D}$. This product term is in the same map locations as $\overline{X}\,\overline{Z}$ in Figure 2-18. The two 1's in the top row are combined with the two 1's in the bottom row to give the term $\overline{B}\,\overline{C}$. The remaining 1 in square, 0110, is combined with its adjacent square, 0010, to give the term $\overline{A}C\overline{D}$. The simplified function is thus

$$F = \overline{B}\,\overline{D} + \overline{B}\,\overline{C} + \overline{A}C\overline{D}$$ ■

2-5 MAP MANIPULATION

When combining squares in a map, it is necessary to ensure that all the minterms of the function are included. At the same time, it is necessary to minimize the number of terms in the simplified function by avoiding any redundant terms whose minterms are already included in other terms. In this section, we consider a manipulative procedure that facilitates the recognition of correct patterns in the map. Other topics to be covered are the simplification of products of sums and the simplification of incompletely specified functions.

Essential Prime Implicants

The procedure for combining squares in a map may be made more systematic if we introduce the terms "implicant," "prime implicant," and "essential prime implicant." A product term is an *implicant* of a function if the function has the value 1 for all minterms of the product term. Clearly, all rectangles on a map made up of squares containing 1's correspond to implicants. If the removal of any literal from an implicant P results in a product term that is not an implicant of the function, then P is a *prime implicant*. On a map for an n-variable function, the set of prime implicants corresponds to the set of all rectangles made up of 2^m squares containing 1's ($m = 0, 1, \ldots n$), with each rectangle containing as many squares as possible.

If a minterm of a function is included in only one prime implicant, that prime implicant is said to be *essential*. Thus, if a square containing a 1 is in only one rectangle representing a prime implicant, then that prime implicant is essential. In Figure 2-15, the terms $\overline{X}Z$ and $X\overline{Z}$ are essential prime implicants, and the terms $X\overline{Y}$ and $\overline{Y}Z$ are nonessential prime implicants.

The prime implicants of a function can be obtained from a map of the function as all possible maximum collections of 2^m squares containing 1's ($m = 0, 1, \ldots n$), that constitute rectangles. This means that a single 1 on a map represents a prime implicant if it is not adjacent to any other 1's. Two adjacent 1's form a rectangle representing a prime implicant, provided that they are not within a rectangle of four or more squares containing 1's. Four 1's form a rectangle representing a prime implicant if they are not within a rectangle of eight or more squares containing 1's, and so on. The essential prime implicants each contain a square in only one rectangle representing a prime implicant.

The systematic procedure for finding the simplified expression from the map requires that we first determine all prime implicants. Then, the simplified expression is obtained from the logical sum of all the essential prime implicants, plus other prime implicants needed to include remaining minterms not included in the essential prime implicants. This procedure will be clarified by a few examples.

Consider the map of Figure 2-21. There are three ways that we can combine four squares into rectangles. The product terms obtained from these combinations are the prime implicants of the function, $\overline{A}D$, $B\overline{D}$, and $\overline{A}B$. The terms $\overline{A}D$ and $B\overline{D}$ are essential prime implicants, but $\overline{A}B$ is not essential. This is because minterms 1 and 3 can be included only in the term $\overline{A}D$, and minterms 12 and 14 can be included only in the term $B\overline{D}$. But minterms 4, 5, 6, and 7 are each included in two prime implicants, one of which is $\overline{A}B$, so the term $\overline{A}B$ is not an essential prime implicant. In fact, once the essential prime implicants are chosen, the third term is not needed because all the minterms are already included in the essential prime implicants. The simplified expression for the function of Figure 2-21 is

$$F = \overline{A}D + B\overline{D}$$

A second example is shown in Figure 2-22. The function plotted in part (a) has seven minterms. If we try to combine squares, we will find that there are six prime implicants. In order to obtain a minimum number of terms for the function, we must first determine the prime implicants that are essential. As shown in part

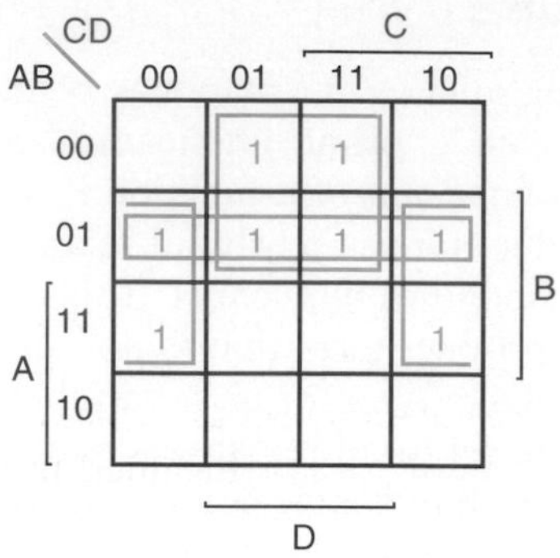

FIGURE 2-21
Prime Implicants: $\overline{A}D$, $B\overline{D}$, and $\overline{A}B$

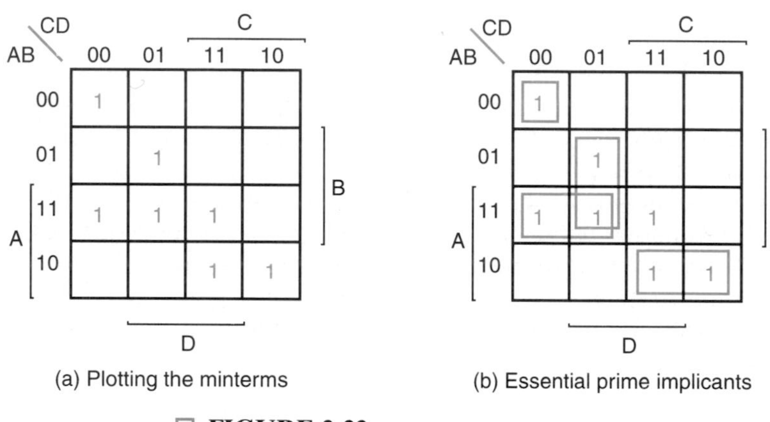

FIGURE 2-22
Simplification with Prime Implicants

(b) of the figure, the function has four essential prime implicants. The product term $\overline{A}\,\overline{B}\,\overline{C}\,\overline{D}$ is essential because it is the only prime implicant that includes minterm 0. Similarly, the product terms $B\overline{C}D$, $AB\overline{C}$, and $A\overline{B}C$ are essential prime implicants because they are the only prime implicants that include minterms 5, 12, and 10, respectively. Minterm 15 is included in two nonessential prime implicants. The simplified expression for the function consists of the logical sum of the four essential prime implicants and one prime implicant that includes minterm 15:

$$F = \overline{A}\,\overline{B}\,\overline{C}\,\overline{D} + B\overline{C}D + AB\overline{C} + A\overline{B}C + \begin{cases} ACD \\ \text{or} \\ ABD \end{cases}$$

The identification of essential prime implicants in the map provides an additional tool which shows the terms that must absolutely appear in every sum-of-products expression for a function and provides a partial structure for a more systematic method for choosing patterns of squares.

Nonessential Prime Implicants

Beyond using all essential prime implicants, the following rule can be applied to include the remaining minterms of the function in nonessential prime implicants:

Selection Rule: Minimize the overlap among prime implicants as much as possible. In particular, in the final solution, make sure that each prime implicant selected includes at least one minterm not included in any other prime implicant selected.

In most cases, this will result in a simplified, although not necessarily minimum, sum-of-products expression. The use of the selection rule is illustrated in the following example:

EXAMPLE 2-7

Find a simplified sum-of-products form for $F(A, B, C, D) = \Sigma m\,(0, 1, 2, 4, 5, 10, 11, 13, 15)$.

The map for F is given in Figure 2-23, with all prime implicants shown. $\overline{A}\,\overline{C}$ is the only essential prime implicant. Using the preceding selection rule, we can choose the remaining prime implicants for the sum-of-products form in the order indicated by the numbers. Note how the prime implicants 1 and 2 are selected in order to include minterms without overlapping. Prime implicant 3 ($\overline{A}\,\overline{B}\,\overline{D}$) and prime implicant $\overline{B}C\overline{D}$ both include the one remaining minterm 0010, and prime implicant 3 is arbitrarily selected to include the minterm and complete the sum-of-products expression:

$$F(A, B, C, D) = \overline{A}\,\overline{C} + ABD + A\overline{B}C + \overline{A}\,\overline{B}\,\overline{D}$$

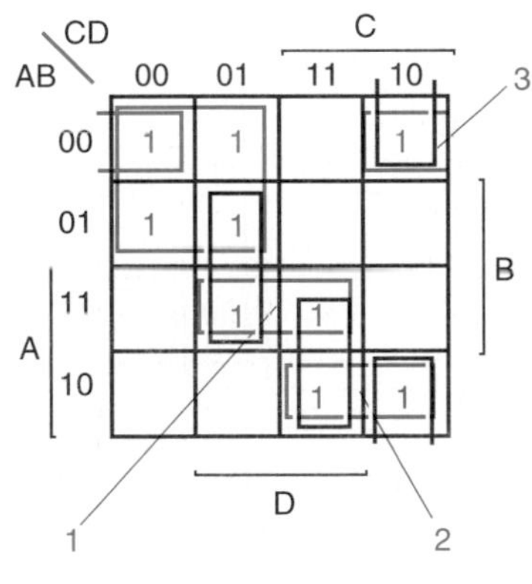

FIGURE 2-23
Map for Example 2-7

Product-of-Sums Simplification

The simplified Boolean functions derived from the maps in all of the previous examples were expressed in sum-of-products form. With only minor modification, the product-of-sums form can be obtained.

The procedure for obtaining a simplified expression in product-of-sums form follows from the basic properties of Boolean functions. The l's placed in the squares of the map represent the minterms of the function. The minterms not included in the function belong to the complement of the function. From this, we see that the complement of a function is represented in the map by the squares not marked by 1's. If we mark the empty squares with 0's and combine them into valid rectangles, we obtain a simplified expression of the complement of the function. We then take the complement of $\overline{F}$ to obtain the function F as a product of sums. This is done by taking the dual and complementing each literal, as described in Example 2-2.

EXAMPLE 2-8

Simplify the following Boolean function in product-of-sums form:

$$F(A,B,C,D) = \Sigma m(0,1,2,5,8,9,10)$$

The 1's marked in the map of Figure 2-24 represent the minterms of the function. The squares marked with 0's represent the minterms not included in F and therefore denote the complement of F. Combining the squares marked with 0's, we obtain the simplified complemented function

$$\overline{F} = AB + CD + B\overline{D}$$

Taking the dual and complementing each literal gives the complement of $\overline{F}$. This is F in product-of-sums form:

$$F = (\overline{A} + \overline{B})(\overline{C} + \overline{D})(\overline{B} + D)$$

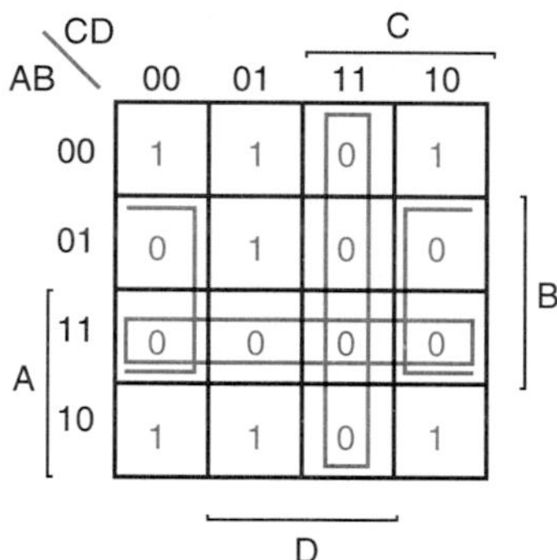

FIGURE 2-24
Map for Example 2-8: $F = (\overline{A} + \overline{B})(\overline{C} + \overline{D})(\overline{B} + D)$

The previous example shows the procedure for obtaining the product-of-sums simplification when the function is originally expressed as a sum of minterms. The procedure is also valid when the function is originally expressed as a product of maxterms or a product of sums. Remember that the maxterm numbers are the same as the minterm numbers of the complemented function, so 0's are entered in the map for the maxterms or for the complement of the function. To enter a function expressed as a product of sums into the map, we take the complement of the function and, from it, find the squares to be marked with 0's. For example, the function

$$F = (\overline{A} + \overline{B} + C)(B + D)$$

can be plotted in the map by first obtaining its complement

$$\overline{F} = AB\overline{C} + \overline{B}\,\overline{D}$$

and then marking 0's in the squares representing the minterms of $\overline{F}$. The remaining squares are marked with 1's. Then, combining the 1's gives the simplified expression in sum-of-products form. Combining the 0's and then complementing gives the simplified expression in product-of-sums form. Thus, for any function plotted on the map, we can derive the simplified function in either one of the two standard forms.

Don't-Care Conditions

The minterms of a Boolean function specify all combinations of variable values for which the function is equal to 1. The function is assumed to be equal to 0 for the rest of the minterms. This assumption, however, is not always valid, since there are applications in which the function is not specified for certain variable value combinations. There are two cases in which this occurs. In the first case, the input combinations never occur. As an example, the four-bit binary code for the decimal digits has six combinations that are not used and not expected to occur. In the second case, the input combinations are expected to occur, but we do not care what the outputs are in response to these combinations. In both cases, the outputs are said to be unspecified for the input combinations. Functions that have unspecified outputs for some input combinations are called *incompletely specified functions.* In most applications, we simply do not care what value is assumed by the function for the unspecified minterms. For this reason, it is customary to call the unspecified minterms of a function *don't-care conditions.* These conditions can be used on a map to provide further simplification of the function.

It should be realized that a don't-care minterm cannot be marked with a 1 on the map, because that would require that the function always be a 1 for such a minterm. Likewise, putting a 0 in the square requires the function to be 0. To distinguish the don't-care condition from 1's and 0's, an X is used. Thus, an X inside a square in the map indicates that we do not care whether the value of 0 or 1 is assigned to F for the particular minterm.

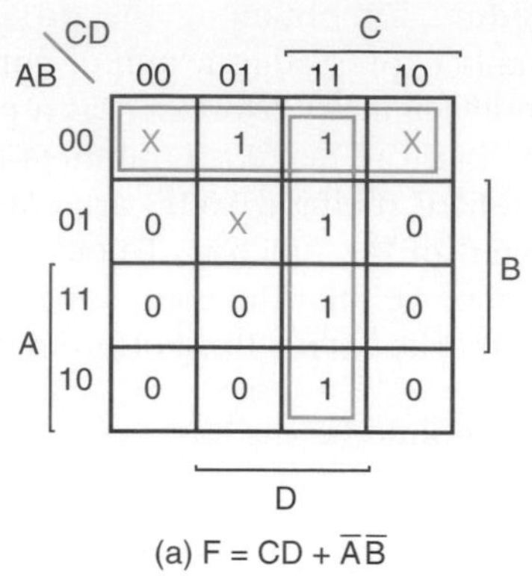

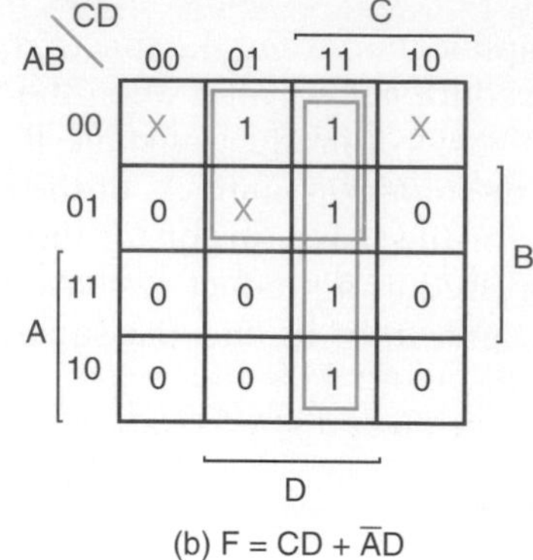

☐ **FIGURE 2-25**
Example with Don't-Care Conditions

In choosing adjacent squares to simplify the function in a map, the don't-care minterms may be assumed to be either 1 or 0. When simplifying the function, we can choose to include each don't-care minterm with either the 1's or the 0's, depending on which combination gives the simplest expression. In addition, a don't-care minterm need not be chosen at all if it does not contribute to producing a larger implicant. The choice depends entirely on the simplification that can be achieved.

To clarify the procedure for handling the don't-care conditions, consider the following incompletely specified function F that has three don't-care minterms d:

$$F(A, B, C, D) = \Sigma m(1, 3, 7, 11, 15)$$

$$d(A, B, C, D) = \Sigma m(0, 2, 5)$$

The minterms of F are the variable combinations that make the function equal to 1. The minterms of d are the don't-care minterms that may be assigned either 0 or 1. The map simplification is shown in Figure 2-25. The minterms of F are marked by 1's, those of d are marked by X's, and the remaining squares are filled with 0's. To get the simplified function in sum-of-products form, we must include all five 1's in the map, but we may or may not include any of the X's, depending on the way the function is simplified. The term CD includes the four minterms in the third column. The remaining minterm in square 0001 can be combined with square 0011 to give a three-literal term. However, by including one or two adjacent X's, we can combine four squares into a rectangle to give a two-literal term. In part (a) of the figure, don't-care minterms 0 and 2 are included with the 1's, which results in the simplified function

$$F = CD + \overline{A}\,\overline{B}$$

In part (b), don't-care minterm 5 is included with the 1's, and the simplified function now is

$$F = CD + \overline{A}D$$

The two expressions represent two functions that are algebraically unequal. Both include the specified minterms of the original incompletely specified function, but each includes different don't-care minterms. As far as the incompletely specified function is concerned, both expressions are acceptable. The only difference is in the value of F for the unspecified minterms.

It is also possible to obtain a simplified product-of-sums expression for the function of Figure 2-25. In this case, the way to combine the 0's is to include don't-care minterms 0 and 2 with the 0's, to give the simplified complemented function

$$\overline{F} = \overline{D} + A\overline{C}$$

Taking the complement of $\overline{F}$ gives the simplified expression in product-of-sums form:

$$F = D(\overline{A} + C)$$

The foregoing examples show that the don't-care minterms in the map are initially considered as representing a choice of 1 or 0. The choice is made depending on the way we want to simplify the incompletely specified function. However, once the choice is made, the simplified function will have specific values for all the minterms of the original function, including those that were initially unspecified. Thus, although the outputs in the initial specification may contain X's, the outputs in the implementation are only 0's and 1's.

2-6 NAND AND NOR GATES

Since Boolean functions are expressed in terms of AND, OR, and NOT operations, it is a straightforward procedure to implement a Boolean function with AND, OR, and NOT gates. We find, however, that the possibility of constructing gates with other logic operations is of practical interest. Factors to be taken into consideration when constructing other types of gates are the feasibility and economy of producing the gate with electronic components, the possibility of extending the gate to more than two inputs, and the ability of the gate to implement Boolean functions alone or in conjunction with other gates.

In addition to AND and OR gates, other logic gates are used extensively in the design of digital circuits. The graphics symbols and truth tables of eight logic gates are shown in Figure 2-26. The gates are shown with two binary input variables, X and Y, and one output binary variable, F. Two graphics symbols are drawn for each gate. The rectangular symbols on the right are recommended by the Institute of Electrical and Electronics Engineers' (IEEE) *Standard Graphic Symbols for Logic Functions* (IEEE Standard 91–1984). The distinctively shaped symbols on the left are considered an acceptable alternative within the standard. The rectangular symbols were more convenient for use when computer graphics capabilities were limited. With contemporary graphics, however, the alternative symbols

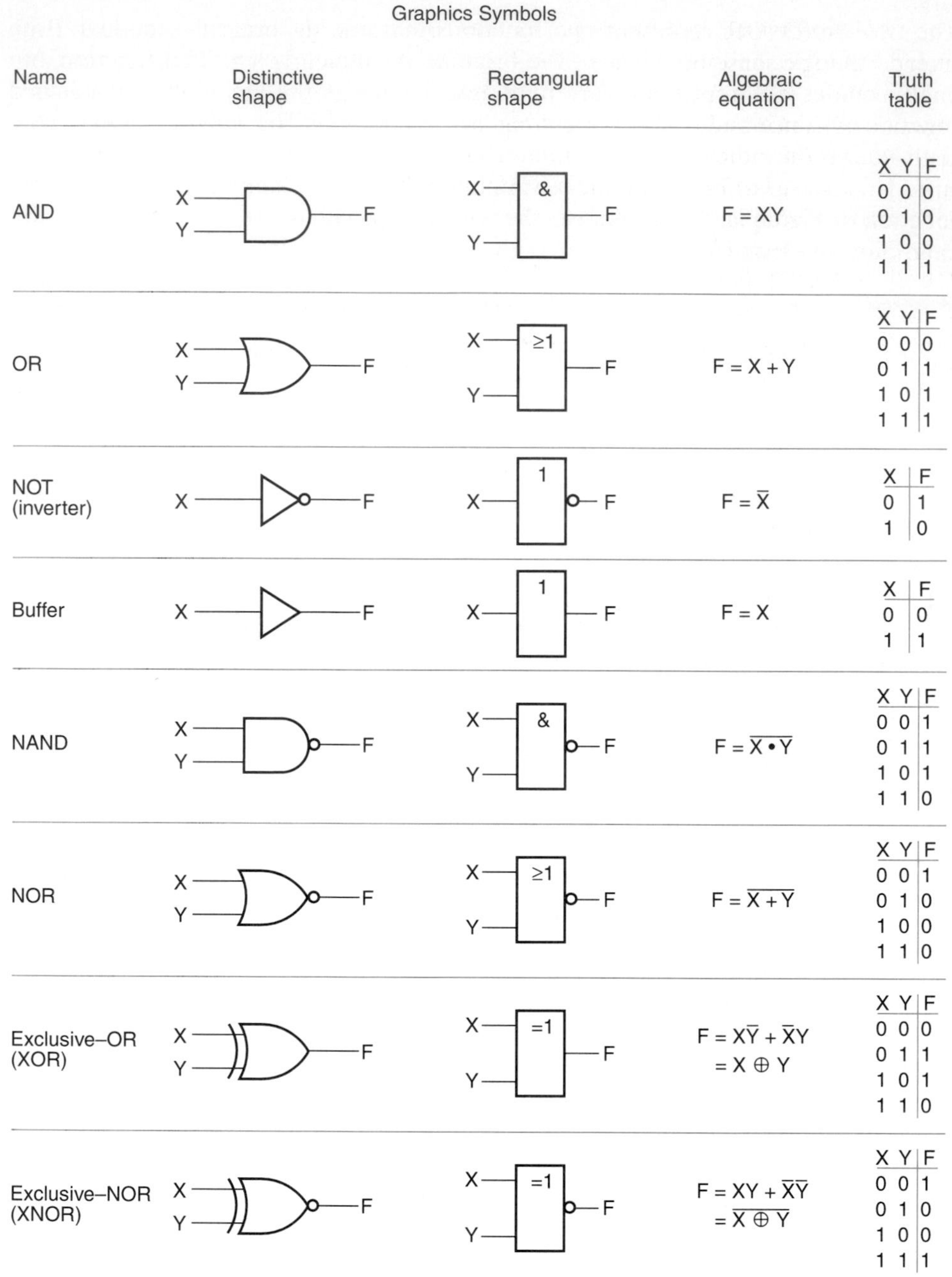

Graphics Symbols

Name	Distinctive shape	Rectangular shape	Algebraic equation	Truth table
AND	X, Y → F	&	$F = XY$	X Y | F 0 0 | 0 0 1 | 0 1 0 | 0 1 1 | 1
OR	X, Y → F	≥1	$F = X + Y$	X Y | F 0 0 | 0 0 1 | 1 1 0 | 1 1 1 | 1
NOT (inverter)	X → F	1	$F = \overline{X}$	X | F 0 | 1 1 | 0
Buffer	X → F	1	$F = X$	X | F 0 | 0 1 | 1
NAND	X, Y → F	&	$F = \overline{X \cdot Y}$	X Y | F 0 0 | 1 0 1 | 1 1 0 | 1 1 1 | 0
NOR	X, Y → F	≥1	$F = \overline{X + Y}$	X Y | F 0 0 | 1 0 1 | 0 1 0 | 0 1 1 | 0
Exclusive–OR (XOR)	X, Y → F	=1	$F = X\overline{Y} + \overline{X}Y$ $= X \oplus Y$	X Y | F 0 0 | 0 0 1 | 1 1 0 | 1 1 1 | 0
Exclusive–NOR (XNOR)	X, Y → F	=1	$F = XY + \overline{X}\overline{Y}$ $= \overline{X \oplus Y}$	X Y | F 0 0 | 1 0 1 | 0 1 0 | 0 1 1 | 1

FIGURE 2-26
Digital Logic Gates

are almost universally used both for computer graphics and in the technical literature.

The AND, OR, and NOT gates were defined previously. The NOT circuit inverts the logic sense of a binary signal to produce the complement operation. The small circle at the output of the graphic symbol of a NOT gate is formally called a *negation indicator* and designates the logical complement. We will informally refer to the negation indicator as a bubble. The triangle symbol by itself designates a buffer circuit. A *buffer* produces the logical function $Z = X$, since the binary value of the output is equal to the binary value of the input. This circuit is used primarily to amplify an electrical signal.

The NAND gate represents the complement of the AND operation. Its name is an abbreviation of NOT AND. The graphics symbol for the NAND gate consists of an AND symbol with a bubble on the output, denoting the complement operation. The NOR gate (an abbreviation of NOT OR) is the complement of the OR operation and is symbolized by an OR graphics symbol with a bubble on the output. NAND and NOR gates are used extensively as standard logic gates and are in fact far more popular than AND and OR gates. This is because NAND and NOR gates are natural functions for the simplest electronic circuits.

The exclusive-OR (XOR) gate is similar to the OR gate, but excludes (has the value 0 for) the combination with both X and Y equal to 1. The graphics symbol for the XOR gate is similar to that for the OR gate, except for the additional curved line on the inputs. The exclusive-OR has the special symbol $\oplus$ to designate its operation. The exclusive-NOR is the complement of the exclusive-OR, as indicated by the bubble at the output of its graphics symbol.

In the rest of this section, we will investigate the implementation of digital circuits with NAND and NOR gates. The XOR gate is discussed in the next section.

NAND Circuits

The NAND gate is said to be a universal gate because any digital system can be implemented with NAND gates alone. To prove that any Boolean function can be implemented with NAND gates, we need only show that the logical operations of AND, OR, and NOT can be obtained with NAND gates only. This is done in Figure 2-27. The complement operation is obtained from a one-input NAND gate that behaves exactly like a NOT gate. In fact, the one-input NAND is an invalid symbol and is replaced by the NOT symbol, as shown in the figure. The AND operation requires two NAND gates. The first produces the NAND operation and the second inverts the logical sense of the signal, so that the combination becomes an AND. The OR operation is achieved using a NAND gate with NOTs on each input. When DeMorgan's theorem is applied as shown in Figure 2-27, the inversions cancel and an OR function results.

A convenient way to implement a Boolean function with NAND gates is to obtain the simplified Boolean function in terms of the Boolean operators AND, OR, and NOT and then convert the function to NAND logic. The conversion of an algebraic expression from AND, OR, and NOT to NAND can be done by simple

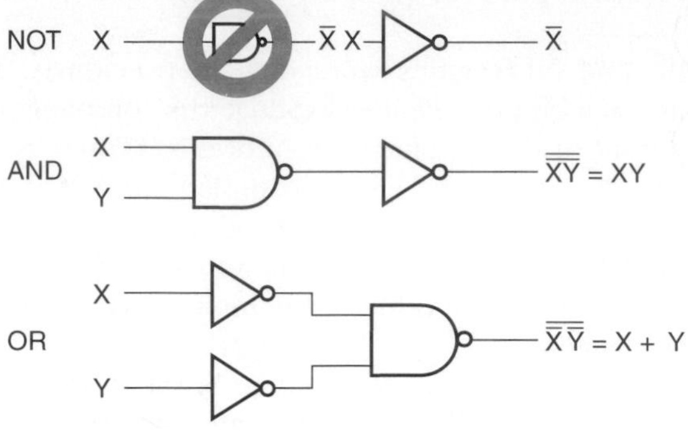

□ **FIGURE 2-27**
Logical Operations with NAND Gates

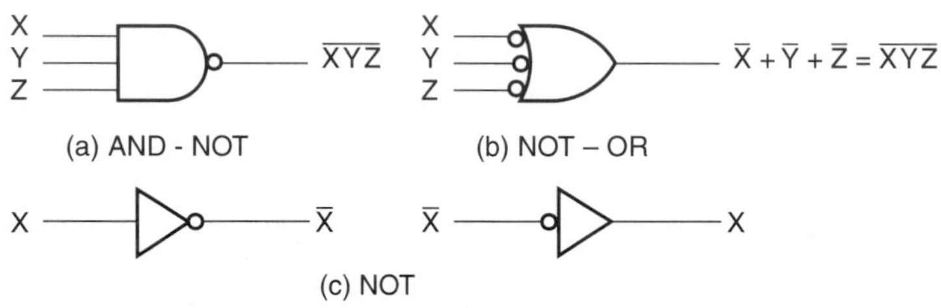

□ **FIGURE 2-28**
Alternative Graphics Symbols for NAND and NOT Gates

circuit manipulation techniques that change AND-OR diagrams to NAND diagrams.

To facilitate the conversion to NAND logic, it is convenient to define alternative graphics symbols for the gate. Two equivalent graphics symbols for the NAND gate are shown in Figure 2-28(a) and Figure 2-28(b). The AND-NOT symbol, given for the NAND previously, consists of an AND graphics symbol with a bubble on its output. Alternatively, it is possible to represent a NAND gate by an OR graphics symbol with a bubble on each input. This NOT-OR symbol for the NAND gate follows DeMorgan's theorem and the convention that negation indicators denote complementation. To deal with the alternative symbols for the one-input NAND case, alternative symbols for the NOT gate are given in Figure 2-28(c). The second symbol is obtained by moving the bubble from the output to the input of the buffer. These alternative graphics symbols are useful in the analysis and design of NAND circuits.

Two-Level Implementation

The implementation of a Boolean function with NAND gates is simplest if the function is in sum-of-products form. This form corresponds to a *two-level circuit*. If AND gates and an OR gate are used to implement the circuit, we say that the

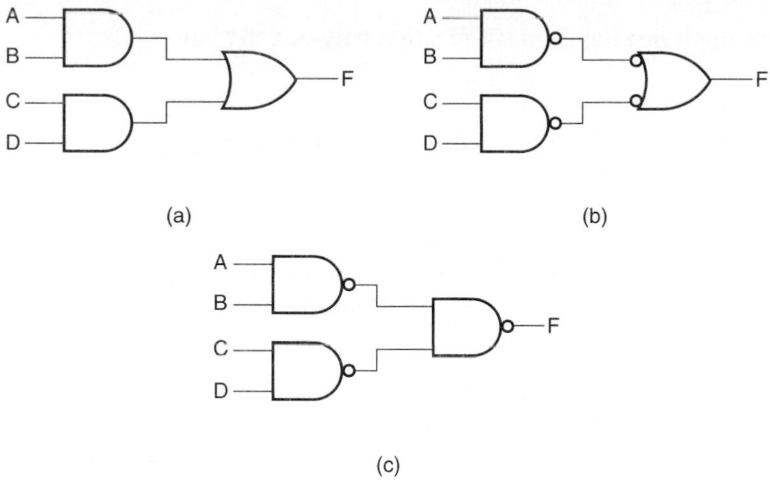

□ **FIGURE 2-29**
Three Ways to Implement $F = AB + CD$

AND gates are in the first level and the OR gate is in the second level. For two-level circuits as defined here, inverters on the inputs to the ANDs and on the output of the OR are not counted as levels. To see the relationship between a sum-of-products expression and its equivalent NAND implementation, consider the logic diagrams drawn in Figure 2-29. All three diagrams are equivalent and implement the function

$$F = AB + CD$$

The function is implemented in (a) with AND and OR gates. In (b), the AND gates are replaced by NAND gates and the OR gate is replaced by a NAND gate with a NOT-OR graphics symbol. Remember that a bubble denotes complementation, and two bubbles along the same line represent double complementation; so both can be removed by applying

$$\overline{\overline{X}} = X$$

Removing the bubble on the gates of (b) produces the circuit of (a). Therefore, the two diagrams implement the same function and are equivalent. In Figure 2-29(c), the output NAND gate is redrawn with the AND-NOT graphics symbol. In drawing NAND logic diagrams, the circuit shown in either (b) or (c) is acceptable. The one in (b) represents a more direct relationship with the Boolean expression it implements. The NAND implementation in Figure 2-29(c) can be verified algebraically. The function it implements can be easily converted to a sum-of-products form by using DeMorgan's theorem:

$$F = \overline{\overline{AB} \cdot \overline{CD}} = AB + CD$$

■ EXAMPLE 2-9

Implement the following Boolean function with NAND gates:

$$F(X, Y, Z) = \Sigma m(1, 2, 3, 4, 5, 7)$$

The first step is to simplify the function into sum-of-products form. This is done by means of the map of Figure 2-30(a), from which the simplified function

$$F = X\overline{Y} + \overline{X}Y + Z$$

is obtained. The two-level NAND implementation is shown in Figure 2-30(b) using the NOT-OR symbol at the second level. Note that input Z must have a NOT gate to compensate for the bubble on the second-level gate. An alternative way of drawing the logic diagram is shown in Figure 2-30(c). Here all the NAND gates are drawn with the same graphics symbol. The inverter with input Z has been removed, but the input variable is complemented and denoted by $\overline{Z}$. ■

The procedure described in the previous example indicates that a Boolean function can be implemented with two levels of NAND gates. The procedure for obtaining the logic diagram from a Boolean function is as follows:

1. Simplify the function and express it in sum-of-products form.
2. Draw a NAND gate for each product term of the expression that has at least two literals. The inputs to each NAND gate are the literals of the term. This constitutes a group of first-level gates.

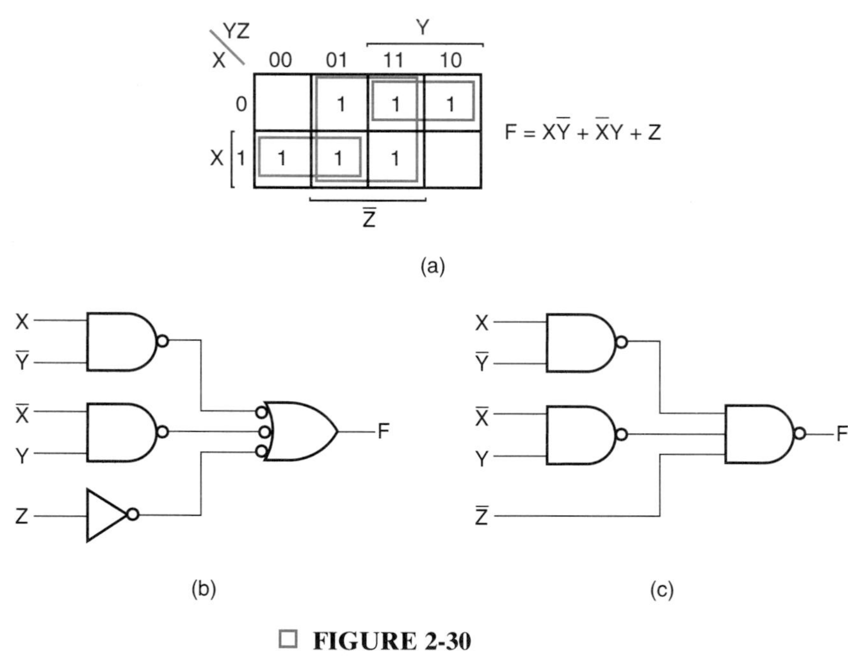

□ **FIGURE 2-30**
Solution to Example 2-9

3. Draw a single gate using the AND-NOT or the NOT-OR graphics symbol at the second level, with inputs coming from outputs of first-level gates.
4. A term with a single literal requires a NOT at the first level. However, if the single literal is complemented from its original appearance it can be connected directly to an input of the second-level NAND gate.

Multilevel NAND Circuits

The standard form of expressing Boolean functions results in a two-level implementation. However, there are occasions when the design of digital systems results in gating structures with three or more levels. The most common procedure in the design of multilevel circuits is to express the Boolean function in terms of AND, OR, and NOT operations. The function can then be implemented directly with AND and OR gates. Then, if necessary, it can be converted into an all-NAND circuit.

For example, consider the Boolean function

$$F = A(CD + B) + B\overline{C}$$

Although it is possible to remove the parentheses and reduce the expression to a standard sum-of-products form, we choose to implement it as a multilevel circuit for illustration. The AND-OR implementation is shown in Figure 2-31(a). There are four levels of gating in the circuit. The first level has two AND gates. The second

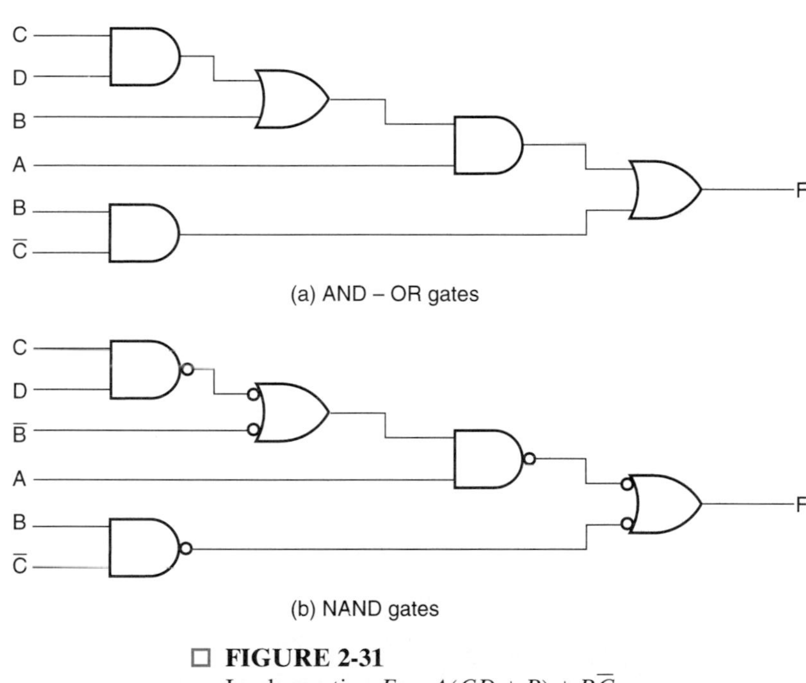

□ **FIGURE 2-31**
Implementing $F = A(CD + B) + B\overline{C}$

level has an OR gate and is followed by an AND gate in the third level and an OR gate in the fourth level. A logic diagram with a pattern of alternate levels of AND and OR gates can be easily converted into a NAND circuit by using the alternative NAND symbols. This is shown in Figure 2-31(b). The procedure is to change every AND gate to an AND-NOT graphics symbol and every OR gate to a NOT-OR graphics symbol. The NAND circuit performs the same logic as the AND-OR circuit, as long as there are zero or two bubbles along each line. The bubble associated with input B to the NOT-OR graphics symbol causes an extra complementation, which must be compensated for by changing the input literal to $\overline{B}$.

The general procedure for converting a multilevel AND-OR diagram into an all-NAND diagram using alternative NAND symbols is as follows:

1. Convert all AND gates to NAND gates with AND-NOT graphic symbols.
2. Convert all OR gates to NAND gates with NOT-OR graphic symbols.
3. Check all the bubbles in the diagram. For every bubble that is not counteracted by another bubble along the same line, insert a NOT gate or complement the input literal from its original appearance.

As another example, consider the multilevel Boolean function

$$F = (A\overline{B} + \overline{A}B)(E(C + \overline{D}))$$

The AND-OR implementation of this function is shown in Figure 2-32(a) with three levels of gating. The conversion into NAND with alternative symbols is pre-

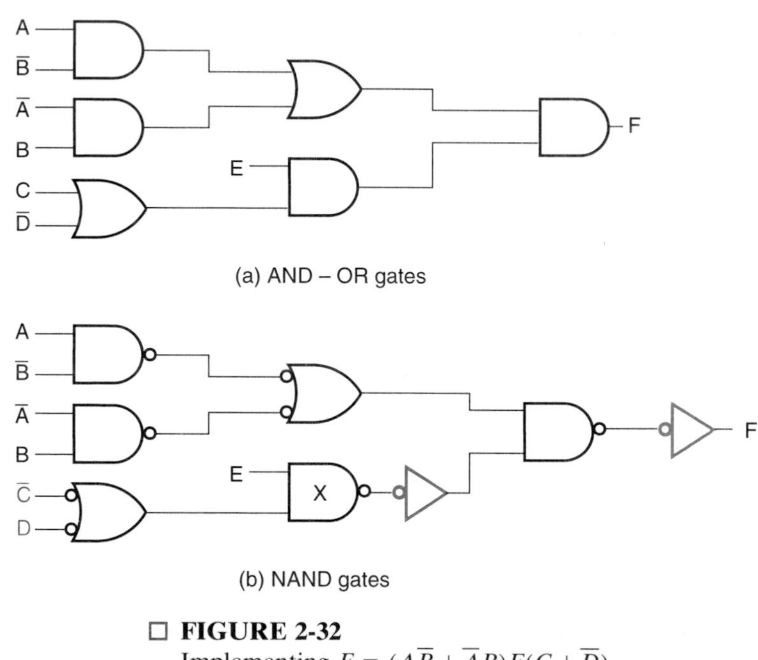

□ **FIGURE 2-32**
Implementing $F = (A\overline{B} + \overline{A}B)E(C + \overline{D})$

sented in part (b) of the diagram. The two additional bubbles associated with inputs C and $\overline{D}$ cause these two literals to be complemented to $\overline{C}$ and D. The bubble on the output NAND gate complements the output value; so we need to insert a NOT gate at the output in order to complement the signal again and get the original value. Likewise, the bubble on the output of NAND gate X is not counteracted, so a NOT gate is inserted.

NOR Circuits

The NOR operation is the dual of the NAND operation. Therefore, all procedures and rules for NOR logic are the dual of the corresponding procedures and rules developed for NAND logic. The NOR gate is another universal gate that can be used to implement any Boolean function. The implementation of the AND, OR, and NOT operations with NOR gates is shown in Figure 2-33. The complement operation is obtained from a one-input NOR gate that, like the NAND, reduces to a NOT gate. The OR operation requires two NOR gates, and the AND operation is obtained with a NOR gate that has NOT gates on each input. The two alternative graphics symbols for the NOR gate are shown in Figure 2-34. The OR-NOT symbol defines the NOR operation as an OR followed by a NOT. The NOT-AND symbol complements each input and then performs an AND operation. These alternative symbols designate the same NOR operation and are logically identical because of DeMorgan's theorem.

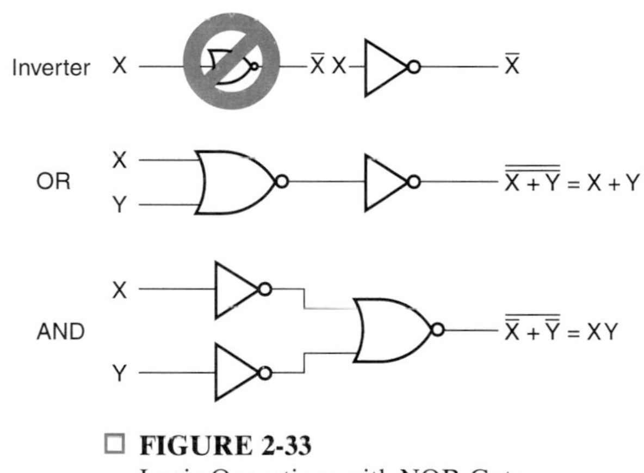

□ **FIGURE 2-33**
Logic Operations with NOR Gates

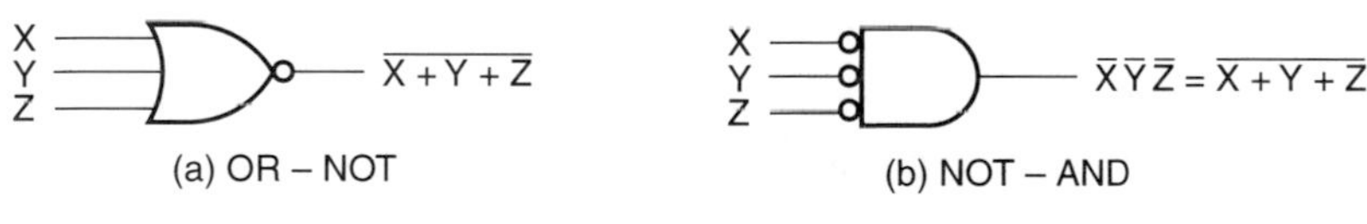

□ **FIGURE 2-34**
Two Graphic Symbols for NOR Gate

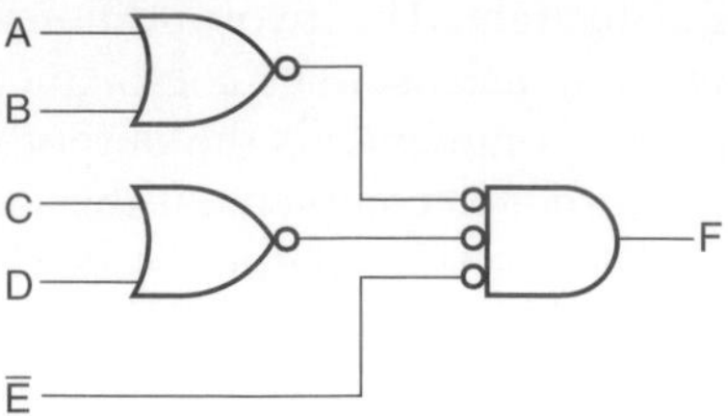

□ **FIGURE 2-35**
Implementing $F = (A + B)(C + D)E$ with NOR Gates

Two-level implementation with NOR gates is simplest if the simplified function is in product-of-sums form. Remember that the simplified product-of-sums expression is obtained from the map by combining the 0's and complementing. A product-of-sums expression is implemented with a first level of OR gates that produce the sum terms, followed by a second-level AND gate to produce the product. The transformation from the OR-AND diagram to a NOR diagram is achieved by changing the OR gates to NOR gates with OR-NOT graphics symbols and the AND gate to a NOR gate with a NOT-AND graphics symbol. A single literal term going into the second-level gate must be complemented from the way it originally appears, or a NOT gate must be inserted. Figure 2-35 shows the NOR implementation of a function expressed in product-of-sums form. The OR-AND pattern can be easily detected by the removal of the bubbles along the same line. The variable E is complemented to compensate for the third bubble at the input of the second-level gate.

The procedure for converting a multilevel AND-OR diagram to an all-NOR diagram is similar to the one presented for NAND gates. For the NOR case, we must convert each OR gate to an OR-NOT symbol and each AND gate to a NOT-AND symbol. Any bubble that is not compensated for by another bubble along the same line needs an inverter or the complementation of the input literal from its original appearance. The transformation of the AND-OR diagram of Figure 2-32(a) into a NOR diagram is shown in Figure 2-36. The Boolean function for this circuit is

$$F = (A\overline{B} + \overline{A}B)E(C + \overline{D})$$

The equivalent AND-OR diagram can be recognized from the NOR diagram by removing all the bubbles. In order for the bubbles to be compensated for, an inverter is added at the lower input of Y and E is complemented.

2-7 Exclusive-OR Gates

The exclusive-OR (XOR), denoted by $\oplus$, is a logical operation that performs the function

$$X \oplus Y = X\overline{Y} + \overline{X}Y$$

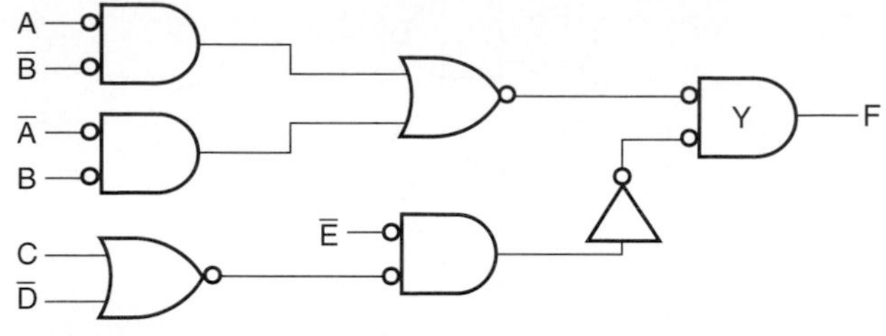

□ **FIGURE 2-36**
Implementing $F = (A\overline{B} + \overline{A}B)E(C + \overline{D})$ with NOR Gates

It is equal to 1 if exactly one variable is equal to 1. The exclusive-NOR, also known as the *equivalence*, is the complement of the exclusive-OR and is expressed by the function

$$\overline{X \oplus Y} = XY + \overline{X}\,\overline{Y}$$

It is equal to 1 if both X and Y are equal to 1 or if both are equal to 0. The two functions can be shown to be the complement of each other, either by means of a truth table or, as follows, by algebraic manipulation:

$$\overline{X \oplus Y} = \overline{X\overline{Y} + \overline{X}Y} = (\overline{X} + Y)(X + \overline{Y}) = XY + \overline{X}\,\overline{Y}$$

The following identities apply to the exclusive-OR operation:

$$X \oplus 0 = X \qquad X \oplus 1 = \overline{X}$$

$$X \oplus X = 0 \qquad X \oplus \overline{X} = 1$$

$$X \oplus \overline{Y} = \overline{X \oplus Y} \qquad \overline{X} \oplus Y = \overline{X \oplus Y}$$

Any of these identities can be verified by using a truth table or by replacing the $\oplus$ operation by its equivalent Boolean expression. It can also be shown that the exclusive-OR operation is both commutative and associative; that is,

$$A \oplus B = B \oplus A$$

$$(A \oplus B) \oplus C = A \oplus (B \oplus C) = A \oplus B \oplus C$$

This means that the two inputs to an exclusive-OR gate can be interchanged without affecting the operation. It also means that we can evaluate a three-variable exclusive-OR operation in any order, and, for this reason, exclusive-ORs with three or more variables can be expressed without parentheses. This implies the possibility of using exclusive-OR gates with three or more inputs.

A two-input exclusive-OR function may be constructed with conventional gates. Two NOT gates, two AND gates, and an OR gate are used. Figure 2-37

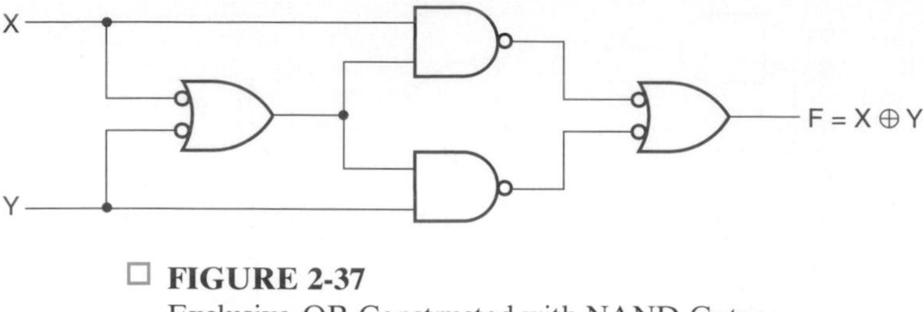

□ **FIGURE 2-37**
Exclusive-OR Constructed with NAND Gates

shows an implementation with four NAND gates. The alterative-symbol NAND diagram performs the operation

$$X(\overline{X}+\overline{Y})+Y(\overline{X}+\overline{Y}) = X\overline{Y}+\overline{X}Y$$

Odd Function

The exclusive-OR operation with three or more variables can be converted into an ordinary Boolean function by replacing the ⊕ symbol with its equivalent Boolean expression. In particular, the three-variable case can be converted to a Boolean expression as follows:

$$\begin{aligned} X \oplus Y \oplus Z &= (X\overline{Y}+\overline{X}Y)\overline{Z}+(XY+\overline{X}\,\overline{Y})Z \\ &= X\overline{Y}\,\overline{Z}+\overline{X}Y\overline{Z}+\overline{X}\,\overline{Y}Z+XYZ \end{aligned}$$

The Boolean expression clearly indicates that the three-variable exclusive-OR is equal to 1 if only one variable is equal to 1 or if all three variables are equal to 1. Hence, whereas in the two-variable function only one variable need be equal to 1, with three or more variables an odd number of variables must be equal to 1. As a consequence, the multiple-variable exclusive-OR operation is defined as the *odd function*. In fact, strictly speaking, this is the correct name for the ⊕ operation with three or more variables; the name "exclusive-OR" is applicable to the case with only two variables.

The Boolean function derived from the odd function is expressed as the logical sum of four minterms whose corresponding binary truth-table values are 001, 010, 100, and 111. Each of these binary numbers has an odd number of 1's. The other four minterms not included in the function are 000, 011, 101, and 110, and they have an even number of 1's. In general, an n-variable odd function is defined as the logical sum of the $2^n/2$ minterms whose corresponding binary truth-table values have an odd number of 1's.

The definition of the odd function can be clarified by plotting the function on a map. Figure 2-38(a) shows the map for the three-variable odd function. The four minterms of the function differ from each other in at least two literals and hence cannot be adjacent on the map. These minterms are said to be *distance two* from each other. The odd function is identified from the four minterms whose binary

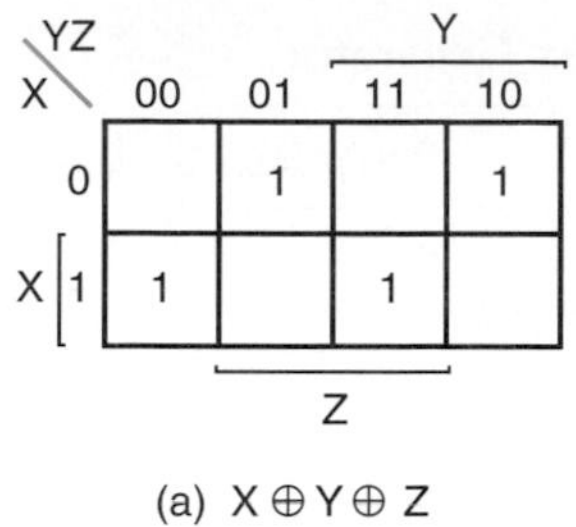

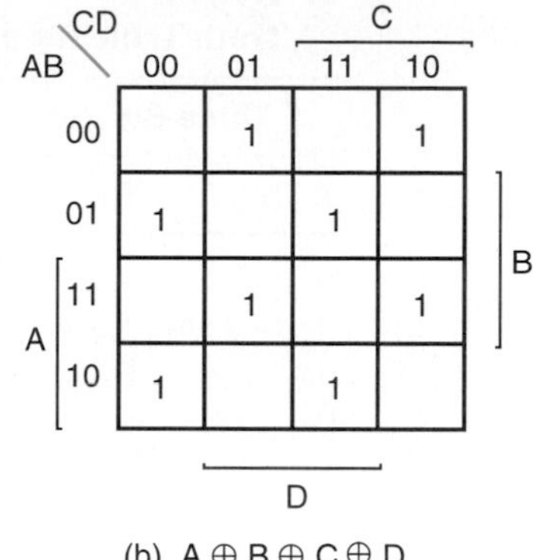

□ **FIGURE 2-38**
Maps for Multiple-Variable Odd Functions

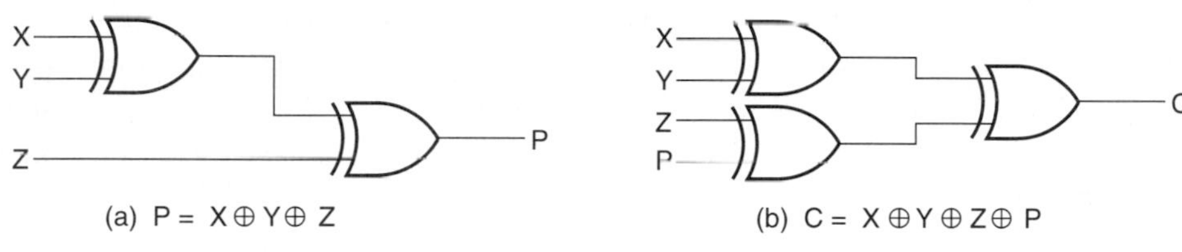

□ **FIGURE 2-39**
Multiple-Input Odd Functions

values have an odd number of 1's. The four-variable case is shown in Figure 2-38(b). The eight minterms marked with l's in the map constitute the odd function. Note the characteristic pattern of the distance between the 1's in the map. It should be mentioned that the minterms not marked with 1's in the map have an even number of 1's and constitute the complement of the odd function, called the *even function*. The odd function is implemented by means of two-input exclusive-OR gates, as shown in Figure 2-39. The even function is obtained by replacing the output gate with an exclusive-NOR gate.

Parity Generation and Checking

Odd or even functions are very useful in systems requiring error detection and correction codes. As discussed in Section 1-5, a parity bit is used in a scheme for detecting errors during the transmission of binary information. A parity bit is an extra bit included with a binary message to make the number of 1's either odd or even. The message, including the parity bit, is transmitted and then checked at the receiving end for errors. An error is detected if the parity of the data bits in the message does not correspond to the parity bit transmitted. The circuit that generates the parity bit in the transmitter is called a *parity generator*. The circuit that checks the parity in the receiver is called a *parity checker*. As an example, consider

□ **TABLE 2-9**
Truth Table for an Even Parity Generator

Three-Bit Message			Parity Bit
X	Y	Z	P
0	0	0	0
0	0	1	1
0	1	0	1
0	1	1	0
1	0	0	1
1	0	1	0
1	1	0	0
1	1	1	1

a three-bit message to be transmitted with an even parity bit. Table 2-9 shows the truth table for this parity generator. The three bits X, Y, and Z constitute the message and are the inputs to the circuit. The parity bit P is the output. For even parity, the bit P must be generated to make the total number of 1's (including P) even. From the truth table, we see that P constitutes an odd function because it is equal to 1 for those minterms whose numerical values have an odd number of 1's. Therefore, P can be expressed as a three-variable odd function:

$$P = X \oplus Y \oplus Z$$

The logic diagram for the parity generator is shown in Figure 2-39(a).

The three bits in the message, together with the parity bit, are transmitted to their destination, where they are applied to a parity-checker circuit to check for possible errors in the transmission. Since the information was transmitted with even parity, the four bits received must have an even number of 1's. An error occurs if these bits have an odd number of 1's, indicating that at least one bit has changed its value during transmission. If such an error occurs, the output of the parity checker,

$$C = X \oplus Y \oplus Z \oplus P$$

will be equal to 1. This is, by definition, an odd function and can be implemented using exclusive-OR gates. The logic diagram of the parity checker is shown in Figure 2-39(b).

It is obvious from the foregoing example that parity generation and checking circuits always have an output function that includes half of the minterms whose numerical values have either an odd or even number of 1's. As a consequence, they can be implemented with exclusive-OR gates. A function with an even number of 1's is the complement of an odd function. It is implemented with exclusive-OR gates, except that the gate associated with the output must be an exclusive-NOR to provide the required complementation.

2-8 Integrated Circuits

Digital circuits are constructed with integrated circuits. An integrated circuit (abbreviated IC) is a small silicon semiconductor crystal, informally called a chip, containing the electronic components for the digital gates. The various gates are interconnected on the chip to form the IC. The chip is mounted in a ceramic or plastic container, and connections are welded from the chip to external pins to form the integrated circuit. The number of pins may range from 14 on a small IC package to several hundred on a larger package. Each IC has a numeric designation printed on the surface of the package for identification. Each vendor publishes a data book or catalog that contains the description and all the necessary information about the ICs that it manufactures.

Levels of Integration

As the technology of ICs has improved, the number of gates that can be put in a single silicon chip has increased considerably. The differentiation between those chips that have a few internal gates and those having thousands to millions of gates is made by a customary reference to a package as being either a small-, medium-, large-, or very large-scale integrated device.

Small-scale integrated (SSI) devices contain several independent gates in a single package. The inputs and outputs of the gates are connected directly to the pins in the package. The number of gates is usually less than 10 and is limited by the number of pins available on the IC.

Medium-scale integrated (MSI) devices have a complexity of approximately 10 to 100 gates in a single package. They usually perform specific elementary digital functions, such as the addition of four bits. MSI digital functions are presented in Chapter 3 and Chapter 5.

Large-scale integrated (LSI) devices contain between 100 and a few thousand gates in a single package. They include digital systems such as small processors, small memories, and programmable modules. LSI logic systems are introduced in Chapter 6.

Very large-scale integrated (VLSI) devices contain several thousand to millions of gates in a single package. Examples are large memory arrays and complex microprocessor and microcomputer chips. Because of their small size and low cost, VLSI devices have revolutionized computer system design technology, giving designers the capability to create complex structures that previously were not economical to manufacture. VLSI-based systems are examined in Chapter 10 through Chapter 12.

Digital Logic Families

Digital integrated circuits are classified not only by their logical operation, but also by the specific circuit technology to which they belong. The circuit technology is referred to as a *digital logic family*. Each such family has its own basic electronic circuit upon which more complex digital circuits and functions are developed. The primitive circuits in each technology are typically NAND, NOR, and NOT gates. The electronic components used in the construction of the basic circuit usually give the name to the technology. Many different logic families of integrated circuits have been introduced commercially. Historically, the following have been the most important.

RTL	Resistor-transistor logic
DTL	Diode-transistor logic
TTL	Transistor-transistor logic
ECL	Emitter-coupled logic
MOS	Metal-oxide semiconductor
CMOS	Complementary metal-oxide semiconductor
BiCMOS	Bipolar complementary metal-oxide semiconductor

RTL and DTL were the earliest logic families and are now obsolete. TTL is a widely used logic family that has been available for decades, but is declining rapidly in use. ECL has an advantage in systems requiring high-speed operation, but is being rapidly overtaken by CMOS. MOS is suitable for circuits that need high component density, and CMOS is preferable in systems requiring low power consumption. In fact, low power consumption is so essential to prevent large, complex, dense devices from overheating that CMOS is now the dominant technology. BiCMOS, which combines CMOS with a bit of TTL, is used selectively in cases in which CMOS alone cannot provide adequate current or the necessary speed.

The characteristics of digital logic families are usually compared by analyzing the circuit of the basic gate in each family. The most important parameters that are evaluated and compared are as follows:

Fan-in specifies the number of inputs available on a gate.

Fan-out specifies the number of standard loads that the output of a typical gate can drive without impairing its performance. A standard load is usually defined as the amount of current needed by an input of another similar gate of the same family.

Noise margin is the maximum external noise voltage superimposed on a normal input value that will not cause an undesirable change in the output of the circuit.

Power dissipation is the power consumed by the gate and made available from the power supply. Much of the power consumed is dissipated as heat, so power dissipation must also be considered in relation to the operating temperature and cooling requirements of the chip.

Propagation delay is the delay time for the change in value of a signal to propagate from input to output. The operating speed is inversely related to the longest propagation delays.

Although all of these parameters are important to the designer, propagation delay is the most important at the logic design level. The determination of propagation delay is illustrated in Figure 2-40. Three propagation delay parameters are defined. The *high-to-low propagation time* t_{PHL} is the delay measured from the ref-

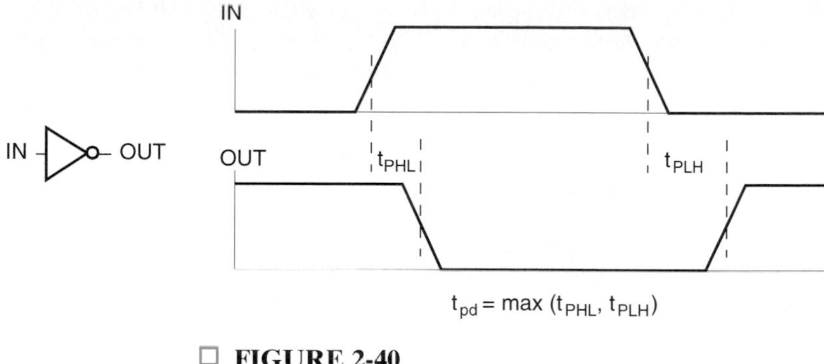

□ **FIGURE 2-40**
Propagation Delay for an Inverter

erence voltage on the input voltage IN to the reference voltage on the output voltage OUT with the output voltage going from H to L. The reference voltage we are using is the 50% point on the voltage signals; other reference voltages may be used, depending on the logic family. The *low-to-high propagation time* t_{PLH} is the delay measured from the reference voltage on the input voltage IN to the reference voltage on the output voltage OUT with the output voltage going from L to H. We define the *propagation delay* t_{pd} as the maximum of these two delays. The reason we have chosen the maximum value is that we will be most concerned about finding the longest time, rather than the shortest time, for a signal to propagate from inputs to outputs. Otherwise the definitions given for t_{pd} may vary, depending on the use of the data. Manufacturers usually specify the maximum and typical values for both t_{PHL} and t_{PLH} or for t_{pd} for their products.

Positive and Negative Logic

Except during transitions, the binary signals at the inputs and outputs of any gate have one of two values: H or L. One value represents logic 1 and the other logic 0. There are two different assignments of signal levels to logic values, as shown in Figure 2-41. Choosing the high level, H, to represent logic 1 defines a *positive-logic* system. Choosing the low level, L, to represent logic 1 defines a *negative-logic* system. The terms "positive" and "negative" are somewhat misleading, since both signals may be positive voltages or both may be negative voltages. It is not the actual signal values that determine the type of logic, but rather the assignment of logic values to the relative amplitudes of the two signal ranges.

Integrated circuit data sheets define digital gates not in terms of logic values, but rather in terms of signal values such as H and L. It is up to the user to decide on a positive or negative logic assignment. Consider, for example, the truth table in Figure 2-42(a). This table is given in a data book for the CMOS gate shown in Figure 2-42(b). The table specifies the physical behavior of the gate when H is 5 volts and L is 0 volts. The truth table of Figure 2-42(c) assumes positive logic, with 1 assigned to H and 0 assigned to L. The table is the same as the truth table for the

Signal value	Logic value
H	1
L	0

(a) Positive logic

Signal value	Logic value
H	0
L	1

(b) Negative logic

□ **FIGURE 2-41**
Signal Assignment and Logic Polarity

X	Y	Z
L	L	L
L	H	L
H	L	L
H	H	H

(a) Truth table with H and L

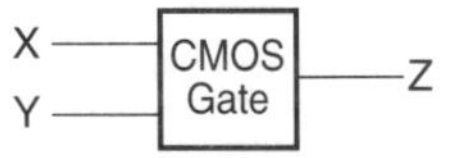

X, Y → CMOS Gate → Z

(b) Gate block diagram

X	Y	Z
0	0	0
0	1	0
1	0	0
1	1	1

(c) Truth table for positive logic

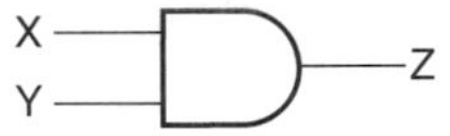

(d) Positive-logic AND gate

X	Y	Z
1	1	1
1	0	1
0	1	1
0	0	0

(e) Truth table for negative logic

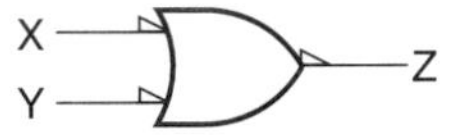

(f) Negative-logic OR gate

□ **FIGURE 2-42**
Demonstration of Positive and Negative Logic

AND operation. The graphics symbol for a positive-logic AND gate is shown in Figure 2-42(d).

Now consider the negative logic assignment for the same physical gate, with 1 assigned to L and 0 assigned to H. The result is the truth table of Figure 2-42(e). This table represents the OR operation, even though the entries are reversed. The graphic symbol for the negative-logic OR gate is shown in Figure 2-42(f). The small triangles on the inputs and output are *polarity indicators.* The presence of a polarity indicator at an input or output signifies that negative logic is assumed for the signal. Thus, the same physical gate can operate either as a positive-logic AND gate or as a negative-logic OR gate.

The conversion from positive logic to negative logic and vice versa is essentially an operation that changes 1's to 0's and 0's to 1's at both the inputs and the output of a gate. Since this interchange of 1's and 0's is a part of taking the dual, the conversion operation produces the dual of the gate function. Thus, the change of all gate inputs and outputs from one polarity to the other results in taking the dual of the function, with all AND operations (or graphics symbols) converted to OR operations (or graphics symbols) and vice versa. In addition, one must not forget to include the polarity indicator in the graphics symbols when negative logic is assumed, and one must also recognize that the polarity definitions for circuit inputs and circuit outputs have been changed. In this book we do not use negative logic, but assume that all gates operate with a positive-logic assignment.

2-9 CMOS CIRCUITS

So far we have dealt with implementing logic circuits in terms of gates. In this section, we briefly explore implementing the gates themselves using CMOS technology. In addition, we study how structures other than primitive logic gates can be implemented directly in terms of electronic elements called transistors. CMOS implementation is important because we often design CMOS logic from Boolean equations directly to the transistor level, skipping the logic gate level.

Switch Models for CMOS Transistors

CMOS technology employs two types of transistor: *n-channel* and *p-channel*. The two differ in the characteristics of the semiconductor materials used in their implementation and in the mechanism governing the conduction of a current through them. Most important to us, however, is the difference in behavior of the two types of transistor. We will model this behavior using switches controlled by voltages corresponding to logic 0 and logic 1. Such a model ignores complex electronic devices and captures only logical behavior.

The symbol for an n-channel transistor is shown in Figure 2-43(a). The transistor has three terminals: the gate (G), the source (S), and the drain (D), as shown in Figure 2-43(b). The voltage applied between G and S determines whether a path for current to flow exists between D and S. If a path exists, we say that the transistor is ON, and if a path does not exist, we say that the transistor is OFF. The n-channel transistor is ON if the applied gate-to-source voltage is H and OFF if the applied voltage is L. Here we will make the usual assumption that a 1 represents the H voltage range and a 0 represents the L voltage range.

The notion of whether a path for current to flow exists is easily modeled by a switch, as shown in Figure 2-43(c). The switch consists of two fixed terminals corresponding to the S and D terminals of the transistor. In addition, there is a movable contact that, depending on its position, determines whether the switch is open or closed. The position of the contact is controlled by the voltage applied to the gate terminal G. Since we are looking at logic behavior, this control voltage is represented on the symbol by the input variable X on the gate terminal. For an n-chan-

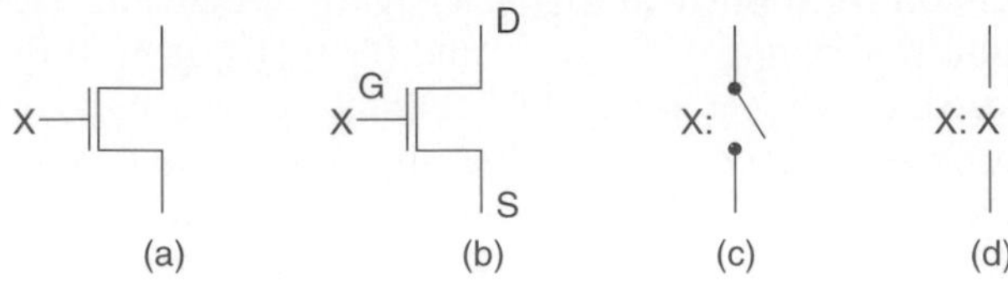

FIGURE 2-43
Symbol and Switch Model for n-Channel Transistor

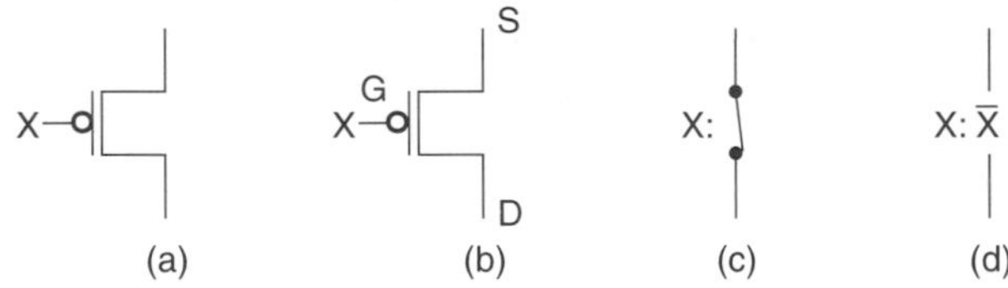

FIGURE 2-44
Symbol and Switch Model for p-Channel Transistor

nel transistor, the contact is open (no path exists) for the input variable X equal to 0 and closed (a path exists) for the input variable X equal to 1. Such a contact is traditionally referred to as being *normally open*, that is, open without a positive voltage applied to activate or close it. Figure 2-43(d) shows a shorthand notation for the n-channel switch model with the variable X applied. This notation represents the fact that a path between S and D exists for X equal to 1 and does not exist for X equal to 0.

The symbol for a p-channel transistor is shown in Figure 2-44(a). In Figure 2-44(b), the positions of the source S and drain D are seen to be interchanged relative to their positions in the n-channel transistor. The voltage applied between the gate G and the source S determines whether a path exists between the drain and source. Note in Figure 2-44(a) that the negation indicator or bubble appears as a part of the symbol. This is because, in contrast to the behavior of an n-channel transistor, a path exists between S and D in the p-channel transistor for input variable X equal to 0 (at value L) and does not exist for input variable X equal to 1 (at value H). This behavior is represented by the model in Figure 2-44(c), which has a *normally closed* contact through which a path exists for X equal to 0. No path exists through the contact for X equal to 1. In addition, the shorthand notation of the p-channel switch model with variable X applied is given in Figure 2-44(d). Since a 0 on input X causes a path to exist through the switch and a 1 on X produces no path, the literal shown on the switch is $\overline{X}$ instead of X.

Networks of Switches

A network made up of switches that model transistors can be used to design CMOS logic. The network implements a function F if there is a path through the network for F equal to 1 and no path through the network for F equal to 0. A sim-

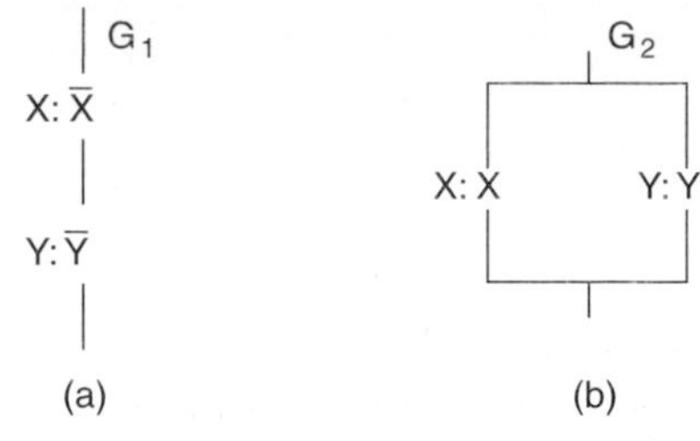

□ **FIGURE 2-45**
Example of Switch Model Networks

ple network of p-channel transistor switch models is shown in Figure 2-45(a). The function G_1 implemented by this network can be determined by finding the input combinations for which a path exists through the network. In order for the path to exist through G_1, both switches must be closed; that is, the path exists for $\overline{X}$ and $\overline{Y}$ both 1. This implies that $X = 0$ *and* $Y = 0$. Thus, the function G_1 of the network is $\overline{X} \cdot \overline{Y} = \overline{X + Y}$, in other words, the NOR function. In Figure 2-45(b), for function G_2, a path exists through the n-channel switch model network if either switch is closed, that is, for $X = 1$ *or* $Y = 1$. Thus, the function G_2 is $X + Y$.

In general, switches in series give an AND function and switches in parallel give an OR function. (The function for the preceding network that models p-channel transistors is a NOR function because of the complementation of the variables and the application of DeMorgan's law.) By using these network functions to produce paths in a circuit that attach logic 1 (H) or logic 0 (L) to an output, we can implement a logic function on the output, as discussed next.

Fully Complementary CMOS Circuits

The subfamily of CMOS circuits that we will now consider has the general structure shown in Figure 2-46(a). Except during transitions, there is a path to the output of the circuit F either from the power supply +V (logic 1) or from ground (logic 0). Such a circuit is called *static* CMOS. In order to have a static circuit, the transistors must implement networks of switches for both function F and function $\overline{F}$. In other words, both the 0's and the 1's of the function F must be implemented with paths through networks. The switch network implementing F is constructed using p-channel transistors and connects the circuit output to logic 1. P-channel transistors are used because they conduct logic-1 values better than logic-0 values. The switch network implementing $\overline{F}$ is constructed using n-channel transistors and connects the circuit output to logic 0. Here n-channel transistors are used because they conduct logic-0 values better than logic-1 values. Note that the same input variables enter both the p-channel and n-channel switch networks.

To illustrate a fully complementary circuit, we use transistors corresponding to the networks G_1 and G_2 from Figure 2-45(a) and (b) as the p-channel implementation of G and the n-channel implementation of $\overline{G}$, respectively, in Figure 2-46(b). A path exists through G_1 for $\overline{X + Y} = 1$, which means that a path exists in

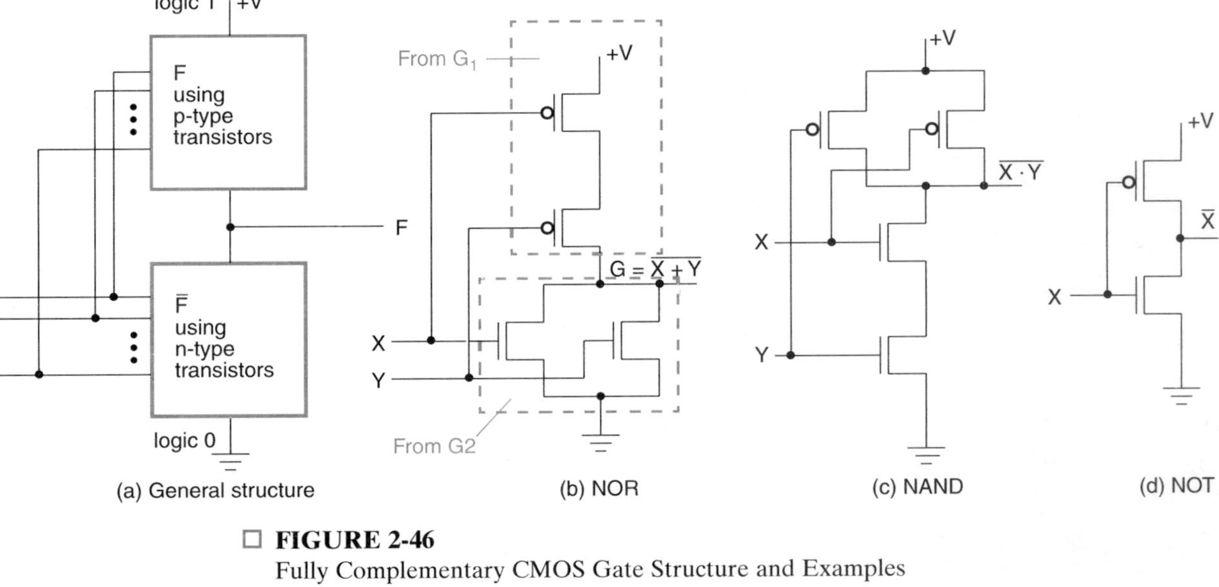

FIGURE 2-46
Fully Complementary CMOS Gate Structure and Examples

Figure 2-46(b) from logic 1 to the circuit output, making $G = 1$ for $\overline{X+Y} = 1$. This provides the 1's on the output for the function G. A path exists through G_2 for $X + Y = 1$, which means that a path exists in Figure 2-46(b) from logic 0 to the output for $X + Y = \overline{\overline{X+Y}} = 1$. This path makes $G = 0$ for the complement of $\overline{X+Y}$. Thus, the n-channel circuit implements $\overline{G}$. This provides the 0's on the output for function G. Since both the 1's and 0's are provided for G, we can say that the circuit output $G = \overline{X+Y}$, which is a NOR gate. This is the standard static CMOS implementation for a NOR.

Since the NAND is just the dual of the NOR, we can implement the CMOS NAND by simply replacing the + by $\cdot$ in the equations for G_1 and G_2. In terms of the switch network, the dual of switches in series is switches in parallel and vice versa. This duality applies to the transistors that are modeled as well, giving the NAND implementation in Figure 2-46(c). The final gate in Figure 2-46(d) is the implementation of the NOT.

Note that all of the circuits in Figure 2-46 implement inverting functions under DeMorgan's laws. This inversion property is characteristic of CMOS gates. In fact, as we look at a general design procedure, we will assume that all functions are implemented as $F = \overline{\overline{F}}$. This avoids working directly with p-channel switches, which involve complementing variables. Thus, we will design the n-channel network for $\overline{F}$ and take the dual to get the p-channel network for F. For functions more complex than NAND, NOR, and NOT, the resulting circuits are called *complex gates* and are designed in accordance with the following procedure for function F:

1. Find and simplify the complement of F, $\overline{F}$.
2. Implement $\overline{F}$ as a switch network using n-channel switch models.
3. Connect the n-channel switch network between ground and output F, and convert the switches to n-channel transistors.
4. Take the dual of the n-channel switch network for $\overline{F}$, and replace the n-channel switch models with p-channel switch models, keeping switch inputs unchanged.
5. Connect the p-channel switch network between +V and the output F, and convert the switches to p-channel transistors.

Step 4 is different than expected: we take the dual of the n-channel switch network to get the p-channel switch network. Recall that the difference between the dual and the complement is that, in taking the complement, all literals in the expression are complemented. We need not do this complementing, however, since the complement of the variables is automatically taken by replacing the n-channel switch models with p-channel switch models. Taking the dual of a function means replacing AND with OR and OR with AND. For a switch network, this corresponds to taking switches or subnetworks that are in parallel and placing them in series and taking switches or subnetworks that are in series and placing them in parallel.

We illustrate this design procedure with an example.

■ EXAMPLE 2-10

Design of a Complex Gate for $F = A\overline{B} + AC + B\overline{C}$

In step 1 of the foregoing procedure, we place F on a map and, from the map, use the 0's to obtain a sum-of-products expression for $\overline{F}$ and the 1's to obtain a product-of-sums expression for $\overline{F}$:

$$\overline{F} = \overline{A}\,\overline{B} + \overline{A}C \qquad\qquad \overline{F} = \overline{A}(\overline{B} + C)$$

Since the product-of-sums expression has fewer literals (three instead of four), we select it for use in the next step.

In step 2, we find an n-channel switch model network for $\overline{F}$. From the product-of-sums expression, $\overline{A}$ is ANDed with the term $\overline{B} + C$. Thus, we place a switch with input $\overline{A}$ in series with a network implementing $\overline{B} + C$, as shown in Figure 2-47(a). We implement $\overline{B} + C$ with a switch having input $\overline{B}$ in parallel with a switch having input C. Checking step 2, the final network in Figure 2-47(a) has a path through it for $A = 0$ and either $B = 0$ or $C = 1$ or both.

The network from Figure 2-47(a) is converted into the corresponding n-channel transistor circuit between ground and the output F, as given in the lower part of Figure 2-47(c), completing step 3.

Next, in step 4, we are to take the dual of the n-channel switch network from step 2. First, we take the switches B and C that are in parallel in Figure 2-47(a) and place them in series. Then we take A, which is in series with the parallel combina-

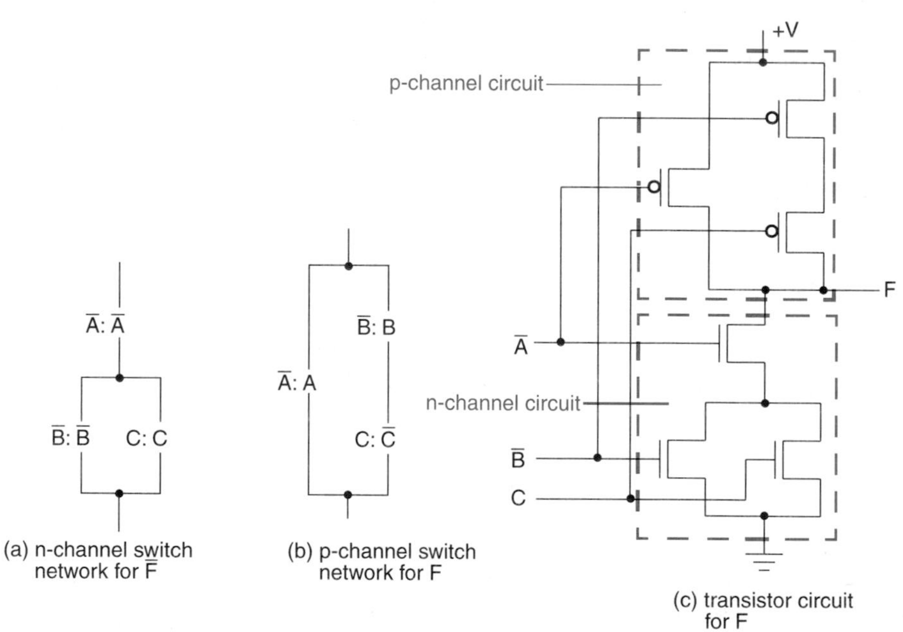

□ **FIGURE 2-47**
Networks and Circuit for Example 2-10, $F = A + B\overline{C}$

tion of B and C and place it in parallel with the series combination of B and C. Finally, we replace the n-channel switch models with p-channel switch models, keeping the complementation on the input values to the switches unchanged. The circuit in Figure 2-47(b) results.

The network in Figure 2-47(b) is converted to the p-channel transistor circuit between +V and F, as given in the upper part of Figure 2-47(c), completing the circuit in step 5. ■

We can use any Boolean expression for $\overline{F}$, although by minimizing the number of literals as much as possible, as is done in this example, we minimize the number of transistors in the circuit. In the actual design of these transistor circuits, it is also necessary to take electronic considerations into account. For example, in most cases, no path through one of the networks can contain more than four or five transistors in series. This clearly limits the functions that can be implemented in a single complex gate.

Transmission Gates

Besides primitive and complex gates, there is one additional transistor circuit frequently used in CMOS logic. This circuit is the transmission gate (TG). It has its own symbol and is often included in gate-level logic circuit diagrams. A transmission gate is used as an electronic switch for making a connection between two points in a circuit. It consists of an n-channel transistor and a p-channel transistor in parallel, as shown in Figure 2-48(a). The two types of transistor are used because the p-channel transistor passes 1 (H) well and the n-channel transistor passes 0 (L) well. Figure 2-48(b) is the switch model for the transmission gate. Here X is the input, Y is the output, and the two terminals C and $\overline{C}$ are control inputs. If $C = 1$ (H) and $\overline{C} = 0$ (L), there is a path between X and Y for the signal to pass through. If $C = 0$ and $\overline{C} = 1$, there is no path, and the circuit behaves like an open switch. The IEEE symbol for the transmission gate is given in Figure 2-48(c). Normally, the control inputs are connected through an inverter, as shown in Figure 2-48(d), so that C and $\overline{C}$ are the complements of each other.

Transmission gates are particularly useful for performing selection functions. A TG-based circuit that selects one of two values A and B to apply to an output F is shown in Figure 2-49(a). If $C = 0$, then a path exists through TG0 connecting F to A, and no path exists through TG1. If $C = 1$, then a path exists through TG1 connecting F to B, and no path exists through TG0. In Chapter 3, we will find that selection circuits, such as this one, are called multiplexers. So we call this circuit a transmission gate–based multiplexer.

By making $B = \overline{A}$ for the selector, an exclusive-OR gate can be constructed with two transmission gates and two inverters, as shown in Figure 2-49(b). Input C controls the paths in the transmission gates, and input A provides the output for F. If input C is equal to 1, a path exists through transmission gate TG1 connecting F to $\overline{A}$, and no path exists through TG0. If input C is equal to 0, a path exists through TG0 connecting F to A, and no path exists through TG1. Thus, the output F is connected to A. This results in the exclusive-OR truth table, as indicated in Figure 2-49(c).

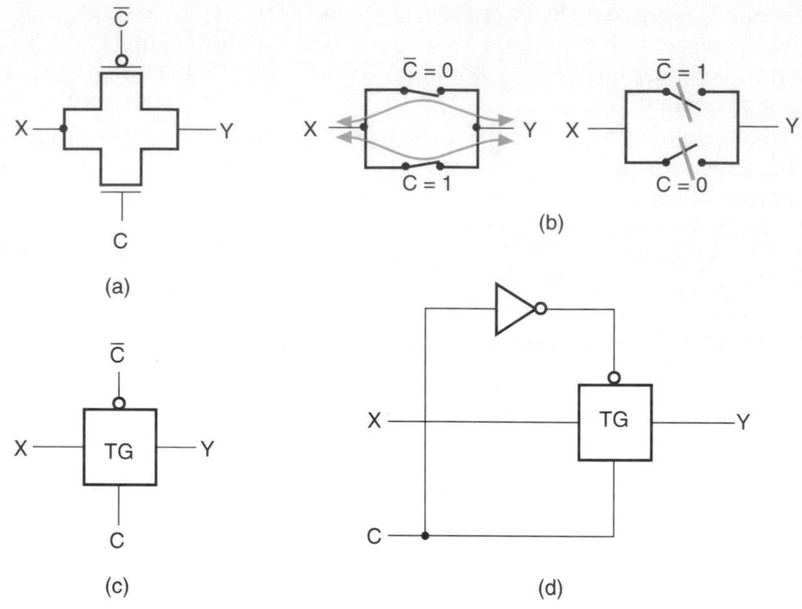

□ **FIGURE 2-48**
Transmission Gate (TG)

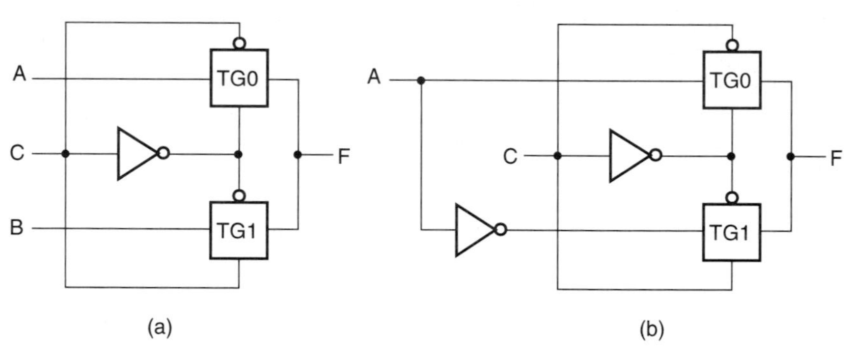

A	C	TG1	TG0	F
0	0	No path	Path	0
0	1	Path	No path	1
1	0	No path	Path	1
1	1	Path	No path	0

(c)

□ **FIGURE 2-49**
Selector and Exclusive-OR Constructed with Transmission Gates

2-10 Chapter Summary

The primitive logic operations AND, OR, and NOT define the primitive logic components called gates, from which digital systems are implemented. A Boolean algebra defined in terms of these operations provides a tool to manipulate Boolean functions in designing digital logic circuits. Minterm and maxterm standard forms correspond directly to truth tables. These standard forms can be manipulated into sum-of-products and product-of-sums standard forms, which correspond to two-level gate circuits. The cost measures to be minimized in simplifying such circuits are the number of gates and the number of input literals to the gates. Maps with two to four variables are an effective alternative to algebraic manipulation in simplifying small circuits. These maps can be used to simplify sum-of-products forms, product-of-sums forms, and incompletely specified functions with don't-care conditions.

The primitive operations AND and OR are not directly implemented by primitive logic elements in the most popular logic families. Instead, these families implement NAND and NOR primitives. The design of NAND and NOR circuits is done by graphical conversion from AND-OR circuits. A more complex primitive, the exclusive-OR, as well as its complement, the exclusive-NOR, is useful for selected applications, such as parity generation and checking.

A number of practical issues associated with integrated circuits and logic families are important to designers. Among these are parameters such as fan-in, fan-out, and propagation delay and the concepts of positive and negative logic. The most prominent logic family, CMOS, can be modeled logically at the transistor level by using switch models. These models provide insight into how transistors are used to implement primitive gates. In addition, the models are a vehicle for designing circuits using a primitive called the transmission gate and for designing gates having complex (as contrasted with primitive) functions.

References

1. Boole, G. *An Investigation of the Laws of Thought.* New York: Dover, 1854.
2. Karnaugh, M. "A Map Method for Synthesis of Combinational Logic Circuits," *Transactions of AIEE, Communication and Electronics*, 72, part I (Nov. 1953), 593-99.
3. Dietmeyer, D. L. *Logic Design of Digital Systems,* 3rd ed. Boston: Allyn Bacon, 1988.
4. Mano, M. M. *Digital Design,* 2nd ed. Englewood Cliffs, NJ: Prentice Hall, 1991.
5. Roth, C. H. *Fundamentals of Logic Design*, 4th ed. St. Paul: West, 1992.
6. Hayes, J. P. *Introduction to Digital Logic Design*. Reading, MA: Addison-Wesley, 1993.
7. Wakerly, J. F. *Digital Design: Principles and Practices,* 2nd ed. Englewood Cliffs, NJ: Prentice Hall, 1994.

8. *The TTL Data Book,* vols. 1, 2, and 3. Dallas: Texas Instruments, 1984.
9. *High-Speed CMOS Logic Data Book.* Dallas: Texas Instruments, 1984.
10. *IEEE Standard Graphic Symbols for Logic Functions.* (Includes IEEE Std 91a-1991 Supplement and IEEE Std 91-1984.) New York: The Institute of Electrical and Electronics Engineers, 1991.
11. WESTE, N. H. E., AND ESHRAGHIAN, K. *Principles Of CMOS VLSI Design: A Systems Perspective,* 2nd ed. Reading, MA: Addison-Wesley, 1993.
12. WOLF, W. *Modern VLSI Design: A Systems Approach.* Englewood Cliffs, NJ: Prentice Hall, 1994.

PROBLEMS

The asterisk (*) indicates a more advanced problem.

2–1. Demonstrate by means of truth tables the validity of the following identities:

(a) DeMorgan's theorem for three variables: $\overline{XYZ} = \overline{X} + \overline{Y} + \overline{Z}$

(b) The second distributive law: $X + YZ = (X + Y)(X + Z)$

(c) $\overline{X}Y + \overline{Y}Z + X\overline{Z} = X\overline{Y} + Y\overline{Z} + \overline{X}Z$

2–2. Prove the identity of each of the following Boolean equations, using algebraic manipulation:

(a) $AB + A\overline{B} + \overline{A}\,\overline{B} = A + \overline{B}$

(b) $\overline{Y}Z + Y\overline{Z} + YZ + \overline{Y}\,\overline{Z} = 1$

(c) $\overline{A} + AB + A\overline{C} + A\overline{B}\,\overline{C} = \overline{A} + B + \overline{C}$

(d) $Y\overline{Z} + \overline{X}\,\overline{Z} + \overline{X}\,\overline{Y} = Y\overline{Z} + \overline{X}\,\overline{Y}$

2–3. *Prove the identity of each of the following Boolean equations, using algebraic manipulation:

(a) $A\overline{B} + \overline{A}\,\overline{C}\,\overline{D} + \overline{A}\,\overline{B}D + \overline{A}\,\overline{B}C\overline{D} = \overline{B} + \overline{A}\,\overline{C}\,\overline{D}$

(b) $XZ + W\overline{Y}\,\overline{Z} + \overline{W}Y\overline{Z} + W\overline{X}\,\overline{Z} = XZ + W\overline{Y}\,\overline{Z} + WX\overline{Y} + \overline{W}XY + \overline{X}Y\overline{Z}$

(c) $CD + A\overline{B} + AC + \overline{A}\,\overline{C} + \overline{A}B + \overline{C}\,\overline{D} = (\overline{A} + \overline{B} + C + \overline{D})(A + B + \overline{C} + D)$

2–4. *A specific Boolean algebra with just two elements 0 and 1 has been used in this chapter. Other Boolean algebras can be defined with more than two elements by using elements that correspond to binary strings. These algebras form the mathematical foundation for bitwise logical operations that we will study in Chapter 7. Suppose that the strings are each a byte of eight bits. Then there are 2^8, or 256, elements in the algebra, where an element I is the eight-bit byte in binary corresponding to I in decimal. Based on bitwise application of the two-element Boolean algebra, define each of the following for the new algebra so that the Boolean identities hold:

(a) The OR operation $A + B$ for any two elements A and B.

(b) The AND operation $A \cdot B$ for any two elements A and B.

(c) The element that acts as the 0 for the algebra.

(d) The element that acts as the 1 for the algebra.

(e) For any element A, the element $\overline{A}$.

2–5. Simplify the following Boolean expressions to a minimum number of literals:

(a) $ABC + AB\overline{C} + \overline{A}B$

(b) $\overline{(A + B)}(\overline{A} + \overline{B})$

(c) $\overline{A}BC + AC$

(d) $BC + B(AD + A\overline{D})$

(e) $(A + \overline{B} + A\overline{B})(AB + \overline{A}C + BC)$

2–6. Reduce the following Boolean expressions to the indicated number of literals:

(a) $\overline{X}\,\overline{Y} + XYZ + \overline{X}Y$ to three literals

(b) $X + Y(Z + \overline{X + Z})$ to two literals

(c) $\overline{W}X(\overline{Z} + \overline{Y}Z) + X(W + \overline{W}YZ)$ to one literal

(d) $(AB + \overline{A}\,\overline{B})(\overline{C}\,\overline{D} + CD) + \overline{AC}$ to four literals

2–7. Using DeMorgan's theorem, express the function

$$F = \overline{A}\,\overline{B} + AB + \overline{B}C$$

(a) with only OR and complement operations.
(b) with only AND and complement operations.

2–8. Find the complement of the following expressions:

(a) $A\overline{B} + \overline{A}B$

(b) $(\overline{V}W + X)Y + \overline{Z}$

(c) $WX(\overline{Y}Z + Y\overline{Z}) + \overline{W}\,\overline{X}(\overline{Y} + Z)(Y + \overline{Z})$

(d) $(A + \overline{B} + C)(\overline{A}\,\overline{B} + C)(A + \overline{B}\,\overline{C})$

2–9. Obtain the truth table of the following functions, and express each function in sum-of-minterms and product-of-maxterms form:

(a) $(XY + Z)(Y + XZ)$

(b) $(\overline{A} + B)(\overline{B} + C)$

(c) $WX\overline{Y} + WX\overline{Z} + WXZ + Y\overline{Z}$

2–10. For the Boolean functions E and F, as given in the following truth table:

X	Y	Z	E	F
0	0	0	1	0
0	0	1	1	0
0	1	0	1	1
0	1	1	0	0
1	0	0	1	0
1	0	1	0	0
1	1	0	0	1
1	1	1	0	1

(a) List the minterms and maxterms of each function.
(b) List the minterms of $\overline{E}$ and $\overline{F}$.
(c) List the minterms of $E + F$ and $E \cdot F$.
(d) Express E and F in sum-of-minterms algebraic form.
(e) Simplify E and F to expressions with a minimum number of literals.

2–11. Convert the following expressions into sum-of-products and product-of-sums forms:

(a) $(AB + C)(B + \overline{C}D)$
(b) $\overline{X} + X(X + \overline{Y})(Y + \overline{Z})\,)$
(c) $(A + B\overline{C} + CD)(\overline{B} + EF)$

2–12. Draw the logic diagram for the following Boolean expressions. The diagram should correspond exactly to the equation.

(a) $B\overline{C} + AB + ACD$
(b) $(AB + \overline{A}\,\overline{B})(C\overline{D} + \overline{C}D)$
(c) $W\overline{X}(\overline{Y} + Z) + (\overline{W} + Y\overline{Z})(X + \overline{Z})$

2–13. Simplify the following Boolean functions by means of a three-variable map:

(a) $F(X, Y, Z) = \Sigma m(1, 3, 6, 7)$
(b) $F(X, Y, Z) = \Sigma m(3, 5, 6, 7)$
(c) $F(A, B, C) = \Sigma m(0, 1, 2, 4, 6)$
(d) $F(A, B, C) = \Sigma m(0, 3, 4, 5, 7)$

2–14. Simplify the following Boolean expressions, using a map:

(a) $\overline{X}\,\overline{Z} + Y\overline{Z} + XYZ$
(b) $\overline{A}B + \overline{B}C + \overline{A}\,\overline{B}\,\overline{C}$
(c) $\overline{A}\,\overline{B} + A\overline{C} + \overline{B}C + \overline{A}B\overline{C}$

2–15. Simplify the following Boolean functions, by means of a four-variable map:

(a) $F(A, B, C, D) = \Sigma m(1, 5, 9, 12, 13, 15)$
(b) $F(W, X, Y, Z) = \Sigma m(1, 3, 9, 11, 12, 13, 14, 15)$
(c) $F(A, B, C, D) = \Sigma m(0, 2, 4, 5, 6, 7, 8, 10, 13, 15)$

2–16. Simplify the following Boolean functions, using a map:

(a) $F(W, X, Y, Z) = \Sigma m(1, 3, 4, 6, 7, 13, 15)$
(b) $F(A, B, C, D) = \Sigma m(0, 1, 2, 5, 8, 10, 11, 13)$

2–17. Find the minterms of the following expressions by first plotting each expression on a map:

(a) $XY + XZ + \overline{X}YZ$
(b) $XZ + \overline{W}X\overline{Y} + WXY + \overline{W}YZ + W\overline{Y}Z$
(c) $\overline{B}\,\overline{D} + ABD + \overline{A}BC$

2–18. Find all the prime implicants for the following Boolean functions, and determine which are essential:

(a) $F(W, X, Y, Z) = \Sigma m(0, 2, 5, 7, 8, 10, 12, 13, 14, 15)$

(b) $F(A,B,C,D) = \Sigma m(0,2,3,5,7,8,10,11,14,15)$

(c) $F(A,B,C,D) = \Sigma m(1,3,4,5,9,10,11,12,13,14,15)$

2–19. Simplify the following Boolean functions by finding all prime implicants and essential prime implicants and applying the selection rule:

(a) $F(A,B,C,D) = BD + AB\overline{C} + \overline{A}\,\overline{C}D + \overline{A}BC + ACD$

(b) $F(A,B,C,D) = \Sigma m(1,3,7,9,12,13,14,15)$

(c) $F(W,X,Y,Z) = \Sigma m(0,1,4,5,6,7,8,9,10,11,14,15)$

2–20. Simplify the following Boolean functions in product-of-sums form:

(a) $F(W,X,Y,Z) = \Sigma m(0,1,2,6,8,9,10,13)$

(b) $F(A,B,C,D) = \Pi M(1,3,5,6,7,9,10,11,14)$

2–21. Simplify the following expressions in (1) sum-of-products and (2) product-of-sums forms:

(a) $A\overline{C} + \overline{B}D + \overline{A}CD + ABCD$

(b) $(\overline{A} + \overline{B} + \overline{D})(A + \overline{B} + \overline{C})(\overline{A} + B + \overline{D})(B + \overline{C} + \overline{D})$

(c) $(\overline{A} + \overline{B} + D)(\overline{A} + \overline{D})(A + B + \overline{D})(A + \overline{B} + C + D))$

2–22. Simplify the following Boolean functions F together with the don't-care conditions d:

(a) $F(X,Y,Z) = \Sigma m(0,1,2,4,5)$, $d(X,Y,Z) = \Sigma m(3,6,7)$

(b) $F(A,B,C,D) = \Sigma m(0,6,8,13,14)$, $d(A,B,C,D) = \Sigma m(2,4,10)$

(c) $F(A,B,C,D) = \Sigma m(1,3,5,7,9,15)$, $d(A,B,C,D) = \Sigma m(4,6,12,13)$

2–23. Simplify the following Boolean functions F together with the don't-care conditions d. Find all prime implicants and essential prime implicants, and apply the selection rule.

(a) $F(A,B,C) = \Sigma m(3,5,6)$, $d(A,B,C) = \Sigma m(0,7)$

(b) $F(W,X,Y,Z) = \Sigma m(0,2,4,5,8,14,15)$, $d(W,X,Y,Z) = \Sigma m(7,10,13)$

(c) $F(A,B,C,D) = \Sigma m(4,6,7,8,12,15)$,
$d(A,B,C,D) = \Sigma m(2,3,5,10,11,14)$

2–24. Simplify the following Boolean functions F together with the don't-care conditions d in (1) sum-of-products and (2) product-of-sums form:

(a) $F(W,X,Y,Z) = \Sigma m(0,1,2,3,7,8,10)$, $d(W,X,Y,Z) = \Sigma m(5,6,11,15)$

(b) $F(A,B,C,D) = \Sigma m(3,4,13,15)$,
$d(A,B,C,D) = \Sigma m(1,2,5,6,8,10,12,14)$

2–25. Simplify each of the following expressions, and implement them with NAND gates. Assume that both true and complement versions of the input variables are available.

(a) $W\overline{X} + WXZ + \overline{W}\,\overline{Y}\,\overline{Z} + \overline{W}X\overline{Y} + WX\overline{Z}$ **(b)** $XZ + XY\overline{Z} + W\overline{X}\,\overline{Y}$

2–26. Implement the following expression with two-input NAND and NOT gates. Assume that only true values of the inputs are available.

$$(AB + \overline{A}\,\overline{B})(C\overline{D} + \overline{C}D)$$

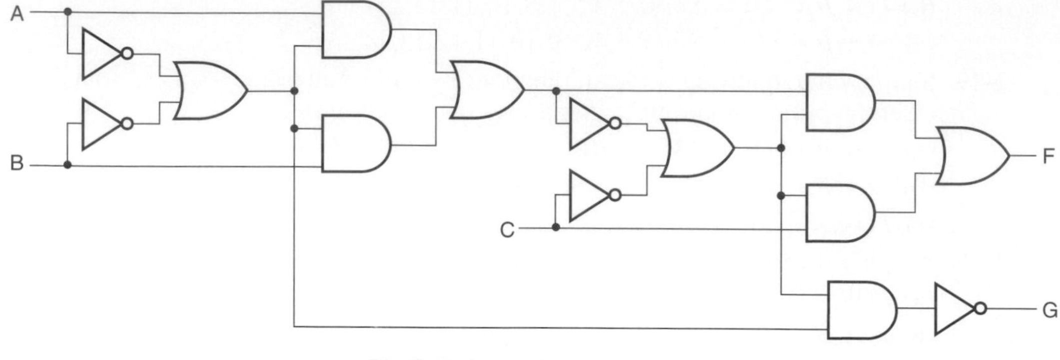

□ **FIGURE 2-50**
Diagram for Problem 2-30

(a) Use a two-level implementation plus NOT gates on the inputs, as needed.

(b) Use a multiple-level implementation to reduce the number of gates. (*Hint*: $X(\overline{XY}) = X(\overline{X} + \overline{Y}) = X\overline{Y}$.)

2–27. Draw the NAND logic diagram for each of the following expressions, using a multiple-level NAND circuit:

(a) $W(X + Y + Z) + XYZ$

(b) $(\overline{A}B + C\overline{D})E + B\overline{D}(A + B)$

2–28. Simplify each of the following expressions, and implement them with NOR gates:

(a) $W\overline{X} + Y\overline{Z} + \overline{W}Y\overline{Z}$

(b) $A\overline{B}\,\overline{C}D + \overline{A}B\overline{C}D + A\overline{B}C\overline{D} + \overline{A}BC\overline{D}$

2–29. Repeat problems 2-25 and 2-27 using NOR gates.

2–30. Convert the AND/ OR/ NOT logic diagram in Figure 2-50 to a) a NAND logic diagram and b) a NOR logic diagram.

2–31. Prove that the dual of the exclusive-OR is also its complement.

2–32. Derive the exclusive-OR/exclusive-NOR circuits for a three-bit parity generator and a four-bit parity checker, using an even parity bit.

2–33. Implement the Boolean function in problem 2-28(b) with exclusive-OR and AND gates.

2–34. An integrated circuit logic family has NAND gates with a fan-out of 5 and buffers with a fan-out of 10. Show how the output signal of a single NAND gate can be applied to 50 other gate inputs using buffers.

2–35. *A NAND gate with seven inputs is required. For each of the following cases, minimize the number of gates used in the multiple-level result:

(a) Design the 7-input NAND gate using 2-input NAND gates and NOT gates.

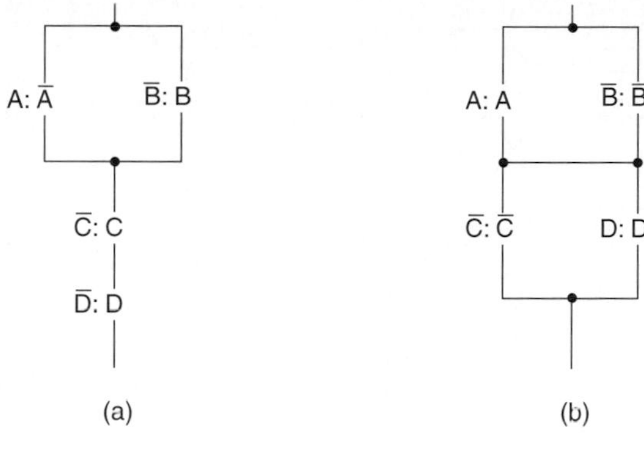

□ **FIGURE 2-51**
Switch Networks for Problem 2-38

(b) Design the 7-input NAND gate using 2-input NAND gates, 2-input NOR gates, and NOT gates.

(c) Compare the number of gates used in (a) and (b).

2–36. The NOR gates in Figure 2-36 have propagation delay $t_{pd} = 7.0$ ns. What is the propagation delay of the longest path through the circuit.

2–37. Show that a positive logic NAND gate is a negative logic NOR gate and vice versa.

2–38. Find the Boolean function that corresponds to the closed paths through each of the given switch model networks in Figure 2-51.

2–39. Implement each of the following Boolean functions as closed paths through (1) an n-channel switch model network and (2) a p-channel switch model network. In each case, use a minimum number of transistors.

(a) $F(X, Y, Z) = YZ + \overline{X}Z + \overline{X}Y\overline{Z}$

(b) $F(A, B, C, D) = A\overline{D} + A\overline{B} + BD + BC$

2–40. Find the CMOS complex gate circuit for each of the following functions:

(a) $F(A, B, C, D) = (A + \overline{C})(\overline{A} + C)(B + \overline{D})(\overline{B} + D)$

(b) $F(W, X, Y, Z) = \Sigma m(4, 7, 9, 11, 12, 13, 14, 15)$, $d(W, X, Y, Z) = \Sigma m(3, 10)$

2–41. Find a multiple-level NAND circuit for F in Problem 2-40(a), and compare the number of transistors used with the number used in that problem. A NAND gate with n inputs uses $2n$ transistors.

2–42. Construct an exclusive-NOR circuit with two NOT gates and two transmission gates.

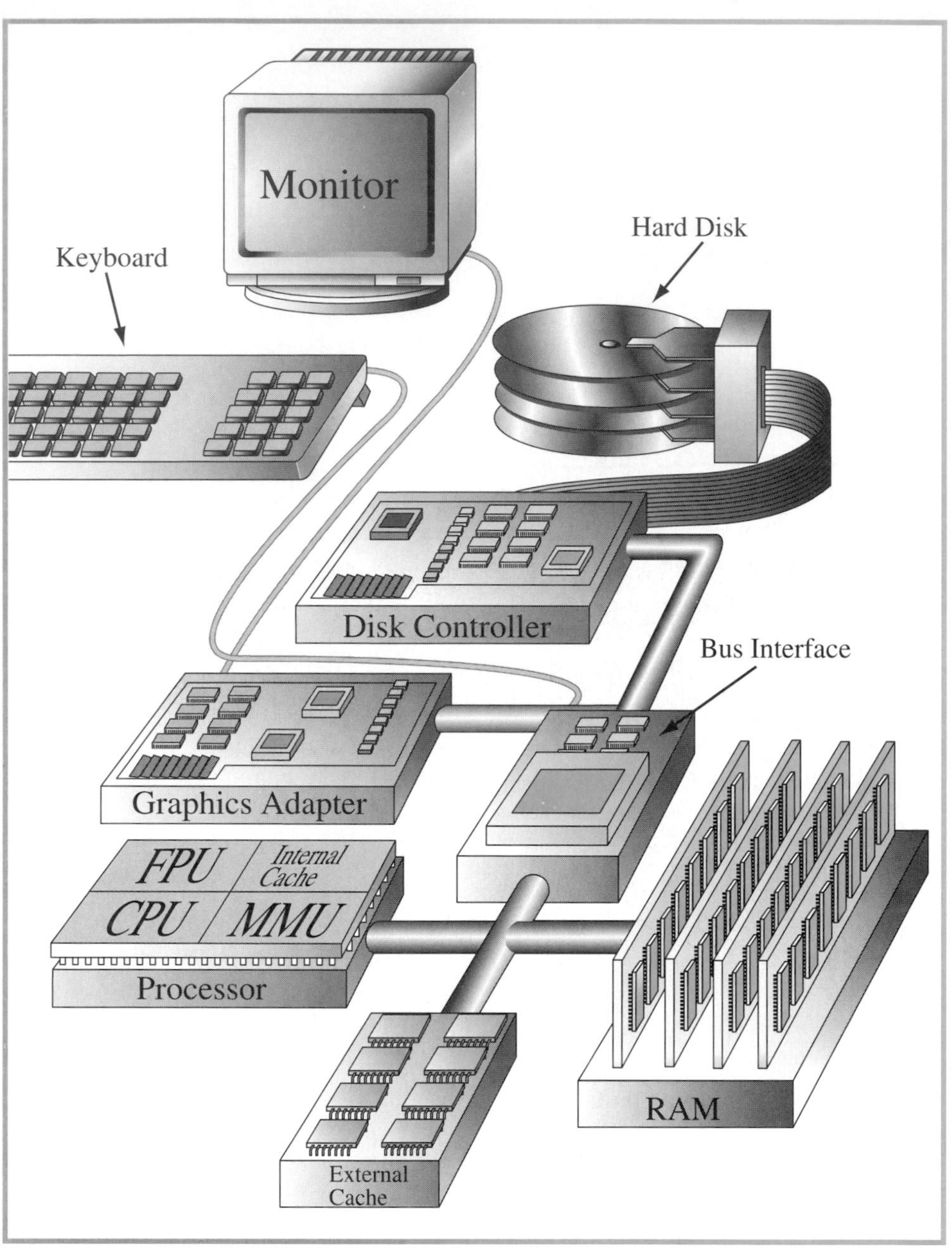
Monitor
Hard Disk
Keyboard
Disk Controller
Bus Interface
Graphics Adapter
FPU
Internal Cache
CPU
MMU
Processor
RAM
External Cache

CHAPTER

3

COMBINATIONAL LOGIC DESIGN

In this chapter, we will learn about the design of combinational circuits and about a number of fundamental circuits that are very useful in designing larger digital circuits. We introduce the use of a hierarchy and the use of computer-aided design tools, both of which are essential in the design of modern, complex circuits. Fundamental, reusable circuits referred to as functional blocks are introduced as well. These circuits include code converters, decoders, encoders, and multiplexers. A special class of functional blocks, including adders, adder-subtractors, and multipliers, performs binary and BCD arithmetic. The simplicity of these circuits comes from using complement representations for numbers and complement-based arithmetic. The chapter concludes with an alternative way of representing functional blocks in diagrams that allows the block function to be completely deduced from the graphics symbol.

Multiplexers and decoders are widely used in the generic computer. Multiplexers are very important for selecting data, in the processor, in memory, and on I/O boards. Decoders are used for selecting boards attached to the input-output bus and to decode instruction codes to determine operations that are to be performed in the processor. Encoders are used in a number of components, such as the keyboard. Adders, adder-subtractors, and multipliers are used in the processor. Since both this design methodology and functional blocks are widely used, we find the computer shaded in blue over most of the electronic portions of its hardware, including memory.

3-1 COMBINATIONAL CIRCUITS

Logic circuits for digital systems may be combinational or sequential. A combinational circuit consists of logic gates whose outputs at any time are determined by combining the values of the applied inputs using logic operations. A combinational

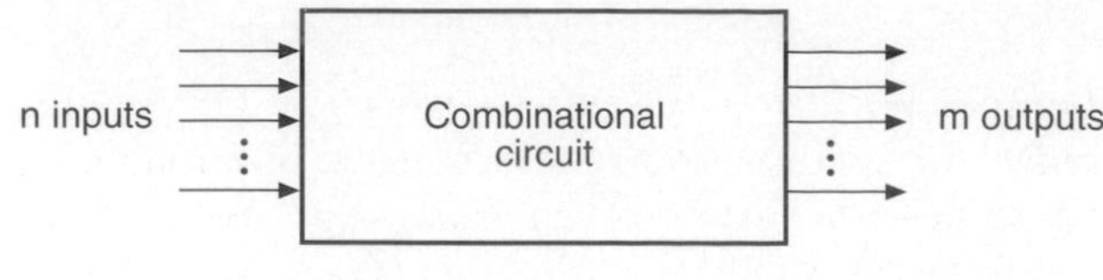

□ **FIGURE 3-1**
Block Diagram of a Combinational Circuit

circuit performs an operation that can be specified logically by a set of Boolean expressions. In addition to using logic gates, sequential circuits employ elements that store bit values. Sequential circuit outputs are a function of the inputs and the bit values in the storage elements. These values, in turn, are a function of previously applied inputs and stored values. As a consequence, the outputs of a sequential circuit depend not only on the presently applied values of the inputs, but also on past inputs, and the behavior of the circuit must be specified by a sequence in time of inputs and internal stored bit values. Sequential circuits are presented in the next chapter.

A combinational circuit consists of input variables, output variables, logic gates, and interconnections. The interconnected logic gates accept signals from the inputs and generate signals at the outputs. A block diagram of a combinational circuit is shown in Figure 3-1. The n input variables come from the environment of the circuit, and the m output variables are available for use by the environment. Each input and output variable exists physically as a binary signal that represents logic 1 or logic 0.

For n input variables, there are 2^n possible binary input combinations. For each binary combination of the input variables, there is one possible binary value on each output. Thus, a combinational circuit can be specified by a truth table that lists the output values for each combination of the input variables. A combinational circuit can also be described by m Boolean functions, one for each output variable. Each such function is expressed as a function of the n input variables.

In Chapter 1, we learned about binary numbers and binary codes that represent discrete quantities of information. In Chapter 2, we introduced the various logic gates and learned how to simplify Boolean functions in order to achieve economical gate implementations. The purpose of this chapter is to use the knowledge acquired in the previous chapters to formulate systematic analysis and design procedures for combinational circuits. The various examples introduced provide some practice in dealing with analysis and design, while treating circuits that are important as basic building blocks for larger circuits.

3-2 Design Topics

Modern digital design involves a number of techniques and tools that are essential to the design of complex circuits and systems. Design hierarchy, computer-aided design tools, and top-down design are among the most important concepts in effective and efficient digital design.

Design Hierarchy

A circuit may be specified by a symbol showing its inputs and outputs and a description defining exactly how it operates. In terms of implementation, however, a circuit is composed of logic gates that are interconnected. A complex digital system may contain millions of such interconnected gates. In fact, a single VLSI processor circuit often contains several million gates. With such complexity, the interconnected gates appear as an incomprehensible maze. Thus, no complex system or circuit can be designed simply by interconnecting gates one at a time. In order to deal with such circuit complexity, a "divide and conquer" approach is used. For design, the circuit is broken up into pieces we call *blocks*. The blocks are interconnected to form the circuit. The functions of the blocks and the interfaces between them are carefully defined, so that the circuit formed by interconnecting the blocks obeys the circuit specification. If a block is still too large and complex to be designed as a single entity, it can be broken into smaller blocks. This process can be repeated as necessary. Note that since we are working primarily with logic circuits, we use the term "circuit" in this discussion, but the ideas apply equally well to the "systems" covered in later chapters.

The "divide and conquer" approach is illustrated in Figure 3-2 by a combinational circuit design based on concepts from Section 2-7. The circuit to be designed is a 9-input odd function useful for checking even parity of a byte plus one parity bit. A symbol for this overall circuit is shown in part (a) of the figure. In part (b), a logic diagram, or *schematic*, is given for the circuit represented by the symbol in part (a). In this schematic, the designer has chosen to break up the circuit into four identical blocks, each of which is a 3-input odd function. The symbol for the 3-input odd function is repeated four times. The four symbols are interconnected to form the 9-input odd-function circuit. In part (c), a 3-input odd function block is shown to consist of two interconnected exclusive-OR gates. Finally, in part (d), the exclusive-OR is implemented by using the circuit from Figure 2-37. Note that in each case, as we move downward from the top level, symbols are replaced by schematics that represent the implementation of the symbol.

This design approach is referred to as *hierarchical design*, and the resulting related symbols and schematics constitute a *hierarchy* representing the circuit designed. The structure of the hierarchy can be represented without the interconnections by starting with the top block and connecting below each block those blocks from which it is made. The hierarchy for the 9-input odd-function circuit is shown in Figure 3-3(a) using this representation. Note that the resulting structure has a form of a tree with the root at the top. The "leaves" of the tree are the NAND gates, in this case 32 of them. In order to provide a more compact representation of the hierarchy, we can reuse blocks as shown in Figure 3-3(b). This diagram corresponds to blocks used in Figure 3-2 with only one copy of each distinct block shown. These diagrams and the circuit in Figure 3-2 are helpful in illustrating a number of useful concepts associated with hierarchies and hierarchical blocks.

First of all, a hierarchy reduces the complexity required to represent the schematic diagram of a circuit. For example, note in Figure 3-3(a) that 32 NAND blocks appear. This means that if a 9-input odd-function circuit was designed

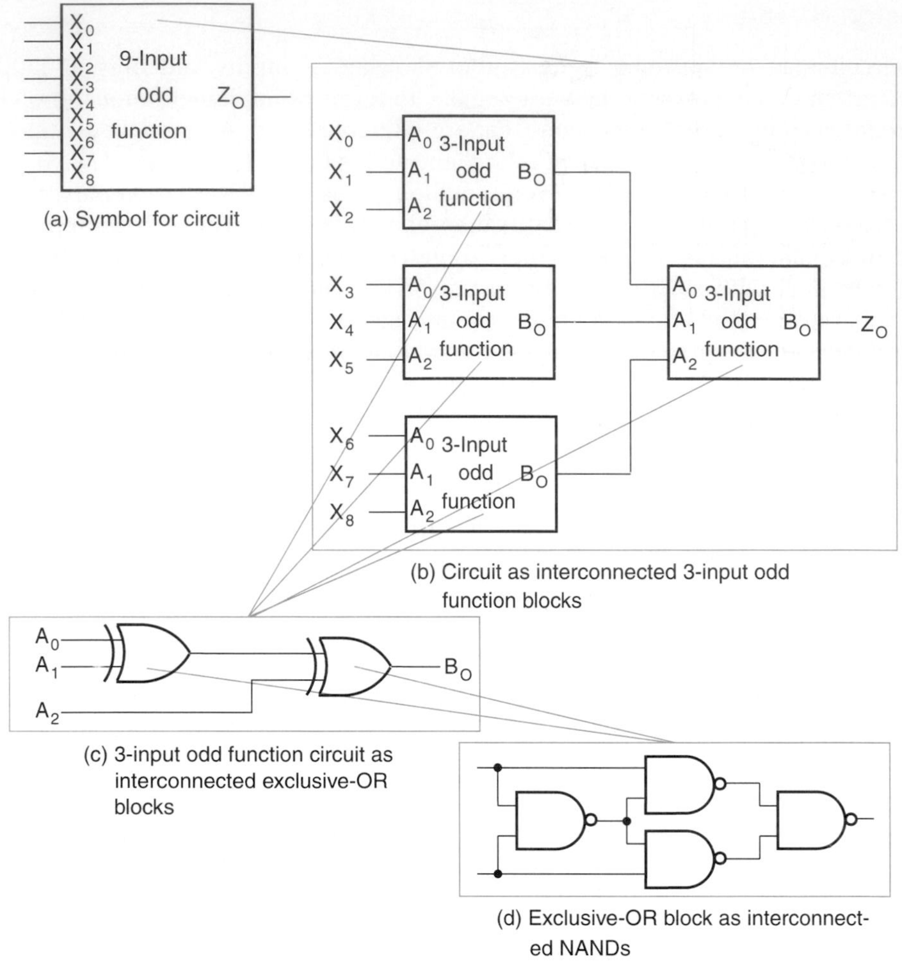

□ **FIGURE 3-2**
Example of Design Hierarchy and Reusable Blocks

directly in terms of NAND gates, the schematic for the circuit would consist of 32 interconnected NAND gate symbols, in contrast to just 10 symbols used to describe the circuit implementation as a hierarchy in Figure 3-2. Thus, a hierarchy gives a simpler representation of a complex circuit.

Second, the hierarchy ends at a set of "leaves" in Figure 3-3. In this case, the leaves consist of NAND gates. Since the NAND gates are electronic circuits, and we are interested here only in designing the logic, the NAND gates are commonly called *primitive blocks*. These are rudimentary blocks, such as gates, that have a symbol, but no logic schematic. Primitive blocks are the rudimentary type of *predefined blocks*, which are more complex structures that likewise have symbols, but no logic schematics. Instead of schematics, their function can be defined by a pro-

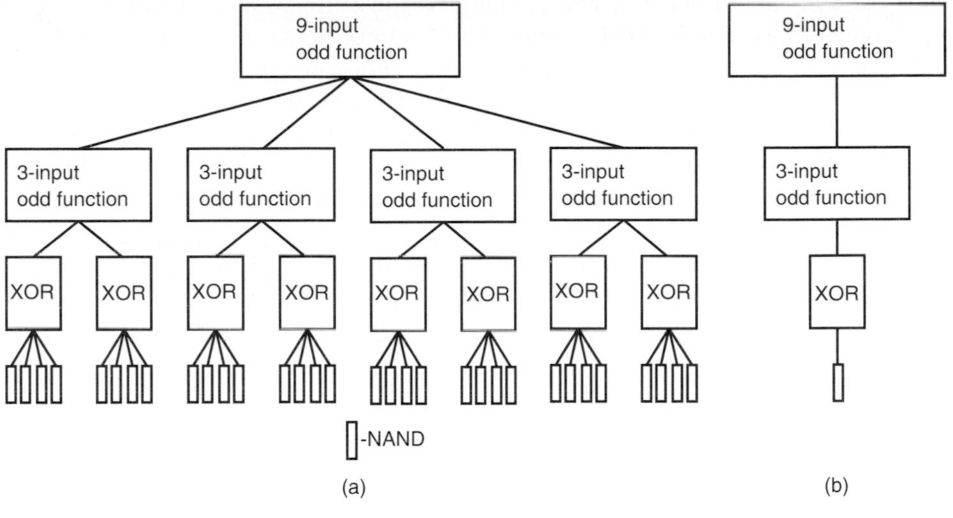

□ **FIGURE 3-3**
Diagrams Representing the Hierarchy for Figure 3-2

gram or description that can serve as a model. For example, in the hierarchy depicted in Figure 3-3, the exclusive-OR gates could have been considered as predefined blocks. In such a case, the diagram describing the exclusive-OR blocks in Figure 3-2(d) would not be necessary. The hierarchy representations in Figure 3-3 would then end with the exclusive-OR blocks. In any hierarchy, the "leaves" consist of predefined blocks, some of which may be primitives.

A third very important property that results from hierarchical design is the reuse of blocks. This is illustrated in Figures 3-3(a) and (b). In part (a), there are four copies of the 3-input odd-function block and eight copies of the exclusive-OR block. In part (b), there is only one copy of the 3-input odd-function block and one copy of the exclusive-OR block. This represents the fact that the designer has to design only one 3-input odd-function block and one exclusive-OR block and can use these blocks four times and eight times, respectively, in designing the 9-input odd-function circuit. In general, suppose that at various levels of the hierarchy, blocks used are carefully defined in such a manner that many of them are identical. For repeated blocks, only one design is necessary. This design can be used in all places where the block is required. The use of a block in each such place is called an *instance* of the block. The block is *reusable* in the sense that it can be used in more than one place in the design of the circuit and, possibly, in the design of other circuits as well. This concept greatly reduces the design required for complex circuits.

In this chapter, we will focus on predefined, reusable blocks that typically lie in the lower levels of logic design hierarchies. These are blocks that provide functions that are broadly useful in digital design. They allow designers to do much of the design process above the primitive block level. We will refer to these particular blocks as *functional blocks*. Thus, a functional block in this chapter is a predefined collection of interconnected gates. Many of these functional blocks have been available for decades in MSI circuits. These same blocks are now in computer-aided design tool libraries used for designing larger integrated circuits. Although the blocks, when used in designing integrated circuits, are no longer MSI circuits, their identity as MSI circuits is so long-standing that many are still often identified by their MSI part numbers. Functional blocks provide a catalog of elementary digital components that are used extensively in the design of digital computers and systems.

Computer-Aided Design

Designing complex systems and integrated circuits would not be feasible without the use of *computer-aided design* (*CAD*) tools. *Schematic capture* tools support the drawing of blocks and interconnections at all levels in the hierarchy. At the level of primitives and functional blocks, *libraries* of graphics symbols are provided. Schematic capture tools facilitate the construction of a hierarchy by supporting the generation of symbols for hierarchical blocks and by permitting block symbols to be easily replicated for reuse.

The primitive blocks and the functional block symbols from libraries have associated models that allow the behavior and the timing of the hierarchical blocks and the circuit to be verified. This verification is performed by applying inputs to the blocks or circuit and using a *logic simulator* to determine the outputs. We will be illustrating logic simulation in a number of examples.

Also available are tools for performing logic synthesis, a process by which a specification of a block is given, along with a library of functional blocks and primitives. The specification may be a truth table or text in the form of a *hardware description language* (*HDL*). With the use of the library, the logic design for the block is automatically produced. The library elements may consist of integrated circuit layouts that can be placed in position and interconnected by automatic chip placement and routing tools. In such a case, the design process, from an HDL description to a final integrated circuit layout, can be achieved automatically. We do not deal in detail with logic or integrated circuit synthesis tools here, but make occasional references to them.

Top-Down Design

Ideally, the design process will be performed *top down*. This means that the circuit function is specified typically by text, plus constraints on cost, performance, and reliability. The circuit is then repeatedly divided into blocks as necessary to complete the design. In fact, reality departs significantly from this ideal view. In order to obtain reusability and to make maximum use of predefined modules, it is often necessary to perform portions of the design *bottom up*. In addition, a particular circuit design for a given specification may violate one of the constraints in the initial specification. In this case, it is necessary to backtrack upward through the hierarchy until a level is reached at which the violation can be eliminated. The design is then started anew at that level.

In this text, since reader familiarity with logic and computer design is probably limited, we need to build a ready set of functional blocks to provide direction in top-down design. Likewise, a sense of how to break up a circuit into blocks that can serve to guide the top-down approach also must be mastered. So the focus in much of the text will be on bottom-up rather than top-down design. In this chapter and in Chapter 5 and Chapter 6, we focus our efforts on the design of frequently used functional blocks to begin building the basis for top-down design. In Chapter 7 and Chapter 8, we illustrate how larger circuits and systems are broken down into blocks and how these blocks are implemented with functional blocks. Finally, beginning with Chapter 9, we apply these ideas to look at design from more of a top-down perspective.

3-3 Analysis Procedure

The analysis of a combinational circuit consists of determining the function that the circuit implements. The analysis starts with a given logic circuit diagram and culminates with a set of Boolean functions or a truth table, together with a possible explanation of the operation of the circuit. If the logic diagram to be analyzed is accompanied by a recognizable function name or a statement of what the diagram is assumed to accomplish, then the analysis problem reduces to a verification of the stated function. We can perform the analysis by manually finding the Boolean equations or truth table or by using logic simulation to apply inputs to the circuit, as mentioned in the last section. Both methods will be covered here.

The first step in the analysis is to make sure that the given circuit is combinational and not sequential. The diagram of a combinational circuit has logic gates with no feedback or storage elements. A feedback path exists if, from the output of a gate, an input of the same gate can be reached via a path of interconnections or gates. Feedback paths or storage elements in a digital circuit may result in a sequential circuit and must be analyzed according to procedures outlined in Chapter 4.

Once the logic diagram is verified to be that of a combinational circuit, one can proceed to obtain the output Boolean functions or the truth table. If the function of the circuit is to be investigated, then it is necessary to interpret how the circuit operates from the derived Boolean functions or truth table. The success of such an investigation is enhanced if one has previous experience and familiarity with a wide variety of digital circuits.

Derivation of Boolean Functions

To obtain the output Boolean functions from a logic diagram, proceed as follows:

1. Label all gate outputs that are a function only of input variables or their complements with arbitrary symbols. Determine the Boolean functions for each gate output.

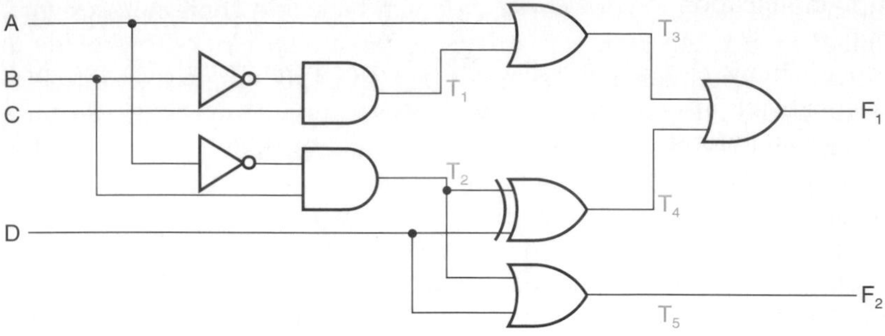

□ **FIGURE 3-4**
Logic Diagram for Analysis Example

2. Label the gates that are a function of input variables and previously labeled gates with different arbitrary symbols. Find the Boolean functions for the outputs of these gates.
3. Repeat the process outlined in step 2 until the outputs of the circuit are obtained in terms of the input variables.

The analysis of the combinational circuit of Figure 3-4 illustrates this procedure.

Note that the circuit has four binary input variables A, B, C, and D and two binary output variables F_1 and F_2. The outputs of the various gates are labeled with intermediate symbols. The outputs of gates that are a function of input variables only are T_1 and T_2. The Boolean functions for these two outputs are

$$T_1 = \overline{B}C$$

$$T_2 = \overline{A}B$$

Next, we consider the outputs of gates that are functions of gates with symbols already defined:

$$T_3 = A + T_1 = A + \overline{B}C$$

$$T_4 = T_2 \oplus D = (\overline{A}B) \oplus D = \overline{A}B\overline{D} + AD + \overline{B}D$$

$$T_5 = T_2 + D = \overline{A}B + D$$

The Boolean functions for the outputs are thus

$$F_2 = T_5 = \overline{A}B + D$$

$$F_1 = T_3 + T_4 = A + \overline{B}C + \overline{A}B\overline{D} + AD + \overline{B}D$$

$$= A + \overline{B}C + B\overline{D} + \overline{B}D$$

The last simplification for F_1 can be verified by algebraic manipulation or by means of a map.

If the circuit to be analyzed is designed hierarchically, the diagrams for the lowest level block instances are analyzed first, those for block instances in the next level up are analyzed next, and so on, until Boolean equations are obtained for all outputs. (See Problem 3–6 at the end of the chapter.)

Derivation of the Truth Table

The derivation of the truth table for a combinational circuit is a straightforward process once the output Boolean functions are known. Alternatively, to obtain the truth table from the logic diagram without going through the derivation of the Boolean functions, proceed as follows:

1. Determine the number of input variables in the circuit. For n inputs, list the binary numbers from 0 to $2^n - 1$ in a table.
2. Break the circuit into small single-output blocks by labeling each block output with an arbitrary symbol.
3. Obtain the truth table for the blocks which have functions that depend on input variables only.
4. Proceed to obtain the truth table for blocks which have functions that depend on previously defined inputs and block outputs, until the columns for all circuit outputs are determined.

The foregoing process is illustrated by the combinational circuit for a binary adder in Figure 3-5. The problem here is to verify that the circuit forms the arithmetic sum of the three bits at inputs X, Y, and Z. The output pair (C, S) ranges in value from binary 00 to 11 (decimal 3), depending on the number of 1's in the inputs. For example, when $(X, Y, Z) = 101$, (C, S) must be equal to binary 10 to indicate that there are two 1's on the inputs.

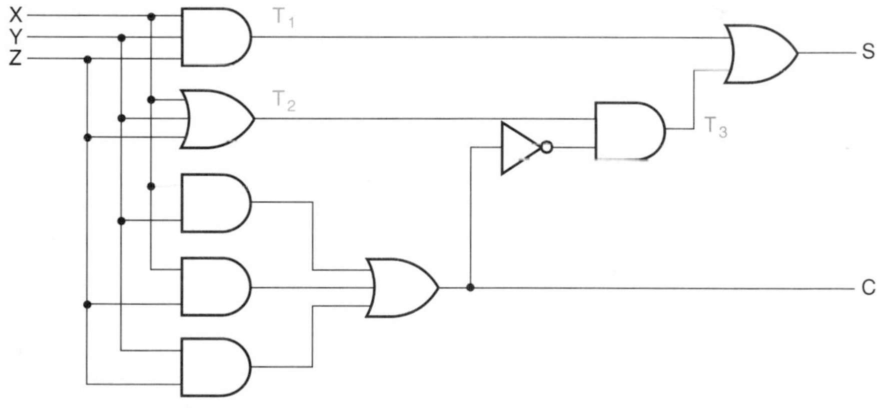

□ **FIGURE 3-5**
Logic Diagram for Binary Adder

□ **TABLE 3-1**
Truth Table for Binary Adder

X	Y	Z	C	$\overline{C}$	T_1	T_2	T_3	S
0	0	0	0	1	0	0	0	0
0	0	1	0	1	0	1	1	1
0	1	0	0	1	0	1	1	1
0	1	1	1	0	0	1	0	0
1	0	0	0	1	0	1	1	1
1	0	1	1	0	0	1	0	0
1	1	0	1	0	0	1	0	0
1	1	1	1	0	1	1	0	1

Table 3-1 shows the procedure for deriving the truth table of the circuit. First, we split up the circuit into blocks by labeling selected gate outputs. Next, we form the eight binary combinations for the three input variables and then find and logically combine truth tables. The truth table for C is determined from the values of the inputs X, Y, and Z. C is equal to 1 when $(X, Y) = 11$ or $(X, Z) = 11$ or $(Y, Z) = 11$. Otherwise, C is equal to 0. The truth table for $\overline{C}$ is the complement of C. The truth tables for T_1 and T_2 are the AND and OR functions of the input variables, respectively. The binary values for T_3 are derived from ANDing T_2 and $\overline{C}$. Thus, T_3 is equal to 1 when both T_2 is equal to 1 and $\overline{C}$ is equal to 1. Otherwise, T_3 is equal to 0. Finally, S is equal to 1 for those combinations in which either T_1 or T_3 or both are equal to 1.

Inspection of the truth table reveals that $(C, S) = 00, 01, 10,$ or 11 when the total number of 1's on the three inputs X, Y, and Z is either zero, one, two, or three, respectively. This verifies the operation of the circuit as a binary adder. The design of the binary adder with a different gate structure is presented in Section 3-8.

If the circuit to be analyzed is hierarchically designed with blocks containing more than a few gates, it is wise to find the Boolean equations for the circuit outputs and evaluate these equations to get the truth table. The alternative, combining truth tables of inputs to complex logical structures to get the output truth table, is difficult to do manually. (See Problem 3–7 at the end of the chapter.)

Logic Simulation

Logic simulation is a fast, accurate method of analyzing a combinational circuit if logic waveforms, a truth table, or a portion of a truth table is the desired result. Simulation, however, does not usually produce Boolean equations for the circuit.

In order to simulate a circuit, it must be described in a form that the simulator can read. This can be a *net list* which specifies using text the inputs, gates, outputs, and their interconnections. More likely, it will be a logic *schematic*, which is a form of logic diagram. To generate a schematic from a logic diagram, we use a *schematic capture* tool. The tool provides symbols for gates, inputs, and outputs that can be taken from a library and included in the schematic. In addition, the

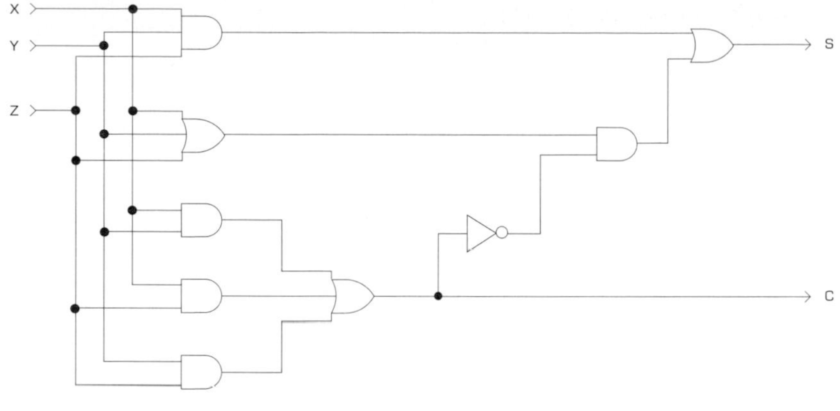

□ **FIGURE 3-6**
ViewDraw Schematic for Binary Adder in Figure 3-5

schematic capture tool provides a wiring tool that permits the gates to be interconnected and a labeling tool to label the schematic. It also often provides a tool for generating symbols that can be used in the schematic to produce a design hierarchy. For use in simulation, underlying software models are associated with the primitive or predefined blocks. A schematic obtained from the logic diagram in Figure 3-5 by using the Viewlogic® ViewDraw® schematic capture tool is shown in Figure 3-6.

Once the circuit has been entered as a net list or schematic, logic simulation can be performed. Typically, inputs to the circuit can be provided in one of two ways: by specifying the inputs as the contents of a file that the simulator can read, or by entering the inputs interactively into the simulator. For the circuit we are considering, if we want all of the truth table information, we need to apply all eight possible binary input combinations. On the other hand, if we are interested only in the response to two or three combinations, we need to apply just those combinations.

For the binary adder, we assume that we want to generate waveforms corresponding to all of the truth table, so apply all eight input combinations. Figure 3-7 shows ViewTrace™ waveforms for the simulation inputs we applied and the corresponding outputs generated by Viewlogic ViewSim®. The eight binary combinations on inputs X, Y, and Z are represented from left to right along with the output values on S and C. The inputs are changed at intervals of 20 nanoseconds (ns) in this simulation. Note that the outputs do not change immediately for each input change. For this simulation, we set all gate delays to 2 ns. Since the circuit has up to five gates between an input and the S output, the S output may change up to 10 ns after the input change. If we use actual gate delays in place of the 2 ns delays, we can obtain useful timing information from the simulation.

The cross-hatch regions in outputs S and C at $T = 0$ are also due to the gate delays. Since we apply the inputs beginning at $T = 0$, the outputs will be unknown until the inputs propagate through to the circuit outputs. The cross-hatched regions

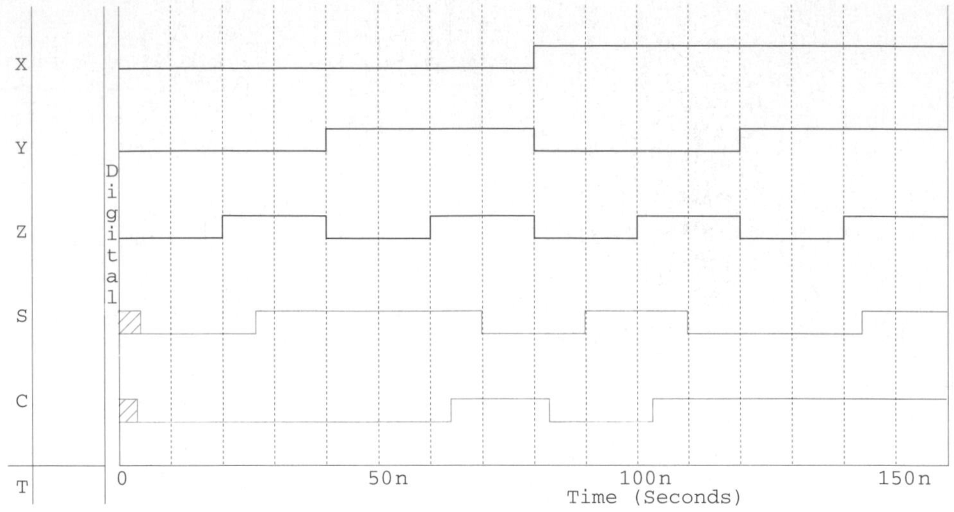

□ **FIGURE 3-7**
ViewTrace Input and Output Waveforms for the Binary Adder Schematic in Figure 3-6

represent an unknown output on S for three gate delays (6 ns) and on C for two gate delays (4 ns). These delays are the lengths of time it takes for the all zero combination applied at $T = 0$ to fix the outputs at a known value.

To verify the correctness of the circuit function from the waveforms, we check the values of the outputs of S and C against the truth table for the full adder in Table 3-1. When an input combination changes, we can read the output values from the waveforms only after the input change has propagated to the outputs. For example, the change in output S due to the input change in Z at $T = 60$ ns occurs at $T = 70$ ns. A good time at which to read off the values for input combination i is just before input combination $i + 1$ is applied. For example, we read the output values for input combination 3 (011) at $T = 79$ ns, just before input combination 4 (100) is applied at $T = 80$ ns. This is late enough in the input interval for the values to have propagated to the outputs. If this is not true, then the interval we used for applying each input is too short and must be lengthened to obtain a valid simulation.

3-4 Design Procedure

The design of a combinational circuit starts from the specification of the problem and culminates in a logic diagram or set of Boolean equations from which the logic diagram can be obtained. The procedure involves the following steps:

1. From the specifications of the circuit, determine the required number of inputs and outputs, and assign a letter symbol to each.

2. Derive the truth table that defines the required relationship between inputs and outputs.
3. Obtain the simplified Boolean functions for each output as a function of the input variables.
4. Draw the logic diagram.
5. Verify the correctness of the design.

A truth table for a combinational circuit consists of input columns and output columns. The input columns are obtained from the 2^n binary numbers for the n input variables. The binary values for the outputs are determined from the stated specifications. The output functions specified in the truth table give the exact definition of the combinational circuit. It is important that the verbal specifications be interpreted correctly in forming the truth table. Often they are incomplete, and any wrong interpretation may result in an incorrect truth table.

The output binary functions listed in the truth table are simplified by any available method, such as algebraic manipulation, the map method, or computer-based simplification programs. Frequently, there is a variety of simplified expressions from which to choose. In a particular application, certain criteria will serve as a guide in the process of choosing an implementation. A practical design must consider constraints such as the number of gates used, number of inputs to a gate, maximum allowable propagation time of a signal through the circuit, number of interconnections, and limitations on the fanout of each gate. Since the importance of each constraint is dictated by the particular application, it is difficult to make a general statement about what constitutes an acceptable implementation. In most cases, the simplification begins by satisfying an elementary objective, such as producing the simplified Boolean functions in a standard form, and then proceeds with further steps to meet other performance criteria. Example 3-1 is the first of a number of examples in this chapter that illustrate the procedure outlined.

■ **EXAMPLE 3-1**

Design a combinational circuit with three inputs and one output. The output must be logic 1 when the binary value of the inputs is less than 011(3) and logic 0 otherwise. Use only NAND gates.

The design of the circuit is undertaken in Figure 3-8. We designate the inputs with the letter symbols $X, Y,$ and Z and the output with F. The truth table is listed in part (a) of the figure. $F = 1$ when the binary inputs correspond to 0, 1, or 2; otherwise, $F = 0$. The simplified Boolean output function is derived from the map in part (b). The logic diagram is drawn in part (c) using alternative symbols for NAND. (See Section 2-6.) The result can be verified by finding the truth table for the resulting NAND circuit and checking it against the initial verbal specification. ■

Code Converters

When a combinational circuit has two or more outputs, each output must be expressed separately as a function of all the input variables. An example of a multi-

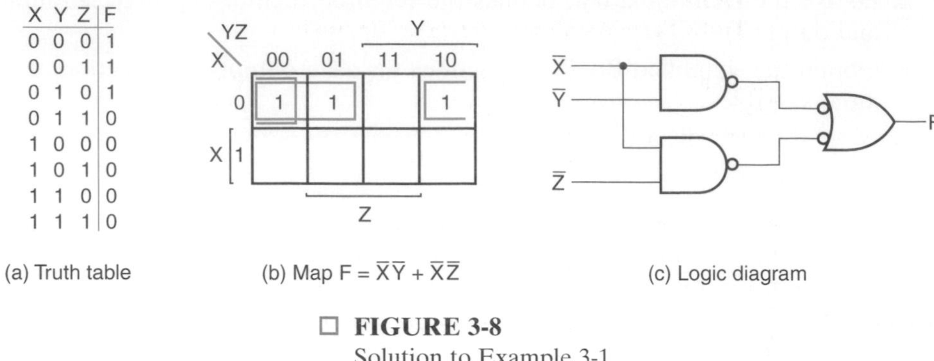

X	Y	Z	F
0	0	0	1
0	0	1	1
0	1	0	1
0	1	1	0
1	0	0	0
1	0	1	0
1	1	0	0
1	1	1	0

(a) Truth table (b) Map $F = \overline{X}\,\overline{Y} + \overline{X}\,\overline{Z}$ (c) Logic diagram

□ **FIGURE 3-8**
Solution to Example 3-1

ple-output circuit is a code converter, which is a circuit that translates information from one binary code to another. The inputs of the circuit provide the bit combination of the elements as specified by the first code, and the outputs generate the corresponding bit combination of the second code. The combinational circuit performs the transformation from one code to the other. The design of code converters will be illustrated by means of two examples. One converts BCD to what we call excess-3 code for the decimal digits. The other converts BCD to the seven signals required to drive a seven-segment light-emitting diode (LED) display.

■ EXAMPLE 3-2
BCD–to–Excess-3 Code Converter

The *excess-3 code* for a decimal digit is the binary combination corresponding to the decimal digit plus 3. For example, the excess-3 code for decimal digit 5 is the binary combination for $5 + 3 = 8$, which is 1000. The excess-3 code has desirable properties with respect to implementing decimal subtraction.

Since both ordinary BCD and excess-3 use four bits to represent each decimal digit, there must be four input variables and four output variables. Designate the inputs by A, B, C, D and the outputs by W, X, Y, Z. The truth table relating the input and output variables is shown in Table 3-2. The excess-3 code word is easily obtained from a BCD code word by adding binary 0011 (3) to it. Note that four binary variables may have 16 bit combinations, but only 10 are listed in the truth table. The 6 combinations 1010 through 1111 are not listed under the inputs. These combinations have no meaning in the BCD code, and we can assume that they will never occur. Hence, it does not matter what binary values we assign to the outputs, and therefore, we can treat them as don't-care conditions. The maps in Figure 3-9 are plotted to obtain the simplified Boolean functions for the outputs. Each one of the four maps represents one of the outputs of the circuit as a function of the four inputs. The l's in the map are obtained directly from the truth table by going over the output columns one at a time. For example, the column under output W has 1's for minterms 5, 6, 7, 8, and 9. Therefore, the map for W must have five 1's in the squares that correspond to these minterms. The six don't-care minterms 10 through

□ **TABLE 3-2**
Truth Table for Code Converter Example

Decimal Digit	Input BCD				Output Excess-3			
	A	B	C	D	W	X	Y	Z
0	0	0	0	0	0	0	1	1
1	0	0	0	1	0	1	0	0
2	0	0	1	0	0	1	0	1
3	0	0	1	1	0	1	1	0
4	0	1	0	0	0	1	1	1
5	0	1	0	1	1	0	0	0
6	0	1	1	0	1	0	0	1
7	0	1	1	1	1	0	1	0
8	1	0	0	0	1	0	1	1
9	1	0	0	1	1	1	0	0

15 are each marked with an × in all the maps. The simplified functions are listed in sum-of-products form under the map of each variable.

The two-level AND-OR logic diagram for the circuit can be obtained directly from the Boolean expressions derived from the maps. There are, however, various other possibilities for a logic diagram that implements the circuit. For example, expressions may be manipulated algebraically for the purpose of sharing common gates between the four functions being implemented. The following manipulation illustrates the flexibility obtained with multiple-output circuits implemented with three levels of gates:

$$W = A + BC + BD = A + B(C + D)$$

$$X = \overline{B}C + \overline{B}D + B\overline{C}\,\overline{D} = \overline{B}(C + D) + B\overline{C}\,\overline{D}$$

$$Y = CD + \overline{C}\,\overline{D} = \overline{C \oplus D}$$

$$Z = \overline{D}$$

The logic diagram that implements these expressions is drawn in Figure 3-10. The manipulation performed allows the gate producing $C + D$ to be shared by the output logic for W and X. Compared to the straightforward sum-of-products implementation, we have saved one gate by using the distributive law to factor out $C + D$ and sharing its gate. ■

■ **EXAMPLE 3-3**
BCD–to–Seven-Segment Decoder

Digital readouts found in electronic calculators and digital watches use LEDs. Each digit of the readout is formed from seven segments, each consisting of one LED that can be illuminated by digital signals. A BCD–to–seven-segment decoder

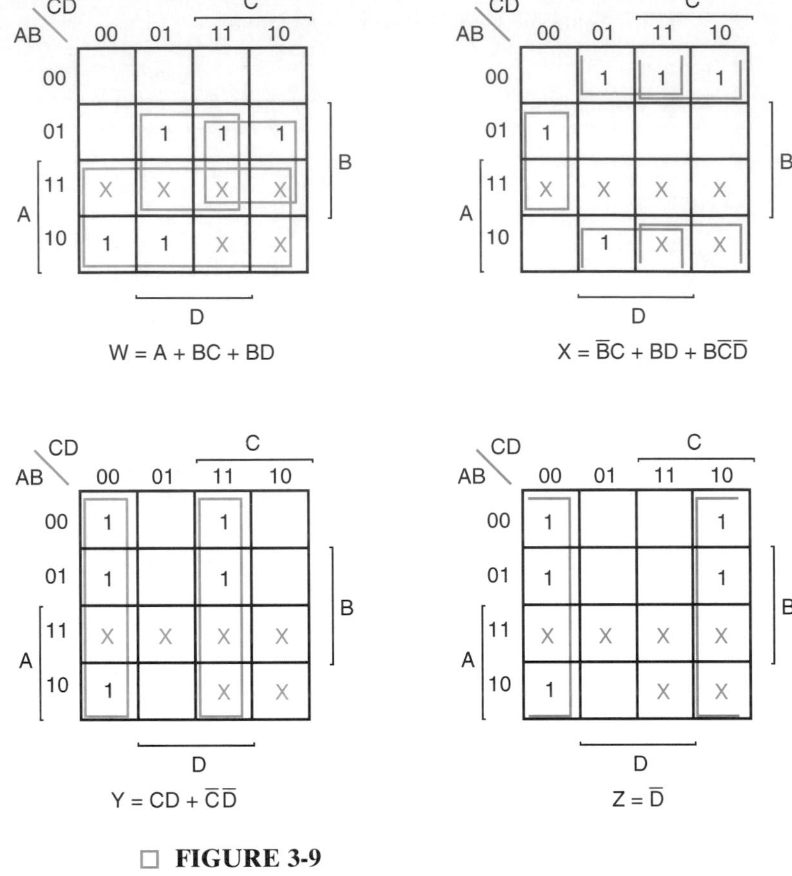

□ **FIGURE 3-9**
Maps for BCD–to–Excess-3 Code Converter

is a combinational circuit that accepts a decimal digit in BCD and generates the appropriate outputs for the selection of segments that display the decimal digit. The seven outputs of the decoder (*a*, *b*, *c*, *d*, *e*, *f*, *g*) select the corresponding segments in the display, as shown in Figure 3-11(a).The numeric designations chosen to represent the decimal digits are shown in Figure 3-11(b). The BCD–to–seven-segment decoder has four inputs for the BCD digit and seven outputs for choosing the segments. The truth table of the combinational circuit is listed in Table 3-3. Each decimal digit illuminates the proper segments for the decimal display. For example, BCD 0011 corresponds to decimal 3, whose display needs segments *a*, *b*, *c*, *d*, and *g*. The truth table assumes that a logic-1 signal illuminates the segment and a logic-0 signal turns the segment off. Some seven-segment displays operate in reverse fashion and are illuminated by a logic-0 signal. For these displays, the seven outputs must be complemented.

The six binary combinations 1010 through 1111 have no meaning in BCD. In the previous example, we assigned these combinations to don't-care conditions. If

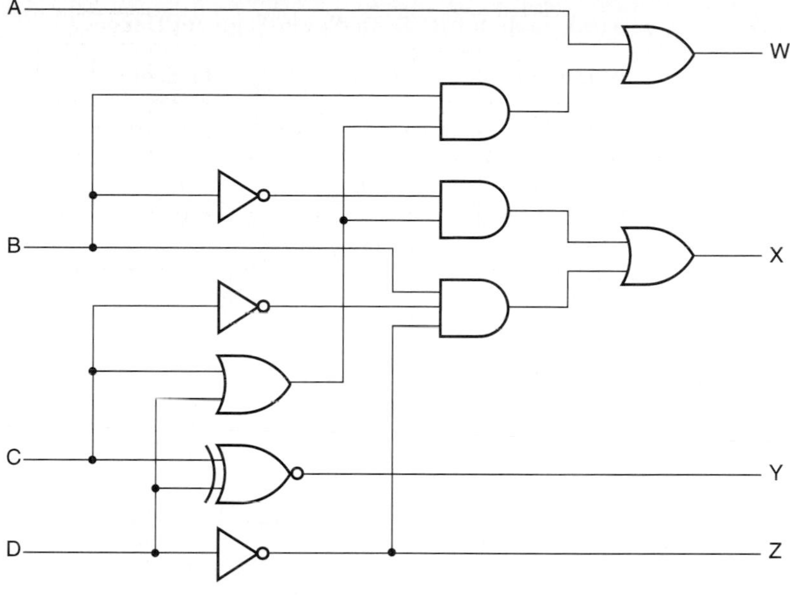

□ **FIGURE 3-10**
Logic Diagram of BCD–to–Excess-3 Code Converter

□ **FIGURE 3-11**
Seven-Segment Display

we do the same here, the design will most likely produce some arbitrary and meaningless displays for the unused combinations. A better choice may be to turn off all the segments when any one of the unused input combinations occur. This can be accomplished by assigning all 0's to minterms 10 through 15.

The information from the truth table can be transferred into seven maps from which the simplified output functions can be derived. The plotting of the seven functions in map form is left as an exercise. (See Problem 12 at the end of this chapter.) One possible way of simplifying the seven functions results in the following Boolean functions:

$$a = \overline{A}C + \overline{A}BD + \overline{B}\,\overline{C}\,\overline{D} + A\overline{B}\,\overline{C}$$

$$b = \overline{A}\,\overline{B} + \overline{A}\,\overline{C}\,\overline{D} + \overline{A}CD + A\overline{B}\,\overline{C}$$

□ **TABLE 3-3**
Truth Table for BCD–to–Seven-Segment Decoder

BCD Input				Seven-Segment Decoder						
A	**B**	**C**	**D**	**a**	**b**	**c**	**d**	**e**	**f**	**g**
0	0	0	0	1	1	1	1	1	1	0
0	0	0	1	0	1	1	0	0	0	0
0	0	1	0	1	1	0	1	1	0	1
0	0	1	1	1	1	1	1	0	0	1
0	1	0	0	0	1	1	0	0	1	1
0	1	0	1	1	0	1	1	0	1	1
0	1	1	0	1	0	1	1	1	1	1
0	1	1	1	1	1	1	0	0	0	0
1	0	0	0	1	1	1	1	1	1	1
1	0	0	1	1	1	1	1	0	1	1
All other inputs				0	0	0	0	0	0	0

$$c = \overline{A}B + \overline{A}D + \overline{B}\,\overline{C}\,\overline{D} + A\overline{B}\,\overline{C}$$

$$d = \overline{A}C\overline{D} + \overline{A}\,\overline{B}C + \overline{B}\,\overline{C}\,\overline{D} + A\overline{B}\,\overline{C} + \overline{A}B\overline{C}D$$

$$e = \overline{A}C\overline{D} + \overline{B}\,\overline{C}\,\overline{D}$$

$$f = \overline{A}B\overline{C} + \overline{A}\,\overline{C}\,\overline{D} + \overline{A}B\overline{D} + A\overline{B}\,\overline{C}$$

$$g = \overline{A}C\overline{D} + \overline{A}\,\overline{B}C + \overline{A}B\overline{C} + A\overline{B}\,\overline{C}$$

The implementation of these seven functions requires 14 AND gates and 7 OR gates. There is a total of 27 product terms in all seven functions, and it would seem that 27 AND gates will be needed. However, there are six common terms whose corresponding AND gates can be shared by two or more outputs. For example, the term $\overline{B}\,\overline{C}\,\overline{D}$ occurs in a, c, d, and e. The output of the AND gate that implements this product term goes to the inputs of the OR gates in all four functions. ■

In general, the total number of gates can be reduced in a multiple-output combinational circuit by using common terms in the output functions. The maps of the output functions may help in finding the common terms by finding identical implicants from two or more maps. Some of the common terms may not be prime implicants of the individual functions. The designer must be inventive and combine squares in the maps in such a way as to create common terms. This can be more done more formally by using a procedure for simplifying multiple-output functions in which prime implicants are defined not only for each individual function, but for all possible combinations of the output functions. These prime implicants are formed by using the AND operator on the output functions. Such a procedure is implemented in various forms in logic simplification software incorporated into logic synthesis tools.

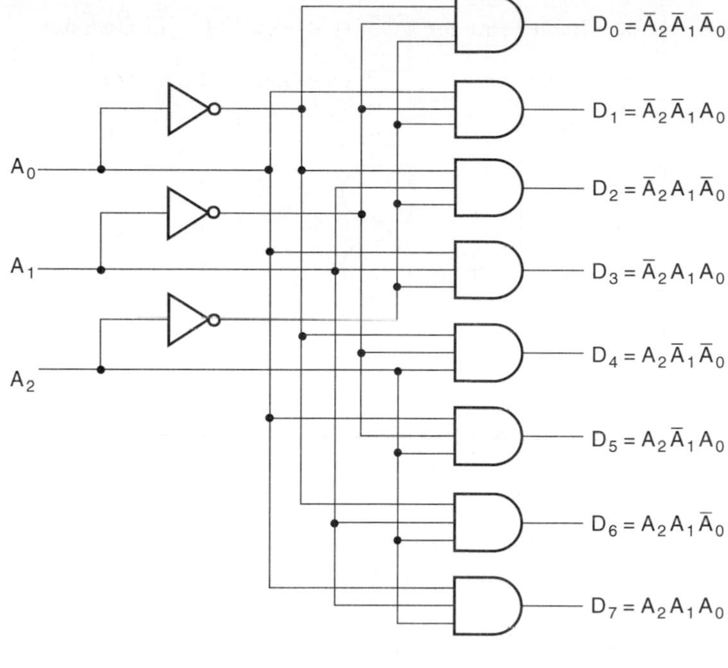

□ **FIGURE 3-12**
3–to–8-Line Decoder

The BCD–to–seven-segment decoder is called a decoder by most manufacturers of integrated circuits because it decodes a binary code for a decimal digit. However, it is actually a code converter that converts a four-bit decimal code to a seven-bit code. The word "decoder" is usually reserved for another type of circuit, presented in the next section.

3-5 Decoders

Discrete quantities of information are represented in digital computers by binary codes. A binary code of n bits is capable of representing up to 2^n distinct elements of coded information. A decoder is a combinational circuit that converts binary information from the n coded inputs to a maximum of 2^n unique outputs. If the n-bit coded information has unused bit combinations, the decoder may have fewer than 2^n outputs.

The decoders presented in this section are called n–to–m-line decoders, where $m \leq 2^n$. Their purpose is to generate the 2^n (or fewer) minterms of n input variables. The logic diagram of a 3–to–8-line decoder is shown in Figure 3-12. The three inputs are decoded into eight outputs, each representing one of the minterms of the three input variables. The three inverters provide the complement of the inputs, and each one of the eight AND gates generates one of the minterms. A par-

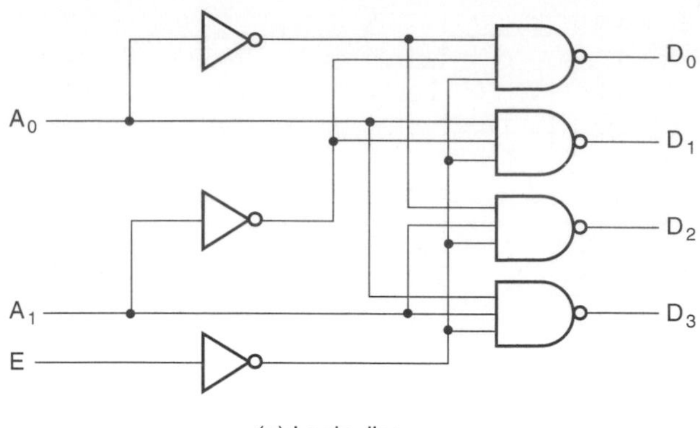

(a) Logic diagram

E	A_1	A_0	D_0	D_1	D_2	D_3
0	0	0	0	1	1	1
0	0	1	1	0	1	1
0	1	0	1	1	0	1
0	1	1	1	1	1	0
1	X	X	1	1	1	1

(b) Truth table

□ **FIGURE 3-13**
A 2–to–4-Line Decoder with Enable Input

ticular application of this decoder is binary-to-octal conversion. The input variables represent a binary number, and each output represents one of the eight digits in the octal number system.

The operation of the decoder may be clarified from the truth table listed in Table 3-4. For each possible input combination, there are seven outputs that are equal to 0 and only one that is equal to 1. The output variable equal to 1 represents the minterm equivalent of the binary number that is applied to the input lines.

Some decoders are constructed with NAND gates directly replacing the AND gates. This is more economical, since in typical logic families, an AND is implemented by a NAND followed by a NOT gate. By using the NAND instead of AND, the cost and delay of the NOT gate are eliminated. The resulting decoder minterms are in their complement form. Furthermore, most commercial decoders include one or more enable inputs to control the operation of the circuit. A 2–to–4-line decoder

□ **TABLE 3-4**
Truth Table for 3–to–8-Line Decoder

Inputs			Outputs							
A_2	A_1	A_0	D_7	D_6	D_5	D_4	D_3	D_2	D_1	D_0
0	0	0	0	0	0	0	0	0	0	1
0	0	1	0	0	0	0	0	0	1	0
0	1	0	0	0	0	0	0	1	0	0
0	1	1	0	0	0	0	1	0	0	0
1	0	0	0	0	0	1	0	0	0	0
1	0	1	0	0	1	0	0	0	0	0
1	1	0	0	1	0	0	0	0	0	0
1	1	1	1	0	0	0	0	0	0	0

with an enable input constructed with NAND gates is shown in Figure 3-13. The circuit operates with complemented outputs and enable input E complemented to match the outputs of a NAND gate decoder. The decoder is enabled when E is equal to 0. As indicated by the truth table, only one output can be equal to 0 at any given time, all other outputs being equal to 1. The output with a value of 0 represents the minterm selected by inputs A_1 and A_0. The circuit is disabled when E is equal to 1, regardless of the values of the other two inputs. When the circuit is disabled, none of the outputs are equal to 0, and none of the minterms are selected. In general, a decoder may operate with complemented or uncomplemented outputs, and the enable input may be activated with a 0 or with a 1 signal. Some decoders have two or more enable inputs that must satisfy a given logic condition in order to enable the circuit.

Decoder Expansion

There are occasions when a certain size of decoder is needed, but only smaller sizes are available. When this occurs, it is possible to combine two or more decoders with enable inputs to form a larger decoder. This is a very effective use of design hierarchy. For example, if a 6–to–64-line decoder is needed, it is possible to design it with four 4–to–16-line decoders and one 2–to–4-line decoder as functional blocks in the logic diagram.

Figure 3-14 shows how decoders with enable inputs can be connected to form a larger decoder. Two 2–to–4-line decoders are combined to achieve a 3–to–8-line decoder. The two least significant bits of the input are connected to both decoders. The most significant bit is connected to the enable input of one decoder and,

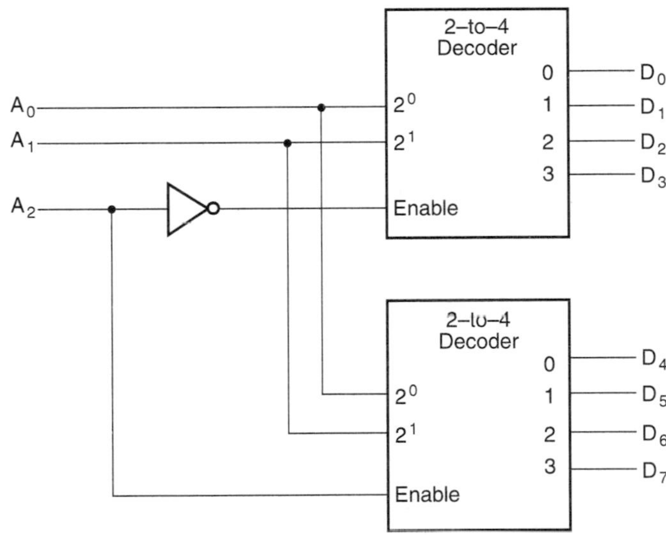

□ **FIGURE 3-14**
A 3–to–8 Decoder Constructed with Two 2–to–4 Decoders

through an inverter, to the enable input of the other decoder. When $A_2 = 0$, the upper decoder is enabled and the lower disabled. The outputs of the lower decoder become inactive and equal to 0. The upper decoder generates the minterms D_0 through D_3, using the values of A_1 and A_0. When $A_2 = 1$, the enable conditions are reversed, and minterms D_4 through D_7 are generated. Note that these decoders are enabled with a logic-1 signal.

The decoder in Figure 3-14 demonstrates the usefulness of the enable input on decoders or any other combinational logic component. Enable inputs are a convenient feature for interconnecting two or more integrated circuits or functional blocks for the purpose of expanding the digital function into a similar function with more inputs and outputs. In this sense, providing an enable input on a functional block makes that block more valuable for use in specific design hierarchies, such as that in Figure 3-14.

Combinational Circuit Implementation

A decoder provides the 2^n minterms of n input variables. Since any Boolean function can be expressed as a sum of minterms, one can use a decoder to generate the minterms and an external OR gate to form their logical sum. In this way, any combinational circuit with n inputs and m outputs can be implemented with an n–to–2^n-line decoder and m OR gates.

The procedure for implementing a combinational circuit by means of a decoder and OR gates requires that the Boolean functions for the circuit be expressed as a sum of minterms. This form can be obtained from the truth table or by plotting each function on a map. A decoder is then chosen that generates all the minterms of the input variables. The inputs to each OR gate are selected from the decoder outputs according to the list of minterms of each function.

■ **EXAMPLE 3-4**

Implement a binary adder circuit with a decoder and OR gates.

From the truth table for the binary adder (Table 3-1), we obtain the functions for the combinational circuit in sum-of-minterms form:

$$S(X, Y, Z) = \Sigma m(1, 2, 4, 7)$$

$$C(X, Y, Z) = \Sigma m(3, 5, 6, 7)$$

Since there are three inputs and a total of eight minterms, we need a 3–to–8-line decoder. The implementation is shown in Figure 3-15. The decoder generates the eight minterms for inputs X, Y, and Z. The OR gate for output S forms the logical sum minterms of 1, 2, 4, and 7. The OR gate for output C forms the logical sum of minterms 3, 5, 6, and 7. ■

A function with a long list of minterms requires an OR gate with a large number of inputs. A function having a list of k minterms can be expressed in its complement form with $2^n - k$ minterms. If the number of minterms in a function

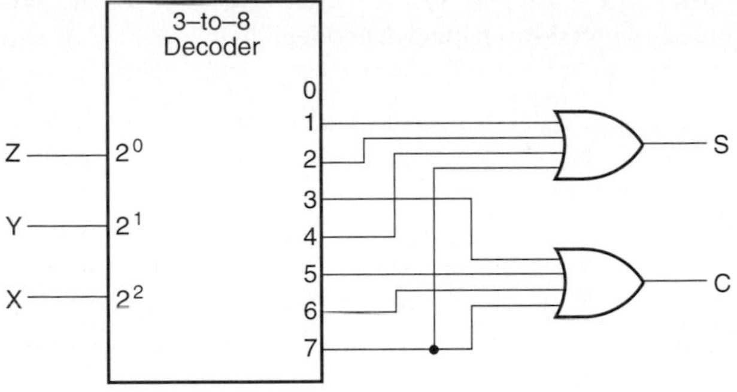

□ **FIGURE 3-15**
Implementing a Binary Adder Using a Decoder

F is greater than $2^n/2$, then the complement of F, $\overline{F}$, can be expressed with fewer minterms. In such a case, it is advantageous to use a NOR gate instead of an OR gate. The OR portion of the NOR gate produces the logical sum of the minterms of $\overline{F}$. The output bubble of the NOR gate complements this sum and generates the normal output F.

If NAND gates are used for the decoder, as in Figure 3-13, the external gates must be NAND gates instead of OR gates. This is because a two-level NAND gate circuit implements a sum-of-minterms function and is equivalent to a two-level AND-OR circuit. (See Section 2-6.)

The decoder method can be used to implement any combinational circuit. However, this implementation must be compared with other possible implementations to determine the best solution. In some cases, the method may provide the best solution, especially if the combinational circuit has many outputs and each output function is expressed with a small number of minterms.

3-6 ENCODERS

An encoder is a digital function that performs the inverse operation of a decoder. An encoder has 2^n (or fewer) input lines and n output lines. The output lines generate the binary code corresponding to the input value. An example of an encoder is the octal-to-binary encoder whose truth table is given in Table 3-5. This encoder has eight inputs, one for each of the octal digits, and three outputs that generate the corresponding binary number. It is assumed that only one input has a value of 1 at any given time.

The encoder can be implemented with OR gates using inputs determined directly from the truth table. Output $A_0 = 1$ if the input octal digit is 1 or 3 or 5 or 7. Similar conditions apply for the other two outputs. These conditions can be expressed by the following Boolean output functions:

$$A_0 = D_1 + D_3 + D_5 + D_7$$

$$A_1 = D_2 + D_3 + D_6 + D_7$$

$$A_2 = D_4 + D_5 + D_6 + D_7$$

□ TABLE 3-5
Truth Table for Octal–to–Binary Encoder

Inputs								Outputs		
D_7	D_6	D_5	D_4	D_3	D_2	D_1	D_0	A_2	A_1	A_0
0	0	0	0	0	0	0	1	0	0	0
0	0	0	0	0	0	1	0	0	0	1
0	0	0	0	0	1	0	0	0	1	0
0	0	0	0	1	0	0	0	0	1	1
0	0	0	1	0	0	0	0	1	0	0
0	0	1	0	0	0	0	0	1	0	1
0	1	0	0	0	0	0	0	1	1	0
1	0	0	0	0	0	0	0	1	1	1

The encoder can be implemented with three 4-input OR gates.

The encoder just defined has the limitation that only one input can be active at any given time: if two inputs are active simultaneously, the output produces an incorrect combination. For example, if D_3 and D_6 are 1 simultaneously, the output of the encoder will be 111 because all the three outputs are equal to 1. This represents neither a binary 3 nor a binary 6. To resolve this ambiguity, some encoder circuits must establish an input priority to ensure that only one input is encoded. If we establish a higher priority for inputs with higher subscript numbers, and if both D_3 and D_6 are 1 at the same time, the output will be 110 because D_6 has higher priority than D_3. Another ambiguity in the octal-to-binary encoder is that an output of all 0's is generated when all the inputs are 0, but this output is the same as when D_0 is equal to 1. This discrepancy can be resolved by providing one more output to indicate that at least one input is equal to 1.

Priority Encoder

A priority encoder is a combinational circuit that implements a priority function. As mentioned in the preceding paragraph, the operation of the priority encoder is such that if two or more inputs are equal to 1 at the same time, the input having the highest priority takes precedence. The truth table of a four-input priority encoder is given in Table 3-6. With the use of X's, this condensed truth table with just five rows represents the same information as the usual 16-row truth table. Whereas X's in output columns represent don't-care conditions, X's in input columns are used to represent product terms that are not minterms. For example, 001X represents the product term $\overline{D}_3\overline{D}_2D_1$. Just as with minterms, each variable is complemented if the corresponding bit in the input combination from the table is 0 and is not complemented if the bit is 1. If the corresponding bit in the input combination is an X, then the variable does not appear in the product term. Thus, for 001X, the variable D_0, corresponding to the position of the X, does not appear in $\overline{D}_3\overline{D}_2D_1$.

□ **TABLE 3-6**
Truth Table of Priority Encoder

Inputs				Outputs		
D_3	D_2	D_1	D_0	A_1	A_0	V
0	0	0	0	X	X	0
0	0	0	1	0	0	1
0	0	1	X	0	1	1
0	1	X	X	1	0	1
1	X	X	X	1	1	1

The number of rows of a full truth table represented by a row in the condensed table is 2^p, where p is the number of X's in the row. For example, in Table 3-6, 1XXX represents $2^3 = 8$ truth table rows, all having the same value for all outputs. In forming a condensed truth table, we must include each minterm in at least one of the rows in the sense that that minterm can be obtained by filling in 1's and 0's for the X's. Also, a minterm must never be included in more than one row such that the rows in which it appears have one or more conflicting output values.

We form Table 3 6 as follows: Input D_3 has the highest priority; so, regardless of the values of the other inputs, when this input is 1, the output for A_1A_0 is 11 (binary 3). From this, we obtain the last row of the table. D_2 has the next priority level. The output is 10 if $D_2 = 1$, provided that $D_3 = 0$, regardless of the values of the other two inputs with lower priority. From this, we obtain the fourth row of the table. The output for D_1 is generated only if all inputs with higher priority are 0, and so on down the priority levels. From this, we obtain the remaining rows of the table. The valid output designated by V is set to 1 only when one or more of the inputs are equal to 1. If all inputs are 0, V is equal to 0, and the other two outputs of the circuit are not used and hence are specified as don't-care conditions in the output part of the table.

The maps for simplifying outputs A_1 and A_0 are shown in Figure 3-16. The minterms for the two functions are derived from Table 3-6. The output values in the table can be transferred directly to the maps by placing them in the squares covered by the corresponding product term represented in the table. The simplified function obtained from each map is listed under that map. The condition for output V is an OR function of all the input variables. The priority encoder is implemented in Figure 3-17 according to the following Boolean functions:

$$A_0 = D_3 + D_1\overline{D}_2$$

$$A_1 = D_2 + D_3$$

$$V = D_0 + D_1 + D_2 + D_3$$

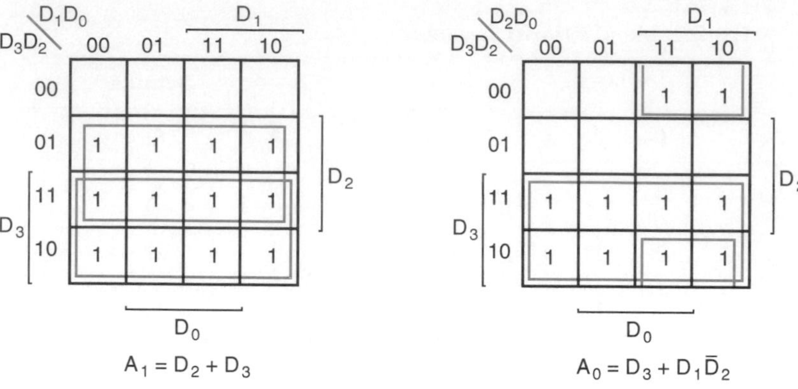

□ **FIGURE 3-16**
Maps for Priority Encoder

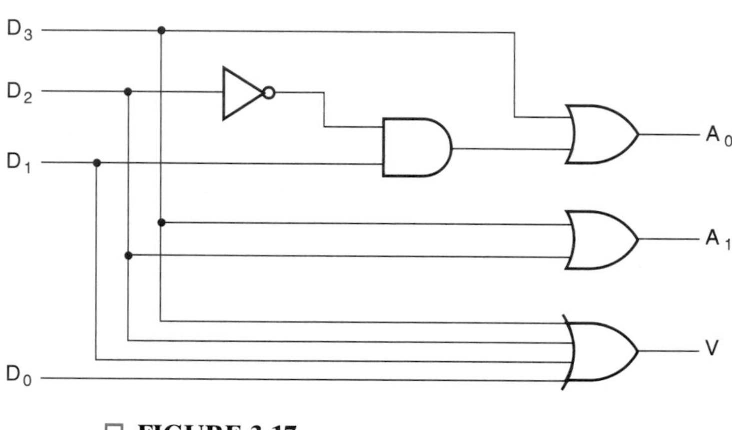

□ **FIGURE 3-17**
Logic Diagram of a 4-Input Priority Encoder

3-7 Multiplexers

A multiplexer is a combinational circuit that selects binary information from one of many input lines and directs the information to a single output line. The selection of a particular input line is controlled by a set of input variables, called *selection* inputs. Normally, there are 2^n input lines and n selection inputs whose bit combinations determine which input is selected. A single 4–to–1-line multiplexer is shown in Figure 3-18. Each of the four inputs D_0 through D_3 is applied to one input of an AND gate. Selection inputs S_1 and S_0 are decoded to select a particular AND gate. The outputs of the AND gates are applied to a single OR gate to provide the 1-line output. To visualize the operation of the circuit, consider the case when $(S_1, S_0) = 10$. The AND gate associated with input D_2 has two of its inputs equal to 1 and the third input connected to D_2. The other three AND gates have at least one

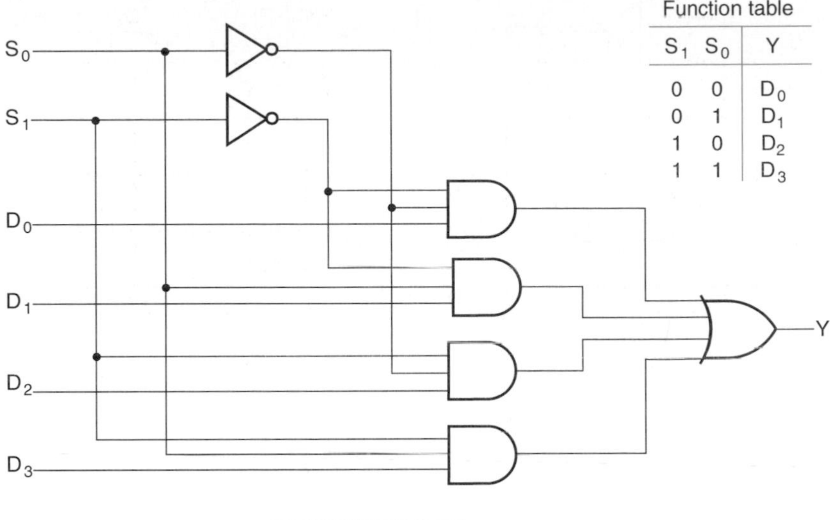

S_1	S_0	Y
0	0	D_0
0	1	D_1
1	0	D_2
1	1	D_3

☐ **FIGURE 3-18**
4–to–1-Line Multiplexer

input equal to 0, which makes their outputs equal to 0. The OR gate output is now equal to the value of D_2, providing a path from the selected input to the output. A multiplexer is also called a data selector, since it selects one of many inputs and steers the binary information to the output line. The function table in the figure lists the input that provides the path to the output for each combination of the binary selection variables. This is yet another form of condensed truth table, in which input literals are listed in the output columns. In this case, the input variable does not appear as an input column of the table. Instead, a single input literal in an output column is used to represent many truth table rows. For example, for the multiplexer in Figure 3-18, the truth table row 00 D_0 represents all rows in which $(S_1, S_0) = 00$ and, for $D_0 = 1$, gives $Y = 1$ and, for $D_0 = 0$, gives $Y = 0$. Since there are six variables, and only S_1 and S_0 are fixed, this single row represents 16 rows of the corresponding full truth table.

The AND gates and inverters in the multiplexer resemble a decoder circuit, and indeed, they decode the selection input lines. In general, a 2^n–to–1-line multiplexer is constructed from an n–to–2^n decoder by adding 2^n input lines to it, one from each data input. The size of the multiplexer is specified by the number 2^n of its data input lines and the single output line. Also, it is implied that the multiplexer contains n selection inputs. The term "multiplexer" is often abbreviated as "MUX."

A multiplexer can be constructed with transmission gates. The 2–to–1-line selector shown in Figure 2-49(a) is a multiplexer. A larger, single-bit 4–to–1-line multiplexer implemented with transmission gates is shown in Figure 3-19. This TG circuit provides a transmission path between the horizontal input and output lines when the two vertical control inputs have the value of 1 on the terminal without a bubble and 0 on the terminal with a bubble. With the opposite values on the con-

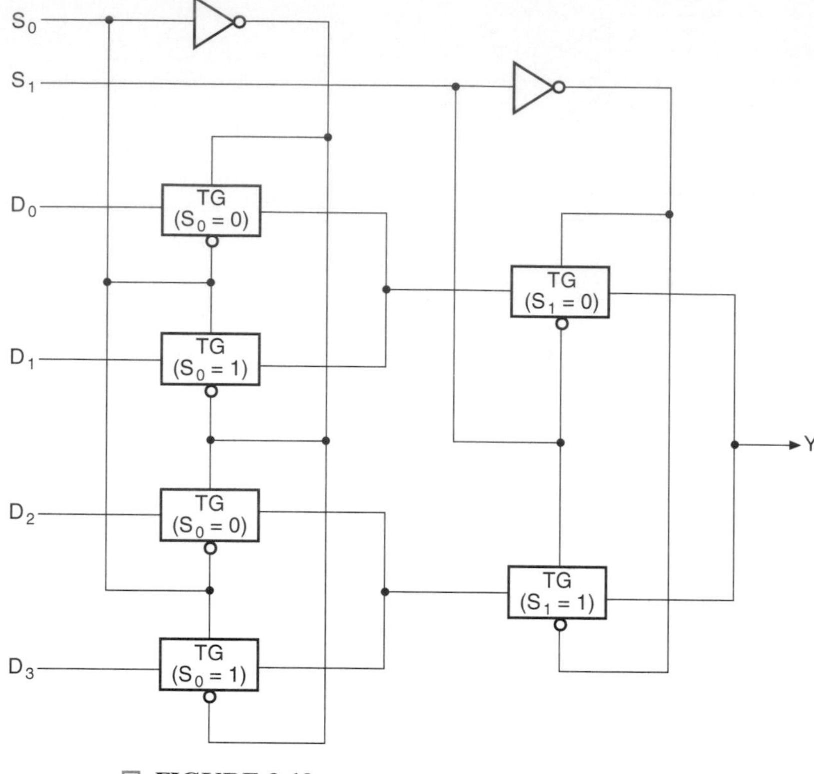

□ **FIGURE 3-19**
4–to–1-Line Multiplexer with Transmission Gates

trol inputs, the path is disconnected, and the circuit behaves like an open switch. The two selection inputs S_1 and S_0 control the transmission paths in the TG circuits. Inside each TG box is marked the condition for a path through the gate. Thus, if $S_0 = 0$ and $S_1 = 0$, there is a closed path from input D_0 to output Y, and the other three inputs are disconnected by one of the other TG circuits.

As in decoders, multiplexers may have an enable input to control the operation of the unit. When the enable input is in the inactive state, the outputs are disabled, and when it is in the active state, the circuit functions as a normal multiplexer. The enable input is useful for expanding two or more multiplexers into a multiplexer with a larger number of inputs.

Multiplexer blocks can be combined in parallel with common selection and enable lines to perform selection on multiple-bit quantities. As an illustration, a quadruple 2–to–1-line multiplexer is shown in Figure 3-20. The circuit has four multiplexers, each capable of selecting one of two input lines. Output Y_0 can be selected to come from either input A_0 or B_0. Similarly, output Y_1 may have the value of A_1 or B_1, and so on. Input selection line S selects one of the lines in each of the four multiplexers. The enable input E must be active for normal operation of the circuit. Although we can think of the circuit as containing four multiplexers, we

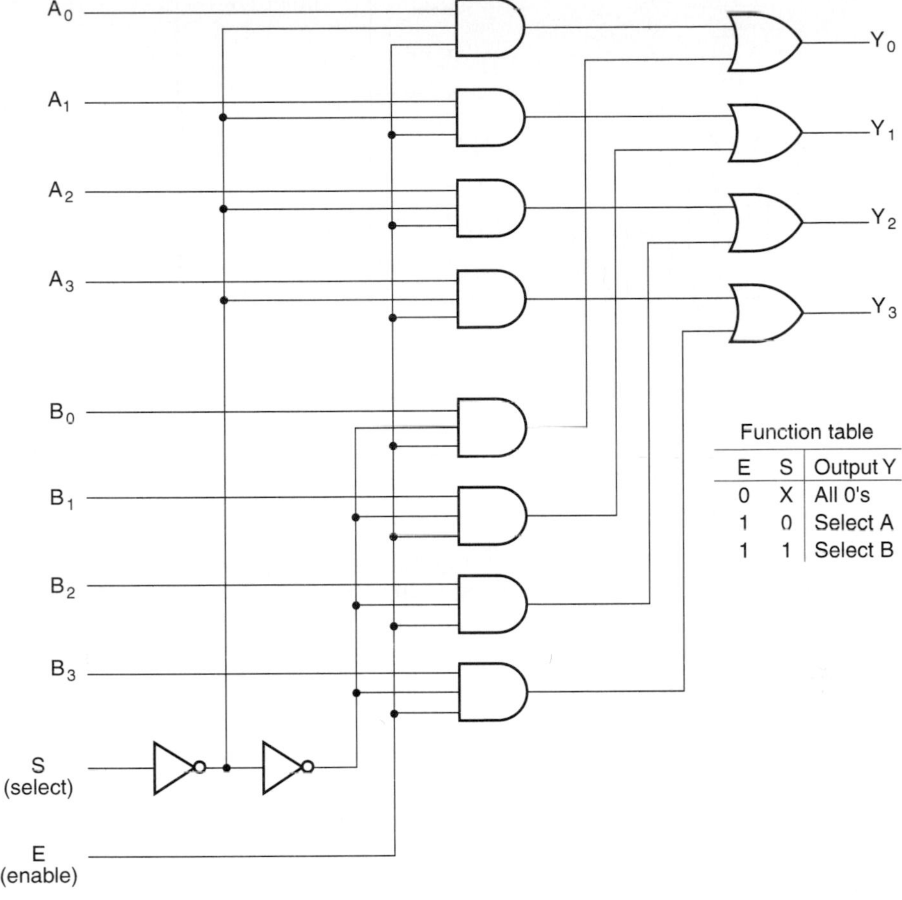

E	S	Output Y
0	X	All 0's
1	0	Select A
1	1	Select B

□ **FIGURE 3-20**
Quadruple 2–to–1-Line Multiplexer

are more likely to view it as a circuit that selects one of two 4-bit sets of data lines. As shown in the function table accompanying the diagram of the circuit, the unit is enabled when $E = 1$. Then, if $S = 0$, the four A inputs have a path to the four outputs. On the other hand, if $S = 1$, the four B inputs are applied to the outputs. The outputs are all 0's when $E = 0$, regardless of the values of S.

Combinational Circuit Implementation

In Section 3-5, we learned a decoder can be used to implement Boolean functions by employing external OR gates. An examination of the logic diagram of a multiplexer reveals that it is essentially a decoder that includes the OR gate within the block. The minterms of a function are generated in a multiplexer by the circuit associated with the selection inputs. The individual minterms can be selected by the

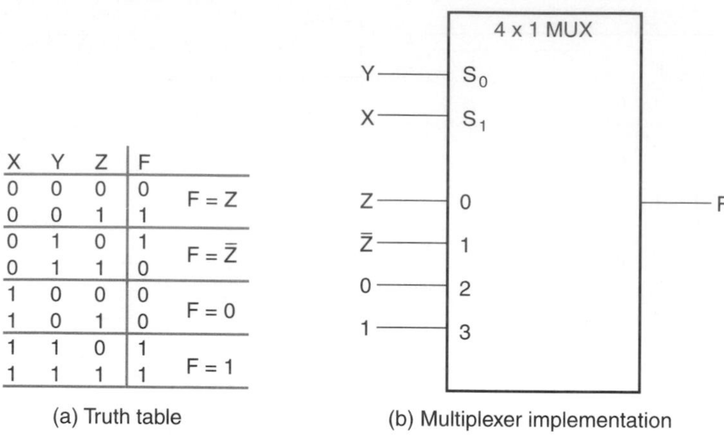

X Y Z F
0 0 0 0
0 0 1 1
F = Z
0 1 0 1
0 1 1 0
F = Z̄
1 0 0 0
1 0 1 0
F = 0
1 1 0 1
1 1 1 1
F = 1
4 x 1 MUX
Y
X
S0
S1
Z
Z̄
0
1
0
1
2
3
F

(a) Truth table

(b) Multiplexer implementation

□ **FIGURE 3-21**
Implementing a Boolean Function with a Multiplexer

data inputs. This provides a method of implementing a Boolean function of n variables with a multiplexer that has n selection inputs and 2^n data inputs, one for each minterm. We will now show a more efficient method for implementing a Boolean function of n variables with a multiplexer that has only $n - 1$ selection inputs. The first $n - 1$ variables of the function are connected to the selection inputs of the multiplexer. The remaining single variable of the function is used for the data inputs. If the single variable is denoted by Z, each data input of the multiplexer will be either Z, $\overline{Z}$, 1, or 0. To demonstrate this procedure, consider the following Boolean function of three variables:

$$F(X, Y, Z) = \Sigma m(1, 2, 6, 7)$$

This function can be implemented with a 4–to–1-line multiplexer, as shown in Figure 3-21. The two variables X and Y are applied to the selection lines in that order; X is connected to the S_1 input and Y to the S_0 input. The values for the data input lines are determined from the truth table of the function. When $(X, Y) = 00$, the output F is equal to Z because $F = 0$ when $Z = 0$ and $F = 1$ when $Z = 1$. This requires that the variable Z be applied to data input 0. The operation of the multiplexer is such that, when $(X, Y) = 00$, data input 0 has a path to the output that makes F equal to Z. In a similar fashion, we can determine the required input to lines 1, 2, and 3 from the value of F when $(X, Y) = 01$, 10, and 11, respectively. This particular example shows all four possibilities that can be obtained for the data inputs.

The general procedure for implementing any Boolean function of n variables with a multiplexer with $n - 1$ selection inputs and 2^{n-1} data inputs follows from the preceding example. The Boolean function is first listed in a truth table. The first $n - 1$ variables in the table are applied to the selection inputs of the multiplexer. For each combination of the selection variables, we evaluate the output as a function of the last variable. This function can be 0, 1, the variable, or the complement

A	B	C	D	F	
0	0	0	0	0	F = D
0	0	0	1	1	
0	0	1	0	0	F = D
0	0	1	1	1	
0	1	0	0	1	$F = \bar{D}$
0	1	0	1	0	
0	1	1	0	0	F = 0
0	1	1	1	0	
1	0	0	0	0	F = 0
1	0	0	1	0	
1	0	1	0	0	F = D
1	0	1	1	1	
1	1	0	0	1	F = 1
1	1	0	1	1	
1	1	1	0	1	F = 1
1	1	1	1	1	

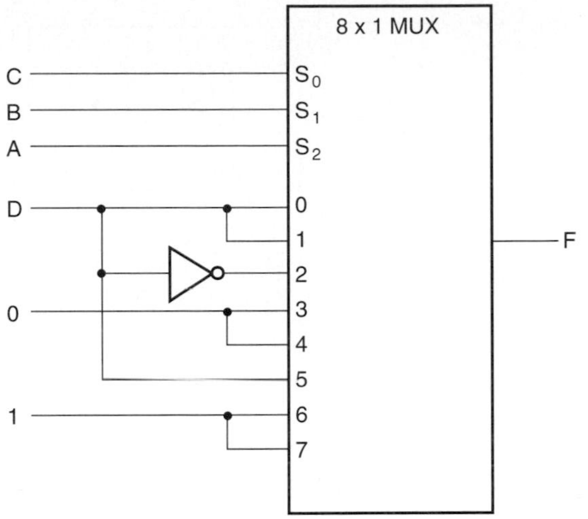

□ **FIGURE 3-22**
Implementing a Four-Input Function with a Multiplexer

of the variable. These values are then applied to the data inputs in the proper order. As a second example, consider the implementation of the following Boolean function:

$$F(A,B,C,D) = \Sigma m(1,3,4,11,12,13,14,15)$$

This function is implemented with a multiplexer with three selection inputs, as shown in Figure 3-22. To correspond to the order of the variables in the associated truth table, variables must be connected to selection inputs such that A, B, and C correspond to selection inputs S_2, S_1, and S_0, respectively. The values for the data inputs are determined from the truth table. The data line number is determined from the binary combination of A, B, and C. For example, when $(A, B, C) = 101$, the truth table shows that $F = D$; so the input variable D is applied to data input 5. The binary constants 0 and 1 correspond to two fixed signal values. Logic 1 and logic 0 are not appropriate schematic symbols; for CMOS- or TTL-type integrated circuits, fixed logic 0 corresponds to signal ground (⏚), and fixed logic 1 is equivalent to a $+V_{DD}$ or $+V_{CC}$ power signal. Hence, in schematics, these are shown instead of the constants 0 and 1, respectively.

Demultiplexer

A demultiplexer is a digital function that performs the inverse of the multiplexing operation: that is, a demultiplexer receives information from a single line and transmits it to one of 2^n possible output lines. The selection of the specific output is controlled by the bit combination of n selection lines. A 1–to 4-line demultiplexer is shown in Figure 3-23. The data input E has a path to all four outputs, but the

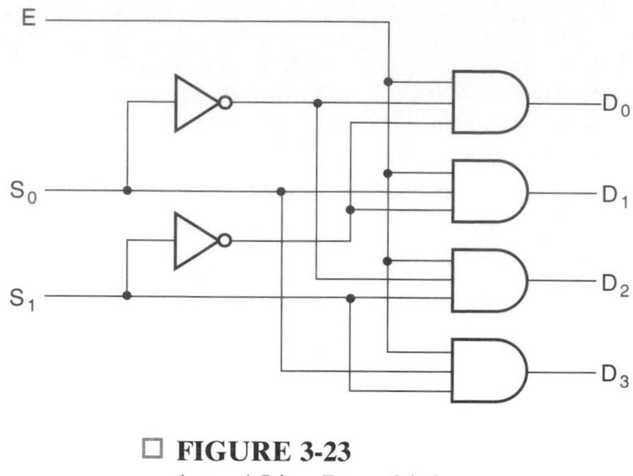

□ **FIGURE 3-23**
1–to–4-Line Demultiplexer

input information is directed to only one of the outputs, as specified by the two selection lines S_1 and S_0. For example, if $(S_1, S_0) = 10$, output D_2 will be the same as the input value of E, while all other outputs remain inactive at logic 0.

A careful inspection of the demultiplexer circuit shows that it is identical to a 2–to–4-line decoder with enable input. For the decoder, the data inputs are S_1 and S_0, and the enable is input E. For the demultiplexer, input E provides the data, while the other inputs accept the selection variables. Although the two circuits have different applications, their logic diagrams are exactly the same. For this reason, a decoder with enable input is referred to as a decoder/demultiplexer.

3-8 Binary Adders

An arithmetic circuit is a combinational circuit that performs arithmetic operations such as addition, subtraction, multiplication, and division with binary numbers or with decimal numbers in a binary code. We will develop arithmetic circuits by means of a hierarchical design. We begin at the lowest level by finding a circuit that performs the addition of two binary digits. This simple addition consists of four possible elementary operations: $0 + 0 = 0$, $0 + 1 = 1$, $1 + 0 = 1$, and $1 + 1 = 10$. The first three operations produce a sum requiring only one bit to represent, but when both the augend and addend are equal to 1, the binary sum requires two bits. Because of this case, the result is always represented by two bits, the carry and the sum. The carry obtained from the addition of two bits is added to the next higher order pair of significant bits. A combinational circuit that performs the addition of two bits is called a *half adder*. One that performs the addition of three bits (two significant bits and a previous carry) is called a *full adder*. The names of the circuits stem from the fact that two half adders can be employed to implement a full adder. The half adder and the full adder are basic arithmetic blocks with which other arithmetic circuits are designed.

Half Adder

A half adder is an arithmetic circuit that generates the sum of two binary digits. The circuit has two inputs and two outputs. The input variables are the augend and addend bits to be added, and the output variables produce the sum and carry. We assign the symbols X and Y to the two inputs and S (for "sum") and C (for "carry") to the outputs. The truth table for the half adder is listed in Table 3-7. The C output is 1 only when both inputs are 1. The S output represents the least significant bit of the sum. The Boolean functions for the two outputs, easily obtained from the truth table, are

$$S = (\overline{X}Y + X\overline{Y}) = X \oplus Y$$

$$C = XY$$

The half adder can be implemented with onc exclusive-OR gate and one AND gate, as shown in Figure 3-24.

□ **TABLE 3-7**
Truth Table of Half Adder

Inputs		Outputs	
X	**Y**	**C**	**S**
0	0	0	0
0	1	0	1
1	0	0	1
1	1	1	0

Full Adder

A full adder is a combinational circuit that forms the arithmetic sum of three input bits. Besides the three inputs, it has two outputs. Two of the input variables, denoted by X and Y, represent the two significant bits to be added. The third input, Z, represents the carry from the previous lower significant position. Two outputs are necessary because the arithmetic sum of three bits ranges in value from 0 to 3, and binary 2 and 3 need two digits for their representation. Again, the two outputs

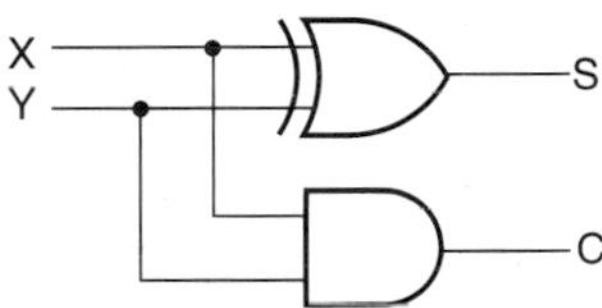

□ **FIGURE 3-24**
Logic Diagram of Half Adder

are designated by the symbols S for "sum" and C for "carry"; the binary variable S gives the value of the bit of the sum, and the binary variable C gives the output carry. The truth table of the full adder is listed in Table 3-8. The values for the outputs are determined from the arithmetic sum of the three input bits. When all the input bits are 0, the outputs are 0. The S output is equal to 1 when only one input is equal to 1 or when all three inputs are equal to 1. The C output has a carry of 1 if two or three inputs are equal to 1. The maps for the two outputs of the full adder are shown in Figure 3-25. The simplified sum-of-product functions for the two outputs are

$$S = \overline{X}\,\overline{Y}Z + \overline{X}Y\overline{Z} + X\overline{Y}\,\overline{Z} + XYZ$$

$$C = XY + XZ + YZ$$

This implementation requires seven AND gates and two OR gates. However, the map for output S is recognized as an odd function, as discussed in Section 2-7. Fur-

□ TABLE 3-8
Truth Table of Full Adder

Inputs			Outputs	
X	**Y**	**Z**	**C**	**S**
0	0	0	0	0
0	0	1	0	1
0	1	0	0	1
0	1	1	1	0
1	0	0	0	1
1	0	1	1	0
1	1	0	1	0
1	1	1	1	1

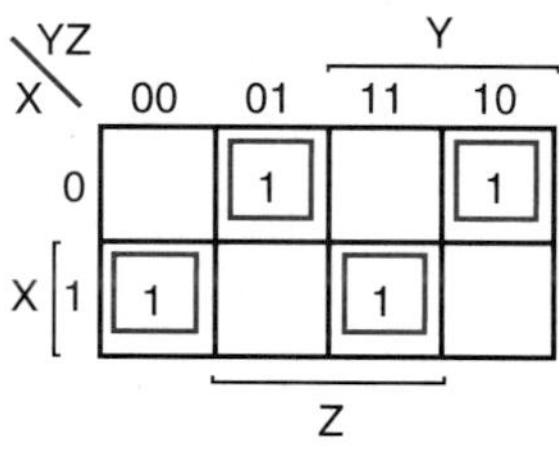

$S = \overline{X}\,\overline{Y}Z + \overline{X}Y\overline{Z} + X\overline{Y}\,\overline{Z} + XYZ$
$= X \oplus Y \oplus Z$

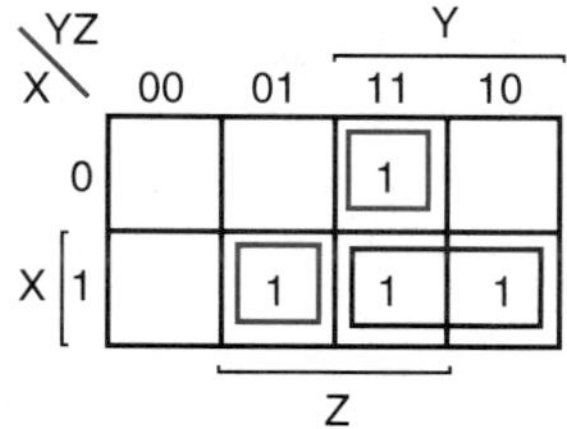

$C = XY + XZ + YZ$
$= XY + Z(X\overline{Y} + \overline{X}Y)$
$= XY + Z(X \oplus Y)$

□ FIGURE 3-25
Maps for Full Adder

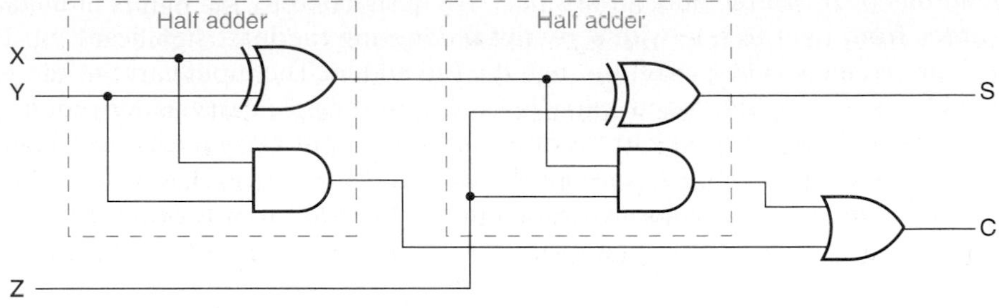

□ **FIGURE 3-26**
Logic Diagram of Full Adder

thermore, the C output function can be manipulated as shown in Figure 3-25 to include the exclusive-OR of X and Y. The Boolean functions for the full adder in terms of exclusive-OR operations can then be expressed as:

$$S = (X \oplus Y) \oplus Z$$

$$C = XY + Z(X \oplus Y)$$

The logic diagram for this implementation is shown in Figure 3-26. It consists of two half adders and an OR gate.

Binary Ripple Carry Adder

A parallel binary adder is a digital circuit that produces the arithmetic sum of two binary numbers using only combinational logic. The parallel adder uses n full adders in parallel, with all input bits applied simultaneously to produce the sum. The full adders are connected in cascade, with the carry output from one full adder connected to the carry input of the next full adder. Since a 1 carry may appear near the least significant bit of the adder and yet propagate through many full adders to the most significant bit, just as a wave ripples outward from a pebble dropped in a pond, the parallel adder is referred to as a *ripple carry adder*. Figure 3-27 shows the interconnection of four full-adder blocks to form a 4-bit ripple carry adder. The

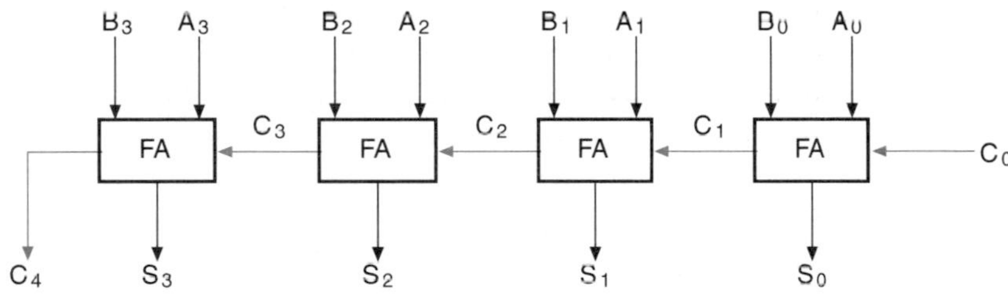

□ **FIGURE 3-27**
4-Bit Ripple Carry Adder

augend bits of A and the addend bits of B are designated by subscripts in increasing order from right to left, with subscript 0 denoting the least significant bit. The carries are connected in a chain through the full adders. The input carry to the parallel adder is C_0, and the output carry is C_4. An n-bit ripple carry adder requires n full adders, with each output carry connected to the input carry of the next-higher-order full adder. For example, consider the two binary numbers $A = 1011$ and $B = 0011$. Their sum $S = 1110$ is formed with a four-bit ripple carry adder as follows:

Input carry	0 1 1 0
Augend A	1 0 1 1
Addend B	0 0 1 1
Sum S	1 1 1 0
Output carry	0 0 1 1

The input carry in the least significant position is 0. Each full adder receives the corresponding bits of A and B and the input carry and generates the sum bit for S and the output carry. The output carry in each position is the input carry of the next-higher-order position, as indicated by the blue lines.

The 4-bit adder is a typical example of a digital component that can be used as a building block in constructing a digital system and thus is defined as a functional block. It can be used in many applications involving arithmetic operations. Observe that the design of this circuit by the usual method would require a truth table with 512 entries, since there are nine inputs to the circuit. By cascading the four instances of the known full adders, it is possible to obtain a simple and straightforward implementation without directly solving this larger problem. This is an example of the power of hierarchy and reuse in design.

Carry Lookahead Adder

The ripple carry adder, although simple in concept, has a long circuit delay due to the many gates in the carry path from the least significant bit to the most significant bit. For a typical design, the longest delay path through an n-bit ripple carry adder is $2n + 2$ gate delays. Thus, for a 16-bit ripple carry adder, the delay is 34 gate delays. This delay tends to be one of the largest in a typical computer design. Accordingly, we find an alternative design, the *carry lookahead adder*, attractive. This adder is a practical design with reduced delay at the price of more complex hardware. The carry lookahead design can be obtained by a transformation of the ripple carry design in which the carry logic over fixed groups of bits of the adder is reduced to two-level logic. The transformation is shown for a 4-bit adder group in Figure 3-28.

First, we construct a new logic hierarchy, separating the parts of the full adders not involving the carry propagation path from those containing the path.

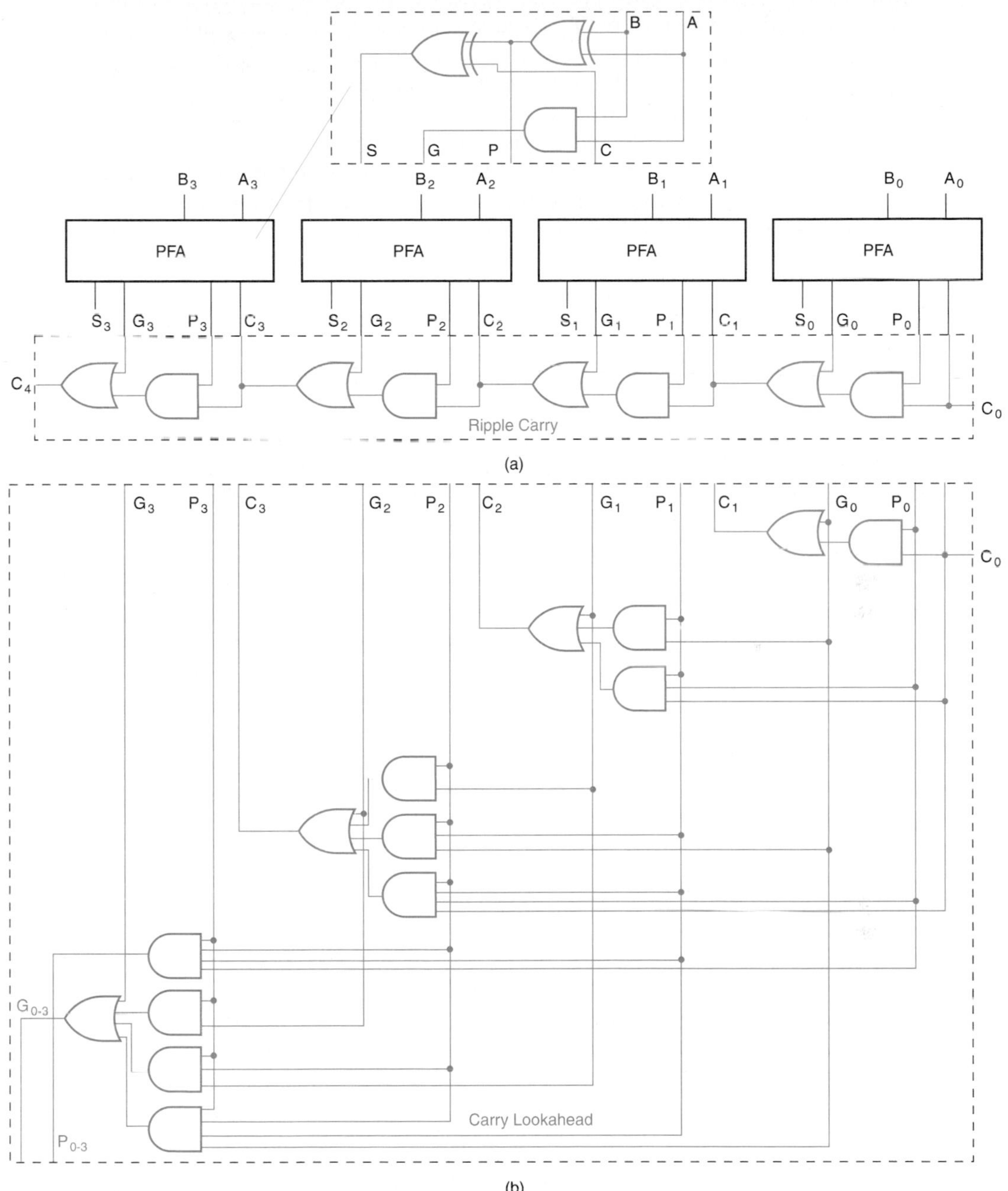

□ **FIGURE 3-28**
Development of a Carry Lookahead Adder

We call the first part of each full adder a *partial full adder* (PFA). This separation is shown in Figure 3-28(a), which presents a diagram of a PFA and a diagram of four PFAs connected to the carry path. We have removed the OR gate and one of the AND gates from each of the full adders to form the ripple carry path.

There are two outputs P_i and G_i from each PFA to the ripple carry path and one input C_i, the carry input, from the carry path to each PFA. The function P_i is the exclusive-OR of A_i and B_i and is called the *propagate* function. Whenever P_i is equal to 1, an incoming carry is propagated through the bit position from C_i to C_{i+1}. For P_i equal to 0, carry propagation through the bit position is blocked. The function G_i is the AND of A_i and B_i and is called the *generate* function. Whenever G_i is equal to 1, the carry output from the position is 1, regardless of the value of P_i, so a carry has been generated in the position. When G_i is 0, a carry is not generated, so that C_{i+1} is 0 if the carry propagated through the position from C_i is also 0. The generate and propagate functions correspond exactly to the half adder and are essential in controlling the values in the ripple carry path. In addition, as in the full adder, the PFA generates the sum function by the exclusive-OR of the incoming carry C_i and the propagate function P_i.

The carry path remaining in the 4-bit ripple carry adder has a total of eight gates in cascade, so the circuit has a delay of eight gate delays. Since only AND and OR gates are involved in the carry path, ideally the delay for each of the four carry signals produced, C_1 through C_4, could be just two gate delays. The basic carry lookahead circuit is simply a circuit in which functions C_1 through C_3 have a delay of only two gate delays. The implementation of C_4 is more complicated in order to allow the 4-bit carry lookahead adder to be extended to multiples of 4 bits, such as 16 bits. The 4-bit carry lookahead circuit is shown in Figure 3-28(b). It is designed to directly replace the ripple carry path in Figure 3-28(a). Since the logic generating C_1 is already two level, it remains unchanged. The logic for C_2, however, has four levels. So to find the carry lookahead logic for C_2, we must reduce the logic to two levels. This can be done by finding the equation for C_2 from the carry path in Figure 3-28(a) and applying the distributive law as follows:

$$\begin{aligned} C_2 &= G_1 + P_1(G_0 + P_0C_0) \\ &= G_1 + P_1G_0 + P_1P_0C_0 \end{aligned}$$

This equation is implemented by the logic with output C_2 in Figure 3-28(b). We obtain the two-level logic for C_3 by finding its equation from the carry path in Figure 3-28(a) and applying the distributive law:

$$\begin{aligned} C_3 &= G_2 + P_2(G_1 + P_1(G_0 + P_0C_0)) \\ &= G_2 + P_2(G_1 + P_1G_0 + P_0C_0) \\ &= G_2 + P_2G_1 + P_2P_1G_0 + P_2P_1P_0C_0 \end{aligned}$$

The two-level logic with output C_3 in Figure 3-28(b) implements this function.

We could implement C_4 using the same method. But some of the gates would have a fan-in of 5, which may increase the delay. Also, we are interested in reusing this same circuit for higher numbered bits, e.g., 4 through 7, 8 through 11, and 12 through 15 of a 16-bit adder. For this adder, in positions 4, 8, and 12 we would like the carry to be produced as fast as possible without using excessive fan-in. Accordingly, we want to repeat the same carry lookahead trick that we used to handle the four bits for *groups of four bits*. This will allow us to reuse the carry lookahead circuit for each group of four bits and also to use the same circuit for the four groups of four bits as if they were individual bits. So instead of generating C_4, we produce generate and propagate functions that apply to 4-bit groups instead of a single bit. To propagate a carry from C_0 to C_4, we need to have all four of the propagate functions equal to 1, giving the *group propagate* function

$$P_{0-3} = P_3P_2P_1P_0$$

To represent the generation of a carry and its propagation to C_4, we need to consider the generation of a carry in each of the positions 0 through 3, as represented by G_0 through G_3, and the propagation of each of these four generated carries to position 4. This gives the *group generate* function

$$G_{0-3} = G_3 + P_3G_2 + P_3P_2G_1 + P_3P_2P_1G_0$$

The group propagate and group generate equations are implemented by the logic in the lower part of Figure 3-28(b). If there are only four bits in the adder, then just the logic used for C_1 can be used to generate C_4 from these two outputs. In a longer adder, a carry lookahead circuit identical to that in the figure, except for labeling, is placed at the second level to generate C_4, C_8, and C_{12}. This concept can be extended with more carry lookahead circuits in the second level and with one carry lookahead circuit in the third level to generate carries for positions 16, 32, and 48 in a 64-bit adder.

Assuming that an exclusive OR contributes 2 gate delays, the longest delay in the 4-bit carry lookahead adder is 6 gate delays, compared to 10 gate delays in the ripple carry adder. The improvement is very modest and perhaps not worth all the extra logic. But applying the carry lookahead circuit to a 16-bit adder using five copies in two levels of lookahead reduces the delay from 34 to just 10 gate delays, improving the performance of the adder by a factor of close to 3. In a 64-bit adder, with the use of 21 carry lookahead circuits in three levels of lookahead, the delay is reduced from 130 gate delays to 14 gate delays, giving more than a factor of 8 in improved performance. In general, for the implementation we have shown, the delay of a carry lookahead adder designed for the best performance is $4L + 2$ gate delays, where L is the number of lookahead levels in the design.

3-9 Binary Subtraction

In Chapter 1, we briefly examined the subtraction of unsigned binary numbers. Although beginning texts cover only signed number addition and subtraction, to the complete exclusion of the unsigned alternative, unsigned number arithmetic plays an important role in computation and computer hardware design. It is used in floating-point units, in sign-magnitude addition and subtraction algorithms, and in extending the precision of fixed-point numbers. For these reasons, we will treat unsigned number addition and subtraction here. We also, however, choose to treat it first so that we can clearly justify, in terms of hardware cost, that which otherwise appears bizarre and often is accepted on faith, namely, the use of complement representations in arithmetic.

In Section 1-3, subtraction is performed by comparing the subtrahend with the minuend and subtracting the smaller from the larger. The use of a method containing this comparison operation results in inefficient and costly circuitry. As an alternative, we can simply subtract the subtrahend from the minuend. Using the same numbers as in a subtraction example from Section 1-3, we have:

Borrows into:	11100
Minuend:	10011
Subtrahend:	−11110
Difference:	10101
Correct Difference:	−01011

If no borrow occurs into the most significant position, then we know that the subtrahend is not larger than the minuend and that the result is positive and correct. If a borrow does occur into the most significant position, as indicated in blue, then we know that the subtrahend is larger than the minuend. The result must then be negative, and so we need to correct its magnitude. We can do this by examining the result of the calculation when a borrow occurs:

$$M - N + 2^n$$

Note that the added 2^n represents the value of the borrow into the most significant position. Instead of this result, the desired magnitude is $N - M$. This can be obtained by subtracting the preceding formula from 2^n:

$$2^n - (M - N + 2^n) = N - M$$

In the foregoing example, 100000 − 10101 = 01011, which is the correct magnitude.

In general, the subtraction of two n-digit numbers $M - N$ in base 2 can be done as follows:

1. Subtract the subtrahend N from the minuend M.
2. If no end borrow occurs, then $M \geq N$, and the result is nonnegative and correct.
3. If an end borrow occurs, then $N > M$, the difference $N - M$ is subtracted from 2^n, and a minus sign is appended to the result.

Subtraction of a binary number from 2^n to obtain an n-digit result is called taking the *2's complement* of the number. So in step 3, we are taking the 2's complement of the difference $N - M$. Use of the 2's complement in subtraction is illustrated by the following example.

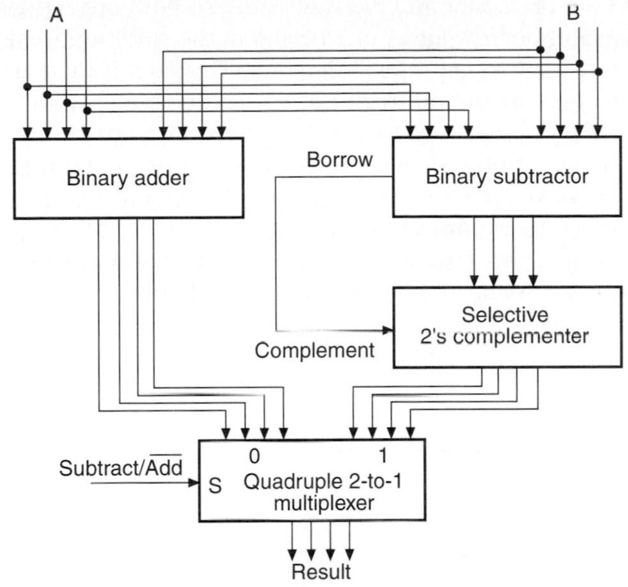

□ **FIGURE 3-29**
Block Diagram of Binary Adder-Subtractor

EXAMPLE 3-5

Perform the binary subtraction 01100100 − 10010110

Borrows into:	1 0011110
Minuend:	01100100
Subtrahend:	− 10010110
Initial Result	11001110

End borrow of 1 implies correction:

2^8	100000000
− Initial Result	− 11001110
Final Result	− 00110010

To perform subtraction using this method requires a subtractor for the initial subtraction. In addition, when necessary, either the subtractor must be used a second time to perform the correction, or a separate 2's complementer circuit must be provided. So, thus far, we require a subtractor, an adder, and possibly a 2's complementer to perform both addition and subtraction. The block diagram for a 4-bit adder-subtractor using these functional blocks is shown in Figure 3-29. The inputs are applied to both the adder and the subtractor, so both operations are performed in parallel. If an end borrow value of 1 occurs in the subtraction, then the selective 2's complementer receives an input value of 1. This circuit then takes the 2's complement of the output of the subtractor. If the end borrow has value of 0, the selective 2's complementer passes the output of the subtractor through unchanged. If subtraction is the operation, then a 1 is applied to S of the multiplexer that selects the output of the complementer. If addition is the operation, then a 0 is applied to S, thereby selecting the output of the adder.

As we will see, this circuit is more complex than necessary. To reduce the amount of hardware, we would like to share logic between the adder and the subtractor. This can also be done using the notion of the complement. So before considering the combined adder/subtractor further, we will take a more careful look at complements.

Complements

There are two types of complements for each base-r system: the *radix complement*, which we saw earlier for base 2, and the *diminished radix complement*. The first is referred to as the *r's complement* and the second as the *(r − 1)'s complement*. When the value of the base r is substituted in the names, the two types are referred to as the 2's and 1's complements for binary numbers and the 10's and 9's complements for decimal numbers, respectively. Since our interest for the present is in binary numbers and operations, we will deal with only 2's and 1's complements.

Given a number N in binary having n digits, the *1's complement of N* is defined as $(2^n - 1) - N$. 2^n is represented by a binary number that consists of a 1 followed by n 0's. $2^n - 1$ is a binary number represented by n 1's. For example, if $n = 4$, we have $2^4 = (10000)_2$ and $2^4 - 1 = (1111)_2$. Thus, the 1's complement of a binary number is obtained by subtracting each digit from 1. When subtracting binary digits from 1, we can have either $1 - 0 = 1$ or $1 - 1 = 0$, which causes the original bit to change from 0 to 1 or from 1 to 0, respectively. Therefore, the 1's complement of a binary number is formed by changing all 1's to 0's and all 0's to 1's—that is, applying the NOT or complement operation to each of the bits. Following are two numerical examples:

The 1's complement of 1011001 is 0100110
The 1's complement of 0001111 is 1110000

In similar fashion, the 9's complement of a decimal number, the 7's complement of an octal number, and the 15's complement of a hexadecimal number are obtained by subtracting each digit from 9, 7, and F (decimal 15), respectively.

Given an n-digit number N in binary, the *2's complement of N* is defined as $2^n - N$ for $N \neq 0$ and 0 for $N = 0$. The reason for the special case of $N = 0$ is that the result must have n bits, and subtraction of 0 from 2^n gives an $(n + 1)$-bit result, 100…0. This special case is achieved by using only an n-bit subtractor or otherwise dropping the 1 in the extra position. Comparing with the 1's complement, we note that the 2's complement can be obtained by adding 1 to the 1's complement, since

$2^n - N = \{[(2^n - 1) - N] + 1\}$. For example, the 2's complement of binary 101100 is 010011 + 1 = 010100 and is obtained by adding 1 to the 1's complement value. Again, for $N = 0$, the result of this addition is 0, achieved by ignoring the carry out of the most significant position of the addition. These concepts hold for other bases as well. As we will see later, they are very useful in simplifying 2's complement and subtraction hardware.

Also, the 2's complement can be formed by leaving all least significant 0's and the first 1 unchanged and then replacing 1's with 0's and 0's with 1's in all other higher significant bits. Thus, the 2's complement of 1101100 is 0010100 and is obtained by leaving the two low-order 0's and the first 1 unchanged and then replacing 1's with 0's and 0's with 1's in the other four most significant bits. In other bases, the first nonzero digit is subtracted from the base r, and the remaining digits to the left are replaced with $r - 1$ minus their values.

It is also worth mentioning that the complement of the complement restores the number to its original value. To see this, note that the 2's complement of N is $2^n - N$, and the complement of the complement is $2^n - (2^n - N) = N$, giving back the original number.

Subtraction with Complements

Earlier, we expressed a desire to simplify hardware by sharing adder and subtractor logic. Armed with complements, we are prepared to define a binary subtraction procedure that uses addition and the corresponding complement logic. The subtraction of two n-digit unsigned numbers, $M - N$, in binary can be done as follows:

1. Add the 2's complement of the subtrahend N to the minuend M. This performs $M + (2^n - N) = M - N + 2^n$.
2. If $M \geq N$, the sum will produce an end carry, 2^n, which is discarded; what is left is the result $M - N$.
3. If $M < N$, the sum does not produce an end carry and is equal to $2^n - (N - M)$, which is the 2's complement of $N - M$. The correction is performed as before, taking the 2's complement of the sum and placing a minus sign in front.

The following examples further illustrate the foregoing procedure. Note that, although we are dealing with unsigned numbers, there is no way to get an unsigned result for the case in step 3. When working with paper and pencil, we recognize, by the absence of the end carry, that the answer must be changed to a negative number. If the minus sign for the result is to be preserved, it must be stored separately from the corrected n-bit result.

EXAMPLE 3-6

Given the two binary numbers $X = 1010100$ and $Y = 1000011$, perform the subtraction $X - Y$ and $Y - X$ using 2's complement operations.

$$\begin{aligned} X &= \quad 1010100 \\ \text{2's complement of } Y &= \quad \underline{0111101} \\ \text{Sum} &= \;\; 10010001 \\ \text{Discard end carry } 2^7 &= -\underline{10000000} \\ \textit{Answer: } X - Y &= \quad 0010001 \end{aligned}$$

$$\begin{aligned} Y &= \quad 1000011 \\ \text{2's complement of } X &= \quad \underline{0101100} \\ \text{Sum} &= \quad 1101111 \end{aligned}$$

There is no end carry.

Answer: $Y - X = -(\text{2's complement of } 1101111) = -0010001.$ ■

Subtraction of unsigned numbers also can be done by means of the 1's complement. Remember that the 1's complement is one less than the 2's complement. Because of this, the result of adding the minuend to the complement of the subtrahend produces a sum that is one less than the correct difference when an end carry occurs. Discarding the end carry and adding one to the sum is referred to as an *end-around carry.*

■ **EXAMPLE 3-7**

Repeat Example 3-6 using 1's complement operations.

$$X - Y = 1010100 - 1000011$$

$$\begin{aligned} X &= \quad 1010100 \\ \text{1's complement of } Y &= + \underline{0111100} \\ \text{Sum} &= \;\; 10010000 \\ \text{End-around carry} & \quad \underline{\hookrightarrow + 1} \\ \textit{Answer: } X - Y &= \quad 0010001 \end{aligned}$$

$$Y - X = 1000011 - 1010100$$

$$\begin{aligned} Y &= \quad 1000011 \\ \text{1's complement of } X &= + \underline{0101011} \\ \text{Sum} &= \quad 1101110 \end{aligned}$$

There is no end carry.

Answer: $Y - X = -(\text{1's complement of } 1101110) = -0010001.$ ■

Note that the negative result is obtained by taking the 1's complement of the sum, since this is the type of complement being used.

3-10 BINARY ADDER-SUBTRACTORS

Using either the 2's or 1's complement, we have eliminated the subtraction operation and need only the appropriate complementer and an adder. When performing a subtraction we complement the subtrahend N, and when performing an addition we do not complement N. These operations can be accomplished by using a selective complementer and adder interconnected to form an adder-subtractor. We have used 2's complement, since it is most prevalent in modern systems. The 2's complement can be obtained by taking the 1's complement and adding 1 to the least significant bit. The 1's complement can be implemented easily with inverter circuits, and we can add 1 to the sum by making the input carry of the parallel adder equal to 1. Thus, by using 1's complement and an unused adder input, the 2's complement is obtained inexpensively. In 2's complement subtraction, as the correction step after adding, we complement the result and append a minus sign if an end carry does not occur. The correction operation is performed by using either the adder-subtractor a second time with $M = 0$ or a selective complementer as in Figure 3-29.

The circuit for subtracting $A - B$ consists of a parallel adder as shown in Figure 3-27, with inverters placed between each B terminal and the corresponding full-adder input. The input carry C_0 must be equal to 1. The operation that is performed becomes A, plus the 1's complement of B, plus 1. This is equal to A plus the 2's complement of B. For unsigned numbers, it gives $A - B$ if $A \geq B$ or the 2's complement of $B - A$ if $A < B$.

The addition and subtraction operations can be combined into one circuit with one common binary adder. This is done by including an exclusive-OR gate with each full adder. A 4-bit adder-subtractor circuit is shown in Figure 3-30. Input S controls the operation. When $S = 0$ the circuit is an adder, and when $S = 1$ the circuit becomes a subtractor. Each exclusive-OR gate receives input S and one of the inputs of B, B_i. When $S = 0$, we have $B_i \oplus 0 = B_i$. If the full adders receive the value of B, and the input carry is 0, the circuit performs A plus B. When $S = 1$, we have $B_i \oplus 1 = \overline{B}_i$ and $C_0 = 1$. The circuit performs the operation A plus the 2's complement of B.

Signed Binary Numbers

In the previous section, we dealt with the addition and subtraction of unsigned numbers. We will now extend this approach to signed numbers, including a further use of complements that eliminates the correction step.

Positive integers and the number zero can be represented as unsigned numbers. To represent negative integers, we need a notation for negative values. In ordinary arithmetic, a negative number is indicated by a minus sign and a positive number by a plus sign. Because of hardware limitations, computers must represent

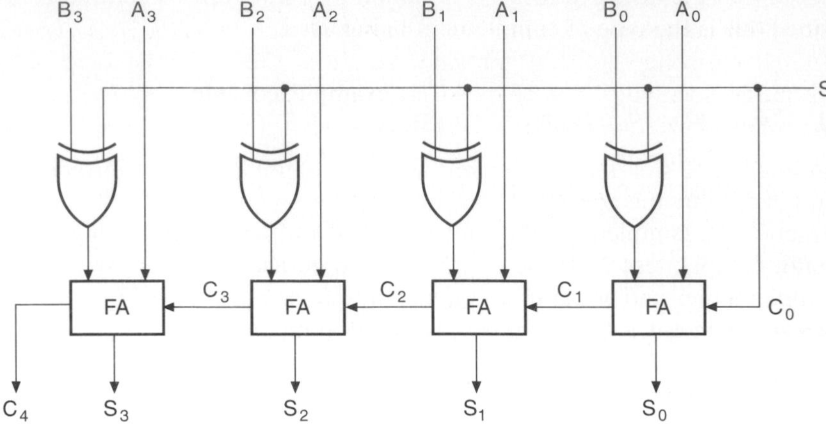

□ **FIGURE 3-30**
Adder-Subtractor Circuit

everything with 1's and 0's, including the sign of a number. As a consequence, it is customary to represent the sign with a bit placed in the most significant position of an n-bit number. The convention is to make the sign bit 0 for positive numbers and 1 for negative numbers.

It is important to realize that both signed and unsigned binary numbers consist of a string of bits when represented in a computer. The user determines whether the number is signed or unsigned. If the binary number is signed, then the leftmost bit represents the sign and the rest of the bits represent the number. If the binary number is assumed to be unsigned, then the leftmost bit is the most significant bit of the number. For example, the string of bits 01001 can be considered as 9 (unsigned binary) or +9 (signed binary), because the leftmost bit is 0. Similarly, the string of bits 11001 represents the binary equivalent of 25 when considered as an unsigned number or −9 when considered as a signed number. The latter is because the 1 in the leftmost position designates a minus sign and the remaining four bits represent binary 9. Usually, there is no confusion in identifying the bits because the type of number representation is known in advance. The representation of signed numbers just discussed is referred to as the *signed-magnitude* system. In this system, the number consists of a magnitude, and a symbol (+ or −) or a bit (0 or 1) indicating the sign. This is the representation of signed numbers used in ordinary arithmetic.

In implementing signed-magnitude addition and subtraction for n-bit numbers, the single sign bit in the leftmost position and the $n - 1$ magnitude bits are processed separately. The magnitude bits are processed as unsigned binary numbers. Thus, subtraction involves the correction step. To avoid this step, we use a different system for representing negative numbers, referred to as a *signed-complement* system. In this system, a negative number is represented by its complement. While the signed-magnitude system negates a number by changing its sign, the signed-complement system negates a number by taking its complement. Since positive numbers always start with 0 (representing a plus sign) in the leftmost position, their complements will always start with a 1, indicating a negative number. The signed-complement system can use either the 1's or the 2's complement, but the latter is the most common. As an example, consider the number 9, represented in binary with eight bits. +9 is represented with a sign bit of 0 in the leftmost position, followed by the binary equivalent of 9, to give 00001001. Note that all eight bits must have a value, and therefore, 0's are inserted between the sign bit and the first 1. Although there is only one way to represent +9, we have three different ways to represent −9 using eight bits:

In signed-magnitude representation:	10001001
In signed-1's complement representation:	11110110
In signed-2's complement representation:	11110111

In signed magnitude, −9 is obtained from +9 by changing the sign bit in the leftmost position from 0 to 1. In signed 1's complement, −9 is obtained by complementing all the bits of +9, including the sign bit. The signed 2's complement representation of −9 is obtained by taking the 2's complement of the positive number, including the 0 sign bit.

Table 3-9 lists all possible 4-bit signed binary numbers in the three representations. The equivalent decimal number is also shown. Note that the positive numbers in all three representations are identical and have 0 in the leftmost position. The signed 2's complement system has only one representation for 0, which is always

□ **TABLE 3-9**
Signed Binary Numbers

Decimal	Signed 2's Complement	Signed 1's Complement	Signed Magnitude
+7	0111	0111	0111
+6	0110	0110	0110
+5	0101	0101	0101
+4	0100	0100	0100
+3	0011	0011	0011
+2	0010	0010	0010
+1	0001	0001	0001
+0	0000	0000	0000
−0	—	1111	1000
−1	1111	1110	1001
−2	1110	1101	1010
−3	1101	1100	1011
−4	1100	1011	1100
−5	1011	1010	1101
−6	1010	1001	1110
−7	1001	1000	1111
−8	1000	—	—

positive. The other two systems have a positive 0 and a negative 0, which is something not encountered in ordinary arithmetic. Note that all negative numbers have a 1 in the leftmost bit position; this is the way we distinguish them from positive numbers. With four bits, we can represent 16 binary numbers. In the signed-magnitude and the 1's complement representations, there are eight positive numbers and eight negative numbers, including two zeros. In the 2's complement representation, there are eight positive numbers, including one zero, and eight negative numbers.

The signed-magnitude system is used in ordinary arithmetic, but is awkward when employed in computer arithmetic due to the separate handling of the sign and the correction step required for subtraction. Therefore, the signed complement is normally used. The 1's complement imposes difficulties because of its two representations of 0 and is seldom used for arithmetic operations. It is useful as a logical operation, since the change from 1 to 0 or 0 to 1 is equivalent to a logical complement operation. The following discussion of signed binary arithmetic deals exclusively with the signed-2's complement representation of negative numbers because it prevails in actual use. By using 1's complementation and the end-around carry, the same procedures as those for signed 2's complement can be applied to signed 1's complement.

Signed Binary Addition and Subtraction

The addition of two numbers $M + N$ in the signed-magnitude system follows the rules of ordinary arithmetic: If the signs are the same, we add the two magnitudes and give the sum the common sign. If the signs are different, we subtract the magnitude of N from the magnitude of M. The absence or presence of an end borrow then determines the sign of the result, based on the sign of M, and determines whether or not a 2's complement correction is performed. For example, since the signs are different, (0 0011001) + (1 0100101) causes 0100101 to be subtracted from 0011001: The result is 1110100, and an end borrow of 1 occurs. The end borrow indicates that the magnitude of M is smaller than the magnitude of N. So the sign of the result is opposite that of M and therefore a minus. The end borrow indicates that the magnitude of the result, 1110100, must be corrected by taking its 2's complement. Combining the sign and the corrected magnitude of the result, we obtain 1 0001100.

In contrast to this signed-magnitude case, the rule for adding numbers in the signed-complement system does not require comparison or subtraction, but only addition. The procedure is very simple and can be stated as follows for binary numbers:

> The addition of two signed binary numbers with negative numbers represented in signed-2's complement form is obtained from the addition of the two numbers, including their sign bits. A carry out of the sign bit position is discarded.

Numerical examples of signed binary addition are given in Example 3-8. Note that negative numbers will already be in 2's complement form and that the sum obtained after the addition, if negative, is left in that same form.

EXAMPLE 3-8

Addition of signed binary numbers using 2's complement.

+ 6	00000110	− 6	11111010	+ 6	00000110	− 6	11111010
+ 13	00001101	+13	00001101	+ 13	11110011	− 13	11110011
+ 19	00010011	+ 7	00000111	− 7	11111001	− 19	11101101

In each of the four cases, the operation performed is addition, including the sign bits. Any carry out of the sign bit position is discarded, and negative results are automatically in 2's complement form. ■

The complement form for representing negative numbers is unfamiliar to people used to the signed-magnitude system. To determine the value of a negative number in signed-2's complement, it is necessary to convert the number to a positive number in order to put it in a more familiar form. For example, the signed binary number 11111001, is negative because the leftmost bit is 1. Its 2's complement is 00000111 which is the binary equivalent of +7. We therefore recognize the original number to be equal to −7.

The subtraction of two signed binary numbers when negative numbers are in 2's complement form is very simple and can be stated as follows:

> Take the 2's complement of the subtrahend (including the sign bit) and add it to the minuend (including the sign bit). A carry out of the sign bit position is discarded.

This procedure stems from the fact that a subtraction operation can be changed to an addition operation if the sign of the subtrahend is changed. That is,

$$(\pm A) - (+B) = (\pm A) + (-B)$$

$$(\pm A) - (-B) = (\pm A) + (+B)$$

But changing a positive number to a negative number is easily done by taking its 2's complement. The reverse is also true, because the complement of a negative number that is already in complement form produces the corresponding positive number.

EXAMPLE 3-9

Subtraction of signed binary numbers using 2's complement

− 6	11111010	11111010	+ 6	00000110	00000110
− (−13)	− 11110011	+ 00001101	− (−13)	− 11110011	+ 00001101
+ 7		00000111	+ 19		00010011

The end carry is discarded. ■

It is worth noting that binary numbers in the signed-complement system are added and subtracted by the same basic addition and subtraction rules as are unsigned numbers. Therefore, computers need only one common hardware circuit to handle both types of arithmetic. The user or programmer must interpret the results of such addition or subtraction differently, depending on whether it is assumed that the numbers are signed or unsigned. Thus, the same adder/subtractor designed for unsigned numbers can be used for signed numbers.

If the signed numbers are in 2's complement representation, then the circuit in Figure 3-30 can be used with no correction step required. For 1's complement, the input from S to C_0 of the adder must be replaced by an input from C_n to C_0.

Overflow

In order to obtain a correct answer when adding and subtracting, we must ensure that the result has a sufficient number of bits to accommodate the sum. If we start with two n-bit numbers, and the sum occupies $n + 1$ bits, we say that an *overflow* occurs. This is true for binary or decimal numbers, whether signed or unsigned. When one performs addition with paper and pencil, an overflow is not a problem, since we are not limited by the width of the page. We just add another 0 to a positive number and another 1 to a negative number, in the most significant position, to extend them to $n + 1$ bits and then perform the addition. Overflow is a problem in computers because the number of bits that hold a number is fixed, and a result that exceeds the number of bits cannot be accommodated. For this reason, computers detect and can signal the occurrence of an overflow. The overflow condition may be handled automatically by interrupting the execution of the program and taking special action. An alternative is to monitor for overflow conditions using software.

The detection of an overflow after the addition of two binary numbers depends on whether the numbers are considered to be signed or unsigned. When two unsigned numbers are added, an overflow is detected from the end carry out of the most significant position. In unsigned subtraction, the magnitude of the result is always equal to or smaller than the larger of the original numbers, making overflow impossible. In the case of signed 2's complement numbers, the most significant bit always represents the sign. When two signed numbers are added, the sign bit is treated as a part of the number, and an end carry of 1 does not necessarily indicate an overflow.

With signed numbers, an overflow cannot occur for an addition if one number is positive and the other is negative: Adding a positive number to a negative number produces a result whose magnitude is equal to or smaller than the larger of the original numbers. An overflow may occur if the two numbers added are both positive or both negative. To see how this can happen, consider the following 2's complement example: Two signed numbers, +70 and +80, are stored in two 8-bit registers. The range of binary numbers, expressed in decimal, that each register can accommodate is from +127 to −128. Since the sum of the two stored numbers is +150, it exceeds the capacity of an 8-bit register. This is also true for −70 and −80.

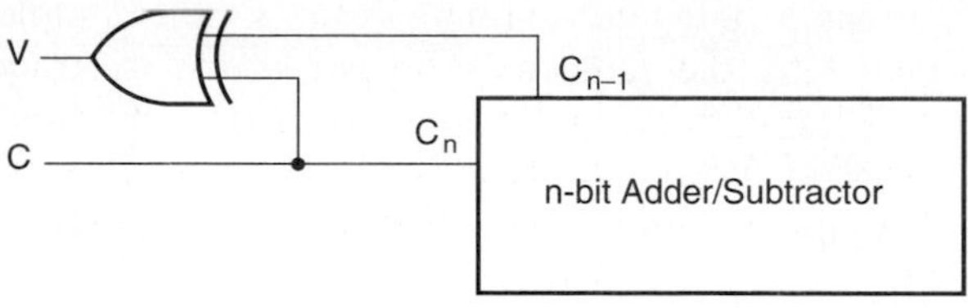

□ **FIGURE 3-31**
Overflow Detection Logic for Addition and Subtraction

These two additions are shown below, together with the two most significant carry bit values.

Carries:	0 1		Carries:	1 0
+ 70	0 1000110		− 70	1 0111010
+ 80	0 1010000		− 80	1 0110000
+ 150	1 0010110		−150	0 1101010

Note that the 8-bit result that should have been positive has a negative sign bit and that the 8-bit result that should have been negative has a positive sign bit. If, however, the carry out of the sign bit position is taken as the sign bit of the result, then the 9-bit answer so obtained will be correct. But since there is no position in the result for the 9th bit, we say that an overflow has occurred.

An overflow condition can be detected by observing the carry into the sign bit position and the carry out of the sign bit position. If these two carries are not equal, an overflow has occurred. This is indicated in the 2's complement example, just completed, where the two carries are explicitly shown. If the two carries are applied to an exclusive-OR gate, an overflow is detected when the output of the gate is equal to 1. For this method to work correctly for 2's complement, it is necessary either to apply the 1's complement of the subtrahend to the adder and add 1 or to have overflow detection on the circuit that forms the 2's complement. The latter condition is due to overflow when complementing the maximum negative number.

Simple logic that provides overflow detection is shown in Figure 3-31. If the numbers are considered unsigned, then the C output being equal to 1 detects a carry (an overflow) for an addition and indicates that no correction step is required for a subtraction. C being equal to 0 detects no carry (no overflow) for an addition and indicates that a correction step is required for a subtraction.

If the numbers are considered signed, then the output V is used to detect an overflow. If $V = 0$ after a signed addition or subtraction, it indicates that no overflow has occurred and the result is correct. If $V = 1$, then the result of the operation contains $n + 1$ bits, but only the rightmost n of those bits fit in the n-bit result, so an overflow has occurred. The $(n+1)$th bit is the actual sign, but it cannot occupy the sign bit position in the result.

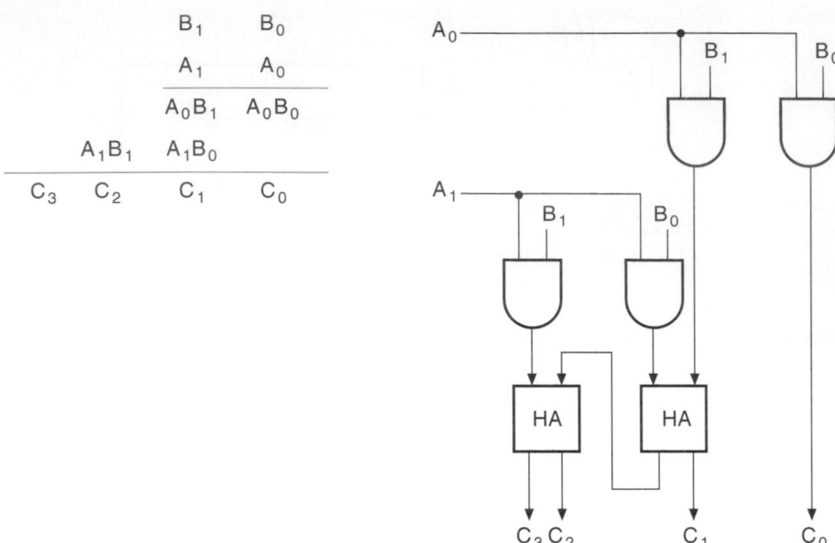

FIGURE 3-32
A 2-Bit by 2-Bit Binary Multiplier

3-11 Binary Multipliers

Multiplication of binary numbers is performed in the same way as with decimal numbers. The multiplicand is multiplied by each bit of the multiplier, starting from the least significant bit. Each such multiplication forms a partial product. Successive partial products are shifted one bit to the left. The final product is obtained from the sum of the partial products.

To see how a binary multiplier can be implemented with a combinational circuit, consider the multiplication of two 2-bit numbers, as shown in Figure 3-32. The multiplicand bits are B_1 and B_0, the multiplier bits are A_1 and A_0, and the product is $C_3C_2C_1C_0$. The first partial product is formed by multiplying B_1B_0 by A_0. The multiplication of two bits such as A_0 and B_0 produces a 1 if both bits are 1; otherwise it produces a 0. This is identical to an AND operation. Therefore, the partial product can be implemented with AND gates as shown in the diagram. The second partial product is formed by multiplying B_1B_0 by A_1 and is shifted one position to the left. The two partial products are added with two half-adder (HA) circuits. Usually there are more bits in the partial products, and it will be necessary to use full adders to produce the sum of the partial products. Note that the least significant bit of the product does not have to go through an adder, since it is formed by the output of the first AND gate.

A combinational circuit binary multiplier with more bits can be constructed in a similar fashion. A bit of the multiplier is ANDed with each bit of the multiplicand in as many levels as there are bits in the multiplier. The binary output in each level of AND gates is added in parallel with the partial product of the previous level to form a new partial product. The last level produces the product. For J mul-

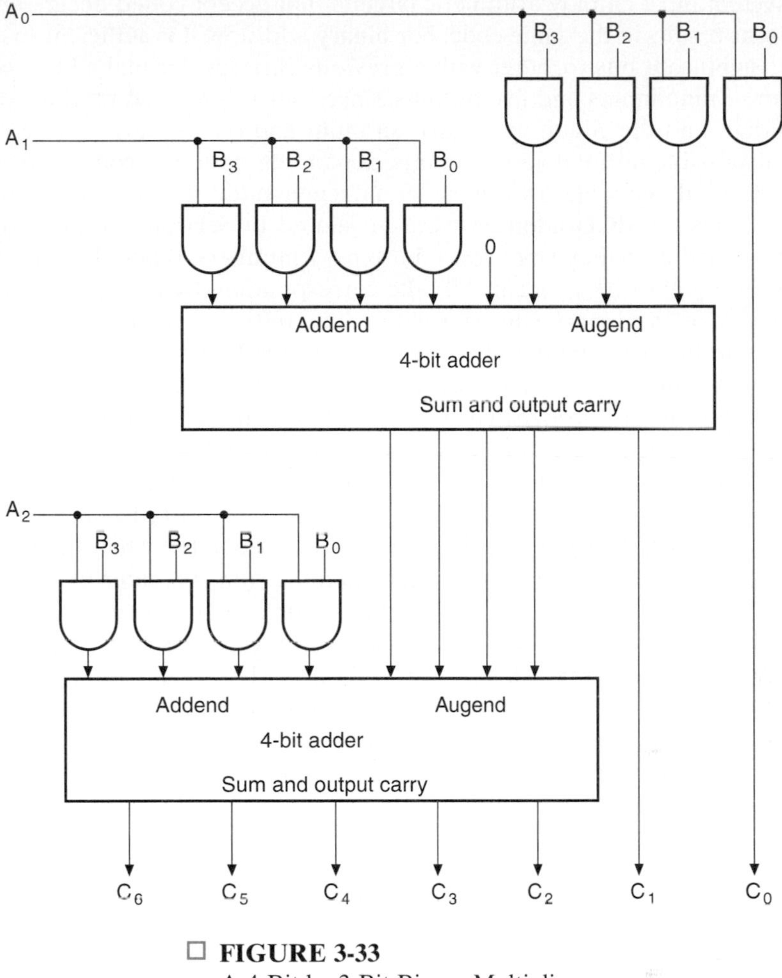

☐ **FIGURE 3-33**
A 4-Bit by 3-Bit Binary Multiplier

tiplier bits and K multiplicand bits, we need $J \times K$ AND gates and $(J - 1)$ K-bit adders to produce a product of $J + K$ bits. As an example of a combinational circuit binary multiplier, consider a circuit that multiplies a binary number of four bits by a number of three bits. Let the multiplicand be represented by $B_3B_2B_1B_0$ and the multiplier by $A_2A_1A_0$. Since $K = 4$ and $J = 3$, we need 12 AND gates and two 4-bit adders to produce a product of seven bits. The logic diagram of this kind of multiplier circuit is shown in Figure 3-33.

3-12 Decimal Arithmetic

Computers or calculators that perform arithmetic operations directly in the decimal number system represent decimal numbers in binary coded form. An adder for such a system must employ arithmetic circuits that accept coded decimal numbers and present results in the same code. For binary addition, it is sufficient to consider a pair of significant bits together with a previous carry. A decimal adder requires a minimum of nine inputs and five outputs, since four bits are required to code each decimal digit and the circuit must have an input and output carry. There is a wide variety of possible decimal adder circuits, depending upon the code used to represent the decimal digits. Here we consider a decimal adder for the BCD code.

The rules for BCD addition were presented in Section 1-4. First, the BCD digits are added as if they were two 4-bit binary numbers. When the binary sum is less than or equal to 1001 (decimal 9), the corresponding BCD digit sum is correct. However, when the binary sum is greater than 1001, we obtain an invalid BCD result. The addition of binary 0110 (decimal 6) to the binary sum converts it to the correct BCD representation and also produces an output carry as required. Consider the addition of two decimal digits in BCD with an input carry. Since each digit does not exceed 9, the sum cannot be greater than $9 + 9 + 1 = 19$.

The logic circuit that checks for the necessary BCD correction can be derived by detecting the occurrence of the binary numbers from 1010 through 10011 (decimal 10 through 19). It is obvious that a correction is needed when the binary sum has an output carry. This condition occurs when the sum is greater than or equal to 16. The other six combinations, from 1010 through 1111, that need a correction have a 1 in the most significant position and a 1 in the second or third significant position. A BCD adder that adds two BCD digits and produces a sum digit in BCD is shown in Figure 3-34. It has two 4-bit binary adders and correction logic. The two decimal digits, together with an input carry, are added in the first 4-bit binary adder, to produce the binary sum. The condition for correction can be expressed by the Boolean function

$$C = K + Z_1Z_3 + Z_2Z_3$$

Here C is the output carry from the BCD adder, and K is the output carry from the first binary adder. The two terms with the Z variables detect the binary outputs from 1010 through 1111. When the BCD carry is equal to 0, nothing is added to the binary sum. This condition occurs if the sum of the two digits plus the input carry is less than or equal to binary 1001. When the output carry is equal to 1, binary 0110 is added to the binary sum through the second 4-bit adder. This condition occurs when the sum is greater than or equal to 1010. Any output carry from the second binary adder can be neglected. A decimal parallel adder that adds two n-digit decimal numbers needs n BCD adders. The output carry from each adder must be connected to the input carry of the adder in the next higher position.

Use of Complements in Decimal

The formation of the 9's and 10's complement was briefly discussed earlier. The following two examples illustrate the formation of complements in decimal:

The 9's complement of 546700 is $999999 - 546700 = 453299$.

The 10's complement of 546700 is $1000000 - 546700 = 453300$.

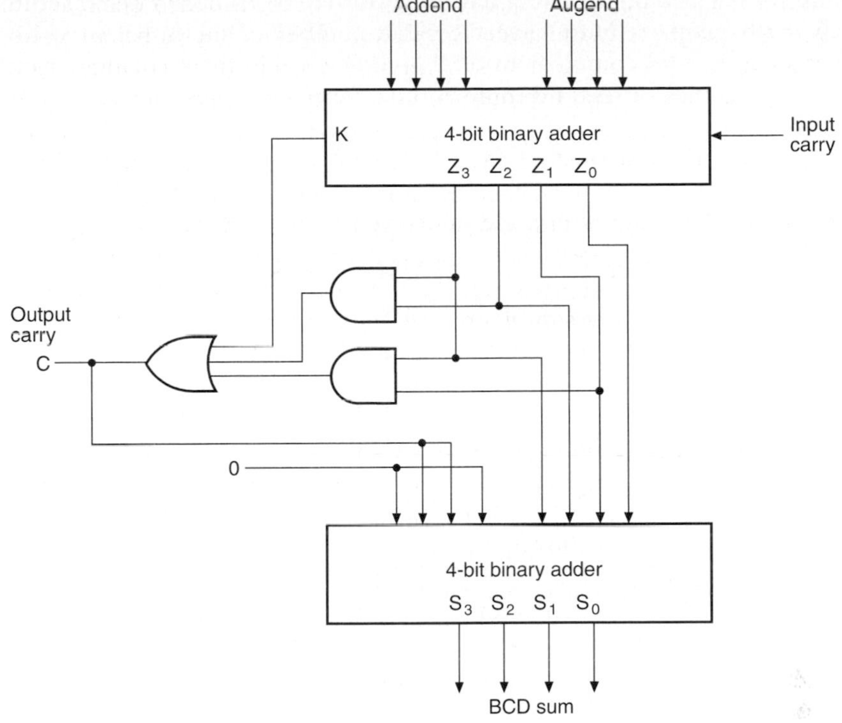

□ **FIGURE 3-34**
Block Diagram of BCD Adder

Thus, as with binary, the decimal 10's complement is obtained by adding 1 to the 9's complement value.

Since 10^n is a number represented by a 1 followed by n 0's, $10^n - N$, which is the 10's complement of N, also can be formed by leaving all least significant 0's unchanged, subtracting the first nonzero least significant digit from 10, and then subtracting all higher significant digits from 9. Accordingly, the 10's complement of 234500 is 765500 and is obtained by leaving the two zeros unchanged, subtracting 5 from 10, and subtracting the other three digits from 9.

The next two examples apply the use of complements to unsigned decimal subtraction.

EXAMPLE 3-10

Using 10's complement, perform the subtraction 72532 − 3250.

$$\begin{aligned} M &= \ 72532 \\ \text{10's complement of } N &= +\ \underline{96750} \\ \text{Sum} &= \ 169282 \\ \text{Discard end carry } 10^5 &= -\ \underline{100000} \\ \textit{Answer} &= \ 69282 \end{aligned}$$

Note that M has five digits and N has only four digits. To perform the subtraction $M - N$, both numbers must have the same number of digits, so we write N as 03250. Taking the 10's complement of N produces a 9 in the most significant position. The occurrence of the end carry in the addition signifies that $M > N$ and the result is positive. ■

■ **EXAMPLE 3-11**

Using 10's complement, perform the subtraction 3250 − 72532.

$$\begin{aligned} M &= \ 03250 \\ \text{10's complement of } N &= +\ \underline{27468} \\ \text{Sum} &= \ 30718 \end{aligned}$$

There is no end carry.

Answer: − (10's complement of 30718) = −69282

Note that since 3250 < 72532, the result is negative. ■

The procedure with end-around carry, as used in binary, is also applicable for subtracting unsigned decimal numbers with 9's complement.

3-13 Standard Graphics Symbols

Standard graphics symbols have been developed for functional blocks and more complex standard blocks so that the user can recognize each function from the unique graphics symbol assigned to it. This standard, known as IEEE Std 91-1984/IEEE Std 91a-1991, has been approved by industry, government, and professional organizations and is consistent with international standards.

The standard uses a rectangular outline to represent each particular logic function. Within the outline, there is a general qualifying symbol denoting the logical operation performed by the unit. For example, the general qualifying symbol for a multiplexer is MUX. The size of the outline is arbitrary, and the shape can be either square or rectangular with an arbitrary ratio of length to width. Input lines are placed on the left, output lines on the right. If the direction of the signal flow is reversed, this must be indicated by arrows. The rectangular symbols for digital gates were introduced in Figure 2-26. An example of a standard graphics symbol is the 4-bit parallel adder shown in Figure 3-35. The qualifying symbol for an adder is the Greek letter Σ. The preferred letters for the arithmetic operands are P and Q. The bit-grouping symbols in the two types of inputs and the sum output are the decimal equivalent of the weights of the bits to the power of 2. Thus, the input labeled 3 corresponds to the value of $2^3 = 8$. The input carry is designated by CI and the output carry by CO. Before introducing the graphics symbols of other components, it is necessary to review some terminology. As mentioned in Section 2-8, a positive-logic system defines the more positive of two signal levels (designated by H) as logic 1 and the more negative (designated by L) as logic 0. Negative

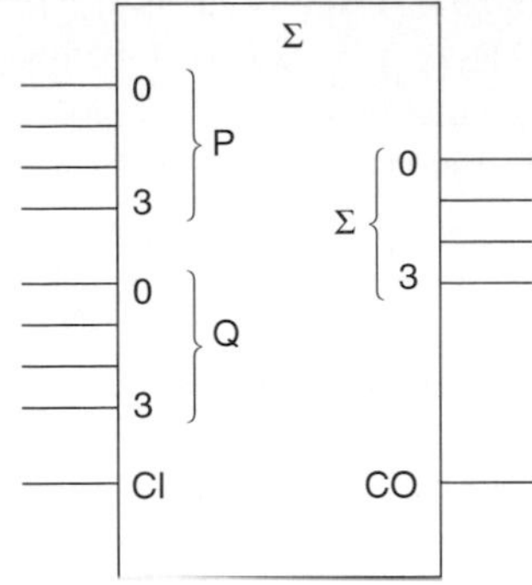

□ **FIGURE 3-35**
Standard Graphics Symbol for a 4-Bit Parallel Adder

logic assumes the opposite assignment. A third alternative is to employ a mixed-logic convention in which the signals are considered entirely in terms of their H and L values. At any point in the circuit, the user is allowed to define the polarity of the logic by assigning logic 1 to either the H or the L value. The mixed-logic notation uses a small right-triangle graphics symbol to designate a negative logic polarity at any input or output terminal. (See Figure 2-42.) Integrated circuit manufacturers specify the operation of their circuits in terms of H and L values. When an input or output is considered in terms of positive logic, it is defined as active high; when it is considered in terms of negative logic, it is assumed to be active low. Active-low inputs or outputs are recognized by the presence of the small-triangle *polarity indicator* graphics symbol. When positive logic is used exclusively throughout the entire system, the small-triangle polarity indicator is equivalent to the small circle (bubble) that designates negation; this circle is the *negation indicator* that distinguishes an inverter from a buffer and a NOR gate from an OR gate. In this book, we usually assume positive logic, and we will employ the bubble when drawing logic diagrams. When an input or output line does not include the small circle, we will define it to be active if it is logic 1. A line that includes the small circle symbol will be considered active if it is in the logic-0 state. However, we will use the small-triangle polarity symbol to indicate an active-low assignment in all drawings that represent standard diagrams. This will conform with integrated circuit data books, in which the polarity symbol is usually employed.

Standard Graphics Symbol for Decoder

Two graphics symbols for the decoder are shown in Figure 3-36. In part (a), we have a 3–to–8-line decoder with an enable input designated by EN. Inputs are on the left and outputs on the right. The identifying symbol X/Y indicates that the circuit converts from code X to code Y. The inputs are assigned binary weights 1, 2, and 4, equivalent to 2^0, 2^1, and 2^2. The outputs are assigned numbers from 0 to 7. The sum of the weights of the inputs determines the output that is active. Thus, if

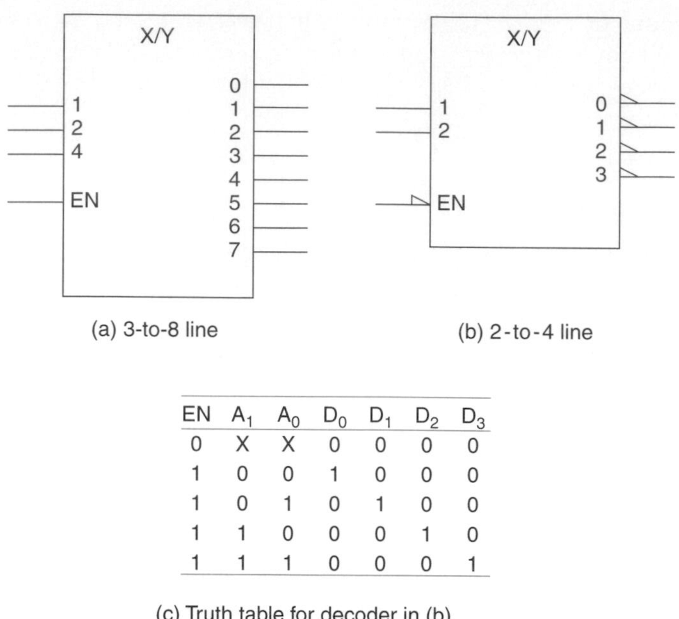

(a) 3-to-8 line

(b) 2-to-4 line

EN	A_1	A_0	D_0	D_1	D_2	D_3
0	X	X	0	0	0	0
1	0	0	1	0	0	0
1	0	1	0	1	0	0
1	1	0	0	0	1	0
1	1	1	0	0	0	1

(c) Truth table for decoder in (b) with EN and outputs negative logic

□ **FIGURE 3-36**
Standard Graphics Symbols for Decoders

the two input lines with weights 1 and 4 are activated, the total weight is $1 + 4 = 5$, and output 5 is activated. Of course, the EN input must be activated for any output to be active. The two-letter symbol EN is reserved to identify an input that enables outputs. If the input labeled EN is active, all outputs are active or inactive according to the input conditions that affect them. If the EN input is inactive, all outputs are inactive. A 2-to-4 line decoder with an active-low enable input and active-low outputs is shown in part(b) of the figure. The circuit is enabled if the enable input is active. This means that the enable signal must be in a low-level state, as indicated by the polarity symbol. If the circuit is not enabled, all outputs are inactive and in the high-level state. When the circuit is enabled, the input weights determine which output will be active. The polarity symbols in the outputs indicate that the selected output will be at a low level. For positive-logic assignments to the inputs and outputs, the inputs and outputs are as specified in the truth table of Figure 3-13(b). For a mixed-logic assignment with the EN input and the four outputs negative logic, the truth table is given in Figure 3-36(c). Note that, although the circuit is the same, the functions represented by the truth tables are quite different due to the use of positive logic for one and mixed logic for the other.

The decoder is a special case of a more general component referred to as a coder. A *coder* is a device that receives a binary code on a number of inputs and produces a different binary code on a number of outputs. Instead of using the qualifying symbol *X/Y*, we can specify the coder by the code name. For example, the 3-

to-8-line decoder of Figure 3-36(a) can be symbolized with the name BIN/OCT, since the circuit converts a 3-bit binary code into an octal (8-bit) output. When the digital component represented by the outline is also a commercial integrated circuit, it is customary to write the IC pin number along each input and output line.

Standard Graphics Symbol for the Multiplexer

Before showing the graphics symbol for a multiplexer, it is necessary to define a notation called *G-* or *AND-dependency.* The letter *G* followed by a number is reserved for specifying AND-dependency. In a block diagram, any input or output labeled with the number associated with *G* is considered to be ANDed with it. For example, if one input in the block diagram has the label *G*1, and another input is labeled with the number 1, then the two inputs labeled *G*1 and 1 are considered to be ANDed together internally.

An example of AND-dependency is shown in Figure 3-37. In part (a), we have a portion of a graphics symbol with two AND-dependency labels *G*1 and *G*2. Also, there are two inputs labeled with the number 1 and one input labeled with the number 2. The equivalent interpretation is shown in part (b) of the figure. Note that the AND gates are drawn with the ampersand (&) inside a rectangular outline to conform with the standard shapes. Input *X* labeled *G*1 is considered to be ANDed with inputs *A* and *B*, which are labeled with a 1. Similarly, input *Y* is ANDed with input *C* to conform with the dependency between *G*2 and 2.

To deal with structures such as multiplexers, AND-dependency is sometimes represented by a shorthand notation that looks like $G\frac{0}{7}$. This symbol stands for eight AND-dependency symbols, from 0 to 7, as follows:

$$G0, G1, G2, G3, G4, G5, G6, G7$$

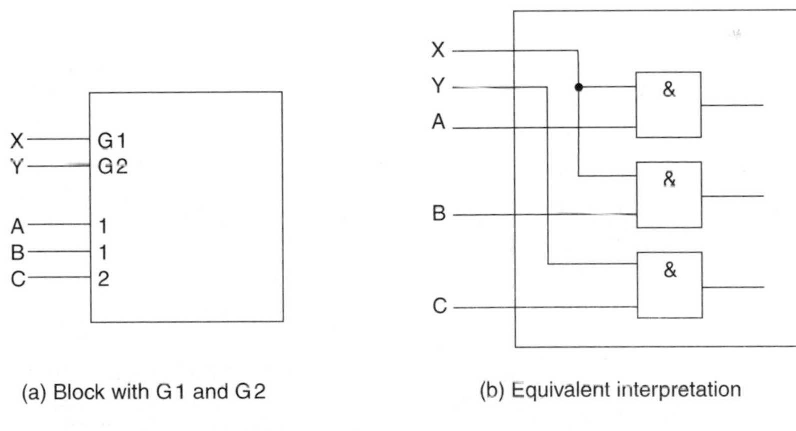

□ **FIGURE 3-37**
Example of a *G*-(AND) Dependency

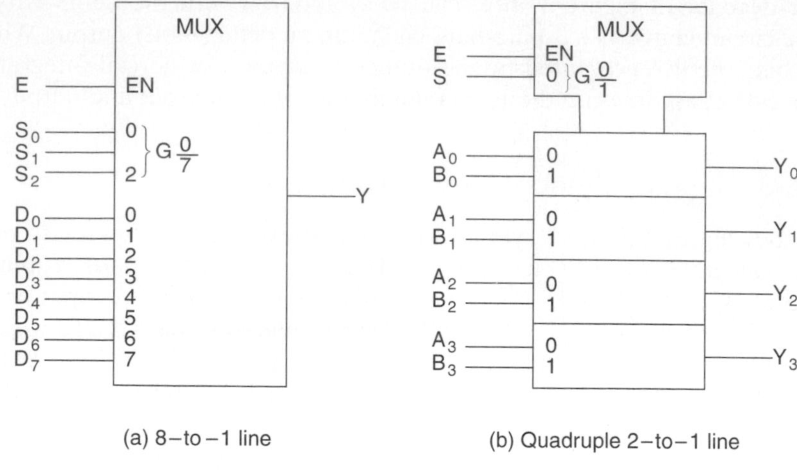

□ **FIGURE 3-38**
Standard Graphics Symbols for Multiplexers

At any given time, only one out of the eight AND gates can be active. The active AND gate is determined from the inputs associated with the *G* symbol. These inputs are marked with weights equal to the powers of 2. For the eight AND gates listed, the weights are 0, 1, and 2, corresponding to the numbers 2^0, 2^1, and 2^2, respectively. The AND gate that is active at any given time is determined from the sum of the weights of the active inputs. Thus, if inputs 0 and 2 are active, the AND gate that is active has the number $2^0 + 2^2 = 5$. This makes *G*5 active and the other seven AND gates inactive.

The standard graphics symbol for the multiplexer is shown in Figure 3-38(a). The label MUX identifies the device as a multiplexer. The symbols inside the block are part of the standard notation, but the symbols marked outside are user defined. In order to understand the standard notation, we use the same external variables here as in the logic diagrams of the multiplexers in Section 3-7.

The diagram in Figure 3-38 (a) represents an 8–to–1-line multiplexer with enable (EN) input. The AND-dependency is marked with $G\frac{0}{7}$ and is associated with the inputs indicated by the brace. These inputs have weights of 0, 1, and 2. They are actually what we earlier called the selection inputs. The eight data inputs are marked with numbers from 0 to 7. The net weight of the active inputs associated with the *G* symbol specify the number in the data input that is active. For example, if selection inputs $(S_2, S_1, S_0) = 110$, then inputs 1 and 2 associated with *G* are active. This gives a numerical value for the AND-dependency of $2^2 + 2^1 = 6$, which makes *G*6 active. Since *G*6 is ANDed with data input number 6, it makes this input active. Thus, the output will be equal to data input *D*6, provided that the enable input is active. The diagram of Figure 3-38(b) represents the quadruple 2–to–1-line multiplexer, whose logic diagram is shown in Figure 3-20. The enable and selection inputs are common to all four multiplexers. This is indicated in the standard notation by the indented box at the top of the diagram, which represents a *common control block*. The inputs to a common control block control all lower sections of the diagram. The common enable input EN is active when in the high-level state. The AND-dependency G_1^0 represents two AND gates, $G0$ and $G1$. When $S = 0$, $G0$ is active, and the inputs marked with 0 are active. When $S = 1$, $G1$ is active, and the inputs marked with 1 are active. The active inputs are applied to the corresponding outputs if EN is active.

Dependency Notation

The most important aspect of the standard logic symbols is the dependency notation. This is used to provide the means of denoting the relationship between different inputs or outputs without actually showing all the elements and interconnections between them. We have already defined the *G*-(AND) dependency in connection with the symbol for the multiplexer. The IEEE standard defines 10 other types of dependencies. Each (except *EN*) is denoted by a letter symbol. The letter appears at the input or output and is followed by a number. Each input or output affected by that dependency is labeled with the same number. The 11 dependencies and their corresponding letter designations are as follows:

G	Denotes an AND (gate) relationship
EN	Specifies an enable action
C	Identifies a control dependency
S	Specifies a setting action
R	Specifies a resetting action
M	Identifies a mode dependency
A	Identifies an address dependency
Z	Indicates an internal interconnection
X	Indicates a controlled transmission
V	Denotes an OR relationship
N	Denotes a negate (exclusive-OR) relationship

The *G*- and EN-dependencies were introduced in this section. The control dependency *C* is used to identify a clock input in a sequential element and to indicate which input is controlled by the clock. The set (*S*) and reset (*R*) dependencies specify internal logic states of an *SR* flip-flop. The *C*-, *S*-, and *R*-dependencies are explained in Section 4-3 in conjunction with the flip-flop circuit. The mode (*M*) dependency is used to identify inputs that select the mode of operation of the unit. This dependency is presented in Chapter 5 in conjunction with registers and counters. The address (*A*) dependency identifies the address input of a memory. The *Z*-dependency indicates interconnections inside the unit; it signifies the existence of internal logic connections between inputs, outputs, internal inputs, and internal outputs, in any combination. The *X* dependency is used to indicate the

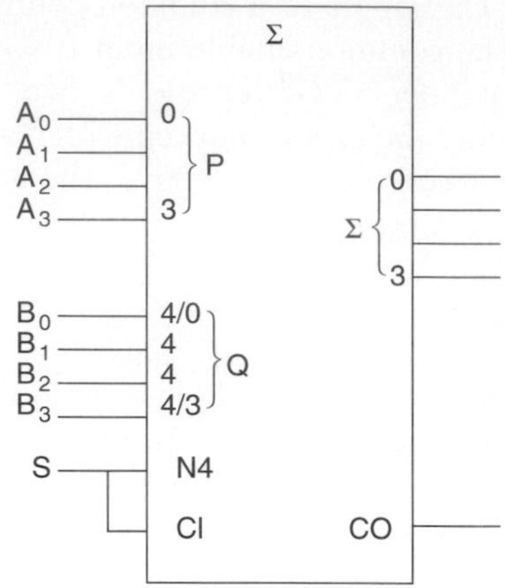

□ **FIGURE 3-39**
Graphics Symbol for an Adder-Subtractor

controlled transmission path in a transmission gate similar to the one shown in Figure 2-48.

The *V*- and *N*-dependencies are used to denote the Boolean relationships of OR and exclusive-OR, respectively. They are similar to the *G*-dependency that denotes the Boolean AND. An example of a graphics symbol that uses the *N*-dependency is shown in Figure 3-39. This is the symbol for the adder-subtractor circuit from Figure 3-30. The arithmetic operands are *P* and *Q*, as in the adder. Input *S* is labeled with *N*4, and the four *Q* inputs have the numerical label 4. The dependency notation here implies the existence of the exclusive-OR relationship between the *S* input and each of the *B* inputs. This means that when $S = 0$, the sum on *Q* uses the *B* inputs, but when $S = 1$, the four inputs are negated by an exclusive-OR gate, and *Q* is based on the 1's complement of *B*.

Qualifying Symbol for Inputs and Outputs

In addition to dependency notation, the IEEE standard graphics symbols for logic functions include qualifying symbols associated with inputs, outputs, and other connections. Among these are the polarity and negation symbols, symbols for internal connections, symbols for inside the outline, and symbols for nonlogical connections. The entire list of symbols is too numerous to be given here. Some of the most commonly used ones, however, are shown in Figure 3-40.

The active-low input or output is the polarity indicator. As mentioned previously, it is equivalent to negation when positive logic is assumed. The EN input is similar to the EN dependency, except that it does not need to be followed by a number. If there is no number following it, the simple EN input has the effect of enabling all the outputs when it is active.

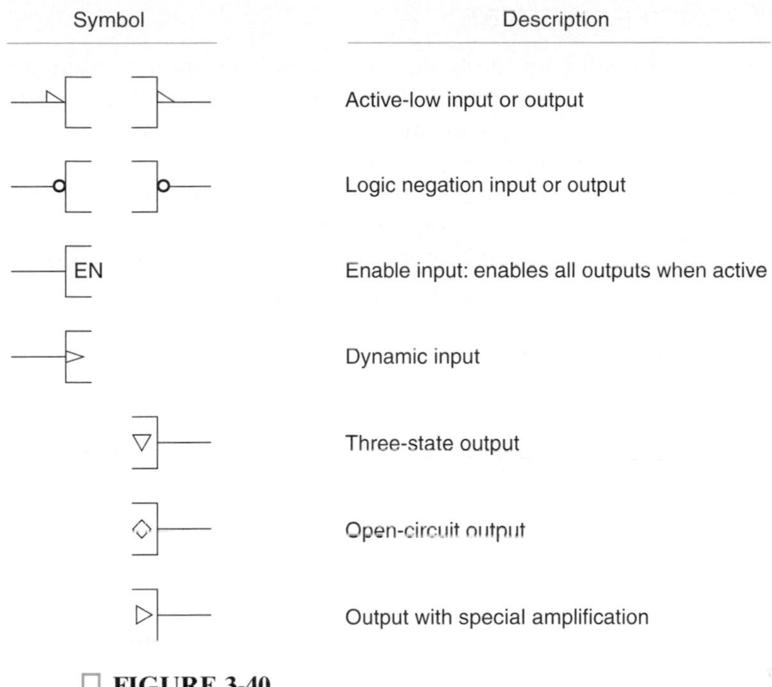

Symbol
Description
Active-low input or output
Logic negation input or output
EN
Enable input: enables all outputs when active
Dynamic input
Three-state output
Open-circuit output
Output with special amplification

□ **FIGURE 3-40**
Qualifying Symbols Associated with Inputs and Outputs

The *dynamic input* is associated with the clock input in flip-flop circuits. It indicates that the input is active on a transition from a low- to a high-level signal. This is explained in more detail in Section 4-3. The three-state output has a third external high-impedance state that has no significance with regard to logic. This is explained in Section 6-3 in conjunction with the discussion of memory data outputs.

The open-circuit output is also referred to as an open-collector or open-drain output. One of the two possible logic states of this type of output corresponds to an external high-impedance condition. An externally connected resistor is sometimes required in order to produce the proper logic level. The diamond-shaped symbol may have a bar on top for high type or on the bottom for low type. The high or low type specifies the logic level when the output is not in the high-impedance state. For example, TTL-type integrated circuits have a special output called open-collector output, recognized by a diamond-shaped symbol with a bar under it. When used as part of a physically distributed implementation of a logic function, two or more open-collector outputs connected to a common resistor to the power supply perform a positive logic AND function or a negative-logic OR function.

The output with special amplification is used in outputs of gates that provide special driving capabilities. These gates are employed in components such as clock drivers or bus-oriented transmitters.

3-14 Chapter Summary

This chapter has dealt with the basic analysis and design of combinational logic circuits. The concepts of a design hierarchy, predefined blocks, and reusable blocks essential to the design of large circuits and systems were introduced. Another essential element of modern design, the use of CAD tools, particularly for schematic capture and logic simulation, was also introduced. Logic circuit analysis by means of truth tables or Boolean equations for the circuit was illustrated, as was is logic simulation for finding waveforms and verifying circuits.

A design procedure for combinational circuits was presented and illustrated with the use of code converters. These converters translate from one code to another and are the first of a number of combinational circuits called functional blocks that are frequently used in designing larger circuits. Another functional block type that was discussed is the decoder, which is used to activate one of a number of lines. In combination with OR or NAND gates, decoders provide a simple minterm-based approach to implementing combinational circuits. Encoders, the opposite of decoders, generate a code associated with the line that is active from among a set of lines. Multiplexers, which take data applied at the input selected and present it at the output, were illustrated. A procedure was given for using an n–to–1 multiplexer and a single inverter to implement any $(n + 1)$-input Boolean function.

The chapter continued with an examination of circuits for performing arithmetic. The implementation of binary adders, including the carry lookahead adder for improved performance, was treated in detail. The subtraction of unsigned binary numbers using 2's and 1's complements was presented, as was the representation of signed binary numbers and their addition and subtraction. The adder-subtractor using 2's complement arithmetic developed for unsigned binary was found to apply directly to the addition and subtraction of signed numbers. A very brief introduction to binary multiplication using combinational circuits made up of AND gates and binary adders was given.

The foundations for BCD addition and subtraction that were presented included the use of 10's complement arithmetic. A BCD digit adder that can be cascaded to form a decimal adder was discussed. The chapter concluded with a discussion of IEEE standard symbols for functional blocks.

References

1. Blakeslee, T. R. *Digital Design with Standard MSI and LSI*, 2nd ed. New York: Wiley, 1979.
2. Mano, M. M. *Digital Design,* 2nd ed. Englewood Cliffs, NJ: Prentice Hall, 1991.
3. Roth, C. H. *Fundamentals of Logic Design*, 4th ed. St. Paul: West, 1992.
4. Wakerly, J. F. *Digital Design: Principles and Practices,* 2nd ed. Englewood Cliffs, NJ: Prentice Hall, 1994.
5. *The TTL Data Book,* vols. 1, 2, and 3. Dallas: Texas Instruments, 1984.

6. *High-Speed CMOS Logic Data Book*. Dallas: Texas Instruments, 1989.
7. *IEEE Standard Graphic Symbols for Logic Functions.* (Includes IEEE Std 91a-1991 Supplement and IEEE Std. 91-1984.) New York: The Institute of Electrical and Electronics Engineers, 1991.
8. KAMPEL, I. *A Practical Introduction to the New Logic Symbols.* Boston: Butterworths, 1985.

PROBLEMS

The asterisk (*) indicates a more advanced problem.

3–1. Determine the Boolean functions for outputs *J* and *K* as a function of the four inputs in the circuit of Figure 3-41.

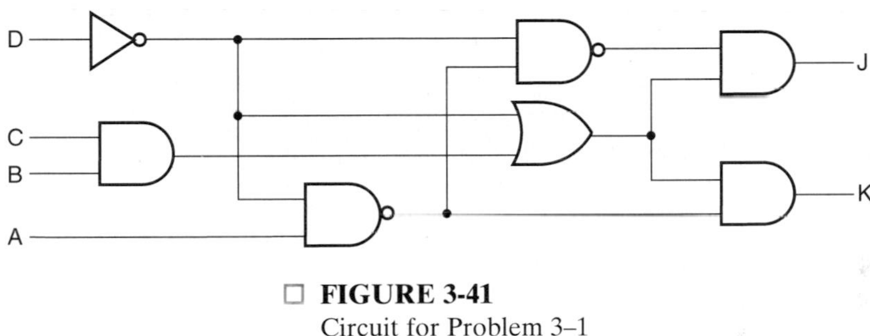

FIGURE 3-41
Circuit for Problem 3–1

3–2. Obtain the truth table for the circuit shown in Figure 3-42. Draw an equivalent circuit for *F* with fewer NAND gates.

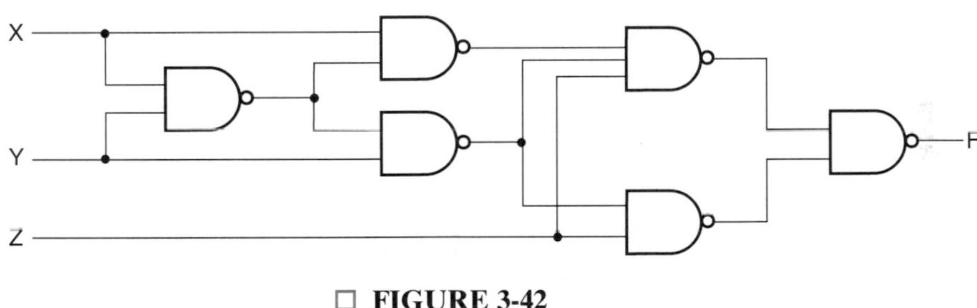

FIGURE 3-42
Circuit for Problem 3–2

3–3. Verify using Boolean algebra that the circuit of Figure 3-43 generates the exclusive-NOR function.

3–4. Find the simplified Boolean functions for outputs *F* and *G* of the circuit in Figure 3-44.

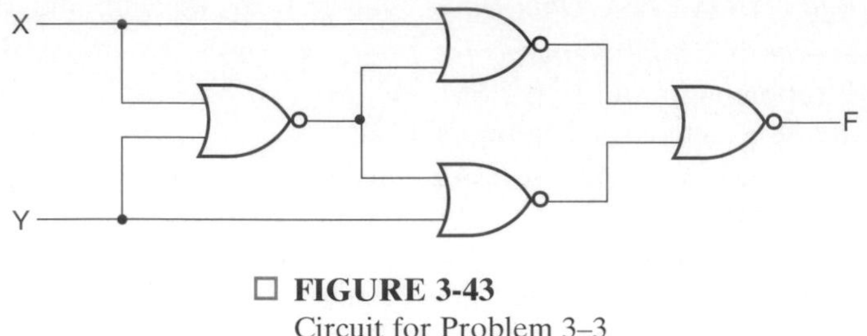

☐ **FIGURE 3-43**
Circuit for Problem 3–3

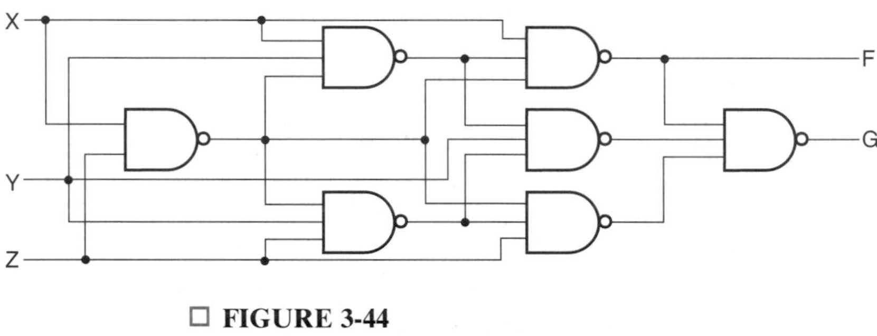

☐ **FIGURE 3-44**
Circuit for Problem 3–4 and Problem 3–5

3–5. Find the truth table for F and G of the circuit in Figure 3-44 by using logic simulation.

3–6. Find simplified Boolean equations for the outputs F and G of the hierarchical circuit in Figure 3-45.

3–7. Find the truth table for the outputs F and G of the hierarchical circuit in Figure 3-46. The symbol shown is for the decoder block in Figure 3-13.

3–8. The logic diagram for a 74HC138 MSI CMOS circuit is given in Figure 3-47. Find the Boolean function for each of the outputs. Describe carefully the circuit function.

3–9. Do Problem 3–8 by using logic simulation to find the output waveforms of the circuit, rather than finding Boolean functions.

3–10. A majority function is generated in a combinational circuit when the output is equal to 1 if the input variables have more 1's than 0's. The output is 0 otherwise. Design a 4-input majority function.

3–11. Design a combinational circuit that detects an error in the representation of a decimal digit in BCD. In other words, obtain a logic diagram whose output is equal to 1 when the inputs contain any one of the six unused bit combinations in the BCD code.

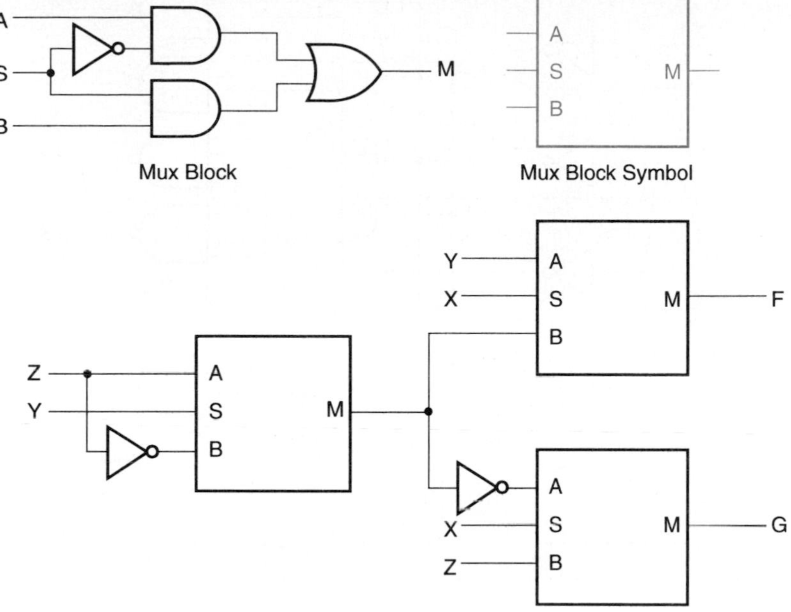

□ **FIGURE 3-45**
Circuit for Problem 3–6

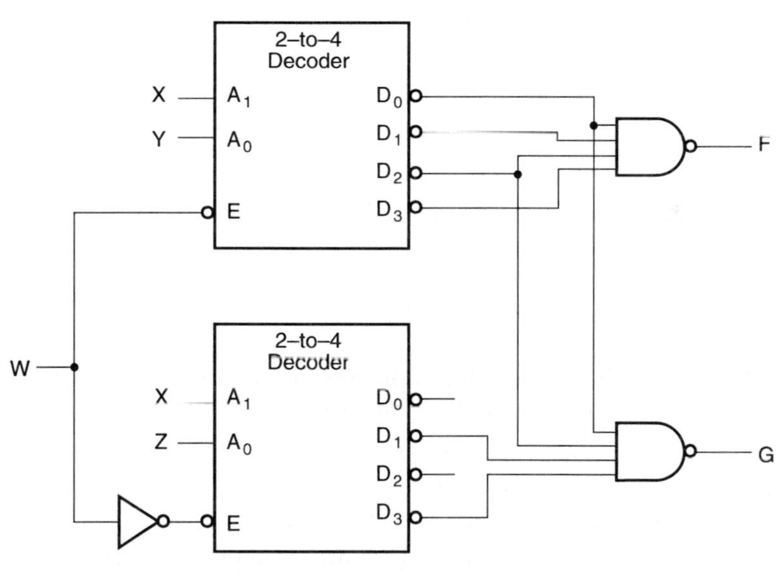

□ **FIGURE 3-46**
Circuit for Problem 3–7

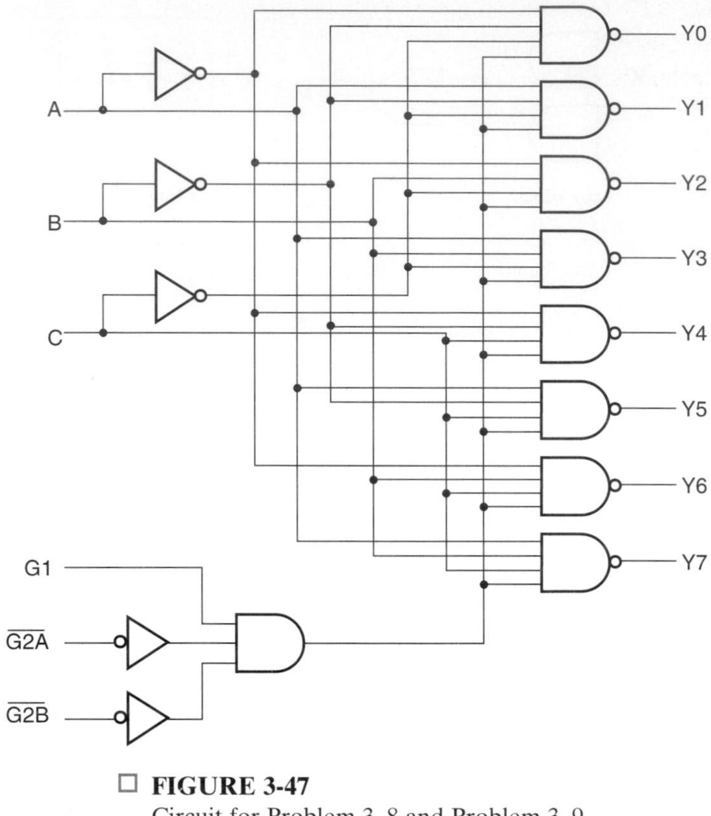

□ **FIGURE 3-47**
Circuit for Problem 3–8 and Problem 3–9

3–12. Complete the design of the BCD–to–seven-segment decoder by performing the following steps:

(a) Plot the seven maps for each of the outputs for the BCD–to–seven-segment decoder specified in Table 3-3.

(b) Simplify the seven output functions in sum–of–products form, and determine the total number of gates that will be needed to implement the decoder.

(c) Verify that the seven output functions listed in the text give a valid simplification. Compare the number of gates with that obtained in part (b).

3–13. Draw the logic diagram of the BCD–to–seven-segment decoder from the Boolean functions listed in Section 3-4 of the text. Use only 21 NAND gates. Assume that both the complemented and uncomplemented inputs are available.

3–14. Construct a 5–to–32-line decoder with four 3–to–8-line decoders with enable input and one 2–to–4-line decoder.

3–15. Design a BCD-to-decimal decoder using the unused combinations of the BCD code as don't-care conditions.

3–16. A combinational circuit is defined by the following three Boolean functions:

$$F_1 = \overline{X}\,\overline{Y} + XY\overline{Z}$$

$$F_2 = \overline{X} + Z$$

$$F_3 = XY + \overline{X}\,\overline{Y}$$

Design the circuit with a decoder and external gates.

3–17. A combinational circuit is specified by the following three Boolean functions:

$$F_1(A, B, C) = \Sigma m(1, 5, 7)$$

$$F_2(A, B, C) = \Sigma m(2, 3, 6)$$

$$F_3(A, B, C) = \Sigma m(0, 4, 7)$$

Implement the circuit with a decoder constructed with NAND gates (similar to Figure 3-13) and external NAND gates.

3–18. Draw the logic diagram of a 2–to–4-line decoder with only NOR and NOT gates. Include an enable input.

3–19. Design a 4-input priority encoder with inputs and outputs as in Table 3-6, but with the truth table representing the case in which input D_0 has the highest priority and input D_3 has the lowest priority.

3–20. Derive the truth table of an octal-to-binary priority encoder.

3–21. Construct an 8–to–1-line multiplexer with enable input using transmission gates.

3–22. Construct a 16–to–1-line multiplexer with two 8–to–1-line multiplexers and one 2–to–1-line multiplexer.

3–23. Construct a quad 9–to–1-line multiplexer with four single 8–to–1-line multiplexers and one quadruple 2–to–1-line multiplexer. The multiplexers should be interconnected and inputs labeled so that the selection codes 0000 through
1000 can be directly applied to the multiplexer selection inputs without added logic.

3–24. Construct a 15–to–1-line multiplexer with two 8–to–1-line multiplexers. Interconnect the two multiplexers and label the inputs such that any added logic required to have selection codes 0000 through 1110 is minimized.

3–25. Construct a dual 7–to–1-line multiplexer from two dual 4–to–1-line multiplexers plus minimum added control logic such that the following selection codes can be used:

Selection Code							Data Input
X_0	X_1	X_2	X_3	X_4	X_5	X_6	Selected
1	0	0	0	0	0	0	D_0
0	1	0	0	0	0	0	D_1
0	0	1	0	0	0	0	D_2
0	0	0	1	0	0	0	D_3
0	0	0	0	1	0	0	D_4
0	0	0	0	0	1	0	D_5
0	0	0	0	0	0	1	D_6

3–26. Implement a binary adder with a dual 4–to–1-line multiplexer.

3–27. Implement the following Boolean function with an 8–to–1-line multiplexer:

$$F(A, B, C, D) = \Sigma m(2, 3, 5, 6, 8, 9, 12, 14)$$

3–28. Implement the Boolean function defined in the truth table of Figure 3-22 with a 4–to–1-line multiplexer and external gates. Connect inputs A and B to the selection lines. The input requirements for the four data lines will be a function of the variables C and D. The values of these variables are obtained by expressing F as a function of C and D for each of the four cases when $AB = 00, 01, 10$, and 11. These functions may have to be implemented with external gates.

3–29. Rearrange the condensed truth table for the circuit of Figure 3-13, and verify that the circuit can function as a demultiplexer.

3–30. Design a combinational circuit that forms the binary sum of two 2-bit numbers A_1A_0 and B_1B_0. Do not use half adders or full adders, but instead use a two-level circuit plus inverters for the input variables, as needed. Design the circuit by starting with a truth table.

3–31. The logic diagram of the first stage of a 4-bit adder, as implemented in integrated circuit type 74283, is shown in Figure 3-48. Verify that the circuit implements a full adder.

3–32. Design a combinational circuit that accepts a 3-bit number and generates as output a binary number equal to the square of the input number.

3–33. Obtain the 1's and 2's complements of the following binary numbers: 10011100, 10010001, 10101010, and 00000000.

3–34. Perform the indicated subtraction with the following unsigned binary numbers by taking the 2's complement of the subtrahend:

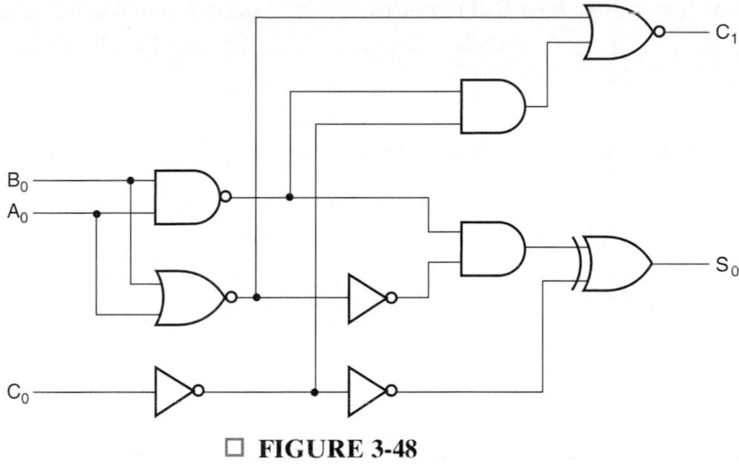

□ **FIGURE 3-48**
Circuit for Problem 3–31

(a) 11011 − 10000 **(c)** 100 − 101000
(b) 10110 − 1011 **(d)** 1011100 − 1011100

3–35. Perform the arithmetic operations (+52) + (−17) and (−52) − (−17) in binary using signed-2's complement representation for negative numbers.

3–36. The following binary numbers have a sign in the leftmost position and, if negative, are in 2's complement form. Perform the indicated arithmetic operations and verify the answers.
(a) 101111 + 111010 **(c)** 110001 − 001010
(b) 001010 + 100010 **(d)** 101010 − 100111

3–37. *Design three versions of the combinational circuit whose input is a 4-bit number and whose output is the 2's complement of the input number such that:
(a) The circuit is a simplified two-level circuit, plus inverters as needed for the input variables.
(b) The circuit is made up of four identical two-input, two-output cells, one for each bit. The cells are connected in cascade, with lines similar to a carry between them. The value applied to the rightmost carry bit is 0.
(c) In order to speed up the circuit in part (b) to enable it to apply to more than four bits, the circuit is to be redesigned with carry lookahead-like logic.

3–38. The adder-subtractor circuit of Figure 3-30 has the following values for input select *S* and data inputs *A* and *B*:

	S	A	B
(a)	0	0111	0110
(b)	0	0100	0101
(c)	1	1100	1010
(d)	1	0101	1010
(e)	1	0000	0010

Determine, in each case, the values of the outputs S_3, S_2, S_1, S_0, and C_4.

3–39. *Design a 5-bit sign-magnitude adder-subtractor. Divide the circuit for design into (1) sign generation and add-subtract control logic, (2) an unsigned number adder-subtractor using 2's complement of the minuend for subtraction, and (3) selective 2's complement result correction logic.

3–40. Design a binary multiplier that multiplies two 4-bit numbers. Use AND gates and binary adders.

3–41. Design a combinational circuit that compares two 4-bit numbers A and B to see whether they are equal. The circuit has one output X, so that $X = 1$ if $A = B$ and $X = 0$ if $A \neq B$.

3–42. *Repeat Problem 3–41 by using three-input, one-output circuits, one for each of the four bits. The four circuits are connected together in cascade by carrylike signals. One of the inputs to each cell is a carry input, and the single output is a carry output.

3–43. Obtain the 9's complement of the following 8-digit decimal numbers: 12983476, 00199500, 90309941, and 00000000.

3–44. Obtain the 10's complement of the following 6-digit decimal numbers: 459100, 007892, 000000, and 100000.

3–45. Perform the indicated subtraction with the following unsigned decimal numbers by taking the 10's complement of the subtrahend:
(a) 5678 − 1234 **(b)** 1995 − 1066 **(c)** 30 − 110 **(d)** 2048 − 512

3–46. Design a simplified combinational circuit that generates the 9's complement of (a) a BCD digit and (b) an excess-3 digit. (c) Compare the gate and literal counts of the two circuits.

3–47. Construct a BCD adder-subtractor using the BCD adder from Figure 3-34 and the 9's complementer of Problem 3–46(a), as well as other logic or functional blocks, as necessary. Use block diagrams for the components, showing only inputs and outputs where possible.

3–48. It is necessary to design a decimal adder for digits represented in the excess-3 code. Show that the correction after adding two digits with a 4-bit binary adder is as follows:
(a) The output carry is equal to the carry from the binary adder.
(b) If the output carry = 1, then add 0011.
(c) If the output carry = 0, then add 1101.
(d) Construct the decimal adder with two 4-bit adders and an inverter.

3–49. Draw the IEEE standard graphics symbol for the circuit in Figure 3-47. (This is the CMOS MSI integrated circuit 74HC138.)

3–50. Draw the IEEE standard graphics symbol diagram of a dual 4–to–1-line multiplexer with separate enable inputs and common selection inputs. (This is the same as the CMOS MSI integrated circuit 74HC153.)

3–51. Define, in your own words: (a) Positive and negative logic. (b) Active high and active low. (c) Polarity indicator. (d) Common control block. (e) Dependency notation.

3–52. Show an example of a graphics symbol that has the three Boolean dependencies G, V, and N. Draw the equivalent interpretation.

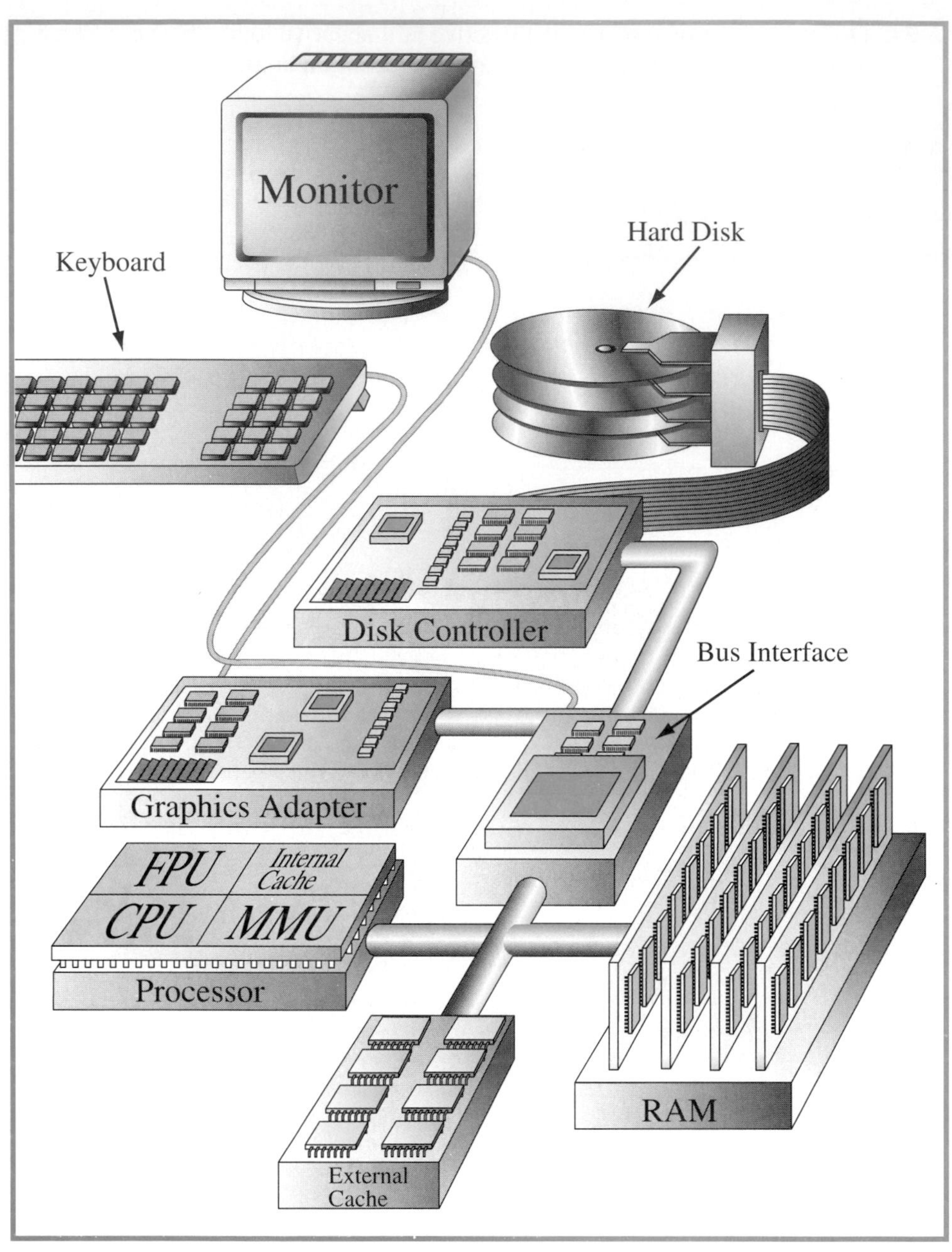
Monitor
Hard Disk
Keyboard
Disk Controller
Bus Interface
Graphics Adapter
FPU
Internal Cache
CPU
MMU
Processor
RAM
External Cache

CHAPTER

4

SEQUENTIAL CIRCUITS

To this point, we have studied only combinational logic. Although such logic is capable of interesting operations, such as addition and subtraction, the performance of useful sequences of operations using combinational logic alone requires cascading many structures together. The hardware to do this, however, is very costly and inflexible. In order to perform useful or flexible sequences of operations that are governed by applied inputs, we need to be able to construct circuits that can store information between the operations performed by combinational circuits. Such circuits are called sequential circuits. This chapter begins with an introduction to sequential circuits, which is followed by a study of the basic elements for storing binary information, called latches and flip-flops. We distinguish flip-flops from latches and study various types of each. We then analyze sequential circuits consisting of both flip-flops and combinational logic. State tables and state diagrams provide a means for describing the behavior of sequential circuits. The remaining sections of the chapter develop the techniques for designing sequential circuits.

Latches, flip-flops, and sequential circuits are fundamental components in the design of almost all digital logic. So we find the generic computer shaded in blue over most of the electronic portions of the hardware. The memory circuits are only lightly shaded, however, since large portions of memory are designed as electronic circuits rather than as logic circuits. What we study in this chapter is widely applied and is used throughout the design of almost all of the computer. The chapter contains fundamental material for any in-depth understanding of computers and digital systems and how they are designed.

4-1 SEQUENTIAL CIRCUIT DEFINITIONS

The digital circuits considered thus far have been combinational. Although every digital system is likely to include a combinational circuit, most systems encoun-

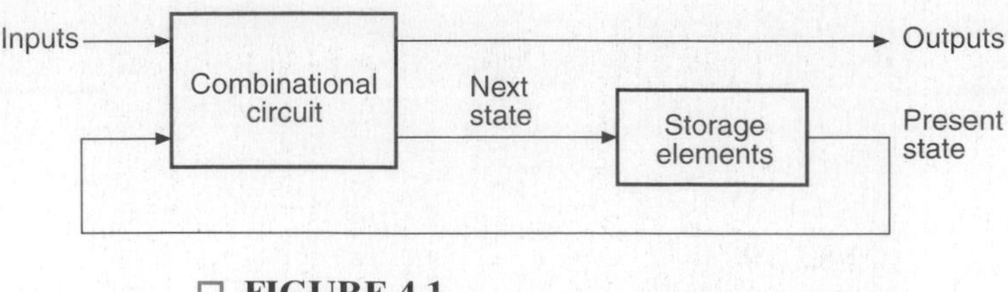

□ **FIGURE 4-1**
Block Diagram of a Sequential Circuit

tered in practice also include storage elements, requiring that the systems be described as sequential circuits.

A block diagram of a sequential circuit is shown in Figure 4-1. A combinational circuit and storage elements are interconnected to form the sequential circuit. The storage elements are circuits that are capable of storing binary information. The binary information stored in these elements at any given time defines the *state* of the sequential circuit at that time. The sequential circuit receives binary information from its environment via the inputs. These inputs, together with the present state of the storage elements, determine the binary value of the outputs. They also determine the values used to specify the next state of the storage elements. The block diagram demonstrates that the outputs in a sequential circuit are a function not only of the inputs, but also of the present state of the storage elements. The next state of the storage elements is also a function of the inputs and the present state. Thus, a sequential circuit is specified by a time sequence of inputs, interna1 states, and outputs.

There are two main types of sequential circuits, and their classification depends on the times at which their inputs are observed and their internal state changes. The behavior of a *synchronous sequential circuit* can be defined from the knowledge of its signals at discrete instants of time. The behavior of an *asynchronous sequential circuit* depends upon the inputs at any instant of time and the order in continuous time in which the inputs change.

Information is stored in digital systems in many ways, including the use of logic circuits. Figure 4-2 (a) shows a buffer. This buffer has a propagation delay t_{pd}. Since information present at the buffer input at time t appears at the buffer output at time $t + t_{pd}$, the information has effectively been stored for time t_{pd}. But, in general, we wish to store information for an indefinite time that is typically much longer than the time delay of one or even many gates. This stored value is to be changed at arbitrary times based on the inputs applied to the circuit and should not depend on the specific time delay of a gate.

Suppose that the output of the buffer in Figure 4-2(a) is connected to its input as shown in Figures 4-2(b) and (c). Suppose further that the value on the input to the buffer in part (b) has been 0 for at least time t_{pd}. Then the output produced by the buffer will be 0 at time $t + t_{pd}$. This output is applied to the input so that the output will also be 0 at time $t + 2\,t_{pd}$. This relationship between input and output holds for all t, so the 0 will be stored indefinitely. The same argument can be made for storing a 1 in the circuit in Figure 4-2(c).

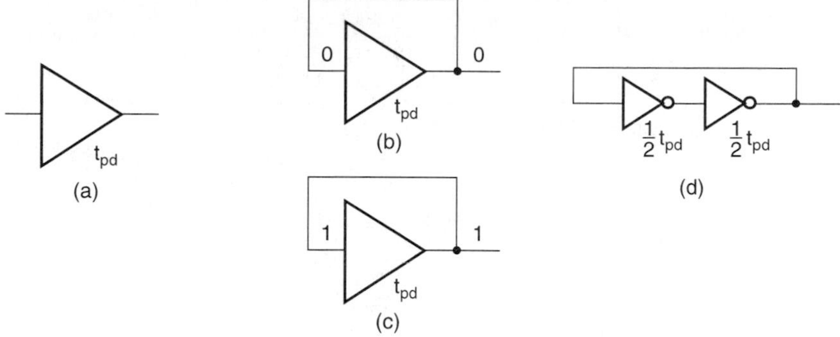

□ **FIGURE 4-2**
Logic Structures for Storing Information

The example of the buffer illustrates that storage can be constructed from logic with delay connected in a closed loop. Any loop that produces such storage must also have a property possessed by the buffer, namely, that there must be no inversion of the signal around the loop. A buffer is usually implemented by using two inverters, as shown in Figure 4-2(d). The signal is inverted twice,

$$\overline{\overline{X}} = X$$

giving no net inversion of the signal around the loop. In fact, this example is an illustration of one of the most popular methods of implementing storage in computer memories. (See Chapter 6.) However, although the circuits in Figures 4-2(b) through (d) are able to store information, there is no way for the information to be changed. By replacing the inverters with NOR or NAND gates, the information can be changed. Asynchronous storage circuits called latches are made in this manner and are discussed in the next section.

In general, more complex asynchronous circuits are difficult to design, since their behavior is highly dependent on the propagation delays of the gates and on the timing of the input changes. Thus, circuits that fit the synchronous model are the choice of most designers. Nevertheless, some asynchronous design is necessary. A very important case is the use of asynchronous latches as blocks to build storage elements, called flip-flops, that store information in synchronous circuits.

A synchronous sequential circuit employs signals that affect the storage elements only at discrete instants of time. Synchronization is achieved by a timing device called a *clock generator* that produces a periodic train of *clock pulses*. The pulses are distributed throughout the system in such a way that synchronous storage elements are affected only in some specified relationship to every pulse. In practice, the clock pulses are applied with other signals that specify the required change in the storage elements. The outputs of storage elements can change their value only in the presence of clock pulses. Synchronous sequential circuits that use clock pulses as inputs to storage elements are called *clocked sequential circuits*. These are the type of circuit most frequently encountered in practice, since they

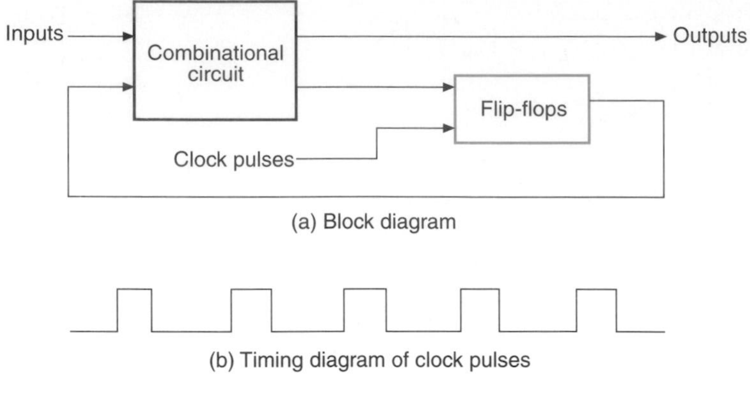

□ **FIGURE 4-3**
Synchronous Clocked Sequential Circuit

operate correctly in spite of wide differences in circuit delays and are relatively easy to design.

The storage elements we use in clocked sequential circuits are called flip-flops. For simplicity, we assume circuits with a single clock signal. A *flip-flop* is a binary storage device that is capable of storing one bit of information and that has the timing characteristics defined in Section 4-3 that simplify design. The block diagram of a synchronous clocked sequential circuit is shown in Figure 4-3.The flip-flops receive their inputs from the combinational circuit and also from a clock signal with pulses that occur at fixed intervals of time, as shown in the timing diagram. The flip-flops can change state only in response to a clock pulse. For synchronous operation, when a clock pulse is absent, the flip-flop outputs cannot change even if the outputs of the combinational circuit driving their inputs change in value. Thus, the feedback loops shown in the figure are broken. As a result, a transition from one state to the other occurs only at fixed time intervals dictated by the clock pulses, giving synchronous operation. The sequential circuit outputs are shown as outputs of the combinational circuit. This is valid even when some sequential circuit outputs are actually the flip-flop outputs. In this case, the combinational circuit part between the flip-flop outputs and the sequential circuit outputs consists of connections only.

A flip-flop has one or two outputs, one for the normal value of the bit stored and an optional one for the complemented value of the bit stored. Binary information can enter a flip-flop in a variety of ways, a fact that gives rise to different types of flip-flops. In the next section, we lay the groundwork for considering the various flip-flops by studying the latches from which they are constructed.

4-2 Latches

A storage element can maintain a binary state indefinitely (as long as power is delivered to the circuit), until directed by an input signal to switch states. The

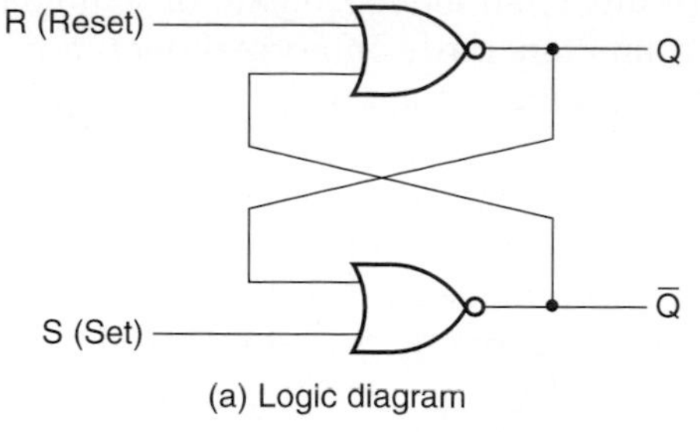

(a) Logic diagram

S	R	Q	$\overline{Q}$	
1	0	1	0	Set state
0	0	1	0	
0	1	0	1	Reset state
0	0	0	1	
1	1	0	0	Undefined

(b) Function table

□ **FIGURE 4-4**
SR Latch with NOR Gates

major differences among the various types of latches and flip-flops are the number of inputs they possess and the manner in which the inputs affect the binary state. The most basic storage elements are latches, from which flip-flops are usually constructed. Although latches are most often used within flip-flops, they can also be used with more complex clocking methods to implement sequential circuits directly. The design of such circuits is, however, beyond the scope of the basic treatment given here. In this section, we focus on latches as basic primitives for constructing storage elements.

SR and $\overline{S}\,\overline{R}$ Latches

The *SR* latch is a circuit constructed from two cross-coupled NOR gates. It is derived from the single-loop storage element in Figure 4-2(d) by simply replacing the inverters with NOR gates, as shown in Figure 4-4(a). This replacement allows the stored value in the latch to be changed. The latch has two inputs, labeled *S* for set and *R* for reset, and two useful states. When output $Q = 1$ and $\overline{Q} = 0$, the latch is said to be in the *set state*. When $Q = 0$ and $\overline{Q} = 1$, it is in the *reset state*. Outputs Q and $\overline{Q}$ are normally the complements of each other. When both inputs are equal to 1 at the same time, an undefined state with both outputs equal to 0 occurs.

Under normal conditions, both inputs of the latch remain at 0 unless the state is to be changed. The application of a 1 to the *S* input causes the latch to go to the set (1) state. The *S* input must go back to 0 before *R* is changed to 1 to avoid occurrence of the undefined state. As shown in the function table in Figure 4-4(b), two input conditions cause the circuit to be in the set state. The initial condition is $S = 1$, $R = 0$, to bring the circuit to the set state. Applying a 0 to *S* with $R = 0$ leaves the circuit in the same state. After both inputs return to 0, it is possible to enter the reset state by applying a 1 to the *R* input. We can then remove the 1 from *R*, and the circuit remains in the reset state. Thus, when both inputs are equal to 0, the latch can be in either the set or the reset state, depending on which input was most recently a 1.

If a 1 is applied to both the inputs of the latch, both outputs go to 0. This produces an undefined state because it violates the requirement that the outputs be the complement of each other. It also results in an indeterminate or unpredictable next state when both inputs return to 0 simultaneously. In normal operation, these problems are avoided by making sure that 1's are not applied to both inputs simultaneously.

The behavior of the *SR* latch described in the preceding paragraph is illustrated by the Viewlogic® logic simulator waveforms shown in Figure 4-5. The propagation delays of the NOR gates are set to a value of 2 nanoseconds, and input changes are separated by 20 nanoseconds. Initially, the state of the latch is unknown, as indicated by the cross-hatching of Q and $\overline{Q}$. Since R is initially 1 and S initially 0, the latch is reset, with Q going to 0 after 2 ns and $\overline{Q}$ going to 1 after 4 ns. When R returns to 0 and S becomes 1, the latch is set. It remains set when S returns to 0. When R becomes 1, the latch is reset. It remains reset when R returns to 0. When both S and R become 1, both Q and $\overline{Q}$ become 0. For input sequence $(S, R) = (1, 1)$ followed by $(1, 0)$, both Q and $\overline{Q}$ go to 0, but then Q changes to 1 in response to the (1,0). These values remain for S and R both 0. When $(S, R) = (1, 1)$ is followed by $(0, 0)$, Q becomes 0 and $\overline{Q}$ becomes 1. This is one form that the indeterminate state behavior can take for that sequence of inputs. Arbitrarily, Q could become 1 and $\overline{Q}$ become 0, or the outputs could oscillate between 0 and 1. The particular behavior that does occur depends on circuit delays and slight differences in the times at which S and R change in the actual circuit. Regardless of the simulation results, since these indeterminate behaviors are viewed as undesirable, the input combination (1,1) is avoided. In general, the latch state changes only in response to input changes and remains unchanged otherwise, unless the oscillating behavior just discussed occurs.

The $\overline{S}\,\overline{R}$ latch with two cross-coupled NAND gates is shown in Figure 4-6(a). It operates with both inputs normally at 1, unless the state of the latch has to be changed. The application of a 0 to the S input causes output Q to go to 1, putting the latch in the set state. When the S input goes back to 1, the circuit remains in the set state. With both inputs at 1, we can change the state of the latch by placing a 0 on the R input. This causes the circuit to go to the reset state and stay there even after both inputs return to 1. The condition that is undefined for this NAND latch is when both inputs are equal to 0 at the same time, an input combination that should be avoided.

Comparing the NAND latch with the NOR latch, we note that the input signals for the NAND require the complement of those values used for the NOR. Because the NAND latch requires a 0 signal to change its state, it is referred to as an $\overline{S}\,\overline{R}$ latch. The bar above the letters designates the fact that the inputs must be in their complement form in order to act upon the circuit state.

The operation of the basic NOR and NAND latches can be modified by providing an additional control input that determines when the state of the latch can be changed. An *SR* latch with a control input is shown in Figure 4-7(a). It consists of the basic $\overline{S}\,\overline{R}$ latch and two additional NAND gates. The control input C acts as an enable signal for the other two inputs. The output of the NAND gates stays at the logic-1 level as long as the control input remains at 0. This is the quiescent condition for the $\overline{S}\,\overline{R}$ latch composed of two NAND gates. When the control input goes to 1, information from the S and R inputs is allowed to affect the $\overline{S}\,\overline{R}$ latch. The set state

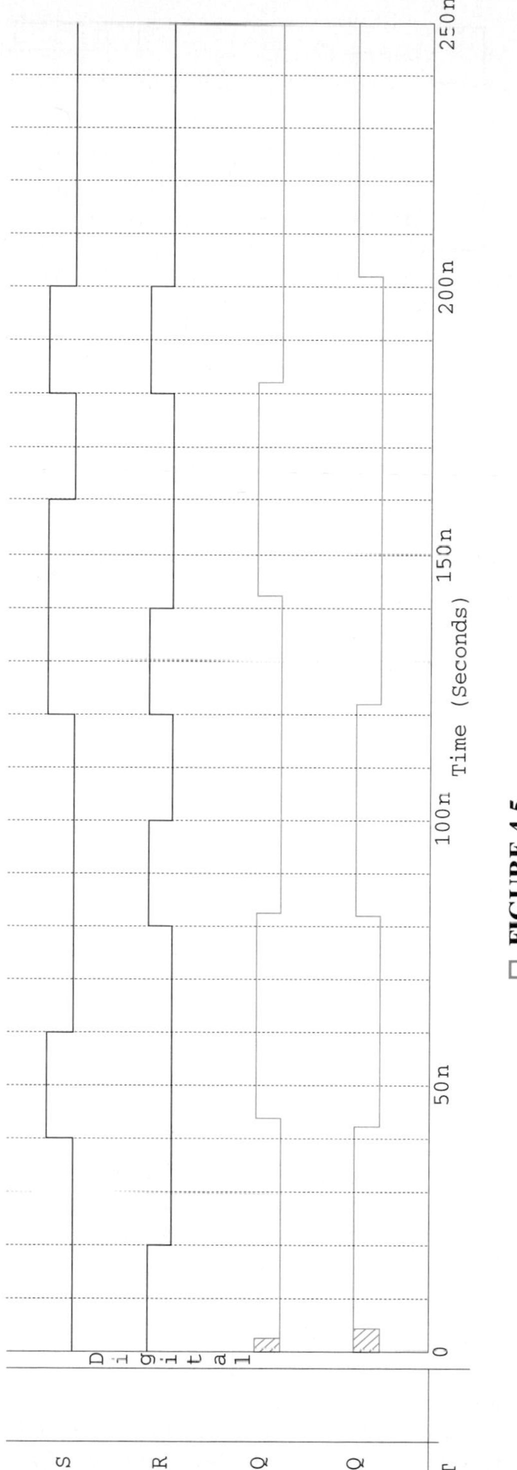

□ **FIGURE 4-5**
Logic Simulation of *SR* Latch Behavior

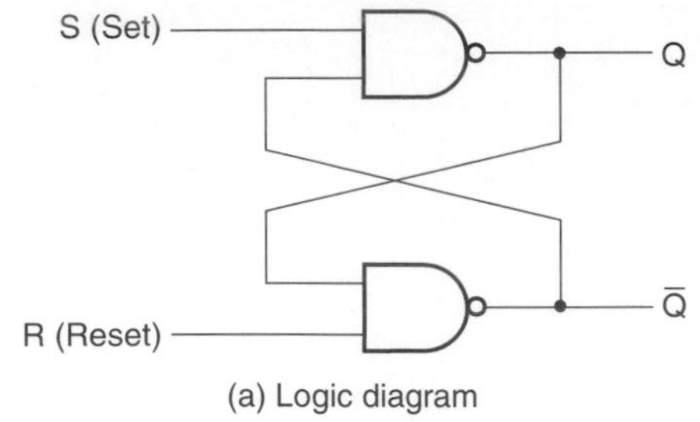

(a) Logic diagram

S	R	Q	$\overline{Q}$	
0	1	1	0	Set state
1	1	1	0	
1	0	0	1	Reset state
1	1	0	1	
0	0	1	1	Undefined

(b) Function table

□ **FIGURE 4-6**
$\overline{S}\,\overline{R}$ Latch with NAND Gates

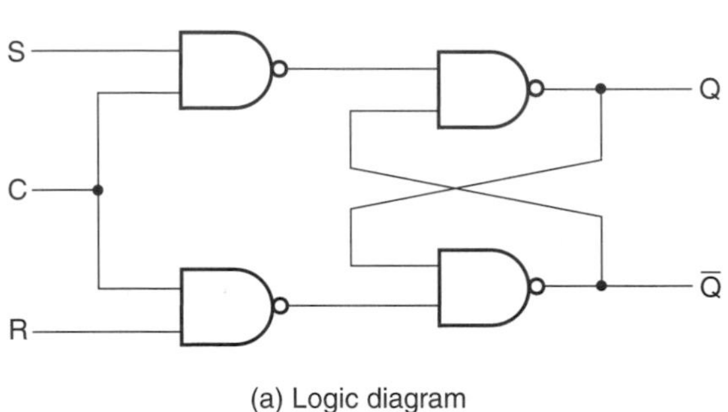

(a) Logic diagram

C	S	R	Next state of Q
0	X	X	No change
1	0	0	No change
1	0	1	Q = 0; Reset state
1	1	0	Q = 1; Set state
1	1	1	Undefined

(b) Function table

□ **FIGURE 4-7**
SR Latch with Control Input

is reached with $S = 1$, $R = 0$, and $C = 1$. To change to the reset state, the inputs must be $S = 0$, $R = 1$, and $C = 1$. In either case, when C returns to 0, the circuit remains in its current state. Control input $C = 0$ disables the circuit so that the state of the output does not change, regardless of the values of S and R. Moreover, when $C = 1$ and both the S and R inputs are equal to 0, the state of the circuit does not change. These conditions are listed in the function table accompanying the diagram.

An undefined state occurs when all three inputs are equal to 1. This condition places 0's on both inputs of the basic $\overline{S}\,\overline{R}$ latch, giving an undefined state. When the control input goes back to 0, we cannot conclusively determine the next state, since the $\overline{S}\,\overline{R}$ latch sees inputs $(0, 0)$ followed by $(1, 1)$. The *SR* latch with control input is an important circuit because other latches and flip-flops are constructed from it. Sometimes the *SR* latch with control input is referred to as an *SR* (or *RS*) flip-flop: however, according to our terminology, it does not qualify as a flip-flop, since the circuit does not fulfill the flip-flop requirements presented in the next section.

D Latch

One way to eliminate the undesirable undefined state in the *SR* latch is to ensure that inputs S and R are never equal to 1 at the same time. This is done in the *D*

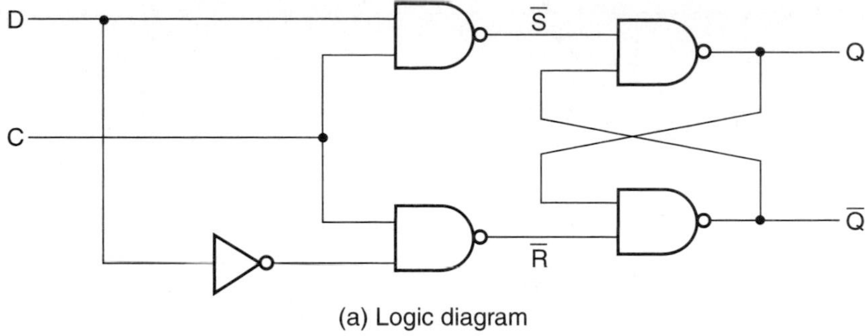

(a) Logic diagram

C	D	Next state of Q
0	X	No change
1	0	Q = 0; Reset state
1	1	Q − 1; Set state

(b) Function table

□ **FIGURE 4-8**
D Latch

latch, shown in Figure 4-8(a). This latch has only two inputs: D (data) and C (control). The complement of D input goes directly to the $\overline{S}$ input, and D is applied to the $\overline{R}$ input. As long as the control input is 0, the $\overline{S}\,\overline{R}$ latch has both inputs at the 1 level, and the circuit cannot change state regardless of the value of D. The D input is sampled when $C = 1$. If D is 1, the Q output goes to 1, placing the circuit in the set state. If D is 0, output Q goes to 0, placing the circuit in the reset state.

The D latch receives its designation from its ability to hold *data* in its internal storage. The binary information present at the data input of the D latch is transferred to the Q output when the control input is enabled (1). The output follows changes in the data input, as long as the control input is enabled. When the control input is disabled (0), the binary information that was present at the data input at the time the transition occurred is retained at the Q output until the control input is enabled again.

The D latch in VLSI circuits is often constructed with transmission gates (TGs), as shown in Figure 4-9. The TG was defined in Figure 2-48. The C input controls two TGs. When $C = 1$, the TG connected to input D conducts, and the TG connected to output Q disconnects. This produces a path from input D through two inverters to output Q. Thus, the output follows the data input as long as C remains active (1). When C changes to 0, the first TG disconnects input D from the circuit, and the second TG connects the two inverters at the output into a loop. Hence, the value that was present at input D at the time that C went from 1 to 0 is retained at the Q output by the loop.

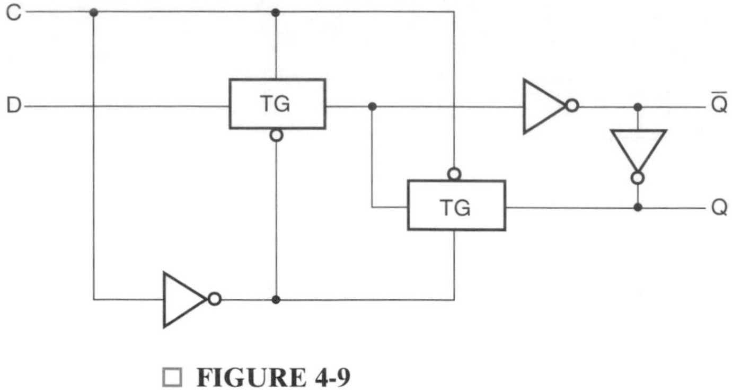

□ **FIGURE 4-9**
D Latch with Transmission Gates

4-3 Flip-Flops

The state of a latch in a flip-flop is allowed to switch by a momentary change in value on the control input. This change is called a *trigger*, and it enables, or triggers the flip-flop. The *D* latch with clock pulses on its control input is triggered every time a pulse to the logic-l level occurs. As long as the pulse remains at the active (1) level, any changes in the data input will change the state of the latch. In this sense, we say that the latch is *transparent*, since its input value can be seen from the outputs.

As the block diagram of Figure 4-3 shows, a sequential circuit has a feedback path from the outputs of the flip-flops to the combination circuit. As a consequence, the data inputs of the flip-flops are derived in part from the outputs of the same and other flip-flops. When latches are used for the storage elements, a serious difficulty arises. The state transitions of the latches start as soon as the clock pulse changes to the logic-1 level. The new state of a latch may appear at its output while the pulse is still active. This output is connected to the inputs of some of the latches through a combinational circuit. If the inputs applied to the latches change while the clock pulse is still in the logic-1 level, the latches will respond to *new state values* of other latches instead of the *original state values*, and a succession of changes of state instead of a single one may occur. The result is an unpredictable situation, since the state may keep changing and continue to change until the clock returns to 0. The final state depends on how long the clock pulse stays at level logic 1. Because of this unreliable operation, the output of a latch cannot be applied directly or through combinational logic to the input of the same or another latch when all the latches are triggered by a single clock signal.

Flip-flop circuits are constructed in such a way as to make them operate properly when they are part of a sequential circuit that employs a single clock. Note that the problem with the latch is that it is transparent: As soon as an input changes, shortly thereafter the corresponding output changes to match it. This transparency is what allows a change on a latch output to produce additional changes at other latch outputs while the clock pulse is at logic 1. The key to the

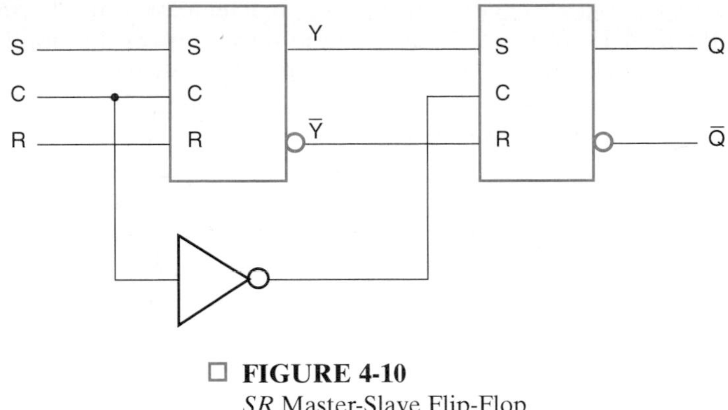

□ **FIGURE 4-10**
SR Master-Slave Flip-Flop

proper operation of flip-flops is to prevent them from being transparent. In a flip-flop, before an output can change, the path from its inputs to its outputs is broken. So a flip-flop cannot "see" the change of its output or of the outputs of other, like flip-flops at its input during the same clock pulse. Thus, the new state of a flip-flop depends only on the immediately preceding state, and the flip-flops do not go through multiple changes of state.

There are two ways that latches are combined to form a flip-flop. One way is to combine two latches such that (1) the inputs presented to the flip-flop when a clock pulse is present control its state and (2) the state of the flip-flop changes only when a clock pulse is not present. Such a circuit is called a *master-slave* flip-flop. Another way is to produce a flip-flop that triggers only during a signal *transition* from 0 to 1 (or from 1 to 0) on the clock and that is disabled at all other times, including for the duration of the clock pulse. This type of circuit is called an *edge-triggered* flip-flop. We next proceed to show the implementation of both types of flip-flops.

Master-Slave Flip-Flop

The master-slave flip-flop consists of two latches and an inverter. A master-slave *SR* flip-flop is shown in Figure 4-10. The symbol with *S*, *C*, and *R* on it is that for the *SR* latch with control input (Figure 4-7), which we will refer to here as a clocked *SR* latch. The left clocked *SR* latch in Figure 4-10 is called the master, the right the slave. When the clock input *C* is 0, the output of the inverter is 1. The slave latch is then enabled, and its output *Q* is equal to the master output *Y*. The master latch is disabled, because *C* is 0. When a logic-1 clock pulse is applied, the values on *S* and *R* control the value stored in the master latch *Y*. The slave, however, is disabled as long as the pulse remains at the 1 level, because its *C* input is equal to 0. Any changes in the external *S* and *R* inputs change the master output *Y*, but cannot affect the slave output *Q*. When the pulse returns to 0, the master is disabled and is isolated from the *S* and *R* inputs. At the same time, the slave is enabled, and the current value of *Y* is transferred to the output of the flip-flop at *Q*.

A Viewlogic® logic simulation illustrating master-slave flip-flop *SR* behavior is shown in Figure 4-11. *R* is initially 1, with *S* at 0 to reset the flip-flop in response to the first clock pulse. Initially, both *Y* and *Q* are unknown, with *Y* going to 0 in response to the clock at 1 and *Q* going to 0 in response to the clock back at 0. After *R* returns to 0 and input *S* becomes 1, the next clock pulse should change the flip-flop to the set state with $Q = 1$. After the clock pulse transition from 0 to 1, the master latch is set and changes *Y* to 1. The slave latch is not affected, because its *C* input is 0. Since the master is an internal circuit, its change of state is not presented at output *Q*. Even though the inputs may be changing, the output of the flip-flop remains in its previous state. When the pulse returns to 0, the information from the master is allowed to pass through to the slave, making the external output $Q = 1$. Note that these changes are delayed from the pulse changes by gate delays. Also, the external inputs *S* and *R* can change anytime after the clock pulse goes through its negative transition. This is because, as the *C* input reaches 0, the master is disabled, and *S* and *R* have no effect until the next clock pulse. For the third clock pulse, both *S* and *R* are 0, and *Y* and *Q* remain unchanged. The results from the simulation in Figure 4-11 continue with a fourth clock pulse that has input $R = 1$. The flip-flop state changes to 0 by first switching the master *Y* to 0 during the clock pulse and the slave *Q* to 0 after the pulse.

A modified version of the *SR* flip-flop which eliminates the undesirable condition that leads to undefined outputs is the *JK* flip-flop. In this type of flip-flop, the condition with both inputs equal to 1 causes the output to complement its value. A master-slave *JK* flip-flop is shown in Figure 4-12(a). The master is an *SR* latch with control input receiving the clock pulses. The slave is an *SR* latch with control input that operates with the complement of the clock pulse. The *J* input behaves like the *S* input to set the flip-flop. The *K* input is similar to the *R* input for resetting the flip-flop. The only difference between the *SR* and *JK* flip-flops is their response to the condition when both inputs are equal to 1. As can be verified from the diagram, this condition complements the outputs in the *JK* flip-flop. This is because of the feedback connection from the outputs to the inputs. When $Q = 1$ and $K = 1$, the *R* input is equal to 1. The *S* input is 0 even if $J = 1$, because $\overline{Q}$, being the complement of *Q*, must be 0. The next clock pulse will reset the flip-flop and change the *Q* output to 0. Similarly, if we start with $Q = 0$, the clock pulse will set the flip-flop and change output *Q* to 1. In either case, the condition $J = 1$ and $K = 1$ causes the outputs of the flip-flop to be complemented in response to a clock pulse. The next state behavior of the *JK* flip-flop is summarized in the table in Figure 4-12(b). The clock input is not explicitly shown, but a clock pulse is assumed to have occurred before the next state of *Q* appears.

Now consider a sequential system containing many master-slave flip-flops, with the outputs of some flip-flops going to inputs of other flip-flops. Assume that the clock pulses to all of the flip-flops are synchronized and occur at the same time. At the beginning of each clock pulse, some of the masters change state, but all the slaves remain in their previous states. This means that the flip-flop slaves are still in their original states, while the flip-flop masters have changed to the new states. After the clock pulse returns to 0, some of the flip-flop slaves change state, but none of the new states have an effect on any of the masters until the next

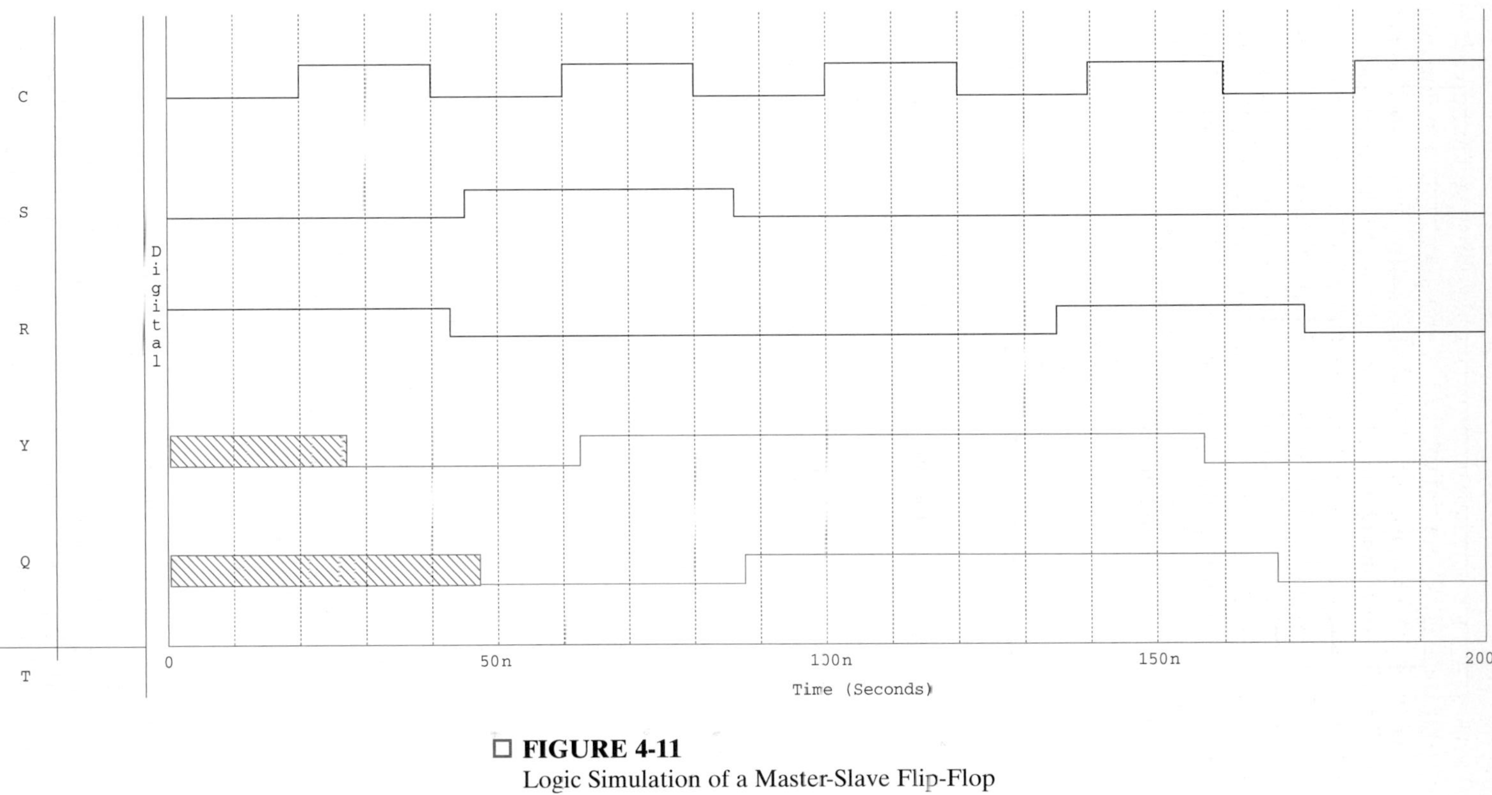

□ **FIGURE 4-11**
Logic Simulation of a Master-Slave Flip-Flop

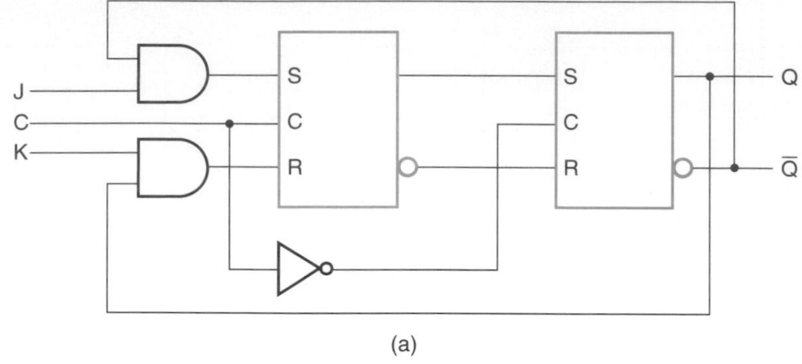

(a)

J	K	Next State of Q
0	0	Q
0	1	0
1	0	1
1	1	$\overline{Q}$

(b)

□ **FIGURE 4-12**
Master-Slave *JK* Flip-Flop

pulse. Thus, the states of flip-flops in a synchronous system can change simultaneously for the same clock pulse, even though outputs of flip-flops are connected to inputs of the same or other flip-flops. This is possible because the inputs affect the state of the flip-flop only while the clock pulse is 1 and the new state appears at the outputs only after the clock pulse has returned to 0, ensuring that the flip-flops are *not* transparent.

For reliable sequential circuit operation, all signals must propagate from the outputs of flip-flops, through the combinational circuit, and back to inputs of master-slave flip-flops, while the clock pulse remains at the logic-0 level. Any changes that occur at the inputs of flip-flops after the clock pulse goes to the logic-1 level, whether intentional or not, are recognized by the outputs. Suppose that the delay in the combinational circuit is such that *S* and *R* are still changing after the clock pulse has gone to the logic-1 level. Suppose also that, as a consequence, the master is set to 1 by the presence of $S = 1$ with $R = 0$. When *S* and *R* finally stop changing, they are both 0, indicating that the state of the flip-flop was *not* to be changed from 0. Thus, the 1 value in the master, which will be transferred to the slave, is in error. There are two consequences of this behavior. First, the master-slave flip-flop is also referred to as a *pulse-triggered* flip-flop, since it can respond to input values that cause a change in state and occur anytime during its clock pulse. Second, the circuit must be designed so that combinational circuit delays are short enough to prevent *S* and *R* from changing during the clock pulse.

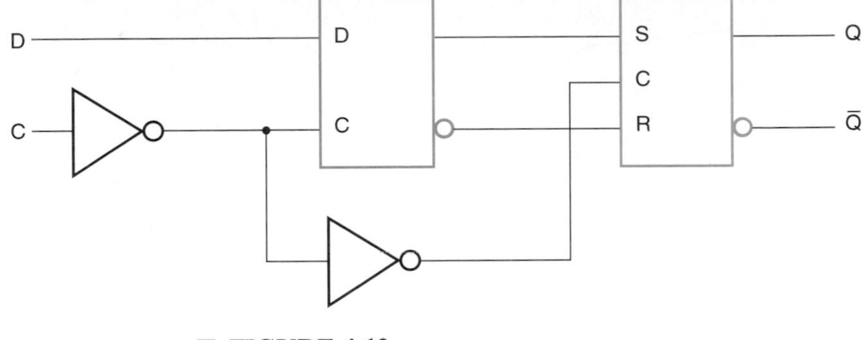

□ **FIGURE 4-13**
D-Type Positive-Edge-Triggered Flip-Flop

Edge-Triggered Flip-Flop

An *edge-triggered* flip-flop ignores the pulse while it is at a constant level and triggers only during a *transition* of the clock signal. Some edge-triggered flip flops trigger on the positive edge (0-to-1 transition), whereas others trigger on the negative edge (1-to-0 transition). The logic diagram of a *D*-type positive-edge-triggered flip-flop is shown in Figure 4-13. This flip-flop takes exactly the form of a master-slave flip-flop, with the master a *D* latch and the slave an *SR* latch or a *D* latch. Also, an inverter is added to the clock input. Because the master latch is a *D* latch, the flip-flop exhibits edge-triggered rather than master-slave or pulse-triggered behavior. For the clock input equal to 0, the master latch is enabled and transparent and follows the *D* input value. The slave latch is disabled and holds the state of the flip-flop fixed. When the positive edge occurs, the clock input changes to 1. This disables the master latch so that its value is fixed and enables the slave latch so that it copies the state of the master latch. The state of the master latch to be copied is the state that is present at the positive edge of the clock. Thus, the behavior appears to be edge triggered. With the clock input equal to 1, the master latch is disabled and cannot change, so the state of both the master and the slave remain unchanged. Finally, when the clock input changes from 1 to 0, the master is enabled and begins following the *D* value. But during the 1–to–0 transition, the slave is disabled before any change in the master can reach it. Thus, the value stored in the slave remains unchanged during this transition. An alternative implementation is given in Problem 4–9 at the end of the chapter.

The timing of the response of a flip-flop to input data *D* and clock *C* must be taken into consideration when one uses an edge-triggered flip-flop. There is a minimum time called the *setup time* for which the *D* input must be maintained at a constant value prior to the occurrence of the clock transition. Otherwise, the master could be changing and be at an intermediate value at the time the slave copies it. Similarly, there is a minimum time called the *hold time* for which the *D* input must not change after the application of the positive transition of the pulse. Otherwise, the master might respond to the input change and be changing at the time the slave latch copies it. The *propagation delay time* of the flip-flop is defined as the interval

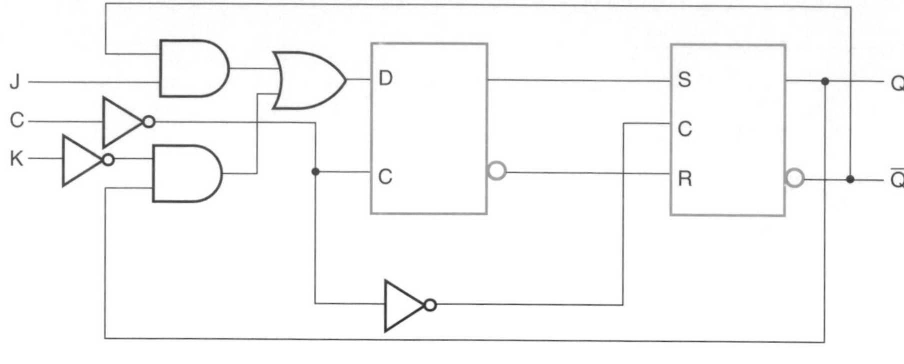

☐ **FIGURE 4-14**
Positive-Edge-Triggered *JK* Flip-Flop

between the trigger edge and the stabilization of the output to the new state. Since the changes of the outputs are to be separated from the control by the inputs, the minimum propagation delay time should be longer than the maximum hold time. These and other parameters are specified in manufacturers' data books for specific logic families.

Earlier, we implemented a master-slave *JK* flip-flop by adding logic to a master-slave *SR* flip-flop. Here, we will implement a positive-edge-triggered *JK* flip-flop by adding logic to the positive-edge-triggered *D* flip-flop. The resulting circuit is shown in Figure 4-14. D is equal to 1 for $J = 1$ and $Q = 0$ or for $K = 0$ and $Q = 1$. So a flip-flop in the 0 state is set for $J = 1$, regardless of the value of K, and remains in the 1 state for $K = 0$, regardless of the value of J. D is equal to 0 for $K = 1$ and $Q = 1$ or for $J = 0$ and $Q = 0$. So a flip-flop in the 1 state is reset to 0 for $K = 1$, regardless of the value of J, and remains in the 0 state for $J = 0$, regardless of the value of K. This is exactly the *JK* behavior. Since the underlying flip-flop is edge triggered, the *JK* flip-flop is also edge triggered, and its value after a positive transition is determined by the values of J and K during the setup time–hold time interval about the transition on clock input C.

Standard Graphics Symbols

The standard graphics symbols for the different types of latches and flip-flops are shown in Figure 4-15. A flip-flop or latch is designated by a rectangular block with inputs on the left and outputs on the right. One output designates the normal state of the flip-flop, and the other, with a bubble, designates the complement output. The graphics symbol for the *SR* latch or *SR* flip-flop has inputs S and R indicated inside the block. In the case of the $\overline{S}\,\overline{R}$ latch, bubbles are added to the inputs to indicate that setting and resetting occur for 0-level inputs. The graphics symbol for the D latch or D flip-flop has inputs D and C indicated inside the block. The graphics symbol for the *JK* flip-flop has inputs J, K, and C inside.

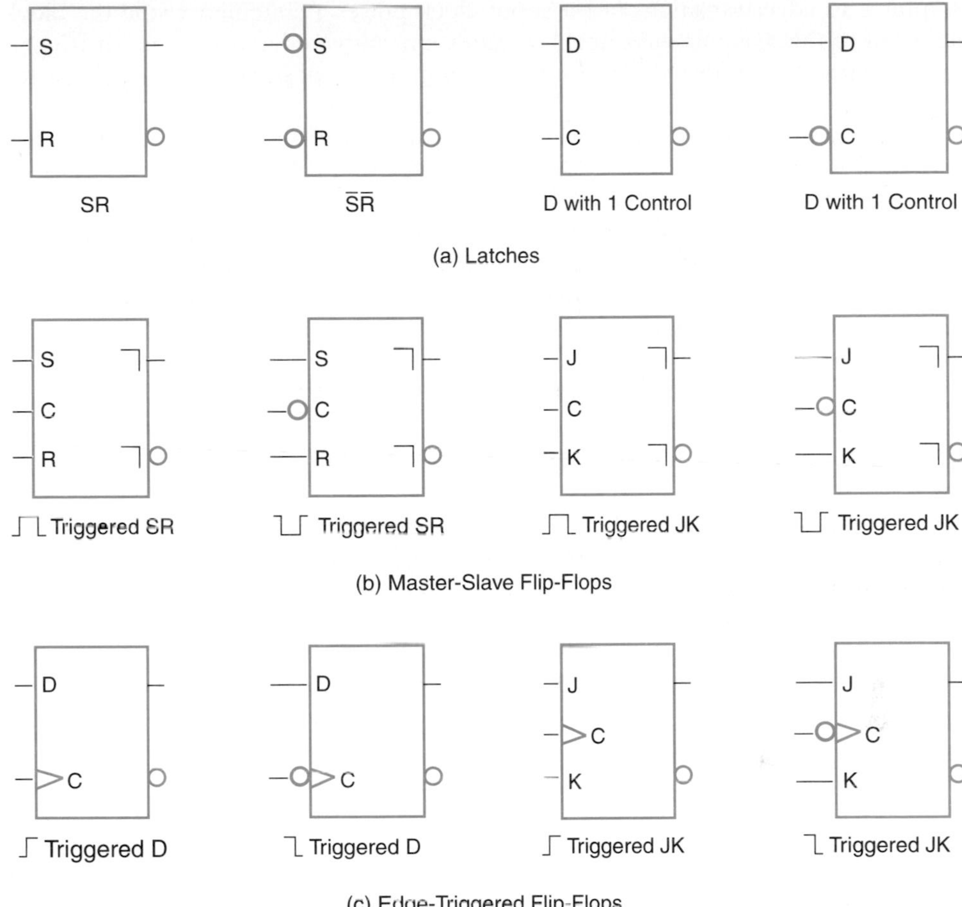

□ **FIGURE 4-15**
Standard Graphics Symbols for Latches and Flip-Flops

Below each symbol, a descriptive title, which is not part of the symbol, is given. In the titles, ⊓ denotes a positive pulse, ⊔ a negative pulse, ↑ a positive edge, and ↓ a negative edge.

Triggering by the 0 level rather than the 1 level is denoted on the latch symbols by adding a bubble at the triggering input. The master-slave is a pulse-triggered flip-flop and is indicated as such with a right-angle symbol called a *postponed output indicator* in front of the outputs. This symbol shows that the output signal changes at the end of the pulse. To denote that the master-slave flip-flop will respond to a negative pulse (i.e., a pulse to 0 with the inactive clock value at 1), a bubble is placed on the *C* input. To denote that the edge-triggered flip-flop responds to an edge, an arrowheadlike symbol in front of the letter *C* designates a *dynamic input*. This *dynamic indicator* symbol denotes the fact that the flip-flop responds to edge transitions of the input clock pulses. A bubble outside the block adjacent to the dynamic indicator designates a negative-edge transition for triggering the circuit. The absence of a bubble designates a positive-edge transition for triggering.

Often, all of the flip-flops used in a circuit are of the same triggering type, such as positive-edge triggered. All of the flip-flops will then change in relation to the same clocking event. When using flip-flops having different triggering in the same sequential circuit, one may still wish to have all of the flip-flop outputs change relative to the same clocking event. Those flip-flops that behave in a manner opposite from the adopted polarity transition can be changed by the addition of inverters to their clock inputs. A preferred procedure is to provide both positive and negative pulses from the master clock generator that are carefully aligned. We apply positive pulses to positive-pulse-triggered (master-slave) and negative-edge-triggered flip-flops and negative pulses to negative-pulse-triggered (master-slave) and positive-edge-triggered flip-flops. In this way, all flip-flop outputs will change at the same time. Finally, to prevent specific timing problems, some designers use flip-flops having different triggering, i.e., both positive and negative edge-triggered flip-flops with a single clock. In these cases, flip-flop outputs are purposely made to change at different times.

In this text, we assume that all flip-flops are of the positive-edge-triggered type, unless otherwise indicated. This provides a uniform graphics symbol for the flip-flops and consistent timing diagrams.

Characteristic Tables

A characteristic table defines the logical properties of a flip-flop by describing its operation in tabular form. The characteristic tables of four types of flip-flops are presented in Table 4-1. They define the next state as a function of the inputs and present state. $Q(t)$ refers to the present state prior to the application of a clock pulse. $Q(t + 1)$ is the state one clock period later, which is referred to as the *next state*. Note that the pulse at input C is not listed in the characteristic table, but is assumed to occur between time t and $t + 1$. The characteristic table for the JK flip-flop shows that the next state is equal to the present state when inputs J and K are both equal to 0. This can be expressed as $Q(t + 1) = Q(t)$, indicating that the clock pulse produces no change of state. When $K = 1$ and $J = 0$, the next clock pulse resets the flip-flop with $Q(t + 1) = 0$. With $J = 1$ and $K = 0$, the flip-flop sets with $Q(t + 1) = 1$. When both J and K are equal to 1, the next state changes to the complement of the present state, which can be expressed as $Q(t + 1) = \overline{Q}(t)$. The SR flip-flop is similar to the JK with J replaced by S and K by R, except for the undefined case. The question mark for the next state when S and R are both equal to 1 indicates an unpredictable next state. The next state of a D flip-flop is dependent only on the D input and is independent of the present state. This can be expressed as $Q(t + 1) = D(t)$.

Note that there is no input to the D flip-flop that produces a "no change" condition. This condition can be accomplished either by disabling the clock pulses on the C input or by leaving the clock pulses undisturbed and connecting the output back into the D input using a multiplexer when the state of the flip-flop must remain the same. The technique that disables clock pulses is referred to as *clock gating*. This technique typically uses fewer gates and saves power, but is often avoided because the gated clock pulses into the flip-flops are delayed. The delay, called *clock skew*, causes gated and ungated flip-flops to change at different times. This can make the circuit unreliable, since the outputs of some flip-flops may reach others while their inputs are still affecting their state.

☐ TABLE 4-1
Flip-Flop Characteristic Tables

(a) *JK* Flip-Flop

J	K	$Q(t+1)$	Operation
0	0	$Q(t)$	No change
0	1	0	Reset
1	0	1	Set
1	1	$\overline{Q}(t)$	Complement

(b) *SR* Flip-Flop

S	R	$Q(t+1)$	Operation
0	0	$Q(t)$	No change
0	1	0	Reset
1	0	1	Set
1	1	?	Undefined

(c) *D* Flip-Flop

D	$Q(t+1)$	Operation
0	0	Reset
1	1	Set

(d) *T* Flip-Flop

T	$Q(t+1)$	Operation
0	$Q(t)$	No change
1	$\overline{Q}(t)$	Complement

The T (toggle) flip-flop listed in the table is obtained from a JK flip-flop by tying inputs J and K together. The graphics symbol for the T flip-flop is like that for the D flip-flop, with the D replaced by T. The characteristic table has only two conditions. When $T = 0$ ($J = K = 0$), a clock pulse does not change the state. When $T = 1$ ($J = K = 1$), a clock pulse complements the state of the flip-flop. The use of the T flip-flop allows complementing behavior to be obtained with logic that drives a single input rather than two inputs. This sometimes produces a simpler circuit.

Direct Inputs

Flip-flops often provide special inputs for setting and resetting them asynchronously, i. e., independently of the clock input C. The inputs that asynchronously set the flip-flop are called *direct set*, or *preset*. The inputs that asynchronously reset the flip-flop are called *direct reset*, or *clear*. Application of a logic 1 (or a logic 0 if a bubble is present) to these inputs affect the flip-flop output without the use of the clock. When power is turned on in a digital system, the states of its flip-flops can be anything. The direct inputs are useful for bringing all flip-flops in a digital system to an initial state prior to the normal clocked operation.

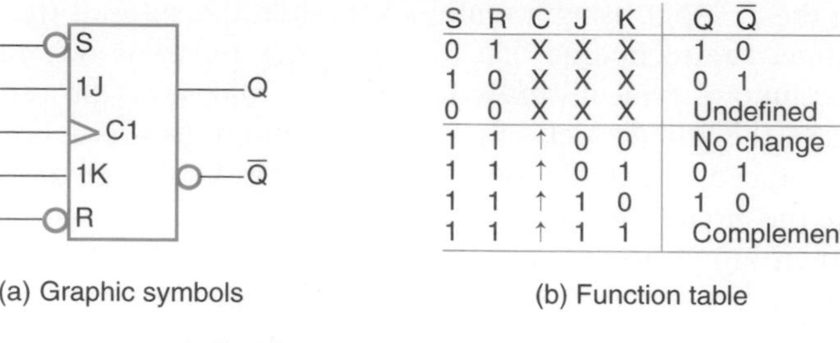

(a) Graphic symbols

S	R	C	J	K	Q $\overline{Q}$
0	1	X	X	X	1 0
1	0	X	X	X	0 1
0	0	X	X	X	Undefined
1	1	↑	0	0	No change
1	1	↑	0	1	0 1
1	1	↑	1	0	1 0
1	1	↑	1	1	Complement

(b) Function table

□ **FIGURE 4-16**
JK Flip-Flop with Direct Set and Reset

The IEEE standard graphics symbol for a *JK* flip-flop with direct set and reset is shown in Figure 4-16. The notations *C*1, 1*J*, and 1*K* are examples of control dependency. An input labeled *Cn*, where *n* is any number, controls all the other inputs starting with the number *n*. In the figure, *C*1 controls inputs 1*J* and 1*K*. *S* and *R* have no 1 in front of them, and therefore, they are not controlled by the clock at *C*1. The *S* and *R* inputs have a circle along the input line to indicate that they are active at the logic-0 level.

The function table specifies the operation of the circuit. The first three rows in the table specify the operation of the direct inputs *S* and *R*. These inputs behave like NAND $\overline{S}\,\overline{R}$ latch inputs (see Figure 4-6) and operate independently of the clock and are therefore asynchronous inputs. The last four rows in the function table specify the clocked operation when both the *S* and *R* inputs are inactive at the logic-1 level. The clock at *C* is shown with an upward arrow to indicate that the flip-flop is a positive-edge-triggered type. The *J* and *K* inputs are controlled by the clock in the usual manner.

4-4 Sequential Circuit Analysis

The behavior of a sequential circuit is determined from the inputs, outputs, and present state of the circuit. The outputs and the next state are a function of the inputs and the present state. The analysis of a sequential circuit consists of obtaining a suitable description that demonstrates the time sequence of inputs, outputs, and states.

A logic diagram is recognized as a synchronous sequential circuit if it includes flip-flops with the clock inputs driven directly or indirectly by a clock signal and the direct sets and resets unused during the normal functioning of the circuit. The flip-flops may be of any type, and the logic diagram may or may not include combinational gates. In this section, we introduce an algebraic representation for specifying the logic diagram of a sequential circuit. We then present a state table and state diagram that describe the behavior of the circuit. Specific examples will be used throughout the discussion to illustrate the various procedures.

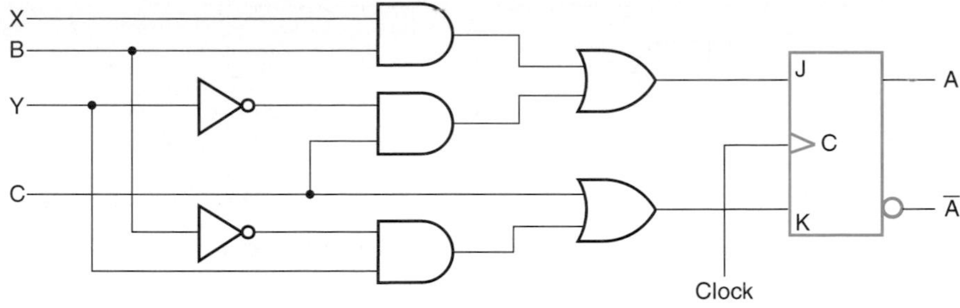

□ **FIGURE 4-17**
Implementing Input Equations

Input Equations

The logic diagram of a sequential circuit consists of flip-flops and, usually, combinational gates. The knowledge of the type of flip-flops used and a list of Boolean functions for the combinational circuit provide all the information needed to draw the logic diagram of the sequential circuit. The part of the combinational circuit that generates the signals for the inputs of flip-flops can be described by a set of Boolean functions called *flip-flop input equations.* We will adopt the convention of using the flip-flop input symbol to denote the flip-flop input equation variable and using the name of the flip-flop output as the subscript for the variable. As an example, consider the following flip-flop input equations:

$$J_A = (XB + \overline{Y}C)$$

$$K_A = (Y\overline{B} + C)$$

Here J_A and K_A are two Boolean variables. The J and K symbols are the inputs of a JK flip-flop. The subscript A is the name of the flip-flop output. The implementation of the two input equations is shown in the logic diagram of Figure 4-17. The JK flip-flop has been assigned an output symbol letter A. The flip-flop has two inputs J and K and a clock input C that is driven by the signal *Clock* and is easily distinguished from circuit input variable C, since they are at different levels of the block hierarchy. The combinational circuit drawn in the diagram is the implementation of the algebraic expression given by the flip-flop input equations. The outputs of the combinational circuit, denoted by J_A and K_A, are then applied to the J and K inputs of flip-flop A.

From this example, we see that a flip-flop input equation is a Boolean expression for a combinational circuit. The subscripted symbol is an output variable of the combinational circuit. This output is always connected to the input of a flip-flop—thus the name "flip-flop input equation."

The flip-flop input equations constitute a convenient algebraic expression for specifying the logic diagram of a sequential circuit. They imply the type of flip-flop from the letter symbol, and they fully specify the combinational circuit that drives

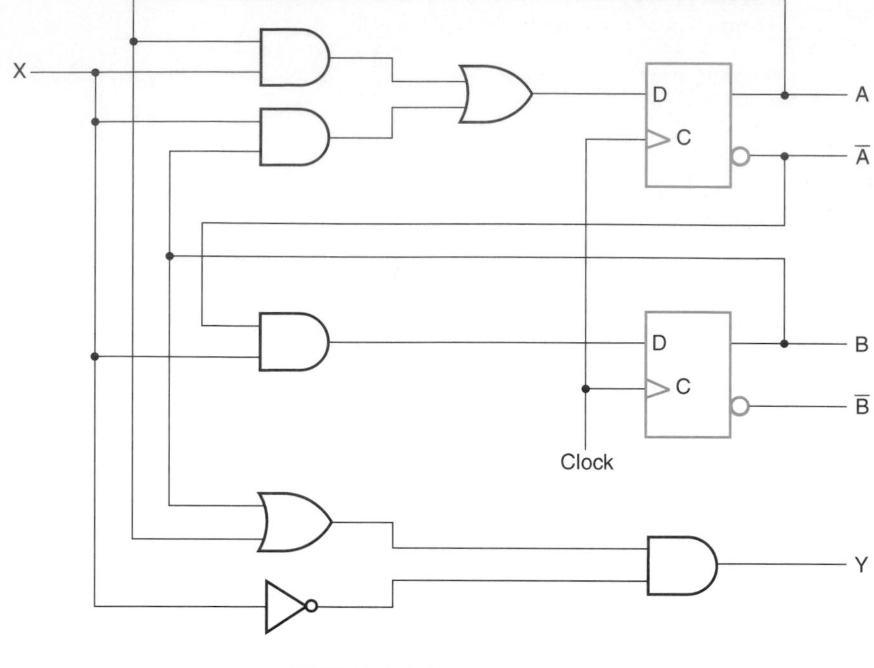

□ **FIGURE 4-18**
Example of a Sequential Circuit

the flip-flops. Time is not included explicitly in these equations, but is implied from the clock at the C input of the flip-flops.

An example of a sequential circuit is given in Figure 4-18. The circuit has two D-type flip-flops, an input X, and an output Y. It can be specified by the following equations:

$$D_A = (AX + BX)$$

$$D_B = \overline{A}X$$

$$Y = (A + B)\overline{X}$$

The first two equations are for flip-flop inputs, and the third equation specifies the output Y. Note that the input equations use the symbol D, which is the same as the input symbol of the flip-flops. The subscripts A and B designate the outputs of the two flip-flops.

State Table

The functional relationships between the inputs, outputs, and flip-flop states of a sequential circuit can be enumerated in a *state table*. The state table for the circuit of Figure 4-18 is shown in Table 4-2. The table consists of four sections, labeled

□ **TABLE 4-2**
State Table for Circuit of Figure 4-18

Present State		Input	Next State		Output
A	B	X	A	B	Y
0	0	0	0	0	0
0	0	1	0	1	0
0	1	0	0	0	1
0	1	1	1	1	0
1	0	0	0	0	1
1	0	1	1	0	0
1	1	0	0	0	1
1	1	1	1	0	0

present state, *input*, *next state*, and *output*. The present-state section shows the states of flip-flops A and B at any given time t. The input section gives each value of X for each possible present state. Note that for each possible input combination, each of the present states is repeated. The next-state section shows the states of the flip-flops one clock period later, at time $t + 1$. The output section gives the value of Y at time t for each combination of present state and input. The derivation of a state table consists of first listing all possible binary combinations of present state and inputs. In Table 4-2, we have eight binary combinations, from 000 to 111. The next-state values are then determined from the logic diagram or from the flip-flop input equations. For a D flip-flop, we have the relationship $A(t + 1) = D_A(t)$. This means that the next state of flip-flop A is equal to the present value of its input D. (See Table 4-1.) The value of the D input is specified in the flip-flop input equation as a function of the present state of A and B and input X. Therefore, the next state of flip-flop A must satisfy the equation

$$A(t+1) = D_A = AX + BX$$

The next-state section in the state table under column A has three 1's where the present state and input value satisfy the conditions $(A,X) = 11$ or $(B,X) = 11$. Similarly, the next state of flip-flop B is derived from the input equation

$$B(t+1) = D_B = \overline{A}X$$

and is equal to 1 when the present state of A is 0 and input X is equal to 1. The output column is derived from the output equation

$$Y = A\overline{X} + B\overline{X}$$

The state table of any sequential circuit with D-type flip-flops is obtained by the procedure just outlined. In general, a sequential circuit with m flip-flops and n inputs needs 2^{m+n} rows in the state table. The binary numbers from 0 through $2^{m+n} - 1$ are listed in the combined present-state and input columns. The next-state section has m columns, one for each flip-flop. The binary values for the next state are derived directly from the D flip-flop input equations. The output section has as many columns as there are output variables. Its binary values are derived from the circuit or from the Boolean functions in the same manner as in a truth table.

Table 4-2 is one dimensional in the sense that the present state and input combinations are combined into a single column of combinations. A two-dimensional state table having the present state tabulated in the left column and the inputs tabulated across the top row is also frequently used. The next-state entries are made in each cell of the table for the present-state and input combination corresponding to the location of the cell. A similar two-dimensional table is used for the outputs if they depend upon the inputs. Such a state table is shown in Table 4-3. Sequential circuits in which the outputs depend on the inputs, as well as on the states, are referred to as *Mealy model* circuits. Otherwise, if the outputs depend only on the states, then a one-dimensional column suffices. In this case, the circuits are referred to as *Moore model* circuits. Each model is named after its originator.

□ **TABLE 4-3**
Two-Dimensional State Table for the Circuit in Figure 4-18

Present state		Next state				Output	
		X = 0		X = 1		X = 0	X = 1
A	B	A	B	A	B	Y	Y
0	0	0	0	0	1	0	0
0	1	0	0	1	1	1	0
1	0	0	0	1	0	1	0
1	1	0	0	1	0	1	0

As an example of a Moore model circuit, suppose we want to obtain the logic diagram and state table of a sequential circuit that is specified by the flip-flop input equation

$$D_A = A \oplus X \oplus Y$$

and output equation

$$Z = A$$

The D_A symbol implies a D-type flip-flop with output designated by the letter A. The X and Y variables are taken as inputs and Z as the output. The logic diagram and state table for this circuit are shown in Figure 4-19. The state table has one column for the present state and one column for the inputs. The next state and output are also in single columns. The next state is derived from the flip-flop input equation which specifies an odd function. (See Section 2-7.) The output column is simply a copy of the column for the present-state variable A.

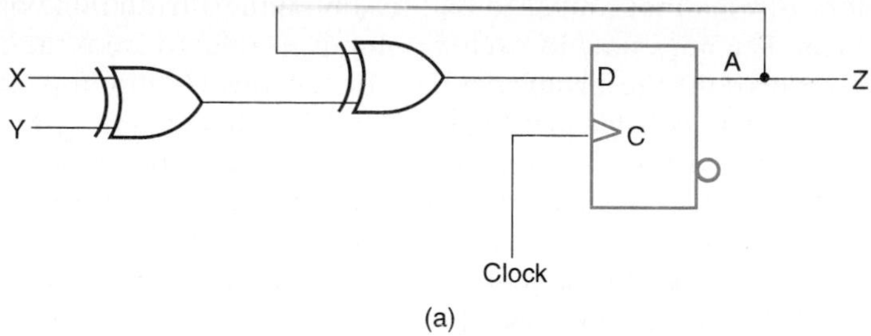

(a)

Present state	Inputs		Next state	Output
A	X	Y	A	Z
0	0	0	0	0
0	0	1	1	0
0	1	0	1	0
0	1	1	0	0
1	0	0	1	1
1	0	1	0	1
1	1	0	0	1
1	1	1	1	1

(b) State table

□ **FIGURE 4-19**
Logic Diagram and State Table for $D_A = A \oplus X \oplus Y$

Analysis with *JK* Flip-Flops

So far we have considered the state table for sequential circuits that employ *D*-type flip-flops, in which case the next-state values are obtained directly from the input equations. For circuits with other types of flip-flops, such as *JK*, the next-state values are obtained by following a two-step procedure:

1. Obtain the binary values of each flip-flop input equation in terms of the present-state and input variables.
2. Use the corresponding flip-flop characteristic from Table 4-1 to determine the next state.

To illustrate this procedure, consider the sequential circuit with two *JK* flip-flops *A* and *B* and one input *X* specified by the following input equations:

$$J_A = B \qquad K_A = B\overline{X}$$

$$J_B = \overline{X} \qquad K_B = A\overline{X} + \overline{A}X$$

The state table for this circuit is shown in Table 4-4. The binary values listed under the columns spanned by the header "Flip-flop inputs" are not part of the state table. They are needed for the purpose of evaluating the next state as specified in step 2 of the procedure. These binary values are obtained directly from the four input equations in a manner similar to that for obtaining a truth table from an algebraic expression. The next state of each flip-flop is evaluated from the corresponding J and K inputs and the characteristic table of the JK flip-flop, Table 4-1(a). There are four cases to consider. When $J = 1$ and $K = 0$, the next state is 1. When $J = 0$ and $K = 1$, the next state is 0. When $J = K = 0$, there is no change of state, and the next-state value is the same as that of the present state. When $J = K = 1$, the next-state bit is the complement of the present-state bit. Examples of the last two cases occur in the table when the present-state and input (A, B, X) are 100. J_A and K_A are both equal to 0, and the present state of A is 1. Therefore, the next state of A remains the same and is equal to 1. In the same row of the table, J_B and K_B are both equal to 1. Since the present state of B is 0, the next state of B is complemented and changes to 1.

☐ **TABLE 4-4**
State Table for Circuit with *JK* Flip-Flops

Present state		Input	Next state		Flip-flop inputs			
A	B	X	A	B	J_A	K_A	J_B	K_B
0	0	0	0	1	0	0	1	0
0	0	1	0	0	0	0	0	1
0	1	0	1	1	1	1	1	0
0	1	1	1	0	1	0	0	1
1	0	0	1	1	0	0	1	1
1	0	1	1	0	0	0	0	0
1	1	0	0	0	1	1	1	1
1	1	1	1	1	1	0	0	0

State Diagram

The information available in a state table may be represented graphically in the form of a state diagram. In this type of diagram, a state is represented by a circle, and transitions between states are indicated by directed lines connecting the circles. Examples of state diagrams are given in Figure 4-20. The diagram in Figure 4-20(a) is for the sequential circuit of Figure 4-18 and the state table of Table 4-2. The state diagram provides the same information as the state table and is obtained directly from it. The binary number inside each circle identifies the state of the flip-flops. For Mealy model circuits, the directed lines are labeled with two binary numbers separated by a slash. The input value during the present state precedes the slash, and the value following the slash gives the output value during the present state with the given input applied. For example, the directed line from state 00 to state 01 is labeled 1/0, meaning that when the sequential circuit is in the present state 00 and the input is 1, the output is 0. After the next clock transition, the circuit goes to the next state, 01. If the input changes to 0, then the output becomes 1, but if the input remains at 1, the output stays at 0. This information is obtained

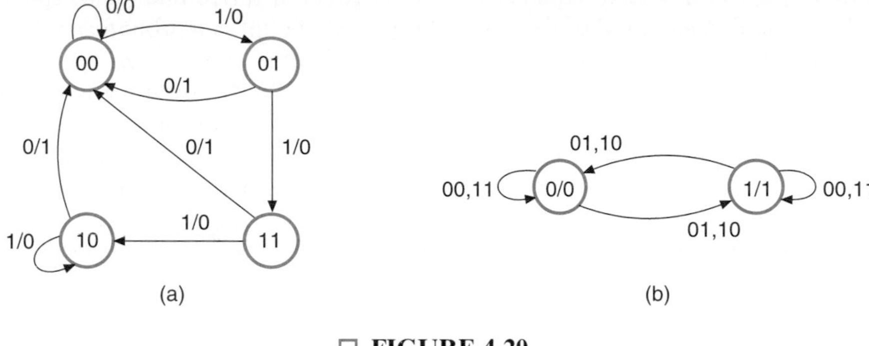

□ **FIGURE 4-20**
State Diagrams

from the state diagram along the two directed lines emanating from the circle with state 01. A directed line connecting a circle with itself indicates that no change of state occurs.

The state diagram of Figure 4-20(b) is for the sequential circuit of Figure 4-19. Here, we have only one flip-flop with two states. There are two binary inputs, and the output depends only on the state of the flip-flop. For such a Moore model circuit, the slash on the directed lines is not included, since the outputs depend only on the state and not on the input values. Instead, the output is included under a slash below the state in a circle. There are two input conditions for each state transition in the diagram, and they are separated by a comma. When there are two input variables, each state may have up to four directed lines coming out of the corresponding circle, depending upon the number of states and the next state for each binary combination of the input values.

There is no difference between a state table and a state diagram, except for their manner of representation. The state table is easier to derive from a given logic diagram and input equations. The state diagram follows directly from the state table. The state diagram gives a pictorial view of state transitions and is the form more suitable for human interpretation of the operation of the circuit. For example, the state diagram of Figure 4-20(a) clearly shows that, starting at state 00, the output is 0 as long as the input stays at 1. The first 0 input after a string of 1's gives an output of 1 and sends the circuit back to the initial state of 00. The state diagram of Figure 4-20(b) shows that the circuit stays at a given state as long as the two inputs have the same value (00 or 11). There is a state transition between the two states only when the two inputs are different (01 or 10).

4-5 Sequential Circuit Design

The design of clocked sequential circuits starts from a set of specifications and culminates in a logic diagram or a list of Boolean functions from which the logic diagram can be obtained. In contrast to a combinational circuit, which is fully specified by a truth table, a sequential circuit requires a state table for its specification. Thus, the first step in the design of a sequential circuit is to obtain a state table or an equivalent representation such as a state diagram.

A synchronous sequential circuit is made up of flip-flops and combinational gates. The design of the circuit consists of choosing the flip-flops and finding a combinational circuit structure which, together with the flip-flops, produces a circuit that fulfills the stated specifications. The number of flip-flops is determined from the number of states in the circuit; n flip-flops can represent up to 2^n binary states. The combinational circuit is derived from the state table by evaluating the flip-flop input equations and output equations. In fact, once the type and number of flip-flops are determined, the design process involves a transformation from a sequential circuit problem into a combinational circuit problem. In this way, the techniques of combinational circuit design can be applied.

Design Procedure

This following is a procedure for the design of sequential circuits:

1. Obtain either the state diagram or the state table from the statement of the problem.
2. If only a state diagram is available from step 1, obtain the state table.
3. Assign binary codes to the states.
4. Derive the flip-flop input equations from the next-state entries in the encoded state table.
5. Derive output equations from the output entries in the state table.
6. Simplify the flip-flop input equations and output equations.
7. Draw the logic diagram with D flip-flops and combinational gates, as specified by the flip-flop input equations and output equations.

Finding State Diagrams and State Tables

The specification for a circuit is often in the form of a verbal description of the behavior of the circuit. This description needs to be interpreted in order to find a state diagram or state table as the first step of the design procedure. In this section, we illustrate the construction of a state diagram for a circuit that recognizes the occurrence of a particular sequence of bits, regardless of where it occurs in a longer sequence.

The sequence-recognizing circuit is to have one input X and one output Z. In addition, it has direct resets on its flip-flops to initialize the state of the circuit to all zeros. The circuit is to recognize the occurrence of the sequence of bits 1101 on X by making Z equal to 1 when the previous three inputs to the circuit were 110 and current input is a 1. Otherwise, Z equals 0.

A key factor in the formulation of any state diagram is to recognize that states are used to "remember" something about the history of past inputs. For example, for the sequence 1101, in order to be able to produce the output value 1

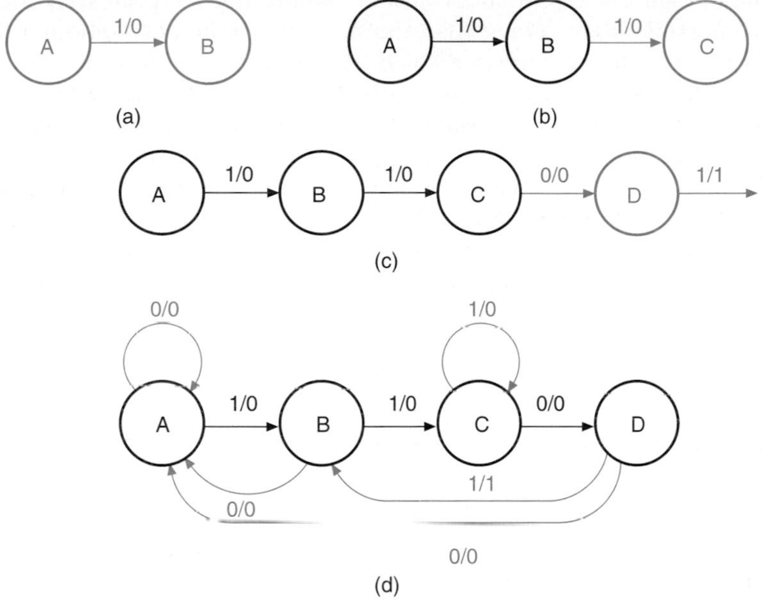

□ **FIGURE 4-21**
Construction of a State Diagram

coincident with the final 1 in the sequence, the circuit must be in a state that "remembers" that the previous three inputs were 110. With this concept in mind, suppose we begin to formulate the state diagram by defining an arbitrary state—say, *A*—as the state in which none of the first portion of the sequence to be recognized has occurred. If a 1 occurs on the input, since 1 is the first bit in the sequence, the event must be "remembered," and the state after the clock pulse cannot be *A*. So we establish a second state, *B*, to represent the occurrence of the first 1 in the sequence. Further, to represent the occurrence of the first 1 in the sequence, we place a transition from *A* to *B* labeled with a 1. Since this is not the final 1 in the sequence 1101, its output is a 0. This initial portion of the state diagram is given in Figure 4-21(a).

The next bit of the sequence is a 1. When this 1 occurs in state *B*, a state is needed to represent the occurrence of two 1's in a row on the input—that is, the occurrence of an additional 1 while in state *B*. So a state *C* and the associated transition are added, as shown in Figure 4-21(b). The next bit of the sequence is a 0. When this 0 occurs in state *C*, a state is needed to represent the occurrence of the two 1's in a row followed by a 0. So the additional state *D* with a transition having a 0 input and 0 output is added. Since state *D* represents the occurrence of 110 as the previous three input bit values on *X*, the occurrence of a 1 in state *D* completes the sequence to be recognized, so the transition for the input value 1 from state *D* has an output value of 1. The resulting partial state diagram, which completely represents the occurrence of the sequence to be recognized, is shown in Figure 4-21(c).

Note in Figure 4-21(c) that a transition is specified for each state for only one of the two possible input values. Also, the state that is the destination of the transition from D for input 1 is not yet defined. The remaining transitions must be based on the idea that the recognizer is to identify the sequence 1101, regardless of where it occurs in a longer sequence. Thus, if we have an initial part of the sequence 1101 anywhere in the state diagram, the transition for an input value that represents an additional bit in the sequence must go to a state such that the 1 output occurs if the sequence is eventually completed.

First, suppose we evaluate where the transition for the 1 input from the D state is to go. Since the transition input is a 1, it could be the first or second bit in the sequence to be recognized. But because the circuit is in state D, we know that the prior input was a 0. So this 1 input is the first 1 in the sequence, since it is not preceded by a 1. The state that represents the occurrence of a first 1 in the sequence is B, so the transition with input 1 from state D is to state B. This transition is shown in the diagram in Figure 4-21(d). Examining state C, we can trace back through states B and A to see that the occurrence of a 1 input at C is at least the second 1 in the sequence. The state representing the occurrence of two 1's in sequence is C, so the transition is to state C. Since the combination of two 1's is not the sequence to be recognized, the output for the transition is 0. Repeating this same analysis for missing transitions from states B and A, we obtain the final state diagram in Figure 4-21(d). The corresponding state table is given in two-dimensional form in Table 4-5.

□ TABLE 4-5
State Table for State Diagram in Figure 4-21

Present State	Next State		Output Z	
	X = 0	X = 1	X = 0	X = 1
A	A	B	0	0
B	A	C	0	0
C	D	C	0	0
D	A	B	0	1

One issue that arises in the formulation of any state diagram is whether excess states have been used. This is not the case in the preceding example, since each state represents input history that is essential for recognition of the stated sequence. If excess states are present, then it is desirable to combine states into the fewest number needed, using state minimization procedures. Due to their complexity, however, particularly in the case in which don't-care entries appear in the state table, such procedures will not be covered here. For the interested student, state minimization procedures are found in the references listed at the end of the chapter.

In contrast to the state diagram in the example we have been considering, the states in the diagram we have constructed have been assigned symbolic names rather than binary codes. It is necessary to replace these symbolic names with binary codes in order to proceed with the design. In general, if there are m states, then the codes must contain n bits, where $2^n \geq m$, and each state must be assigned a unique code. So for the circuit under consideration, the codes assigned to the states require two bits.

We begin by assigning a code to the initial reset state. If 1101 occurs as the initial four inputs to the circuit, it should be recognized. But if 101, 01, or 1 occurs as the first input sequence, it should not be recognized. The only initial state that can provide this property is state A. So if direct resets are used on the flip-flops, the code 00 should be assigned to state A. As a basis for encoding the remaining states, extensive work on the assignment of codes to states exists, but it is too complex for our treatment here. These methods have focused primarily on attempting to select codes in such a way that the logic required to implement the flip-flop input equations and output equations is minimized. In our example, we will simply assign the codes to the states in Gray code order, beginning with state A. The Gray code is selected in this case simply because it makes it easier for the next-state and output functions to be placed on a Karnaugh map. The state table with the codes assigned is shown in Table 4-6.

□ **TABLE 4-6**
Table 4-5 with Names Replaced by Binary Codes

	Next State		Output Z	
Present State	**X = 0**	**X = 1**	**X = 0**	**X = 1**
00	00	01	0	0
01	00	11	0	0
11	10	11	0	0
10	00	01	0	1

4-6 Designing with *D* Flip-Flops

The remainder of the sequential circuit design procedure will be illustrated by means of the following example. We wish to design a clocked sequential circuit that operates according to the state diagram, including binary codes, shown in Figure 4-22. The state diagram specifies four states, two input values, and two output values. Two D flip-flops are needed to represent the four states. We label the flip-flop outputs with the letters A and B, the input with X, and the output with Y.

The state table of the circuit is listed in Table 4-7. It is derived directly from the state diagram. Since the states are already represented by binary codes, step 3 of the design procedure can be omitted. The flip-flop input equations are obtained from the next-state values as listed in the table. The output equation is obtained from the binary values of Y in the table. The three Boolean equations for the combinational gates can be expressed as a sum of minterms of the present-state variables A and B and the input variable X:

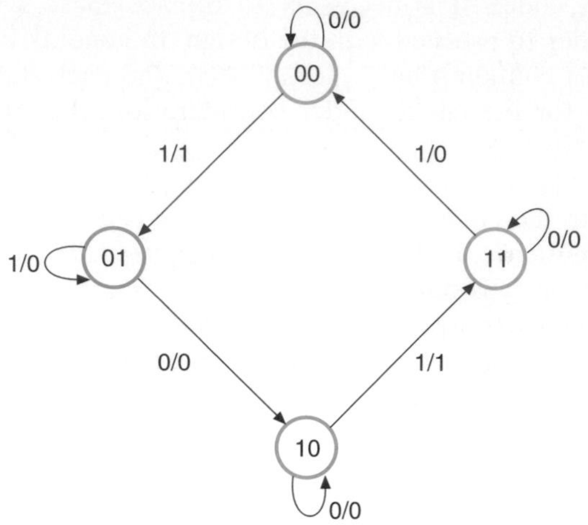

□ **FIGURE 4-22**
State Diagram for Design Example

□ **TABLE 4-7**
State Table for Design Example

Present State		Input	Next State		Output
A	**B**	**X**	**A**	**B**	**Y**
0	0	0	0	0	0
0	0	1	0	1	1
0	1	0	1	0	0
0	1	1	0	1	0
1	0	0	1	0	0
1	0	1	1	1	1
1	1	0	1	1	0
1	1	1	0	0	0

$$A(t+1) = D_A(A, B, X) = \Sigma\, m(2,4,5,6)$$

$$B(t+1) = D_B(A, B, X) = \Sigma\, m(1,3,5,6)$$

$$Y(A, B, X) = \Sigma\, m(1,5)$$

The Boolean functions are simplified by means of the maps plotted in Figure 4-23.

The simplified functions are

$$D_A = A\overline{B} + B\overline{X}$$

$$D_B = \overline{A}X + \overline{B}X + AB\overline{X}$$

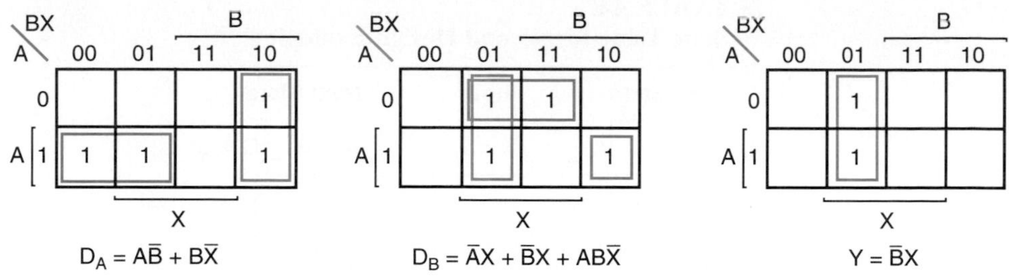

FIGURE 4-23
Maps for Input Equations and Output Y

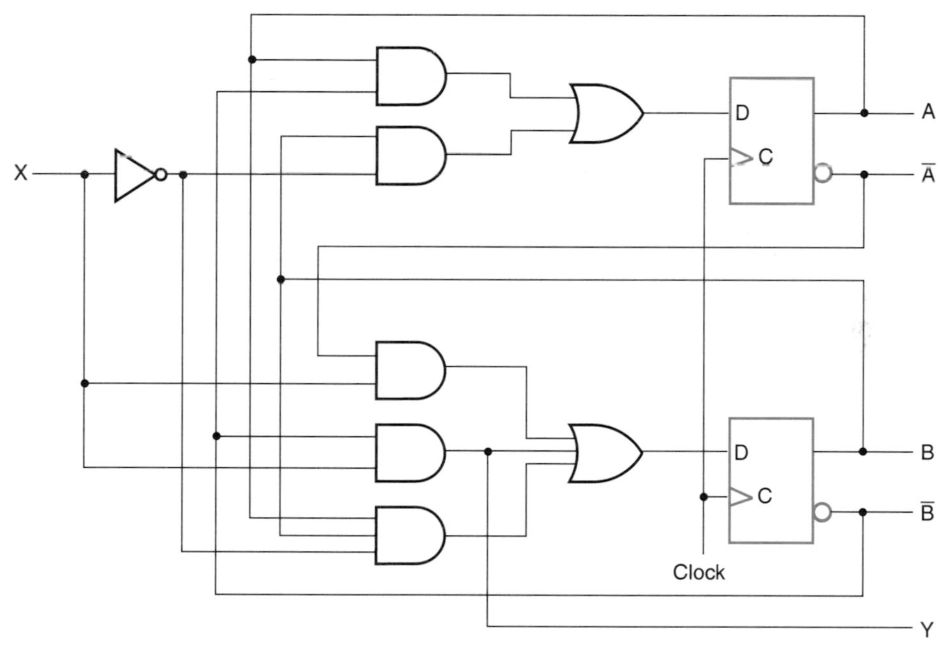

FIGURE 4-24
Logic Diagram for Sequential Circuit with D Flip-Flops

$$Y = \overline{B}X$$

The logic diagram of the sequential circuit is shown in Figure 4-24.

Designing with Unused States

A circuit with n flip-flops has 2^n binary states. The state table from which the circuit was originally derived, however, may have any number of states $m \leq 2^n$. States that are not used in specifying the sequential circuit are not listed in the state table. In simplifying the input equations, the unused states can be treated as don't-care

☐ **TABLE 4-8**
State Table for Second Design Example

Present State			Input	Next State		
A	B	C	X	A	B	C
0	0	1	0	0	0	1
0	0	1	1	0	1	0
0	1	0	0	0	1	1
0	1	0	1	1	0	0
0	1	1	0	0	0	1
0	1	1	1	1	0	0
1	0	0	0	1	0	1
1	0	0	1	1	0	0
1	0	1	0	0	0	1
1	0	1	1	1	0	0

conditions. Consider the state table listed in Table 4-8. The table defines three flip-flops *A*, *B*, and *C*, and one input *X*. There is no output column, which means that the flip-flops serve as outputs of the circuit. With three flip-flops, it is possible to specify eight states, but the state table lists only five. Thus, there are three unused states that are not included in the table: 000, 110, and 111. When an input of 0 or 1 is included with the unused present-state values, we obtain six unused combinations for the present-state and input columns: 0000, 0001, 1100, 1101, 1110, and 1111. These six combinations are not listed in the state table and hence may be treated as don't-care minterms.

The three input equations for the *D* flip-flops are derived from the next-state values and are simplified in the maps of Figure 4-25. Each map has six don't-care

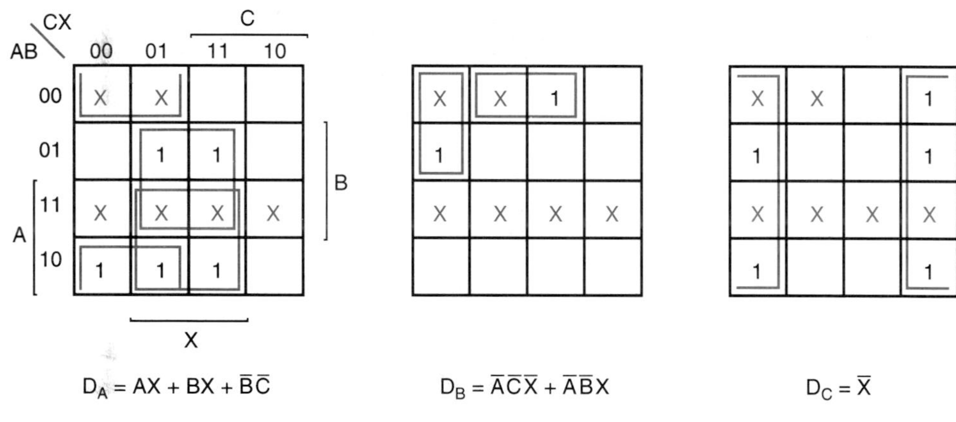

☐ **FIGURE 4-25**
Maps for Simplifying Input Equations

minterms in the squares corresponding to binary 0, 1, 12, 13, 14, and 15. The simplified equations are

$$D_A = AX + BX + \overline{B}\,\overline{C}$$

$$D_B = \overline{A}\,\overline{C}\,\overline{X} + \overline{A}\,\overline{B}X$$

$$D_C = \overline{X}$$

The logic diagram can be obtained directly from the input equations and will not be drawn here.

One factor touched upon only briefly to this point is the initial state of a sequential circuit. Since one does not know in what state the flip-flops will be when the power in a digital system is first turned on, it is customary to provide a *master reset* signal to initialize the state of the flip-flops in the circuit. This avoids the circuit starting in an unused state. Typically, the master reset signal is applied to the flip-flops at their asynchronous (direct) inputs (see Figure 4-16) before the clocked operations start. In most cases flip-flops are reset to 0, but some may be set to 1, depending on the initial state desired. If there is a large number of flip-flops, based on the circuit operation only a portion of them may be initialized.

It is possible that outside interference or a malfunction will cause the circuit to enter one of the unused states. Thus, it is sometimes desirable to specify fully or at least partially the next-state values or output values for the unused states. Depending on the function and application of the circuit, a number of ideas may be applied. First, the outputs for the unused states may be specified so that any actions that result from entry into and transitions between the unused states are not harmful. Second, an additional output may be provided or an unused output code employed which indicates that the circuit has entered an incorrect state. Third, to ensure that a return to normal operation is possible without resetting the entire system, the next-state behavior for the unused states may be specified. Typically, next states are selected such that one of the normally occurring states is reached regardless of the input values within a few clock cycles. The decision as to which of the three options to apply, either individually or in combination, is based on the application of the circuit or the policies of a particular design group.

4-7 Designing with *JK* Flip-Flops

The design of a sequential circuit with flip-flops other than the *D* type is complicated by the fact that the flip-flop input equations for the circuit must be derived indirectly from the state table. When *D*-type flip-flops are employed, the input equations are obtained directly from the next state. This is not the case for the *JK* and other types of flip-flops. In order to determine the input equations for these flip-flops, it is necessary to derive a functional relationship between the state table and the input equations.

Flip-Flop Excitation Tables

The flip-flop characteristic tables presented in Table 4-1 provide the value of the next state when the values of the inputs and the present state are known. These tables are used to analyze sequential circuits and to define the operation of the flip-flops. During the design process, we usually know the transition from the present state to the next state and wish to find the flip-flop input conditions that will cause this transition. Accordingly, we need a table that lists the required inputs for a given change of state. Such a table is called an *excitation table.* Table 4-9 presents the excitation tables for four different types of flip-flops. Each table has a column for the present state $Q(t)$, a column for the next state $Q(t + 1)$, and a column for each flip-flop input to show how the required transition is achieved. There are four possible transitions from the present state to the next state. The required input conditions for each of these transitions are derived from the information available in the characteristic table. The symbol X in the table represents a don't-care condition, which means, again, that it does not matter whether the input is 0 or 1.

□ **TABLE 4-9**
Flip-Flop Excitation Tables

(a) *JK* Flip-Flop

$Q(t)$	$Q(t+1)$	J	K
0	0	0	X
0	1	1	X
1	0	X	1
1	1	X	0

(b) *SR* Flip-Flop

$Q(t)$	$Q(t+1)$	S	R
0	0	0	X
0	1	1	0
1	0	0	1
1	1	X	0

(c) *D* Flip-Flop

$Q(t)$	$Q(t+1)$	D
0	0	0
0	1	1
1	0	0
1	1	1

(d) *T* Flip-Flop

$Q(t)$	$Q(t+1)$	T
0	0	0
0	1	1
1	0	1
1	1	0

The excitation table for the *JK* flip-flop can be derived from the knowledge of how the flip-flop operates. Consider the first entry in the table. The transition from a present state of 0 to a next state of 0 can be accomplished in two ways. If $J = 0$ and $K = 0$, there is no change of state, and the flip-flop stays at 0. If $J = 0$ and $K = 1$, the flip-flop resets to 0. This dictates that J must be equal to 0, but K can be either 0 or 1, and in either case, we obtain the required transition. This is indicated in the first row of the table by a 0 under J and a don't-care symbol X under K. The transformation from a present state of 0 to a next state of 1 can also be done in two ways. Letting $J = 1$ and $K = 0$ sets the flip-flop to 1. Letting $J = 1$ and $K = 1$ complements the flip-flop from 0 to 1. For this case it is necessary that J be equal to 1, but it does not matter whether K is 0 or 1. Either way, we get the proper transition. This information is listed in the second row of the table. In a similar manner, it is possible to derive the rest of the entries in the excitation tables for the JK flip-flop and the other three flip-flops. The excitation tables for the T and D flip-flops have no don't-care conditions and can be specified with an excitation function. The excitation table for the D flip-flop shows that the next state is always equal to the D input and is independent of the present state. This can be represented algebraically by the excitation function

$$D = Q(t+1)$$

Therefore, the values for input D can be taken directly from the values in the next-state column, as was done in the previous section. The excitation table for the T flip-flop shows that the T input is equal to the exclusive-OR of the present state and the next state. This can be expressed by the excitation function

$$T = Q(t) \oplus Q(t+1)$$

Design Procedure

The design procedure for sequential circuits with JK flip-flops is the same as that for sequential circuits with D flip-flops, except that the input equations must be evaluated from the present-state to next-state transition derived from the excitation table. To illustrate the procedure, we will design the sequential circuit specified by Table 4-10. This table is the same as Table 4-7, but without the output section shown. In addition to having columns for the present state, input, and next state, as in a conventional state table, the table also shows the flip-flop input conditions, from which we obtain the input equations. The flip-flop inputs in the table are derived from the state table in conjunction with the excitation table for the JK flip-flop. For example, in the first row of Table 4-10 we have a transition for flip-flop A

□ **TABLE 4-10**
State Table with *JK* Flip-Flop Inputs

Present State		Input	Next State		Flip-Flop Inputs			
A	**B**	**X**	**A**	**B**	J_A	K_A	J_B	K_B
0	0	0	0	0	0	X	0	X
0	0	1	0	1	0	X	1	X
0	1	0	1	0	1	X	X	1
0	1	1	0	1	0	X	X	0
1	0	0	1	0	X	0	0	X
1	0	1	1	1	X	0	1	X
1	1	0	1	1	X	0	X	0
1	1	1	0	0	X	1	X	1

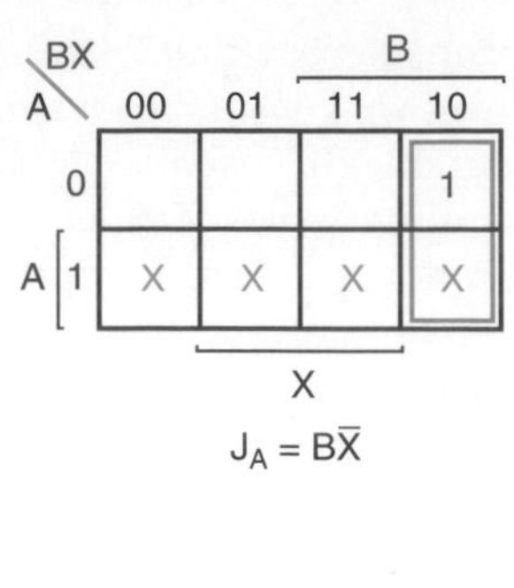

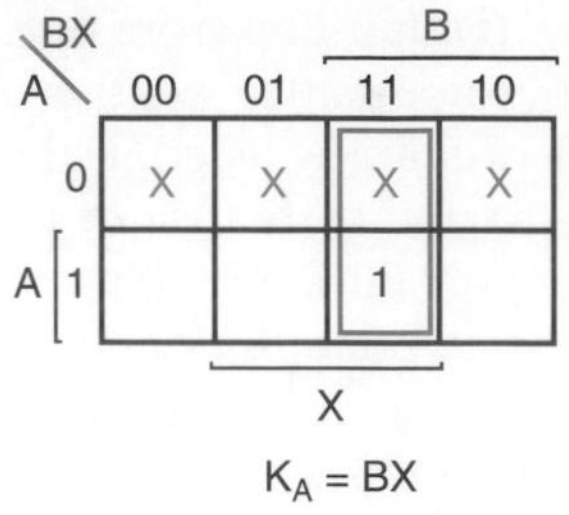

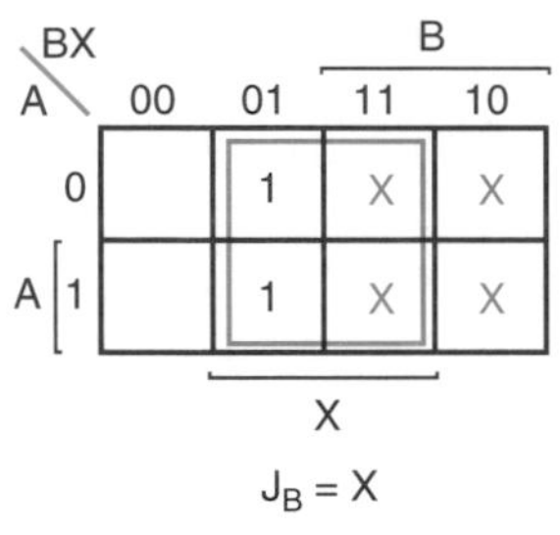

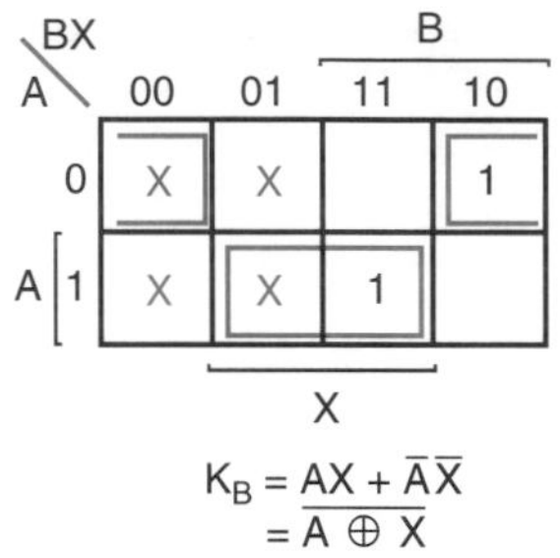

☐ **FIGURE 4-26**
Maps for *J* and *K* Input Equations

from 0 in the present state to 0 in the next state. In Table 4-9(a) for the *JK* flip-flop, we find that a transition of states from 0 to 0 requires that input *J* be 0 and input *K* be a don't-care condition. So 0 and X are entered in the first row under J_A and K_A. Since the first row also shows a transition for flip-flop *B* from 0 in the present state to 0 in the next state, 0 and X are inserted in the first row under J_B and K_B, respectively. The second row of the table shows a transition for flip-flop *B* from 0 in the present state to 1 in the next state. From the excitation table, we find that a transition from 0 to 1 requires that *J* be 1 and *K* be a don't-care condition, so 1 and X are copied in the second row under J_B and K_B, respectively. This process is continued for each row in the table and for each flip-flop, with the input conditions from the excitation table copied into the proper row of the particular flip-flop being considered.

The flip-flop inputs in Table 4-10 specify the truth table for the flip-flop input equations as a function of the present-state variables *A* and *B* and input *X*. The input equations are simplified in the maps of Figure 4-26. The next-state values are not used during the simplification, since the input equations are a function of the present state and input only. Note the advantage of using *JK*-type flip-flops when designing sequential circuits. The fact that there are so many don't-care entries indicates that the combinational circuit for the input equations is likely to be simpler, because don't-care minterms usually help in obtaining simpler expressions. If there are unused states in the state table, there may be additional don't-care conditions in the map. The four input equations for the two *JK* flip-flops are listed

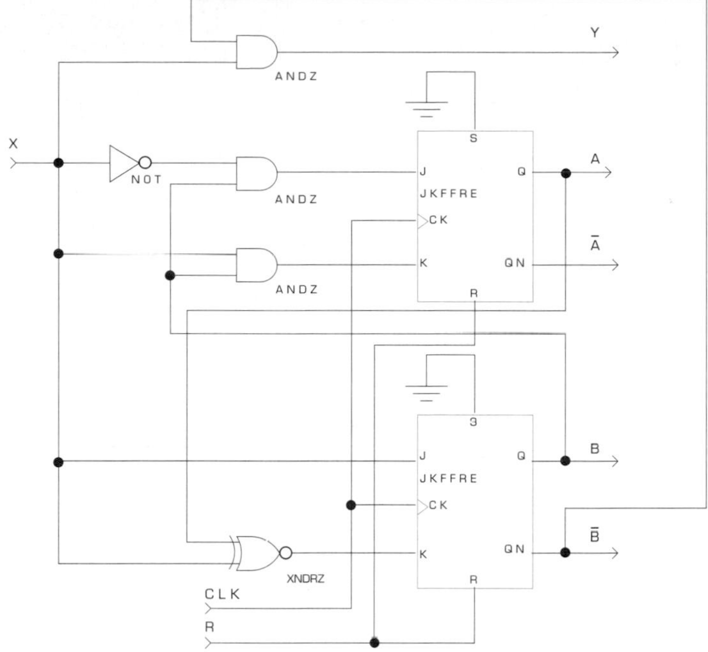

☐ **FIGURE 4-27**
Logic Diagram for Sequential Circuit with *JK* Flip-Flops

under the maps of Figure 4-26. The logic diagram of the sequential circuit is drawn in Figure 4-27, with output logic from Figure 4-24. Since we are planning to simulate this circuit, we have used the ViewDraw™ schematic capture tool to draw the diagram. Note that direct reset inputs reset the flip-flops for a 1 applied to master reset input *R*.

Thus far, we have not verified correctness of any of the circuits designed. To illustrate this process, we will verify the correctness of the logic circuit in Figure 4-27 with respect to the initial state diagram in Figure 4-22. We do this by simulating the logic with the Viewlogic® simulator ViewSim® and comparing its response to that obtained from the state diagram for the same input sequence. First we determine the sequence to apply to verify the circuit thoroughly. We devise a short sequence that passes through each of the transitions in the state diagram in Figure 4-22 to verify that the next state and the output are correct. In addition, at the end of the sequence, we make sure that the master reset *R* resets both of the flip-flops in the circuit from 1 to 0. The sequence, shown in Figure 4-28(a), applies the values listed for *R* and *X* at successive clock cycles. The respective states and output values, as determined from the state diagram, are also shown.

In order to perform the simulation, we need to have a clock, as well as the input signals *R* and *X*. In doing the simulation of any sequential circuit, sufficient time must be provided in the clock period for each of the following:

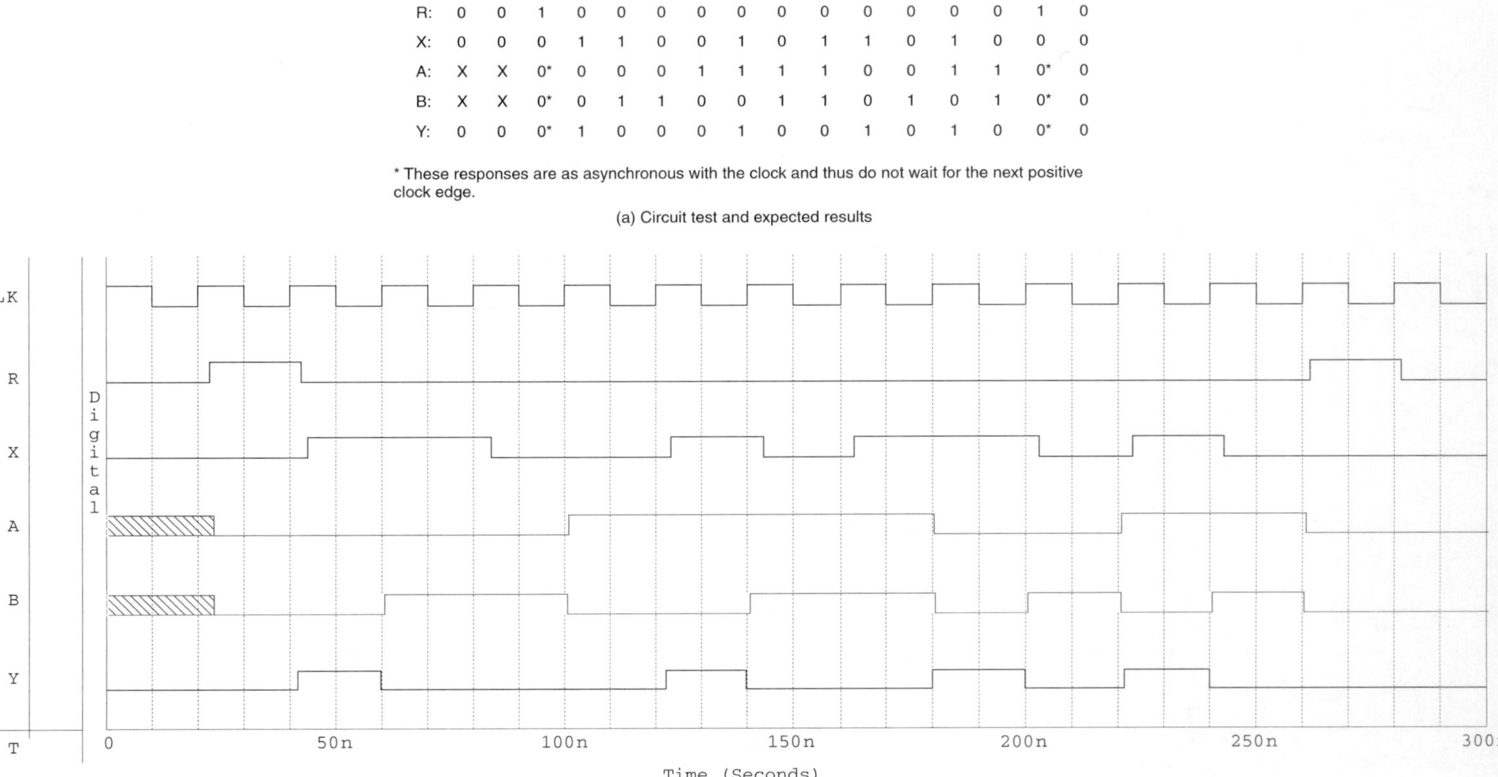

R:	0	0	1	0	0	0	0	0	0	0	0	0	0	0	1	0
X:	0	0	0	1	1	0	0	1	0	1	1	0	1	0	0	0
A:	X	X	0*	0	0	0	1	1	1	1	0	0	1	1	0*	0
B:	X	X	0*	0	1	1	0	0	1	1	0	1	0	1	0*	0
Y:	0	0	0*	1	0	0	0	1	0	0	1	0	1	0	0*	0

* These responses are as asynchronous with the clock and thus do not wait for the next positive clock edge.

(a) Circuit test and expected results

(b) Simulation results

□ **FIGURE 4-28**
Logic Simulation Verification for the Circuit in Figure 4-27

1. all flip-flops and inputs to change;
2. the effects of these changes to propagate through the combinational logic of the circuit to the flip-flop inputs; and
3. the setup of the flip-flops for the next clock edge to occur.

For our simulation, we will use a clock period of 20 ns, which is adequate for the delays and setup times present. Note that if circuit inputs change such that flip-flop inputs change during the setup/hold time interval, the state may become incorrect or the simulator may signal a setup/hold time violation. Thus, we apply the inputs early in the clock period. If the inputs come from combinational logic driven by flip-flops, in a careful simulation the delay of this logic must be considered in timing the input changes so that any problems with setup times are detected.

The results of the simulation appear in Figure 4-28(b). Note when the inputs change relative to the clock edge. Also, there is a time delay for the flip-flops to change after each clock, although it is short enough that it is not visible in the waveforms. The values just prior to each positive clock edge are to be compared with those given in the sequence derived from the state diagram. This comparison indicates that the values are identical. Hence, since the circuit has the state diagram originally given, it meets the circuit specification.

As a final example, consider the design of a sequential circuit with T flip-flops. Using the state table portion of Table 4-10, we obtain the binary values for the T flip-flop inputs from the excitation functions for A and B:

$$T_A = A(t) \oplus A(t+1)$$

$$T_B = B(t) \oplus B(t+1)$$

Going over the present-state and next-state values for A and B in the table, we determine the binary values for T_A and T_B in the following manner. For each present- to next-state transition from 0 to 1 or from 1 to 0, we place a 1 on the corresponding T input; for each case where there is no change (from 0 to 0 or from 1 to 1), we place a 0 on the corresponding T input. The input equations can be obtained from the minterms that produce a 1 in the flip-flop input columns and are

$$T_A(A,B,X) = \Sigma\, m(2,7) = \overline{A}B\overline{X} + ABX$$

$$T_B(A,B,X) = \Sigma\, m(1,2,5,7) = \overline{A}B\overline{X} + ABX + \overline{B}X$$

The algebraic expressions are the simplified input equations from which one can draw the gates of the sequential circuit. The circuit consists of two T flip-flops, three AND gates, two OR gates, and an inverter. Remember that a T flip-flop can be constructed from a JK flip-flop with inputs J and K tied together to form a single input T.

4-8 Chapter Summary

Sequential circuits are the foundation on which most digital design is based. Flip-flops are the basic storage elements for synchronous sequential circuits. Flip-flops are constructed of more fundamental elements called latches. By themselves, latches are transparent and, as a consequence, unsuitable for use in synchronous sequential circuits using a single clock. But when latches are combined to form flip-flops, nontransparent storage elements suitable for such use result. Two triggering methods are used for flip-flops: master-slave and edge triggering. In addition, flip-flops are of a number of types, including D, SR, JK, and T.

Sequential circuits are formed using these flip-flops and combinational logic. Sequential circuits can be analyzed to find state tables and state diagrams that represent the behavior of the circuits. In order to handle the analysis for different types of flip-flops, we use a characteristic table for each type. Analysis can also be performed by using logic simulation.

These same state diagrams and state tables can be formulated from verbal specifications of digital circuits. By assigning binary codes to the states and finding flip-flop input equations for the type or types of flip-flops selected for the circuit, sequential circuits can be designed. In order to handle the various types of flip-flops, excitation tables are used in the formation of the flip-flop input equations. Sequential circuit design also includes issues such as finding logic for the circuit outputs, resetting the state at power-up, and controlling the behavior of the circuit in the event that it enters states unused in the original specification. Finally, logic simulation plays an important role in verifying that the circuits which have been designed meet the original specification.

References

1. Mano, M. M. *Digital Design,* 2nd ed. Englewood Cliffs, NJ: Prentice Hall, 1991.
2. Hayes, J. P. *Introduction to Digital Logic Design*. Reading, MA: Addison-Wesley, 1993.
3. Roth, C. H. *Fundamentals of Logic Design*, 4th ed. St. Paul: West, 1992.
4. Wakerly, J. F. *Digital Design: Principles and Practices,* 2nd ed. Englewood Cliffs, NJ: Prentice Hall, 1994.

Problems

The asterisk (*) indicates a more advanced problem.

4–1. Perform a manual or computer-based logic simulation similar to that given in Figure 4-5 for the $\overline{S}\,\overline{R}$ latch shown in Figure 4-6. Construct the input sequence, keeping in mind that changes in state for this type of latch occur in response to 0 rather than 1.

4–2. Perform a manual or computer-based logic simulation similar to that given in Figure 4-5 for the SR latch with control input C in Figure 4-7. In particular, examine the behavior of the circuit when S and R are changed while C has the value 1.

4–3. The D latch shown in Figure 4-8 can be constructed with only four NAND gates. This can be done by removing the inverter and connecting the output of the upper NAND gate to the input of the lower gate. Use manual or computer-based logic simulation to show that the new circuit is functionally the same as the original one.

4–4. Obtain the logic diagram of the D latch shown in Figure 4-8, using NOR gates only.

4–5. Obtain the logic diagram of the SR master-slave flip-flop in Figure 4-10, using NAND gates only.

4–6. Obtain the logic diagram of a D-type master-slave flip-flop using transmission gates and inverters. Which of the symbols in Figure 4-15 corresponds to your circuit? (*Hint*: more than one symbol may correspond.)

4–7. Obtain a timing diagram similar to Figure 4-11 for a JK master-slave flip-flop during four clock pulses. Show the timing signals for $C, J, K, Y,$ and Q. Assume that initially the output Q is equal to 1, with the first pulse $J = 0$ and $K = 1$. Then, for successive pulses, J goes to 1, followed by K going to 0 and then J going back to 0. Assume that each input changes after the negative edge of the pulse.

4–8. Repeat Problem 4–7, using a positive-edge-triggered JK flip-flop. Show the timing diagrams for $C, J, K,$ and Q.

4–9. A popular alternative design for a positive-edge-triggered D flip-flop is shown in Figure 4-29. Simulate the circuit to determine whether its functional behavior is identical to that of the circuit in Figure 4-13.

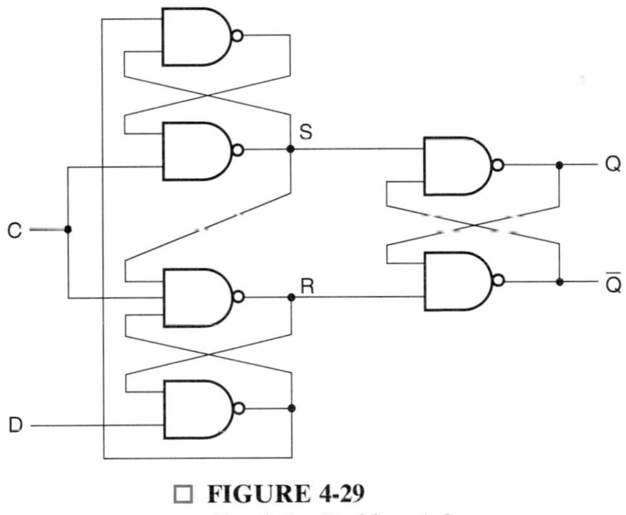

□ **FIGURE 4-29**
Circuit for Problem 4–9

4–10. *Use logic simulation to compare the master-slave *JK* flip-flop in Figure 4-12 with a negative-edge-triggered *JK* flip-flop which is identical to the circuit in Figure 4-14, except that the inverter on the clock input *C* to the circuit is removed. Apply waveforms that demonstrate the difference in behavior in these two circuits for changes in the *J* and *K* inputs while $C = 1$.

4–11. Write *characteristic equations* for each type of flip-flop, using the information in Table 4-1. A characteristic equation gives the function $Q(t + 1)$ in terms of $Q(t)$ and the input variables to the flip-flop. Use the characteristic equation for the *JK* flip-flop to find equations $A(t + 1)$ and $B(t + 1)$ from the flip-flop input equations corresponding to Table 4-4.

4–12. A sequential circuit with two *D* flip-flops *A* and *B*, two inputs *X* and *Y*, and one output *Z* is specified by the following input equations:

$$D_A = \overline{X}Y + XA \quad D_B = \overline{X}B + XA \quad Z = B$$

(a) Draw the logic diagram of the circuit.
(b) Derive the state table.
(c) Derive the state diagram.

4–13. A sequential circuit has three *D* flip-flops *A*, *B*, and *C*, and one input *X*. The circuit is described by the following input equations:

$$D_A = (B\overline{C} + \overline{B}C)X + (BC + \overline{B}\,\overline{C})\overline{X}$$

$$D_B = A$$

$$D_C = B$$

(a) Derive the state table for the circuit.
(b) Draw two state diagrams, one for $X = 0$ and the other for $X = 1$.

4–14. A sequential circuit has one flip-flop *Q*, two inputs *X* and *Y*, and one output *S*. The circuit consists of a full adder circuit connected to a *D* flip-flop, as shown in Figure 4-30. Derive the state table and state diagram of the sequential circuit.

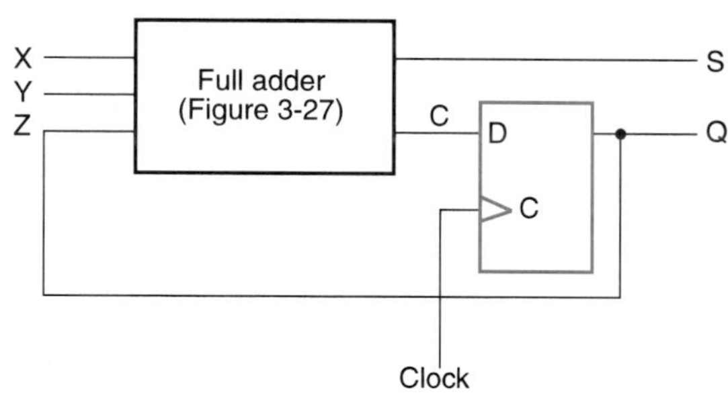

□ **FIGURE 4-30**
Circuit for Problem 4–14

4–15. Starting from state 00 in the state diagram of Figure 4-20(a), determine the state transitions and output sequence that will be generated when an input sequence of 010110111011110 is applied.

4–16. Draw the state diagram of the sequential circuit specified by the state table in Table 4-4.

4–17. Draw the logic diagram of a sequential circuit with two *JK* flip-flops and one input. The circuit is specified by the input equations associated with the flip-flop inputs in Table 4-4.

4–18. A sequential circuit has two *JK* flip-flops, one input *X*, and one output *Y*. The logic diagram of the circuit is shown in Figure 4-31. Derive the state table and state diagram of the circuit.

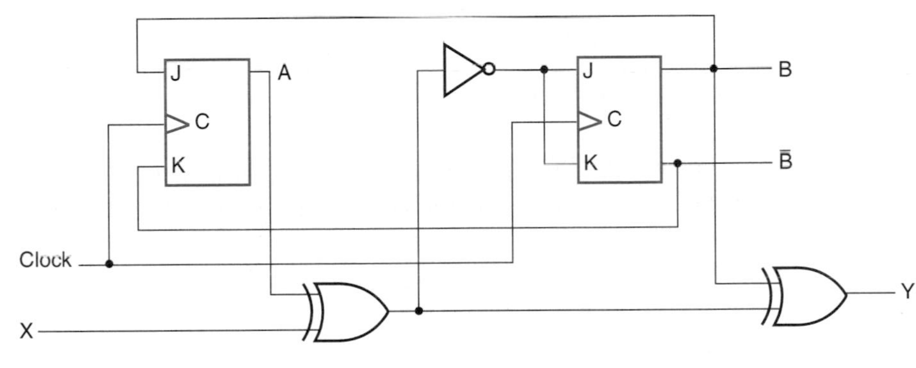

□ **FIGURE 4-31**
Circuit for Problem 4–18

4–19. A sequential circuit has two *JK* flip-flops *A* and *B*, two inputs *X* and *Y*, and one output *Z*. The flip-flop input equations and output function are as follows:

$$J_A = BX + \overline{B}\,\overline{Y} \qquad K_A = \overline{B}X\overline{Y} \quad Z = AXY + B\overline{X}\,\overline{Y}$$

$$J_B = \overline{A}X \qquad K_B = A + X\overline{Y}$$

(a) Draw the logic diagram of the circuit.
(b) Derive the state table and state diagram of the circuit.

4–20. Design a sequential circuit with two *D* flip-flops *A* and *B* and one input *X*. When $X = 0$, the state of the circuit remains the same. When $X = 1$, the circuit goes through the state transitions from 00 to 01 to 11 to 10, back to 00, and then repeats.

4–21. A sequential circuit has two flip-flops *A* and *B*, one input *X* and one output *Y*. The state diagram is shown in Figure 4-32. Design the circuit with *D* flip-flops.

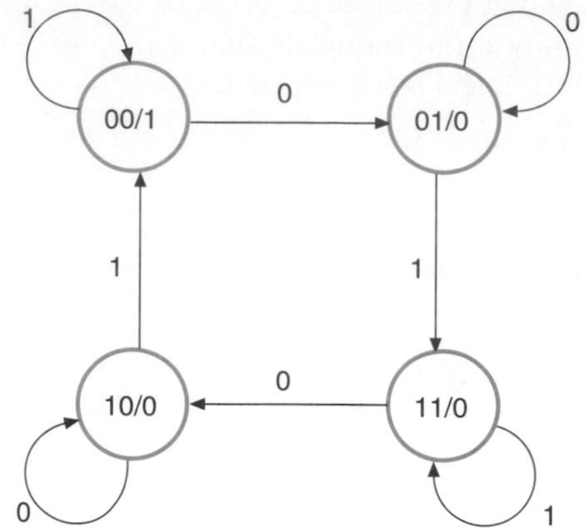

□ **FIGURE 4-32**
State Diagram for Problem 4–21 and Problem 4–27

4–22. Convert a D-type flip-flop into a JK flip-flop, using external gates. The gates can be derived by means of a sequential circuit design procedure starting from a state table with the D flip-flop output as the present and next state and with J and K as inputs.

4–23. A set-dominant flip-flop has set and reset inputs. It differs from a conventional SR flip-flop in that, when both S and R are equal to 1, the flip-flop is set. Obtain the characteristic table of the set-dominant flip-flop.

4–24. A JN flip-flop has two inputs J and N. Input J behaves like the J input of a JK flip-flop, and input N behaves like the complement of the K input of a JK flip-flop (that is $N = \overline{K}$).

(a) Obtain the characteristic table of the flip-flop.

(b) Show that, by connecting the two inputs together, one obtains a D-type flip-flop.

4–25. Derive an excitation table for the JN flip-flop defined in Problem 4–24.

4–26. Derive the excitation table of the set-dominant flip-flop defined in Problem 4–23.

4–27. Design a sequential circuit for the state diagram given in Figure 4-32, using JK flip-flops.

4–28. Find the logic diagram for the sequence recognizer circuit having the state table given in Table 4-6. Use D flip-flops.

4–29. Do a logic simulation of your design in Problem 4–28, verifying that only the correct sequence causes the output to go to 1. The input sequence used in the simulation should include all transitions in the state diagram of Figure 4-21. The simulation output should include both the state variables and the output Z.

4–30. Design a sequential circuit with two *JK* flip-flops *A* and *B* and two inputs *E* and *X*. If $E = 0$, the circuit remains in the same state, regardless of the value of *X*. When $E = 1$ and $X = 1$, the circuit goes through the state transitions from 00 to 01 to 10 to 11, back to 00, and then repeats. When $E = 1$ and $X = 0$, the circuit goes through the state transitions from 00 to 11 to 10 to 01, back to 00, and then repeats.

4–31. *The state table for a circuit called a twisted ring counter is given in Table 4-11. This circuit has no inputs, and its outputs are the uncomplemented

☐ TABLE 4-11
State Table for Problem 4–31

Present State	Next State
ABC	ABC
000	100
100	110
110	111
111	011
011	001
001	000

outputs of the flip-flops. Since it has no inputs, it simply goes from state to state whenever a clock pulse occurs.

(a) Design the circuit using *D* flip-flops and assuming that the unspecified next states are don't-care conditions.

(b) Add the necessary logic to the circuit to initialize it to state 000 on power-up master reset.

(c) In the last paragraph of Section 4-6, three techniques for dealing with situations in which a circuit accidentally enters an unused state are discussed. If the circuit you designed in parts (a) and (b) is used in a child's toy, which of the three techniques given would you apply? Justify your decision.

(d) Based on your decision in part (c), redesign the circuit if necessary.

(e) Repeat part (c) for the case in which the circuit is used to control engines on a commercial airliner. Justify your decision.

(f) Repeat part (d) based on your decision in part (e).

4–32. Design the sequential circuit specified by the state table of Table 4-10, using *SR* flip-flops.

4–33. Using the state table part of Table 4-10 (remove the flip-flop input list for *J* and *K*), derive a new list of binary values for the inputs of two *T* flip-flops T_A and T_B. Verify the corresponding *T*-type flip-flop input equations listed in Section 4-7, and draw the logic diagram of the circuit with *T* flip-flops.

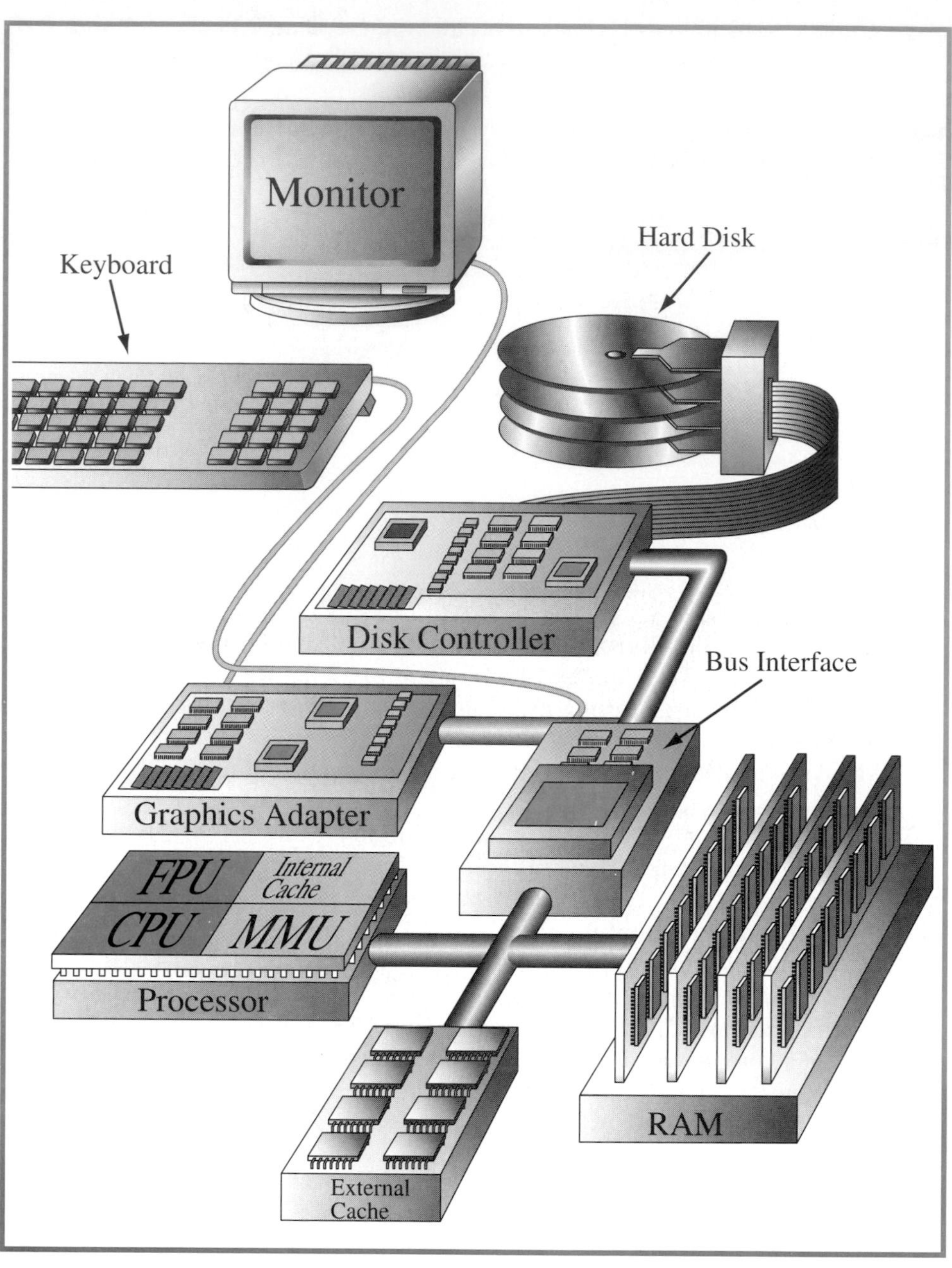

Monitor
Hard Disk
Keyboard
Disk Controller
Bus Interface
Graphics Adapter
FPU
Internal Cache
CPU
MMU
Processor
RAM
External Cache

CHAPTER

5

REGISTERS AND COUNTERS

In Chapter 3, we studied combinational functional blocks. In Chapter 4, we examined sequential circuits. In this chapter, we bring together the two ideas and present sequential functional blocks, generally referred to as registers and counters. In Chapter 4, the circuits that were analyzed or designed did not have any particular structure, and the number of flip-flops was quite small. In contrast, the circuits we shall consider here have more structure, with multiple stages or cells that are identical or close to identical. Also, because of this structure, it is easy to add more stages to produce circuits with many flip-flops compared to the circuits in Chapter 4. Registers are particularly useful for storing information during the processing of data; counters assist us in sequencing the processing.

In the generic computer, registers are used extensively for temporary storage of data in areas other than memory. Registers of this kind are often large, having at least 32 bits. Special registers called shift registers are used less frequently, appearing primary in the input-output parts of the system. Counters are used in the various parts of the computer to control or keep track of the sequencing of activities. Overall, sequential functional blocks are used widely in the generic computer, but more or less so in some areas. As a consequence, the computer is lightly shaded in blue over the electronic portions. Also, memory areas (covered in the next chapter) are unshaded, and shading is heavier in the CPU and FPU parts of the processor, each of which contains large numbers of registers.

5-1 DEFINITION OF REGISTER AND COUNTER

A register includes a set of flip-flops. Since each flip-flop is capable of storing one bit of information, an n-bit register, including n flip-flops, is capable of storing n bits of binary information. By the broadest definition, a *register* consists of a set of flip-flops, together with gates that implement their state transitions. This broad definition includes the various sequential circuits considered in Chapter 4. In a narrower sense, the term *register* is applied to a set of flip-flops, possibly with added combinational gates, that perform data-processing tasks. The flip-flops hold data, and the gates determine the new or transformed data to be transferred into the flip-flops.

A *counter* is a register that goes through a predetermined sequence of states upon the application of clock pulses. The gates in the counter are connected in such a way as to produce the prescribed sequence of binary states. Although counters are special types of registers, it is common to differentiate them from registers.

Registers and counters are sequential functional blocks, which are used extensively in the design of digital systems in general and digital computers in particular. Registers are useful for storing and manipulating information; counters are employed in circuits that sequence and control operations in a digital system.

5-2 Registers

The simplest register is a register that consists of only flip-flops without external gates. Figure 5-1(a) shows such a register constructed with four *D*-type flip-flops. The common *Clock* input triggers all flip-flops on the rising edge of each pulse, and the binary data available at the four *D* inputs are transferred into the 4-bit register. The four *Q* outputs can be sampled to obtain the binary information stored in the register. The $\overline{Clear}$ input goes to the $\overline{R}$ inputs of all four flip-flops and is useful for clearing the register to all 0's prior to its clocked operation. This input is labeled $\overline{Clear}$ rather than *Clear*, since a 0 must be applied to it to cause all flip-flops to reset asynchronously. Activation of the asynchronous $\overline{R}$ inputs to flip-flops during normal clocked operation can lead to circuit designs that are highly delay dependent for their correct operation and that can therefore easily malfunction. Thus, we maintain $\overline{Clear}$ at logic 1 during normal clocked operation, allowing it to be logic 0 only when a system reset is desired. We note that the ability to clear a register to all 0's is optional; whether a clear operation is provided depends on the use of the register in the system.

The transfer of new information into a register is referred to as *loading* the register. If all the bits of the register are loaded simultaneously with a common clock pulse, we say that the loading is done in parallel. A positive clock transition applied to the *Clock* input of thc register of Figure 5-1(a)will load all four *D* inputs into the flip-flops in parallel.

Figure 5-1(b) shows a symbol for the register in Figure 5-1(a). This symbol permits the use of the register in a design hierarchy. The symbol has all of the inputs to the logic circuit on its left and all of the outputs on the right. The inputs include the clock input with the dynamic indicator to represent positive-edge triggering of the flip-flops. We note that the name *Clear* appears inside the symbol, with a bubble in the signal line on the outside of the symbol. This notation indicates that application of a logic 0 to the signal line activates the clear operation on the register flip-flops. If the signal line were labeled outside the symbol, the label would be $\overline{Clear}$.

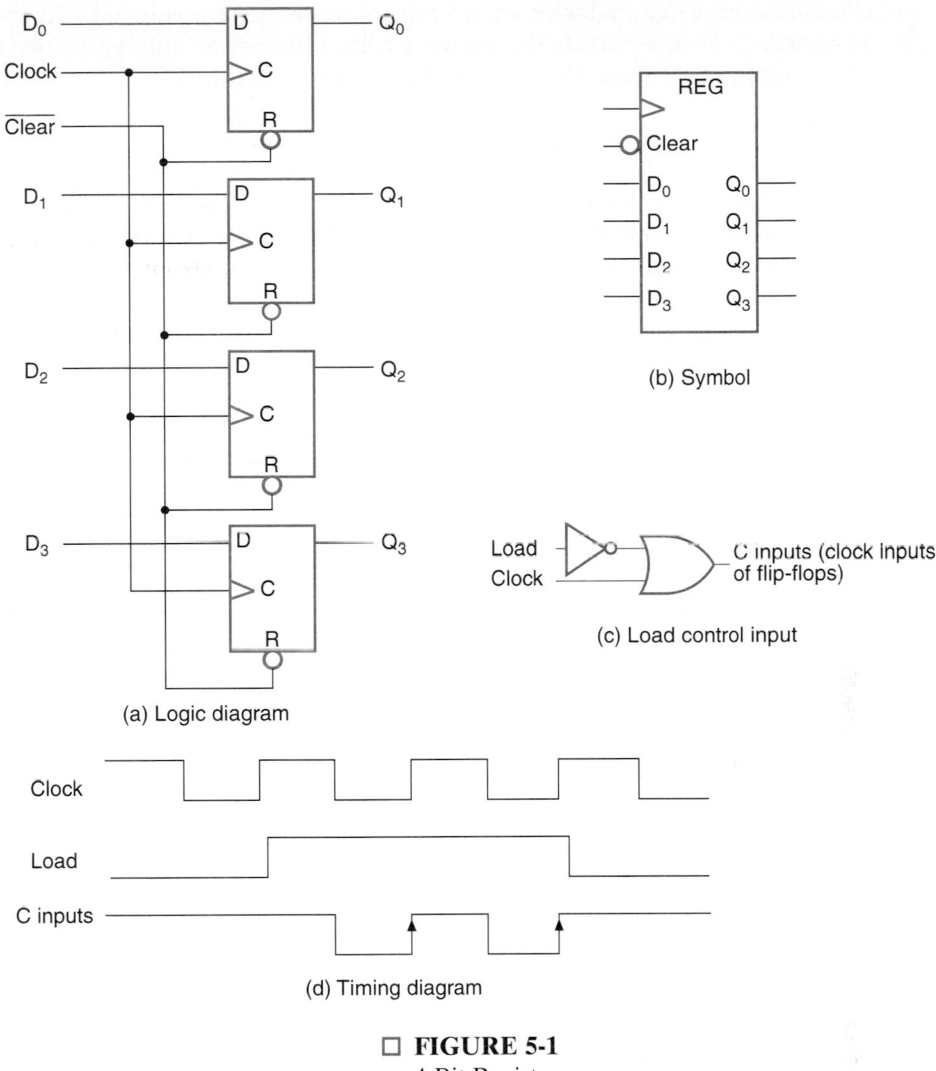

FIGURE 5-1
4-Bit Register

Register with Parallel Load

Most digital systems have a master clock generator that supplies a continuous train of clock pulses. The pulses are applied to all flip-flops and registers in the system. In effect, the master clock acts like a heart that supplies a constant beat to all parts of the system. For the design in Figure 5-1(a), the clock must be prevented from reaching the clock input to the circuit if the contents of the register are to be left unchanged. Thus, a separate control signal is used to govern those clock cycles during which clock pulses will have an effect on the register. The clock pulses are prevented from reaching the register when its contents are not to be changed. This approach can be implemented with a load control input *Load* combined with the clock, as shown in Figure 5-1(c). The output of the OR gate is applied to the *C* inputs of the register flip-flops. The equation for the logic shown is

$$C\ inputs = \overline{Load} + Clock$$

When the *Load* signal is 1, *C inputs* = *Clock*, so the register is clocked normally, and new information can be transferred into the register on the positive transitions of the clock. When the *Load* signal is 0, *C inputs* = 1. With this constant input applied, there are no positive transitions on *C inputs*, so the contents of the register remain unchanged. The effect of the *Load* signal on the signal *C inputs* is shown in Figure 5-1(d). Note that the clock pulses that appear on *C inputs* are pulses to 0 which end with the positive edge that triggers the flip-flops. These pulses and edges appear when *Load* is 1 and are replaced by a constant 1 when *Load* is 0. In order for this circuit to work correctly, *Load* must be constant at the correct value, either 0 or 1, throughout the interval when *Clock* is 0. One condition in which this occurs is if *Load* comes from a flip-flop that is triggered on a positive edge of *Clock*, a normal circumstance if all flip-flops in the system are positive-edge triggered. Since the clock is turned on and off at the register *C* inputs by the use of a logic gate, the technique is referred to as *clock gating*.

Inserting gates in the clock pulse path produces different propagation delays between *Clock* and the inputs of flip-flops having and not having clock gating attached. If clock signals arrive at the flip-flops or registers at different times, *clock skew* is said to exist. But to have a truly synchronous system, we must ensure that all clock pulses arrive simultaneously throughout the system so that all flip-flops trigger at the same time. For this reason, it is advisable to control the operation of the register without using clock gating if possible. Otherwise, control of delays must be employed to drive the clock skew as close to zero as possible.

A 4-bit register with a control input *Load* that is directed through gates into the *D* inputs of the flip-flops, instead of through clock gating on the *C* inputs, is shown in Figure 5-2. The *Load* input to the register determines the action to be taken with each clock pulse. When *Load* is 1, the data on the four inputs is transferred into the register with the next positive transition of a clock pulse. When *Load* is 0, the data inputs are blocked, and the *D* inputs of the flip-flops are connected to their outputs. The feedback connection from output to input of the flip-flop is necessary because the *D* flip-flop does not have a "no change" input condition: With each clock pulse, the *D* input determines the next state of the output. To leave the output unchanged, it is necessary to make the *D* input equal to the present value of the output.

Note that the clock pulses are applied continuously to the *C* inputs. *Load* determines whether the next pulse will accept new information or leave the information in the register intact. The transfer of information from inputs to register is done simultaneously for all four bits during a single positive pulse transition. This method of transfer is usually preferred over clock gating, since it avoids clock skew and the potential for malfunctions of the circuit.

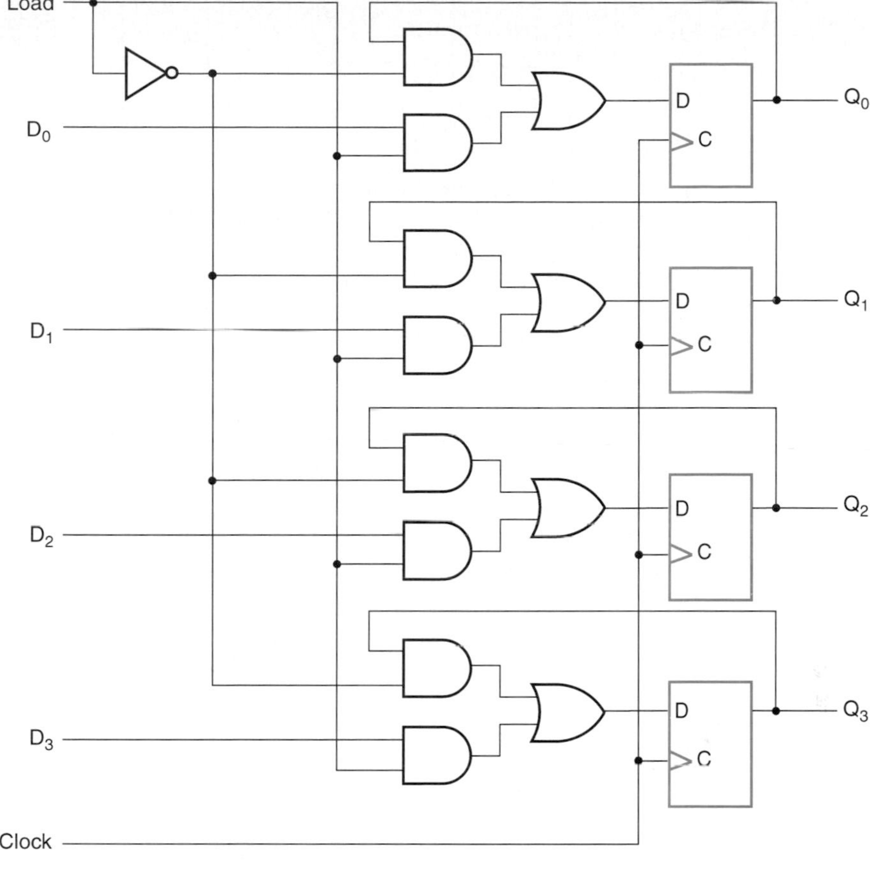

□ **FIGURE 5-2**
4-Bit Register with Parallel Load

5-3 Shift Registers

A register capable of shifting its stored bits laterally in one or both directions is called a *shift register*. The logical configuration of a shift register consists of a chain of flip-flops in cascade, with the output of one flip-flop connected to the input of the next flip-flop. All flip-flops receive a common clock pulse, which activates the shift from each stage to the next.

The simplest possible shift register is one that uses only flip-flops, as shown in Figure 5-3(a). The output of a given flip-flop is connected to the D input of the flip-flop at its right. The clock is common to all flip-flops. The *serial input SI* is the input to the leftmost flip-flop during the shift. The *serial output SO* is taken from the output of the rightmost flip-flop. A symbol for the shift register is given in Figure 5-3(b).

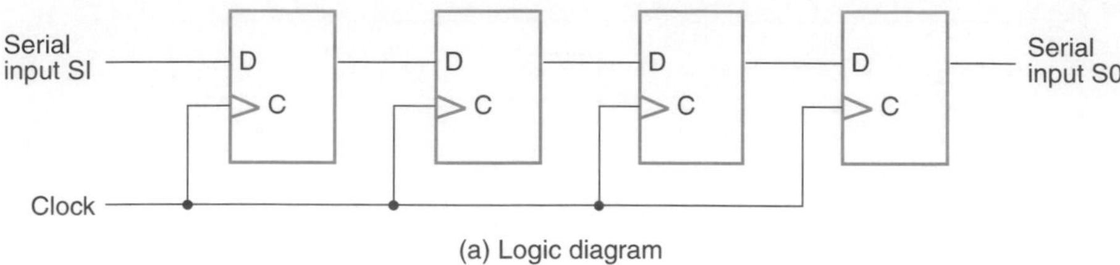

(a) Logic diagram

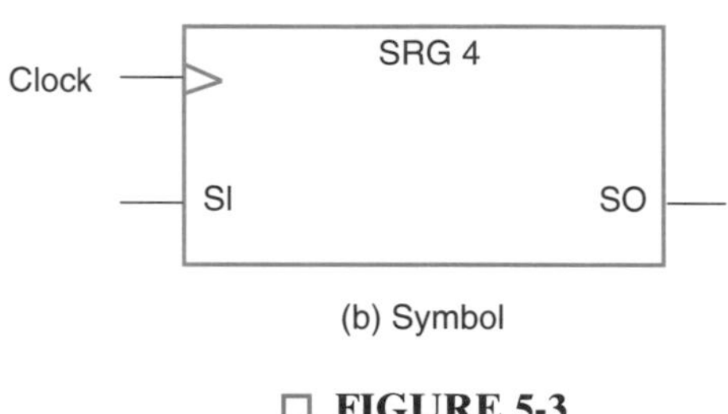

(b) Symbol

□ **FIGURE 5-3**
4-Bit Shift Register

Sometimes it is necessary to control the shift so that it occurs only with certain pulses, but not with others. For the shift register in Figure 5-3, the shift can be controlled by connecting the clock through the logic shown in Figure 5-1(c), with *Shift* replacing *Load*. Again, due to clock skew, this is usually not the most desirable approach. Thus, we learn later that the shift operation can be controlled through the *D* inputs of the flip-flops rather than through the clock inputs *C*.

Serial Transfer

A digital system is said to operate in a serial mode when information in the system is transferred or manipulated one bit at a time. Information is transferred one bit at a time by shifting the bits out of one register and into a second register. This transfer method is in contrast to parallel transfer, in which all the bits of the register are transferred at the same time.

The serial transfer of information from register *A* to register *B* is done with shift registers, as shown in the block diagram of Figure 5-4(a). The serial output of register *A* is connected to the serial input of register *B*. The serial input of register *A* receives 0's while its data are transferred to register *B*. It is also possible for register *A* to receive other binary information, or if we want to maintain the data in register *A*, we can connect its serial output to its serial input so that the information is circulated back into the register. The initial content of register *B* is shifted out through its serial output and is lost unless it is transferred back into register *A*, to a third shift register, or to other storage. The shift control input *Shift* determines when and how many times the registers are shifted. The registers using *Shift* are

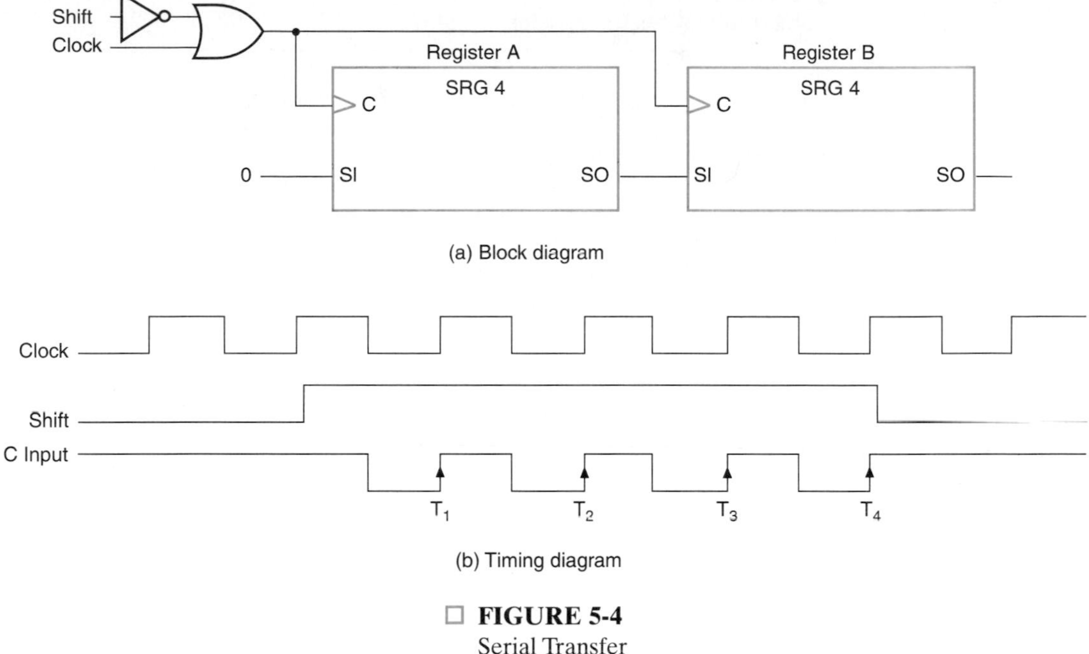

□ **FIGURE 5-4**
Serial Transfer

controlled by means of the logic from Figure 5-1(c), which allows the clock pulses to pass to the shift register clock inputs only when *Shift* has the value logic 1.

In the figure, each shift register has four stages. The logic that supervises the transfer must be designed to enable the shift registers, through the *Shift* signal, for a fixed time of four clock pulses. Shift register enabling is shown in the timing diagram for the clock gating logic in Figure 5-4(b). Four pulses find *Shift* in the active state, so that the output of the logic connected to the clock inputs of the registers produces four pulses: T_1, T_2, T_3, and T_4. Each positive transition of these pulses causes a shift in both registers. After the fourth pulse, *Shift* changes back to 0 and the shift registers are disabled. We note again that, for positive-edge triggering, the pulses on the clock inputs are 0, and the inactive level when no pulses are present is a 1 rather than a 0.

Now suppose that the binary content of register A before the shift is 1011, that of register B is 0010, and the *SI* of register A is logic 0. Then the serial transfer from A to B occurs in four steps, as shown in Table 5-1. With the first pulse T_1, the rightmost bit of A is shifted into the leftmost bit of B, the leftmost bit of A receives a 0 from the serial input, and at the same time, all other bits of A and B are shifted one position to the right. The next three pulses perform identical operations, shifting the bits of A into B one at a time while transferring 0's to A. After the fourth shift, the logic supervising the transfer changes the *Shift* signal to 0 and the shifts stop. Register B contains 1011, which is the previous value of A. Register A contains all 0's.

□ **TABLE 5-1**
Example of Serial Transfer

Timing pulse	Shift Register A				Shift Register B			
Initial value	1	0	1	1	0	0	1	0
After T_1	0	1	0	1	1	0	0	1
After T_2	0	0	1	0	1	1	0	0
After T_3	0	0	0	1	0	1	1	0
After T_4	0	0	0	0	1	0	1	1

The difference between serial and parallel modes of operation should be apparent from this example. In the parallel mode, information is available from all bits of a register, and all bits can be transferred simultaneously during one clock pulse. In the serial mode, the registers have a single serial input and a single serial output, and information is transferred one bit at a time.

Serial Addition

Operations in digital computers are usually done in parallel because of the faster speed attainable. Serial operations are slower, but have the advantage of requiring less hardware. To demonstrate the serial mode of operation, we will show the operation of a serial adder. Also, we compare the serial adder to the parallel counterpart presented in Section 3-8 to illustrate the time-space trade-off in design.

The two binary numbers to be added serially are stored in two shift registers. Bits are added one pair at a time through a single full-adder (FA) circuit, as shown in Figure 5-5. The carry out of the full adder is transferred into a D flip-flop. The output of this carry flip-flop is then used as the carry input for the next pair of significant bits. The sum bit on the S output of the full adder could be transferred into a third shift register, but we have chosen to transfer the sum bits into register A as the contents of the register are shifted out. The serial input of register B can receive a new binary number as its contents are shifted out during the addition.

The operation of the serial adder is as follows: Register A holds the augend, register B holds the addend, and the carry flip-flop has been reset to 0. The serial outputs of A and B provide a pair of significant bits for the full adder at X and Y. The output of the carry flip-flop provides the carry input at Z. When *Shift* is set to 1, the OR gate enables the clock for both registers and the flip-flop. Each clock pulse shifts both registers once to the right, transfers the sum bit from S into the leftmost flip-flop of A, and transfers the carry output into the carry flip-flop. Shift control logic enables the registers for as many clock pulses as there are bits in the registers (four pulses in this example). For each pulse, a new sum bit is transferred to A, a new carry is transferred to the flip-flop, and both registers are shifted once to the right. This process continues until the shift control logic changes *Shift* to 0. Thus, the addition is accomplished by passing each pair of bits and the previous

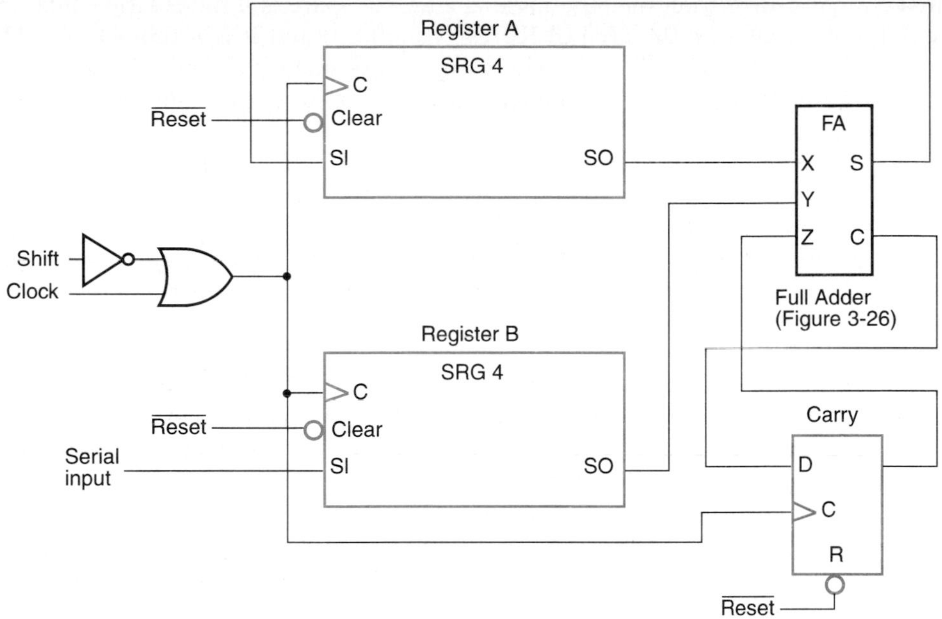

□ **FIGURE 5-5**
Serial Addition

carry through a single full-adder circuit and transferring the sum, one bit at a time, back into register *A*.

Initially we can reset register *A*, register *B*, and the *Carry* flip-flop to 0. Then we shift the first number into *B*. Next, the first number from *B* is added to the 0 in *A*. While *B* is being shifted through the full adder, we can transfer a second number to it through its serial input. The second number can be added to the contents of register *A* at the same time that a third number is transferred serially into register *B*. Serial addition may be repeated to form the addition of two, three, or more numbers, with their sum accumulated in register *A*.

A comparison of the serial adder with the parallel adder described in Section 3-8 provides an example of space-time trade-off. The parallel adder has *n* full adders for *n*-bit operands, whereas the serial adder requires only one full adder. Excluding the registers from both, the parallel adder is a combinational circuit, whereas the serial adder is a sequential circuit because it includes the carry flip-flop. The serial circuit also takes *n* clock cycles to complete an addition. Identical circuits, such as the *n* full adders in the parallel adder, connected together in a chain constitute an example of an *iterative logic array*. If the values on the carries between the full adders are regarded as state variables, then the states from the least significant end to the most significant end are the same as the states appearing in sequence on the flip-flop output in the serial adder. Note that in the iterative logic array the states appear in space, but in the sequential circuit the states appear in time. By converting from one of these implementations to the other, one can make a space-time trade-off. The parallel adder in space is n times larger than the serial adder (ignoring the area of the carry flip-flop), but it is n times faster. The serial adder, although it is n times slower, is n times smaller in space. This gives the designer a significant choice in emphasizing speed or area, where more area translates into more cost.

Shift Register with Parallel Load

If all the flip-flop outputs of a shift register are accessible, then information entered serially by shifting can be taken out in parallel from the outputs of all of the flip-flops. If a parallel load capability is also added to a shift register, then data entered in parallel can be taken out in serial fashion by shifting out the data in the register. Thus, a shift register with accessible flip-flop outputs and parallel load can be used for converting incoming parallel data to outgoing serial data and vice versa.

The logic diagram of a 4-bit shift register with parallel load and the symbol for this register are shown in Figure 5-6. There are two control inputs, one for the shift and the other for the load. Each stage of the register consists of a D flip-flop, an OR gate, and three AND gates. The first AND gate enables the shift operation. The second AND gate enables the input data. The third AND gate restores the contents of the register when no operation is required.

The operation of this register is specified in Table 5-2. When both the shift and load control inputs are 0, the third AND gate in each stage is enabled, and the output of each flip-flop is applied to its D input. A positive transition of the clock restores the contents of the register, and no change in output occurs. When the shift input is 0 and the load input is 1, the second AND gate in each stage is enabled, and the input D is applied to the corresponding D input of the flip-flop. The next positive clock transition transfers the input data into the register. When the shift input is equal to 1, the first AND gate in each stage is enabled and the other two are disabled. Since the Load input is disabled by $\overline{\text{Shift}}$ into the second AND gate, we mark it with a don't-care condition in the table. The shift operation causes the data from the serial input SI to be transferred to flip-flop Q_0, the output of Q_0 to be transferred to flip-flop Q_1, and so on down the line on a positive transition of the clock. Note that because of the way the circuit is drawn, the shifting occurs in the downward direction. If we rotate the page a quarter turn counterclockwise, the register will be shifting from left to right.

Shift registers are often used to interface digital systems that are situated remotely from each other. For example, suppose it is necessary to transmit an n-bit

□ **TABLE 5-2**
Function Table for the Register of Figure 5-6

Shift	Load	Operation
0	0	No change
0	1	Load parallel data
1	×	Shift down from Q_0 to Q_3

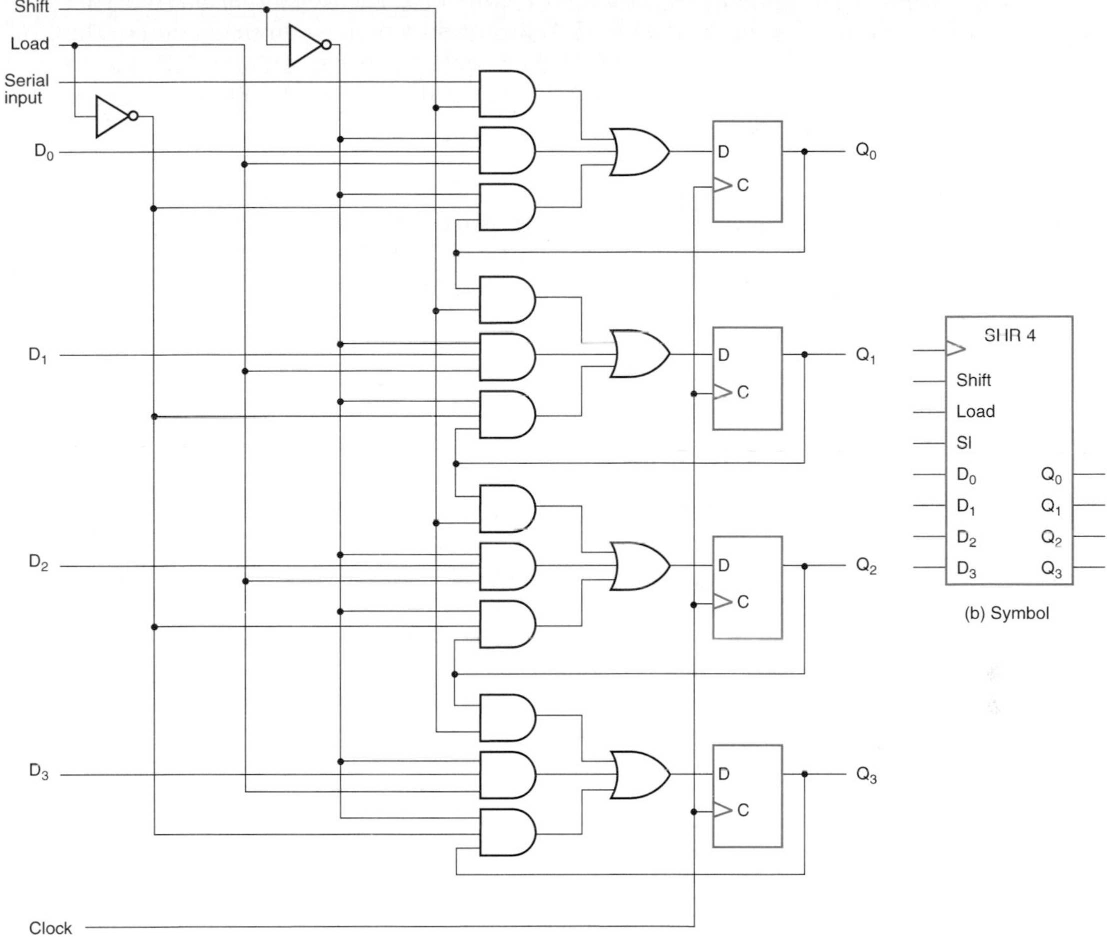

□ **FIGURE 5-6**
Shift Register with Parallel Load

quantity between two points. If the distance is far, it will be expensive to use n lines to transmit the n bits in parallel. It may be more economical to use a single line and transmit the information serially, one bit at a time. The transmitter accepts the n-bit data in parallel into a shift register and then transmits the data serially along the common line. The receiver accepts the data serially into a shift register. When all n bits are accumulated, they can be taken in parallel from the outputs of the register. Thus, the transmitter performs a parallel-to-serial conversion of data, and the receiver does a serial-to-parallel conversion.

Bidirectional Shift Register

A register capable of shifting in one direction only is called a *unidirectional shift register*. A register that can shift in both directions is called a *bidirectional shift reg-*

ister. It is possible to modify the circuit of Figure 5-6 by adding a fourth AND gate in each stage, for shifting the data in the upward direction. An investigation of the resultant circuit will reveal that the four AND gates, together with the OR gate in each stage, constitute a multiplexer with the selection inputs controlling the operation of the register.

□ **TABLE 5-3**
Function Table for the Register of Figure 5-7

Mode control		Register Operation
S_1	S_0	
0	0	No change
0	1	Shift down
1	0	Shift up
1	1	Parallel load

One stage of a bidirectional shift register with parallel load is shown in Figure 5-7(a). Each stage consists of a D flip-flop and a 4–to–1-line multiplexer. The two selection inputs S_1 and S_0 select one of the multiplexer inputs for the D flip-flop. The selection lines control the mode of operation of the register according to the function table of Table 5-3. When the mode control $S_1S_0 = 00$, input 0 of the multiplexer is selected. This forms a path from the output of each flip-flop into its own input. The next clock transition transfers into each flip-flop the binary value it held previously, and no change of state occurs. When $S_1S_0 = 01$, the terminal marked 1 on the multiplexer has a path to the D input of each flip-flop. These paths cause a shift-down operation. The serial input is transferred into the first stage, and the contents of each stage Q_{i-1}, are transferred into stage Q_i. When $S_1S_0 = 10$, a shift-up operation results in another serial input going into the last stage. In addition, the value in each stage Q_{i+1} is transferred into stage Q_i. Finally, when $S_1S_0 = 11$, the binary information on each parallel input line is transferred into the corresponding flip-flop, resulting in a parallel load.

Figure 5-7(b) shows a symbol for the bidirectional shift register in Figure 5-7(a). Note that both a left serial input (LSI) and a right serial input (RSI) are provided. If serial outputs are desired, Q_3 is used as the left one and Q_0 as the right one.

5-4 Ripple Counter

A register that goes through a prescribed sequence of states upon the application of input pulses is called a *counter*. The input pulses may be clock pulses or may originate from some other source, and they may occur at fixed intervals of time or at random intervals. In our discussion of counters, we assume clock pulses, but these other signals can be substituted for the clock. The sequence of states may fol-

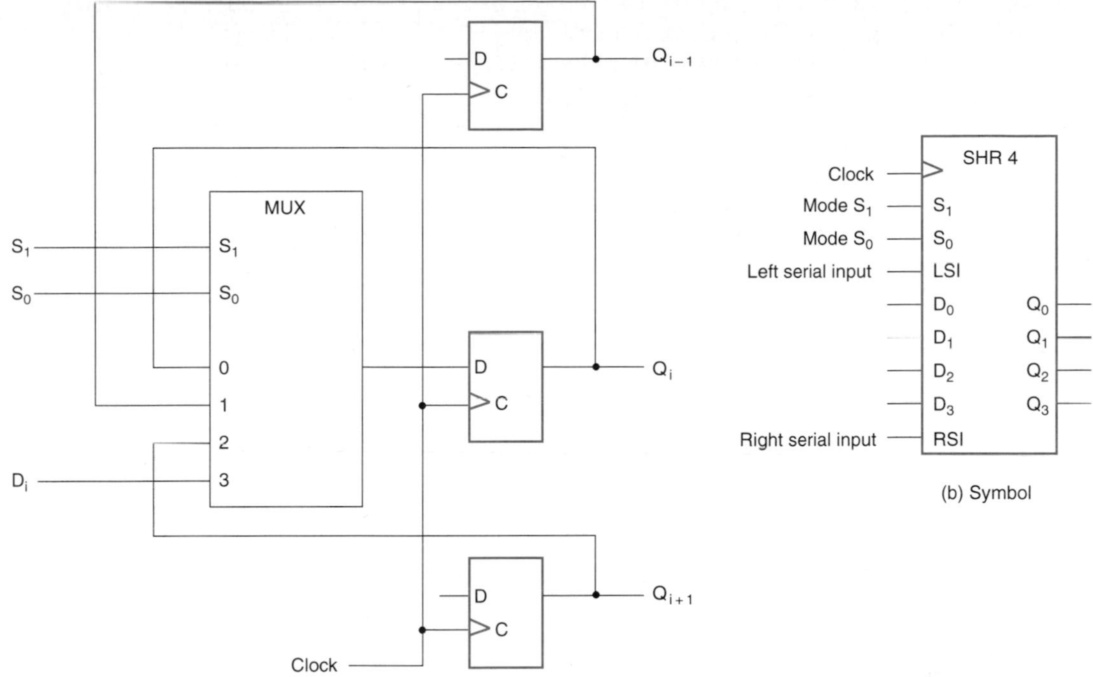

(a) Logic diagram of one typical stage

(b) Symbol

□ **FIGURE 5-7**
Bidirectional Shift Register with Parallel Load

low the binary number sequence or any other sequence of states. A counter that follows the binary number sequence is called a *binary counter*. An n-bit binary counter consists of n flip-flops and can count in binary from 0 through $2^n - 1$.

Counters are available in two categories: ripple counters and synchronous counters. In a ripple counter, the flip-flop output transition serves as a source for triggering other flip-flops. In other words, the C input of some or all of the flip-flops is triggered not by the common clock pulses, but rather by the transition that occurs in other flip-flop outputs. In a synchronous counter, the C inputs of all of the flip-flops receive the common clock pulse, and the change of state is determined from the present state of the counter. Synchronous counters are discussed in the next two sections. Here we present the binary ripple counter and explain its operation.

The logic diagram of a 4-bit binary ripple counter is shown in Figure 5-8. The counter is constructed with flip-flops capable of complementing their contents, such as JK. The output of each flip-flop is connected to the C input of the next flip-flop in sequence. The flip-flop holding the least significant bit receives the incoming clock pulses. The J and K inputs of all the flip-flops are connected to a permanent logic 1. Also, the bubble in front of the dynamic indicator symbol next to the C

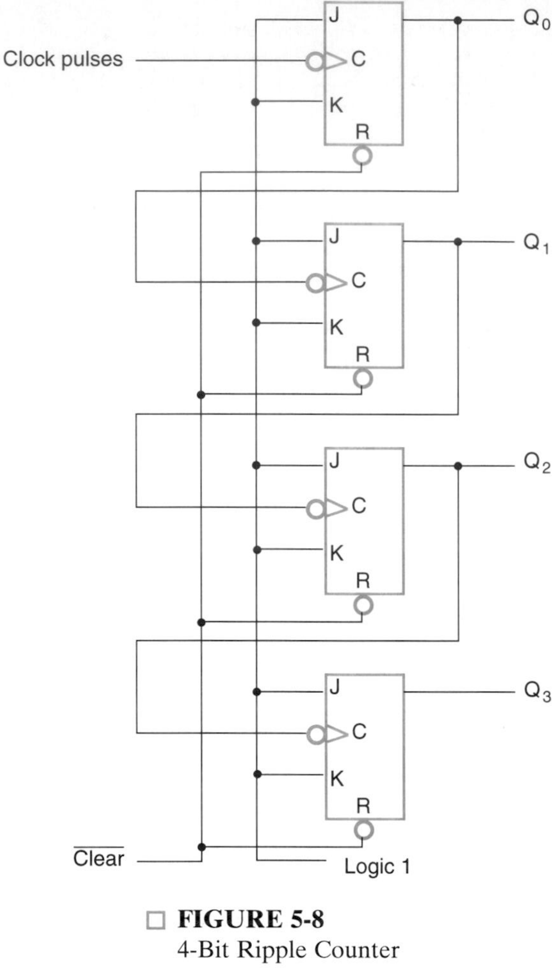

FIGURE 5-8
4-Bit Ripple Counter

label indicates that the flip-flops respond to the negative-edge transition of the input. J and K's being connected to logic 1 and negative-edge triggering make each flip-flop complement its value if the signal on its C input goes through a negative transition. The negative transition occurs when the output of the previous flip-flop, to which C is connected, goes from 1 to 0. A 0-level signal on $\overline{Clear}$ driving the $\overline{R}$ inputs clears the register to 0 asynchronously.

To understand the operation of a binary counter, let us examine the upward counting sequence given in the left half of Table 5-4. The count starts with binary 0 and increments by one with each count pulse input. After the count of 15, the counter goes back to 0 to repeat the count. The least significant bit (Q_0) is complemented with each count pulse input. Every time that Q_0 goes from 1 to 0, it complements Q_1. Every time that Q_1 goes from 1 to 0, it complements Q_2. Every time that Q_2 goes from 1 to 0, it complements Q_3, and so on for any higher order bits in

□ **TABLE 5-4**
Counting Sequence of Binary Counter

Upward Counting Sequence				Downward Counting Sequence			
Q_3	Q_2	Q_1	Q_0	Q_3	Q_2	Q_1	Q_0
0	0	0	0	1	1	1	1
0	0	0	1	1	1	1	0
0	0	1	0	1	1	0	1
0	0	1	1	1	1	0	0
0	1	0	0	1	0	1	1
0	1	0	1	1	0	1	0
0	1	1	0	1	0	0	1
0	1	1	1	1	0	0	0
1	0	0	0	0	1	1	1
1	0	0	1	0	1	1	0
1	0	1	0	0	1	0	1
1	0	1	1	0	1	0	0
1	1	0	0	0	0	1	1
1	1	0	1	0	0	1	0
1	1	1	0	0	0	0	1
1	1	1	1	0	0	0	0

the ripple counter. For example, consider the transition from count 0011 to 0100. Q_0 is complemented with the count pulse. Since Q_0 goes from 1 to 0, it triggers Q_1 and complements it. As a result, Q_1, goes from 1 to 0, which complements Q_2, changing it from 0 to 1. Q_2 does not trigger Q_3, because Q_2 produces a positive transition, and the flip-flops respond only to negative transitions. Thus, the count from 0011 to 0100 is achieved by changing the bits one at a time so that the counter goes from 0011 to 0010 (Q_0 from 1 to 0), then to 0000 (Q_1 from 1 to 0), and finally to 0100 (Q_2 from 0 to 1). The flip-flops change one at a time in succession, and the signal propagates through the counter in a ripple fashion from one stage to the next.

A ripple counter that counts downward gives the sequence in the right half of Table 5-4. Downward counting can be accomplished by connecting the complement output of each flip-flop to the C input of the next flip-flop or by using positive-edge-triggered flip-flops. (See Problem 5–14 at the end of the chapter.)

The advantage of ripple counters is their simple hardware. But they are asynchronous circuits and, with added logic, can be unreliable and delay dependent. This is particularly true for logic that provides feedback paths from counter outputs to counter inputs. Also, due to the length of time required for the ripple to occur, large ripple counters are slow circuits. These properties are incompatible with modern system design, and as a consequence, synchronous binary counters are favored in most designs.

5-5 Synchronous Binary Counters

Synchronous counters are different from ripple counters in that, in synchronous counters, clock pulses are applied to the inputs of all of the flip-flops. Thus, the common clock pulse triggers all flip-flops simultaneously rather than one at a time, as happens in a ripple counter. For *JK* flip-flops, the decision whether the flip-flop is complemented or not is determined by the *J* and *K* data inputs. If these are both equal to 0, the flip-flop does not change state, even in the presence of a pulse. If they are both equal to 1, the flip-flop complements its value on the clock transition.

Design of Binary Counter

The design procedure for a synchronous counter is the same as with any other synchronous sequential circuit. A counter may operate without an external input, except for the clock pulses. The output of the counter is taken from the outputs of the flip-flops without any additional outputs from gates. In the absence of inputs, the state table of a counter will consist of columns for the present state and next state only. The outputs will be implicitly represented by the present-state column.

The state table of a 4-bit binary counter is listed in the present-state and next-state columns of Table 5-5. The present-state column shows the 16 states from 0000 to 1111. The next state is equal to the value of the present state plus one. When the circuit reaches the state 1111, it goes to the next state, 0000, which causes the counter to repeat its cycle.

TABLE 5-5
State Table and Flip-Flop Inputs for Binary Counter

Present state				Next state				Flip-flop inputs							
Q_3	Q_2	Q_1	Q_0	Q_3	Q_2	Q_1	Q_0	J_{Q3}	K_{Q3}	J_{Q2}	K_{Q2}	J_{Q1}	K_{Q1}	J_{Q0}	K_{Q0}
0	0	0	0	0	0	0	1	0	×	0	×	0	×	1	×
0	0	0	1	0	0	1	0	0	×	0	×	1	×	×	1
0	0	1	0	0	0	1	1	0	×	0	×	×	0	1	×
0	0	1	1	0	1	0	0	0	×	1	×	×	1	×	1
0	1	0	0	0	1	0	1	0	×	×	0	0	×	1	×
0	1	0	1	0	1	1	0	0	×	×	0	1	×	×	1
0	1	1	0	0	1	1	1	0	×	×	0	×	0	1	×
0	1	1	1	1	0	0	0	1	×	×	1	×	1	×	1
1	0	0	0	1	0	0	1	×	0	0	×	0	×	1	×
1	0	0	1	1	0	1	0	×	0	0	×	1	×	×	1
1	0	1	0	1	0	1	1	×	0	0	×	×	0	1	×
1	0	1	1	1	1	0	0	×	0	1	×	×	1	×	1
1	1	0	0	1	1	0	1	×	0	×	0	0	×	1	×
1	1	0	1	1	1	1	0	×	0	×	0	1	×	×	1
1	1	1	0	1	1	1	1	×	0	×	0	×	0	1	×
1	1	1	1	0	0	0	0	×	1	×	1	×	1	×	1

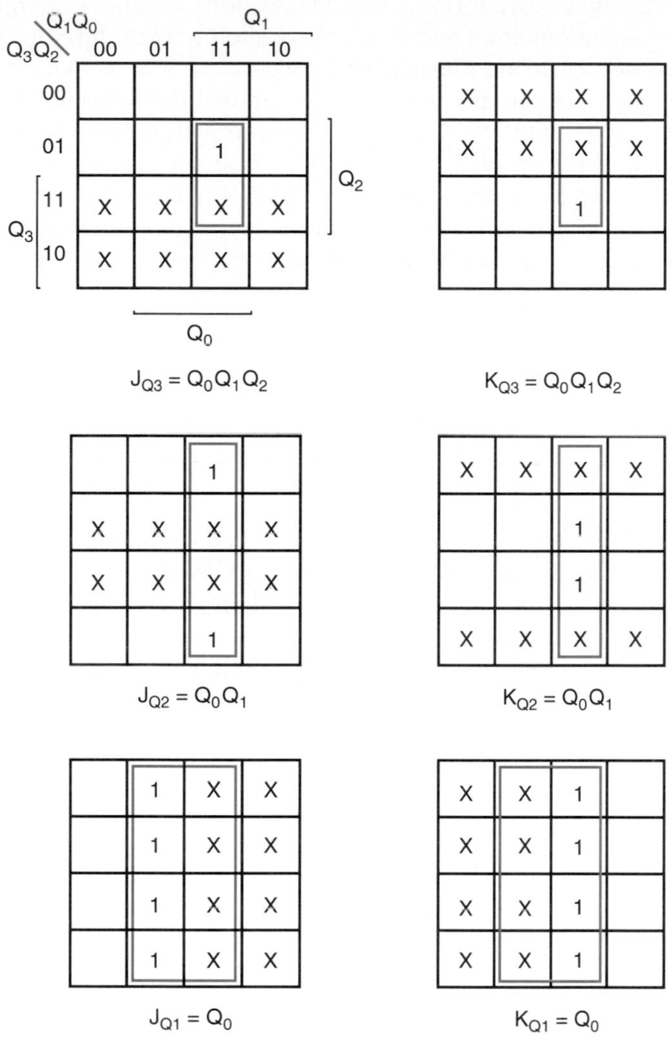

□ **FIGURE 5-9**
Maps for Input Equations of a Binary Counter

Binary counters are most efficiently constructed with complementing T or JK flip-flops. They also can be designed with D flip-flops. We will design a counter with JK flip-flops first and then will repeat the design with D flip-flops.

The procedure for designing a sequential circuit with JK flip-flops was presented in Section 4-7. We use the same procedure here to obtain the circuit of the binary counter. First we obtain the flip-flop input values for each J and K and list them in Table 5-5. These values come from the excitation conditions given in Table 4-9. Then we find simplified input equations by means of maps, as shown in Figure 5-9. The maps for the least significant flip-flop are not drawn. These two maps contain only 1's and don't-care terms, which makes both J_{Q0} and K_{Q0} equal to 1. Note that the input equations for J and K are the same for each flip-flop. This indicates that T flip-flops could be used instead of JK's.

In many applications, it is necessary to control the operation of the counter with a count-enable input. We designate this input with the variable EN. Note that EN is a variable made up of two symbols. When such a variable is used in a Boolean equation, it may be difficult to distinguish EN from E AND N. To denote the latter, we will use $E \cdot N$. With EN as the enable input, the flip-flop input equations for the binary counter can be expressed as follows:

$$J_{Q0} = K_{Q0} = EN$$

$$J_{Q1} = K_{Q1} = Q_0 \cdot EN$$

$$J_{Q2} = K_{Q2} = Q_0 \cdot Q_1 \cdot EN$$

$$J_{Q3} = K_{Q3} = Q_0 \cdot Q_1 \cdot Q_2 \cdot EN$$

When $EN = 0$, all J and K inputs are equal to 0, and the flip-flops remain in the same state, even in the presence of clock pulses. When $EN = 1$, the first input equation becomes $J_{Q0} = K_{Q0} = 1$, and the other input equations reduce to the equations derived in the maps of Figure 5-9. The flip-flop in the least significant position of a synchronous binary counter is complemented with every clock pulse transition. A flip-flop in any other position is complemented with a clock pulse transition if all previous least significant bits are equal to 1. In an n-bit binary counter, the input equation for flip-flop Q_i at any stage for $i = 1, 2, 3, \ldots, n$, is

$$J_{Qi} = K_{Qi} = Q_0 \cdot Q_1 \cdot Q_2 \cdot \ldots \cdot Q_{i-1} \cdot EN$$

Synchronous binary counters have a regular pattern, as can be seen from the 4-bit binary counter shown in Figure 5-10. The C inputs of all flip-flops receive the common clock pulses. The first stage Q_0 is complemented when the counter is enabled, with input EN being a 1. The other J and K inputs are equal to 1 if all previous least significant stages are equal to 1 and the count is enabled. The chain of AND gates generates the required logic for the J and K inputs. The carry output CO can be used to extend the counter to more stages, with each stage having an additional flip-flop and an AND gate.

Note that the flip-flops trigger on the positive-edge transition of the clock. The polarity of the clock is not essential here, as it was with the ripple counter. The synchronous counter can be triggered with either the positive or the negative clock transition. The symbol for the 4-bit counter using positive-edge triggering is shown in Figure 5-10(b).

Counter with *D* Flip-Flops

A binary counter can be designed with D flip-flops by following the design procedure outlined in Section 4-6. The input equations for the D flip-flops are obtained

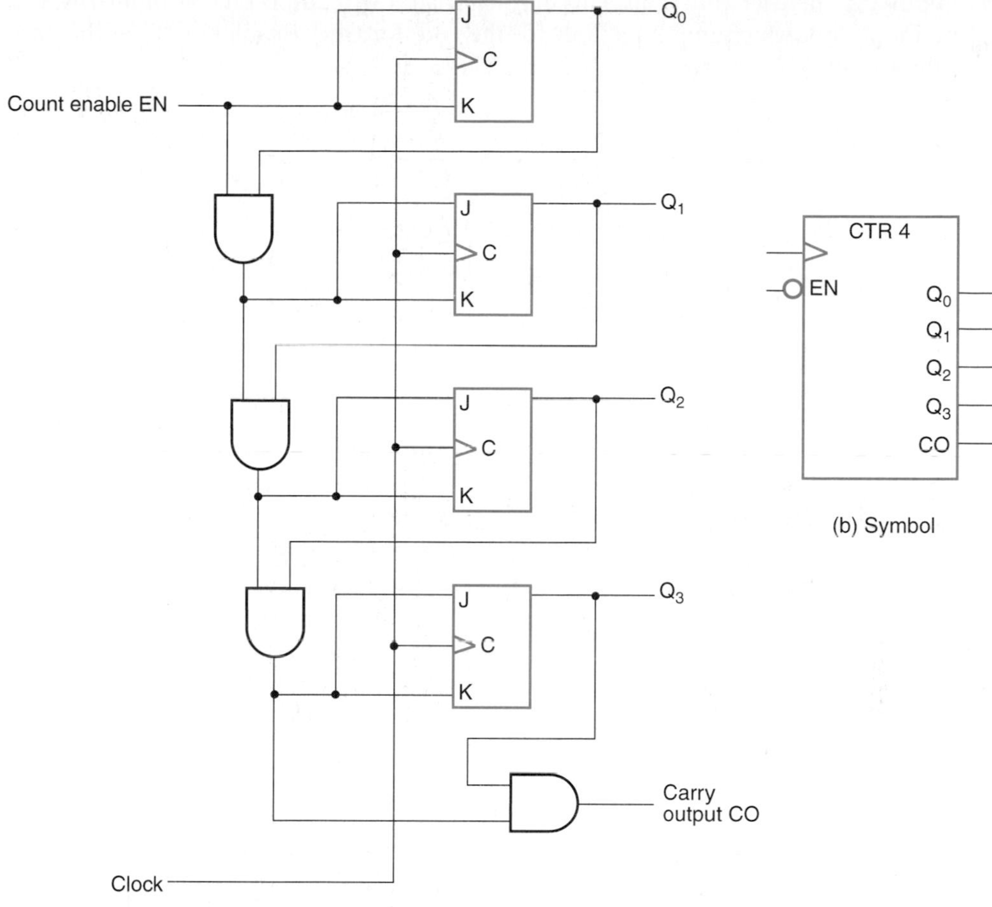

(b) Symbol

(a) Logic diagram

□ **FIGURE 5-10**
4-Bit Synchronous Binary Counter

directly from the next-state values in Table 5-5. The equations can be expressed in sum-of-minterms form as a function of the present state:

$$D_{Q0}(Q_3, Q_2, Q_1, Q_0) = \Sigma\, m(0, 2, 4, 6, 8, 10, 12, 14)$$

$$D_{Q1}(Q_3, Q_2, Q_1, Q_0) = \Sigma\, m(1, 2, 5, 6, 9, 10, 13, 14)$$

$$D_{Q2}(Q_3, Q_2, Q_1, Q_0) = \Sigma\, m(3, 4, 5, 6, 11, 12, 13, 14)$$

$$D_{Q3}(Q_3, Q_2, Q_1, Q_0) = \Sigma\, m(7, 8, 9, 10, 11, 12, 13, 14)$$

Simplifying the four functions with maps and adding a count enable input EN, we obtain the following input equations for the counter (see Problem 5–15 at the end of the chapter):

$$D_{Q0} = Q_0 \oplus EN$$

$$D_{Q1} = Q_1 \oplus (Q_0 \cdot EN)$$

$$D_{Q2} = Q_2 \oplus (Q_0 \cdot Q_1 \cdot EN)$$

$$D_{Q3} = Q_3 \oplus (Q_0 \cdot Q_1 \cdot Q_2 \cdot EN)$$

The input equation for any flip-flop Q_i in state i can be expressed as

$$D_{Qi} = Q_i \oplus (Q_0 \cdot Q_1 \cdot Q_2 \cdot \ldots \cdot Q_{i-1} \cdot EN)$$

The logic diagram of a 4-bit binary counter with D flip-flop is shown in Figure 5-11(a). Note that the circuit has the same chain of AND gates for the carry, as before. The only difference between the circuit of Figure 5-11(a) and the counter with JK flip-flops shown in Figure 5-10 is the exclusive-OR gate on each flip-flop input. The exclusive-OR indicates that the D flip-flop, together with the exclusive-OR gate, corresponds to a T-type flip-flop. One input of the exclusive-OR gate comes from the output of the D flip-flop, and the other input of the gate is equivalent to an input for a T or JK flip-flop. (See Problem 5–16 at the end of the chapter.)

Serial and Parallel Counters

We will use the synchronous counter in Figure 5-11 to demonstrate two alternative designs for binary counters. In Figure 5-11(a), a chain of 2-input AND gates is used to provide information to each stage about the state of prior stages of the counter. This is analogous to the carry logic in the ripple carry adder. A counter that uses such logic is said to have *serial gating* and is referred to as a *serial counter*. The analogy to the ripple carry adder suggests that there might be logic for counters analogous to carry lookahead. Such logic is shown in Figure 5-11(b). This logic can simply replace that in the blue box in Figure 5-11(a) to produce a counter having *parallel gating*, called a *parallel counter*. The advantage of parallel gating logic is that, in going from state 1111 to state 0000, only one AND gate delay occurs instead of the four AND gate delays that occur in the serial counter. This reduction in delay allows the counter to operate much faster. The equations for parallel gating can be taken directly from the AND terms in the flip-flop input equations given previously for the synchronous counter with D flip-flops. The same technique can be used with counters implemented with T or JK flip-flops.

If we connect two 4-bit parallel counters together by connecting the CO output of one to the EN input of the other, the 8-bit counter that results is a serial-parallel counter. This counter has two 4-bit parallel parts connected in series with each other. The idea can be extended to counters of any length. Again, employing the

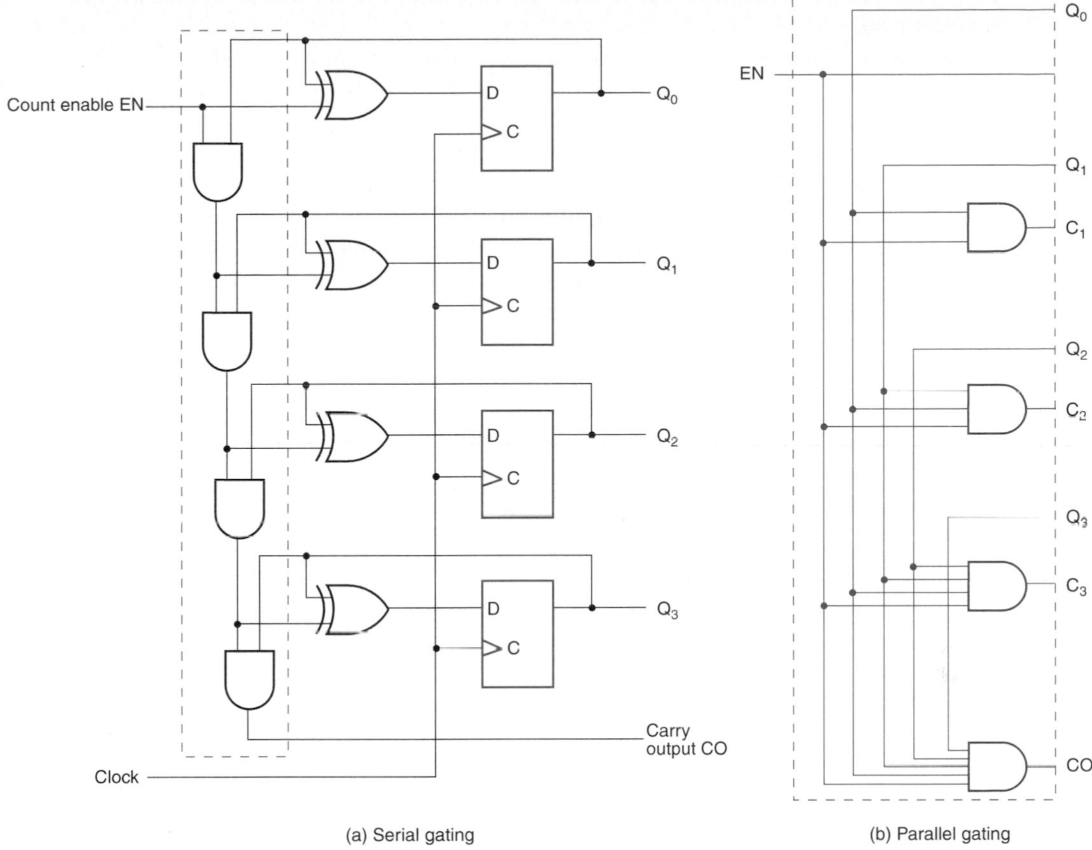

□ **FIGURE 5-11**
4-Bit Binary Counter with *D* Flip-Flops

analogy to carry lookahead adders, additional levels of gating logic can be introduced to replace the serial connections between the 4-bit segments. The further reduction in delay that results is useful for constructing large, fast counters.

Up-Down Binary Counter

A synchronous count-down binary counter goes through the binary states in reverse order from 1111 to 0000 and back to 1111 to repeat the count. It is possible to design a count-down counter in the usual manner, but the result is predictable from an inspection of the binary count-down. The bit in the least significant position is complemented with each count pulse. A bit in any other position is complemented if all lower significant bits are equal to 0. For example, the next state after the present state of 0100 is 0011. The least significant bit is always complemented. The second significant bit is complemented because the first bit is 0. The third significant bit is complemented because the first two bits are equal to 0. But the fourth bit does not change, because it is not the case that all lower significant bits are equal to 0.

The logic diagram of a synchronous count-down binary counter is similar to the circuit of the binary up-counter, except that the inputs to the AND gates must come from the complement outputs of the flip-flops. The two operations can be combined to form a counter that can count both up and down, which is referred to as an up-down binary counter. The up-down counter needs a mode input to select between the two operations. We designate this mode select input by S, with $S = 1$ for up-counting and $S = 0$ for down-counting. Let variable EN be a count enable input, with $EN = 1$ for normal up- or down-counting and $EN = 0$ for disabling both counts. If we use T-type flip-flops (that is, $T = J = K$), a 4-bit, up-down binary counter can be described by the following input equations:

$$T_{A0} = EN$$

$$T_{A1} = Q_0 \cdot S \cdot EN + \overline{Q}_0 \cdot \overline{S} \cdot EN$$

$$T_{A2} = Q_0 \cdot Q_1 \cdot S \cdot EN + \overline{Q}_0 \cdot \overline{Q}_1 \cdot \overline{S} \cdot EN$$

$$T_{A3} = Q_0 \cdot Q_1 \cdot Q_2 \cdot S \cdot EN + \overline{Q}_0 \cdot \overline{Q}_1 \cdot \overline{Q}_2 \cdot \overline{S} \cdot EN$$

The carry outputs for the next stage are:

$$C_{up} = Q_0 \cdot Q_1 \cdot Q_2 \cdot Q_3 \cdot S \cdot EN \qquad \text{For upward counting}$$

$$C_{dn} = \overline{Q}_0 \cdot \overline{Q}_1 \cdot \overline{Q}_2 \cdot \overline{Q}_3 \cdot \overline{S} \cdot EN \qquad \text{For downward counting}$$

The carry outputs supply the input for complementing the next flip-flop if the counter is extended to a fifth stage. The logic diagram of the circuit can be easily obtained from the input equations, but is not included here. (See, however, Problem 5–19 at the end of the chapter.)

Binary Counter with Parallel Load

Counters employed in digital systems quite often require a parallel-load capability for transferring an initial binary number into the counter prior to the count operation. Figure 5-12(a) shows the logic diagram of a register that has a parallel-load capability and that can also operate as a counter. Figure 5-12(b) shows the corresponding symbol. When equal to 1, the input load control disables the count operation and causes a transfer of data from the four parallel inputs into the four flip-flops. If the load input is 0 and the count input is 1, the circuit operates as a binary counter. The carry output CO becomes a 1 if all four flip-flops are equal to 1 while the count input is enabled. This is the condition for complementing the flip-flop that contains the next significant bit. The output is useful for expanding the counter to more stages.

The operation of the circuit is summarized in Table 5-6. With the load and count inputs both at 0, the outputs do not change, even when pulses are applied to the C inputs. If the load input is maintained at logic 0, the count input controls the operation of the counter, and the outputs change to the next binary count for each

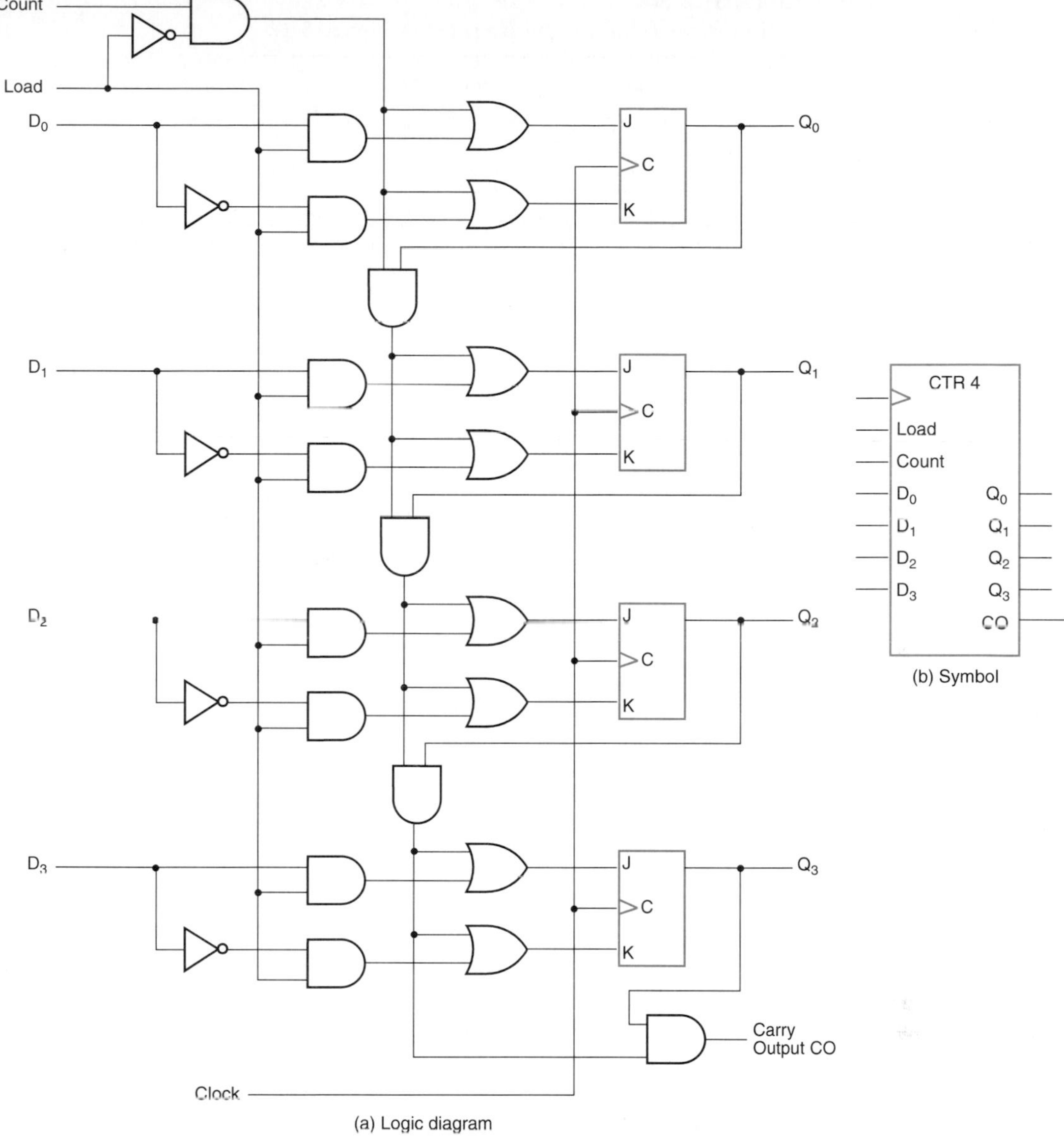

□ **FIGURE 5-12**
4-Bit Binary Counter with Parallel Load

positive transition of the clock. The input data are loaded into the flip-flops when the load control input is equal to 1, regardless of the value of the count input, because the count input is inhibited when the load input is active.

Counters with parallel load are very useful in the design of digital computers. In subsequent chapters, we will refer to them as registers with load and increment

☐ **TABLE 5-6**
Function Table for the Register of Figure 5-12

Load	Count	Operation
0	0	No change
0	1	Count next binary state
1	×	Load inputs

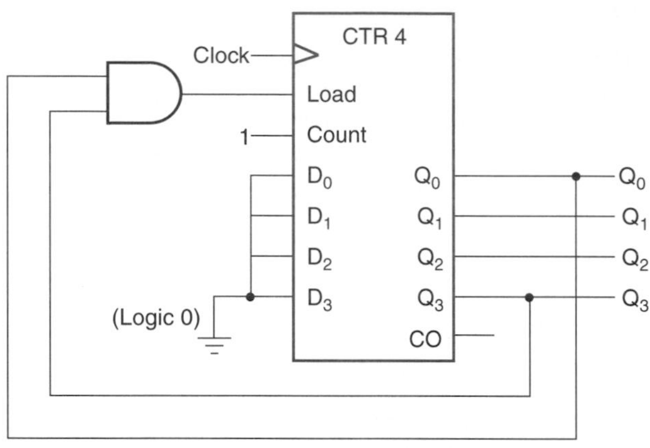

☐ **FIGURE 5-13**
BCD Counter

operations. The *increment* operation adds one to the contents of a register. By enabling the count input during one clock period, the content of the register can be incremented by one.

The binary counter with parallel load can be converted into a synchronous BCD counter (without load input) by connecting an external AND gate to it, as shown in Figure 5-13. The counter starts with an all-zero output, and the count input is active at all times. As long as the output of the AND gate is 0, each positive clock pulse transition increments the counter by one. When the output reaches the count of 1001, both Q_0 and Q_3 become 1, making the output of the AND gate equal to 1. This condition makes *Load* active; so on the next clock transition, the counter does not count, but is loaded from its four inputs. Since all four inputs are connected to logic 0, 0000 is loaded into the counter following the count of 1001. Thus, the circuit counts from 0000 through 1001 and then back to 0000, as required for a BCD counter.

5-6 Other Counters

Counters can be designed to generate any desired number of states in sequence. A *divide-by-N counter* (also known as a *modulo-N counter*) is a counter that goes through a repeated sequence of N states. The sequence may follow the binary count or may be any other arbitrary sequence. In either case, the design of the counter follows the procedure presented in Chapter 4 for the design of synchronous sequential circuits. To demonstrate this procedure, we will present the design of two counters: a BCD counter and a counter with an arbitrary sequence of states.

BCD Counter

As shown in the previous section, a BCD counter can be obtained from a binary counter with parallel load. It is also possible to design a BCD counter using individual flip-flops and gates. Assuming T-type flip-flops for the counter, we derive the state table and input conditions as shown in Table 5-7. The T input in each case is equal to 1 if the flip-flop is complemented during the transition from present state to next state. The T input is equal to 0 if the next state is the same as the present state. An output Y is included in the table. This output is equal to 1 when the present state is 1001. In this way, Y can enable the count of the next decade while its own decade switches from 1001 to 0000.

□ **TABLE 5-7**
State Table and Flip-Flop Inputs for BCD Counter

Present State				Next State				Output	Flip-Flop Inputs			
Q_8	Q_4	Q_2	Q_1	Q_8	Q_4	Q_2	Q_1	Y	T_{Q8}	T_{Q4}	T_{Q2}	T_{Q1}
0	0	0	0	0	0	0	1	0	0	0	0	1
0	0	0	1	0	0	1	0	0	0	0	1	1
0	0	1	0	0	0	1	1	0	0	0	0	1
0	0	1	1	0	1	0	0	0	0	1	1	1
0	1	0	0	0	1	0	1	0	0	0	0	1
0	1	0	1	0	1	1	0	0	0	0	1	1
0	1	1	0	0	1	1	1	0	0	0	0	1
0	1	1	1	1	0	0	0	0	1	1	1	1
1	0	0	0	1	0	0	1	0	0	0	0	1
1	0	0	1	0	0	0	0	1	1	0	0	1

The flip-flop input equations are obtained from the flip-flop inputs listed in the table and can be simplified by means of maps. The unused states for minterms 1010 through 1111 are taken as don't-care conditions. The simplified input equations for the BCD counter are:

$$T_{Q1} = 1$$

$$T_{Q2} = Q_1\overline{Q}_8$$

$$T_{Q4} = Q_1Q_2$$

$$T_{Q8} = Q_1Q_8 + Q_1Q_2Q_4$$

$$Y = Q_1Q_8$$

The circuit can be drawn with four T flip-flops, four AND gates, and one OR gate. Synchronous BCD counters can be cascaded to form counters for decimal numbers of any length. The cascading is done by connecting output Y to T_{Q1} and to inputs of each of the AND gates in the next-higher-order decade.

Arbitrary Count Sequence

Suppose we wish to design a counter that has a repeated sequence of six states, as listed in Table 5-8. In this sequence, flip-flops B and C repeat the binary count 00, 01, 10, while flip-flop A alternates between 0 and 1 every three counts. Thus, the count sequence for the counter is not straight binary, and two states, 011 and 111, are not included in the count. The choice of JK-type flip-flops results in the flip-flop input conditions listed in the table. Inputs K_B and K_C have only 1's and X's in their 1 columns, so these inputs are always equal to 1. The other flip-flop input equations can be simplified using minterms 3 and 7 as don't-care conditions. The simplified functions are:

$$J_A = B \qquad K_A = B$$

$$J_B = C \qquad K_B = 1$$

$$J_C = \overline{B} \qquad K_C = 1$$

☐ TABLE 5-8
State Table and Flip-Flop Inputs for Counter

Present State			Next State			Flip-Flop Inputs					
A	B	C	A	B	C	J_A	K_A	J_B	K_B	J_C	K_C
0	0	0	0	0	1	0	×	0	×	1	×
0	0	1	0	1	0	0	×	1	×	×	1
0	1	0	1	0	0	1	×	×	1	0	×
1	0	0	1	0	1	×	0	0	×	1	×
1	0	1	1	1	0	×	0	1	×	×	1
1	1	0	0	0	0	×	1	×	1	0	×

The logic diagram of the counter is shown in Figure 5-14(a). Since there are two unused states, we analyze the circuit to determine their effect. The state diagram so obtained is drawn in Figure 5-14(b). This diagram indicates that if the circuit ever goes to one of the unused states, the next count pulse transfers it to one of the valid states, and the circuit then continues to count correctly.

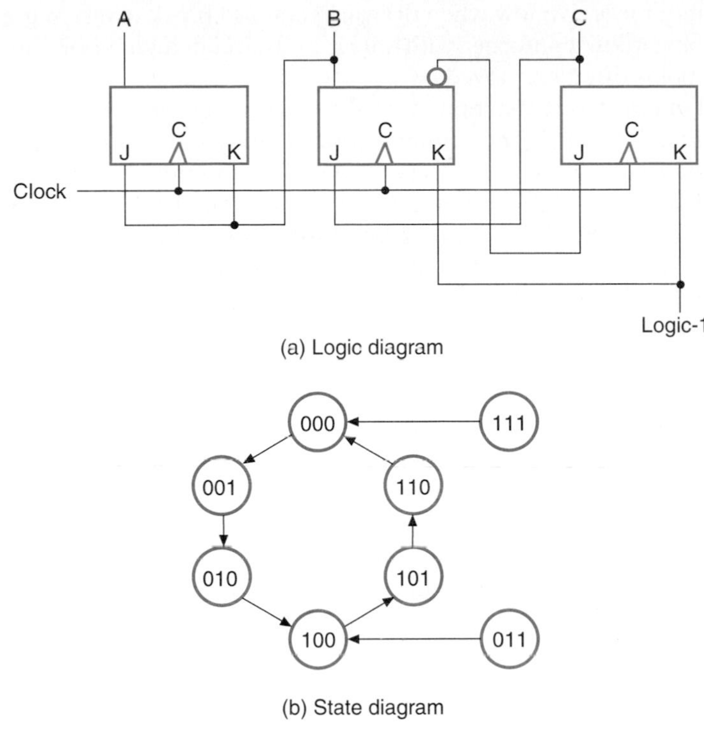

□ **FIGURE 5-14**
Counter with Arbitrary Count

5-7 Standard Graphics Symbols

The standard graphics symbols for logic functions were introduced in Section 3-13. In the current section, we give standard graphics symbols for selected counters and registers from this chapter.

The standard graphics symbol for a register is equivalent to the symbol used for a group of flip-flops with common clock input. Figure 5-15(a) shows the standard block diagram symbol for a group of four D-type flip-flops with common clock and clear inputs. The clock input $C1$ and the clear input R appear in the upper block, referred to as a *common control block*. The inputs to a common control block are considered to be inputs to each of the elements in the lower sections of the diagram. The notation $C1$ represents a control dependency that controls all the $1D$ inputs. Thus, the $1D$ input in each flip-flop is positive-edge triggered, as specified by the dynamic indicator next to the common $C1$ input signal. The common R input resets all flip-flops. The $1D$ symbol is used only once in the upper section, instead of repeating it in each section.

The small right triangle along the R input is a *polarity indicator*. It signifies that the input is active low. When positive logic is used, the low signal is equivalent to logic 0, and therefore, a bubble can be used instead. However, we will show the polarity symbol for active low when drawing standard block diagram graphics symbols. This convention conforms with integrated circuit data books in which the polarity symbol is often employed.

The standard block diagram symbol for a register with parallel load control is shown in Figure 5-15(b). The common control block shows input $M1$ for the load and $C2$ for the clock. The letter M is used to indicate a *mode dependency*. The notation "[Load]" is optional. Each flip-flop section has the notation 1,2D at the input. The 1 refers to $M1$, the 2 refers to $C2$, and D is the input type of the flip-flop. The symbol 1,2D designates the fact that the D input is triggered with clock $C2$, provided that the load input $M1$ is active. Note that the symbol 1,2D is written only once in the top section and is assumed to be repeated in the lower sections.

The convention used in the dependency notation dictates that the number following C determines which input is affected by it. In Figure 5-15(a), we chose $C1$ and 1D. In Figure 5-15(b), the identifying number 1 was chosen for the mode dependency M and the number 2 for the control dependency C. The D input is affected by both, so we place both identifying numbers in front of the D symbol. We could have chosen $M2$ and $C1$ instead, and it does not make a difference in this case.

Figure 5-16 shows the graphics symbol for the unidirectional shift register with parallel load whose detailed diagram is shown in Figure 5-6. The common control block has the qualifying symbol $SRG4$, which stands for a 4-bit shift register. The block has three dependency symbols: $M1$ and $M2$ are mode dependencies for the shift and load operations respectively; the control dependency $C3$ is for the clock input, which, based on the dynamic indicator present, is positive-edge triggered. The symbol " /1→" following $C3$ indicates that the register shifts to the right, or in the downward direction, when $M1$ is active.

The four blocks below the common block represent the four flip-flops. Flip-flop Q_0 has two inputs. The serial input has a label 1,3D. This notation means that the D input of the flip-flop is active when $M1$ (shift) is active and $C3$ goes through a positive clock transition. The other input of Q_0 and the inputs of the other blocks are for the parallel data. This is denoted by the label $\bar{1}$,2,3D. The bar over the 1 specifies that $M1$ must be inactive, the 2 is for $M2$ (load), and 3 is for the clock $C3$.

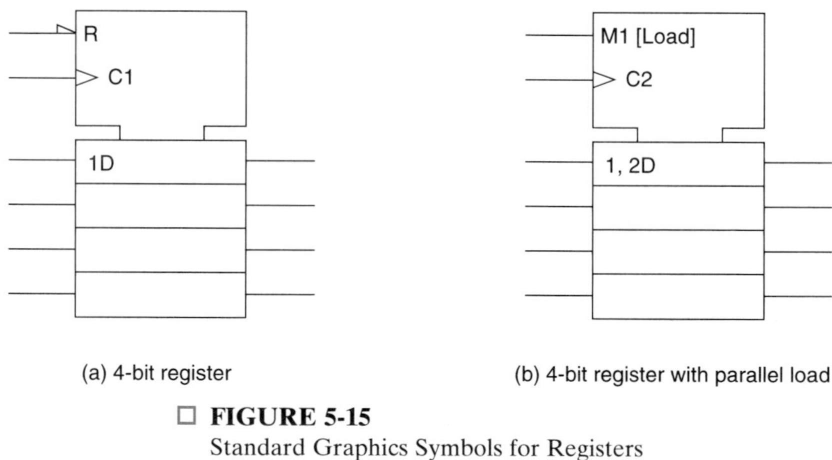

FIGURE 5-15
Standard Graphics Symbols for Registers

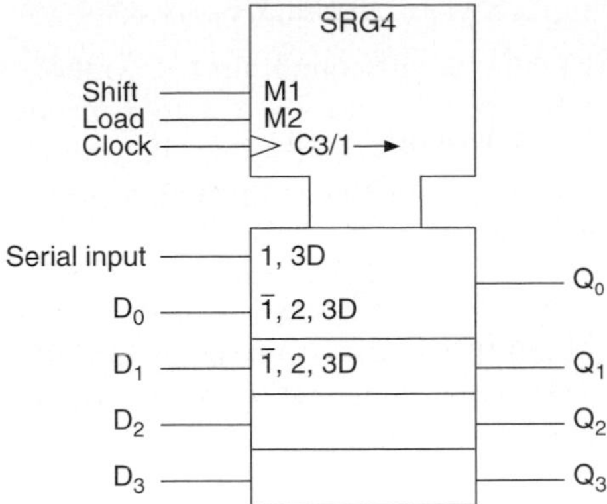

□ **FIGURE 5-16**
Graphics Symbol for Shift Register with Parallel Load

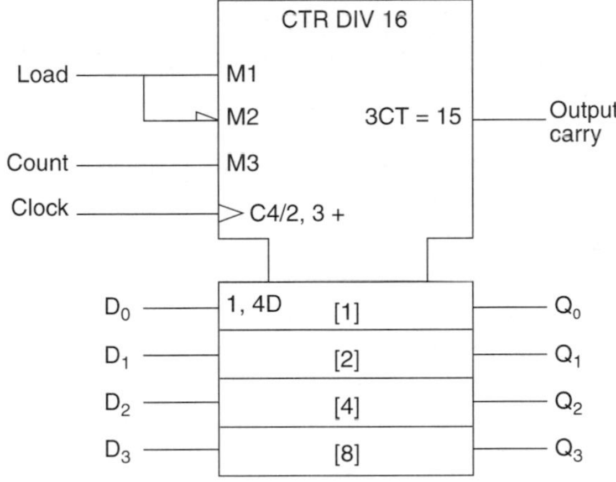

□ **FIGURE 5-17**
Graphics Symbol for a 4-Bit Binary Counter with Parallel Load

If $M1$ and $M2$ are inactive (both equal to 0), there is no active input, and therefore, the outputs are not affected by the clock transition. When there are identical blocks under the control of a common block, the inputs must be labeled in the top-most block, although they may be labeled in all blocks if desired. Note that the parallel-load input is labeled in the first and second blocks in the figure. This is in accordance with the standard: the two blocks are labeled because the first block is different from the remaining blocks.

The standard graphics symbol for the 4-bit counter with parallel load is shown in Figure 5-17. The single-load input is split into two modes labeled $M1$ and $M2$. When the load input is high $M1$ is active, and when the load input is low $M2$ is active, as indicated by the polarity symbol. The condition for counting is designated by the label associated with the clock input:

$$>C4/2,3\ +$$

This notation means that the circuit counts up (+ symbol) when $M2$ and $M3$ are active (load = 0, count = 1) and the clock in $C4$ goes through a positive transition. The parallel inputs have the label 1,4D, meaning that the D inputs are active when $M1$ is active (load = 1) and the clock goes through a positive transition. The carry output is designated by the label

$$3CT = 15$$

This equation is interpreted to mean that the carry output is active (equal to 1) if $M3$ is active (count = 1) and the content (CT) of the counter is binary 1111.

5-8 Chapter Summary

Registers, shift registers, and counters are structured sequential logic blocks. Registers are sets of flip-flops or interconnected sets of flip-flops and combinational logic. The simplest registers are flip-flops that are loaded with new contents from their inputs on every clock cycle. Slightly more complex are registers in which the flip-flops can be loaded with new contents under the control of a signal on only selected clock cycles.

Shift registers add a new dimension, since they are designed to move information laterally one or more bit positions at a time. When combined with the ability to be loaded with data, a shift register can be used to convert data presented in parallel into data presented serially. Likewise, if the outputs of the register are accessible, a shift register can be used to convert data presented serially into data presented in parallel. This lateral movement of data can also be used in hardware structures that perform serial arithmetic operations.

Counters are used to provide a sequence of values, often in binary counting order. The simplest of counters has no inputs other than an asynchronous reset for initialization to zero. This kind of counter simply counts clock pulses. More complex versions can also be loaded with data and have input signals that enable them to count.

References

1. Dietmeyer, D. L. *Logic Design of Digital Systems,* 3rd ed. Boston: Allyn-Bacon, 1988.
2. Mano, M. M. *Digital Design,* 2nd ed. Englewood Cliffs, NJ: Prentice Hall, 1991.
3. Roth, C. H. *Fundamentals of Logic Design,* 4th ed. St. Paul: West, 1992.
4. Wakerly, J. F. *Digital Design: Principles and Practices,* 2nd ed. Englewood Cliffs, NJ: Prentice Hall, 1994.
5. *Graphic Symbols for Logic Functions.* (Includes IEEE Std 91-1984 and IEEE Std 91a-1991.) New York: The Institute of Electrical and Electronics Engineers, 1991.

6. MANN, F. A. *Overview of IEEE Standard 91-1984*. Dallas: Texas Instruments, 1984.

7. *High-Speed CMOS Logic Data Book*. Dallas: Texas Instruments, 1989.

PROBLEMS

The asterisk (*) indicates a more advanced problem.

5–1. Use manual or computer-based simulation of the operation of the register in Figure 5-1(a) with the load control logic in Figure 5-1(c) to demonstrate that the clock gating functions correctly. Use a positive-edge-triggered flip-flop with *Clock* as its clock input to generate *Load*. Be sure to use nonzero gate and flip-flop delays.

5–2. *Change the OR gate in Figure 5-1(c) into an AND gate, and remove the inverter on *Load*.

(a) Perform the same simulation as in Problem 5–1 to demonstrate that the new clock gating circuitry does not work correctly. Explain what goes wrong.

(b) Will the circuit work correctly if the flip-flop generating *Load* is triggered by the negative rather than the positive edge of *Clock*?

5–3. The content of a 4-bit register is initially 0100. The register is shifted six times to the right, with the serial input being the sequence 010110 with the left most bit applied first. What is the content of the register after each shift?

5–4. What is the difference between serial and parallel transfer? Explain how to convert serial data to parallel and parallel data to serial. What type of register is needed?

5–5. The serial adder of Figure 5-5 uses two 4-bit registers. Register A holds the binary number 0101 and register B holds 0011. The carry flip-flop is initially reset to 0. List the binary values in register A and the carry flip-flop after each of four shifts.

5–6. What changes are needed in Figure 5-5 to convert it to a serial subtractor that subtracts the contents of register B from the contents of register A. Explain how it is possible to detect whether $A < B$. What will be the result of the subtraction if $A < B$?

5–7. Modify the register of Figure 5-6 so that it will operate according to the following function table:

Shift	Load	Register Operation
0	0	No change
0	1	Load parallel data
1	0	Shift down
1	1	Clear register to 0

5–8. Draw the logic diagram of a 4-bit register with mode selection inputs S_1 and S_0. The register is to be operated according to the following function table:

S_1	S_0	Register Operation
0	0	No change
0	1	Complement output
1	0	Clear register to 0
1	1	Load parallel data

5–9. Draw the four stages of the bidirectional shift register with parallel load shown in Figure 5-7.

5–10. Using external connections and the shift register from Figure 5-16, construct a bidirectional shift register (without parallel load). This can be done by connecting the outputs into the inputs in the proper order and using the load input to control the shift left operation.

5–11. **(a)** A ring counter is a shift register, as in Figure 5-3, with the serial output connected to the serial input. Starting from an initial state of 1000, list the sequence of states of the four flip-flops after each shift.

(b) A switch-tail ring counter uses the complement of the serial output as the serial input. Starting from an initial state of 0000, list the sequence of states after each shift until the register returns to 0000.

(c) Beginning in state 00…0, how many states are there in the count sequence of an n-bit switch-tail counter?

5–12. A negative-edge-triggered flip-flop has a 4-ns delay between the time its C input goes from 1 to 0 and the time the output is complemented. What is the maximum delay in a 12-bit binary ripple counter that uses these flip-flops? What is the maximum frequency at which the counter can operate reliably?

5–13. How many flip-flops will be complemented in a 12-bit binary ripple counter to reach the next count after

(a) 100110011111. **(b)** 000111111111.

5–14. Draw the logic diagram of a 4-bit ripple binary down-counter using:

(a) Flip-flops that trigger on the positive transition of the clock.

(b) Flip-flops that trigger on the negative transition of the clock.

5–15. Starting from the state table of Table 5-5, obtain the simplified flip-flop input equations for the synchronous binary counter with D flip-flops. Include a count enable input EN.

5–16. Using sequential circuit design procedure, convert a D-type flip-flop to a T-type flip-flop. Show that the logic needed is an exclusive-OR gate.

5–17. Construct a 12-bit serial-parallel counter, using three 4-bit parallel counters. What is the maximum number of AND gates in a chain that a signal must propagate through in the 12-bit counter.

5–18. *A 64-bit synchronous parallel counter is to be designed.

(a) Draw the logic diagram of a 64-bit parallel counter, using 4-bit parallel counter blocks and two levels of parallel gating connections between the blocks. In these blocks, *CO* is not driven by *EN*.

(b) What is the ratio of the maximum frequency of operation of this counter to that of a 64-bit serial-parallel counter? Assume that the flip-flop propagation time is twice the delay of an AND gate and that the flip-flop setup time is equal to the delay of an AND gate.

5–19. Draw the logic diagram of a 4-bit synchronous up-down binary counter with an enable input *EN* and a mode input *S*. Use *JK* flip-flops.

5–20. Using the synchronous binary counter of Figure 5-12 and an AND gate, construct a counter that counts from 0000 through 1011. Repeat for a count from 0000 to 0101.

5–21. Using two binary counters of the type shown in Figure 5-12 and logic gates, construct a binary counter that counts from 0 through binary 64.

5–22. Verify the flip-flop input equations of the synchronous BCD counter specified in Table 5–7. Draw the logic diagram of the BCD counter and include a count enable input.

5–23. Design a synchronous BCD counter with JK flip-flops.

5–24. Use *JK*-type flip-flops to design a binary counter with the following repeated binary sequence:

(a) 0,1,2,3,4 **(b)** 0,1,2,3,4,5,6

5–25. Use *T*-type flip-flops to design a counter with the following repeated binary sequence: 0, 1, 3, 2, 4, 6.

5–26. Use *D*-type flip-flops to design a counter with the following repeated binary sequence: 0, 1, 2, 4, 8.

5–27. Explain the purpose of the common control block when used with the standard graphics symbology.

5–28. Draw the graphics symbol of a 4-bit register with parallel load, using the label *C*1 for the clock and *M*2 for the load input.

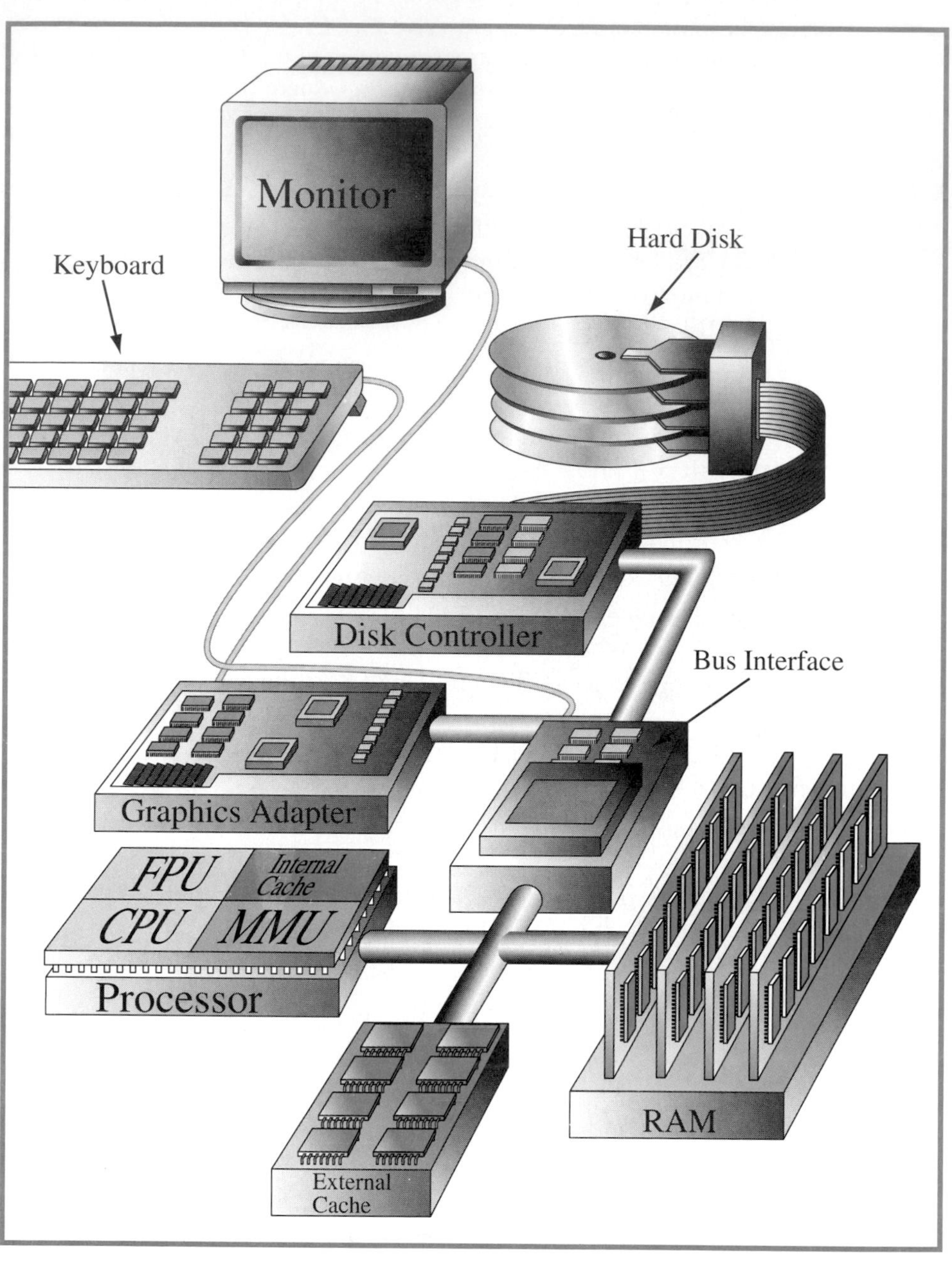

Monitor
Hard Disk
Keyboard
Disk Controller
Bus Interface
Graphics Adapter
FPU
Internal Cache
CPU
MMU
Processor
RAM
External Cache

CHAPTER 6

MEMORY AND PROGRAMMABLE LOGIC DEVICES

Memory is a major component of a digital computer and is present in a large proportion of all digital systems. Random-access memory (RAM) stores data temporarily, and read-only memory (ROM) stores data permanently. ROM is one form of a variety of components called programmable logic devices (PLDs) that use stored information to define logic circuits.

Our study of RAM begins by looking at it in terms of a model with inputs, outputs, and signal timing. We then use equivalent logical models to understand the internal workings of RAM chips. Finally, we put RAM chips together to build simple RAM systems and study error-detecting and error-correcting codes for handling errors in stored data. Our study of ROM examines the device's basic logical structure and takes an alternative view of ROM as an implementation of a combinational circuit. We then examine a variety of PLDs, including programmable logic array (PLAs) devices, programmable array logic (PAL®) devices, complex programmable logic devices (CPLDs), and field-programmable gate arrays (FPGAs).

In many of the previous chapters, the concepts presented were broad, pertaining to much of the generic computer. In this chapter, for the first time, we can be more precise and point to specific uses of memory and related components. Beginning with the microprocessor, we find that the internal cache is largely very fast RAM, so it is shaded heavily. In the microprocessor, there are several PLAs and a large ROM controlling the execution of instructions in the CPU. The MMU also contains very fast RAM. Outside the CPU, the heavily shaded external cache is largely fast RAM. The heavily shaded RAM subsystem, by its very name, is a type of memory. Not shown is the basic input/output system (BIOS) ROM, which permanently stores a portion of the programs associated with I/O. In the I/O area, we find substantial memory for storing information about the screen image in the video adapter. RAM appears in disk cache in the disk controller, to speed up disk access. Many monitors have a processor that controls the display; such monitors contain ROM to hold programs for the processor. Even the keyboard may contain a small processor and a ROM for storing programs. Aside from the highly central role of the RAM subsystem in storing data and programs, we find memory in various forms applied in most subsystems.

6-1 Memory and Programmable Logic Device Definitions

In digital systems, memory is a collection of cells capable of storing binary information. In addition to these cells, memory contains electronic circuits for storing and retrieving the information. As indicated in the discussion of the generic computer, memory is used in many different parts of a modern computer, providing temporary or permanent storage for substantial amounts of binary information. In order to process this information, it is sent from the memory to processing hardware consisting of registers and combinational logic. The processed information is then returned to the same or a different memory. Input and output devices also interact with memory. Information from an input device is placed in memory so that it can be used in processing. Output information from processing is placed in memory and from there is sent to an output device.

Two types of memories are used in various parts of a computer: *random-access memory* (RAM) and *read-only memory* (ROM). RAM accepts new information for storage to be available later for use. The process of storing new information in memory is referred to as a memory *write* operation. The process of transferring the stored information out of memory is referred to as a memory *read* operation. RAM can perform both the write and the read operations.

ROM is a programmable logic device (PLD). The binary information that is stored within such a device must be specified in some fashion and then embedded within the hardware of the computer. This process is referred to as *programming* the device, where the word "programming" refers to a hardware procedure that specifies the bits that are inserted into the hardware configuration of the device. Unlike RAM, ROM can perform only the memory read operation. This means that suitable information is stored inside the memory and can be retrieved or read at any time; however, the information cannot be altered in any way.

ROM is just one example of a PLD. Other such devices are the programmable logic array (PLA), the programmable array logic (PAL®) device, the complex programmable logic device (CPLD), and the field-programmable gate array (FPGA). A PLD is an integrated circuit with internal logic gates and/or connections that can in some way be changed by a programming process. One of the simplest technologies employs fuses. In the original state of the device, all the fuses are intact. Programming the device involves blowing those fuses along the paths that must be removed in order to obtain the particular configuration of the desired logic function. In this chapter, we introduce four different types of PLD and outline procedures for their use in the design of digital systems.

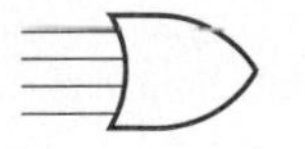

(a) Conventional symbol

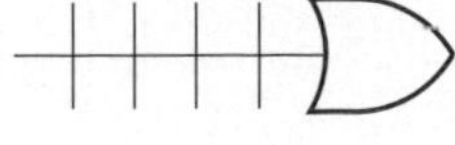

(b) Array logic symbol

□ **FIGURE 6-1**
Conventional and Array Logic Symbols for OR Gate

A typical PLD may have hundreds to thousands of gates. Some, but not all, programmable logic technologies have high fan-in gates. In order to show the internal logic diagram for such technologies in a concise form, it is necessary to employ a special gate symbology applicable to array logic. Figure 6-1 shows the conventional and array logic symbols for a multiple-input OR gate. Instead of having multiple input lines to the gate, we draw a single line to the gate. The input lines are drawn perpendicular to this line and are selectively connected to the gate. If an x is present at the intersection of two lines, there is a connection. If an x is not present, then there is no connection. In a similar fashion, we can draw the array logic for an AND gate. Since this was first done for a fuse-based technology, the graphics representation, when marked with the selected connections, is referred to as a *fuse map*. We will use the same graphics representation and terminology even when the programming technology is not fuses. This type of graphics representation for the inputs of gates will be used subsequently in drawing logic diagrams.

6-2 Random-access Memory

Memory is a collection of binary storage cells together with associated circuits needed to transfer information into and out of the cells. Memory cells can be accessed to transfer information to or from any desired location, with the access taking the same time regardless of the location, hence, the name *random-access memory*. In contrast, *serial memory*, such as is exhibited by a magnetic disk or tape unit, takes different lengths of time to access information, depending on where the desired location is relative to the current physical position of the disk or tape.

Binary information is stored in memory in groups of bits, each group of which is called a *word*. A word is an entity of bits that moves in and out of memory as a unit—a group of 1's and 0's that represents a number, an instruction, one or more alphanumeric characters, or other binary coded information. A group of eight bits is called a *byte*. Most computer memories use words that are multiples of eight bits in length. Thus, a 16-bit word contains two bytes, and a 32-bit word is made up of four bytes. The capacity of a memory unit is usually stated as the total number of bytes that it can store. Communication between a memory and its environment is achieved through data input and output lines, address selection lines, and control lines that specify the direction of transfer of information. A block diagram of a memory is shown in Figure 6-2. The n data input lines provide the information to be stored in memory, and the n data output lines supply the information coming out of memory. The k address lines specify the particular

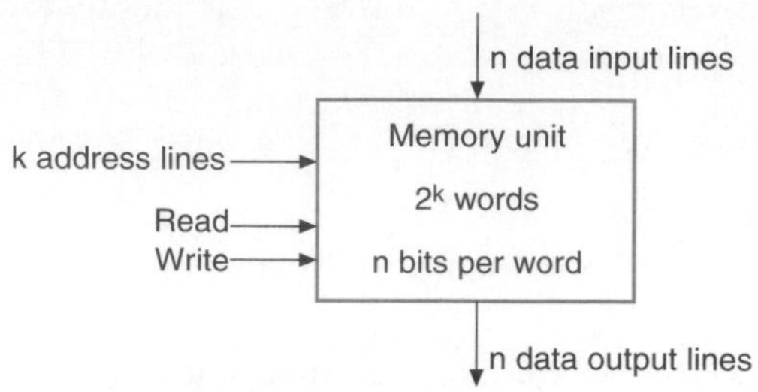

□ **FIGURE 6-2**
Block Diagram of Memory

Memory address		Memory contents
Binary	Decimal	
0000000000	0	10110101 01011100
0000000001	1	10101011 10001001
0000000010	2	00001101 01000110
	• • • • •	• • • • •
1111111101	1021	10011101 00010101
1111111110	1022	00001101 00011110
1111111111	1023	11011110 00100100

□ **FIGURE 6-3**
Contents of a 1024×16 Memory

word chosen among the many available. The two control inputs specify the direction of transfer desired: the Write input causes binary data to be transferred into memory, and the Read input causes binary data to be transferred out of memory.

The memory unit is specified by the number of words it contains and the number of bits in each word. The address lines select one particular word. Each word in memory is assigned an identification number called an *address*. Addresses range from 0 to $2^k - 1$, where k is the number of address lines. The selection of a specific word inside memory is done by applying the k-bit binary address to the address lines. A decoder accepts this address and opens the paths needed to select the word specified. Computer memory varies greatly in size. It is customary to refer to the number of words (or bytes) in memory with one of the letters K (kilo), M (mega), or G (giga). K is equal to 2^{10}, M is equal to 2^{20}, and G is equal to 2^{30}. Thus, $64K = 2^{16}$, $2M = 2^{21}$, and $4G = 2^{32}$.

Consider, for example, a memory with a capacity of 1K words of 16 bits each. Since $1K = 1024 = 2^{10}$, and 16 bits constitute two bytes, we can say that the memory can accommodate 2048, or 2K, bytes. Figure 6-3 shows the possible contents of the first three and the last three words of this size of memory. Each word contains 16 bits that can be divided into two bytes. The words are recognized by their decimal addresses from 0 to 1023. An equivalent binary address consists of 10 bits. The first address is specified using ten 0's, and the last address is specified with ten 1's. This is because 1023 in binary is equal to 1111111111. A word in memory is selected by its binary address. When a word is read or written, the memory operates on all 16 bits as a single unit.

The 1K $\times$ 16 memory of the figure has 10 bits in the address and 16 bits in each word. If, instead, we have a 64K $\times$ 10 memory, it is necessary to include 16 bits in the address, and each word will consist of 10 bits. The number of address bits needed in memory is dependent on the total number of words that can be stored there and is independent of the number of bits in each word. The number of bits in the address for a word is determined from the relationship $2^k \geq m$, where m is the total number of words and k is the minimum number of address bits satisfying the relationship.

Write and Read Operations

The two operations that a random-access memory can perform are write and read. A *write* is a transfer into memory of a new word to be stored. A *read* is a transfer of a copy of a stored word out of memory. A Write signal specifies the transfer-in operation, and a Read signal specifies the transfer-out operation. On accepting one of these control signals, the internal circuits inside memory provide the desired function.

The steps that must be taken for a write are as follows:

1. Apply the binary address of the desired word to the address lines.
2. Apply the data bits that must be stored in memory to the data input lines.
3. Activate the Write input.

The memory unit will then take the bits from the data input lines and store them in the word specified by the address lines.

The steps that must be taken for a read are as follows:

1. Apply the binary address of the desired word to the address lines.
2. Activate the Read input.

The memory will then take the bits from the word that has been selected by the address and apply them to the data output lines. The contents of the selected word are not changed by reading them.

Memory is made up of RAM integrated circuits (chips), plus additional logic circuits. RAM chips usually provide the two control inputs for the read and write operations in a somewhat different configuration from that just described. Instead of having separate Read and Write inputs to control the two operations, most integrated circuits provide at least a Chip Select that selects the chip to be read from or written to and a Read/$\overline{\text{Write}}$ that determines the particular operation. The memory operations that result from these control inputs are shown in Table 6-1.

□ **TABLE 6-1**
Control Inputs to a Memory Chip

Chip select CS	Read/$\overline{\text{Write}}$ R/$\overline{\text{W}}$	Memory operation
0	×	None
1	0	Write to selected word
1	1	Read from selected word

The Chip Select is used to enable the particular RAM chip or chips containing the word to be accessed. When Chip Select is inactive, the memory chip or chips are not selected, and no operation is performed. When Chip Select is active, the Read/$\overline{\text{Write}}$ input determines the operation to be performed. While Chip Select accesses chips, a signal is also provided that accesses the entire memory. We will call this signal the Memory Enable.

Timing Waveforms

The operation of the memory unit is controlled by an external device, such as a CPU. The CPU is synchronized by its own clock pulses. The memory, however, does not employ the CPU clock, nor does it have its own clock. Instead, its read and write operations are timed by changes in values on the control inputs. The *access time* of a memory read operation is the maximum time from the application of the address to the appearance of the data at the Data Output. Similarly, the *write cycle time* is the maximum time from the application of the address to the completion of all internal memory operations required to store a word. Memory writes may be performed one after the other at the intervals of the cycle time. The CPU must provide the memory control signals in such a way as to synchronize its own internal clocked operations with the read and write operations of memory. This means that the access time and the write cycle time of the memory must be related within the CPU to a period equal to a fixed number of CPU clock pulse periods.

Assume, as an example, that a CPU operates with a clock frequency of 50MHz, giving a period of 20ns (1 ns $= 10^{-9}$ s) for one clock pulse. Suppose now that the CPU communicates with a memory with an access time of 65 ns and a write cycle time of 75 ns. The number of clock pulses required for a memory request is the integer value greater than or equal to the maximum of the access time and the write cycle time, divided by the clock period. Since the period of the CPU clock is 20 ns, and the maximum of the access time and write cycle time is 75 ns, it will be necessary to devote at least four clock pulses to each memory request.

The memory cycle timing shown in Figure 6-4 is for a CPU with a 50MHz clock and memory with a 75-ns write cycle time and a 65-ns access time. The write cycle in part (a) shows four pulses $T1$, $T2$, $T3$, and $T4$ with a cycle of 20 ns. For a write operation, the CPU must provide the address and input data to the memory. The address is applied, and Memory Enable is set to the high level at the positive

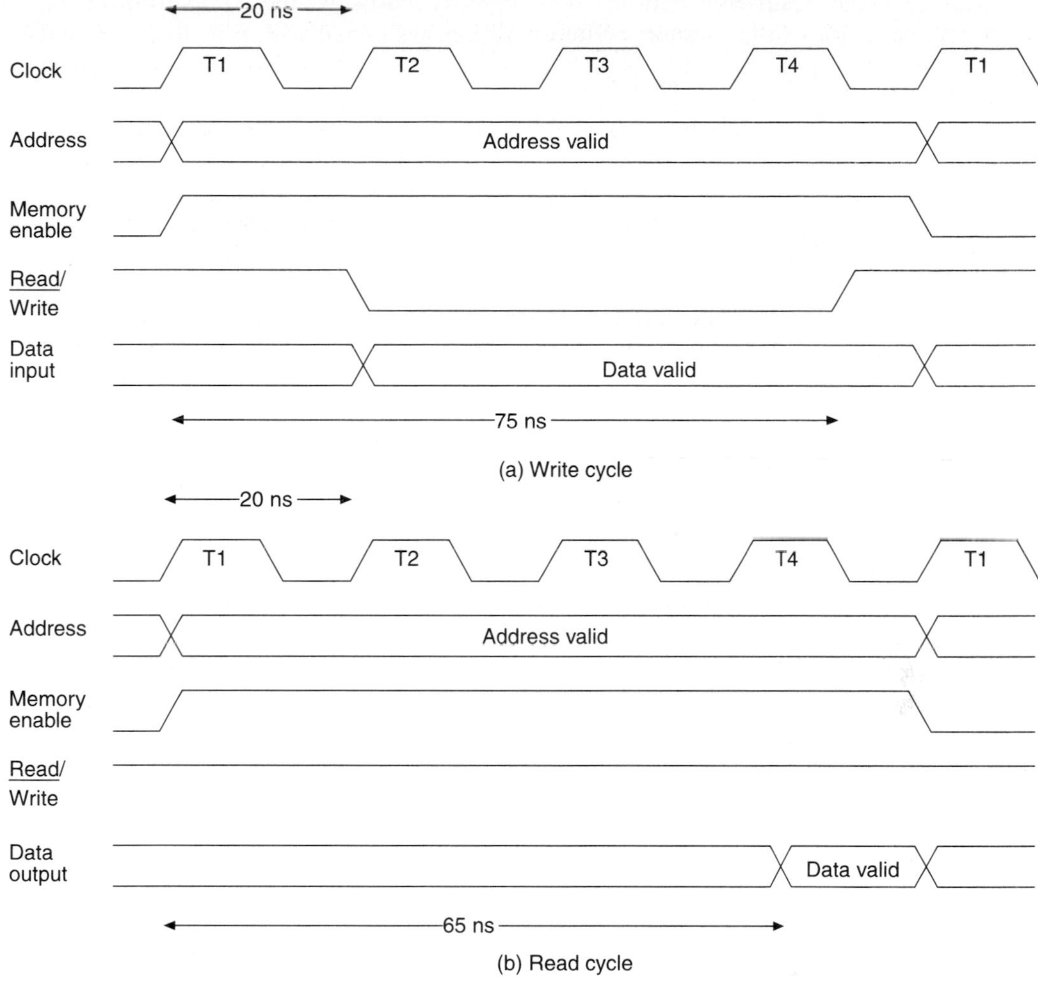

□ **FIGURE 6-4**
Memory Cycle Timing Waveforms

edge of the $T1$ pulse. The data, needed somewhat later in the write cycle, is applied at the positive edge of $T2$. The two lines that cross each other in the address and data waveforms designate a possible change in value of the multiple lines. A change of the Read/$\overline{\text{Write}}$ signal to 0 to designate the write operation is also at the positive edge of $T2$. To avoid destroying data in other memory words, it is important that this change occur after the signals on the address lines have become fixed at the desired values. Otherwise, one or more other words might be momentarily addressed and accidentally written over with different data. The Read/$\overline{\text{Write}}$ signal must stay at 0 long enough after application of the address and Memory Enable to allow the write operation to complete. Finally, the address and data signals must remain stable for a short time after the Read/$\overline{\text{Write}}$ goes to 1, again to avoid destroying data in other memory words. At the completion of the fourth clock pulse, the memory write operation has ended with 5 ns to spare, and the CPU can apply the address and control signals for another memory request with the next *T*1 pulse.

The read cycle shown in Figure 6-4(b) has an address for the memory that is provided by the CPU. The CPU applies the address, sets the Memory Enable to 1, and sets Read/$\overline{\text{Write}}$ to 1 to designate a read operation, all at the positive edge of *T*1. The memory places the data of the word selected by the address onto the data output lines within 65 ns from the time that the address is applied and the memory enable is activated. Then, the CPU transfers the data into one of its internal registers during the positive transition of the next *T*1 pulse, which can also change the address and controls for the next memory request.

Properties of Memory

Integrated circuit RAM may be either static or dynamic. *Static* RAM (SRAM) consists of internal latches that store the binary information. The stored information remains valid as long as power is applied to the RAM. *Dynamic* RAM (DRAM) stores the binary information in the form of electric charges on capacitors. The capacitors are accessed inside the chip by *n*-channel MOS transistors. The stored charge on the capacitors tends to discharge with time, and the capacitors must be periodically recharged by *refreshing* the DRAM. This is done by cycling through the words every few milliseconds, reading and rewriting them to restore the decaying charge. DRAM offers reduced power consumption and larger storage capacity in a single memory chip, but SRAM is easier to use and has shorter read and write cycles. Also, no refresh is required for SRAM.

Memory units that lose stored information when power is turned off are said to be *volatile*. Integrated circuit RAMs, both static and dynamic, are of this category, since the binary cells need external power to maintain the stored information. In contrast, a *nonvolatile memory*, such as magnetic disk, retains its stored information after the removal of power. This is because the data stored on magnetic components is represented by the direction of magnetization, which is retained after power is turned off. Another nonvolatile memory is ROM, discussed in Section 6–7.

6-3 RAM Integrated Circuits

As indicated earlier, memory consists of RAM chips plus additional logic. We will consider the internal structure of the RAM chip first. Then we will study combinations of RAM chips and additional logic used to construct memory. The internal structure of a RAM chip of *m* words with *n* bits per word consists of an array of *mn* binary storage cells and associated circuitry. The circuity is made of decoders to select the word to be read or written, read circuits, write circuits, and output logic. The *RAM cell* is the basic binary storage cell used in the RAM chip, which is typically designed as an electronic circuit rather than a logic circuit. Nevertheless, it is possible and convenient to model the RAM chip using a logic model.

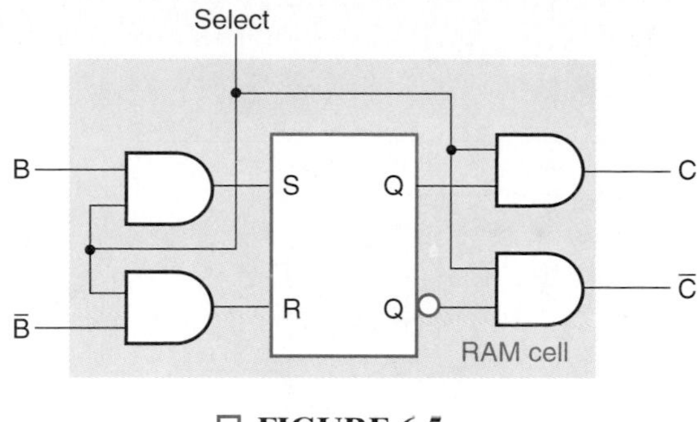

□ **FIGURE 6-5**
Static RAM Cell

A static RAM chip serves as the basis for our discussion. We first present RAM cell logic for storing a single bit and then use the cell in a hierarchy to describe the RAM chip. Figure 6-5 shows the logic model of the RAM cell. The storage part of the cell is modeled by an *SR* latch. The inputs to the latch are enabled by a Select signal. For Select equal to 0, the stored content is held. For Select equal to 1, the stored content is determined by the values on B and $\overline{B}$. The outputs from the latch are gated by Select to produce cell outputs C and $\overline{C}$. For Select equal to 0, both C and $\overline{C}$ are 0, and for Select equal to 1, C is the stored value and $\overline{C}$ is its complement.

To obtain simplified static RAM diagrams, we interconnect a set of RAM cells and read and write circuits to form a *RAM bit slice* that contains all of the circuitry associated with a single bit position of a set of RAM words. The logic diagram for a RAM bit slice is shown in Figure 6-6(a). The portion of the model representing each RAM cell is highlighted in blue. The loading of a cell latch is now controlled by a Word Select input. If this is 0, then both S and R are 0, and the cell latch contents remain unchanged. If the Word Select input is 1, then the value to be loaded into the latch is controlled by two signals B and $\overline{B}$ from the Write Logic. In order for either of these signals to be 1 and potentially change the stored value, Read/$\overline{\text{Write}}$ must be 0 and Bit Select must be 1. Then the Data In value and its complement are applied to B and $\overline{B}$, respectively, to set or reset the latch in the RAM cell selected. If Data In is 1 the latch is set to 1, and if Data In is 0 the latch is reset to 0, completing the write operation.

Only one word is written at a time. That is, only one Word Select line is 1, and all other Word Select lines are 0. Thus, only one RAM cell attached to B and $\overline{B}$ is written. The Word Select also controls the reading of the RAM cells, using shared Read Logic. If Word Select is 0, then the stored value in the *SR* latch is prevented by the AND gates from reaching the pair of OR gates in the Read Logic. But if Word Select is 1, the stored value passes through to the OR gates and is captured in the Read Logic *SR* latch. If Bit Select is also 1, the captured value appears on the Data Out line of the RAM bit slice. Note that for this particular Read Logic design, the read occurs regardless of the value of Read/$\overline{\text{Write}}$.

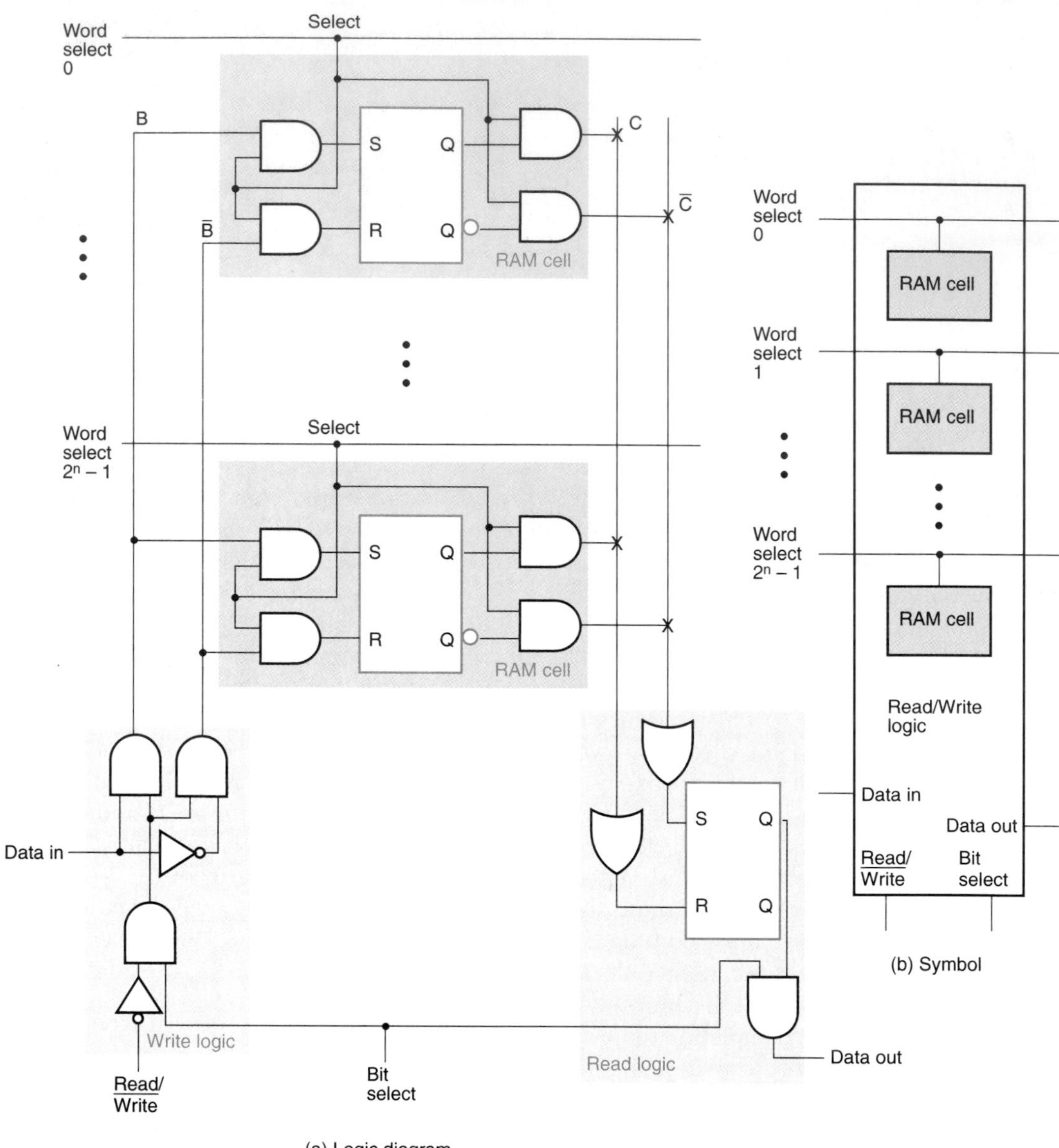

(a) Logic diagram

(b) Symbol

□ **FIGURE 6-6**
RAM Bit Slice Model

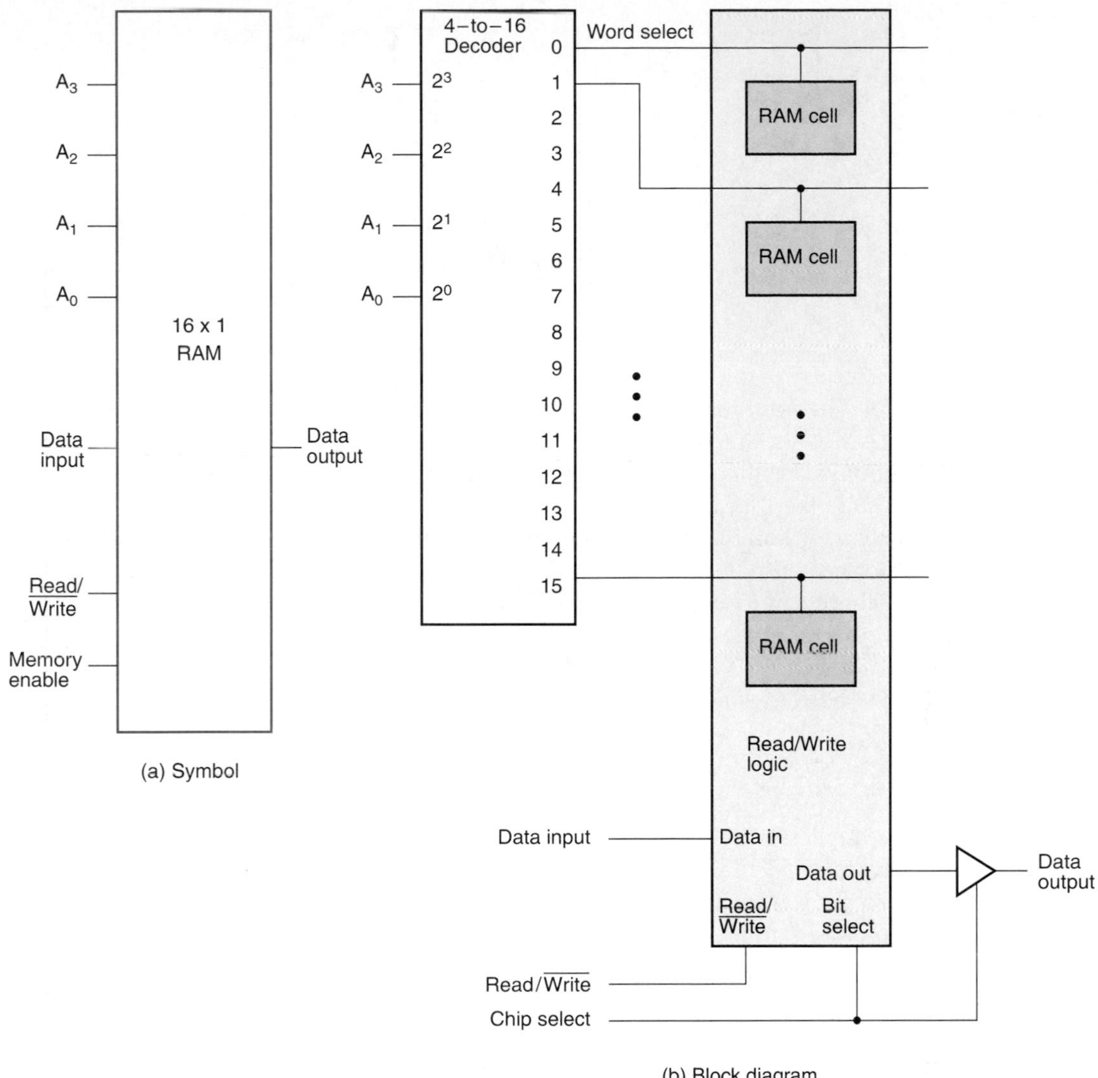

□ **FIGURE 6-7**
16-Word by 1-Bit RAM Chip

The symbol for the RAM bit slice given in Figure 6-6(b) is used to represent the internal structure of RAM chips. Each Word Select line extends beyond the bit slice, so that when multiple RAM bit slices are placed side by side, corresponding Word Select lines connect. The other signals in the lower portion of the symbol may be connected in various ways, depending on the structure of the RAM chip.

The symbol and block diagram for a 16×1 RAM chip are shown in Figure 6-7. Both have four address inputs for the 16 one-bit words stored in RAM. There are also Data Input, Data Output, and Read/$\overline{\text{Write}}$ signals. The Chip Select at the

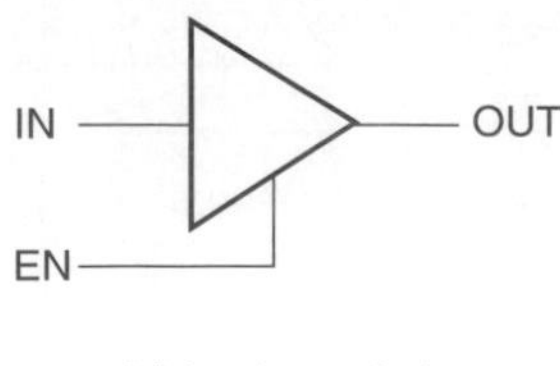

(a) Logic symbol

EN	IN	OUT
0	X	Hi-Z
1	0	0
1	1	1

(b) Truth table

□ **FIGURE 6-8**
Three-state Buffer

chip level corresponds to the Memory Enable at the level of a RAM consisting of multiple chips. The internal structure of the RAM chip consists of a RAM bit slice having 16 RAM cells. Since there are 16 Word Select lines to be controlled such that one and only one has the value logic 1 at a given time, a 4–to–16-line decoder is used to decode the four address bits into 16 Word Select bits.

The only additional logic in the figure is a triangular symbol with one normal input, one normal output, and a second input on the bottom of the symbol. This symbol is a three-state buffer that allows construction of a multiplexer with an arbitrary number of inputs. Three-state outputs are connected together and properly controlled using the Chip Select inputs. We digress at this point to understand the three-state buffers, which are critical to the use of RAM chips.

Three-State Buffers

As the name implies, a three-state logic output exhibits three distinct states. Two of the states are the logic 1 and logic 0 of conventional logic. The third state is the *high-impedance* (Hi-*Z)* state. The high-impedance state behaves like an open circuit, which means that, looking back into the logic circuit, we would find that the output appears to be disconnected. Three-state outputs may appear on any logic block.

The graphics symbol and truth table for a three-state buffer are given in Figure 6-8. The symbol in Figure 6-8(a) is distinguished from that of a normal buffer by the ENABLE input, *EN,* entering the bottom of the symbol. From the truth table in Figure 6-8(b), if $EN = 1$, *OUT* is equal to *IN,* giving normal buffer behavior. But for $EN = 0$, the output value is high impedance (Hi-*Z*), regardless of the value of *IN.*

Three-state buffer outputs can be connected together to form a multiplexed output line. Figure 6-9(a) shows two three-state buffers with their outputs connected to form output line *OL*. We are interested in the output of this structure in terms of the four inputs *EN*1, *EN*0, *IN*1, and *IN*0. The output behavior is given by the truth table in Figure 6-9(b). For *EN*1 and *EN*0 equal to 0, both buffer outputs are Hi-*Z*. Since both appear as open circuits, *OL* is also an open circuit, or Hi-*Z*. For $EN1 = 0$ and $EN0 = 1$, the output of the top buffer is *IN*0 and the output of bottom buffer is Hi-*Z*. Since the value of *IN*0 combined with an open circuit is just

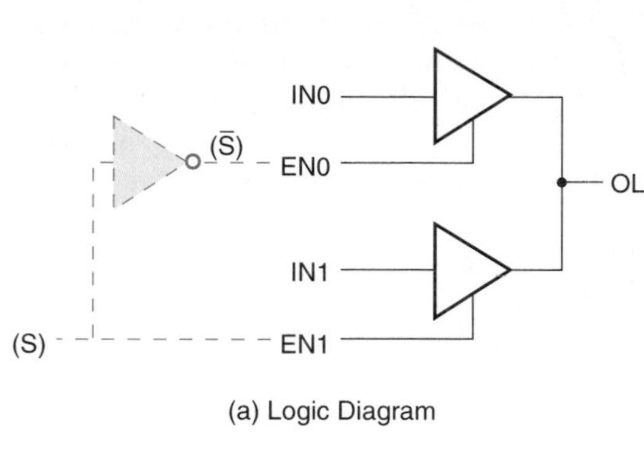

(a) Logic Diagram

EN1	EN0	IN1	IN0	OL
0	0	X	X	Hi-Z
(S) 0	($\bar{S}$)1	X	0	0
0	1	X	1	1
1	0	0	X	0
1	0	1	X	1
1	1	0	0	0
1	1	1	1	1
1	1	0	1	
1	1	1	0	

(b) Truth table

□ **FIGURE 6-9**
Three-state Buffers Forming a Multiplexed Line OL

$IN0$, OL is $IN0$, giving the second and third rows of the truth table. A corresponding, but opposite, case occurs for $EN1 = 1$ and $EN0 = 0$, so OL is $IN1$, giving the fourth and fifth rows of the truth table. For $EN1$ and $EN0$ both 1, the situation is more complicated. If $IN1 = IN0$, then their mutual value appears at OL. But if $IN1 \neq IN0$, then their values conflict at the output. The conflict results in an electrical current flowing from the buffer output that is at 1 into the buffer output that is at 0. This current is often large enough to cause heating and may even damage the circuit, as symbolized by "smoke" in the truth table. Such a situation clearly must be avoided. Since the $IN0$ and $IN1$ values are unpredictable, the only way it can be avoided for sure is to avoid having $EN1 = EN0 = 1$. In the general case, for n three-state buffers attached to a bus line, EN can equal 1 for only one of the buffers and must be 0 for the rest. One way to ensure this is to use a decoder to generate the EN signals. For the two-buffer case, the decoder degenerates to an inverter with select input S, as shown in dotted lines in Figure 6-9(a).

It is interesting to examine the truth table with the inverter in place. It consists of the shaded area of the table in Figure 6-9(b). The constrained truth table is exactly that of a two-way multiplexer with select input S. Thus, a set of n three-state buffers with their outputs connected together and EN inputs driven by a decoder provides n–to–1 selection, just as an n–to–1-line multiplexer does.

By using three-state buffers on the outputs of RAM chips, these outputs can be connected together to provide the word from the chip being read on the bit lines attached to the RAM outputs. The enable signals in the preceding discussion correspond to the Chip Select inputs on the RAM chips. To read a word from a particular RAM chip, the Chip Select value for that chip must be 1, and for all other chips attached to the same output bit lines, the Chip Select must be 0. These combinations containing a single 1 can be obtained from a decoder.

Coincident Selection

Inside a RAM chip, the decoder with k inputs and 2^k outputs requires 2^k AND gates with k inputs per gate if a straightforward design approach is used. In addition, if the number of words is large, and all bits for one bit position in the word are contained in a single RAM bit slice, the number of RAM cells sharing the read and write circuits is also large. The electrical properties resulting from both of these situations cause the access and write cycle times of the RAM to become long, which is undesirable.

The total number of decoder gates, the number of inputs per gate, and the number of RAM cells per bit slice can all be reduced by employing two decoders with a *coincident selection* scheme. In one possible configuration, two $k/2$-input decoders are used instead of one k-input decoder. One decoder controls the word select lines and the other controls the bit select lines. The result is a two-dimensional matrix selection scheme. If the RAM chip has m words with 1 bit per word, then the scheme selects the RAM cell at the intersection of the Word Select row and the Bit Select column. Since the Word Select is no longer strictly selecting words, its name is changed to *Row Select*. An output from the added decoder that selects one or more bit slices is referred to as a *Column Select*.

Coincident selection is illustrated for the 16×1 RAM chip with the structure shown in Figure 6-10. The chip consists of four RAM bit slices of four bits each and has a total of 16 RAM cells in a two-dimensional array. The two most significant address inputs go through the 2–to–4-line row decoder to select one of the four rows of the array. The two least significant address inputs go through the 2–to–4-line column decoder to select one of the four columns (RAM bit slices) of the array. The column decoder is enabled with the Chip Select input. When the Chip Select is 0, all outputs of the decoder are 0 and none of the cells is selected. This prevents writing into any RAM cell in the array. With Chip Select at 1, a single bit in the RAM is accessed. For example, for the address 1001, the first two address bits are decoded to select row 10 (2_{10}) of the RAM cell array. The second two address bits are decoded to select column 01(1_{10}) of the array. The RAM cell accessed, in row 2 and column 1 of the array, is cell 9 ($10_2\ 01_2$). With a row and column selected, the Read/$\overline{\text{Write}}$ input determines the operation. During the read operation (Read/$\overline{\text{Write}}$ = 1), the selected bit of the selected row goes through the OR gate to the three-state buffer. Note that the gate is drawn according to the array logic established in Figure 6-1. Since the buffer is enabled by Chip Select, the value read appears at the Data Output. During the write operation (Read/$\overline{\text{Write}}$ = 0), the bit available on the Data Input line is transferred into the selected RAM cell. Those RAM cells not selected are disabled, and their previous binary values remain unchanged.

The same RAM cell array is used in Figure 6-11 to produce an 8×2 RAM chip (eight words of two bits each). The row decoding is unchanged from that in Figure 6-10; the only changes are in the column and output logic. Since there are just three address bits, and two are handled by the row decoder, the column decoder has only one address bit and Chip Select as inputs and produces just two Column Select lines. Since two bits at a time are to be written or read, the Column Select lines go to adjacent pairs of RAM bit slices. Two input lines, Data Input 0

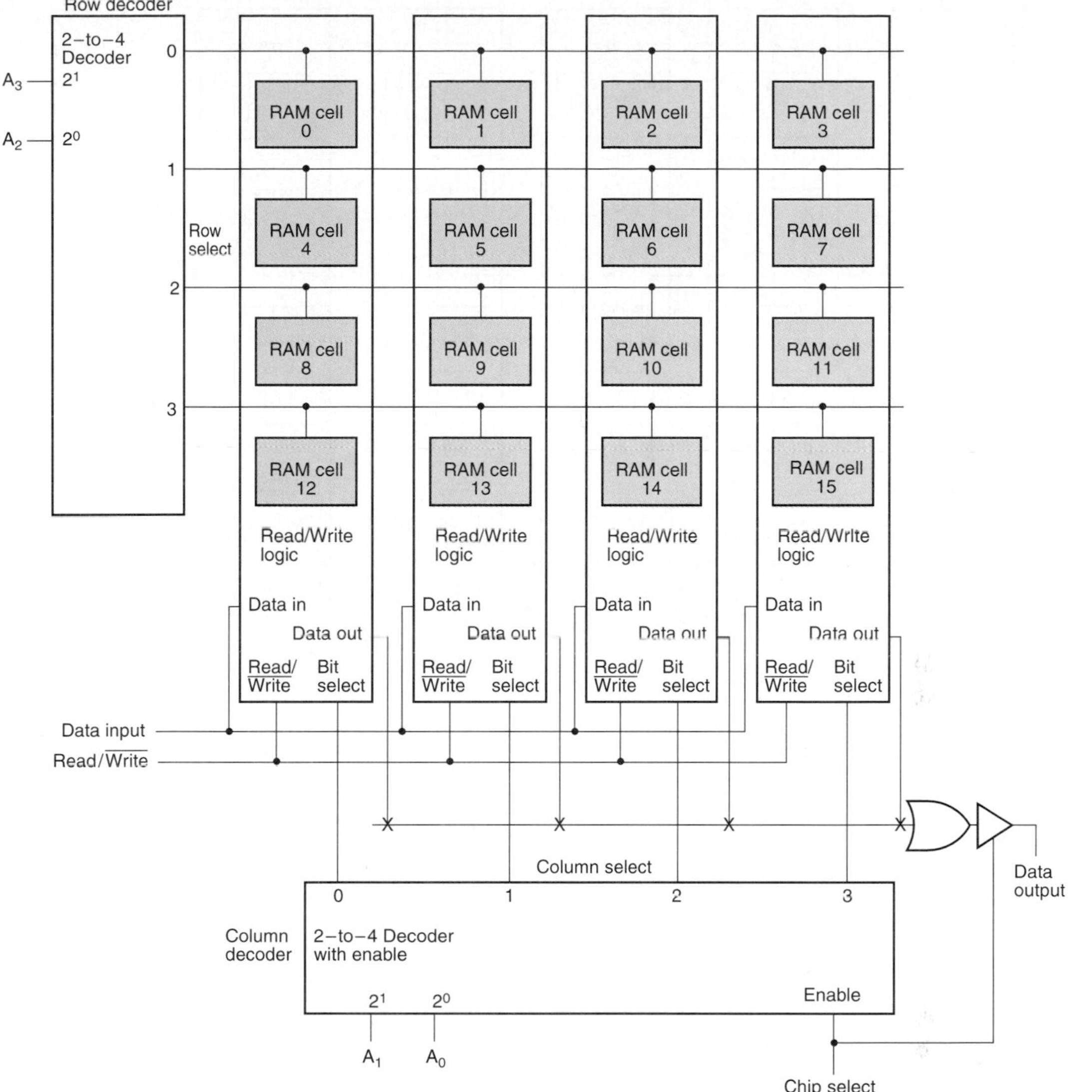

□ **FIGURE 6-10**
Diagram of a 16 × 1 RAM Using a 4 × 4 RAM Cell Array

and Data Input 1, each go to a different bit in all of the pairs. Finally, corresponding bits of the pairs share output OR gates and three-state buffers, giving output lines Data Output 0 and Data Output 1. The operation of this structure can be illustrated by the application of the address 3 (011_2). The first two bits of the address, 01, access row 1 of the array. The final bit, 1, accesses column 1, which con-

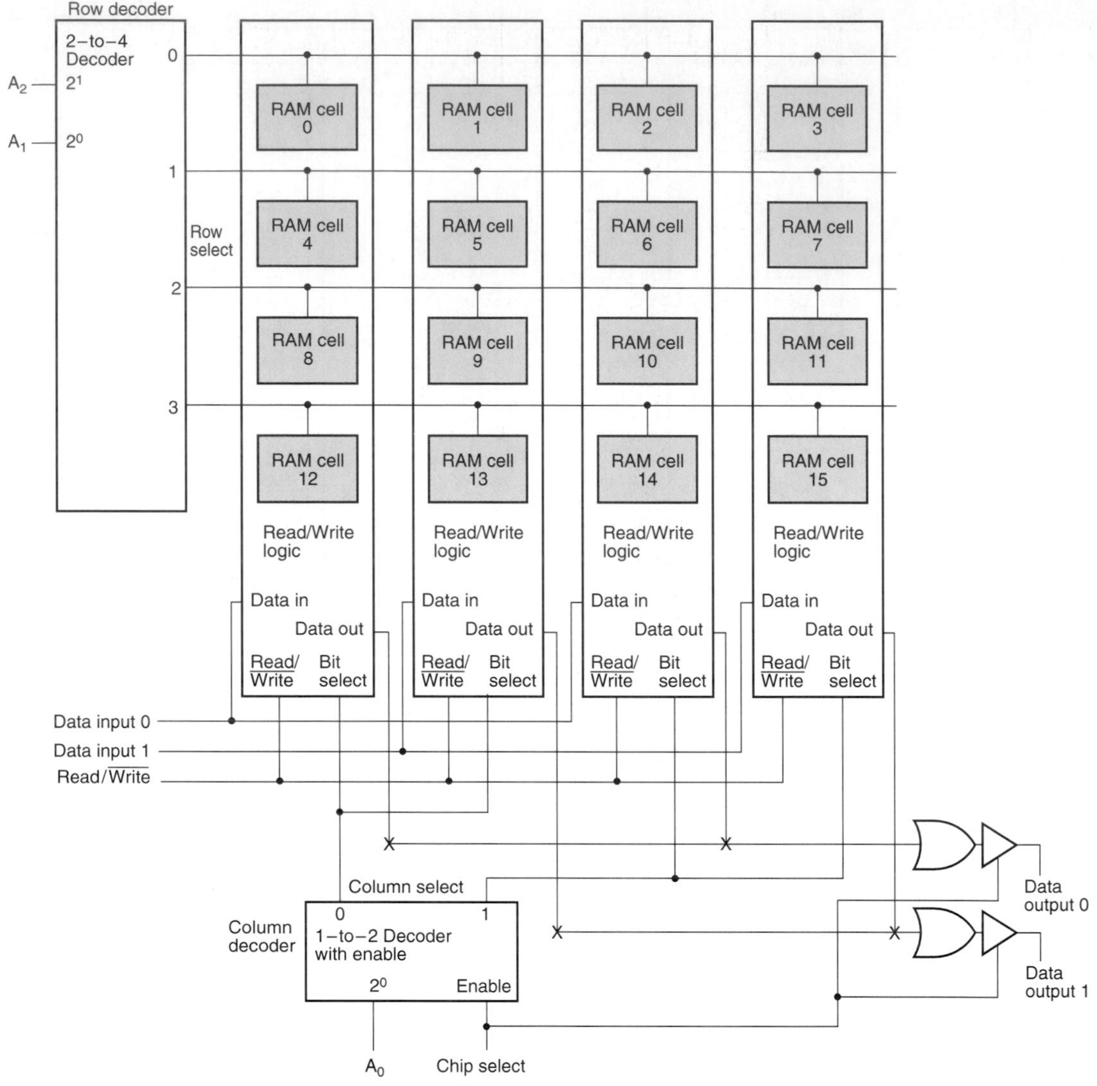

☐ **FIGURE 6-11**
Block Diagram of an 8×2 RAM Using a 4×4 RAM Cell Array

sists of bit slices 2 (10_2) and 3 (11_2). So the word to be written or read lies in RAM cells 6 and 7 ($011\ 0_2$ and $011\ 1_2$), which contain bits 0 and 1, respectively, of word 3.

We can demonstrate the savings of the coincident selection scheme by considering a more realistic static RAM size, $32K \times 8$. This RAM chip contains a total of 256K bits. To make the number of rows and columns in the array equal, we take the square root of 256K, giving $512 = 2^9$. So the first nine bits of the address are fed to the row decoder and the remaining six bits to the column decoder. Without

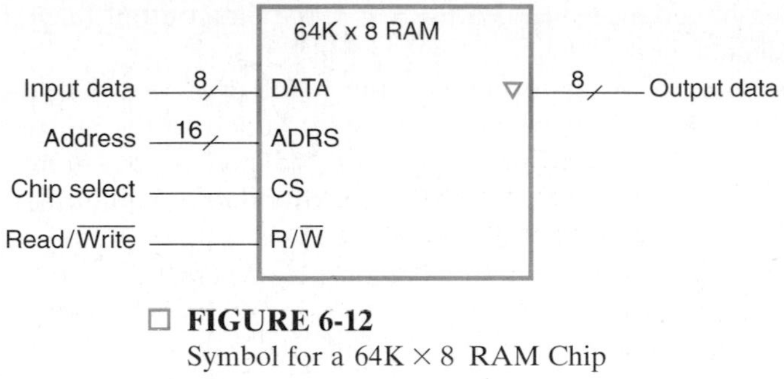

□ **FIGURE 6-12**
Symbol for a 64K × 8 RAM Chip

coincident selection, the single decoder would have 15 inputs and 32,768 outputs. With coincident selection, there is one 9–to–512 line decoder and one 6–to–64 line decoder. The number of gates for a straightforward design of the single decoder would be 32,800. For the two coincident decoders, the number of gates is 608, reducing the gate count by a factor of more than 50. In addition, although it appears that there are 64 times as many Read/Write circuits, the column selection can be done between the RAM cells and the Read/Write circuits, so that only the original eight circuits are required. Because of the reduced number of RAM cells attached to each Read/Write circuit at any time, the access time of the chip is also improved.

6-4 Array of RAM ICs

Integrated circuit RAM chips are available in a variety of sizes. If the memory unit needed for an application is larger than the capacity of one chip, it is necessary to combine a number of chips in an array to form the required size of memory. The capacity of the memory depends on two parameters: the number of words and the number of bits per word. An increase in the number of words requires that we increase the address length. Every bit added to the length of the address doubles the number of words in memory. An increase in the number of bits per word requires that we increase the number of data input and output lines, but the address length remains the same.

To illustrate an array of RAM ICs, let us first introduce a RAM chip using the condensed representation for inputs and outputs shown in Figure 6-12. The capacity of this chip is 64K words of eight bits each. The chip requires a 16-bit address and eight input and output lines. Instead of 16 lines for the address and eight lines each for data input and data output, each is shown in the block diagram by a single line. Each line has a slash across it with a number indicating the number of lines represented. The CS (Chip Select) input selects the particular RAM chip, and the $R/\overline{W}$ (Read/$\overline{\text{Write}}$) input specifies the read or write operation when the chip is selected. The small triangle shown at the outputs is the standard graphics symbol for three-state outputs. The CS input of the RAM controls the behavior of the data output lines. When $CS = 0$, the chip is not selected, and all its data outputs are in the high-impedance state. With $CS = 1$, the data output lines carry the eight bits of the selected word.

Suppose that we want to increase the number of words in the memory by using two or more RAM chips. Since every bit added to the address doubles the binary number that can be formed, it is natural to increase the number of words in factors of two. For example, two RAM chips will double the number of words and add one bit to the composite address. Four RAM chips multiply the number of words by four and add two bits to the composite address.

Consider the possibility of constructing a $256K \times 8$ RAM with four $64K \times 8$ RAM chips, as shown in Figure 6-13. The eight data input lines go to all the chips. The three-state outputs can be connected together to form the eight common data output lines. This type of output connection is possible only with three-state outputs. Just one chip select input will be active at any time, while the other three chips will be disabled. The eight outputs of the selected chip will contain 1's and 0's, and the other three will be in a high-impedance state presenting only open circuits to the binary output signals of the selected chip.

The 256K-word memory requires an 18-bit address. The 16 least significant bits of the address are applied to the address inputs of all four chips. The 2 most significant bits are applied to a 2×4 decoder. The four outputs of the decoder are applied to the CS inputs of the four chips. The memory is disabled when the EN input of the decoder, Memory Enable, is equal to 0. All four outputs of the decoder are then 0, and none of the chips is selected. When the decoder is enabled, address bits 17 and 16 determine the particular chip that is selected. If these bits are equal to 00, the first RAM chip is selected. The remaining 16 address bits then select a word within the chip in the range from 0 to 65,535. The next 65,536 words are selected from the second RAM chip with an 18-bit address that starts with 01 followed by the 16 bits from the common address lines. The address range for each chip is listed in decimal under its symbol in the figure.

It is also possible to combine two chips to form a composite memory containing the same number of words, but with twice as many bits in each word. Figure 6-14 shows the interconnection of two $64K \times 8$ chips to form a $64K \times 16$ memory. The 16 data input and data output lines are split between the two chips. Both receive the same 16-bit address and the common CS and $R/\overline{W}$ control inputs.

The two techniques just described may be combined to assemble an array of identical chips into a large-capacity memory. The composite memory will have a number of bits per word that is a multiple of that for one chip. The total number of words will increase in factors of two times the word capacity of one chip. An external decoder is needed to select the individual chips based on the additional address bits of the composite memory.

To reduce the number of pins on the chip package, many RAM ICs provide common terminals for the data input and data output. The common terminals are said to be *bidirectional*, which means that for the read operation they act as outputs, and for the write operation they act as inputs. Bidirectional lines are constructed with three-state buffers and are discussed further in Section 7-5. The use of bidirectional signals requires control of the three-state buffers by both Chip Select and Read/$\overline{\text{Write}}$.

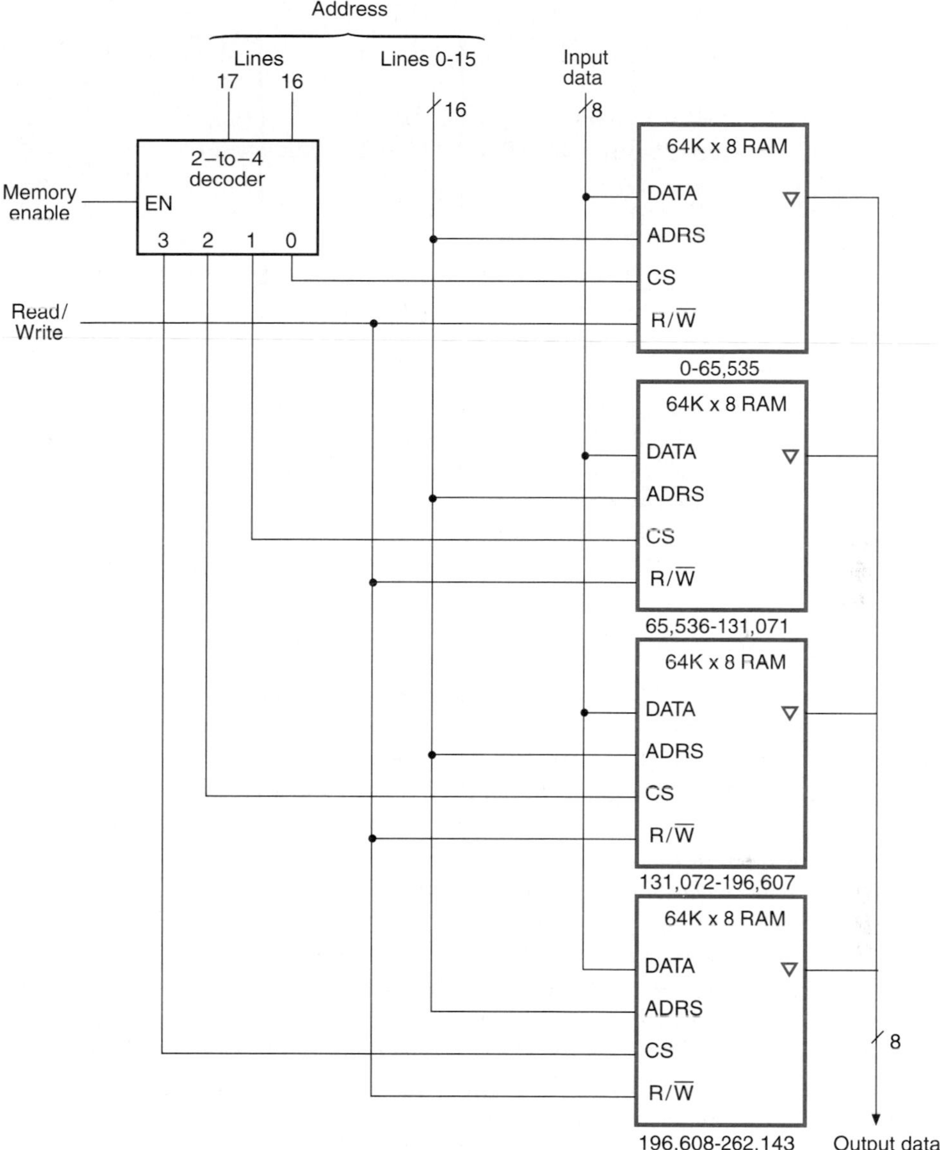

□ **FIGURE 6-13**
Block Diagram of a 256K × 8 RAM

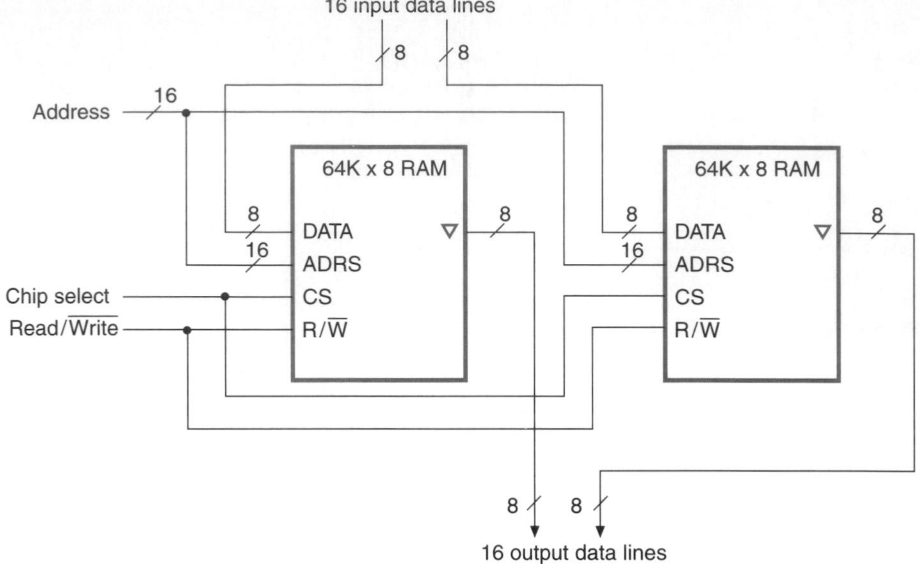

□ **FIGURE 6-14**
Block Diagram of a 64K × 16 RAM

6-5 Error Detection and Correction

The small size of the transistors or capacitors, combined with cosmic ray effects, causes occasional errors in stored information in large, dense RAM chips, particularly those that are dynamic. These errors can be detected and corrected by employing error-detecting and -correcting codes in RAMs. The most common error detection scheme is the parity bit. (See Section 2-7.) A parity bit is generated and stored along with the data word in memory. The parity of the word is checked after reading the word from memory. The word is accepted if the parity of the bits read out is correct. If the parity of the bits read is incorrect, an error is detected, but it cannot be corrected.

An error-correcting code uses multiple parity check bits that are stored with the data word in memory. Each check bit is a parity bit for a group of bits in the data word. When the word is read back from memory, the parity of each group, including the check bit, is evaluated. If the parity is correct for all groups, it signifies that no detectable error has occurred. If one or more of the newly generated parity values is incorrect, a unique pattern called a syndrome results that may be able to identify which bit is in error. A single error occurs when a bit changes in value from 1 to 0 or from 0 to 1 while stored or if it erroneously changes during a write or read operation. If the specific bit in error is identified, then the error can be corrected by complementing the erroneous bit.

Hamming Codes

The most common types of error-correcting codes used in RAM are based on the codes devised by R. W. Hamming. In the Hamming code, k parity bits are added to an n-bit data word, forming a new word of $n + k$ bits. The bit positions are numbered in sequence from 1 to $n + k$. Those positions numbered with powers of two are reserved for the parity bits. The remaining bits are the data bits. The code can be used with words of any length.

Before giving the general characteristics of the Hamming code, we will illustrate its operation with a data word of eight bits. Consider, for example, the 8-bit data word 11000100. We include four parity bits with this word and arrange the 12 bits as follows:

Bit position	1	2	3	4	5	6	7	8	9	10	11	12
	P_1	P_2	1	P_4	1	0	0	P_8	0	1	0	0

The 4 parity bits P_1 through P_8 are in positions 1, 2, 4, and 8, respectively. The 8 bits of the data word are in the remaining positions. Each parity bit is calculated as follows:

$$P_1 = \text{XOR of bits } (3, 5, 7, 9, 11) = 1 \oplus 1 \oplus 0 \oplus 0 \oplus 0 = 0$$

$$P_2 = \text{XOR of bits } (3, 6, 7, 10, 11) = 1 \oplus 0 \oplus 0 \oplus 1 \oplus 0 = 0$$

$$P_4 = \text{XOR of bits } (5, 6, 7, 12) = 1 \oplus 0 \oplus 0 \oplus 0 = 1$$

$$P_8 = \text{XOR of bits } (9, 10, 11, 12) = 0 \oplus 1 \oplus 0 \oplus 0 = 1$$

Recall that the exclusive-OR operation performs the odd function. It is equal to 1 for an odd number of l's among the variables and to 0 for an even number of 1's. Thus, each parity bit is set so that the total number of l's in the checked positions, including the parity bit, is always even.

The 8-bit data word is written into the memory together with the 4 parity bits as a 12-bit composite word. Substituting the 4 parity bits in their proper positions, we obtain the 12-bit composite word written into memory:

Bit position	1	2	3	4	5	6	7	8	9	10	11	12
	0	0	1	1	1	0	0	1	0	1	0	0

When the 12 bits are read from memory, they are checked again for errors. The parity of the word is checked over the same groups of bits, including their parity bits. The four check bits are evaluated as follows:

$$C_1 = \text{XOR of bits } (1, 3, 5, 7, 9, 11)$$

$$C_2 = \text{XOR of bits } (2, 3, 6, 7, 10, 11)$$

$$C_4 = \text{XOR of bits } (4, 5, 6, 7, 12)$$

$$C_8 = \text{XOR of bits } (8, 9, 10, 11, 12)$$

A 0 check bit designates an even parity over the checked bits, and a 1 designates an odd parity. Since the bits were written with even parity, the result, $C = C_8C_4C_2C_1 = 0000$, indicates that no error has occurred. However, if $C \neq 0$, the 4-bit binary number formed by the check bits gives the position of the erroneous bit if only a single bit is in error. For example, consider the following three cases:

Bit position	1	2	3	4	5	6	7	8	9	10	11	12	
	0	0	1	1	1	0	0	1	0	1	0	0	No error
	1	0	1	1	1	0	0	1	0	1	0	0	Error in bit 1
	0	0	1	1	0	0	0	1	0	1	0	0	Error in bit 5

In the first case, there is no error in the 12-bit word. In the second case, there is an error in bit position number 1 because it changed from 0 to 1. The third case shows an error in bit position 5 with a change from 1 to 0. Evaluating the XOR of the corresponding bits, we determine the four check bits to be as follows:

	C_8	C_4	C_2	C_1
No error	0	0	0	0
Error in bit 1	0	0	0	1
Error in bit 5	0	1	0	1

Thus, for no error, we have $C = 0000$; with an error in bit 1, we obtain $C = 0001$; and with an error in bit 5, we get $C = 0101$. Hence, when C is not equal to 0, the decimal value of C gives the position of the bit in error. The error can then be corrected by complementing the corresponding bit. Note that an error can occur in the data or in one of the parity bits.

The Hamming code can be used for data words of any length. In general, for k check bits and n data bits, the total number of bits, $n + k$, that can be in a coded word is at most $2^k - 1$. In other words, the relationship $n + k \leq 2^k - 1$ must hold. This relationship gives $n \leq 2^k - 1 - k$ as the number of bits for the data word. For example, when $k = 3$, the total number of bits in the coded word is $n + k \leq 2^3 - 1 = 7$, giving $n \leq 7 - 3 = 4$. For $k = 4$, we have $n + k \leq 15$, giving $n \leq 11$. Thus, the data word may be less than 11 bits, but must have at least five bits; otherwise, only three check bits will be needed. The relationships for $k = 3$ and $k = 4$ justify the use of four check bits for the eight data bits in the previous example.

The grouping of bits for parity generation and checking can be determined from a list of the binary numbers from 0 through $2^k - 1$. The least significant bit is a 1 in the binary numbers 1, 3, 5, 7, and so on. The second significant bit is a 1 in the binary numbers 2, 3, 6, 7, and so on. Comparing these numbers with the bit positions used in generating and checking parity bits in the Hamming code, we note the relationship between the bit groupings in the code and the position of the 1-bits in the binary count sequence. Each group of bits starts with a number that is a power of 2—for example, 1, 2, 4, 8, 16, and so forth. These numbers are also the position numbers for the parity bits.

The basic Hamming code can detect and correct an error in only a single bit. Some multiple-bit errors are detected, but they may be corrected erroneously, as if they were single-bit errors. By adding another parity bit to the coded word, the Hamming code can be used to correct a single error and detect double errors. If we include this additional parity bit, the previous 12-bit coded word becomes 001110010100P_{13}, where P_{13} is evaluated from the exclusive-OR of the other 12 bits. This produces the 13-bit word 0011100101001 (even parity). When this word is read from memory, the check bits and also the parity bit P are evaluated over the entire 13 bits. If $P = 0$, the parity is correct (even parity), but if $P = 1$, the parity over the 13 bits is incorrect (odd parity). The following four cases can occur:

If $C = 0$ and $P = 0$	No error occurred.
If $C \neq 0$ and $P = 1$	A single error occurred that can be corrected.
If $C \neq 0$ and $P = 0$	A double error occurred that is detected but cannot be corrected.
If $C = 0$ and $P = 1$	An error occurred in the P_{13} bit.

Note that this scheme will detect more than two erroneous bits in many cases, but is not guaranteed to detect all such errors.

A modified Hamming code to generate and check parity bits for a single-error-correction, double-error-detection scheme is most often used in real systems. The modified code uses a different parity check bit scheme that balances the number of inputs to the logic for each check bit and thus the number of inputs to each circuit that does the checking. The balancing minimizes the delay through the error correction and detection circuits. These circuits can be used in a RAM subsystem to add check bits during write operations and to correct single errors and detect double errors during read operations. (See Problem 6–14 at the end of the chapter.)

6-6 Programmable Logic Technologies

The next four sections deal with five types of programmable logic devices (PLDs): the read-only memory (ROM), the programmable logic array (PLA), the programmable array logic (PAL®) device, the complex programmable logic device (CPLD), and the field-programmable gate array (FPGA). Before treating these devices, we deal with the supporting programming technologies. In PLDs, programming technologies are applied to establish or break interconnections, build lookup tables, and control transistor switching. We will relate the technologies to these three applications.

The oldest of the programming technologies is the use of a *fuse*. Each of the programmable points in the PDL consists of a connection formed by a fuse. When a voltage considerably higher than the normal power supply voltage is applied across the fuse, the high current breaks the connection by blowing out the fuse. The two connection states, CLOSED and OPEN, are represented by an intact and blown fuse, respectively.

A second programming technology that also controls connections is *mask programming*, which is done by the semiconductor manufacturer during the last steps of the chip fabrication process. Connections are made or not made in the metal layers serving as conductors in the chip. Depending on the desired function for the chip, the structure of these layers is determined by the fabrication process. The procedure is costly because the vendor charges the customer a special fee for custom masking the particular device. For this reason, mask programming is economical only if a large quantity of the same PLD configuration is to be ordered.

A third programming technology that controls connections is the *antifuse*. As the name suggests, the antifuse is just the opposite of a fuse. In contrast to a fuse, an antifuse consists of a small area in which two conductors are separated by a material having a high resistance. The antifuse acts as an OPEN path before programming. By applying a voltage somewhat higher than the normal power supply voltage across the two conductors, the material separating the two conductors is melted or otherwise changed to a low resistance. The low-resistance material conducts, causing the connection to be a CLOSED path.

All three of the preceding connection technologies are permanent: the devices cannot be reprogrammed, because irreversible physical changes have occurred as a result of the previous programming. Thus, if the programming is incorrect or needs to be changed, the device must be discarded.

The final technology that can be applied to connection control is a static RAM (SRAM) bit driving the gate of an MOS *n*-channel transistor at the programming point. If the SRAM bit stores a 1, then the transistor is turned ON, and the connection between its source and drain forms a CLOSED path. For the SRAM bit equal to 0, the transistor is OFF, and the connection between its source and drain is an OPEN path. Since the SRAM bit contents can be changed electronically, the device can be easily reprogrammed. But in order to store the SRAM bit, the power supply must be available. Thus, we find that the SRAM-based technology is volatile; that is, the programmed logic is lost in the absence of the power supply voltage. In addition to controlling connections, SRAM technology can be used to implement logic by using lookup tables. In this case, the logic inputs are address inputs for reading the SRAM, and the logic outputs are the stored values for the addressed word that appear on the SRAM data outputs. Thus, the logic can be implemented simply by storing the truth table in the SRAM—hence the term *lookup table*.

The third use of programming technologies is control of transistor switching. The most popular technology is based on storing charge on a floating gate. The latter is located below the regular gate within an MOS transistor and is completely isolated by an insulating dielectric. Stored negative charge on the floating gate makes the transistor impossible to turn ON. The absence of stored negative charge makes it possible for the transistor to turn ON if a HIGH is applied to its regular gate. Since it is possible to remove the stored charge, these technologies permit erasure and reprogramming. The two technologies are called *erasable* and *electrically erasable*. Programming applies combinations of voltage higher than normal

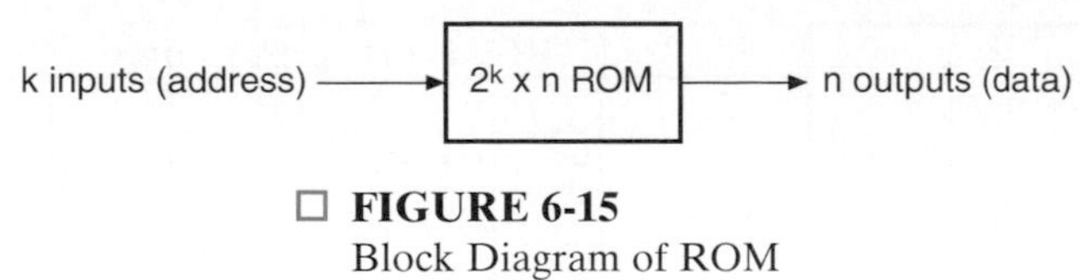

□ **FIGURE 6-15**
Block Diagram of ROM

power supply voltages to the transistor. Erasure uses exposure to a strong ultraviolet light source for a specified amount of time. Once this type of chip has been erased, it can be reprogrammed. An electrically erasable device can be erased by a process somewhat similar to the programming process, using voltages higher than the normal power supply value. Since transistor control prevents or allows a connection to be established between the source and the drain, it is really a form of connection control, giving a choice between (1) always OPEN or (2) OPEN or CLOSED, depending on an applied HIGH or LOW, respectively, on the regular transistor gate.

We next consider four distinct programmable device structures. We will describe each of the structures and indicate which of the technologies is typically used in its implementation.

6-7 Read-Only Memory

A read-only memory (ROM) is essentially a device in which permanent binary information is stored. The information must be specified by the designer and is then embedded into the ROM to form the required interconnection or electronic device pattern. Once the pattern is established, it stays within the ROM even when power is turned off and on again; that is, ROM is nonvolatile.

A block diagram of a ROM device is shown in Figure 6-15. There are k inputs and n outputs. The inputs provide the address for the memory, and the outputs give the data bits of the stored word that is selected by the address. The number of words in a ROM device is determined from the fact that k address input lines can specify 2^k words. Note that ROM does not have data inputs, because it does not have a write operation. Integrated circuit ROM chips have one or more enable inputs and come with three-state outputs to facilitate the construction of large arrays of ROM.

Consider, for example, a 32×8 ROM. The unit consists of 32 words of 8 bits each. There are five input lines that form the binary numbers from 0 through 31 for the address. Figure 6-16 shows the internal logic construction of this ROM. The five inputs are decoded into 32 distinct outputs by means of a 5–to–32-line decoder. Each output of the decoder represents a memory address. The 32 outputs are connected through programmable connections to each of the eight OR gates. The diagram shows the array logic convention used in complex circuits. (See Figure 6-1.) Each OR gate must be considered as having 32 inputs. Each output of the decoder is connected through a fuse to one of the inputs of each OR gate. Since each OR gate has 32 internal programmable connections, and since there are eight OR gates, the ROM contains $32 \times 8 = 256$ programmable connections. In general, a

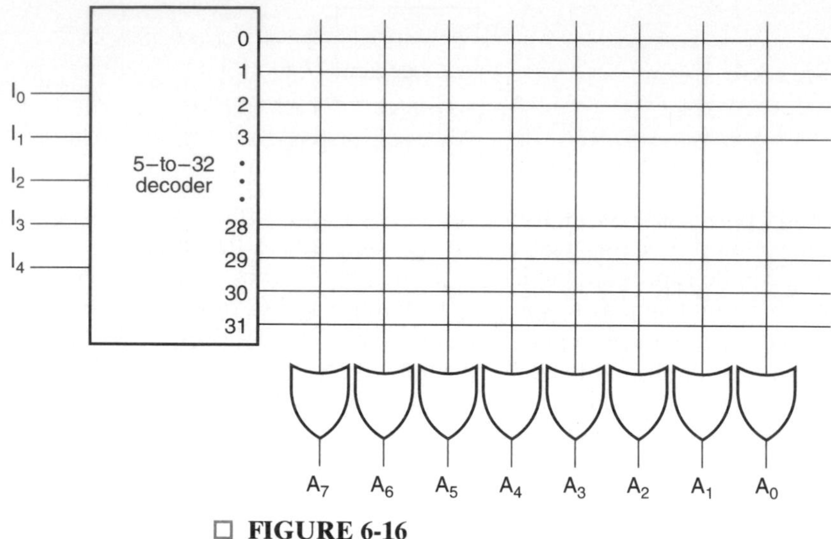

□ **FIGURE 6-16**
Internal Logic of a 32×8 ROM

$2^k \times n$ ROM will have an internal k–to–2^k line decoder and n OR gates. Each OR gate has 2^k inputs, which are connected through programmable connections to each of the outputs of the decoder.

The internal binary storage of a ROM is specified by a truth table that shows the contents of a word in each address. For example, the contents of a 32×8 ROM may be specified with a truth table similar to the one shown in Table 6-2. The truth table shows the five inputs under which are listed all 32 addresses. Each input combination specifies the address of a word of 8 bits whose value is listed under

□ **TABLE 6-2**
ROM Truth Table (Partial)

Inputs					Outputs							
I_4	I_3	I_2	I_1	I_0	A_7	A_6	A_5	A_4	A_3	A_2	A_1	A_0
0	0	0	0	0	1	0	1	1	0	1	1	0
0	0	0	0	1	0	0	0	1	1	1	0	1
0	0	0	1	0	1	1	0	0	0	1	0	1
0	0	0	1	1	1	0	1	1	0	0	1	0
		⋮							⋮			
1	1	1	0	0	0	0	0	0	1	0	0	1
1	1	1	0	1	1	1	1	0	0	0	1	0
1	1	1	1	0	0	1	0	0	1	0	1	0
1	1	1	1	1	0	0	1	1	0	0	1	1

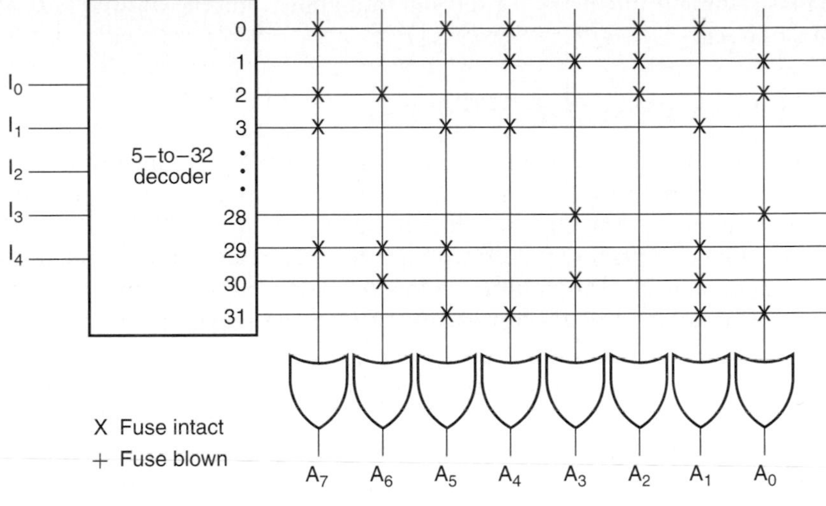

□ **FIGURE 6-17**
Programming the ROM According to Table 6-2

the output columns. Table 6-2 shows only the first four and the last four words in the ROM. The complete table must include the list of all 32 words.

The hardware procedure programs the ROM with connections that follow the truth table. For example, programming the ROM according to the truth table given in Table 6-2 results in the configuration shown in Figure 6-17. Every 0 listed in the truth table specifies an OPEN circuit, and every 1 listed specifies a CLOSED circuit. As an example, the table specifies the 8-bit word 10110010 for permanent storage at input address 00011. The four 0's in the word are programmed by open circuiting the connection between output 3 of the decoder and the inputs of the OR gates associated with outputs A_6, A_3, A_2, and A_0. The four 1's in the word are marked with a cross in the diagram, to designate a closed circuit. When the input of the ROM is 00011, all the outputs of the decoder are 0 except output 3, which is at logic 1. The signal that is equivalent to logic 1 at decoder output 3 then propagates through closed circuits and the OR gates to outputs A_7, A_5, A_4, and A_1. The other four outputs remain at 0. The result is that the stored word 10110010 is applied to the eight data outputs.

Four technologies are used for ROM programming. If mask programming is used, then the ROM is called simply a ROM. If fuses are used, the ROM can be programmed by the user having the proper programming equipment. In this case, the ROM is referred to as a *programmable ROM*, or PROM. If the ROM uses the erasable floating-gate technology, then the ROM is referred to as an *erasable, programmable ROM*, or EPROM. Finally, if the electrically erasable technology is used, the ROM is called an *electrically erasable, programmable ROM*, or EEPROM or E^2PROM. The choice of programming technology depends on many factors, including the number of identical ROMS to be produced, the desired permanence of the programming, the desire for reprogrammability, and the desired performance in terms of delay.

Combinational Circuit Implementation

It was shown in Section 3-5 that a decoder generates the 2^k minterms of the k input variables. By inserting OR gates to sum the minterms of Boolean functions, we were able to generate any desired combinational circuit. The ROM is essentially a device that includes both the decoder and the OR gates within a single unit. By closing connections for those minterms that are included in the function, the ROM outputs can be programmed to represent the Boolean functions of the output variables in a combinational circuit.

In terms of its internal operation, a ROM can be interpreted in two ways: as a memory device that contains a fixed pattern of stored words and as a circuit that implements a combinational function. From the latter point of view, each output terminal is considered separately as the output of a Boolean function expressed as a sum of minterms. For example, the ROM of Figure 6-17 may be considered as a combinational circuit with eight outputs, each a function of the five input variables. Output A_7 can be expressed as a sum of minterms as follows (the three dots represent minterms 4 through 27, which are not specified in the figure):

$$A_7(I_4, I_3, I_2, I_1, I_0) = \Sigma m(0, 2, 3, \ldots, 29)$$

A CLOSED circuit connection includes a minterm for the sum, and an OPEN circuit connection removes the minterm from the sum.

ROM devices are widely used to implement complex combinational circuits directly from their truth tables. They are useful for converting from one code, such as Gray code, to another, such as BCD. They can generate complex arithmetic operations, such as multiplication or division, and in general, they are used in applications requiring a moderate number of inputs and large number of outputs.

In practice, when a combinational circuit is designed by means of a ROM, it is not necessary to design the logic or to show the internal connections inside the unit. All that the designer has to do is specify the particular ROM by its name or its IC number and provide the truth table for the ROM. The truth table gives all the information for programming the ROM. No internal logic diagram is needed to accompany the table.

EXAMPLE 6-1

Design a combinational circuit using a ROM. The circuit accepts a 3-bit number and generates an output binary number equal to the square of the input number.

The first step in designing the circuit is to derive the truth table of the combinational circuit. In most cases, this is all that is needed. In other cases, we can use a partial truth table for the ROM by utilizing certain properties in the output variables. Table 6-3 is the truth table for the combinational circuit. Three inputs and six outputs are needed to accommodate all of the possible binary numbers. We note that output B_0 is always equal to input A_0, so there is no need to generate B_0 with a

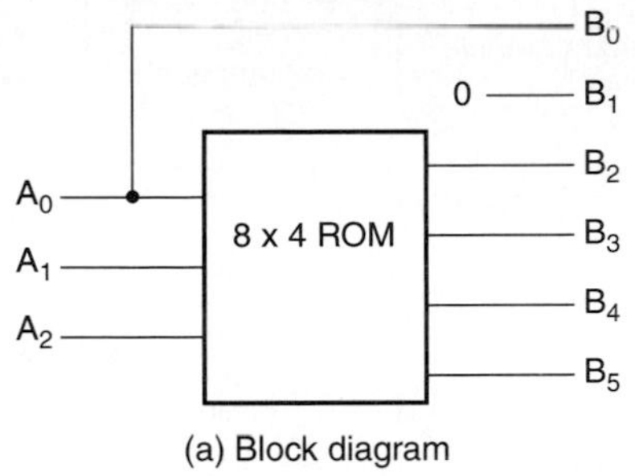

(a) Block diagram

A_2	A_1	A_0	B_5	B_4	B_3	B_2
0	0	0	0	0	0	0
0	0	1	0	0	0	0
0	1	0	0	0	0	1
0	1	1	0	0	1	0
1	0	0	0	1	0	0
1	0	1	0	1	1	0
1	1	0	1	0	0	1
1	1	1	1	1	0	0

(b) ROM truth table

□ **FIGURE 6-18**
ROM Implementation of Example 6-1

□ **TABLE 6-3**
Truth Table for Circuit of Example 6-1

Inputs			**Outputs**						
A_2	A_1	A_0	B_5	B_4	B_3	B_2	B_1	B_0	**Decimal**
0	0	0	0	0	0	0	0	0	0
0	0	1	0	0	0	0	0	1	1
0	1	0	0	0	0	1	0	0	4
0	1	1	0	0	1	0	0	1	9
1	0	0	0	1	0	0	0	0	16
1	0	1	0	1	1	0	0	1	25
1	1	0	1	0	0	1	0	0	36
1	1	1	1	1	0	0	0	1	49

ROM. Moreover, output B_1 is always 0, so this output is a known constant. Thus, we actually need to generate only four outputs with the ROM; the other two are readily obtained. The minimum size of ROM needed must have three inputs and four outputs. Three inputs specify eight words; so the ROM must be of size 8×4. The implementation of the ROM is shown in Figure 6-18. The three inputs specify eight words of four bits each. The block diagram of Figure 6-18(a) shows the required connections of the combinational circuit. The truth table in Figure 6-18(b) specifies the information needed for programming the ROM. ■

The PROM represents one of three major types of PLDs. These types of PLDs differ in the placement of programmable connections in the AND-OR array. Figure 6-19 shows the locations of the connections for the three types. Programmable read-only memory (PROM) has a fixed AND array constructed as a decoder and programmable connections for the output OR gates. The PROM implements Boolean functions in sum-of-minterms form. The programmable array logic (PAL®) device has a programmable connection AND array and a fixed OR array.

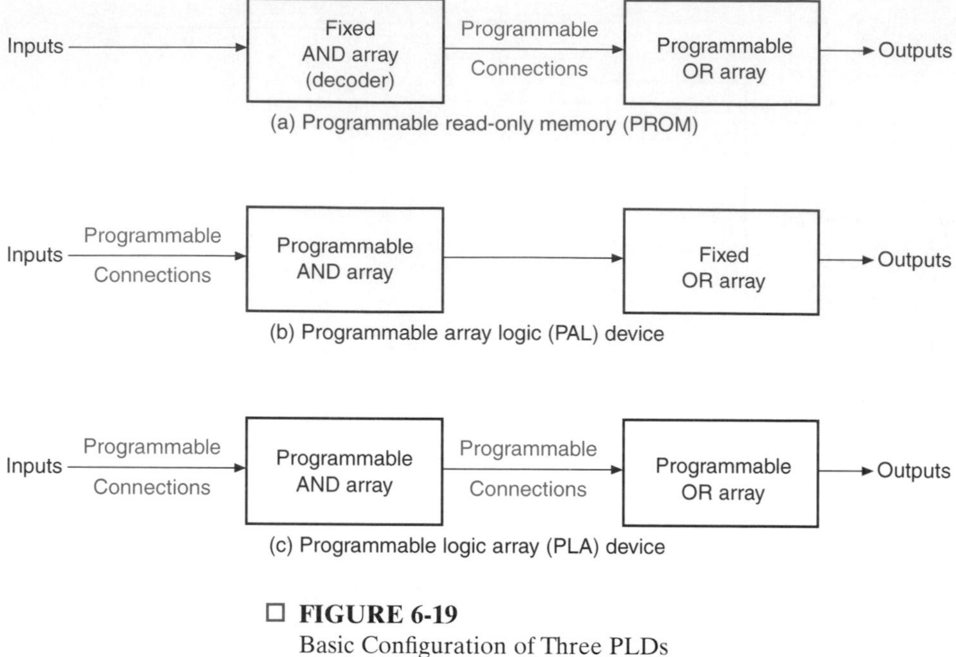

□ **FIGURE 6-19**
Basic Configuration of Three PLDs

The AND gates are programmed to provide the product terms for the Boolean functions, which are logically summed in each OR gate. The most flexible of the three types of PLD is the programmable logic array (PLA), which has programmable connections for both AND and OR arrays. The product terms in the AND array may be shared by any OR gate to provide the required sum-of-products implementation. The names PLA and PAL® emerged for devices from different vendors during the development of PLDs. The implementation of combinational circuits with a PROM was illustrated in the previous section. The design of combinational circuits with PLA and PAL devices is presented in the next two sections.

6-8 Programmable Logic Array

The PLA is similar in concept to the PROM, except that the PLA does not provide full decoding of the variables and does not generate all the minterms. The decoder is replaced by an array of AND gates that can be programmed to generate product terms of the input variables. The product terms are then selectively connected to OR gates to provide the sum of products for the required Boolean functions.

The internal logic of a PLA with three inputs and two outputs is shown in Figure 6-20. Such a circuit is too small to be cost effective, but is presented here to demonstrate the typical logic configuration of a PLA. The diagram uses the array logic graphics symbols for complex circuits. Each input goes through a buffer and an inverter, represented in the diagram by a composite graphics symbol that has

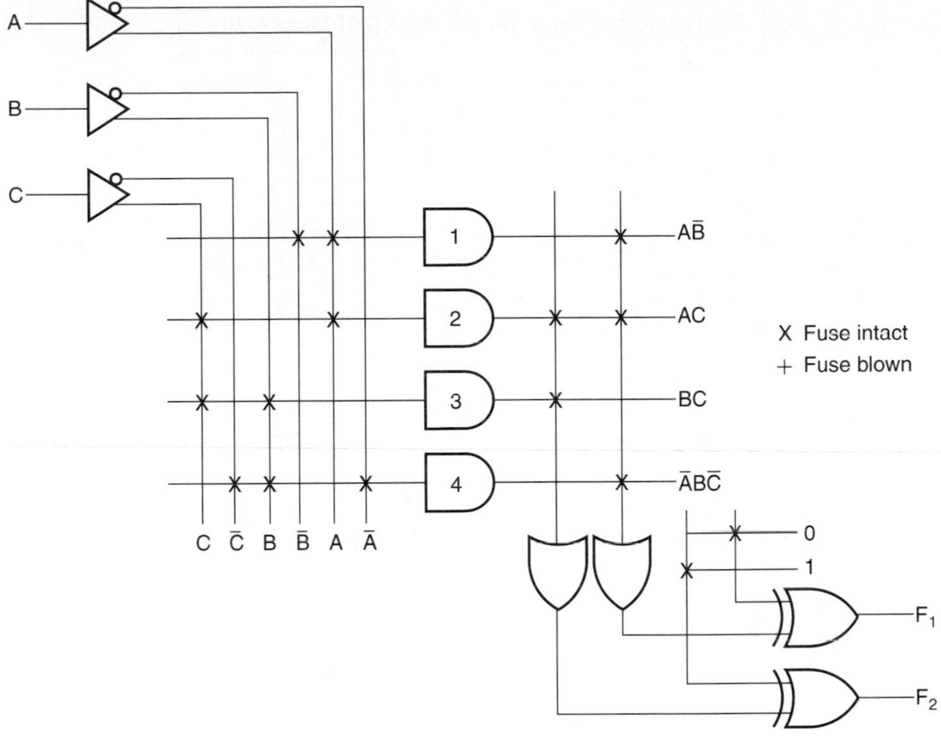

FIGURE 6-20
PLA with Three Inputs, Four Product Terms, and Two Outputs

both the true and the complement outputs. Programmable connections run from each input and its complement to the inputs of each AND gate, as indicated by the intersections between the vertical and horizontal lines. The outputs of the AND gates have programmable connections to the inputs of each OR gate. The output of the OR gate goes to an XOR gate, where the other input can be programmed to receive a signal equal to either logic 1 or logic 0. The output is inverted when the XOR input is connected to 1 (since $X \oplus 1 = \overline{X}$). The output does not change when the XOR input is connected to 0 (since $X \oplus 0 = X$). The particular Boolean functions implemented in the PLA of the figure are

$$F_1 = A\overline{B} + AC + \overline{A}B\overline{C}$$

$$F_2 = \overline{AC + BC}$$

The product terms generated in each AND gate are listed by the output of the gate in the diagram. The product term is determined from the inputs with CLOSED circuit connections. The output of an OR gate gives the logic sum of the selected

TABLE 6-4
Programming Table for the PLA in Figure 6-20

	Product term	Inputs			Outputs	
		A	B	C	(T) F_1	(C) F_2
$A\overline{B}$	1	1	0	—	1	—
AC	2	1	—	1	1	1
BC	3	—	1	1	—	1
$\overline{A}B\overline{C}$	4	0	1	0	1	—

product terms. The output may be complemented or left in its true form, depending on the programming of the connection associated with the XOR gate.

The fuse map of a PLA can be specified in tabular form. For example, the programming table that specifies the PLA of Figure 6-20 is listed in Table 6-4. The table consists of three sections. The first section lists the product term numbers. The second section specifies the required paths between inputs and AND gates. The third section specifies the paths between the AND and OR gates. For each output variable, we may have a T (for true) or C (for complement) for controlling the output exclusive-OR gate. The product terms listed on the left are not part of the table; they are included for reference only. For each product term, the inputs are marked with 1, 0, or — (dash). If a variable in the product term appears in its true form, the corresponding input variable is marked with a 1. If the variable in the product term appears complemented, the corresponding input variable is marked with a 0. If the variable is absent in the product term, it is marked with a dash.

The paths between the inputs and the AND gates are specified under the column heading "Inputs" in the table. A 1 in the input column specifies a CLOSED circuit from the input variable to the AND gate. A 0 in the input column specifies a CLOSED circuit from the complement of the variable to the input of the AND gate. A dash specifies OPEN circuits for both the input variable and its complement. It is assumed that an OPEN terminal on the input of an AND gate behaves like a 1.

The paths between the AND and OR gates are specified under the column heading "Outputs." The output variables are marked with 1's for those product terms that are included in the function. Each product term that has a 1 in the output column requires a CLOSED path from the output of the AND gate to the input of the OR gate. Those product terms marked with a dash specify an OPEN circuit. It is assumed that an open terminal on the input of an OR gate behaves like a 0. Finally, a T (true) output dictates that the other input of the corresponding XOR gate be connected to 0, and a C (complement) specifies a connection to 1.

The size of a PLA is specified by the number of inputs, the number of product terms, and the number of outputs. A typical PLA has 16 inputs, 48 product terms, and eight outputs. For n inputs, k product terms, and m outputs, the internal logic of the PLA consists of n buffer-inverter gates, k AND gates, m OR gates, and m XOR gates. There are $2n \times k$ programmable connections between the inputs and the AND array, $k \times m$ programmable connections between the AND and OR arrays, and m programmable connections associated with the XOR gates.

In designing a digital system with a PLA, there is no need to show the internal connections of the unit, as was done in Figure 6-20. All that is needed is a PLA programming table from which the PLA can be programmed to supply the required logic. As with a ROM, the PLA may be mask programmable or field programmable. With mask programming, the customer submits a PLA program table to the manufacturer. The table is used by the vendor to produce a custom-made PLA that has the internal logic specified by the customer. Field programming uses a PLA called a *field-programmable logic array*, or FPLA. This device can be programmed by the user by means of a commercial hardware programming unit.

In implementing a combinational circuit with a PLA, a careful investigation must be undertaken in order to reduce the number of distinct product terms so that the complexity of the circuit may be reduced. Fewer product terms can be achieved by simplifying the Boolean function to a minimum number of terms. The number of literals in a term is less important, since all the input variables are available anyway. It is wise, however, to avoid extra literals, as these may cause problems in testing the circuit and may reduce the speed of the circuit. Both the true and complement forms of each function should be simplified to see which one can be expressed with fewer product terms and which one provides product terms that are common to other functions.

EXAMPLE 6-2

Implement the following two Boolean functions with a PLA:

$$F_1(A,B,C) = \Sigma m0,1,2,4)$$

$$F_2(A,B,C) = \Sigma m0,5,6,7$$

The two functions are simplified in the maps of Figure 6-21. Both the true and complement outputs of the functions are simplified in sum-of-products form. The combination that gives a minimum number of product terms is

$$F_1 = \overline{AB + AC + BC}$$

$$F_2 = AB + AC + \overline{A}\,\overline{B}\,\overline{C}$$

The simplification gives four distinct product terms: AB, AC, BC, and $\overline{A}\,\overline{B}\,\overline{C}$. The PLA programming table for this combination is shown in the figure. Note that output F_1 is the true output, even though a C is marked over it in the table. This is because $\overline{F_1}$ is generated with an AND-OR circuit and is available at the output, of the OR gate. The XOR gate complements the function $\overline{F_1}$ to produce the true F_1 output.

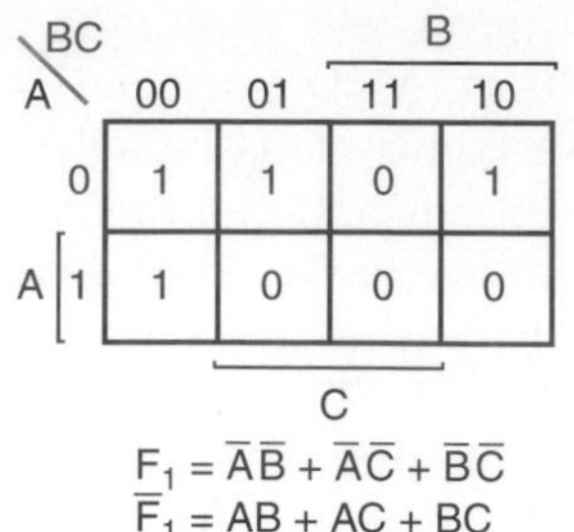

A \ BC	00	01	11	10
0	1	0	0	0
1	0	1	1	1

$F_2 = AB + AC + \overline{A}\overline{B}\overline{C}$
$\overline{F}_2 = \overline{A}C + \overline{A}B + A\overline{B}\overline{C}$

PLA programming table

	Product term	Inputs A B C	Outputs (C) F_1	Outputs (T) F_2
AB	1	1 1 –	1	1
AC	2	1 – 1	1	1
BC	3	– 1 1	1	–
$\overline{A}\overline{B}\overline{C}$	4	0 0 0	–	1

□ **FIGURE 6-21**
Solution to Example 6-2

The combinational circuit used in Example 6-2 is too simple to implement with a PLA. It was presented merely for purposes of illustration. A typical PLA has a sizable number of inputs and product terms. The simplification of complex Boolean functions with so many variables should be carried out by means of computer-assisted simplification procedures. The computer-aided design program simplifies the functions and their complements to a minimum number of terms. The program then selects a minimum number of product terms that cover all functions, each in the true or complement form. The PLA programming table is then generated, and the required connection map is obtained from the table. The connection map is applied to an FPLA programmer that executes the procedure to include or remove the connections in the integrated circuit.

6-9 Programmable Array Logic Devices

The PAL® device is a PLD with a fixed OR array and programmable AND array. Because only the AND gates are programmable, the PAL device is easier to program than, but is not as flexible as, the PLA. Figure 6-22 presents the logic configuration of a typical programmable logic device. The particular device shown has four inputs and four outputs. Each input has a buffer-inverter gate, and each output is generated by a fixed OR gate. The device has four sections, each composed of a three-wide AND-OR array, meaning that there are three programmable AND

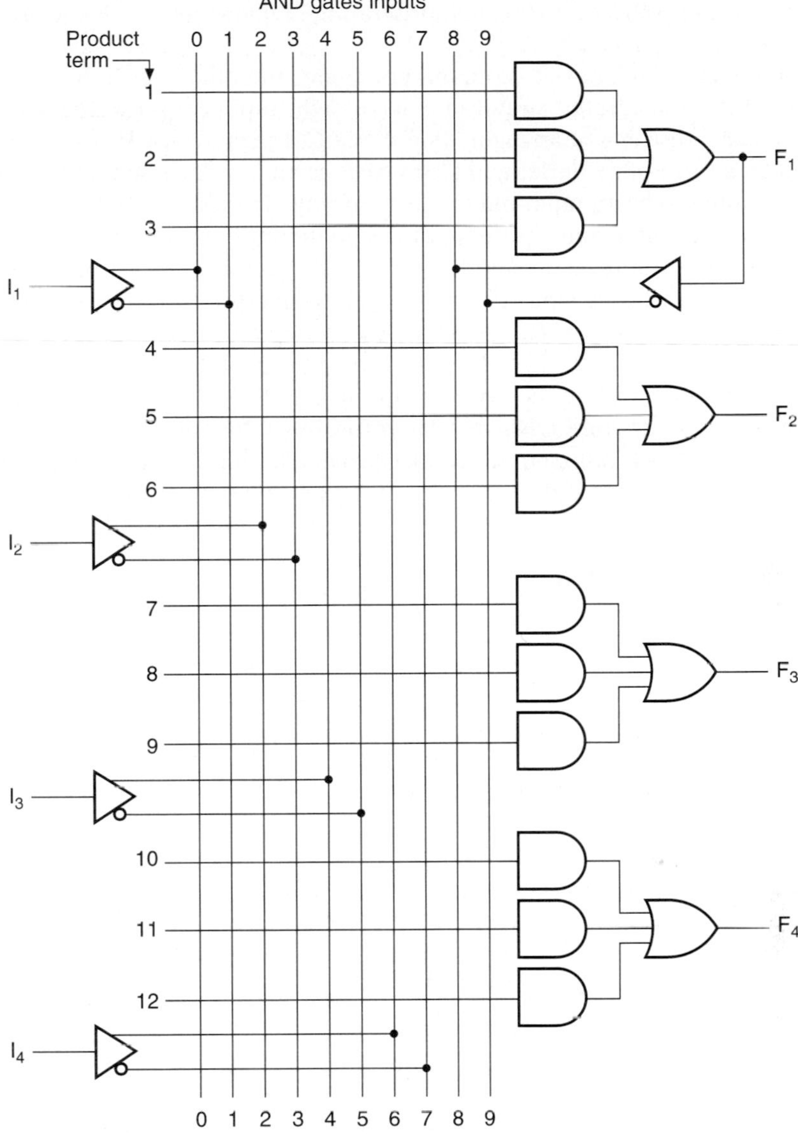

☐ **FIGURE 6-22**
PAL® device with Four Inputs, Four Outputs, and a Three-wide AND-OR Structure

gates in each section. Each AND gate has 10 programmable input connections, indicated in the diagram by 10 vertical lines intersecting each horizontal line. The horizontal line symbolizes the multiple-input configuration of an AND gate. One of the outputs shown is connected to a buffer-inverter gate and then fed back into the inputs of the AND gates through programmed connections. This is often done with all device outputs.

Commercial PAL devices contain more gates than the one shown in Figure 6-22. A small PAL integrated circuit may have eight inputs, eight outputs, and eight sections, each consisting of an eight-wide AND-OR array. Each PAL device output is driven by a three-state buffer and also serves as an input. These input/outputs can be programmed to be an input only, an output only, or bidirectional with a variable signal driving the three-state buffer enable signal. Flip-flops are often included in a PAL device between the array and the three-state buffer at the outputs. Since each output is fed back as an input through a buffer-inverter gate into the AND programmed array, a sequential circuit can be easily implemented.

In designing with a PAL device, the Boolean functions must be simplified to fit into each section. Unlike the arrangement in the PLA, a product term cannot be shared among two or more OR gates. Therefore, each function can be simplified by itself, without regard to common product terms. The number of product terms in each section is fixed, and if the number of terms in the function is too large, it may be necessary to use two or more sections to implement one Boolean function. In such a case, common terms may be useful.

As an example of a PAL device incorporated into the design of a combinational circuit, consider the following Boolean functions, given in sum-of-minterms form:

$$W(A,B,C,D) = \Sigma\, m(2,12,13)$$

$$X(A,B,C,D) = \Sigma\, m(7,8,9,10,11,12,13,14,15)$$

$$Y(A,B,C,D) = \Sigma\, m(0,2,3,4,5,6,7,8,10,11,15)$$

$$Z(A,B,C,D) = \Sigma\, m(1,2,8,12,13)$$

Simplifying the four functions to a minimum number of terms results in the following Boolean functions:

$$W = AB\overline{C} + \overline{A}\,\overline{B}C\overline{D}$$

$$X = A + BCD$$

$$Y = \overline{A}B + CD + \overline{B}\,\overline{D}$$

$$Z = AB\overline{C} + \overline{A}\,\overline{B}C\overline{D} + A\overline{C}\,\overline{D} + \overline{A}\,\overline{B}\,\overline{C}D$$

$$= W + A\overline{C}\,\overline{D} + \overline{A}\,\overline{B}\,\overline{C}D$$

Note that the function for Z has four product terms. The logical sum of two of these terms is equal to W. Thus, by using W, it is possible to reduce the number of terms for Z from four to three, so that the functions can fit into the PAL device in Figure 6-22.

The PAL programming table is similar to the table used for the PLA, except that only the inputs of the AND gates need to be programmed. Table 6-5 lists the PAL programming table for the preceding four Boolean functions. The table is divided into four sections with three product terms in each, to conform with the PAL device of Figure 6-22. The first two sections need only two product terms to implement the Boolean function. By placing W in the first section of the device, the feedback into the array is available to reduce the function Z to three terms.

☐ TABLE 6-5
PAL® Programming Table

Product term	AND Inputs					Outputs
	A	B	C	D	W	
1	1	1	0	—	—	$W = AB\overline{C}$
2	0	0	1	0	—	$+\overline{A}\,\overline{B}C\overline{D}$
3	—	—	—	—	—	
4	1	—	—	—	—	$X = A$
5	—	1	1	1	—	$+BCD$
6	—	—	—	—	—	
7	0	1	—	—	—	$Y = \overline{A}B$
8	—	—	1	1	—	$+CD$
9	—	0	—	0	—	$+\overline{B}\,\overline{D}$
10	—	—	—	—	1	$Z = W$
11	1	—	0	0	—	$+A\overline{C}\,\overline{D}$
12	0	0	0	1	—	$+\overline{A}\,\overline{B}\,\overline{C}D$

The connection map for the PAL device, as specified in the programming table, is shown in Figure 6-23. For each 1 or 0 in the table, we mark the corresponding intersection in the diagram with the symbol for a CLOSED circuit connection. For each dash, we mark the diagram with OPEN circuit connections for both the true and complement inputs. If the AND gate is not used, we leave all of its inputs as CLOSED circuits. Since the corresponding input receives both the true and the complement of each input variable, we have $A \cdot \overline{A} = 0$, and the output of the AND gate is always 0.

As with all PLDs, the design of PAL devices is facilitated by using computer-aided techniques. The programming is accomplished with a commercial hardware programming unit.

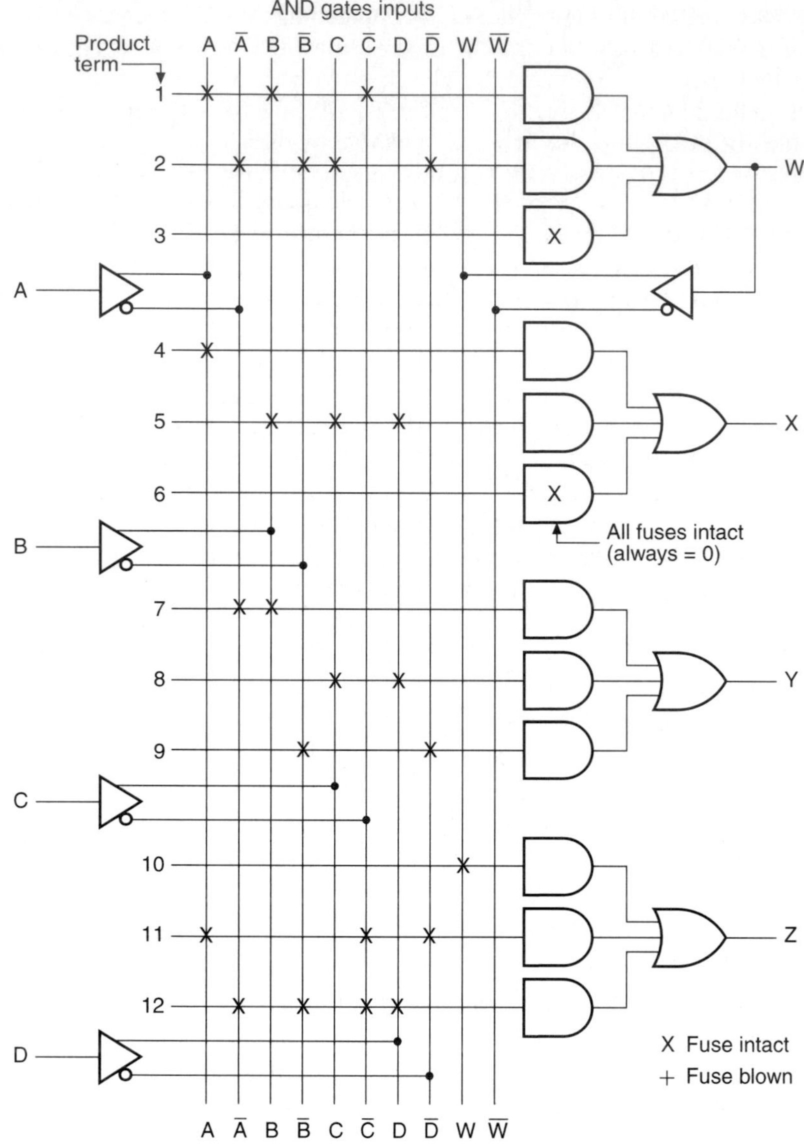

FIGURE 6-23
Connection Map for PAL® Device as Specified in Table 6-5

6-10 VLSI Programmable Logic Devices

The advantage of using the PLD in the design of digital systems is that it can be programmed to incorporate a complex logic function within a single IC. But for larger or more complex functions, VLSI technology is appropriate. VLSI design refers to the design of digital systems that contain thousands to millions of gates within a single IC chip.

There are three primary ways of designing VLSI circuits. In *full custom design*, an entire design of the chip, down to the smallest detail of the layout, is performed. Since this process is very expensive, custom design can be justified only for dense, fast ICs that are likely to be sold in sizable quantities. A closely related technique is *standard cell design*, in which large parts of the design have been performed ahead of time or, possibly, used in previous designs. The predesigned parts are connected to form the IC design. An intermediate-cost procedure, standard cell design gives lower density and lower performance than full custom design.

The third approach to VLSI design is by the use of a *gate array*, which is a pattern of gates fabricated in silicon that is repeated thousands of times, so that the entire chip contains identical gates. Arrays of 1,000 to 100,000 gates can be fabricated within a single IC, depending on the technology used. The application of a gate array requires that the design specify how the gates are interconnected and how the interconnections are routed. Many steps of the fabrication process are common and independent of the final logic function. These steps are economical, since they can be used for numerous different designs. Additional fabrication steps are required to interconnect the gates in order to customize the gate array to the particular design. As with all complex designs, the design process is highly automated with the use of CAD.

More recently, VLSI approaches have been developed for PLDs to handle designs that in the past were implemented by many small chips or with gate arrays having from 1,000 to 50,000 gates. The new approaches yield high-capacity programmable logic devices typically called complex programmable logic devices (CPLDs) or field-programmable gate arrays (FPGAs). These two structures typically share the following properties:

1. substantial amounts of uncommitted combinational logic;
2. preimplemented flip-flops; and
3. programmable interconnections between the combinational logic, flip-flops, and the chip input/outputs.

Aside from these properties common to all VLSI PLDs, the devices differ significantly from vendor to vendor. To illustrate the variety of structures, we will give an overview of a version of a CPLD and two versions of an FPGA.

Altera MAX 7000 CPLDs

One family of CPLDs manufactured by Altera® is based on an EEPROM floating-gate technology. The overall structure of an Altera® MAX 7000™ CPLD is shown

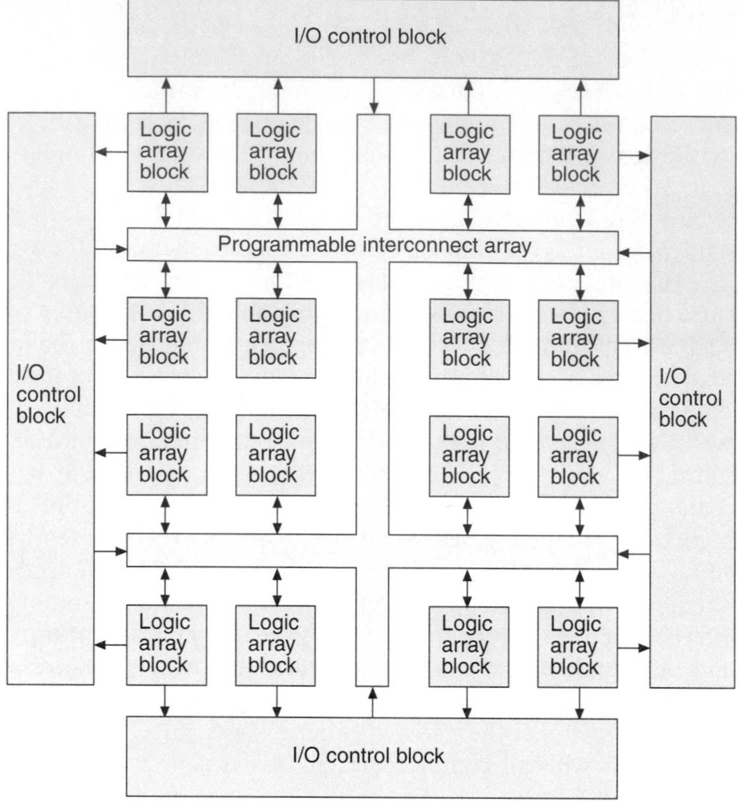

□ **FIGURE 6-24**
Altera® MAX 7000™ Structure (Reprinted with Permission of Altera Corporation,© Altera Corp., 1991)

in Figure 6-24. In the particular device illustrated, there are 16 identical logic array blocks, all of whose outputs feed into the programmable interconnect array. The latter also receives inputs from the I/O control blocks, thereby providing inputs from the outside to the interior of the circuit. Connections can be programmed as needed from all signals within the programmable interconnection array to the inputs of the logic array block.

Each logic array block contains 16 macrocells, each with a flip-flop in addition to a basic combinational logic structure. Some of the AND gates in the macrocell are used for flip-flop control, such as Preset, Clear, and Clock Enable. The flip-flop itself can be programmed to act as a *D*, *T*, *JK*, or *SR* flip-flop. In addition, special connection features allow AND gates from other macrocells to be used by a given macrocell, and NAND gates can be shared at the inputs to all macrocells. These features deal with the problem of functions requiring more AND terms than the five provided in the basic combinational part of each macrocell.

With respect to I/O, as is apparent in Figure 6-24, each macrocell in the logic array blocks around the outer edge of the Altera® structure is connected to I/O

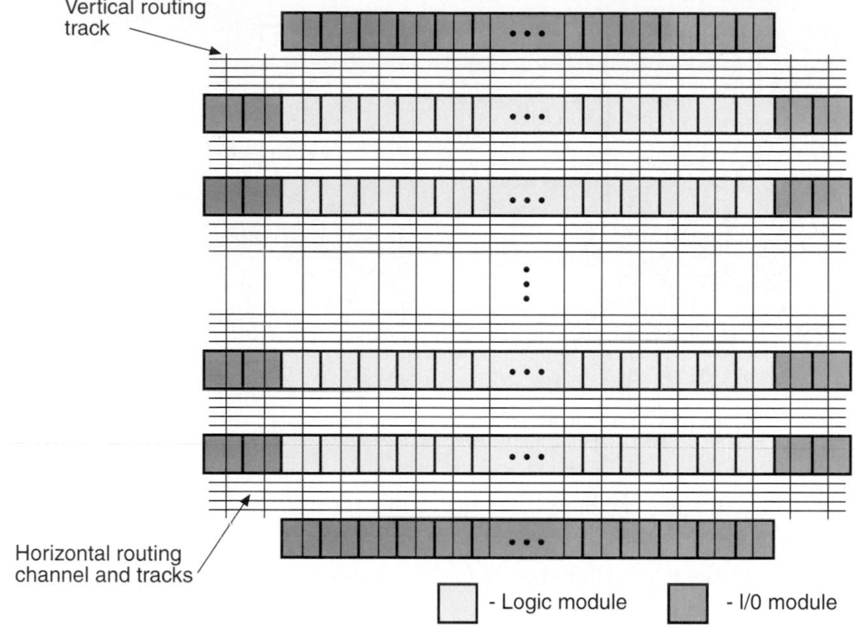

□ **FIGURE 6-25**
Actel® ACT 3® Structure (Reprinted Courtesy of Actel Corporation, © Actel Corp., 1993)

control blocks. These consist of an I/O pin driven by a three-state buffer having the output of the macrocell as its input. Buffer control comes from an AND in the macrocell. Two input paths—one from the input and one from the output of the three-state buffer—enter the programmable interconnect array.

Actel ACT 3 FPGAs

Actel® FPGAs use a gate-array-like structure as their foundation. The structure of an Actel® ACT® 3 chip is shown in Figure 6-25. The logic in the array is provided in rows of logic modules, shown in light blue. At the top and bottom of the array, the rows consisting of I/O modules that attach to the adjacent input/output pins are shown in blue. Likewise, the two modules on each end of the remaining rows are I/O modules that attach to the adjacent I/O pins. The interconnection structure consists of horizontal and vertical interconnection tracks containing wiring segments. Many horizontal tracks lie in each channel between the rows of logic. The vertical tracks are fewer in number and pass over both the horizontal channels and the logic modules.

The Actel FPGA uses antifuse technology that provides permanent and nonvolatile programming. The device has many antifuses for attaching the logic and I/O module inputs and outputs to wiring segments in the channels. It also has anti-

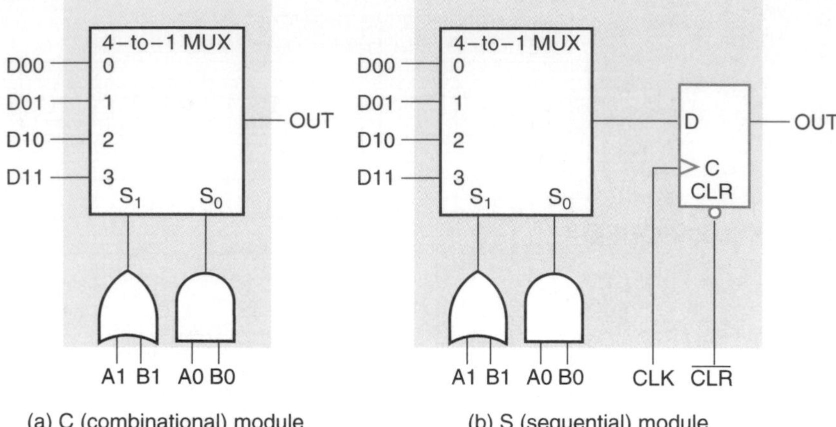

□ **FIGURE 6-26**
Actel® Logic Modules (Reprinted Courtesy of Actel Corporation, © Actel Corp., 1993)

fuses which connect together the wiring segments that lie in the tracks to provide connections of varying lengths.

The two types of Actel logic modules are shown in Figure 6-26. These modules lie in a fixed pattern within the rows of the array. The *C* module is based on a multiplexer implementation of combinational logic functions, as treated in Section 3-7. Functions of three variables in which one of the variables is not complemented can be implemented by a *C* module. With the AND gate and the OR gate on the multiplexer select inputs, some functions of up to eight variables can be implemented. The *S* module contains the same combinational logic structure, with a positive-edge-triggered *D* flip-flop added at its output. The flip-flop has a dedicated Clear input CLR as well. The Actel I/O module contains *D* flip-flops for storage on both the output and the input. The output has a three-state buffer. The storage flip-flops can either hold or load, depending on applied control signals. There is a choice on both the output and the input between presenting combinational or stored data.

The Actel structure is distinguished by the fact that no programming is available within the logic modules. Instead, all programming is done using antifuses in the wiring tracks. Also, *C* modules do not contain flip-flops, so that the chip area wasted on unused flip-flops is limited. If more flip-flops are needed, they can be constructed by wiring together *C* modules. The Actel structure is compact and fast, due to the very small size and low resistance of the antifuses.

Xilinx XC4000 Structure

The Xilinx® XC4000™ FPGA structure is shown in Figure 6-27. The logic within the FPGA is implemented in an array of programmable blocks of logic called configurable logic blocks (CLBs). Input to and output from the array is handled by input/output blocks (IOBs) along the edges of the array. The CLBs and IOBs are

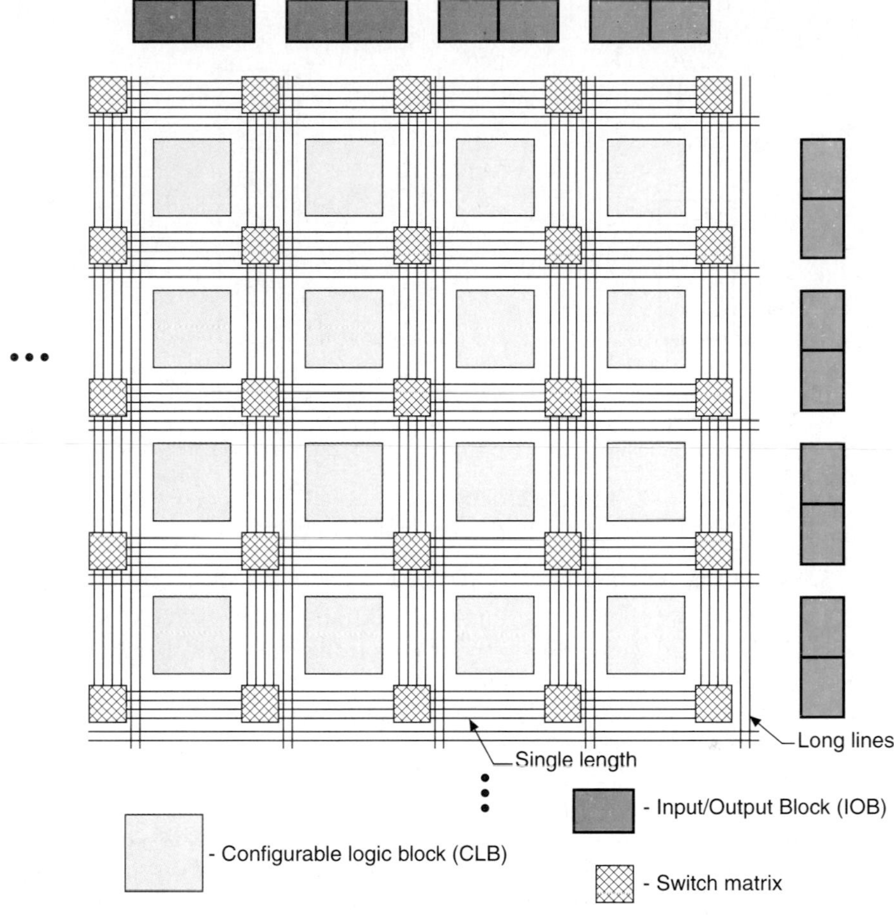

□ **FIGURE 6-27**
Xilinx® XC4000™ FPGA Structure (Adapted with Permission of Xilinx, Inc.)

interconnected by a variety of programmable interconnection structures. Connections to and from CLBs and IOBs can be programmed, and wire segments can be interconnected, to form paths from one block to another by using an array of programmable connection blocks called switch matrices.

Xilinx uses SRAM technology to store the programming information. After power is applied to the circuit, the program data defining the logic configuration must be loaded into the SRAM. There are a number of different ways of loading the information. In fact, FPGA actually contains logic to load the information into itself from a PROM. Once the programming information is loaded, the FPGA switches from programming mode to the operational mode in which the logic is available. The logic remains until either the FPGA is reprogrammed or the power is turned off. The ability to reprogram the FPGA allows different logic to be implemented in a system by the same FPGA at different times.

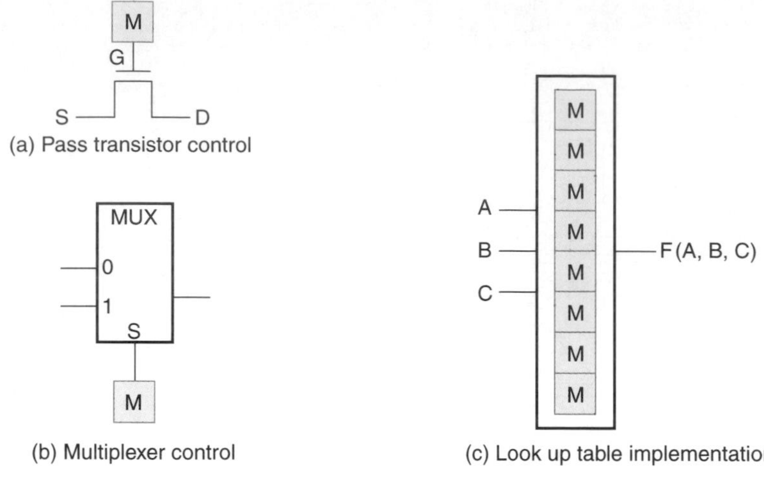

□ **FIGURE 6-28**
SRAM Bit Use in Xilinx® FPGAs

SRAM bits control the logic implemented in a Xilinx FPGA by the use of three techniques, illustrated in Figure 6-28: pass transistor control, multiplexer control, and lookup table implementation.

Figure 6-28(a) shows an SRAM cell driving the gate (G) terminal of an n-channel MOS transistor. (See Section 2-9.) When such a transistor is to make a bidirectional connection for the passage of a signal between two wiring segments, it is called a *pass transistor*. One wiring segment is attached to the source S of the transistor and the other segment to the drain D. When the SRAM cell contains a 0, the transistor is OFF, the path between the wiring segments is OPEN, and no signal is passed. When the SRAM cell contains a 1, the transistor is ON and the path between the wiring segments is CLOSED, allowing a signal to pass between the segments. An XC4000 series part typically contains tens of thousands of such transistors in its interconnection structure.

In Figure 6-28(b), an SRAM cell is attached to the select input S of a 2–to–1 multiplexer. If the SRAM cell contains a 0, then the value on the 0 input of the multiplexer is passed to the multiplexer output. If the SRAM cell contains a 1, then the value on the 1 input is passed to the multiplexer output. The structure is used to make selections between two signals. Sometimes there are two SRAM cells driving a 4–to–1 multiplexer. Finally, in cases where the data inputs to the multiplexer are X and $\overline{X}$, the multiplexer symbol is replaced by an XOR gate with X on one input and the SRAM cell on the other.

The final use of SRAM cells is to build a lookup table, as in Figure 6-28(c). In the figure, a lookup table for a three-variable function $F(A, B, C)$ is illustrated. The SRAM cells in the table store the actual truth table of the function, so each cell contains the value of function F for the corresponding minterm. The lookup table is functionally equivalent to a multiplexer with the SRAM bits applied to the data

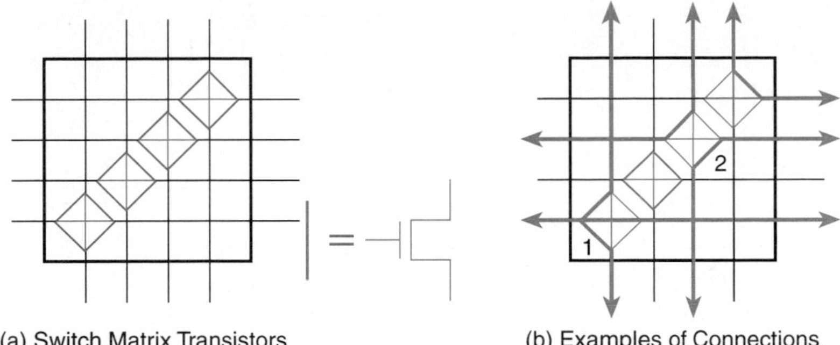

□ **FIGURE 6-29**
Example of Xilinx® Switch Matrix (Adapted with Permission of Xilinx®, Inc.)

inputs and the input variables *A*, *B*, and C on the selection inputs. For example, if (*A*, *B*, *C*) = 0 1 0, the value in SRAM cell 2 (binary 010) appears on the output of the circuit. So the lookup table is actually a multiplexer implementation of combinational logic, as discussed in Section 3-7, with the SRAM cells providing the data inputs.

Xilinx Interconnections

Connections between CLBs and between CLBs and IOBs are made using wiring segments in both horizontal and vertical channels lying between the various blocks. Some of the segments are very long, spanning the entire length or width of the array. Two such segments, referred to as *long lines*, are shown in the routing channels in Figure 6-27. Other segments are just long enough to span a single CLB. These can be interconnected using the switch matrices shown in the figure.

An example of a switch matrix appears in Figure 6-29(a). Where four segments meet, there are six pass transistors—one vertical, one horizontal, and four on the diagonals. Each pass transistor is represented by a blue line. The connection between two segments is CLOSED for a 1 stored in the SRAM cell driving the gate of the transistor. The connection between two segments is OPEN for a 0 stored in the SRAM cell. Several connections are shown in Figure 6-29(b). Note that at point 1 all four segments are joined together by closing three transistors. In this case, all six transistors could be closed to make a connection with less electrical resistance. At point 2, two distinct signal paths pass through a single set of pass transistors. In order to connect to CLBs and IOBs, wiring segments from the block inputs and outputs extend across adjacent wiring channels. SRAM-controlled pass transistors lie at selected intersections between these segments and perpendicular wiring segments in the channels.

In addition to the long-line and single-length wiring segments in Figure 6-27, there are double-length segments, special interconnections for clocks, three-state buffers as inputs to some of the long lines, and special hardware for implementing large AND gates in the wiring channels.

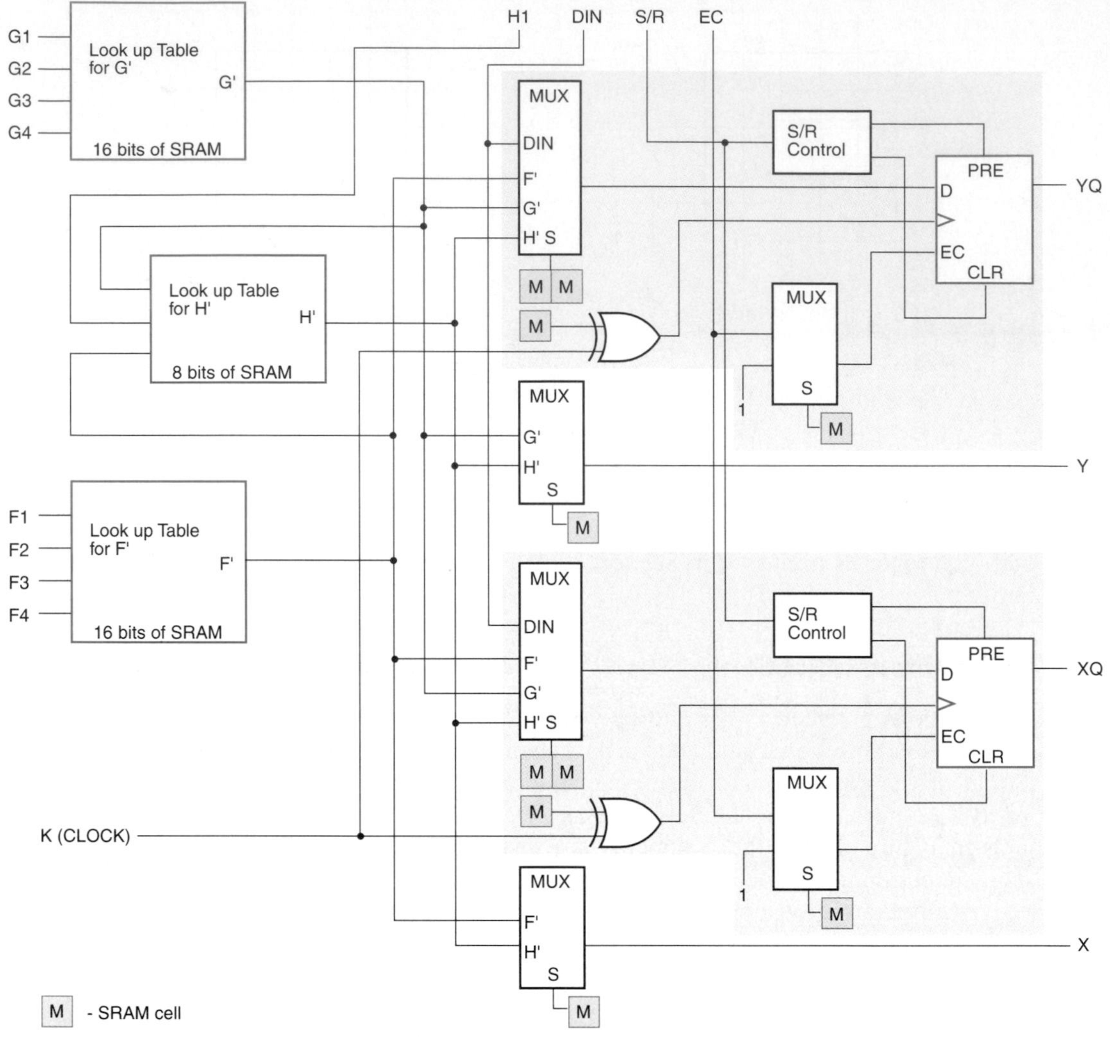

□ **FIGURE 6-30**
Simplified Diagram of a Xilinx® Configurable Logic Block (Adapted with permission of Xilinx®, Inc.)

Xilinx Logic

Most of the logic circuits in a Xilinx FPGA lie within the CLBs and the IOBs. Both of these structures are internally programmable and fairly complex. We will look in detail at the CLB and then sketch the main features of the IOB.

A simplified diagram of a Xilinx CLB is shown in Figure 6-30. First of all, there are 13 inputs to the block, including the *Clock K*. Not shown is a *Global Set/Reset* signal that goes to all of the flip-flops in the FPGA. Two flip-flops and

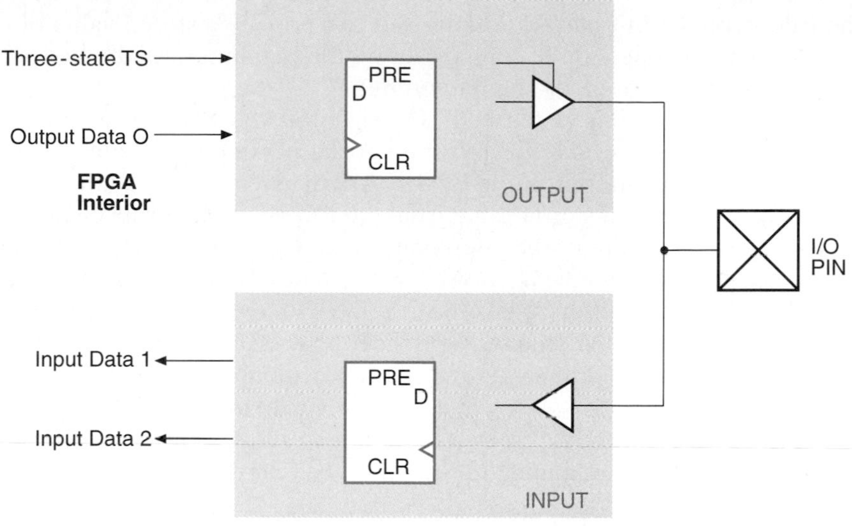

□ **FIGURE 6-31**
Sketch of Xilinx® IOB Structure (Adapted with Permission of Xilinx®, Inc.)

their associated logic are highlighted in blue. The remaining part of the CLB is used to implement combinational logic.

In the figure, three lookup tables implement combinational functions. Two 4-input tables implement two functions labeled F' and G'. These two blocks feed a third lookup table, which implements a three-variable combinational function of F', G', and input $H1$. By properly setting the two 2-to-1 multiplexers, any pair of F', G', and H' can be assigned to the two combinational outputs, X and Y, of the CLB. Two functions of up to four variables and selected functions of up to nine variables can be implemented.

Two D flip-flops directly drive outputs XQ and YQ. Each of the D inputs is selected by one of two multiplexers. Each multiplexer can select from F', G', H', and input DIN. The multiplexers allow any one of the combinational functions that are implemented to drive a flip-flop. In addition, in cases where the combinational functions are used as outputs and do not drive the flip-flops, DIN can be selected as a direct path to the flip-flop, so that it can be used for storage of values from other sources. The two XORs allow each flip-flop to be individually selected to be positive- or negative-edge triggered. The two S*R* controls allow the signal *S/R* to be designated as either an asynchronous Set or Reset for each of the flip-flops. This setting also governs the effect of the *Global Set/Reset* on the state of the flip-flop. Finally, two multiplexers allow the input *EC* to optionally act as an ENABLE signal for each flip-flop.

The Xilinx IOB is also programmable and offers the designer a number of choices. We will briefly sketch its primary features. To simplify the explanation, we consider the output and input portions of the IOB separately, as shown in Figure 6-31.

The output portion of the IOB can provide output data O from the interior of the FPGA on the I/O pin. Alternatively, it can provide a stored value of output data O from a flip-flop. A three-state driver on the output allows the I/O pin to be used as an input, an output, or an input/output.

In the input portion of the IOB, the signal at the I/O pin enters an input buffer. The signal can be fed directly to input data 1 and input data 2, which are two input lines to the interior of the FPGA. Alternatively, one or both of the input data lines can be fed by a stored value from the I/O pin signal or its complement.

The overall structure of the interconnections, CLBs, and IOBs is clearly quite complicated. A designer having to deal with hundreds of CLBs and IOBs and thousands of interconnection points in such an FPGA would have a very difficult job. As a consequence, CAD tools are provided that take a design in the form of a schematic or textual description, automatically partition the design into pieces that fit within a CLB, place the pieces into specific CLBs and route the connections between the CLBs, and to and from IOBs. The end result of this process is tens of thousands of bits of programming information that can be loaded into the FPGA to implement the desired logic.

6-11 Chapter Summary

Memory is of two types: random-access memory (RAM) and read-only memory (ROM). For both types, we apply an address to read from or write into a data word. Read and write operations have specific steps and associated timing parameters, including access time and write cycle time. Memory can be static or dynamic and volatile or nonvolatile. Internally, a RAM chip consists of an array of RAM cells, decoders, write circuits, read circuits, and output circuits. A combination of a write circuit, read circuit, and the associated RAM cells can be logically modeled as a RAM bit slice. RAM bit slices, in turn, can be combined to form two-dimensional RAM cell arrays, which, with decoders and output circuits added, form the basis for a RAM chip. Output circuits use three-state buffers in order to facilitate connecting together an array of RAM chips without significant additional logic. Error detection and correction codes, often based on Hamming codes, are used to detect or correct errors in stored RAM data.

ROMs that implement a combinational circuit are the most straightforward of the programmable logic devices (PLDs), since they simply store a truth table. Programmable logic arrays (PLAs) and programmable array logic (PAL®) devices are other forms of basic PLDs. A ROM can be viewed as having a programmable OR array, a PAL device as having a programmable AND array, and a PLA as having both programmable AND and OR arrays. The primary tasks of the logic designer are to understand each PLD and its limitations and to fit the logic that is to be designed into the PLD.

For more complex logic, very large-scale integrated (VLSI) PLDs are appropriate. A VLSI PLD can be characterized as having substantial uncommitted combinational logic, many flip-flops, and complex programmable interconnection structures. Otherwise, VLSI PLD structures and programming technologies vary significantly from vendor to vendor. Design using VLSI PLDs as well as simple PLDs makes heavy use of CAD tools to isolate the designer as much as possible from complex details.

REFERENCES

1. WESTE, N. H. E., AND ESHRAGHIAN, K. *Principles of CMOS VLSI Design: A Systems Perspective*, 2nd ed. Reading, MA: Addison-Wesley, 1993.
2. HODGES, D. A., AND JACKSON, H. G. *Analysis and Design of Digital Integrated Circuits*, 2nd ed. New York: McGraw-Hill, 1988.
3. HAZNEDAR, H. *Digital Microelectronics.* Redwood City, CA: Benjamin/Cummings, 1991.
4. HAMMING, R. W. "Error Detecting and Error Correcting Codes." *Bell System Tech. Jour.*, 29 (1950): 147–160.
5. LIN, S., AND COSTELLO, D. J., JR. *Error Control Coding.* Englewood Cliffs, NJ: Prentice Hall, 1983.
6. WAKERLY, J. F. *Digital Design: Principles and Practices,* 2nd ed. Englewood Cliffs, NJ: Prentice Hall, 1994.
7. NELSON, V. P., NAGLE, H. T., CARROLL, B. D., AND IRWIN, J. D. *Digital Logic Circuit Analysis and Design.* Englewood Cliffs, NJ: Prentice Hall, 1995.
8. KITSON, B. *Programmable Array Logic Handbook.* Sunnyvale, CA: Advanced Micro Devices, Inc., 1984.
9. TRIMBERGER, S. M., ED. *Field-Programmable Gate Array Technology.* Boston: Kluwer Academic Publishers, 1994.
10. *Actel FPGA Data Book and Design Guide 1994.* Sunnyvale, CA: Actel Corporation, 1993.
11. *Altera Data Book.* San Jose: Altera Corporation, 1995.
12. *The Programmable Logic Data Book*, 2nd ed. San Jose: Xilinx, Inc., 1994.

PROBLEMS

The asterisk (*) indicates a more advanced problem.

6–1. The following memories are specified by the number of words times the number of bits per word. How many address lines and input-output data lines are needed in each case? (a) $4K \times 16$; (b) $128K \times 8$; (c) $16M \times 32$; (d) $2G \times 8$.

6–2. Give the number of bytes stored in the memories listed in Problem 6–1.

6–3. Word number $(723)_{10}$ in the memory shown in Figure 6-3 contains the binary equivalent of $(3451)_{10}$. List the 10-bit address and the 16-bit memory contents of the word.

6–4. Show memory timing waveforms for write and read operations. Assume a CPU clock of 33 MHz and memory access and write cycle times of 70 ns.

6–5. A $16K \times 4$ RAM chip uses coincident decoding by splitting the internal decoder into row select and column select. (a) Assuming that the RAM cell array is square, what is the size of each decoder, and how many AND gates are required for decoding an address? (b) Determine the row and column selection lines that are enabled when the input address is the binary equivalent of $(9000)_{10}$.

6–6. *Assume that the largest decoder that can be used in an $m \times 1$ RAM chip has 11 address inputs and that coincident decoding is employed. In order to construct RAM chips that contain more 1-bit words than m, multiple RAM cell arrays with decoders and read/write circuits are included in the chip.

(a) With the decoder restrictions given, how many RAM cell arrays are required to construct a $16M \times 1$ RAM chip?

(b) Show the decoder required to select from among the different RAM arrays in the chip and its connections to address bits and column decoders.

6–7. **(a)** How many $32K \times 8$ RAM chips are needed to provide a memory capacity of 256K bytes?

(b) How many lines of the address must be used to access 256K bytes? How many of these lines are connected to the address inputs of all chips?

(c) How many lines must be decoded for the chip select inputs? Specify the size of the decoder.

6–8. Using the $64K \times 8$ RAM chip in Figure 6-12 plus a decoder, construct the block diagram for a $256K \times 16$ RAM.

6–9. Given the 8-bit data word 10110101, generate the 13-bit composite word for the Hamming code that corrects single errors and detects double errors.

6–10. Given the 11-bit data word 00100111010, generate the corresponding 15-bit Hamming code word.

6–11. A 12-bit Hamming code word containing 8 bits of data and 4 parity bits is read from memory. What was the original 8-bit data word that was written into memory if the 12-bit word read out is

(a) 000010101010 **(b)**111110010110 **(c)** 100111110100

6–12. How many parity check bits must be included with the data word to achieve single error correction and double error detection when the data word contains (a) 16 bits; (b) 32 bits; (c) 64 bits?

6–13. It is necessary to formulate the Hamming code for four data bits D_3, D_5, D_6. and D_7, together with three parity bits P_1, P_2, and P_4.

(a) Evaluate the 7-bit composite code word for the data word 0101.

(b) Evaluate the three check bits C_1, C_2, and C_4, assuming no error.

(c) Assume an error in bit D_5 during storage into memory. Show how the error in the bit is detected and corrected.

(d) Add a parity bit P to include double error detection in the code. Assume that errors occurred in bits P_2 and D_5. Show how this double error is detected.

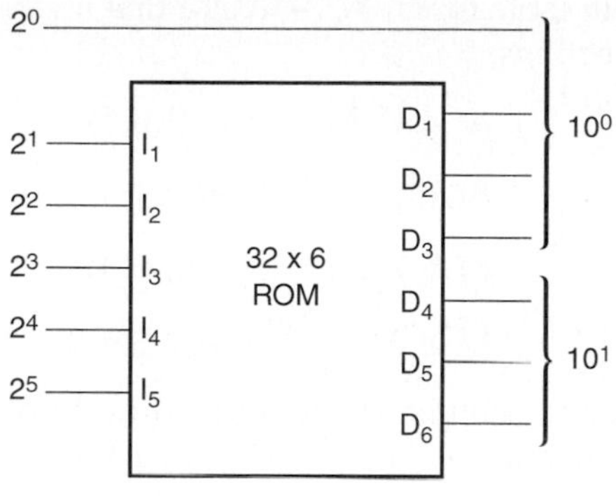

□ **FIGURE 6-32**
Binary-to-Decimal ROM Converter

6–14. *A modified single-error-correcting, double-error-detecting Hamming code for four bits of data D_3, D_5, D_6, and D_7 has the following parity bit equations:

$$P_1 = D_3 \oplus D_5 \oplus D_6$$

$$P_2 = D_3 \oplus D_5 \oplus D_7$$

$$P_4 = D_3 \oplus D_6 \oplus D_7$$

$$P_8 = D_5 \oplus D_6 \oplus D_7$$

(a) Find the binary values of the four check bits for a single error in each of the eight bit positions of the code.
(b) Assuming that either a single or double error has occurred, indicate the type of error for each of the following words read from memory:
(1) 10100011 (2) 11001110 (3) 00011101
(c) Give the correct data bit values for the single-error cases in (b).

6–15. Given a 32 × 8 ROM chip with an enable input, show the external connections necessary to construct a 128 × 8 ROM with four chips and a decoder.

6–16. The 32 × 6 ROM, together with the 2^0 line, as shown in Figure 6-32, converts a 6-bit binary number to its corresponding two-digit BCD number. For example, binary 100001 converts to BCD 011 0011 (decimal 33). Specify the truth table for the ROM.

6–17. Specify the size of a ROM (number of words and number of bits per word) that will accommodate the truth table for the following combinational circuit components:
(a) A 4-bit adder-subtractor.
(b) A binary multiplier that multiplies two 6-bit numbers.
(c) A code converter from a 4-digit BCD number to a binary number.

6–18. Tabulate the truth table for an 8×4 ROM that implements the following four Boolean functions:

$$A(X,Y,Z) = \Sigma\, m(3,6,7)$$

$$B(X,Y,Z) = \Sigma\, m(0,1,4,5,6)$$

$$C(X,Y,Z) = \Sigma\, m(2,3,4)$$

$$D(X,Y,Z) = \Sigma\, m(2,3,4,7)$$

6–19. Obtain the PLA programming table for the four Boolean functions listed in Problem 6–18. Minimize the number of product terms. Be sure to attempt to share product terms between functions and to consider the use of complemented (*C*) outputs.

6–20. Derive the PLA programming table for the combinational circuit that squares a 3-bit number. Minimize the number of product terms. (See Table 6-3 and Figure 6-18 for the equivalent ROM implementation.)

6–21. List the PLA programming table for the BCD–to–excess-3 code converter whose Boolean functions are simplified in Figure 3-10.

6–22. Repeat Problem 6–21, using a PAL device.

6–23. The following is the truth table of a three-input, four-output combinational circuit. Obtain the PAL programming table for the circuit, and mark the fuses to be blown in a PAL diagram similar to the one shown in Figure 6-22.

Inputs			Outputs			
X	**Y**	**Z**	**A**	**B**	**C**	**D**
0	0	0	0	1	0	0
0	0	1	1	1	1	1
0	1	0	1	0	1	1
0	1	1	0	1	0	1
1	0	0	1	0	1	0
1	0	1	0	0	0	1
1	1	0	1	1	1	0
1	1	1	0	1	1	1

6–24. Give the inputs attached to $F1$ through $F4$, $G1$ through $G4$, and $H1$, as well as the truth tables for F', G', and H', to implement the following pair of functions in a single Xilinx CLB:

$$X = AK + BK + \overline{C}\,\overline{D}K + AEJL$$

$$Y = \overline{A}\,\overline{B}(C + D)$$

6–25. *Give the inputs to $F1$ through $F4$, $G1$ through $G4$, and $H1$, as well as the truth tables for F', G', and H' and the values for all SRAM bits in Figure 6-30, to implement the sequential circuit having the state diagram in Figure 4-22. Assume that the flip-flops are to be positive-edge triggered and that the synchronous SR inputs to the flip-flops are both to be Reset. Is it possible to implement this whole circuit, including the logic for output Y, in a single CLB?

6–26. *Specify inputs and outputs for as few Actel S-modules and C-modules as possible to implement the sequential circuit specified by the following equations:

$$D_A(X,A,B,C) = \Sigma\, m(2,3,5,7)$$

$$D_B(X,A,B,C) = \Sigma\, m(2,3,4,5,6,7)$$

$$D_C(X,A,B,C) = \Sigma\, m(3,7,9,13)$$

$$Z(X,A,B,C) = \Sigma\, m(4,5,6,7,11,13,14,15)$$

Draw the resulting logic diagram.

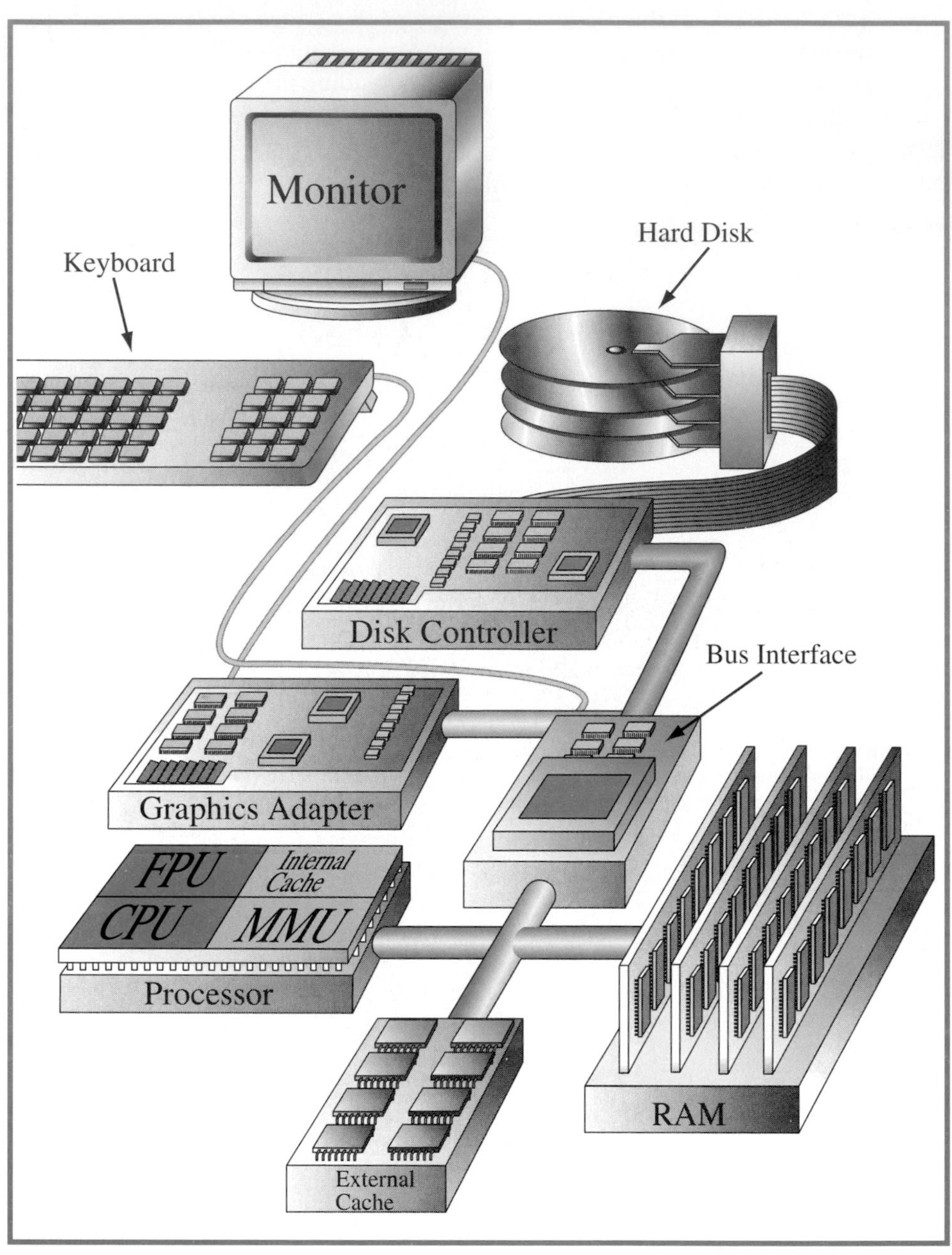

Monitor
Hard Disk
Keyboard
Disk Controller
Bus Interface
Graphics Adapter
FPU
Internal Cache
CPU
MMU
Processor
RAM
External Cache

CHAPTER

7

REGISTER TRANSFERS AND DATAPATHS

In a digital system, a datapath and a control unit are frequently present at the top levels of the design hierarchy. A *datapath* consists of processing logic and a collection of registers that performs data processing. A *control unit* is made up of logic that determines the sequence of data-processing operations performed by the datapath. In this chapter, we see how designers use the various functional blocks discussed in previous chapters to build a datapath. In addition, we introduce the concept of a register transfer for representing data-processing actions performed by a datapath. Register transfer notation describes elementary data-processing actions referred to as *microoperations*. Register transfers move information between registers, between registers and memory, and through processing logic. Dedicated transfer hardware using multiplexers and shared transfer hardware called *buses* implement these movements of data.

We define a sample computer datapath that illustrates data-processing actions and serves as a framework for the design of detailed processing logic. The concept of a control word provides a tie between the datapath and its control unit. Using this concept, the sample datapath serves as the foundation for illustrating control unit design in Chapter 8. In addition, the datapath provides a basis for the design of computers in Chapter 10. Finally, we give an example of a pipelined datapath that provides higher processing rates and allows a faster clock. This example appears in Chapter 8 as a vehicle for demonstrating a pipelined control unit and in Chapter 10 as support for the design of a pipelined computer.

In the current chapter, the main topics are register transfers, microoperations, buses, data-paths, datapath components, and control words. These are quite general topics, so we find the generic computer shaded lightly over many of the electronic portions of the hardware. Note that the CPU and FPU in the processor chip are heavily shaded. This is because these components contain major pipelined datapaths that perform processing. It is in the CPU and the FPU that additions, subtractions, and other operations take place. The connections between various electronic parts of the computer are also shaded heavily. These connections are buses, which we discuss for the first time in this chapter.

7-1 Datapaths and Operations

A digital system is a sequential circuit made up of interconnected flip-flops and gates. In Chapter 4, we learned that sequential circuits can be specified by means of state tables. To specify a large digital system with state tables is very difficult, if not impossible, because the number of states is prohibitively large. To overcome this difficulty, digital systems are designed using a modular, hierarchical approach. The system is partitioned into modular subsystems, each of which performs some functional task. The modules are constructed hierarchically from functional blocks such as registers, counters, decoders, multiplexers, buses, arithmetic elements, flip-flops, and primitive gates. Interconnecting the various subsystems through data and control signals results in a digital system.

In most digital system designs, we partition the system into two types of modules: a *datapath*, which performs data-processing operations, and a *control unit*, which determines the sequence of those operations. Figure 7-1 shows the general relationship between a datapath and a control unit. *Control signals* are binary signals that activate the various data-processing operations. To activate a sequence of such operations, the control unit sends the proper sequence of control signals to the datapath. The control unit, in turn, receives status bits from the datapath. These variables describe aspects of the state of the datapath. The control unit uses the variables in defining the specific sequence of the operations to be performed. Note that the datapath and control unit may also interact with other parts of a digital system, such as memory and input-output logic, through the paths labeled data inputs, data outputs, control inputs, and control outputs.

Datapaths are best defined by their registers and the operations that are performed on binary data stored in the registers. Examples of register operations are

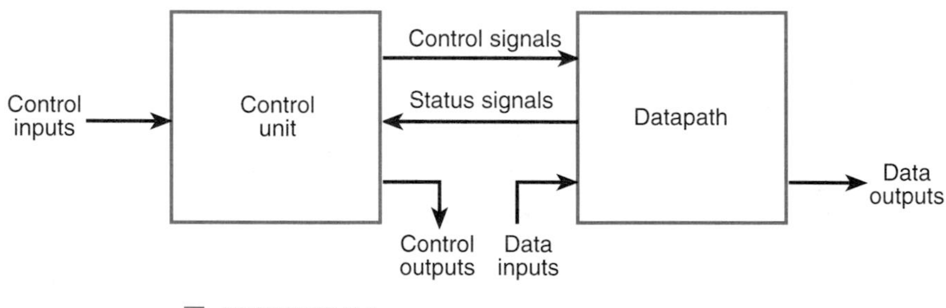

□ **FIGURE 7-1**
Interaction between Datapath and Control Unit

shift, count, clear, and load. The registers are assumed to be basic components of the digital system. The movement of the data stored in registers and the processing performed on the data are referred to as *register transfer operations*. The register transfer operations of digital systems are specified by the following three basic components:

1. the set of registers in the system,
2. the operations that are performed on the data stored in the registers, and
3. the control that supervises the sequence of operations in the system.

A register has the capability to perform one or more *elementary operations* such as load, count, add, subtract, and shift. For example, a right-shift register is a register that can shift data to the right. A counter is a register that increments a number by one. A single flip-flop is a 1-bit register that can be set or cleared. In fact, by this definition, the flip-flops and closely associated gates of any sequential circuit can be called registers.

The elementary operations performed on the data stored in registers are called *microoperations*. Examples of microoperations are loading the contents of one register into another, adding the contents of two registers, and incrementing the contents of a register. A microoperation is usually, but not always, performed in parallel on a string of bits during one clock cycle. As in the case of a counter, the result of the microoperation may replace the previous binary data in the register. Alternatively, the result may be transferred to another register, leaving the previous data unchanged. The sequential functional blocks introduced in Chapter 5 are registers that implement microoperations. A counter with parallel load is able to perform two microoperations—increment and load. A bidirectional shift register is able to perform shift right and shift left microoperations.

The control unit provides signals that sequence the microoperations in a prescribed manner. The results of a currently executing microoperation may determine both the sequence of control signals and the sequence of future microoperations to be executed. Note that the term "microoperation," as used here, does not refer to any particular way of producing the control signals: specifically, it does not imply that the control signals are generated by a control unit based on microprogramming, as will be discussed in Chapter 8.

This chapter introduces the components of register transfer with a symbolic notation for representing registers and specifying the operations on their contents. The register transfer method uses a set of expressions and statements that resemble statements used in programming languages. This notation can concisely specify part or all of a complex digital system such as a computer. The specification then serves as a basis for a more detailed design of the system.

In the sections that follow, we focus on concepts used in designing datapaths. In the last half of the chapter, we illustrate datapaths with two common types of datapath appearing in computers. In the last half of Chapter 8, we add three types of control units to the datapaths. Then, in Chapter 10, we present two complete CPUs made up of combined datapaths and control units. Note that although we

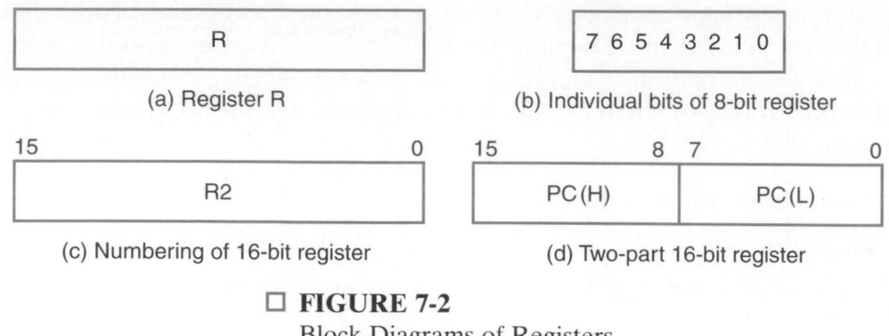

□ **FIGURE 7-2**
Block Diagrams of Registers

use computer datapaths and control units for purposes of illustration, the concepts in this chapter and Chapter 8 apply equally well to other digital systems.

7-2 Register Transfer Operations

We will denote the registers in a digital system by uppercase letters (sometimes followed by numerals) that indicate the function of the register. For example, a register that holds an address for the memory unit is usually called an address register and can be designated by the name *AR*. Other designations for registers are *PC* for program counter, *IR* for instruction register, and *R*2 for register 2. The individual flip-flops in an *n*-bit register are typically numbered in sequence from 0 to $n-1$, starting with 0 in the least significant (often the rightmost) position and increasing toward the most significant position. Since the 0 bit is on the right, this order can be referred to as *little-endian*, in the same manner as for bytes in Chapter 1. The reverse order, with bit 0 on the left, is referred to as *big-endian*. Figure 7-2 shows representations of registers in block diagram form. The most common way to represent a register is by a rectangular box with the name of the register inside, as in part (a) of the figure. The individual bits can be identified as in part (b). The numbering of bits represented by just the leftmost and rightmost values at the top of a register box is illustrated by a 16-bit register *R*2 in part (c). A 16-bit program counter, *PC*, is partitioned into two sections in part (d) of the figure. In this case, bits 0 through 7 are assigned the symbol *L* (for low-order byte), and bits 8 through 15 are assigned the symbol *H* (for high-order byte). The label *PC*(*L*), which may also be written *PC*(7:0), refers to the low-order byte of the register, and *PC*(*H*) or *PC*(15:8) refers to the high-order byte.

Data transfer from one register to another is designated in symbolic form by means of the replacement operator (←). Thus, the statement

$$R2 \leftarrow R1$$

denotes a transfer of the contents of register *R*1 into register *R*2. In other words, the statement designates the copying of the contents of *R*2 into *R*1. The register *R*1 is referred to as the *source* of the transfer and the register *R*2 as the *destination*. By

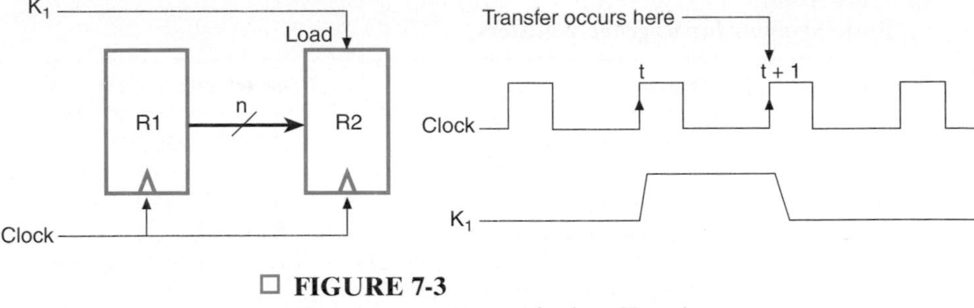

□ **FIGURE 7-3**
Transfer from $R1$ to $R2$ when $K_1 = 1$

definition, the contents of the source register do not change as a result of the transfer. Only the contents of $R2$ as the destination register change.

A statement that specifies a register transfer implies that datapath circuits are available from the outputs of the source register to the inputs of the destination register and that the destination register has a parallel load capability. Normally, we want a given transfer to occur not for every clock pulse, but only for specific values of the control signals. This can be specified by a *conditional statement*, symbolized by the *if-then* form

$$\text{If}(K_1 = 1)\text{then}(R2 \leftarrow R1)$$

where K_1 is a control signal generated in the control unit. In fact, K_1 can be any Boolean function that evaluates to 0 or 1. A more concise way of writing the if-then form is

$$K_1\text{: } R2 \leftarrow R1$$

This control condition, terminated with a colon, symbolizes the requirement that the transfer operation be executed by the hardware only if $K_1 = 1$.

Every statement written in register transfer notation presupposes a hardware construct for implementing the transfer. Figure 7-3 shows a block diagram that depicts the transfer from $R1$ to $R2$. The n outputs of register $R1$ are connected to the n inputs of register $R2$. The letter n is used to indicate the number of bits in the register transfer path from $R1$ to $R2$. When the width of the path is known, n is replaced by an actual number. Register $R2$ has a load control input that is activated by the control signal K_1. It is assumed that the signal is synchronized with the same clock as the one applied to the register. The flip-flops are assumed to be positive-edge triggered by this clock. As shown in the timing diagram, K_1 is set to 1 on the rising edge of a clock pulse at time t. The next positive transition of the clock at time $t + 1$ finds $K_1 = 1$, and the inputs of $R2$ are loaded into the register in parallel. In this case, K_1 returns to 0 on the positive clock transition at time $t + 1$, so that only a single transfer from $R1$ to $R2$ occurs.

Note that the clock is not included as a variable in the register transfer statements. It is assumed that all transfers occur in response to a clock transition. Even though the control condition K_1 becomes active at time t, the actual transfer does

□ **TABLE 7-1**
Basic Symbols for Register Transfers

Symbol	Description	Examples
Letters (and numerals)	Denotes a register	*AR, R2, DR, IR*
Parentheses	Denotes a part of a register	$R2(1), R2(7{:}0), AR(L)$
Arrow	Denotes transfer of data	$R1 \leftarrow R2$
Comma	Separates simultaneous transfers	$R1 \leftarrow R2,\ R2 \leftarrow R1$
Square brackets	Specifies an address for memory	$DR \leftarrow M[AR]$

not occur until the register is triggered by the next positive transition of the clock, at time $t + 1$.

The basic symbols we use in register transfer notation are listed in Table 7-1. Registers are denoted by an uppercase letter, possibly followed by one or more uppercase letters and numerals. Parentheses are used to denote a part of a register by specifying the range of bits in the register or by giving a symbolic name to a portion of the register. The left-pointing arrow denotes a transfer of data and the direction of transfer. A comma is used to separate two or more register transfers that are executed at the same time. For example, the statement

$$K_3\colon R2 \leftarrow R1,\ R1 \leftarrow R2$$

denotes an operation that exchanges the contents of two registers simultaneously for a positive clock edge at which $K_3 = 1$. Such an exchange is possible with registers made of flip-flops, but presents a difficult timing problem with registers made of latches. Square brackets are used in conjunction with a memory transfer. The letter M designates a memory word, and the register enclosed inside the square brackets provides the address of the word in memory. This is explained in more detail in Section 7-5.

7-3 MICROOPERATIONS

A microoperation is an elementary operation performed on data stored in registers or in memory. The microoperations most often encountered in digital systems are of four types:

1. *Transfer* microoperations, which transfer binary data from one register to another.
2. *Arithmetic* microoperations, which perform arithmetic on data in registers.
3. *Logic* microoperations, which perform bit manipulation on data in registers.
4. *Shift* microoperations, which shift data in registers.

A given microoperation may be of more than one type. For example, a 1's complement operation can be both an arithmetic microoperation and a logic microoperation.

Transfer microoperations were introduced in the previous section. This type of microoperation does not change the binary data as it moves from the source register to the destination register. The other three types of microoperations can produce new binary data and, hence, new information. In digital systems, basic sets of operations are used to form sequences that implement more complicated operations. In this section, we define a basic set of microoperations, symbolic notation for these microoperations, and descriptions of the digital hardware that implements them.

Arithmetic Microoperations

We define the basic arithmetic microoperations as add, subtract, increment, decrement, and complement. The statement

$$R0 \leftarrow R1 + R2$$

specifies an add operation. It states that the contents of register $R2$ are to be added to the contents of register $R1$ and the sum transferred to register $R0$. To implement this statement with hardware, we need three registers and a combinational component that performs the addition, such as a parallel adder. The other basic arithmetic operations are listed in Table 7-2. Subtraction is most often implemented through complementation and addition. Instead of using the minus operator, we can specify 2's complement subtraction by the statement

$$R0 \leftarrow R1 + \overline{R2} + 1$$

where $\overline{R2}$ specifies the 1's complement of $R2$. Adding 1 to $\overline{R2}$ gives the 2's complement of $R2$. Finally, adding the 2's complement of $R2$ to the contents of $R1$ is equivalent to $R1 - R2$.

The increment and decrement microoperations are symbolized by a plus-one and minus-one operation, respectively. These operations are implemented by using a special combinational circuit, an adder-subtractor, or a binary up-down counter with parallel load.

□ **TABLE 7-2**
Arithmetic Microoperations

Symbolic designation	Description
$R0 \leftarrow R1 + R2$	Contents of $R1$ plus $R2$ transferred to $R0$
$R2 \leftarrow \overline{R2}$	Complement of the contents of $R2$ (1's complement)
$R2 \leftarrow \overline{R2} + 1$	2's complement of the contents of $R2$
$R0 \leftarrow R1 + \overline{R2} + 1$	$R1$ plus 2's complement of $R2$ transferred to $R0$ (subtraction)
$R1 \leftarrow R1 + 1$	Increment the contents of $R1$ (count up)
$R1 \leftarrow R1 - 1$	Decrement the contents of $R1$ (count down)

Multiplication and division are not listed in Table 7-2. Multiplication can be represented by the symbol $*$ and division by /. These two operations are not included in the basic set of arithmetic microoperations because they are assumed to be implemented by sequences of basic microoperations. In contrast, multiplication can be considered as a microoperation if implemented by a combinational circuit as illustrated in Section 3-11. In such a case, the result is transferred into a destination register at the clock edge after all signals have propagated through the entire combinational circuit.

There is a direct relationship between the statements written in register transfer notation and the registers and digital functions required for their implementation. To illustrate, consider the following two statements:

$$\overline{X}K_1{:}R1 \leftarrow R1 + R2$$

$$XK_1{:}R1 \leftarrow R1 + \overline{R2} + 1$$

Control variable K_1 activates an operation to add or subtract. If, at the same time, control variable X is equal to 0, then $\overline{X}K_1 = 1$, and the contents of $R2$ are added to the contents of $R1$. If X is equal to 1, then $XK_1 = 1$, and the contents of $R2$ are subtracted from the contents of $R1$. Note that the two control conditions are Boolean functions and reduce to 0 when $K_1 = 0$, a condition that inhibits the execution of both operations simultaneously.

A block diagram showing the implementation of the preceding two statements is given in Figure 7-4. An n-bit adder-subtractor, similar to the one shown in

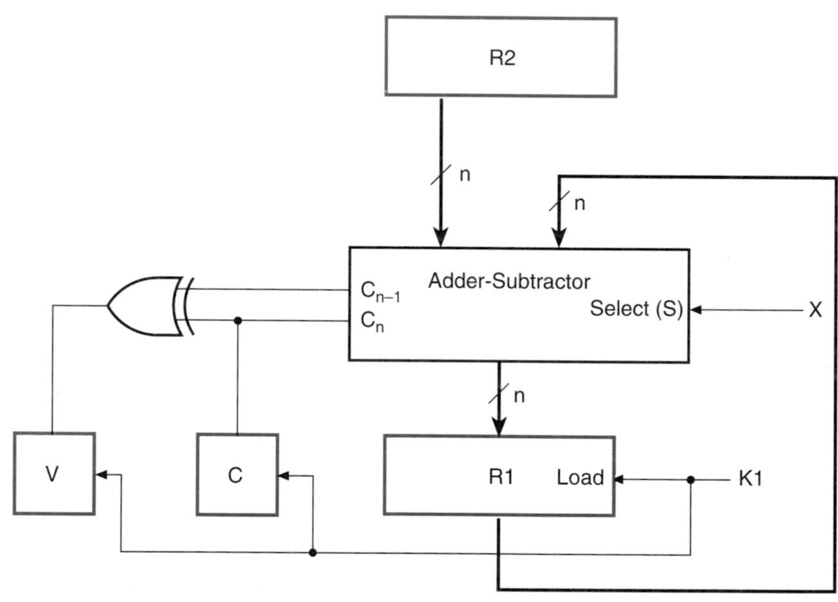

□ **FIGURE 7-4**
Implementation of Add and Subtract Microoperations

Figure 3-30, receives its input data from registers $R1$ and $R2$. The sum or difference is applied to the inputs of $R1$. The Select input S of the adder-subtractor selects the operation in the circuit. When $S = 0$, the two inputs are added, and when $S = 1$, $R2$ is subtracted from $R1$. Applying the control variable X to the S input activates the required operation. The output of the adder-subtractor is loaded into $R1$ on any positive clock edge at which $\overline{X}K_1 = 1$ or $XK_1 = 1$. We can simplify this to just K_1, since

$$\overline{X}K_1 + XK_1 = (\overline{X} + X)K_1 = K_1$$

Thus, the control variable X selects the operation, and the control variable K_1 loads the result into $R1$.

Based on the discussion of overflow in Section 3-10, the overflow output is transferred to flip-flop V, and the output carry from the most significant bit of the adder-subtractor is transferred to flip-flop C, as shown in Figure 7-4. These transfers occur when $K_1 = 1$ and are not represented in the register transfer statements; if desired, we could show them as additional simultaneous transfers.

Logic Microoperations

Logic microoperations are useful in manipulating the bits stored in a register. These operations consider each bit in the register separately and treat it as a binary variable. The symbols for the four basic logic operations are shown in Table 7-3. The NOT microoperation, represented by a bar over the source register name, complements all bits and thus is the same as the 1's complement. The symbol $\wedge$ is used to denote the AND microoperation and the symbol $\vee$ to denote the OR microoperation. By using these special symbols, it is possible to distinguish between the add microoperation represented by a + and the OR microoperation. Although the + symbol has two meanings, one can distinguish between them by noting where the symbol occurs. If the + occurs in a microoperation, it denotes addition. If the + occurs in a control or Boolean function, it denotes OR. The OR microoperation will always use the $\vee$ symbol. For example, in the statement

$$(K_1 + K_2): R1 \leftarrow R2 + R3,\ R4 \leftarrow R5 \vee R6$$

the + between K_1 and K_2 is an OR operation between two variables in a control condition. The + between $R2$ and $R3$ specifies an add microoperation. The OR microoperation is designated by the symbol $\vee$ between registers $R5$ and $R6$. The logic microoperations can be easily implemented with a group of gates, one for each bit position. The NOT of a register of n bits is obtained with n NOT gates in parallel. The AND microoperation is obtained using a group of n AND gates, each of which receives a pair of corresponding inputs from the two source registers. The outputs of the AND gates are applied to the corresponding inputs of the destination register. The OR and exclusive-OR microoperations require a similar arrangement of gates.

□ **TABLE 7-3**
Logic Microoperations

Symbolic designation	Description
$R0 \leftarrow \overline{R1}$	Logical bitwise NOT (1's complement)
$R0 \leftarrow R1 \wedge R2$	Logical bitwise AND (clears bits)
$R0 \leftarrow R1 \vee R2$	Logical bitwise OR (sets bits)
$R0 \leftarrow R1 \oplus R2$	Logical bitwise XOR (complements bits)

The logic microoperations can change bit values, clear a group of bits, or insert new bit values into a register. The following examples show how the bits stored in the 16-bit register $R1$ can be selectively changed by using a logic microoperation and a logic operand stored in the 16-bit register $R2$.

The AND microoperation can be used for clearing one or more bits in a register to 0. The Boolean equations $X \cdot 0 = 0$ and $X \cdot 1 = X$ dictate that, when ANDed with 0, a binary variable X produces a 0, but when ANDed with 1, the variable remains unchanged. A given bit or group of bits in a register can be cleared to 0 if ANDed with 0. Consider the following example:

10101101 10101011	$R1$	(data)
00000000 11111111	$R2$	(mask)
00000000 10101011	$R1 \leftarrow R1 \wedge R2$	

The 16-bit logic operand in $R2$ has 0's in the high-order byte and 1's in the low-order byte. By ANDing the contents of $R2$ with the contents of $R1$, it is possible to clear the high-order byte of $R1$ and leave the bits in the low-order byte unchanged. Thus, the AND operation can be used to selectively clear bits of a register. This operation is sometimes called *masking out* the bits, because it masks or deletes all 1's in the *data* in $R1$, based on bit positions that are 0 in the *mask* provided in $R2$.

The OR microoperation is used to set one or more bits in a register. The Boolean equations $X + 1 = 1$ and $X + 0 = X$ dictates that, when ORed with 1, the binary variable X produces a 1, but when ORed with 0, the variable remains unchanged. A given bit or group of bits in a register can be set to 1 if ORed with 1. Consider the following example:

10101101 10101011	$R1$	(data)
11111111 00000000	$R2$	(mask)
11111111 10101011	$R1 \leftarrow R1 \vee R2$	

The high-order byte of $R1$ is set to all 1's by ORing it with all 1's in the $R2$ operand. The low-order byte remains unchanged because it is ORed with 0's.

The XOR (exclusive-OR) microoperation can be used to complement one or more bits in a register. The Boolean equations $X \oplus 1 = \overline{X}$ and $X \oplus 0 = X$ dictate that, when a binary variable X is XORed with 1, it is complemented, but when XORed with 0, the variable remains unchanged. By XORing a bit or group of bits in register $R1$ with 1's in selected positions in $R2$, it is possible to complement the bits in the selected positions in $R1$. Consider the following example:

10101101 10101011	$R1$	(data)
11111111 00000000	$R2$	(mask)
01010010 10101011	$R1 \leftarrow R1 \oplus R2$	

The high-order byte in $R1$ is complemented after the XOR operation with $R2$, and the low-order byte is unchanged

Shift Microoperations

Shift microoperations are used for lateral movement of data. The contents of a source register can be shifted either right or left. A *left shift* is toward the most significant bit, and a *right shift* is toward the least significant bit. Shift microoperations are used in the serial transfer of data. They are also used for manipulating the contents of registers in arithmetic, logical, and control operations. The destination register for a shift microoperation may be the same as or different from the source register. There are no standard symbols for shift microoperations. We use strings of letters to represent the two shift microoperations defined in Table 7-4. For example,

$$R0 \leftarrow \text{sr } R0, R1 \leftarrow \text{sl } R2$$

are two microoperations that respectively specify a 1-bit shift to the right of the contents of register $R0$ and a transfer of the contents of $R2$ shifted one bit to the left into register $R1$. The contents of $R2$ are not changed by these shifts.

For a left-shift microoperation, we call the rightmost bit of the destination register the *incoming bit*. For a right-shift microoperation, we define the leftmost bit of the destination register as the incoming bit. The incoming bit may have different values, depending upon the type of shift microoperation. Here we assume that, for sr and sl, the incoming bit is 0, as shown in the examples in Table 7-4. The *outgoing bit* is the leftmost bit of the source register for the left-shift operation and the rightmost bit of the source register for the righ-shift operation. For the left and right shifts shown, the outgoing bit value is simply discarded. In chapter 9, we will explore other types of shifts differing in the way incoming and outgoing bits are treated.

□ TABLE 7-4
Examples of Shifts

		Eight-bit examples	
Type	**Symbolic designation**	**Source $R2$**	**After shift: Destination $R1$**
shift left	$R1 \leftarrow \text{sl } R2$	10011110	00111100
shift right	$R1 \leftarrow \text{sr } R2$	11100101	01110010

7-4 Multiplexer-Based Transfer

There are occasions when a register receives data from two or more different sources at different times. Consider the following conditional statement having an *if-then-else* form:

$$\text{If}(K_1 = 1)\text{then}(R0 \leftarrow R1) \text{ else if } (K_2 = 1) \text{ then } (R0 \leftarrow R2)$$

The value in register $R1$ is transferred to register $R0$ when control signal K_1 equals 1; otherwise, the value in register $R2$ is transferred to $R0$ when K_2 equals 1. The conditional statement may be broken into two parts using the following control conditions:

$$K_1\text{: } R0 \leftarrow R1, \overline{K}_1 K_2\text{: } R0 \leftarrow R2$$

This specifies hardware connections from two registers, $R1$ and $R2$, to one common destination register $R0$. In addition, making a selection between two source registers must be based on values of the control variables K_1 and K_2.

The block diagram of a circuit with 4-bit registers that implements the conditional register transfer statements by using a multiplexer is shown in Figure 7-5(a). The quad 2–to–1 multiplexer selects between the two source registers. For $K_1 = 1$, $R1$ is loaded into $R0$, irrespective of the value of K_2. For $K_1 = 0$ and $K_2 = 1$, $R2$ is loaded into $R0$. When both K_1 and K_2 are equal to 0, the multiplexer does select $R2$ as the input to $R0$, but the contents of $R2$ are not loaded into register $R0$ because the control function, $K_2 + K_1$, into the LOAD input of $R0$ is equal to 0.

The detailed logic diagram of the hardware implementation is shown in Figure 7-5(b). The diagram uses functional block symbols based on detailed logic for the registers in Figure 5-2 and for the multiplexer in Figure 3-20. Note that since the detailed diagram represents just a part of a system, there are inputs and outputs that are not yet connected. Also, the clock is not shown in the block diagram, but is shown in the detailed diagram. It is important to relate the information given in a block diagram such as Figure 7-5(a) with the detailed wiring connections in the corresponding logic diagram (Figure 7-5(b)). In order to save space, we often omit such detailed logic diagrams in designs. However, it is possible to obtain a logic diagram with detailed wiring from the corresponding block diagram and a library of functional blocks. In fact, such a procedure is performed by computer programs used for automated logic synthesis.

7-5 Bus-Based Transfer

A typical digital system has many registers. Paths must be provided to transfer data from one register to another. The amount of logic and number of interconnections may be excessive if each register has its own dedicated set of multiplexers. A more efficient scheme for transferring data between registers is a system that uses a shared transfer path called a *bus*. A bus is characterized by a set of common lines usually driven by selection logic. Control signals for the logic select a single source and one or more destinations on any clock cycle for which a transfer occurs.

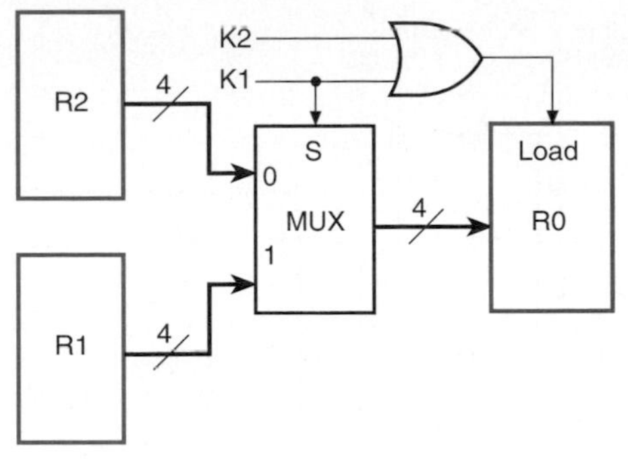

(a) Block diagram

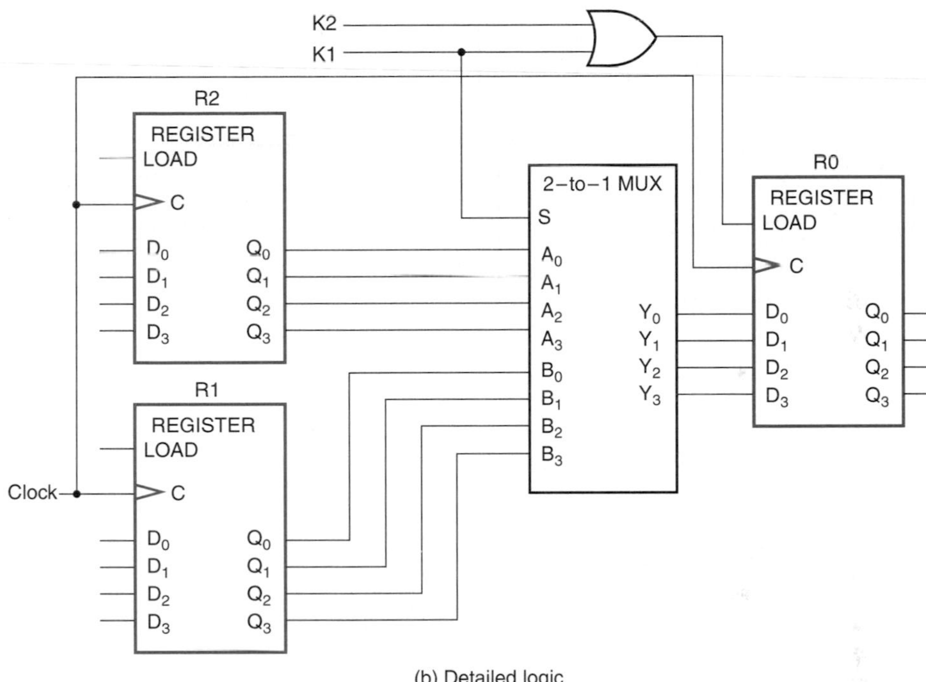

(b) Detailed logic

□ **FIGURE 7-5**
Use of Multiplexers to Select between Two Registers

In Section 7-4, we saw that multiplexers and parallel load registers can be used to implement dedicated transfers from multiple sources. A block diagram for such transfers between three registers is shown in Figure 7-6(a). There are three n-bit 2–to–1 multiplexers, each with its own select signal. Each register has its own load signal. The same system based on a bus can be implemented by using a single n-bit 3–to–1 multiplexer and parallel load registers. If a set of multiplexer outputs is shared as a common path, these output lines are a bus. Such a system with a single bus for transfers between three registers is shown in Figure 7-6(b). The control

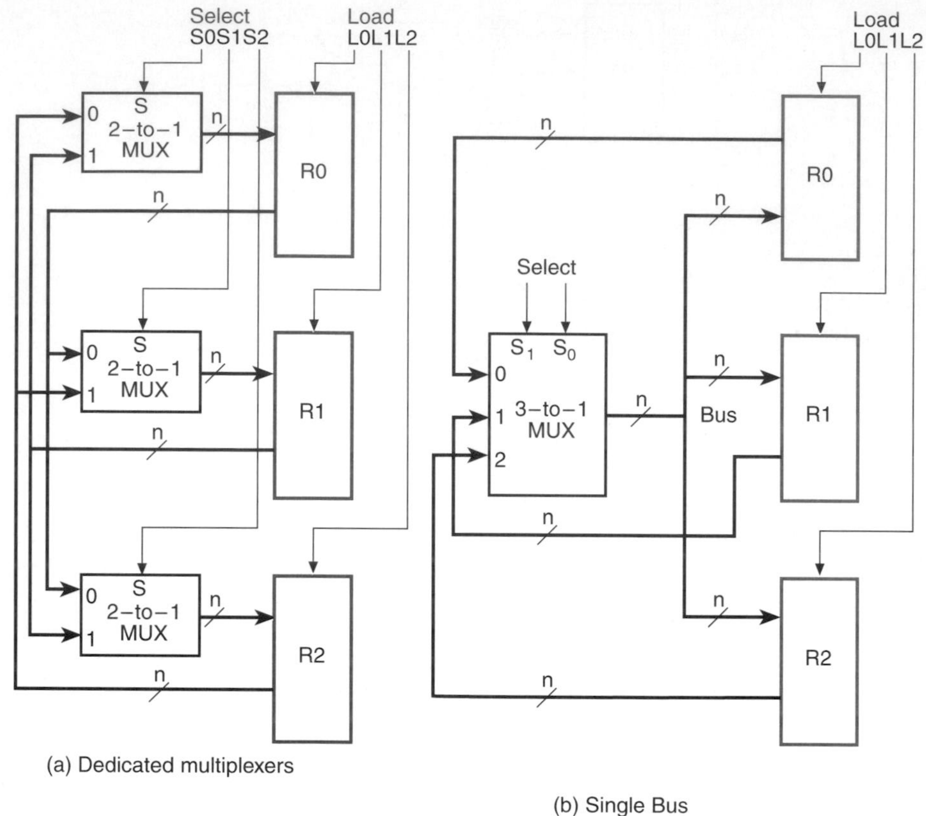

□ **FIGURE 7-6**
Single Bus versus Dedicated Multiplexers

input pair, Select, determines the contents of the single source register that will appear on the multiplexer outputs, i.e., on the bus. The load inputs determine the destination register or registers to be loaded with the bus data.

In Table 7-5, transfers using the single-bus implementation of Figure 7-6(b) are illustrated. The first transfer is from *R*2 to *R*0. Select equals 10, selecting input *R*2 to the multiplexer. Load signal *L*0 for register *R*0 is 1, with all other loads at 0 causing the contents of *R*2 on the bus to be loaded into *R*0 on the next positive clock transition. The second transfer in the table illustrates the loading of the contents of *R*1 into both *R*0 and *R*2. The source *R*1 is selected because Select is equal to 01. In this case, *L*2 and *L*0 are both 1, causing the contents of *R*1 on the bus to be loaded into registers *R*0 and *R*2. The third transfer, an exchange between *R*0 and *R*1, is impossible in a single clock cycle, since it requires two simultaneous sources, *R*0 and *R*1, on the single bus. Thus, this transfer requires at least two buses or a bus combined with a dedicated path from one of the registers to the other. Note that such a transfer can be executed on the dedicated multiplexers in Figure 7-6(a). So, for a single-bus system, simultaneous transfers with different sources in

☐ **TABLE 7-5**
Examples of Register Transfers Using the Single Bus in Figure 7-6(b)

	Select		Load		
Register Transfer	**S1**	**S0**	**L2**	**L1**	**L0**
$R0 \leftarrow R2$	1	0	0	0	1
$R0 \leftarrow R1,\ R2 \leftarrow R1$	0	1	1	0	1
$R0 \leftarrow R1,\ R1 \leftarrow R0$			Impossible		

a single clock cycle are impossible, whereas for the dedicated multiplexers, any combination of transfers is possible. Hence, the reduction in hardware that occurs for a single bus in place of dedicated multiplexers results in limitations on simultaneous transfers.

If we assume that only single-source transfers are needed, then we can use Figure 7-6 to compare the complexity of the hardware in dedicated versus bus-based systems. First of all, assume a multiplexer design, as in Figure 3-20. In Figure 7-6(a), there are $2n$ AND gates and n OR gates per multiplexer (not counting inverters), for a total of $9n$ gates. In contrast, in Figure 7-6(b), the bus multiplexer requires only $3n$ AND gates and n OR gates, for a total of $4n$ gates. Also, the data input connections to the multiplexers are reduced from $6n$ to $3n$. Thus, the cost of the selection hardware is reduced by about half.

Three-State Bus

A bus can be constructed with the three-state buffers introduced in Section 6-3 instead of multiplexers. This has the potential for additional reductions in the number of connections. But why use three-state buffers instead of a multiplexer, particularly for implementing buses? The reason is that many three-state buffer outputs can be connected together to form a bit line of a bus, and this bus is implemented using only one level of logic gates. On the other hand, in a multiplexer, such a large number of sources means a high fan-in OR, which requires multiple levels of OR gates, introducing more logic and increasing delay. In contrast, three-state buffers provide a practical way to construct fast buses with many sources, so they are often preferred in such cases. More important, however, is that signals can travel in two directions on a three-state bus. Thus, the three-state bus can use the same interconnection to carry signals into and out of a logic circuit. This feature, which is most important when crossing chip boundaries, is illustrated in Figure 7-7(a). The figure shows a register with n lines that serve as both inputs and outputs lying across the boundary of the shaded area. If the three-state buffers are enabled, then the lines are outputs; if the three-state buffers are disabled, then the lines can be inputs. The symbol for this structure is also given in the figure. Note that the bidirectional bus lines are represented by a two-headed arrow. Also, a small inverted triangle denotes the three-state outputs of the register.

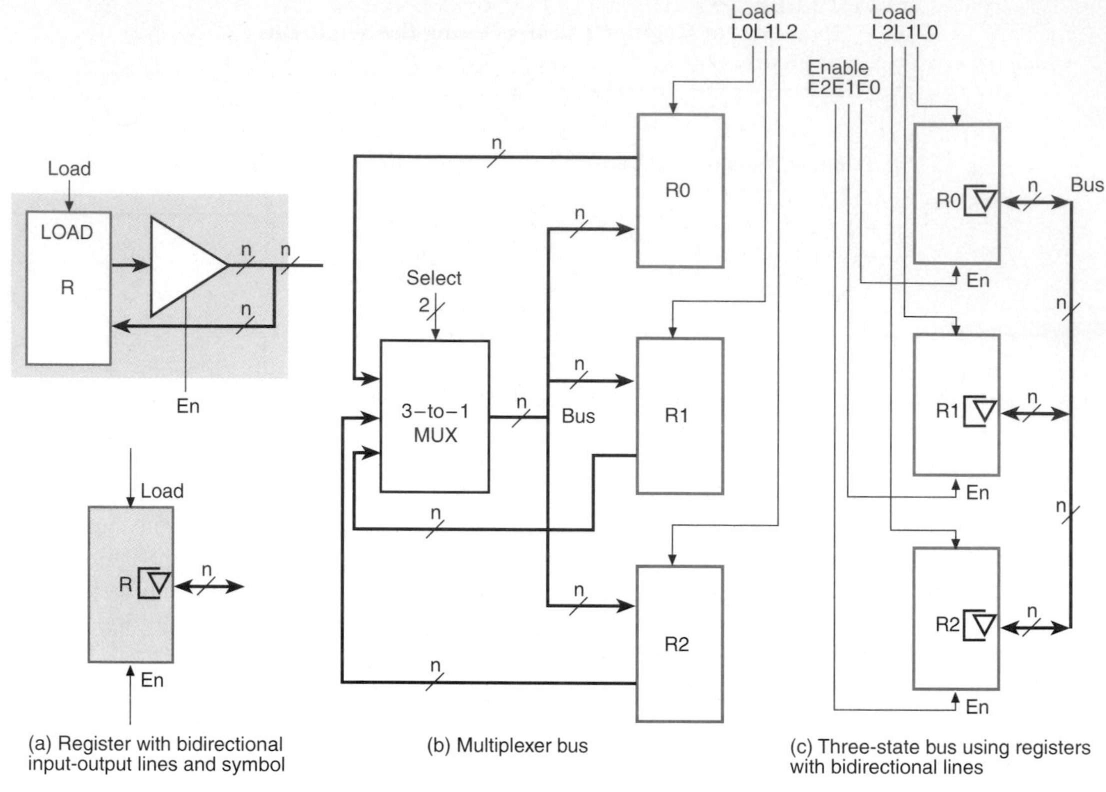

□ **FIGURE 7-7**
Three-State Bus versus Multiplexer Bus

Figure 7-7(b) and Figure 7-7(c) show a multiplexer-implemented bus and a three-state bus, respectively, for comparison. The symbol from Figure 7-7(a) for a register with bidirectional input-output lines is used in Figure 7-7(c). In contrast to the situation in Figure 7-6, where dedicated multiplexers were replaced by a bus, these two implementations are identical in terms of their register transfer capability. Note that in the three-state bus, there are only three data connections to the set of register blocks for each bit of the bus. The multiplexer-implemented bus has six data connections per bit to the set of register blocks. This reduction in the number of data connections by half, along with the ability to easily construct a bus with many sources, makes the three-state bus an attractive alternative. The use of such bidirectional input-output lines is particularly effective between logic circuits in different physical packages.

Memory Transfer

The operation of a memory unit was described in Section 6-4. Recall that the transfer of a data word within memory to the outside of memory is called a *read* operation, while the transfer of new data into a memory word is called a *write* operation. For memory transfers, we denote the memory by a letter representing the array of memory words, usually M if there is a single memory. The particular memory word is selected from the array of words by providing the *address* as an index to a location in memory. For a memory transfer, the address is enclosed in brackets following the letter denoting the memory. For example, $M[63]$ specifies the word in address $(63)_{10}$ of memory M.

Consider a memory unit that receives an address from address register AR. The data in memory is transferred to or from the data register DR. A read operation can then be stated as

$$Read\colon DR \leftarrow M[AR]$$

This causes a transfer of the data word from the portion of memory specified by the address in register AR into register DR.

The write operation is a transfer of data from register DR into the memory word specified by the address in register AR. Symbolically,

$$Write\colon M[AR] \leftarrow DR$$

There are often buses in a system that are associated with the connections for writes to and reads from memory, since multiple units need to access the memory. An example of this use of buses is depicted in Figure 7-8. Since the address bus and data bus shown have multiple sources and bridge across physical packages, they are both implemented using registers with three-state buffers. In this system, there are three source registers—$A0$, $A1$, and $A2$—for the address bus. The source to be used is selected by the address bus decoder. The data bus is bidirectional. For a read, the memory Read signal is activated by the data bus source decoder. The contents of the selected memory word can go to one of three registers, $D0$, $D1$, and $D2$, selected by the data bus destination decoder. For a write, the data word for the memory comes from one of those three registers, selected by the data bus source decoder. The memory Write signal is activated by the data bus destination decoder to write the contents of the selected register into memory. Transfers from one data register to another are also possible with the memory inactive, i.e., neither reading nor writing data.

A memory transfer statement in this system must specify the address register and the data register used. For example, the transfer of data from data register $D2$ to a memory word selected by register $A1$ is specified by

$$Write\colon M[A1] \leftarrow D2$$

This is a write operation with $A1$ supplying the address. Note that the statement does not specify the buses explicitly. Nevertheless, it implies that required select inputs for the address bus decoder are 01 (for $A1$), the select inputs for the data bus source decoder are 10 (for $D2$), and the select inputs for the data bus destination decoder are 11 (for Memory Write).

The read operation in a memory with buses can be specified similarly. The statement

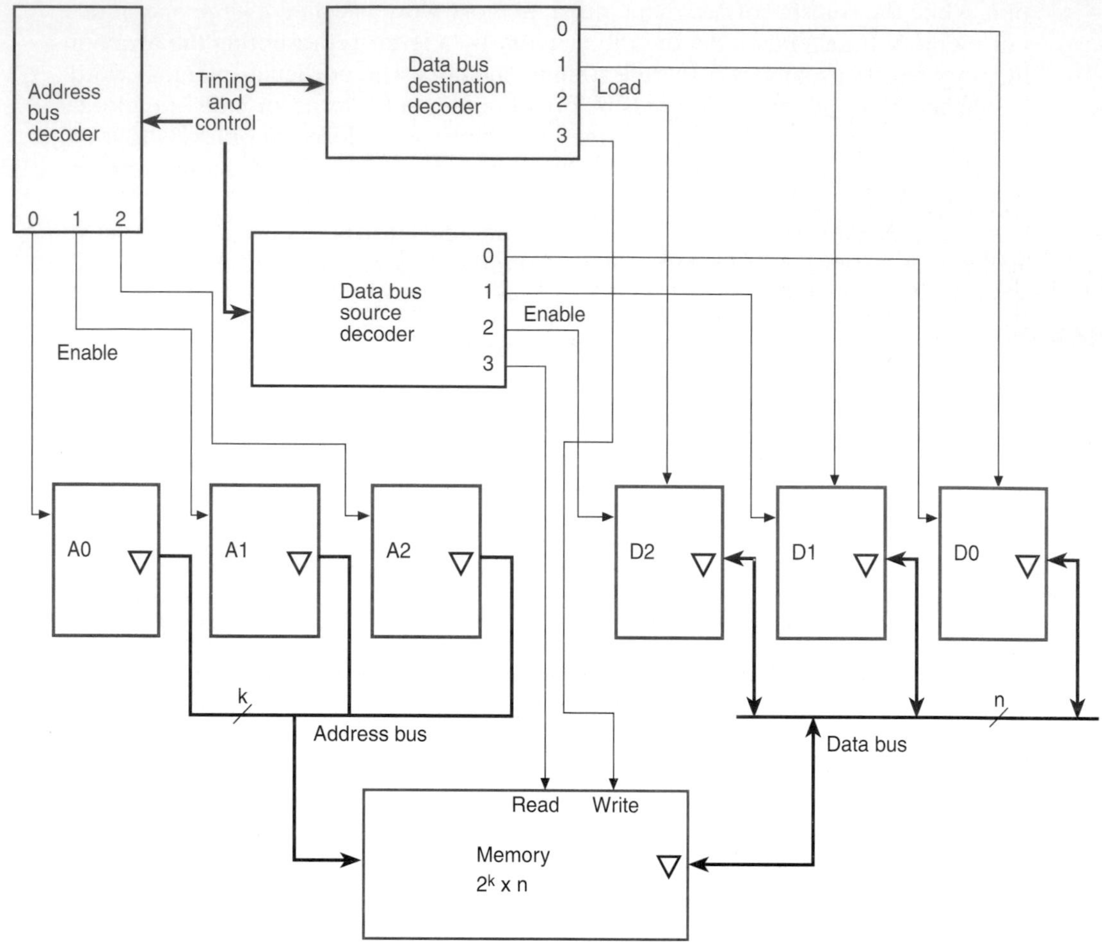

□ **FIGURE 7-8**
Memory Unit Connected to Address and Data Buses

$$Read\text{: } D1 \leftarrow M[A2]$$

specifies a read operation from the memory word having the address given in register *A*2. The data from memory is transferred to register *D*1. The statement implies that the select inputs for the address bus decoder are 10 (for *A*2), the select inputs for the data bus source decoder are 11 (for Memory Read), and the select inputs for the data bus destination decoder are 01 (for *D*1).

In the preceding write operation, if the memory is slow compared to other logic, the address and write data may need to be held on the buses for multiple clock cycles. Likewise, for the read operation, the address may need to be held for multiple clock cycles, and the data captured in register *D*1 several clock cycles after the read is started. Also, note that timing signals may be needed for proper memory operation, adding some complexity to the data bus decoders.

7-6 DATAPATHS

Instead of having each individual register perform its microoperations directly, computer systems often employ a number of storage registers in conjunction with a shared operation unit called an *arithmetic/logic unit*, abbreviated ALU. To perform a microoperation, the contents of specified source registers are applied to the inputs of the shared ALU. The ALU performs an operation, and the result of this operation is transferred to a destination register. With the ALU as a combinational circuit, the entire register transfer operation from the source registers, through the ALU, and into the destination register is performed during one clock cycle. The shift operations are often performed in a separate unit, but sometimes it is considered a part of the ALU.

Recall that the combination of a set of registers with a shared ALU and interconnecting paths is the datapath for the system. The rest of this chapter is concerned with the organization and design of datapaths. The design of a particular ALU is undertaken to show the process involved in implementing a complex combinational circuit. We also design a shifter, combine control signals into control words, and introduce a fast datapath called a pipeline.

The datapath and the control unit are the two parts of the processor, or CPU, of a computer. In addition to the registers, the datapath contains the digital logic that implements the various microoperations. This digital logic consists of buses, multiplexers, decoders, and processing circuits. When a large number of registers are included in a datapath, they are most conveniently connected through one or more buses. Registers in a datapath interact by the direct transfer of data, as well as in the performance of the various types of microoperations. A simple bus-based datapath with four registers, an ALU, and a shifter is shown in Figure 7-9. The shading and blue signal names relate to Figure 7-18 and will be discussed in Section 7-9. The black signal names are used here to describe the details in Figure 7-9. Each register is connected to two multiplexers to form ALU input buses A and B. The select inputs on each multiplexer select one register for the corresponding bus. For Bus B, there is an additional multiplexer, MUX B, so that constants can be brought into the datapath from outside. Bus B also connects to Data out, to send data outside the datapath to other components of the system, such as memory or input-output. Likewise, Bus A connects to Address out, to send address information outside of the datapath for memory or input-output.

Arithmetic and logic microoperations are performed on the operands on the A and B buses by the ALU. The G select inputs select the microoperation to be performed by the ALU. The shift microoperations are performed on data on Bus A by the shifter. The H select input selects a shift microoperation from Table 7-4. MUX F selects the output of the ALU or the output of the shifter. MUX D selects the output of MUX F or external data applied to Data in to be applied to Bus D. The latter is connected to the inputs of all the registers. The destination select inputs determine which register is loaded with the data on Bus D. Since the select inputs are decoded, only one register Load signal is active for any transfer of data into a register from Bus D. A Load Enable signal that can force all Load signals to

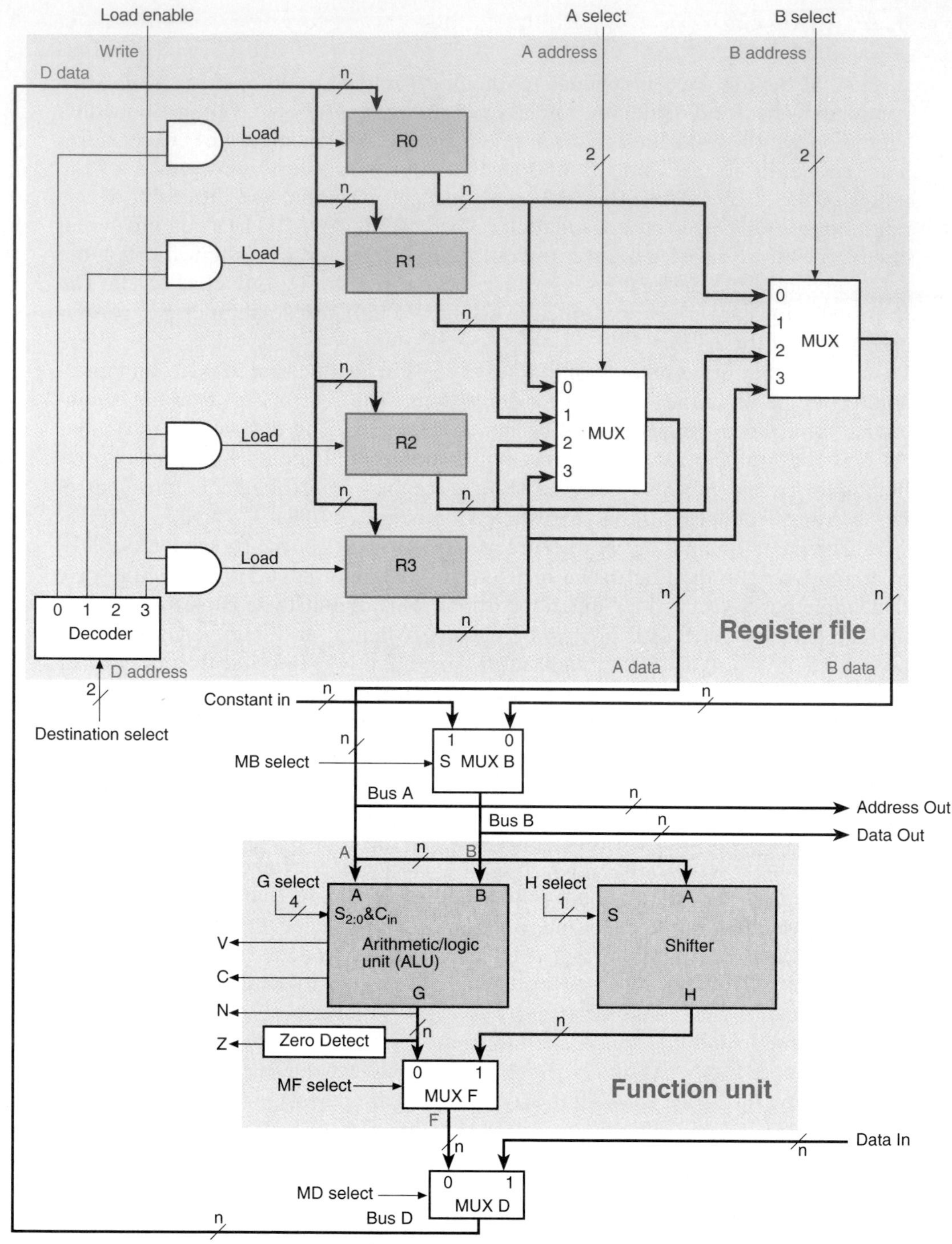

□ **FIGURE 7-9**
Block Diagram of a Datapath

0 using AND gates is present for transfers that are not to change the contents of any of the four registers.

It is useful to have certain information based on the results of an ALU operation available for use by the control unit of the CPU to make decisions. Four status bits are shown with the ALU in Figure 7-9. The status bits, carry C and overflow V, were explained in conjunction with Figure 7-4. The zero status bit Z is 1 if the output of the ALU contains all zeros and is 0 otherwise. Thus, $Z = 1$ if the result of an operation is zero, and $Z = 0$ if the result is nonzero. The sign status bit N (for negative) is the leftmost bit of the ALU output, which is the sign bit for the result in signed-number representations.

The control unit for the datapath directs the information flow through the buses, the ALU, the shifter, and the registers by applying signals to the select inputs. For example, to perform the microoperation

$$R1 \leftarrow R2 + R3$$

the control unit must provide binary selection values to the following sets of control inputs:

1. A select, to place the contents of $R2$ onto A data and, hence, Bus A.
2. B select, to place the contents of $R3$ onto the 0 input of MUX B; and MB select, to put the 0 input of MUX B onto Bus B.
3. G select, to provide the arithmetic operation $A + B$.
4. MF select, to place the ALU output on the MUX F output.
5. MD select, to place the MUX F output onto Bus D.
6. Destination select, to select $R1$ as the destination of the data on Bus D.
7. Load enable, to enable a register—in this case, $R1$—to be loaded.

The sets of values must be generated and must become available on the corresponding control lines early in the clock cycle. The binary data from the two source registers must propagate through the multiplexers and the ALU and on into the inputs of the destination register, all during the remainder of the same clock cycle. Then, when the next positive clock edge arrives, the binary data on Bus D is loaded into the destination register. To achieve fast operation, the ALU and shifter are constructed with combinational logic having a limited number of levels, such as a carry lookahead adder.

7-7 The Arithmetic/Logic Unit

The ALU is a combinational circuit that performs a set of basic arithmetic and logic microoperations. The ALU has a number of selection lines used to determine the operation to be performed. The selection lines are decoded within the ALU, so that k selection lines can specify up to 2^k distinct operations.

Figure 7-10 shows the block diagram of a typical n-bit ALU. The n data inputs from A are combined with the n data inputs from B to generate the result of an operation at the G outputs. The mode-select input S_2 distinguishes between arith-

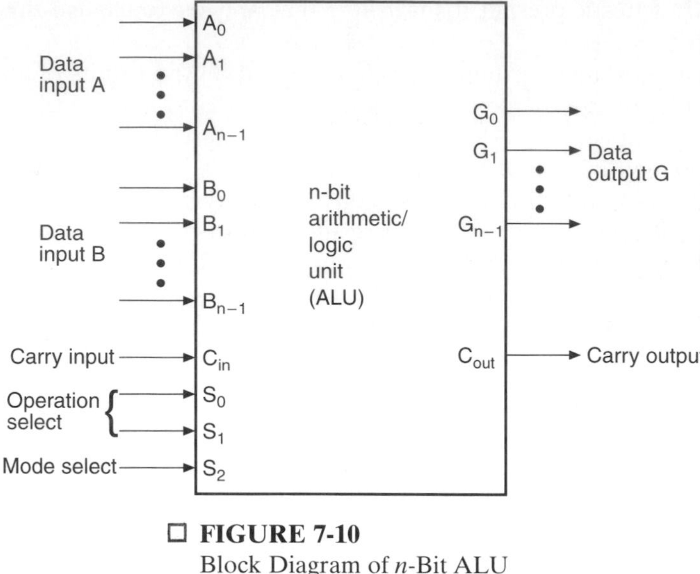

□ **FIGURE 7-10**
Block Diagram of *n*-Bit ALU

metic and logic operations. The two function-select inputs S_1 and S_0 specify the particular arithmetic or logical operation to be performed. With three selection lines, it is possible to specify four arithmetic operations with S_2 at 0 and four logic operations with S_2 at 1. The input and output carries have meaning only during an arithmetic operation.

We perform the design of this ALU in three stages. First, we design the arithmetic section. Then we design the logic section, and finally, we combine the two sections to form the ALU.

Arithmetic Circuit

The basic component of an arithmetic circuit is a parallel adder, which is constructed with a number of full-adder circuits connected in cascade, as in Figure 3-27. By controlling the data inputs to the parallel adder, it is possible to obtain different types of arithmetic operations. The block diagram in Figure 7-11 demonstrates a configuration in which one set of inputs to the parallel adder is controlled by the select lines S_1 and S_0. There are n bits in the arithmetic circuit, with two inputs A and B and output G. The n inputs from B go through the B input logic to the Y inputs of the parallel adder. The input carry C_{in} goes in the carry input of the full adder in the least-significant-bit position. The output carry C_{out} is from the full adder in the most-significant-bit position. The output of the parallel adder is calculated from the arithmetic sum

$$G = X + Y + C_{in}$$

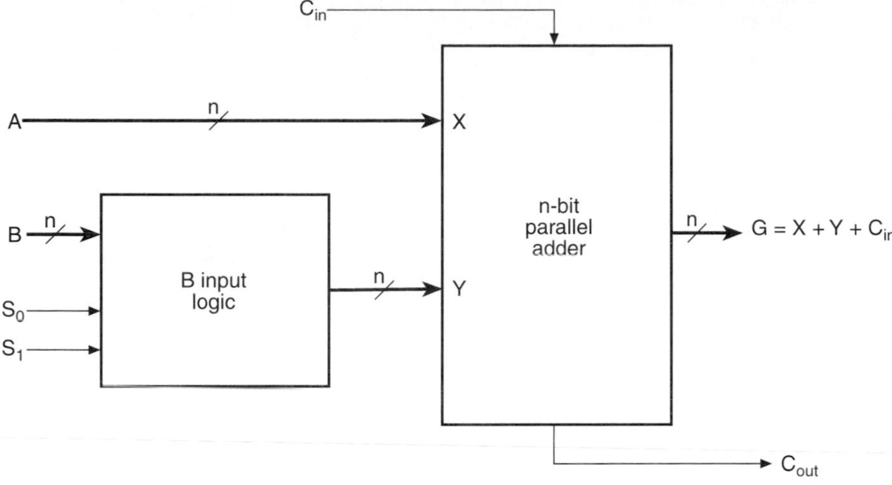

□ **FIGURE 7-11**
Block Diagram of an Arithmetic Circuit

where X is the n-bit binary number from the inputs and Y is the n-bit binary number from the B input logic. C_{in} is the input carry, which equals 0 or 1. Note that the symbol + in the equation denotes arithmetic addition.

Table 7-6 shows the arithmetic operations that are obtainable by controlling the value of Y with the two selection inputs S_1 and S_0. If the inputs from B are ignored and we insert all 0's at the Y inputs, the output sum becomes $G = A + 0 + C_{in}$. This gives $G = A$ when $C_{in} = 0$ and $G = A + 1$ when $C_{in} = 1$. In the first case, we have a direct transfer from input A to output G. In the second case, the value of A is incremented by 1. For a straight arithmetic addition, it is necessary to apply the B inputs to the Y inputs of the parallel adder. This gives $G = A + B$ when $C_{in} = 0$. Arithmetic subtraction is achieved by applying the complement of inputs B to the Y inputs of the parallel adder, to obtain $G = A + \overline{B} + 1$ when $C_{in} = 1$. This gives A plus the 2's complement of B, which is equivalent to 2's complement subtraction. All 1's is the 2's complement representation for –1. Thus, applying all 1's to the Y inputs with $C_{in} = 0$ produces the decrement operation $G = A - 1$.

The B input logic in Figure 7-11 can be implemented with n multiplexers. The data inputs to each multiplexer in stage i for $i = 0, 1, \ldots, n-1$ are 0, B_i, $\overline{B}_i$, and 1, corresponding to selection values S_1S_0: 00, 01, 10, and 11, respectively. Thus, the arithmetic circuit can be constructed with n full adders and n 4–to–1 multiplexers.

The number of gates in the B input logic can be reduced if, instead of using 4–to–1 multiplexers, we go through the logic design of one stage (one bit) of the B input logic. This can be done as shown in Figure 7-12. The truth table for one typical stage i of the logic is given in Figure 7-12(a). The inputs are S_1, S_0, and B_i, and the output is Y_i. Following the requirements specified in Table 7-6, we let $Y_i = 0$ when $S_1S_0 = 00$, and similarly assign the other three values of Y_i for each of the

Inputs			Output	
S_1	S_0	B_i	Y_i	
0	0	0	0	$Y_i = 0$
0	0	1	0	
0	1	0	0	$Y_i = B_i$
0	1	1	1	
1	0	0	1	$Y_i = \overline{B}_i$
1	0	1	0	
1	1	0	1	$Y_i = 1$
1	1	1	1	

(a) Truth table

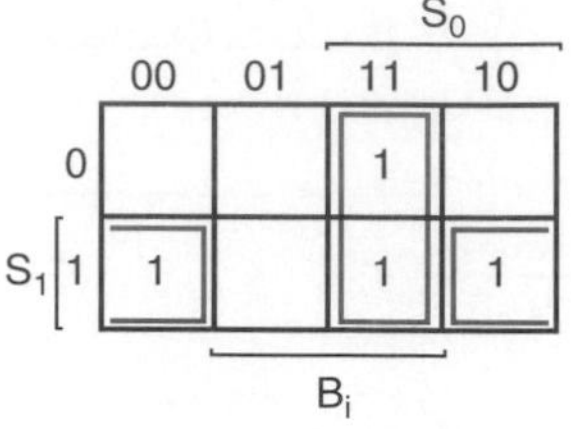

(b) Map Simplification:
$Y_i = B_iS_0 + \overline{B}_iS_1$

☐ **FIGURE 7-12**
B Input Logic for One Stage of Arithmetic Circuit

☐ **TABLE 7-6**
Function Table for Arithmetic Circuit

Select		Input	$G = A + Y + C_{in}$	
S_1	S_0	Y	$C_{in} = 0$	$C_{in} = 1$
0	0	all 0's	$G = A$ (transfer)	$G = A + 1$ (increment)
0	1	B	$G = A + B$ (add)	$G = A + B + 1$
1	0	$\overline{B}$	$G = A + \overline{B}$	$G = A + \overline{B} + 1$ (subtract)
1	1	all 1's	$G = A - 1$(decrement)	$G = A$ (transfer)

combinations of the selection variables. Output Y_i is simplified in the map in Figure 7-12(b), to give

$$Y_i = B_iS_0 + \overline{B_i}S_1$$

S_1 and S_0 are common to all n stages. Each stage i is associated with input B_i and output Y_i for $i = 0, 1, 2, \ldots, n-1$. This logic corresponds to a 2–to–1 multiplexer with B_i on the select input and S_1 and S_0 on the data inputs.

Figure 7-13 shows the logic diagram of an arithmetic circuit for $n = 4$. The four full-adder (FA) circuits constitute the parallel adder. The carry into the first stage is the input carry C_{in}. All other carries are connected internally from one stage to the next. The selection variables are S_1, S_0, and C_{in}. Variables S_1 and S_0 control all Y inputs of the full adders according to the Boolean function derived in Figure 7-12(b). Whenever C_{in} is 1, $Y + B$ has 1 added. The eight arithmetic operations for the circuit as a function of S_1, S_0, and C_{in} are listed in Table 7-6. It is interesting to note that the operation G $= A$ appears twice in the table. This is a harmless by-product of using C_{in} as one of the control variables while implementing both increment and decrement instructions.

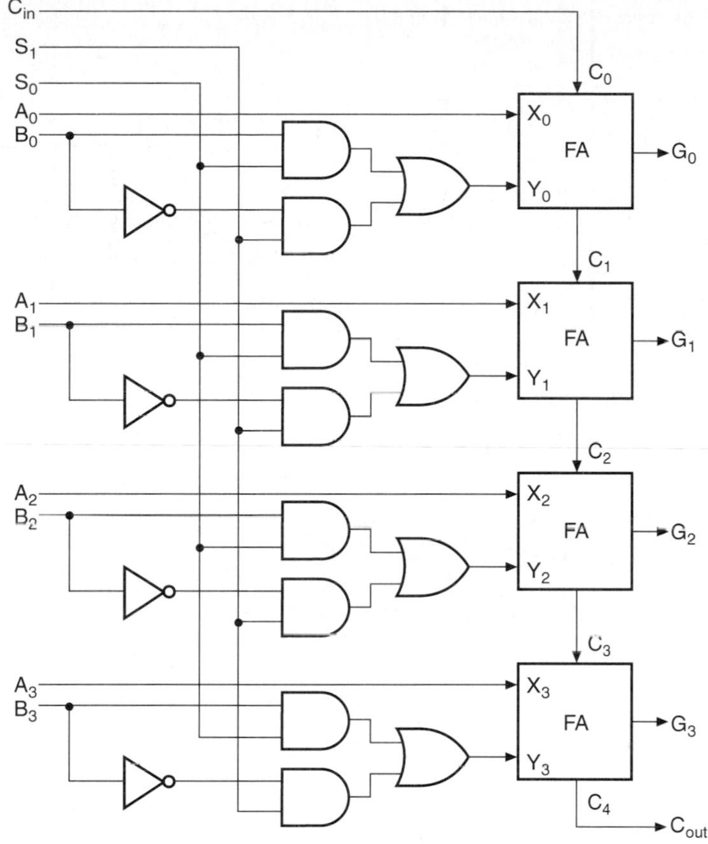

□ **FIGURE 7-13**
Logic Diagram of a 4-bit Arithmetic Circuit

Logic Circuit

The logic microoperations manipulate the bits of the operands by treating each bit in a register as a binary variable, giving bitwise operations. There are four commonly used logic operations—AND, OR, XOR (exclusive-OR), and NOT—from which others can be conveniently derived.

Figure 7-14(a) shows one stage of the logic circuit. It consists of four gates and a 4–to–1 multiplexer, although simplification could yield less complex logic. Each of the four logic operations is generated through a gate that performs the required logic. The outputs of the gates are applied to the inputs of the multiplexer with two selection variables S_1 and S_0. These choose one of the data inputs of the multiplexer and direct its value to the output. The diagram shows a typical stage with subscript i. For the logic circuit with n bits, the diagram must be repeated n times, for $i = 0, 1, 2,\ldots, n-1$. The selection variables are applied to all stages. The

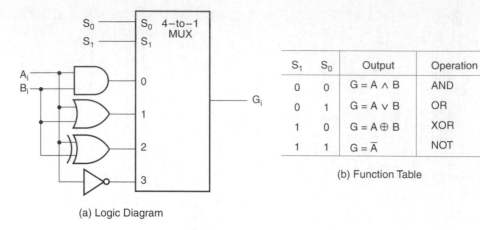

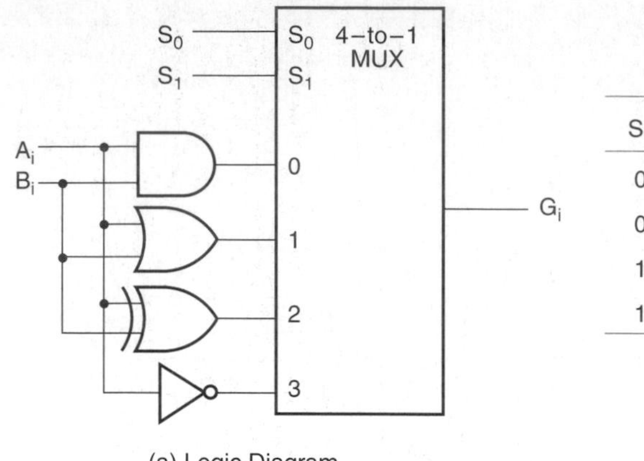

(a) Logic Diagram

S_1	S_0	Output	Operation
0	0	$G = A \wedge B$	AND
0	1	$G = A \vee B$	OR
1	0	$G = A \oplus B$	XOR
1	1	$G = \overline{A}$	NOT

(b) Function Table

☐ **FIGURE 7-14**
One Stage of Logic Circuit

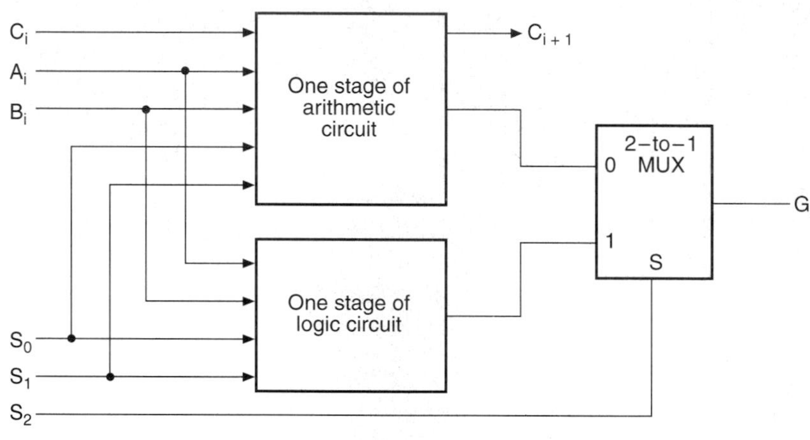

☐ **FIGURE 7-15**
One Stage of ALU

function table in Figure 7-14(b) lists the logic operations obtained for each combination of the selection values.

Arithmetic/Logic Unit

The logic circuit can be combined with the arithmetic circuit to produce an ALU. Selection variables S_1 and S_0 can be common to both circuits, provided that we use a third selection variable to differentiate between the two. The configuration for one stage of the ALU is illustrated in Figure 7-15. The outputs of the arithmetic and logic circuits in each stage are applied to a 2–to–1 multiplexer with selection variable S_2. When $S_2 = 0$, the arithmetic output is selected, and when $S_2 = 1$, the logic output is selected. Note that the diagram shows just one typical stage of the ALU; the circuit must be repeated n times for an n-bit ALU. The output carry C_{i+1} of a given arithmetic stage must be connected to the input carry C_i of the next stage in sequence. The input carry to the first stage is the input carry C_{in}, which also acts as a selection variable for the arithmetic operations.

The ALU specified in Figure 7-15 provides eight arithmetic and four logic operations. Each operation is selected through the variables S_2, S_1, S_0, and C_{in}. Table 7-7 lists the 12 ALU operations. The first 8 are arithmetic operations and are selected with $S_2 = 0$. The next 4 are logic operations and are selected with $S_2 = 1$. The input carry C_{in} has no effect during the logic operations and is marked with X to indicate that its value may be either 0 or 1.

☐ TABLE 7-7
Function Table for ALU

Operation Select					
S_2	S_1	S_0	C_{in}	Operation	Function
0	0	0	0	$G = A$	Transfer A
0	0	0	1	$G = A + 1$	Increment A
0	0	1	0	$G = A + B$	Addition
0	0	1	1	$G = A + B + 1$	Add with carry input of 1
0	1	0	0	$G = A + \overline{B}$	A plus 1's complement of B
0	1	0	1	$G = A + \overline{B} + 1$	Subtraction
0	1	1	0	$G = A - 1$	Decrement A
0	1	1	1	$G = A$	Transfer A
1	0	0	X	$G = A \wedge B$	AND
1	0	1	X	$G = A \vee B$	OR
1	1	0	X	$G = A \oplus B$	XOR
1	1	1	X	$G = \overline{A}$	NOT (1's complement)

The ALU logic we have designed is not as simple as it could be and has a fairly high number of logic levels, contributing to propagation delay in the circuit. With the use of logic simplification software, we can simplify this logic and reduce the delay. For example, it is quite easy to simplify the logic for a single stage of the ALU . Also, for realistic n, a means of further reducing the carry propagation delay in the ALU, such as the carry lookahead adder from Section 3-8, is usually necessary.

7-8 THE SHIFTER

The shifter shifts the value on Bus A, placing the result on an input of MUX F. The basic shifter performs one of two main types of transformations on the data: right shift and left shift.

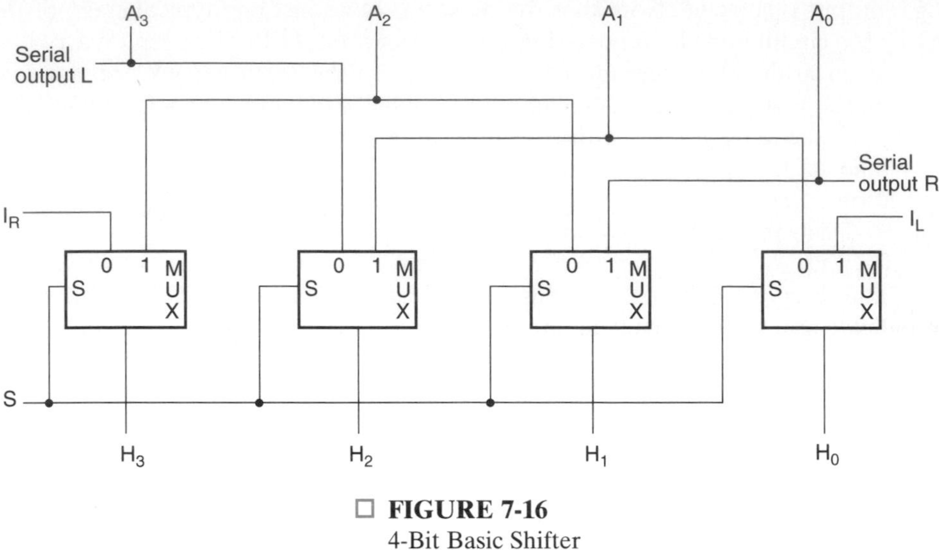

□ **FIGURE 7-16**
4-Bit Basic Shifter

A seemingly obvious choice for a shifter would be a bidirectional shift register with parallel load. Data from Bus *A* can be transferred to the register in parallel and then shifted to the right, the left, or not at all. A clock pulse loads the output of Bus *A* into the shift register, and a second clock pulse performs the shift. Finally, a third clock pulse transfers the data from the shift register to the selected destination register.

Alternatively, the transfer from a source register to a destination register can be done using only one clock pulse if the shifter is implemented as a combinational circuit. Because of the faster operation that results from the use of one clock pulse instead of three, this is the preferred method. In a combinational shifter, the signals propagate through the gates without the need for a clock pulse. Hence, the only clock needed for a shift in the datapath is for loading the data from Bus *H* into the selected destination register. It should be noted, however, that the period of the clock in this case may be longer due to combinational logic delay than when three clocks are used. In the datapath we are designing, this is of no consequence, since there is already a longer delay path through the ALU.

A combinational shifter can be constructed with multiplexers as shown in Figure 7-16. The selection variable *S* is applied to all four multiplexers to select the type of operation within the shifter. $S = 0$ causes a right-shift operation and S = 1 causes a left-shift operation. The right shift fills the position on the left with the value on serial input I_R. The left shift fills the position on the right with the value on serial input I_L. Serial outputs are available from serial output *R* and serial output *L* for right and left shifts, respectively. The right- and left-shift operations are illustrated in Table 7-4.

The diagram of Figure 7-16 shows only four stages of the shifter, which will have *n* stages in a system with *n*-bit operands. Additional selection variables may be employed to specify what goes into I_R and I_L during a single bit-position shift, as

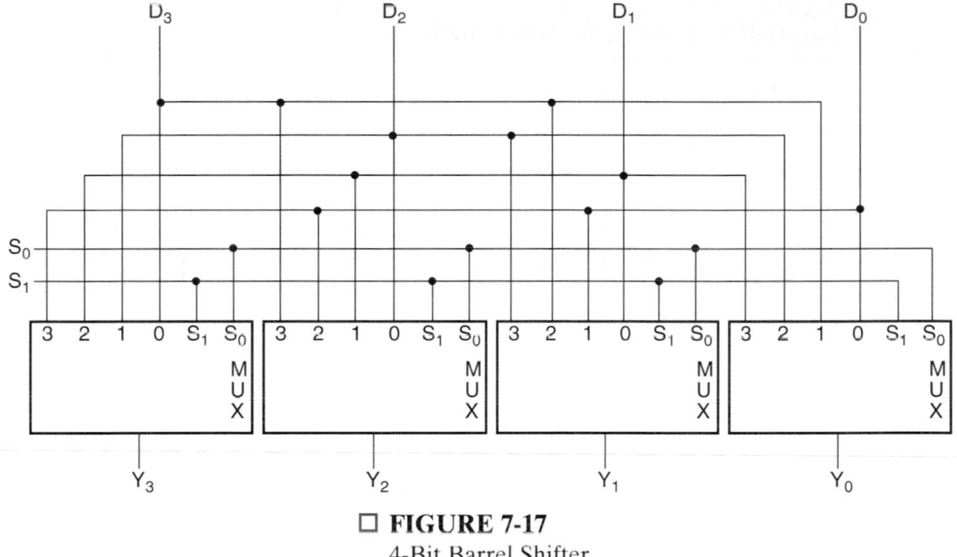

□ **FIGURE 7-17**
4-Bit Barrel Shifter

discussed in Section 10-2. Note that to shift an operand by $m > 1$ bit positions, this shifter must perform m 1-bit position shifts taking m clock cycles.

Barrel Shifter

In some datapath applications, the data must be shifted more than one bit position in a single clock cycle. A *barrel shifter* is one form of combinational circuit that shifts or rotates the input data bits by the number of bit positions specified by a binary value on a set of selection lines. The shift we consider here is a rotation to the left, which means that the binary data is shifted to the left, with the bits coming from the most significant part of the register rotated back into the least significant part of the register.

A 4-bit version of this kind of barrel shifter is shown in Figure 7-17. It consists of four multiplexers with common select lines S_1 and S_0. The selection variables determine the number of positions that the input data will be shifted to the left by rotation. When $S_1S_0 = 00$, no shift occurs, and the input data has a direct path to the outputs. When $S_1S_0 = 01$, the input data is rotated one position, with D_0 going to Y_1, D_1 going to Y_2, D_2 going to Y_3, and D_3 going to Y_0. When $S_1S_0 =$ 10, the input is rotated two positions, and when $S_1S_0 = 11$, the rotation is by three bit positions. Table 7-8 gives the function table for the 4-bit barrel shifter. For each binary value of the selection variables, the table lists the inputs that go to the corresponding output. Thus, to rotate three positions, S_1S_0 must be equal to 11, causing D_0 to go to Y_3, D_1 to go to Y_0, D_2 to go to Y_1 and D_3 to go to Y_2. Note that by using this left-rotation barrel shifter, one can generate all desired right rotations as well. For example, a left rotation by three positions is the same as a right rotation by one position in this 4-bit barrel shifter. In general, in a 2^n-bit barrel shifter, i positions of left rotation is the same as $2^n - i$ bits of right rotation.

□ **TABLE 7-8**
Function Table for 4-Bit Barrel Shifter

Select		Output				
S_1	S_0	Y_3	Y_2	Y_1	Y_0	Operation
0	0	D_3	D_2	D_1	D_0	No rotation
0	1	D_2	D_1	D_0	D_3	Rotate one position
1	0	D_1	D_0	D_3	D_2	Rotate two positions
1	1	D_0	D_3	D_2	D_1	Rotate three positions

A barrel shifter with 2^n input and output lines requires 2^n multiplexers, each having 2^n data inputs and n selection inputs. The number of positions for the data to be rotated is specified by the selection variables and can be from 0 to $2^n - 1$ positions. For a large n, the fan-in to gates is too large, so larger barrel shifters consist of layers of multiplexers, as shown in Section 10-3, or of special structures designed at the transistor level.

7-9 Datapath Representation

The datapath in Figure 7-9 includes the registers, selection logic for the registers, the ALU, the shifter, and three additional multiplexers. With a hierarchical structure, we can reduce the apparent complexity of the datapath. This reduction is important, since we will frequently use this datapath. Also, as illustrated by the register file to be discussed next, the use of a hierarchy allows one implementation of a module to be replaced with another, so that we are not tied to specific logic implementations.

A typical datapath has more than 4 registers. Indeed, computers with 32 or even more registers are common. The construction of a bus system with a large number of registers requires different techniques. A set of registers having common microoperations performed on them may be organized into a *register file*. The typical register file is a special type of fast memory that permits one or more words to be read out and one or more words to be written in, all simultaneously. Functionally, a simple register file contains the equivalent of the logic shaded in blue in Figure 7-9. Due to the memorylike nature of register files, the A select, B select, and Destination select inputs in the figure, become three addresses. As shown in Figure 7-9 in blue and on the register file symbol in Figure 7-18, the A address accesses a word to be read onto A data, the B address accesses a second word to be read onto B data, and the D address accesses a word to be written into from D data. All of these accesses occur in the same clock cycle. A Write input corresponding to the Load Enable signal is also provided. When at 0, the Write signal prevents the registers from being changed during that clock cycle. The size of the register file is $2^m \times n$, where m is the number of register address bits and n is the number of bits per register. For the datapath in Figure 7-9, $m = 2$, giving four registers, and n is unspecified.

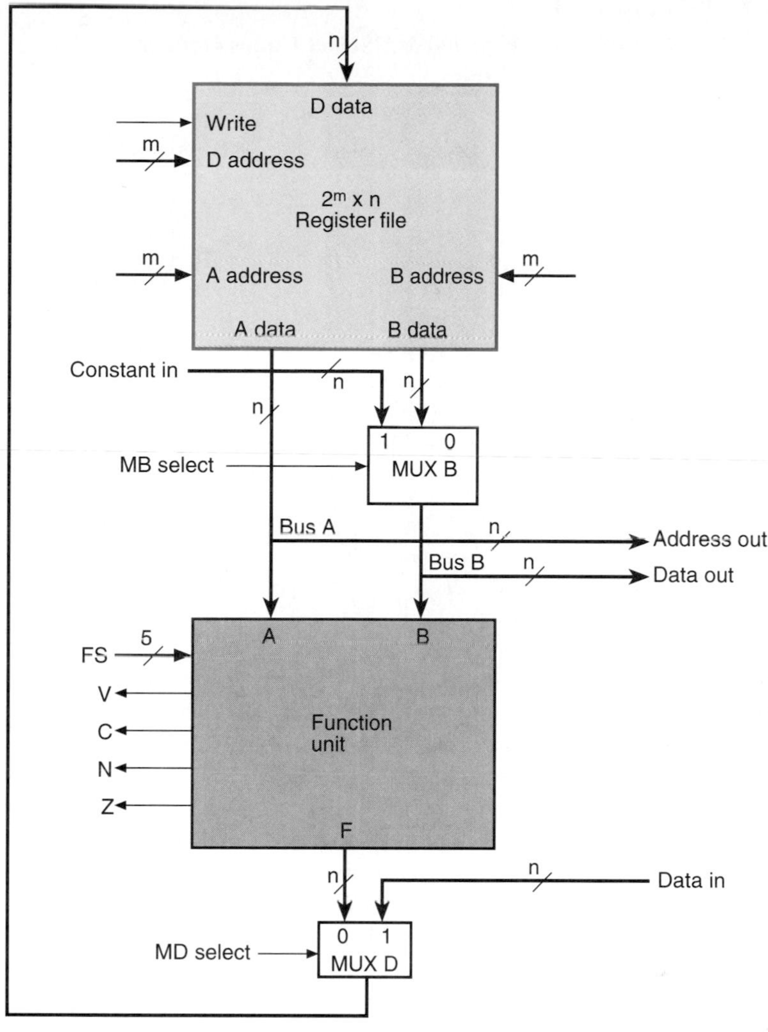

□ **FIGURE 7-18**
Block Diagram of Datapath Using the Register File and Function Unit

Since both the ALU and the shifter are shared processing units with outputs that are selected by MUX *F*, it is convenient to group the two units and the MUX together to form a shared function unit. Gray shading in Figure 7-9 highlights the function unit, which can be represented by the symbol given in Figure 7-18. The inputs to the function unit are from Bus *A* and Bus *B*, and the output of the function unit goes to MUX *D*. The function unit also has the four status bits *C*, *V*, *N*, and *Z* as added outputs.

In Figure 7-9, there are three sets of select inputs: the *G* select, *H* select, and *MF* select. In Figure 7-18, there is a single set of select inputs labeled *FS*, for "function select." To fully specify the function unit symbol in the figure, all of the

☐ **TABLE 7-9**
***G* Select, *H* Select, and *MF* Select Codes Defined in Terms of *FS* Codes**

FS	MF Select	G Select	H Select	Microoperation
00000	0	0000	0	$F = A$
00001	0	0001	1	$F = A + 1$
00010	0	0010	0	$F = A + B$
00011	0	0011	1	$F = A + B + 1$
00100	0	0100	0	$F = A + \overline{B}$
00101	0	0101	1	$F = A + \overline{B} + 1$
00110	0	0110	0	$F = A - 1$
00111	0	0111	1	$F = A$
01000	0	1000	0	$F = A \wedge B$
01010	0	1010	0	$F = A \vee B$
01100	0	1100	0	$F = A \oplus B$
01110	0	1110	0	$F = \overline{A}$
10000	1	0000	0	$F = \text{sr } A$
10001	1	0001	1	$F = \text{sl } A$

codes for *MF* select, *G* select, and *H* select must be defined in terms of the codes for *FS*. Table 7-9 defines these code transformations. The codes for *FS* are given in the left column. The codes for *MF* select are in the leftmost bit of the *FS* codes, the codes for *G* select are in the second through fifth bits of the *FS* codes, and the codes for *H* select are in the fifth bit of the *FS* codes. If *MF* select = 0, then the *G* select codes determine the function on the output of the function unit. If *MF* select = 1, then the *H* select codes determine the function on the output of the function unit. To show this dependency, the codes that determine the function unit outputs are highlighted in blue in the table, which also contains the micro-operation for each *FS* code. The status bits are assumed to be meaningless when the shifter is selected.

7-10 The Control Word

The selection variables for the datapath control the microoperations executed within the datapath for any given clock pulse. For the datapath in Section 7-9, the selection variables control the addresses for the data reads from the register file, the function performed by the function unit, and the data loaded into the register file, as well as the selection of external data. We will now demonstrate how these control variables select the microoperations for a particular datapath to be used in later chapters. The choice of control variable values for typical microoperations will be discussed and a simulation of the datapath illustrated.

A block diagram of a datapath that is a specific version of the datapath in Figure 7-18 is shown in Figure 7-19(a). It has a register file with eight registers, *R*0

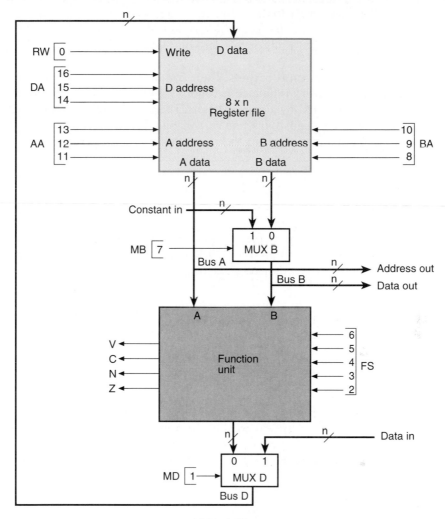

(a) Block Diagram

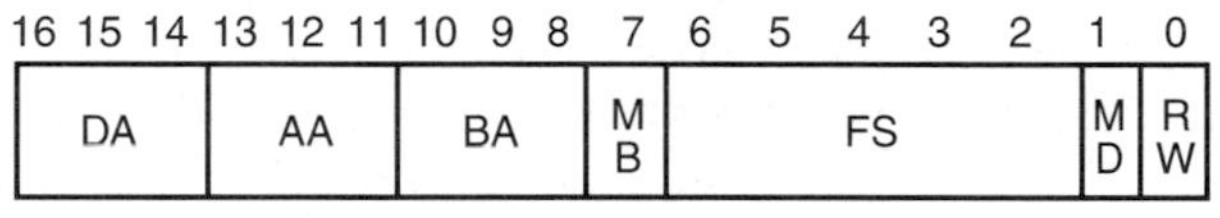

(b) Control word

FIGURE 7-19
Datapath with Control Variables

through $R7$. The register file provides the inputs to the function unit through Bus A and Bus B. MUX B selects between constant values and register values on B data. The ALU and zero-detection logic within the function unit generate the binary data for the four status bits: C (carry), V (overflow), Z (zero), and N (sign). MUX D selects the function unit output or the data on Data in as input for the register file.

There are 17 binary control inputs. Their combined values specify a *control word*. The 17-bit control word is defined in Figure 7-19(b). It consists of seven parts called *fields*, each designated by a pair of letters. The three register fields are three bits each. The remaining fields have one or five bits. The three bits of DA select one of eight destination registers for the result of the microoperation. The three bits of AA select one of eight source registers for the Bus A input of the ALU. The three bits of BA select a source register for the 0 input of the MUX B. The single MB bit determines whether Bus B carries the contents of a register or a constant. The 5-bit FS field controls the operation of the function unit. The FS field contains one of the 14 codes from Table 7-9. The single bit of MD selects the function unit output or the data on Data in as the inputs to Bus D. The final field, RW, determines whether a register is written or not. When applied to the control inputs, the 17-bit control word specifies a particular microoperation.

☐ TABLE 7-10
Encoding of Control Word for the Datapath

DA, AA, BA		MB		FS		MD		RW	
Function	**Code**	**Function**	**Code**	**Function**	**Code**	**Function**	**Code**	**Function**	**Code**
$R0$	000	Register	0	$F = A$	00000	Function	0	No write	0
$R1$	001	Constant	1	$F = A + 1$	00001	Data In	1	Write	1
$R2$	010			$F = A + B$	00010				
$R3$	011			$F = A + B + 1$	00011				
$R4$	100			$F = A + \overline{B}$	00100				
$R5$	101			$F = A + \overline{B} + 1$	00101				
$R6$	110			$F = A - 1$	00110				
$R7$	111			$F = A$	00111				
				$F = A \wedge B$	01000				
				$F = A \vee B$	01010				
				$F = A \oplus B$	01100				
				$F = \overline{A}$	01110				
				$F = \text{sr } A$	10000				
				$F = \text{sl } A$	10001				

The functions of all meaningful control codes are specified in Table 7-10. For each of the fields, a symbolic name and a binary code for each of the functions are given. The register selected by each of the fields DA, AA, and BA is the one with the decimal equivalent equal to the binary number for the code. MB selects either the register selected by the BA field or a constant from outside of the datapath on

Constant in. The ALU operations, the shifter operations, and the selection of the ALU or shifter outputs are all specified by the FS field. The field MD controls the information to be loaded into the register file. The final field, RW, has the functions No Write, to prevent writing to any registers, and Write, to signify writing to a register.

The control word for a given microoperation can be derived by specifying the value of each of the control variables. For example, a subtraction given by the statement

$$R1 \leftarrow R2 + \overline{R3} + 1$$

specifies $R2$ for the A input of the ALU and $R3$ for the B input of the ALU. It also specifies function unit operation $F = A + \overline{B} + 1$ and selection of the function unit output for input into the register file. Finally, the microoperation selects $R1$ as the destination register and sets RW to 1 to cause $R1$ to be written. The control word for this microinstruction is specified by its seven fields, with the binary value for each field obtained from the encoding listed in Table 7-10. The binary control word for this subtraction microoperation is 00101001100010101 and is obtained as follows:

Field:	DA	AA	BA	MB	FS	MD	RW
Symbolic:	$R1$	$R2$	$R3$	Register	$F = A + \overline{B} + 1$	Function	Write
Binary:	001	010	011	0	00101	0	1

The control word for the microoperation and those for several other microoperations are given in Table 7-11 using symbolic notation and in Table 7-12 using binary codes.

The second example in Table 7-11 is a shift microoperation given by the statement

$$R4 \leftarrow \text{sl } R6$$

This statement specifies a shift left for the shifter. The contents of register $R6$, shifted to the left, are transferred to $R4$. From the knowledge of the symbols in

☐ TABLE 7-11
Examples of Microoperations for the Datapath, Using Symbolic Notation

Micro-operation	DA	AA	BA	MB	FS	MD	RW
$R1 \leftarrow R2 + \overline{R3} + 1$	$R1$	$R2$	$R3$	Register	$F = A + \overline{B} + 1$	Function	Write
$R4 \leftarrow$ sl R6	$R4$	$R6$	—	Register	$F = $ sl A	Function	Write
$R7 \leftarrow R7 + 1$	$R7$	$R7$	—	Register	$F = A + 1$	Function	Write
$R1 \leftarrow R0 + 2$	$R1$	—	—	Constant	$F = A + B$	Function	Write
Data out $\leftarrow R3$	—	—	$R3$	Register	—	—	No Write
$R4 \leftarrow$ Data in	$R4$	—	—	—	—	Data in	Write
$R5 \leftarrow 0$	$R5$	$R0$	$R0$	Register	$F = A \oplus B$	Function	Write

□ **TABLE 7-12**
Examples of Microoperations from Table 7-11, Using Binary Control Words

Micro-operation	DA	AA	BA	MB	FS	MD	RW
$R1 \leftarrow R2 - R3$	001	010	011	0	00101	0	1
$R4 \leftarrow$ sl R6	100	110	000	0	10001	0	1
$R7 \leftarrow R7 + 1$	111	111	000	0	00001	0	1
$R1 \leftarrow R0 + 2$	001	000	000	1	00010	0	1
Data out $\leftarrow R3$	000	000	011	0	00000	0	0
$R4 \leftarrow$ Data in	100	000	000	0	00000	1	1
$R5 \leftarrow 0$	101	000	000	0	01100	0	1

each field, the control word in binary is derived as shown in Table 7-12. For many microoperations, neither the A data nor the B data from the register file is used. In these cases, the respective symbolic field is marked with a dash. For convenience, we assign 0's to any unused field when formulating the binary control word. Continuing with the last three examples in Table 7-11, to make the contents of a register available to an external destination only, we place the contents of the register on the B data output of the register file, with RW = No Write (0) to prevent the register file from being written. To place a constant in a register, we place the constant on Constant in, set MB to Constant, and pass the value through the ALU and Bus D to the destination register. To clear a register to 0, Bus D is set to all 0's by using the same register for both A data and B data with FS = 01100 and MD = 0. The DA field is set to the code for the destination register, and RW is Write (1).

It is apparent from these examples that many additional microoperations can be performed by the same datapath. Sequences of such microoperations can be realized by providing a control unit that produces the appropriate sequences of control words. Different approaches to control unit design are the focus of Chapter 8.

To complete this section, we will perform a simulation of the datapath in Figure 7-19 with n, the number of bits in each register, equal to 8. We assume that the microoperations in Table 7-12, executed in sequence, provide the inputs to the datapath and that the initial contents of each register are its number in hexadecimal (e.g., $R5$ contains $(0000\ 0101)_2 = (05)_{16}$). Figure 7-20 gives the result of this simulation. The first value displayed is the Clock. The next eight values are inputs to the datapath. Following these are the eight registers. Finally, the values for the status bits (Z, N, C, V), Address out, and Data out are displayed. Note that these values change at the same time as the register values, indicating that the combinational delay is small compared to the clock period. Binary variables are displayed as logical waveforms, all other variables with hexadecimal values.

Of note in the simulation results is that we find that changes in registers as a result of a particular microoperation appear in the clock cycle *after* that in which the microoperation is specified. For example, the result of the subtraction in the first clock cycle appears in register $R1$ in the second cycle. This is because the result is loaded into flip-flops on the positive edge of the clock at the end of the first clock

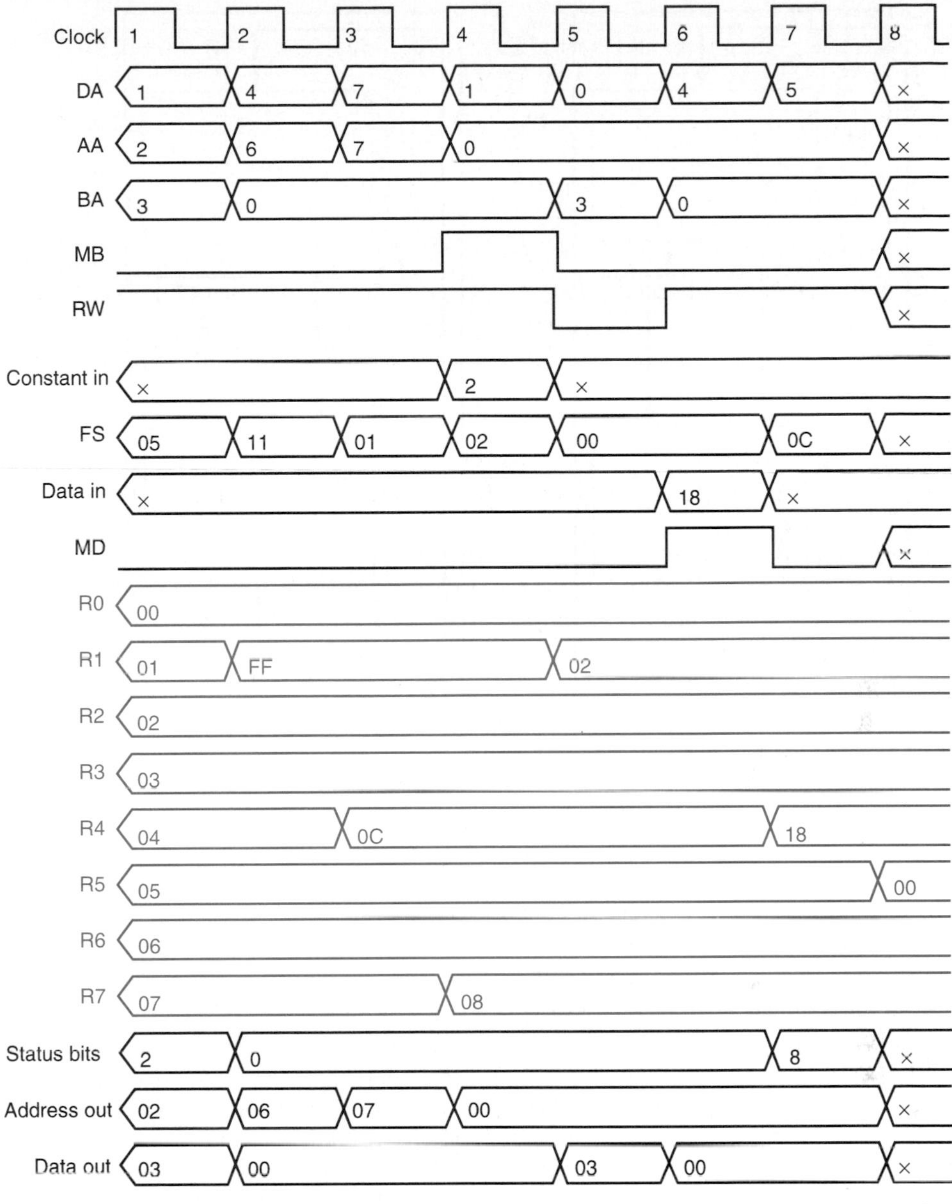

□ **FIGURE 7-20**
Simulation of the Microoperation Sequence in Table 7-12

cycle. On the other hand, the values on the status bits, Address out, and Data out appear in the same clock cycle, since they do not depend on a positive clock edge occurring. We place a specified value on Constant in only in the clock cycle when it is necessary to add 2 to *R*0; otherwise, its value is unspecified and denoted by ×. Likewise, we place the value 18 on Data in only on the clock cycle when it is neces-

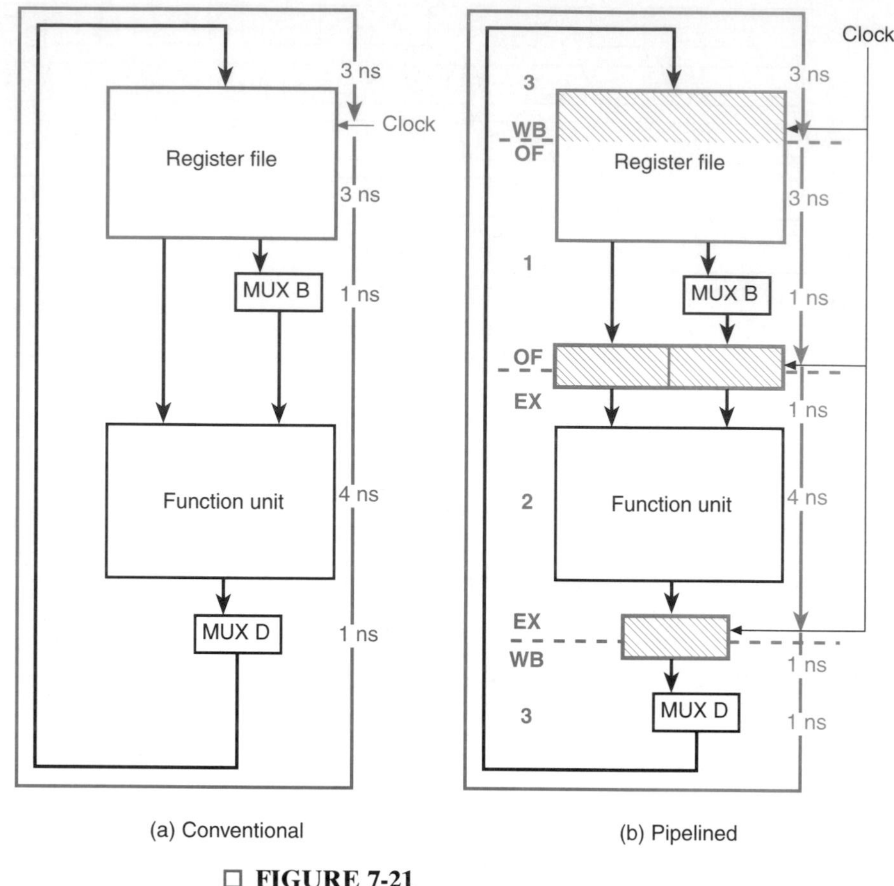

□ **FIGURE 7-21**
Datapath Timing

sary to load 18 into *R*4. Note that the values in the registers in the sixth clock cycle are unchanged from the previous cycle. This is because RW = 0 in the fifth clock cycle, preventing the registers from being written to. Finally, note that eight clock cycles of simulation are used for seven microoperations so that the values in the registers that result from the last microoperation executed can be observed.

7-11 Pipelined Datapath

In addition to providing a datapath that performs the necessary register transfer microoperations, we need to be concerned about the speed or rate at which the microoperations are performed. Figure 7-21(a) illustrates maximum delay values for each of the components of a typical datapath. A maximum of 4 ns (3 ns + 1 ns) is required to read two operands from the register file or to read one operand from the register file and obtain a constant from MUX *B*. A maximum of 4 ns is also required to execute an operation in the functional unit. Finally, a maximum of 4 ns

□ **FIGURE 7-22**
Assembly Line Analogy to Datapath Pipeline

(1 ns + 3 ns) is required to write the result back into the register file, including the delay of MUX *D*. Adding these delays, we find that 12 ns is required to perform a single microoperation. The rate at which the microoperations can be performed is the inverse of 12 ns, which is 83.3 megahertz (MHz; 1 megahertz equals 10^6 clock cycles per second). This is the maximum frequency at which the clock can be operated, since 12 ns is the smallest clock period that will allow each microoperation to be completed with certainty.

Now suppose that the rate at which microoperations are performed is not adequate for a particular application and that there are no faster components available with which to reduce the 12 ns required to complete a microoperation. Still, it may be possible to reduce the clock period and increase the clock frequency. This can be done by breaking up the 12-ns delay path with registers. The resulting datapath, sketched in Figure 7-21(b), is referred to as a *pipelined datapath,* or just a *pipeline.*

Three sets of registers break the delay of the original datapath into three parts. These registers are shown crosshatched in blue. The register file contains the first set of registers. Cross-hatching covers only the top half of the register file, since the lower half is viewed as the combinational logic that selects the two registers to be read. The two registers that store the *A* data from the register file and the output of MUX *B* constitute the second set of registers. The third set of registers stores the inputs to MUX *D*.

The term "pipeline," unfortunately, does not provide the best analogy for the corresponding datapath structure. A better analogy for the datapath pipeline is an assembly line, as shown in Figure 7-22. A custom product being built may pass down the assembly line many times before it is completed. A conveyor belt moves components from stage to stage by proceeding forward periodically the length of one stage. Components and partially completed assemblies are stored in bins. The person at the first stage of the assembly line takes one or more components to be included in the custom product from storage bins and places them on the conveyor. The person at the second stage assembles the components. The person at the last stage puts the assembly on the conveyor belt into the storage bins to be used for further assembly later on. Note that each person has a simple task to perform. Moreover, as soon as all of the tasks in a particular stage are done, the conveyor belt can move forward so that the same tasks can be performed on the next items on the belt.

Observe that one assembly operation is completed at each advance of the belt. At any given time, three assembly operations are in some stage of completion. Is this any faster than having only one stage, as in the case of a conventional datapath? Suppose that a single-stage line with one person doing everything takes one minute to do the operation and that 20 seconds is devoted to each of the three tasks. How many assemblies come from the single-stage line in one minute? Just one. How many assemblies come from the three-stage line in one minute? Since an assembly operation is completed every 20 seconds at the last stage of the line, three assemblies emerge! So the *throughput*, or rate of producing assemblies, of the three-stage line is three times that of the single-stage line. Although both of these assembly lines take one minute to complete one assembly, the three-stage assembly line completes three times as many assemblies in a given amount of time. We say that the *throughput* of the three stage line is three times that of the one stage line.

Now for the analogy: The "take from storage bins" stage corresponds to obtaining or *fetching* operands from the register file, with the bins corresponding to the register file locations. The "assembly" stage corresponds to execution of a function unit operation. Finally, the "put into storage bins" operation corresponds to writing the result back into a register file location. We will refer to these stages in the execution of a microoperation as *operand fetch*, *execution*, and *write-back*, respectively. The analogy suggests that if we can set up a structure with these stages and structures that correspond to the periodic movement of the conveyor belt, then we can process three times as many microoperations in a given time as the conventional datapath does.

The desired structure based on the conventional datapath is sketched in Figure 7-21(b). The operand fetch (OF) is stage 1, the execution (EX) is stage 2, and the write-back (WB) is stage 3. These stages are labeled at their boundaries with appropriate abbreviations. To provide the mechanism corresponding to the conveyor belt, we place registers between the stages. Just as a conveyor belt provides temporary storage for moving assemblies between stages, the registers provide temporary storage for moving data between stages. The resulting structure is called a *pipeline*, and the registers between the stages are called *pipeline platforms.* Stage 1 of the pipeline has the delay required for reading the register file followed by selection by MUX *B*. This delay is 3 plus 1 ns, or 4 ns. Stage 2 of the pipeline has the 1 ns delay of the platform plus the 4 ns delay of the functional unit, giving 5 ns. Stage 3 has the 1 ns delay of the platform, the delay for the selection by MUX *D*, and the delay for writing back into the register file. This delay is 1 + 1 + 3, for a total of 5 ns. Thus, all flip-flop–to–flip-flop delays are at most 5 ns, allowing a minimum clock period of 5 ns (assuming that the setup times for the flip-flops are zero) and a maximum clock frequency of 200 MHz.

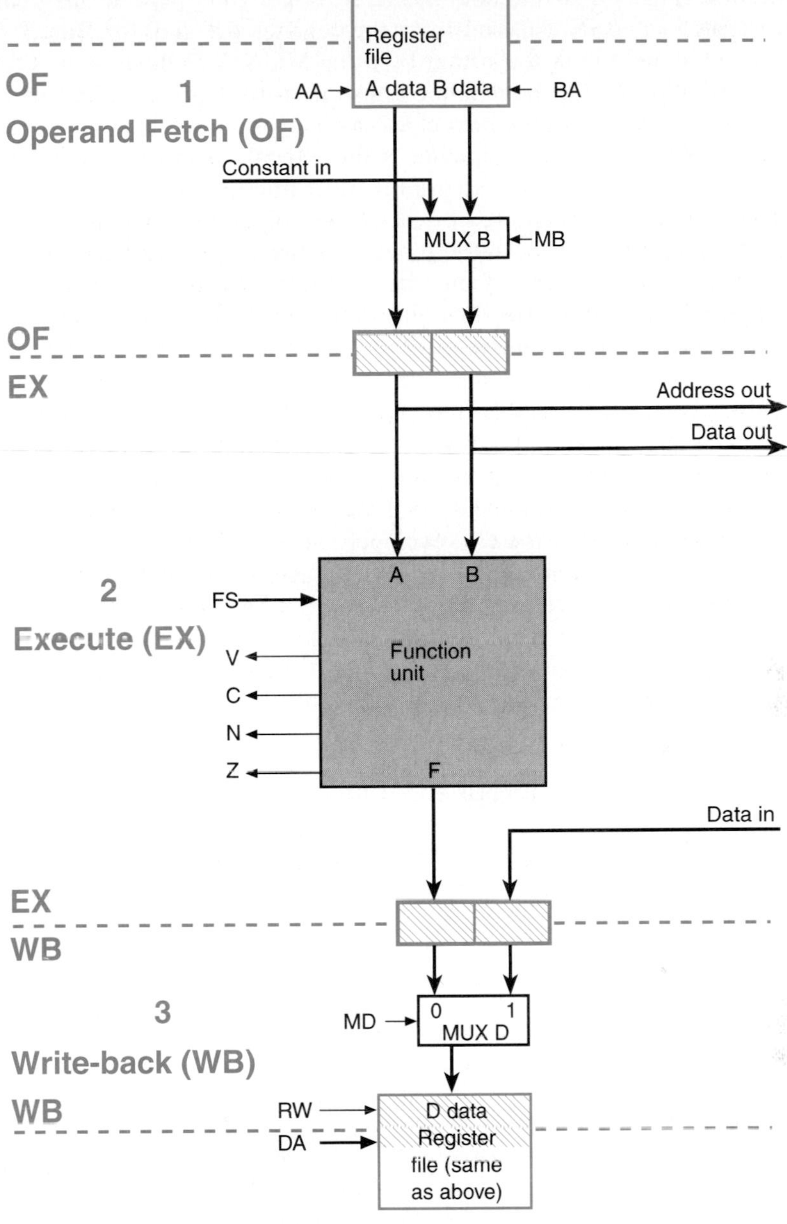

☐ **FIGURE 7-23**
Block Diagram of Pipelined Datapath

A more detailed diagram of the pipelined datapath appears in Figure 7-23. In this diagram, rather than showing the path from the output of MUX D to the register file input, the register file is shown *twice*—once in the OF stage, where it is read, and once in the WB stage, where it is written.

The first stage, OF, is the operand fetch stage. The operand fetch consists of reading register values to be used from the register file and, for Bus B, selecting between a register value or a constant by using MUX *B*. Following the OF stage is the first pipeline platform. The pipeline registers store the operand or operands for use in the next stage during the next clock cycle.

The second stage of the pipeline is the execute stage, denoted EX. In this stage, a function unit operation occurs for most microoperations. The results produced from this stage are captured by the second pipeline platform.

The third and final stage of the pipeline is the write-back stage, denoted WB. In this stage, the result saved from the EX stage, or the value on Data in, is selected by MUX *D* and written back into the register file at the end of the stage. In this case, the write part of the register file is the pipeline platform. The WB stage completes the execution of each microoperation that requires writing to a register.

Before leaving the assembly-line–pipelined-datapath analogy, we examine the cost of the single-stage assembly line versus that of the three-stage assembly line. First, even though the three-stage assembly line produces assemblies three times as fast as the single-stage line does, it costs three times as much in terms of personnel. Plus, it has the overhead of the conveyor belt. So it appears that it is not very cost effective compared to having three single-stage assembly stations operating in parallel. Nevertheless, from an industrial standpoint, it has proven to be cost effective. In terms of an automobile assembly line, can you figure out why? In contrast, for the pipelined datapath, pipeline platforms cut a single datapath into three pieces. Thus, the increased cost is mainly that of the pipeline platforms.

Execution of Pipeline Microoperations

There are up to three operations at some stage of completion in the assembly line at any given time. By analogy, we should be able to have three microoperations at some stage of completion in the pipeline at any given time.

We now examine the execution of this sequence of microoperations with respect to the stages of the pipeline in Figure 7-23. In clock period 1, microoperation 1 is in the OF stage. In clock period 2, microoperation 1 is in the EX stage, and microoperation 2 is in the OF stage. In clock period 3, microoperation 1 is in the WB stage, microoperation 2 is in the EX stage, and microoperation 3 is in the OF stage. So at the end of the third clock period, microoperation 1 has completed execution, microoperation 2 is two-thirds finished, and microoperation 3 is one-third finished. So we have completed $1 + 2/3 + 1/3 = 2.0$ microoperations in three clock periods, or 15 ns. In the conventional datapath, we would have completed execution of microoperation 1 only. So, indeed, the pipelined datapath is faster in this example.

The procedure we have been using to analyze the sequence of microoperations so far is somewhat tedious. So to finish the analysis of the timing of the sequence, we will use a *pipeline execution pattern* diagram, as shown in Figure 7-24. Each vertical position in this diagram represents a microoperation to be performed, and each horizontal position represents a clock cycle. An entry in the diagram represents the stage of processing of the microoperation. So, for example, the

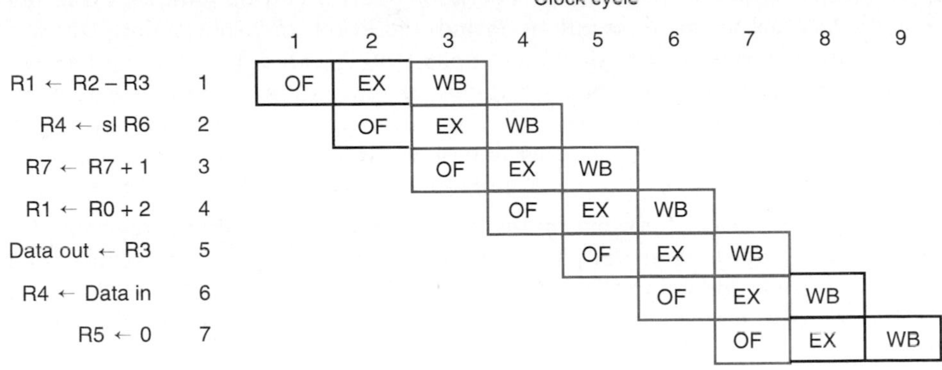

□ **FIGURE 7-24**
Pipeline Execution Pattern for Microoperation Sequence in Table 7-12

execution (EX) stage of microoperation 4, which adds the constant 2 to *R*0, occurs in clock cycle 5.

We can see from the overall diagram that the sequence of seven microoperations requires nine clock cycles to execute completely. The time required is $9 \times 5 = 45$ ns, compared to $7 \times 15 = 105$ ns for the conventional datapath. Thus, the sequence of microoperations is executed about 2.3 times faster.

Now let us examine the pipeline execution pattern carefully. In the first two clock cycles, not all of the pipeline stages are active, since the pipeline is *filling*. In the next five clock cycles, all stages of the pipeline are active, as indicated in blue, and the pipeline is fully utilized. In the last two clock cycles, not all stages of the pipeline are active, since the pipeline is *emptying*. If we want to find the maximum possible improvement of the pipelined datapath over the conventional one, we compare the two when the pipeline is fully utilized. Over these five clock cycles, 3 through 7, the pipeline executes $(5 \times 3) \div 3 = 5$ operations in 25 ns. In the same time, the conventional datapath executes $5 \times 5 = 25 \div 15 = 1.67$ microoperations. So the pipelined datapath executes at best $5 \div 1.67 = 3.0$ times as many microoperations in a given time as the conventional datapath. In this ideal situation, we say that the throughput of the pipelined datapath is three times that of the conventional one. Note that filling and emptying reduce the pipeline speed below the maximum of 3.0. Additional topics associated with pipelines—in particular, providing a control unit for pipelined datapaths and dealing with pipeline hazards—are covered in Chapter 8 and Chapter 10, respectively.

7-12 CHAPTER SUMMARY

In this chapter, we introduced register transfers as a means of representing and specifying elementary processing operations. Register transfers can be related to corresponding digital system hardware, both at the block diagram and at the detailed logic level.

Microoperations are elementary operations performed on data stored in registers. Arithmetic microoperations include addition and subtraction, which are described as register transfers and implemented with corresponding hardware. Logic microoperations—that is, the bitwise application of logic primitives such as AND, OR, and XOR, combined with a binary word—provide masking and selective complementing on other binary words. Left- and right-shift microoperations move data laterally one or more bit positions at a time.

Multiplexers select between multiple transfer paths entering a register. Buses are shared register transfer paths for multiple registers and offer reduced hardware in trade for limitations on possible simultaneous transfers. In addition to multiplexers, three-state buffers enhance the implementation of buses by providing bidirectional transfer paths and reduced connections. Register transfers to and from memory illustrate the interaction of memory reads and writes with registers when using buses.

A datapath is often used for information processing in digital systems. Among the major components of datapaths are register files, buses, arithmetic/logic units (ALUs), and shifters. The control word provides a means of organizing the control of the microoperations performed by the datapath. These concepts were all combined into the concept of a datapath, which will serve as a basis for exploring computers in the remainder of the text.

Finally, the addition of registers forms a *pipelined* datapath, which enables operations to be performed with higher clock frequencies and throughput than are achievable by using the same processing components in a conventional datapath.

REFERENCES

1. MANO, M. M. *Computer Engineering: Hardware Design:* Englewood Cliffs, NJ: Prentice Hall, 1988.
2. DIETMEYER, D. L. *Logic Design of Digital Systems,* 3rd ed. Boston, MA: Allyn-Bacon, 1988.
3. HAMACHER, V. C., VRANESIC, Z. G., AND ZAKY, S. G. *Computer Organization,* 3rd ed. New York, NY: McGraw-Hill, 1990.
4. MANO, M. M. *Computer System Architecture,* 3rd Ed. Englewood Cliffs, NY: Prentice Hall, 1993.
5. PATTERSON, D. A., AND HENNESSY, J. L., *Computer Organization and Design: The Hardware/Software Interface.* San Mateo, CA: Morgan Kaufmann, 1994.
6. HENNESSY, J. L., AND PATTERSON, D. A. *Computer Architecture: A Quantitative Approach,* 2nd ed. San Francisco, CA: Morgan Kaufmann, 1996.

PROBLEMS

7–1. Show the diagram of the hardware that implements the register transfer statement

$$C_3\text{: } R2 \leftarrow R1, R1 \leftarrow R2$$

7–2. The outputs of registers $R0$, $R1$, $R2$, and $R3$ are connected through 4–to–1 multiplexers to the inputs of a fourth register R4. Each register is eight bits long. The required transfers, as dictated by four control variables, are:

$$C_0: R4 \leftarrow R0$$

$$C_1: R4 \leftarrow R1$$

$$C_2: R4 \leftarrow R2$$

$$C_3: R4 \leftarrow R3$$

The control variables are mutually exclusive; i.e., only one variable can be equal to 1 at any time, while the other three are equal to 0. Also, no transfer into $R4$ is to occur for all control variables equal to 0. (a) Using registers and a multiplexer, draw a detailed logic diagram of the hardware implementing a single bit of these register transfers. (b) Draw a logic diagram of the simple logic that converts the control variables as inputs to outputs that are the select variables for the multiplexers and the load signals for registers.

7–3. Using two 4-bit registers $R1$ and $R2$, a quadruple 2–to–1-line multiplexer with enable, and four inverters, draw the detailed logic diagram that implements all of the following statements:

$C_1: R2 \leftarrow R1$	Transfer $R1$ to $R2$
$C_2: R2 \leftarrow 0$	Clear R2 synchronously with the clock
$C_3: R2 \leftarrow \overline{R2}$	Complement $R2$

Assume that the control variables obey the assumptions in Problem 7–2.

7–4. Draw the block diagram for the hardware that implements the statement

$$XC_1 + YC_2: AR \leftarrow AR + BR$$

where AR and BR are n-bit registers and C_1, C_2, X, and Y are control variables. Include the logic gates for the control function. (Remember that the symbol + designates an OR operation in a control or Boolean function, but it represents an arithmetic plus in a microoperation.)

7–5. Logic to implement transfers among three registers $R0$, $R1$, and $R2$ is to be implemented. Use the control variable assumptions given in Problem 7–2. The register transfers are as follows:

$$C_A: R1 \leftarrow R2, R0 \leftarrow R2$$

$$C_B: R2 \leftarrow R0$$

$$C_C: R0 \leftarrow R1, R2 \leftarrow R1$$

(a) Using registers and dedicated multiplexers, draw a detailed logic diagram of the hardware implementing a single bit of these register transfers.

(b) Draw a logic diagram of simple logic that converts the control variables as inputs to outputs that are the select variables for the multiplexers and load signals for the registers.

7–6. Using a 4-bit counter with parallel load as in Figure 5-12 and a 4-bit adder as in Figure 3-27, draw the logic diagram that implements the following statements:

$$C_1\text{: } R1 \leftarrow R1 + R2 \qquad \text{Add } R2 \text{ to } R1$$

$$\overline{C_1}C_2\text{: } R1 \leftarrow R1 + 1 \qquad \text{Increment } R1$$

7–7. *A digital system has three registers: *AR*, *BR*, and *PR*. Three flip-flops provide the control variables for the system: *S* is a flip-flop that is set by an external Start signal to start the operation, and *F* and *R* are two flip-flops used in sequencing the microoperations when the system is in operation. A fourth flip-flop *D* is set by the system when the operation is completed. The operation of the digital system is described by the following register transfer statements:

$$S\text{: } PR \leftarrow 0, S \leftarrow 0, F \leftarrow 1, D \leftarrow 0$$

$$F\text{: } F \leftarrow 0, \text{if } AR = 0 \text{ then } D \leftarrow 1 \text{ else } R \leftarrow 1$$

$$R\text{: } PR \leftarrow PR + BR, AR \leftarrow AR - 1, R \leftarrow 0, F \leftarrow 1$$

(a) Show that the digital system multiplies the contents of registers *AR* and *BR* and places the product in register *PR*.

(b) Draw a block diagram of the hardware implementation. Include a Start input to set flip-flop *S* and a Done output signal from flip-flop *D*.

7–8. Assume that registers *R*1 and *R*2 in Figure 7-4 hold two unsigned numbers. When select input *X* is equal to 1, the adder-subtractor circuit performs the arithmetic operation "*R*1 + 2's complement of *R*2." This sum and the output carry C_n are transferred into *R*1 and *C* when $K_1 = 1$ and a positive edge occurs on the clock.

(a) Show that if $C = 1$, then the value transferred to *R*1 is equal to $R1 - R2$, but if $C = 0$, the value transferred to *R*1 is the 2's complement of $R2 - R1$. (See Section 3-9 and Example 3-6 for the procedure for subtracting unsigned numbers with 2's complement.)

(b) Indicate how the value in the *C* bit can be used to detect a borrow after the subtraction of two unsigned numbers.

7–9. Perform the bitwise logic AND, OR, and XOR with the two 8-bit operands 10100101 and 00111001.

7–10. Given the 16-bit operand 00011011 11000110, what operation must be performed and what operand must be used

(a) to clear all even bit positions (assume bit positions are 15 through 0 from left to right) to 0?

(b) to set the middle 8 bits to 1?

(c) to complement the leftmost 4 and the rightmost 4 bits?

7–11. Starting from the 8-bit operand 101001101, show the values obtained after each one of the shift microoperations given in Table 7-4.

7–12. Repeat Problem 7–5 using a single multiplexer-based bus instead of dedicated multiplexers.

7–13. Draw a logic diagram of a bus system similar to the one shown in Figure 7-5, but use three-state buffers and a decoder instead of the multiplexers.

7–14. A system is to have the following set of register transfers, implemented using buses:

$$C_a\text{: } R0 \leftarrow R4$$

$$C_b\text{: } R3 \leftarrow R0, R1 \leftarrow R4, R4 \leftarrow R0$$

$$C_c\text{: } R2 \leftarrow R3, R0 \leftarrow R3$$

$$C_d\text{: } R2 \leftarrow R4, R4 \leftarrow R2$$

(a) What is the minimum number of buses that can be used to implement the set of transfers?

(b) Draw a block diagram of the system, showing the registers and buses and the connections between them.

7–15. The following memory transfers are to be specified for the system of Figure 7-8:

(a) $M[A2] \leftarrow D1$ **(b)** $D2 \leftarrow M[A1]$ **(c)** $D1 \leftarrow D2$

Specify the memory operation and determine the binary selection variables for all decoders.

7–16. A datapath similar to the one in Figure 7-9 has 48 registers. How many selection lines are needed for each set of multiplexers and for the decoder?

7–17. Given an 8-bit ALU with outputs F_7 through F_0 and carries C_8 and C_7, show the logic circuit for generating the signals for the four status bits N (sign), Z (zero), V (overflow), and C (carry).

7–18. Most contemporary systems use a high-speed addition method such as carry lookahead instead of a ripple carry adder. Modify the arithmetic unit in Figure 7-13 to use a carry lookahead adder.

7–19. Design an arithmetic circuit with one selection variable S and two n-bit data inputs A and B. The circuit generates the following four arithmetic operations in conjunction with carry C_{in}:

S	$C_{in} = 0$	$C_{in} = 1$
0	$F = A + B$ (add)	$F = A + 1$ (increment)
1	$F = A - 1$ (decrement)	$F = A + \overline{B} + 1$ (subtract)

Draw the logic diagram for the two least significant stages of the arithmetic circuit.

7–20. Design a 4-bit arithmetic circuit with two selection variables S_1 and S_0 that generates the following arithmetic operations:

$S_1\ S_0$	$C_{in} = 0$	$C_{in} = 1$
0 0	$F = A + B$ (add)	$F = A + B + 1$
0 1	$F = A$ (transfer)	$F = A + 1$ (increment)
1 0	$F = \overline{B}$ (complement)	$F = \overline{B} + 1$ (negate)
1 1	$F = A + \overline{B}$	$F = A + \overline{B} + 1$ (subtract)

Draw the logic diagram for a single bit stage.

7–21. Inputs X_i and Y_i of each full adder in an arithmetic circuit have digital logic specified by the Boolean functions

$$X_i = A_i \qquad Y_i = \overline{B}_i S + B_i \overline{C}_{\text{in}}$$

where S is a selection variable, C_{in} is the input carry, and A_i and B_i are input data for stage i.

(a) Draw the logic diagram for the 4-bit circuit, using full adders and multiplexers.

(b) Determine the arithmetic operation performed for each of the four combinations of S and C_{in}: 00, 01, 10, and 11.

7–22. Design a digital circuit that performs the four logic operations of exclusive-OR, exclusive-NOR, NOR, and NAND. Use two selection variables. (a) Using a Karnaugh map, design minimum logic for one typical stage, and show the logic diagram. (b) Repeat (a), trying different assignments of the selection codes to the four operations to see whether the logic for the stage can be simplified further.

7–23. *Design an ALU that performs the following operations; give the result of your design as the logic diagram for a single stage of the ALU. Your design should have only a single carry line between stages. If you have access to logic simplification software, apply it to the design to obtain reduced logic.

$A + B$	sl A
$A - B$	$A \vee B$
$\overline{A}$	$A \oplus B$
$\overline{A} + 1$	$A \wedge B$

7–24. Find the output Y of the 4-bit barrel shifter in Figure 7-17 for each of the following bit patterns applied to S_1, S_0, D_3, D_2, D_1, and D_0:

(a) 100110 **(b)** 001110

(c) 011010 **(d)** 110001

7–25. Specify the 17-bit control word that must be applied to the processor of Figure 7-19 to implement each of the following microoperations:

(a) $R1 \leftarrow R2 - R3$ **(b)** $R4 \leftarrow \text{sr } R3$

(c) $R2 \leftarrow \text{Data in}$ **(d)** $R7 \leftarrow R5 + \text{Constant in}$

(e) $R6 \leftarrow 0$ **(f)** $R3 \leftarrow R2 \oplus R7$

(g) $R1 \leftarrow R0 + 1$ **(h)** $R2 \leftarrow \text{sl } R3$

7–26. Given the following 17-bit control words for the datapath of Figure 7-19, determine (a) the microoperation that is executed and (b) the change in the contents of the register for each control word (assume that the registers are 8-bit registers and that, before the execution of a control word, they contain the value of their number (e.g., register $R5$ contains 05 in hexadecimal)):

(a) 110 100 101 0 01000 0 1 **(d)** 101 000 000 0 00000 1 1
(b) 110 001 100 0 00101 0 1 **(e)** 111 111 000 1 01100 0 1
(c) 101 010 000 0 10000 0 1 **(f)** 010 000 000 0 01100 0 1

7–27. Given the following sequence of 17-bit control words for the processor unit in Figure 7-19, and beginning ASCII character codes in registers in 8-bit-wide datapaths, simulate the datapath to determine the alphanumeric characters in the registers after the execution of the sequence (the result is a scrambled word: what is it?):

011 011 001 0 00010 0 1	$R0$	00000000
100 100 001 0 01010 0 1	$R1$	00100000
101 101 001 0 01100 0 1	$R2$	01000100
001 001 000 0 01110 0 1	$R3$	01000111
001 001 000 0 00001 0 1	$R4$	01010100
110 110 001 0 00101 0 1	$R5$	01001100
111 111 001 0 00101 0 1	$R6$	01000001
001 111 000 0 00000 0 1	$R7$	01001001

7–28. A datapath has five major components, A through E, attached in a loop similar to that in Figure 7-21(a). The maximum delay of each of the components is A, 3 ns; B, 3 ns; C, 4 ns; D, 6 ns; and E, 5 ns.

(a) What is the maximum clock frequency that can be used for the datapath?
(b) The datapath is to be changed to one that is pipelined using three stages. How should the components be combined into stages, and what is the maximum clock frequency that can be achieved?
(c) Repeat (b) for four pipeline stages.

7–29. Perform a simulation similar to that of Figure 7-20 for the sequence of microoperations in Figure 7-24 executing on the pipelined datapath of Figure 7-23. Assume that the pipeline registers in Figure 7-23 are labeled OF/EX:A, OF/EX:B, EX/WB:F, and EX/WB:DI. Assume also that the registers in the register file contain a value equal to their number, the value used on Constant in is 4F, and the value used on Data in is 0E.

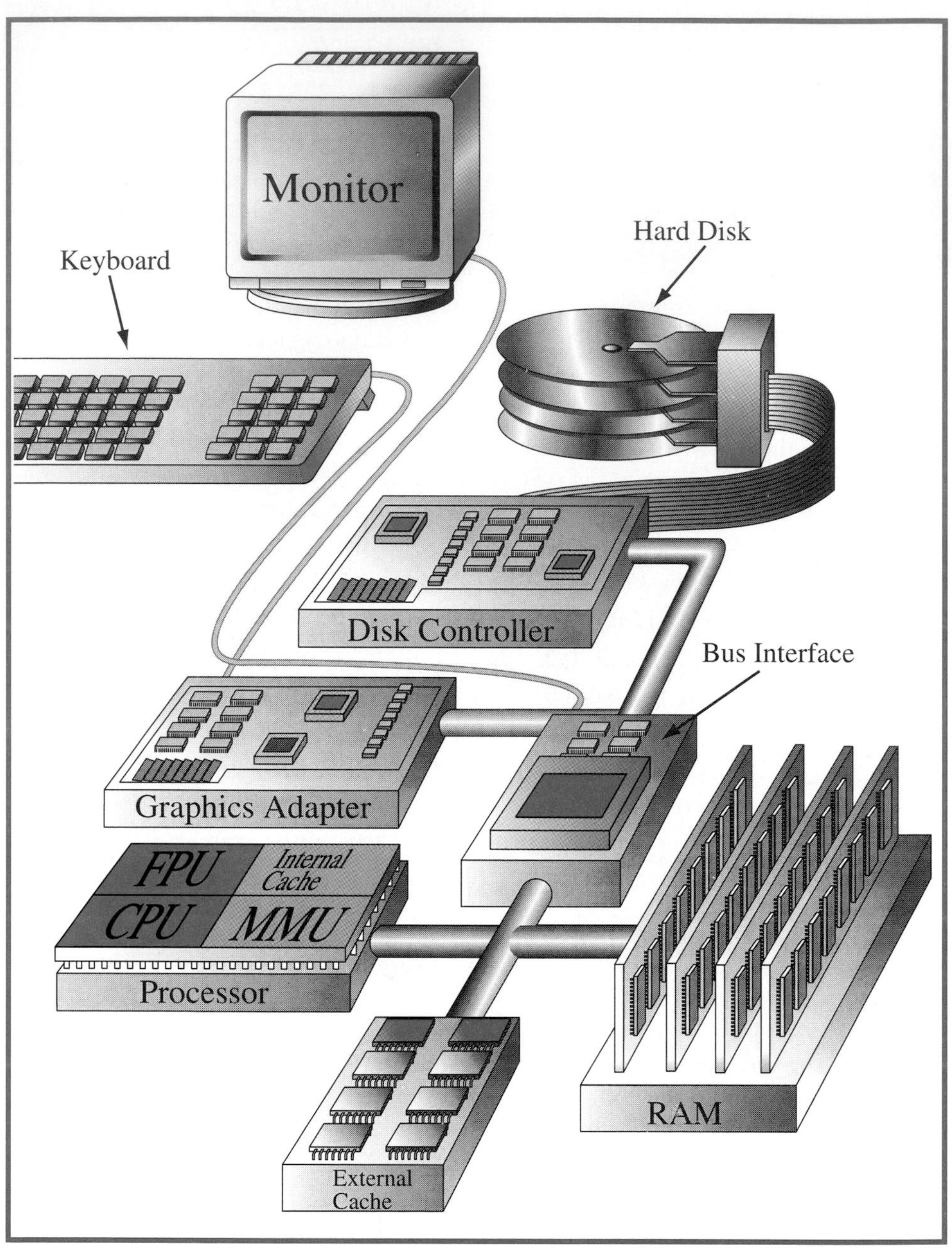
Monitor
Hard Disk
Keyboard
Disk Controller
Bus Interface
Graphics Adapter
FPU
Internal Cache
CPU
MMU
Processor
RAM
External Cache

CHAPTER

8

SEQUENCING AND CONTROL

In Chapter 7, we introduced datapaths for processing data. In the current chapter, our focus is the control of datapaths. In this regard, digital systems are classified into programmable and nonprogrammable systems. Programmable systems include computers, so the combination of a datapath and control unit to form the CPU of a computer is presented. The chapter concludes our examination of general digital systems design and is a prelude to computer design, discussed in Chapters 9 through 12.

The algorithmic state machine (ASM), a more flexible version of the state diagram of a sequential circuit, provides a representation of the behavior of the control unit. By using register transfers in the ASM, combined control unit and datapath behavior is represented. The implementation of a control unit represented by an ASM uses either hardwired or a microprogrammed approaches. There are many specialized techniques for hardwired control design. Here we consider only sequential circuit design as covered in Chapter 4 and a sample of two specialized design approaches based on it. Alternatively, the microprogrammed approach offers a comparatively easy, more structured, and more flexible approach for complex designs.

In this chapter, the main topics are algorithmic state machines, hardwired control, microprogrammed control, and the application of all of these to the design of computer CPUs. Since the design techniques are quite general, we find the generic computer shaded lightly over much of the electronic portion of the hardware. Note, however, that the CPU and FPU in the processor chip are shaded heavily. These components contain significant controls for sequencing and activating processing operations. Both the CPU and the FPU decode instructions and sequence the control words for their respective datapaths. Nevertheless, like the material in previous chapters, what we study in this chapter is broad ranging.

8-1 THE CONTROL UNIT

The binary information stored in a digital computer can be classified as either data or control information. As we saw in the previous chapter, data is manipulated in a datapath to perform arithmetic, logic, shifting, and other data-processing tasks. These operations are implemented with ALUs, registers, multiplexers, and buses. The control unit provides signals that activate the various microoperations in the datapath to perform the specified data-processing tasks. The control unit also determines the sequence in which the various actions are performed. Because the logic design of a digital system is often treated in two distinct parts, the design of the datapath was covered in Chapter 7 and the design of the control unit is covered in this chapter.

The timing of all registers in a synchronous digital system is controlled by a master clock generator. The clock pulses are applied to all flip-flops and registers in the system, including those in the control unit. To prevent continuous clock pulses from changing the state of all registers on every clock cycle, some registers have a load control signal that enables and disables the loading of the register. The binary variables that control the selection inputs of multiplexers, buses, and ALUs and the load control inputs of registers are generated by the control unit.

The control unit that generates the signals for sequencing the operations in the datapath is a sequential circuit with states that dictate the control signals for the system. At any given time, the state of the sequential circuit activates a prescribed set of microoperations. Using status conditions and control inputs, the sequential control unit determines the next state in which additional microoperations are activated. The digital circuit that acts as the control unit provides a sequence of signals for activating the microoperations in the datapath and also determines its own next state.

Based on the overall system design, there are two distinct types of control units used in digital systems, one for a *programmable system* and the other for a *nonprogrammable system*.

In a programmable system, a portion of the input to the processor consists of a sequence of *instructions*. Each instruction specifies the operation the system is to perform, which operands to use for the operation, where to place the results of the operation, and, in some cases, which instruction to execute next. For the programmable system, the instructions are usually stored in memory, which is either RAM or ROM. To execute the instructions in sequence, it is necessary to provide the address in memory of the instruction to be executed. This address comes from a register called the *program counter* (*PC*). As the name implies, the *PC* has logic that permits it to count. In addition, in order to change the sequence of operations using decisions based on status information, the *PC* needs parallel load capability. So, in the case of a programmable system, the control unit contains a *PC* and associated decision logic, as well as the necessary logic to interpret the instruction. *Executing* an instruction means activating the necessary sequence of microoperations in the datapath that are required to perform the operation specified by the instruction.

For a nonprogrammable system, the control unit is not responsible for obtaining instructions from memory, nor is it responsible for sequencing the execution of those instructions. There is no *PC* or similar register in such a system. Instead, the control unit determines the operations to be performed and the sequence of those operations, based on only its inputs and the status bits.

Initially, we consider a binary multiplier as an example of a nonprogrammable digital system. This system illustrates the use of algorithmic state machines (ASMs) for control unit design, plus both hardwired and microprogrammed control logic. We follow this study with consideration of three control approaches to programmable systems, using a simple computer CPU as an example. Note that, although the use of the microprogrammed control logic approach involves programming, it does not necessarily make the overall system programmable from the user's viewpoint, but is simply an alternative way to implement the control unit.

8-2 ALGORITHMIC STATE MACHINES

A data-processing task can be defined by register transfer operations controlled by a sequencing mechanism. Such a task can be specified as a hardware algorithm that consists of a finite number of procedural steps which perform the data processing task. The most challenging and creative part of digital design is the formulation of hardware algorithms for achieving required objectives. A hardware algorithm can be used as a basis for defining both the datapath and the control unit of a system.

A flowchart is a convenient way to specify a sequence of procedural steps and decision paths for an algorithm. A flowchart for a hardware algorithm must have special characteristics that tie it closely to the hardware that implements the algorithm. As a consequence, we use a special flowchart called an *algorithmic state machine* (ASM) chart to define digital hardware algorithms. A *state machine* is just another term for a sequential circuit.

The ASM chart resembles a conventional flowchart, but is interpreted somewhat differently. A conventional flowchart describes procedural steps and decision paths without any concern for their relationship to time. By contrast, the ASM chart is distinguished by the fact that it describes a sequence of events, as well as the timing relationship between the states, of the control unit and the actions that occur in the states in response to clock pulses.

In the next subsection, we specify the blocks that make up an ASM chart and explain the timing relationship. In the following section, we present an example of a binary multiplier in order to illustrate the formulation and implementation of ASM-based control.

The ASM Chart

The ASM chart contains three basic elements: the state box, the decision box, and the conditional output box, as illustrated in Figures 8-1(a) through (c).

A state in the control sequence is indicated by a state box, as shown in Figure 8-1(a). The *state box* is a rectangle containing register transfer operations or output

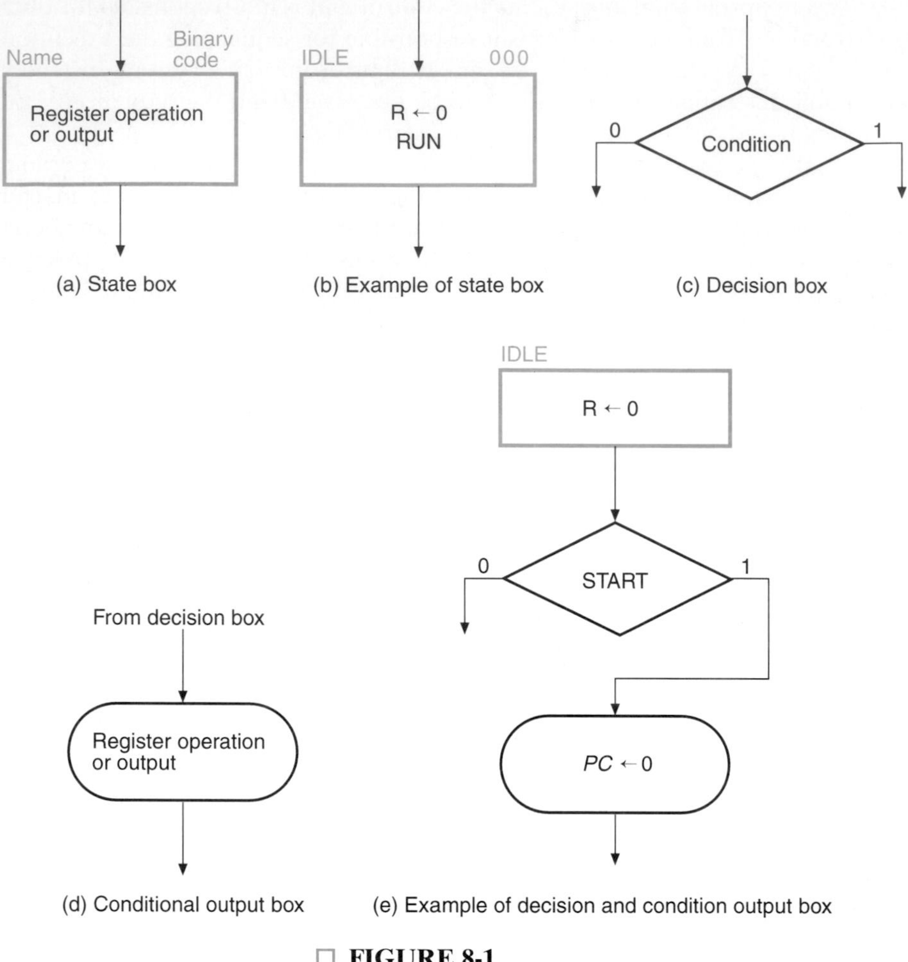

□ **FIGURE 8-1**
ASM Chart Elements

signals that are activated while the control unit is in the state. The symbolic name for the state is placed at the upper left corner of the box, and the binary code for the state, if assigned, is placed at the upper right corner of the box.

Figure 8-1(b) shows a specific example of a state box. The state has the symbolic name IDLE, and the binary code assigned to it is 000. Inside the box is the register transfer $R \leftarrow 0$ and the output RUN. The register transfer indicates that the register R is to be reset to 0 on any clock pulse that occurs with the control in state IDLE. RUN indicates that the output signal RUN is to be 1 during the time that the control is in state IDLE. RUN is 1 for any state box in which it appears and is 0 for any state box in which it does not appear.

The *decision box* describes the effect of inputs on the control. It is a diamond-shaped box with two exit paths, as shown in Figure 8-1(c). The input condition is a single variable or a Boolean expression within the box. One exit path is taken if

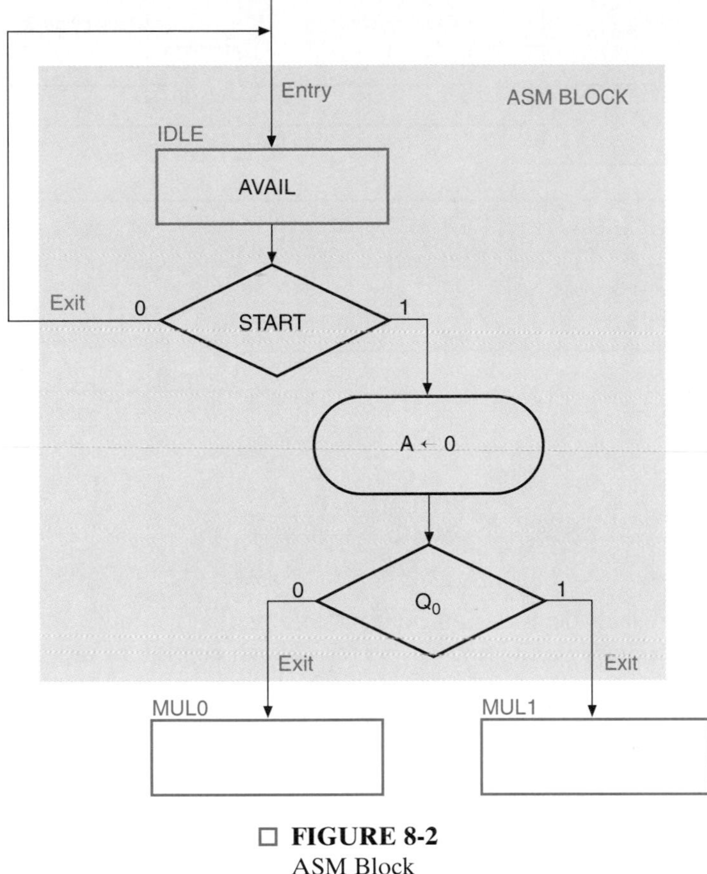

□ **FIGURE 8-2**
ASM Block

the input condition is true (1), and the other is taken if the input condition is false (0).

The third element, the *conditional output box*, is unique to the ASM chart. The oval shape of the box is shown in Figure 8-1(d). The rounded corners differentiate it from the state box. The entry path to a conditional output box from a state box must pass through one or more decision boxes. If the conditions specified on the path through the decision boxes leading from the state box to a conditional output box are satisfied, the register transfers or outputs listed inside the conditional output box are activated.

An example of a decision and conditional output box is given in Figure 8-1(e). If the ASM is in state IDLE, register *R* is reset to 0 for every clock pulse. If the ASM is in state IDLE, register *PC* is cleared to 0 as well, but only if the signal START is equal to 1.

An *ASM block* consists of one state box and all of the decision and conditional output boxes connected between the state box exit and entry paths to the same or other state boxes. An example of an ASM block is shown in Figure 8-2.

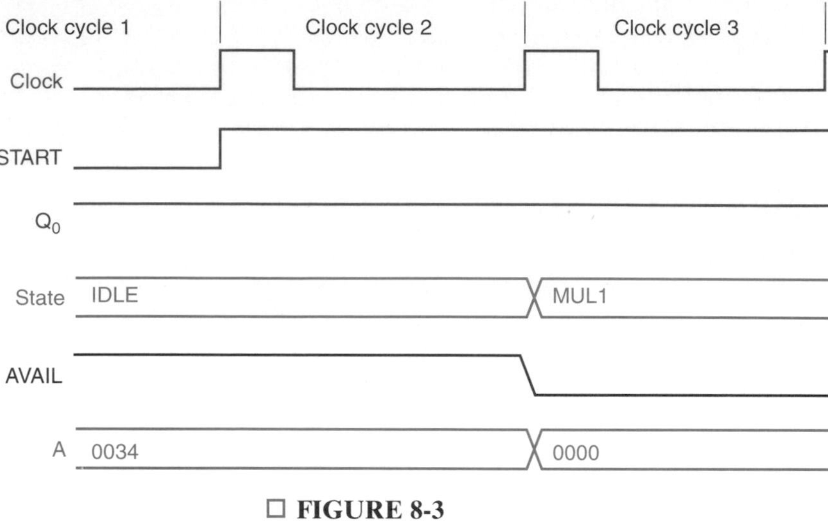

□ **FIGURE 8-3**
ASM Timing Behavior

The block represents decisions and output actions that can take place in the state appearing at its entry point. Any outputs for which conditions are satisfied within the ASM block are activated in the block state. Any register transfers for which conditions are satisfied within the ASM block will be executed when the clock event occurs. This same clock event will also transfer control to one of the next states, as specified by decisions within the ASM block. For the block in Figure 8-2, the state is IDLE. While in the state IDLE, the output AVAIL is equal to 1. If START is 0, then the next state is IDLE. If START is 1, then at the clock event, A is cleared to all 0's, and, if Q_0 is 0, the next state is MUL0. If Q_0 is 1, then the next state is MUL1. In the figure, the entry path and the three exit paths for the ASM block are labeled.

The ASM chart is really form of state diagram for the sequential circuit part of the control unit. Each state box is equivalent to a node in the state diagram. The decision boxes are equivalent to input values on the lines that connect nodes in the diagram. The register transfers and outputs in the state boxes and the conditional output boxes correspond to the outputs of the sequential circuit. Outputs in a state box are those that would be specified on a state node in the state diagram. Outputs in a conditional output box are those corresponding to the input values on the lines connecting states in the state diagram.

Timing Considerations

In order to make clear the timing considerations for the ASM, we use the sample ASM block in Figure 8-2. The timing of the events related to state IDLE is illustrated in Figure 8-3. In considering the timing of these events, recall that all flip-flops used are positive-edge triggered. During clock cycle 1, the control unit is in present state IDLE, output AVAIL is 1, and input START is 0. Based on the ASM block, when a positive clock edge occurs between clock cycle 1 and clock cycle 2, the state remains at IDLE, and AVAIL remains at 1 in clock cycle 2. Also, the contents of register A remain unchanged. In clock cycle 2, START becomes 1. So when the next positive clock edge occurs, between clock cycle 2 and clock cycle 3, register A is cleared to 0. With START at 1, Q_0 is examined and found to be 1. When the clock edge occurs between clock cycle 2 and clock cycle 3, the next state becomes MUL1. The new state MUL1 and the new value of A both appear at the beginning of clock cycle 3. At this clock cycle, the value of AVAIL becomes 0, since AVAIL does not appear in the state box for state MUL1. Finally, note that the output AVAIL = 1 appears concurrently with the present state IDLE, but the result of the register transfer for A appears concurrently with the next state MUL1. This is because outputs occur immediately in response to state and input values, but register transfers and state changes both wait until the next positive clock edge.

8-3 Design Example: Binary Multiplier

In this section, we introduce a hardware algorithm for binary multiplication, propose a simple datapath for its implementation, and then describe its register transfers and control by use of an ASM. We consider two basic approaches to implementing the control unit—hardwired control in Section 8-4 and microprogrammed control in Section 8-5. Hardwired control is less costly for small control units and offers faster operation speeds for high-performance systems. On the other hand, for systems that have many complex instructions to be interpreted, microprogrammed control offers an effective approach, provided that it can be made fast enough to meet the demands of the application.

The system we will examine multiplies two unsigned binary numbers. In Section 3-11, a particular hardware algorithm to execute this multiplication without using storage elements resulted in a combinational multiplier with many adders and AND gates. In contrast, in this chapter, the hardware algorithm is to result in a sequential multiplier that uses only one adder and a long shift register. In this section, the algorithm is illustrated, the datapath proposed, and the ASM chart formulated.

Binary Multiplier

The multiplication of two unsigned binary numbers is done with paper and pencil by successive shifts of copies of the multiplicand to the left and an addition. The process is best illustrated with an actual example. Let us multiply the two binary numbers 10111 and 10011, as shown in Figure 8-4. To carry out the multiplication, we look at successive bits of the multiplier, least significant bit first. If the multiplier bit is 1, the multiplicand is copied down to enter into the addition to be performed later. Otherwise 0's are copied down. The numbers copied in successive lines are shifted one position to the left from the previous number copied, to be in proper alignment with the multiplier bit being processed. Finally, the numbers are

23	10111	Multiplicand
19	10011	Multiplier
	10111	
	10111	
	00000	
	00000	
	10111	
437	110110101	Product

□ **FIGURE 8-4**
Hand Multiplication Example

added and their sum forms the product. Note that the product obtained upon multiplying two n-bit binary numbers can have up to $2n$ bits.

When the multiplication procedure is implemented with digital hardware, it is useful to change the process slightly. First, instead of having a digital circuit that adds n binary numbers simultaneously, it is less expensive to provide a circuit that sums only two numbers. So each time a copy of the multiplicand or 0's are determined to enter into the addition, they are immediately added to a *partial product.* The partial product is stored in a register in preparation for the shift action to follow. Second, instead of shifting the copies of the multiplicand to the left, the partial product being formed is shifted to the right. This leaves the partial product and the copy of the multiplicand in the same relative position as the left shift of the multiplicand did. But, more important, it means that an adder is needed for only n bit positions instead of $2n$ bit positions. The addition always takes place in the same n positions, instead of moving to the left one bit position each time. Third, when the corresponding bit in the multiplier is 0, there is no need to add all 0's to the partial product, since this does not alter its resulting value.

The multiplication example is repeated in Figure 8-5 with these changes. Note that the initial partial product is 0. Each time that the multiplier bit being processed is 1, an addition of the multiplicand, followed by a right shift, is performed. Each time that the multiplier bit is a 0, only a right shift is performed. One of these two actions is performed for each bit of the multiplier, so in this case, five such actions occur. An unsigned overflow occurring during an addition is indicated in blue. This overflow is no problem, however, since a right shift immediately follows that brings the extra partial product bit into the most significant regular bit position.

Multiplier Datapath

The block diagram for the binary multiplier is shown in Figure 8-6. The multiplier datapath is first constructed from components covered in previous chapters. All but counter P are expanded to n bits; counter P requires $\lceil \log_2 n \rceil$ bits in order to count the processing of the n bits of the multiplier. ($\lceil x \rceil$ denotes the smallest inte-

23	10111	Multiplicand
19	10011	Multiplier
	00000	Initial partial product
	10111	Add multiplicand, since multiplier bit is 1
	10111	Partial product after add and before shift
	010111	Partial product after shift
	10111	Add multiplicand, since multiplier bit is 1
	1000101	Partial product after add and before shift[a]
	1000101	Partial product after shift
	01000101	Partial product after shift
	001000101	Partial product after shift
	10111	Add multiplicand, since multiplier bit is 1
	110110101	Partial product after add and before shift
437	0110110101	Product after final shift

a. Note that overflow temporarily occurred.

FIGURE 8-5
Hardware Multiplication Example

ger greater than or equal to x.) We use the parallel adder from Figure 3-27, parallel-load register B from Figure 5-2, and parallel-load shift registers A and Q from Figure 5-6. Counter P is a version of the parallel-load counter in Figure 5-12 modified to count down instead of up, and C is a flip-flop that can be either synchronously cleared to 0 or loaded from C_{out}. These datapath components are connected as shown in Figure 8-6.

The multiplicand is loaded into register B from IN, the multiplier is loaded into register Q from IN, and the partial product is formed in register A and stored in registers A and Q. This dual use of Q is possible because we use a right shift of the multiplier in Q to examine each successive multiplier bit. The right shift vacates space one bit at a time in register Q. This space accepts the lower part of the partial product from A as it is generated. The n-bit binary adder is used for adding B to A. The C flip-flop stores the carry C_{out}, whether 0 or 1, from the addition and is reset to 0 during the right shift. In order to count the number of add-shift or shift actions that are to occur, counter P is provided. It is initially set to $n - 1$ and counted down after the formation of each partial product. The value in P is checked just before it is counted down. So n operations occur, one operation for each value in P, $n - 1$ down through 0. Each operation is either an add and shift or just a shift. When P contains 0, the final product is in the double register A and Q, and processing stops.

The control unit stays in an initial state until the Go signal G becomes 1. Then the system performs multiplication. The sum of A and B forms the n most significant bits of the partial product, which is transferred to A. C_{out} from the addition is transferred to C. Both the partial product and the multiplier in A and Q are

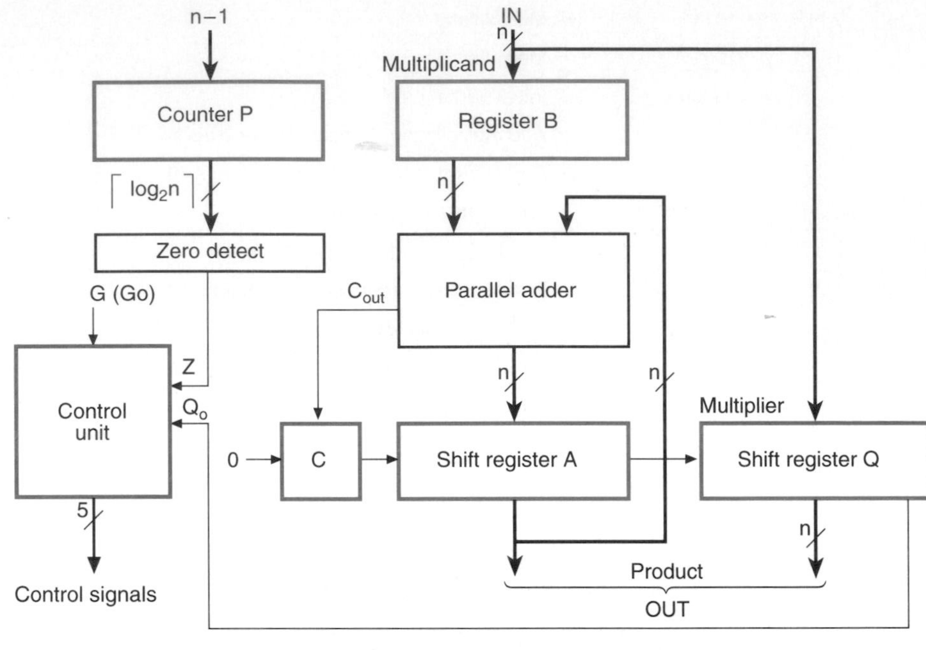

□ **FIGURE 8-6**
Block Diagram for Binary Multiplier

shifted to the right. The carry from C is shifted into the most significant bit of A, the least significant bit of A is shifted into the most significant bit of Q, and the least significant bit of Q is discarded. After this right-shift operation, one (additional) bit of the partial product has been transferred into Q, and the multiplier bits have been shifted one position to the right. In this manner, the least significant bit of Q, Q_0, always holds the bit of the multiplier that must be considered next by the control unit. The control unit "decides" whether to add or not, based on the value of this bit. It also checks signal Z for P equal to zero to determine whether the multiplication is finished. Q_0 and Z are the status inputs for the control unit, while the input G is the only external control input. The control signals from the control unit to the datapath activate the required microoperations.

ASM Chart for Multiplier

An ASM chart giving the sequence of operations in the binary multiplier is shown in Figure 8-7. Initially, the multiplicand is in B and the multiplier in Q. The loading of these two registers is not handled by the multiplier control unit. As long as the ASM is in state IDLE and G is 0, no actions occur, and the ASM remains in IDLE. The multiplication process starts when G becomes 1. As the ASM moves from state IDLE to state MUL0, registers C and A are cleared to 0, and the counter P is loaded with the constant $n - 1$. In state MUL0, a decision is made based on Q_0, the least significant bit of Q. If Q_0 is 1, the contents of B are added to those of A, with

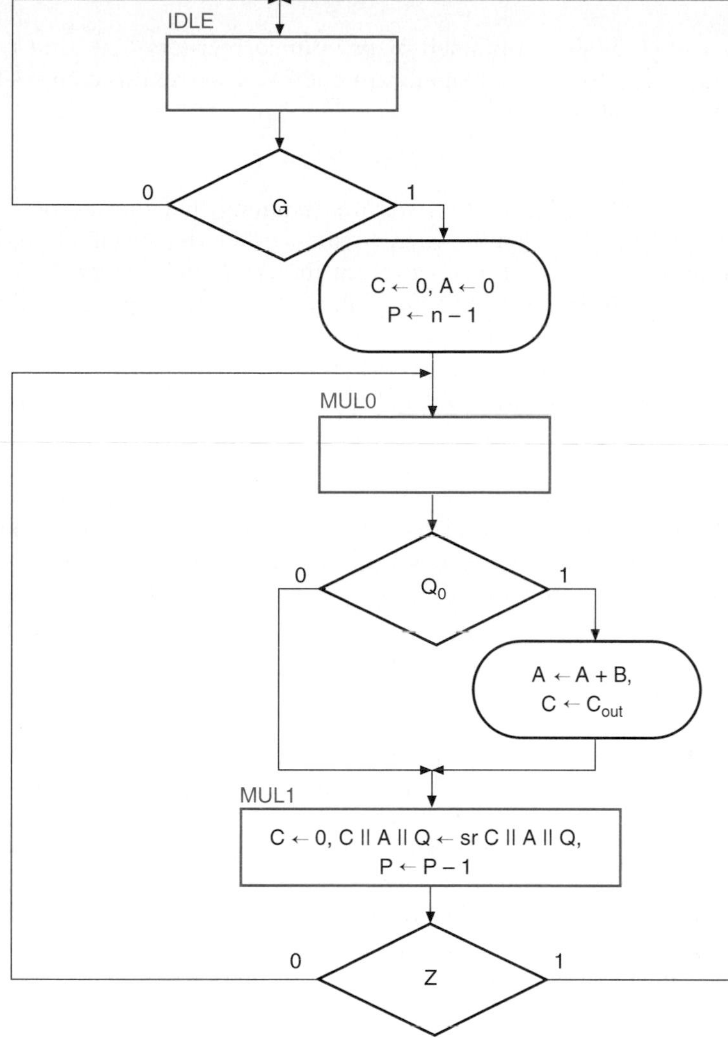

□ **FIGURE 8-7**
ASM Chart for Binary Multiplier

the result transferred to A and the carry transferred to C. If Q_0 is 0, register A and bit C are left unchanged. In both cases, the next state is MUL1.

In state MUL1, a right shift is performed on the combined contents of C, A, and Q. This shift can be expressed by the somewhat messy list of five simultaneous register transfers:

$$C \leftarrow 0,\ A(n-1) \leftarrow C, A \leftarrow \text{sr } A,\ \ Q(n-1) \leftarrow A(0),\ Q \leftarrow \text{sr } Q$$

Instead of this formulation, we will add a bit of notation and use $\|$ to define a composite register made up of other registers or pieces of other registers. For example,

$$C \| A \| Q$$

represents a single register obtained by combining registers C, A, and Q from the most significant end to the least significant end. We can use this composite register to represent the right shift

$$C \| A \| Q \leftarrow \text{sr } C \| A \| Q$$

as shown in Figure 8-7. Recall that we are assuming that the leftmost bit of the result for a right shift takes on the value 0 unless otherwise specified, so C becomes 0. This is represented explicitly, however, in the ASM chart, since C is set to 0 in another state as well. The explicit listing allows $C \leftarrow 0$ to be performed by using a single control signal for both states.

Counter P is decremented in MUL1. The value in P is checked after the formation of each partial product, but before P is decremented. For the first $n - 1$ times that P is checked, its content is nonzero, so status bit Z is 0, and the loop consisting of states MUL0 and MUL1 is executed again. For the nth time P is checked, the content of P is zero, so status bit Z is 1. This indicates that the multiplication is complete, so that the ASM returns to state IDLE. The final product is available in $|A \| Q$, with A holding the n most significant bits and Q the n least significant bits of the product. It is worthwhile to reexamine the hardware multiplication example for $n = 5$ in Figure 8-5, this time considering the relationship to the datapath and the flow of the ASM chart.

The type of registers originally selected for the datapath correspond to the microoperations listed in the ASM chart. Register A is a shift register with parallel load to accept the sum from the adder; it also needs a synchronous clear to reset the register to 0. Register Q is a shift register. The C flip-flop needs to accept the input carry and needs a synchronous clear. Registers B and Q also need parallel load in order to load the multiplicand and the multiplier prior to the start of the multiplication process.

In the next section, we consider implementing the binary multiplier control unit using the hardwired control approach and, in the following section, using microprogrammed control.

8-4 Hardwired Control

In implementing a control unit, there are two distinct aspects with which to deal: the control of the microoperations and the sequencing of the control unit and microoperations. Very simply put, the first has to do with the part of the control that generates the control signals, and the second has to do with the part of the control that determines what happens next. Here, we separate these two aspects by dividing the original ASM specification into two parts: a table that defines the control signals in terms of states and inputs, and a simplified ASM chart that represents only transitions from state to state. Although we are separating these two aspects for design purposes, they can share logic.

□ **TABLE 8-1**
Control Signals for Binary Multiplier

Block Diagram Module	Microoperation	Control Signal Name	Control Expression
Register A:	$A \leftarrow 0$	Initialize	$\text{IDLE} \cdot G$
	$A \leftarrow A + B$	Load	$\text{MUL0} \cdot Q_0$
	$C\|A\|Q \leftarrow \text{sr } C\|A\|Q$	Shift_dec	MUL1
Register B:	$B \leftarrow IN$	Load_B	LOADB
Flip-Flop C:	$C \leftarrow 0$	Clear_C	$\text{IDLE} \cdot G + \text{MUL1}$
	$C \leftarrow C_{\text{out}}$	Load	—
Register Q:	$Q \leftarrow IN$	Load_Q	LOADQ
	$C\|A\|Q \leftarrow \text{sr } C\|A\|Q$	Shift_dec	—
Counter P:	$P \leftarrow n - 1$	Initialize	—
	$P \leftarrow P - 1$	Shift_dec	—

The control signals are based on the ASM chart. The control signals needed for the multiplier datapath are listed in Table 8-1, where we have chosen to examine the datapath registers and tabulate the microoperations for each register. Next, control signals are defined. If a control signal can be used for activating microoperations in more than one register, we do so. This is reasonable, since the datapath is dedicated to only one operation, multiplication. Thus, the control signals do not need to be separated to provide the generality required for implementing additional, potentially unknown operations. Finally, the Boolean expression for each control signal is derived from the location or locations of the microoperation in the ASM chart. For example, for register A, there are three microoperations shown in Table 8-1: clear, add and load, and right shift. Since the clear operation always occurs at the same time as the clear for flip-flop C and the loading of counter P, all of these microoperations can be activated by the same control signal, named Initialize. Because C is cleared in state MUL1 as well, however, we choose to separate its control signal. So Initialize is used for clearing A and loading P. In the last column for Initialize, the Boolean expression for which Initialize is to be active, as determined from the ASM chart, is given in terms of the state IDLE and input G. Since Initialize is to be 1 when G is 1 in state IDLE, IDLE and G are ANDed. At this point, the name for the state is treated as a Boolean variable. Depending on the implementation, there may be such a signal representing the state, or the state may need to be expressed as a function of the state variables. The signal for clearing C, Clear_C, is to be active in state IDLE for G equal to 1, as well as in state MUL1. Thus, G is ANDed with IDLE, and the result is ORed with MUL1. The other two internal multiplier control signals, Load and Shift_Dec, are defined in a similar manner. The final two signals, Load_B and Load_Q, load the multiplicand and multiplier from outside the multiplier system. These signals will not be considered explicitly in the remainder of the design.

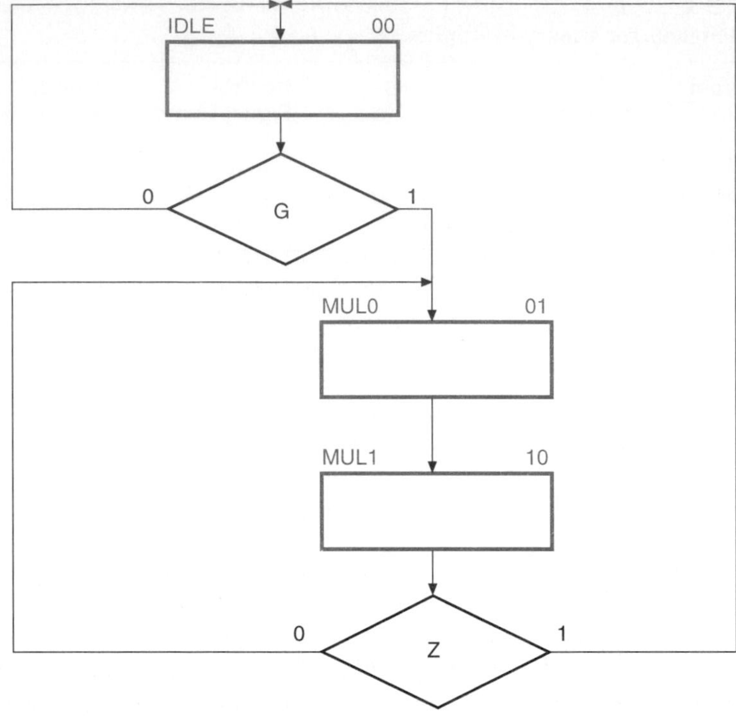

□ **FIGURE 8-8**
Sequencing Part of ASM Chart for the Binary Multiplier

With the information on microoperations removed, we can redraw the ASM chart so that only the information on sequencing is represented. This modified ASM chart for the binary multiplier appears in Figure 8-8. Note that all of the conditional output boxes have been removed. In addition, any decision box not affecting the next state is removed. In particular, in Figure 8-7, the decision box Q_0 affected only a conditional output box. Once that conditional output box is removed, the two exit paths from decision box Q_0 clearly go to the same state. So this decision box has no effect on the next state and is removed.

From this modified ASM chart, we can design the sequencing part of the control unit—i.e., the part that represents the next-state behavior. The division of control into next-state behavior in the form of the modified ASM chart and output behavior in the form of the control signal table shows how the ASM corresponds to the next-state and output parts of a sequential circuit. Figure 8-8 corresponds to the state diagram of a sequential circuit without the outputs specified, except that the representations used in the diagram for states and transitions are different. Because of this correspondence, we can treat the ASM chart as a state diagram and form a state table for the sequencing part of the control unit. Then the control unit can be designed by the sequential logic design procedure, as outlined in Chapter 4. However, in many cases, this method is difficult to carry out because of the large number of states for a typical control unit. As a consequence, we use specialized methods for control unit design that are variations of the classical sequential logic methods. We next present and illustrate two such design methods.

Sequence Register and Decoder

The sequence register and decoder method, as the name implies, uses a sequence register for the control states and a decoder to provide an output signal corresponding to each of the states. A register with n flip-flops can have up to 2^n states and an n-to-2^n decoder has up to 2^n outputs, one for each of the states. An n-bit sequence register is essentially n flip-flops, together with the associated gates that effect their state transitions. Additional logic may be required to produce the necessary control signal outputs.

The sequencing part of the ASM chart for the binary multiplier has three states and two inputs. To implement the ASM chart with a sequence register and decoder, we need two flip-flops for the register and a 2–to–4-line decoder. Since there are three states, only three of the four decoder outputs are used. Although this is a simple example, the procedure to be outlined applies to more complex situations as well.

The state table for the sequencing part of the control unit is shown in Table 8-2; it is derived directly from the ASM chart in Figure 8-8. We designate the two flip-flops as M_1 and M_0 and assign the binary states 00, 01, and 10 to IDLE, MUL0, and MUL1, respectively. Note that the input columns have unspecified entries ($\times$) whenever an input variable is not used to determine the next state. The outputs of the sequencing part of the control are designated by the state names. The binary code for the present state determines the particular output variable that is equal to 1 at any given time. Thus, when the present state is $M_1M_0 = 00$, output IDLE equals 1, while the other outputs equal 0. Since these outputs depend on the present state only, they can be generated with the 2–to–4-line decoder having inputs $M_1$and M_0 and outputs IDLE, MUL0, and MUL1.

□ **TABLE 8-2**
State Table for Sequence Register and Decoder Part of Multiplier Control Unit

Present state			Inputs		Next state		Decoder Outputs		
Name	**M_1**	**M_0**	**G**	**Z**	**M_1**	**M_0**	**IDLE**	**MUL0**	**MUL1**
IDLE	0	0	0	×	0	0	1	0	0
	0	0	1	×	0	1	1	0	0
MUL0	0	1	×	×	1	0	0	1	0
MUL1	1	0	×	0	0	1	0	0	1
	1	0	×	1	0	0	0	0	1
—	1	1	×	×	×	×	×	×	×

As mentioned earlier, the sequential circuit can be designed from the state table by means of the sequential logic design procedure presented in Chapter 4. This example has a small number of states and inputs, so we could use maps to simplify the Boolean functions. In most control logic applications, however, the number of states is much larger. The application of the conventional method requires excessive work to obtain the simplified input equations for the flip-flops. Here, the design can be simplified if we take into consideration the fact that the decoder outputs are available for use in the design. Instead of using flip-flop outputs as the present state conditions, we might as well use the outputs of the decoder to obtain this information. These outputs supply a single signal representing each of the possible present states of the circuit. Moreover, instead of using maps to simplify the flip-flop equations, we can obtain them directly by inspection of the state table. For example, from the next-state conditions in the table, we find that the next state of M_0 is equal to 1 when the present state is IDLE and input G is equal to 1 or when the present state is MUL1 and input Z is equal to 0. These conditions give

$$D_{M_0} = IDLE \cdot G + MUL1 \cdot \overline{Z}$$

for the D input of the M_0 flip-flop. Similarly, the D input of the M_1 flip-flop is

$$D_{M_1} = MUL0$$

Note that these equations derived by inspection from the state table use the state names rather than the state variable names, since the decoder producing the state symbols is present. In some cases, it may be possible to find simpler D flip-flop input equations by using the state variables directly instead of the states. We can remove redundancy and reduce cost by writing the Boolean equations for the decoder and applying a simplification program to the entire set of control equations.

The logic diagram for the control is drawn in Figure 8-9. It consists of a two-bit register with flip-flops M_1 and M_0 and a 2–to–4-line decoder. The three outputs of the decoder are used to generate the control outputs, as well as inputs to the next-state logic. The outputs Initialize, Clear_C, Shift_dec, and Load are determined from Table 8-1. Initialize and Shift_dec are already available as signals, so that only labeled output lines are added. However, as shown in the figure, we must add logic gates for Clear_C and Load. We complete the binary multiplier design by connecting the outputs of the control unit to the control inputs of the datapath.

One Flip-Flop per State

Another possible method of control logic design is the use of one flip-flop per state. A flip-flop is assigned to each of the states, and at any time, only one of the flip-flops contains a 1, with all the rest containing 0. When the 1 is in the flip-flop assigned to a particular state, the sequential circuit is in that same state. The single 1 propagates from one flip-flop to another under the control of decision logic. In such a configuration, each flip-flop represents a state that is present only when the single 1 is stored in the flip-flop.

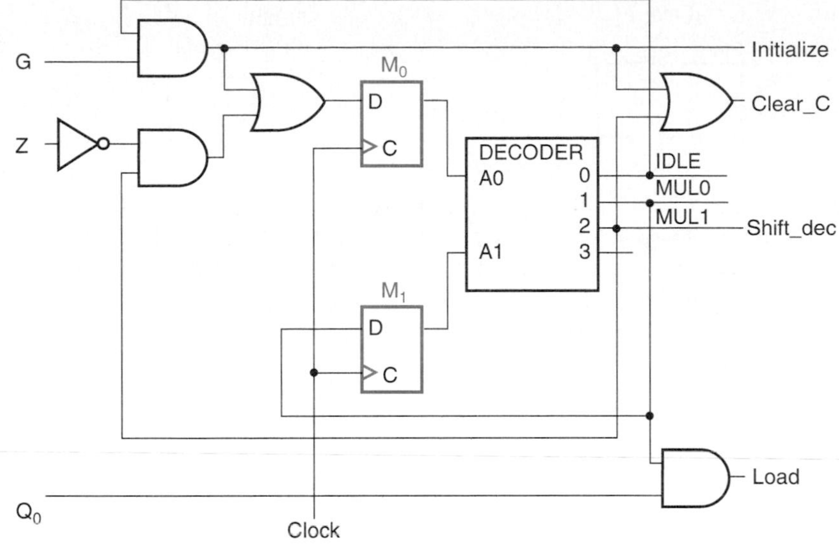

□ **FIGURE 8-9**
Control Unit for Binary Multiplier Using a Sequence Register and a Decoder

It is obvious that, short of some error detection or correction techniques, this method uses the maximum number of flip-flops for the sequential circuit. For example, a sequential circuit with 12 states using minimum variable encoding needs four flip-flops. With one flip-flop per state, the circuit requires 12 flip-flops, one for each state. At first glance, it may seem that this method would increase the cost of the system, since more flip-flops are used. But the method offers some cost advantages that may not be apparent. One advantage is the simplicity with which the logic can be designed—merely by inspection of the ASM chart or state diagram. No state or excitation tables are needed if D flip-flops are employed. This offers a savings in design effort, an increase in logic simplicity.

Figure 8-10 shows the symbol replacement rules for transforming an ASM chart into a sequential circuit with one flip-flop per state. These rules are most easily applied to an ASM chart representing only sequencing information, such as that of Figure 8-8. Each rule specifies the replacement of a component of an ASM chart with a logic circuit. As shown in Figure 8-10(a), the state box is replaced by a D flip-flop labeled with the name of the state. The entry to the state box corresponds to the D input to the flip-flop. The exit of the state box corresponds to the output of the flip-flop.

In Figure 8-10(b), the decision box is replaced by a demultiplexer. The signal corresponding to the entry to the decision box is sent to one of two circuit lines, depending on the value of signal X. If X is 0, the signal is sent to the exit 0 line; if X is 1, the signal is sent to the exit 1 line. So, for example, if the single 1 in the circuit is on the entry to the decision box, and X is 0, the 1 is passed to the exit 0 line. The demultiplexer acts like a switch that directs the 1 through the paths in the circuit corresponding to paths in the ASM chart.

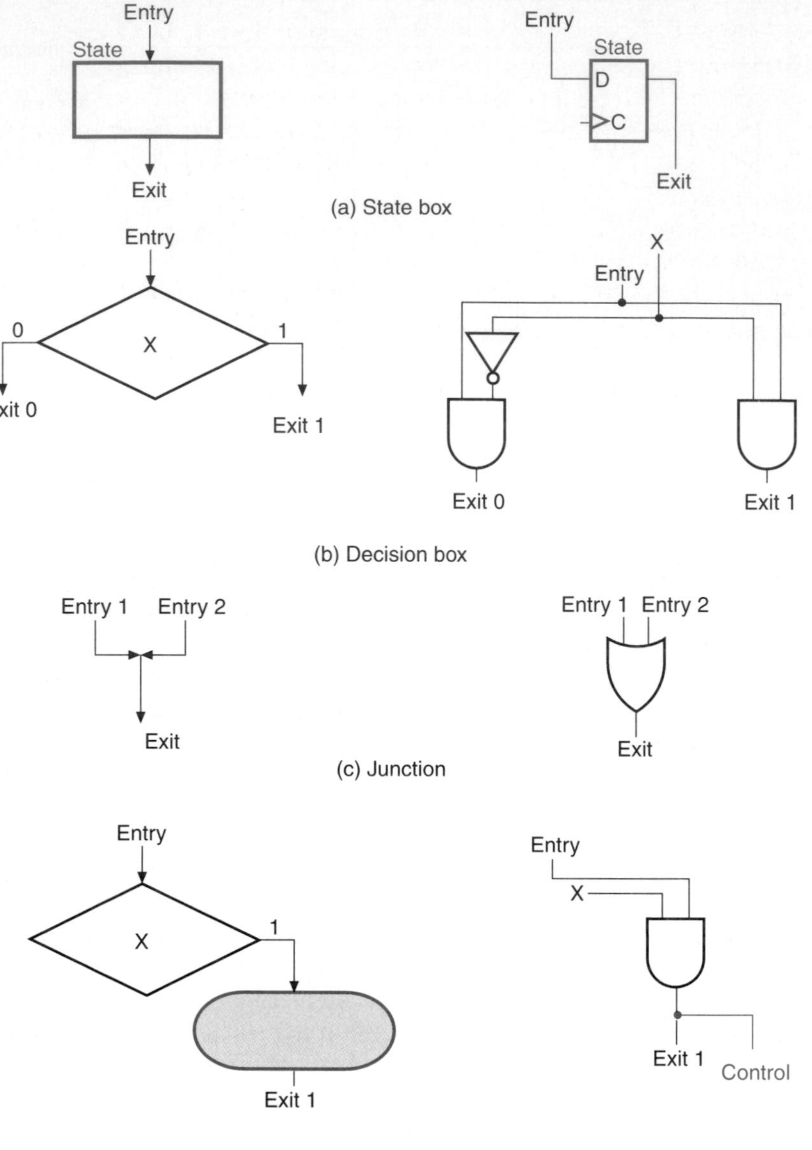

□ **FIGURE 8-10**
Transformation Rules for Control Unit with One Flip-Flop per State

The junction in Figure 8-10(c) is any point at which two or more directed lines in the ASM chart join together. If a 1 is present in the circuit on any line corresponding to one of the entry paths, then it must appear on the line corresponding to the exit path, giving that line the value 1. If none of the lines corresponding to entry paths into the junction have the value 1, then the exit line must have the value 0. Thus, the junction is replaced by an OR gate.

With these three transformations, the sequencing part of the ASM chart can be replaced by a circuit with one flip-flop per state just by inspection. In order to handle outputs, it is merely a matter of attaching control lines to the proper locations in the circuit or adding output logic. The outputs are based on the original ASM chart or the control signal table derived from the chart. Attaching a control line based on an ASM chart is illustrated by the conditional output box in Figure 8-10(d). The box in the ASM chart is just replaced by a connection in the circuit. But to cause the output actions to happen, a control line is tapped from the connection. The transformation is shown in blue for clarity.

We now use these transformations to find the control unit with one flip-flop per state for the binary multiplier. The ASM chart in Figure 8-8 will be used for the sequencing part of the design; note that the binary codes given are ignored, since they were for the former design approach. The resulting logic diagram is shown in Figure 8-11.

First, we replace each of the three state boxes by a *D* flip-flop labeled with the name of the state, as indicated by the circled 1's in the figure. Second, each of the decision boxes is replaced by a demultiplexer with the decision variable as its selection input, as indicated by the circled 2's in the figure. Third, each junction is replaced by an OR gate, as indicated by the circled 3's. Finally, the connections represented by the directed lines in the ASM chart are added from the outputs to the inputs of the corresponding components.

To handle the control outputs, we can use either Table 8-1 or the original ASM chart in Figure 8-7. From the table, we see that the Boolean function for Initialize is already available in the logic diagram, so we simply add the output labeled Initialize. Likewise, the output for Shift_dec can be added. For Clear_C and Load, however, logic gates are added. All of the output connections and logic added are designated by the circled 4's in Figure 8-11.

One final issue in the design of the control logic with one flip-flop per state is initialization to the state having a 1 in the IDLE flip-flop and a 0 in all of the others. This can be done by using an asynchronous PRESET input on the IDLE flip-flop and an asynchronous CLEAR on the other flip-flops. If only an asynchronous CLEAR is available, rather than both PRESET and CLEAR, a NOT gate can be placed just before the *D* input and another NOT gate just after the output of the IDLE flip-flop. Then the IDLE flip-flop will actually contain a 0 when in state IDLE and a 1 at all other times. This permits the asynchronous CLEAR to be used to initialize all three flip-flops in the circuit. It should be noted that, other than for resetting the circuit, the use of asynchronous flip-flop inputs for implementing ASMs or other sequential circuits is generally poor design practice.

Once the basic design of the control logic with one flip-flip per state is completed, it may be desirable to refine the design. For example, if there are a number of junctions connected together by lines, the OR gates that resulted from the transformation may be combined. Also, demultiplexers cascaded with each other may be combined. Other logic reduction or conversion of gates to NANDs or NORs may also be applied to the design.

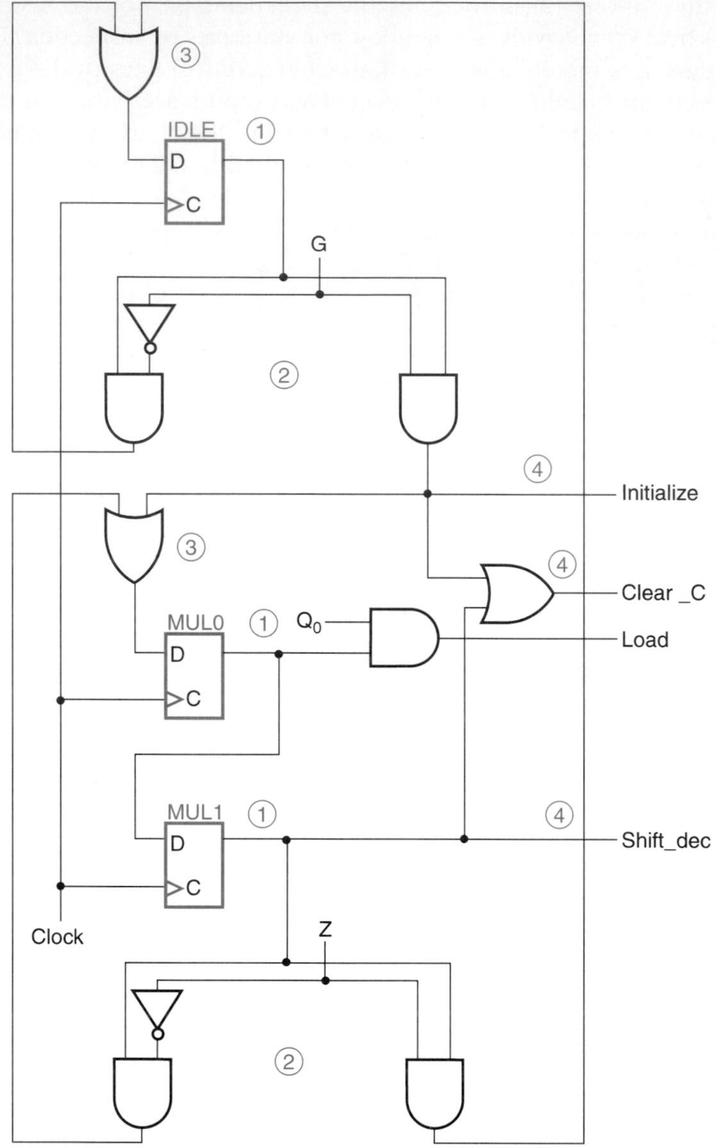

☐ **FIGURE 8-11**
Control Unit with One Flip-Flop per State for the Binary Multiplier

8-5 Microprogrammed Control

A control unit with its binary control values stored as words in memory is called a *microprogrammed control*. Each word in the control memory contains a *microinstruction* that specifies one or more microoperations for the system. A sequence of microinstructions constitutes a *microprogram*. The latter is often fixed at the time

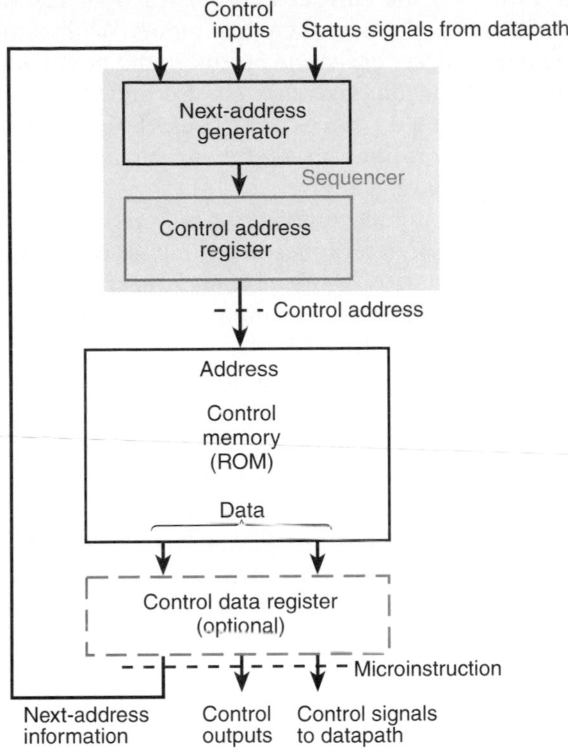

☐ **FIGURE 8-12**
Microprogrammed Control Unit Organization

of the system design and so is usually stored in ROM. Microprogramming involves placing some representation for combinations of values of control variables in words of ROM for use by the rest of the control logic via successive read operations. The contents of a word in ROM at a given address specify the microoperations to be performed for both the datapath and the control unit. A microprogram can also be stored in RAM. In this case, it is loaded initially at system startup from the computer console or from some form of nonvolatile storage such as a magnetic disk. With either ROM or RAM, the memory in the control unit is called *control memory*; if RAM is used, the memory is referred to as *writable control memory*.

Figure 8-12 shows the general configuration of a microprogrammed control. The control memory is assumed to be a ROM within which all control information is permanently stored. The *control address register* (*CAR*) specifies the address of the microinstruction. The *control data register (CDR)*, which is optional, may hold the microinstruction currently being executed by the datapath and the control unit. One of the functions of the control word is to determine the address of the next microinstruction to be executed. This microinstruction may be the next one in sequence, or it may be located somewhere else in the control memory. Therefore, one or more bits that specify how to determine the address of the next microinstruction must be present in the current microinstruction. The next address may also be a function of status and external control inputs. While a microinstruction is being executed, the *next-address generator* produces the next address. This address is transferred to the *CAR* on the next clock pulse and is used to read the next microinstruction to be executed from ROM. Thus, the microinstructions contain bits for activating microoperations in the datapath and bits that specify the sequence of microinstructions executed.

The next-address generator, in combination with the *CAR*, is sometimes called a microprogram *sequencer,* as it determines the sequence of instructions that is read from control memory. The address of the next microinstruction can be specified in several ways, depending on the sequencer inputs. Typical functions of a microprogram sequencer are incrementing the *CAR* by one and loading the *CAR*. Possible sources for the load operation include an address from control memory, an externally provided address, and an initial address to start control unit operation.

The *CDR* holds the present microinstruction while the next address is being computed and the next microinstruction is being read from memory. The *CDR* breaks up a long combinational delay path through the control memory and the datapath. Insertion of this register is just like inserting a pipeline platform, as in Section 7-11; it allows the system to use a higher clock frequency and hence perform processing faster. The inclusion of a *CDR* in a system, however, complicates the sequencing of microinstructions, particularly when decision making based on status bits is involved. Hence, for simplicity, we omit the *CDR* and take the microinstructions directly from the ROM outputs. The ROM operates as a combinational circuit, with the address as the input and the corresponding microinstruction as the output. The contents of the specified word in ROM remain on the output lines of the ROM as long as the address value is applied to the inputs. No read/write signal is needed, as it is with RAM. Each clock pulse executes the microoperations specified by the microinstruction and also transfers a new address to the *CAR*, which, in this case, is the only component in the control that receives clock pulses and stores state information. The next-address generator and the control memory are combinational circuits. Thus, the state of the control unit is given by the contents of the *CAR*.

The status bits enter the next-address generator and affect the determination of the next state. Unless the status bits bypass the control unit and directly control the microoperations being executed in the datapath, they can do no more than influence the next microoperation by affecting the address generated by the next-address generator. This has a profound effect on the structure of the ASM charts for microprogrammed controls. The sequential circuits must be Moore-type sequential circuits, and as a consequence, conditional output boxes are *not* permitted in the ASM charts. This often means that more states will be required in the ASM for a given hardware algorithm. An ASM chart for the binary multiplier, developed under the restriction that the system contain no conditional output boxes, is given in Figure 8-13. Compared to the ASM chart in Figure 8-7, this chart has two more states, INIT and ADD, that have been added where originally conditional output boxes were used. Besides being a Moore-type circuit, this ASM has only single decision boxes determining the sequencing between states. Although next-state

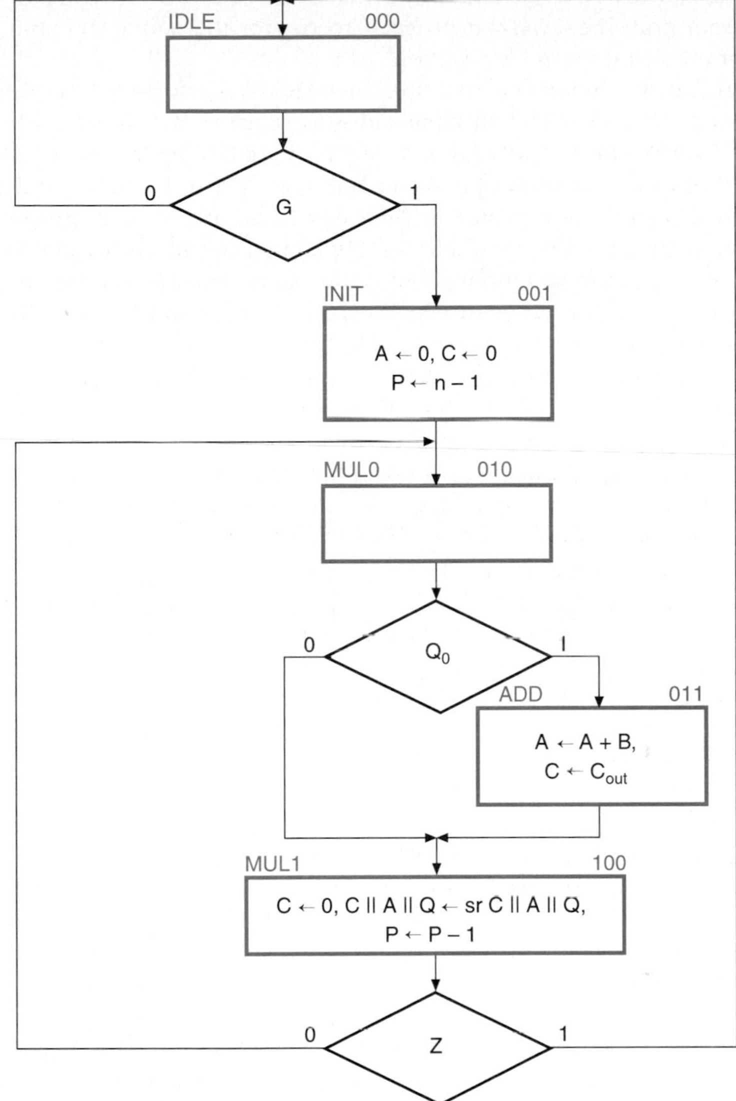

□ **FIGURE 8-13**
ASM Chart for Microprogrammed Binary Multiplier Control Unit

decisions based on multiple values are possible, they are often excluded in simpler next-address generator designs.

Binary Multiplier Example

To illustrate microprogrammed control design, we consider again the binary multiplier. We need to determine three things: the bits in the control word for the microinstructions, the sizes of the ROM and the *CAR*, and the structure of the next-address generator. Then we can proceed to design the sequencer and write the microprogram for binary multiplication.

We focus on the microoperations needed to perform the multiplications, ignoring those that load the multiplicand into register B and the multiplier into register Q. From Table 8-1, we find that only four control signals are needed for the datapath to perform the multiplication: Initialize, Load, Clear_C, and Shift_dec. These control signals are repeated in Table 8-3. In addition, the corresponding register transfers are copied from Table 8-1. By comparing these transfers with those in the states in the microprogrammed control ASM chart for the multiplier in Figure 8-13, we can list the states in which each control signal is active. These states appear in the third column of Table 8-3. This information forms the foundation for designing the part of the microinstruction that controls the datapath.

□ TABLE 8-3
Control Signals for Microprogrammed Multiplier Control

Control Signal	Register Transfers	States in Which Signal is Active	Micro-instruction Bit Position	Symbolic Notation
Initialize	$A \leftarrow 0, P \leftarrow n-1$	INIT	0	IT
Load	$A \leftarrow A + B, C \leftarrow C_{out}$	ADD	1	LD
Clear_C	$C \leftarrow 0$	INIT, MUL1	2	CC
Shift_dec	$C \| A \| Q \leftarrow \text{sr } C \| A \| Q, P \leftarrow P - 1$	MUL1	3	SD

The four control signals can be used as given or can be encoded to reduce the number of bits in the microinstruction. If the control signals are not encoded, four bits, one for each of the control signals, are required in the DATAPATH field of the microinstruction. If the control signals are to be encoded in some fashion, there must be sufficient flexibility to encode all possible distinct combinations of the control signals needed. Suppose that we wish to use a single code word for each such combination. From Table 8-3, the three control signal combinations used are (Initialize, Clear_C) in state INIT, (Load) in state ADD and (Clear_C, Shift_dec) in state MUL1. In addition, for states IDLE and MUL1, we need the combination with no control signals active, giving a total of four distinct combinations. With encoding, the number of bits in the field of the microinstruction for controlling the datapath can be reduced to two, since four combinations can be represented using two bits. If this is done, then a decoder will be required at the ROM output to regenerate the original control signals. In that case, since only two bits are saved and the number of states is small, the saving in the size of ROM is unlikely to be sufficient to justify adding the decoder. As a consequence, we will not encode the control signals.

The bit position in the microinstruction format to be occupied by each control signal is given in the fourth column of Table 8-3. The fifth column lists the

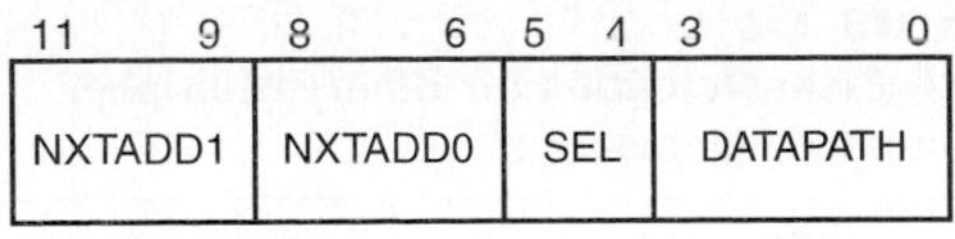

□ **FIGURE 8-14**
Microinstruction Control Word Format

symbolic designation for each of the microoperations for use in writing microprograms. The format for the microinstruction control word for the binary multiplier control is given in Figure 8-14. The 4-bit codes based on Table 8-3 are used in the DATAPATH field of the word.

The remainder of the microinstruction control word is devoted to the sequencing of the control unit. There are many different ways to design the sequencer. The design method determines the fields needed in the microinstruction for sequencing. As a first step in this design, we consider the sequencing requirement defined by the ASM chart in Figure 8-13. In states INIT and ADD, the next state does not depend on status signals or inputs. In state IDLE, the next state depends on the value of G. In state MUL0, the next state depends on the value of Q_0. Finally, in state MUL1, the next state depends on the value of Z. For the cases where the next state depends on a status or input value, a pair of address values is needed, one for the input value equal to 0 and one for the input value equal to 1.

The approach used to define the addresses is a major decision in the sequencer design. There are many possible approaches, but two are most typical. One method includes the two addresses in the microinstruction controlling the decision. Based on the value of the decision variable, one of the two address values is loaded into the *CAR*. This method permits the arbitrary assignment of addresses to states and ensures that no states need to be added to provide the desired sequencing. But it requires two addresses in each microinstruction, potentially resulting in a long microinstruction word and wide ROM. The other method uses a counter with parallel load as the *CAR*. One of the two addresses is obtained from the microinstruction, but the other is obtained by simply counting up the *CAR*. This method requires at most one address per microinstruction word. But the assignment of the addresses to the states can be problematic, and states may have to be added to provide the desired sequencing. These states can slow the operation of the system due to the added clock cycles needed to pass through them.

The choice of method is based on a complex relationship between the speed and the cost of the system. We choose the two-address-per-microinstruction method, which yields the simplest design for purposes of illustration. Later in this chapter and in Chapter 10, we will use the parallel-load counter method for more complex control units. Since we have five states in the multiplier control, the addresses are three bits. Thus, we place two 3-bit next address fields, NXTADD0 and NXTADD1, in the microinstruction control word format in Figure 8-14.

In addition, we add a 2-bit select field, SEL, to select whether to make a decision and, if so, to select which one of the three decision variables, G, Q_0, or Z, is to control the decision. Table 8-4 gives the definition of the four binary values for the

□ **TABLE 8-4**
SEL Field Definition for Binary Multiplier Control Sequencing

SEL		
Symbolic notation	**Binary Code**	**Sequencing Microoperations**
NXT	00	$CAR \leftarrow NXTADD0$
DG	01	$\overline{G}: CAR \leftarrow NXTADD0$ $G: CAR \leftarrow NXTADD1$
DQ	10	$\overline{Q_0}: CAR \leftarrow NXTADD0$ $Q_0: CAR \leftarrow NXTADD1$
DZ	11	$\overline{Z}: CAR \leftarrow NXTADD0$ $Z: CAR \leftarrow NXTADD1$

SEL field. If SEL is 00, symbolically denoted by NXT, then the next address is obtained from the NXTADD0 field in the instruction. If SEL is 01, denoted by DG, then a decision is made based on *G*. If *G* is 0, the next address loaded into the *CAR* is from the NXTADD0 field; if *G* is 1, the next address loaded into the *CAR* is from the NXTADD1 field. Similar behavior applies to a decision on Q_0, denoted by DQ, and a decision on *Z*, denoted by DZ. The information in the table completes the specification for the microprogrammed control hardware. The overall microinstruction contains 12 bits: 4 for controlling the datapath and 8 for sequencing microinstructions.

Using the information in Table 8-3, Table 8-4, and Figure 8-13, we now design the control unit. Based on the length of the microinstruction, the length of the ROM control words is 12 bits. Since there are only five states in Figure 8-13, the ROM contains 5 words. We need a 3-bit address to address 5 words and use a simple 3-bit parallel load register as the *CAR*. The addresses to be loaded into the *CAR* come from NXTADD0 and NXTADD1 in the microinstruction on the ROM output. A quad 2–to–1 multiplexer on the data input to the *CAR* selects between these two address sources. The select signal *S* for this address multiplexer is determined on the basis of Table 8-4. A 4–to–1 multiplexer can be used to select the constant or decision variable required for each of the four SEL codes.

The resulting multiplier block diagram containing the components discussed, as well as the datapath, appears in Figure 8-15. The control memory is a ROM with a capacity of five 12-bit microinstructions. Four of the output bits of the ROM go to the datapath control inputs, while the remaining 8 bits determine the next address for the *CAR*. The SEL bits control the 4–to–1 multiplexer MUX2. For SEL at 00, MUX2 selects input 0, which has the value 0. A 0 on *S* of MUX1 selects NXTADD0 as the next address, as specified in Table 8-4. For SEL at 01, MUX 2 selects input 1, which is *G*. If *G* is 0, then the *S* on MUX1 is 0, selecting NXTADD0

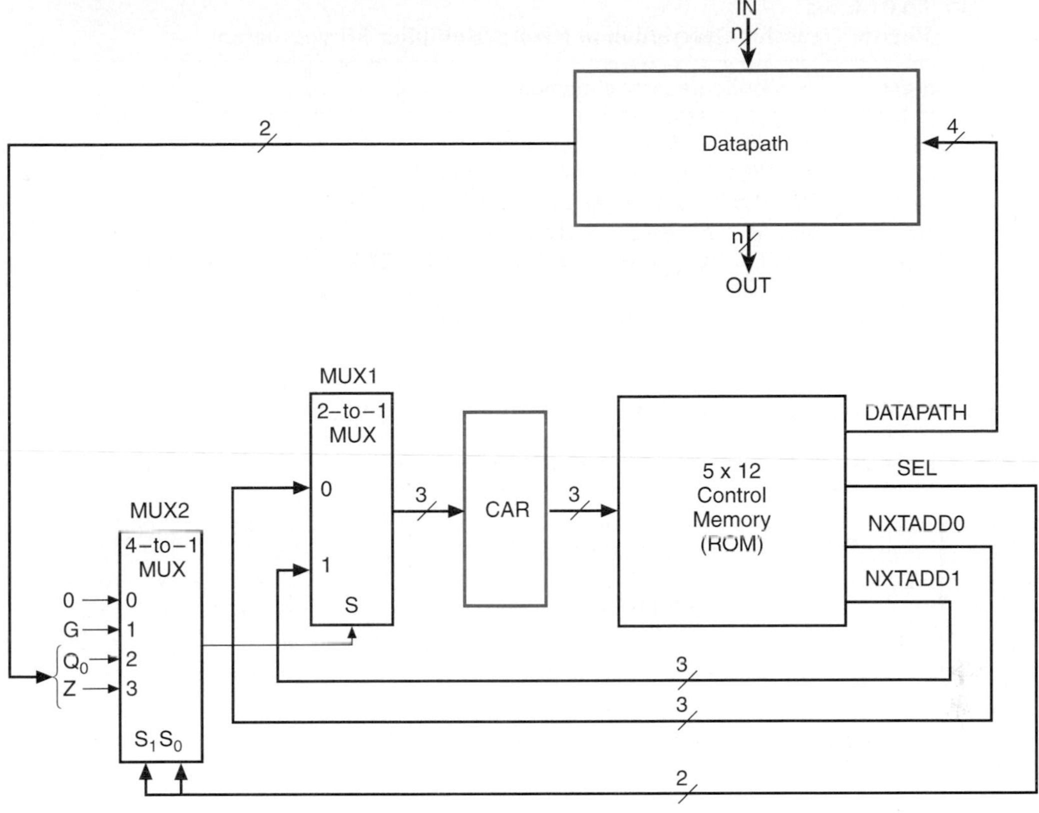

□ **FIGURE 8-15**
Microprogrammed Control Unit for Multiplier

as the address; if G is 1, the address becomes NXTADD1. The other two decision variables, Q_0 and Z, operate in a similar manner.

Table 8-5 gives a microprogram for the binary multiplier in register transfer notation. This description corresponds to the ASM in Figure 8-13. Note that there is a microinstruction in the microprogram that corresponds to each of the states in the ASM chart. On the chart, the binary code for each state is the contents of the *CAR* for that state. In the left half of Table 8-6, the register transfer microprogram is converted to a symbolic microprogram by replacing each register transfer with the symbolic names for the operation and using symbolic (state) names for the addresses. In the right half of the table, the symbolic microprogram is converted to a binary microprogram by replacing symbolic names with the corresponding binary codes from Figure 8-13, Table 8-3, and Table 8-4. In the binary microprogram, we have chosen to change unspecified symbolic entries to all zeros.

With the preceding introduction to hardwired and microprogrammed control unit design, we are now prepared to consider more complex control units for pro-

□ **TABLE 8-5**
Register Transfer Description of Binary Multiplier Microprogram

Address	Symbolic transfer statement
IDLE	G: $CAR \leftarrow \text{INIT}, \overline{G}$: $CAR \leftarrow \text{IDLE}$
INIT	$C \leftarrow 0, A \leftarrow 0, P \leftarrow n-1, CAR \leftarrow \text{MUL0}$
MUL0	Q_0: $CAR \leftarrow \text{ADD}, \overline{Q_0}$: $CAR \leftarrow \text{MUL1}$
ADD	$A \leftarrow A + B, C \leftarrow C_{out}, CAR \leftarrow \text{MUL1}$
MUL1	$C \leftarrow 0, C \Vert A \Vert Q \leftarrow \text{sr } C \Vert A \Vert Q$, Z: $CAR \leftarrow \text{IDLE}, \overline{Z}$: $CAR \leftarrow \text{MUL0}$, $P \leftarrow P - 1$

□ **TABLE 8-6**
Symbolic Microprogram and Binary Microprogram for Multiplier

Address	NXTADD1	NXTADD0	SEL	DATAPATH	Address	NXTADD1	NXTADD0	SEL	DATAPATH
IDLE	INIT	IDLE	DS	None	000	001	000	01	0000
INIT	—	MUL0	NXT	IT, CC	001	000	010	00	0101
MUL0	ADD	MUL1	DQ	None	010	011	100	10	0000
ADD	—	MUL1	NXT	LD	011	000	100	00	0010
MUL1	IDLE	MUL0	DZ	CC, SD	100	010	000	11	1100

grammable digital systems. Our specific focus is simple computers, thereby building a basis for studying CPU designs in Chapters 9 through 12.

8-6 A Simple Computer Architecture

We introduce a simple computer architecture as a foundation for studying control designs for programmable systems. We show how the operations specified by instructions can be implemented by register transfers in the datapath, plus movement of information between the datapath and memory. We also show how the different control structures can be designed to implement the sequences of operations necessary for controlling program execution. The purpose here is to illustrate the different approaches to control design and the effects that such approaches have on datapath design and system performance. A more extensive study of the concepts associated with instruction sets for digital computers is presented in detail in the next chapter, and more complete CPU designs are undertaken in Chapter 10.

Instructions

The user instructs the computer as to the operations to be performed and their sequence by the use of a *program*, which is a list of instructions that specifies the operations, the operands, and the sequence in which processing is to occur. The data processing performed by a computer can be altered by specifying a new program with different instructions or by specifying the same instructions with different data. Instructions and data are usually stored together in the same memory. By means of the techniques discussed in Chapter 12, however, they may appear to be coming from different memories. The control unit reads an instruction from memory and decodes and executes the instruction by issuing a sequence of one or more microoperations. The particular instructions available differ from one computer to another. The ability to execute a program from memory is the most important single property of a general-purpose computer. Execution of a program from memory is in sharp contrast to the multiplier control unit considered earlier, which executes only a single, fixed operation.

An *instruction* is a collection of bits that instructs the computer to perform a specific operation. The *operation code* of an instruction, often shortened to "opcode," is a group of bits in the instruction that specifies an operation, such as add, subtract, shift, or complement. We call the collection of instructions for a computer its *instruction set* and a thorough description of the instruction set its *instruction set architecture.*

The number of bits required for the opcode of an instruction is a function of the total number of operations in the instruction set. It must consist of at least m bits for up to 2^m distinct operations. The designer assigns a bit combination (a code) to each operation. The computer is designed to accept this bit configuration at the proper time in the sequence of activities and to supply the proper control word sequence to execute the specified operation. As a specific example, consider a computer with a maximum of 128 distinct operations, one of them an addition operation. The opcode assigned to this operation consists of seven bits in the bit configuration 0000010. When the opcode 0000010 is detected by the control unit, a sequence of control words is applied to the datapath to perform the intended addition.

The opcode of an instruction specifies the operation to be performed. The operation must be performed using data stored in computer registers or in memory. An instruction, therefore, must specify not only the operation, but also the registers or memory words in which the operands are to be found and the result is to be placed. The operands may be specified by an instruction in two ways. An operand is said to be specified *explicitly* if the instruction contains special bits for its identification. For example, the instruction performing an addition may contain three binary numbers specifying the registers containing the two operands and the register that will receive the result. An operand is said to be defined *implicitly* if it is included as a part of the definition of the operation itself, rather than being given in the instruction. For example, in an Increment Register operation, one of the operands is implicitly +1.

Instruction Formats

The format of an instruction is usually depicted by a rectangular box symbolizing the bits of the instruction, as they appear in memory words or in a control register. The bits are divided into groups or parts called *fields*. Each field is assigned a specific item, such as the opcode, a constant value, or a register file address. The vari-

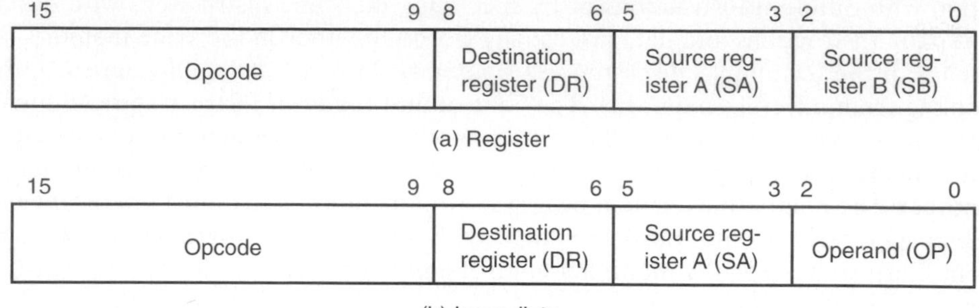

□ **FIGURE 8-16**
Two Instruction Formats

ous fields specify different functions for the instruction and, when shown together, constitute an instruction format.

Consider, for example, the two instruction formats depicted in Figure 8-16. Suppose that the computer has a register file consisting of eight registers, $R0$ through $R7$. The instruction format in Figure 8-16(a) consists of an opcode that specifies the use of three or fewer registers, as needed. One of the registers is designated a destination for the result and two of the registers sources for operands. For convenience, the field names are abbreviated DR, for "Destination Register," SA for "Source Register A," and SB for "Source Register B." The number of register fields and registers actually used are determined by the specific opcode. The opcode also specifies the use of the registers. For example, for a subtraction operation, suppose that the three bits in SA are 010, specifying $R2$, the three bits in SB are 011, specifying $R3$, and the three bits in DR are 001, specifying $R1$. Then the contents of $R3$ will be subtracted from the contents of $R2$, and the result will be placed in $R1$. As an additional example, suppose that the operation is a store (to memory). Suppose further, that the three bits in SA specify $R4$ and the three bits in SB specify $R5$. For this particular operation, it is assumed that the register specified in SA contains the address and the register specified in SB contains the operand to be stored. So the value in $R5$ is stored in the memory location given by the value in $R4$. The DR field has no effect, since the store operation prevents the register file from being written to.

The instruction format in Figure 8-16(b) has an opcode, two register fields, and an operand. The operand is a constant called an *immediate operand*, since it is immediately available in the instruction. For example, for an add immediate operation with SA specified as $R7$, DR specified as $R2$, and operand OP equal to 011, the value 3 is added to the contents of $R7$, and the result of the addition is placed in $R2$. The two formats in Figure 8-16 are used for the simple computer to be discussed in this chapter. In Chapter 9, we present and discuss more generally other instruction types and formats.

Now suppose that we have a memory with 16 bits per word and that the opcodes contain 7 bits, as in either format in Figure 8-16. Then instructions and

Decimal address	Memory contents	Decimal opcode	Other specified fields	Operation
25	0000101 001 010 011	5 (Subtract)	DR:1, SA:2 SB:3	R1 ← R2 – R3
35	0100000 000 100 101	32 (Store)	SA:4 SB:5	M [R4] ← R5
45	1000010 010 111 011	66 (Add Immediate)	DR:2 SA:7 OP:3	R2 ← R7 + 3
70	0000000 011 000 000	Data = 192. After execution of instruction in 35, Data = 80.		

□ **FIGURE 8-17**
Memory Representation of Instructions and Data

data, in binary, are placed in memory as shown in Figure 8-17. This stored information represents the three instructions used to illustrate formats in the previous three paragraphs. At address 25, we have a register format instruction that specifies an operation to subtract *R*3 from *R*2 and load the difference into *R*1. This operation is represented symbolically in the rightmost column of Figure 8-17. Note that the 7-bit opcode for subtraction is 0000101, or decimal 5. The remaining bits of the instruction specify the three registers: 001 specifies the destination register as *R*1, 010 specifies the source register *A* as *R*2, and 011 specifies the source register *B* as *R*3.

In memory location 35 is a register format instruction to store the contents of *R*5 in the memory location specified by *R*4. The opcode is 0100000, or decimal 32, and the operation is given symbolically, again, in the rightmost column of the figure. Suppose *R*4 contains 70 and *R*5 contains 80. Then the execution of this instruction will store the value 80 in memory location 70, replacing the original value of 192 stored there. (See Figure 8-17.)

At address 45, an immediate format instruction appears that adds 3 to the contents of *R*7 and loads the result into *R*2. The opcode for this instruction is 66, and the operand to be added is the value 3 (011) in the OP field, the last three bits of the instruction.

The placement of instructions in memory as shown in Figure 8-17 is quite arbitrary. In many computers, the word length is from 32 to 64 bits, so the instruction formats can hold much larger immediate operands than those we have given. Depending on the computer architecture, some of the instruction formats may occupy two or more consecutive memory words. Also, the number of registers is often larger, so the register fields in the instructions must contain more bits.

At this point, it is vital to recognize the difference between a computer *operation* and a hardware *microoperation*. An operation is specified by an instruction

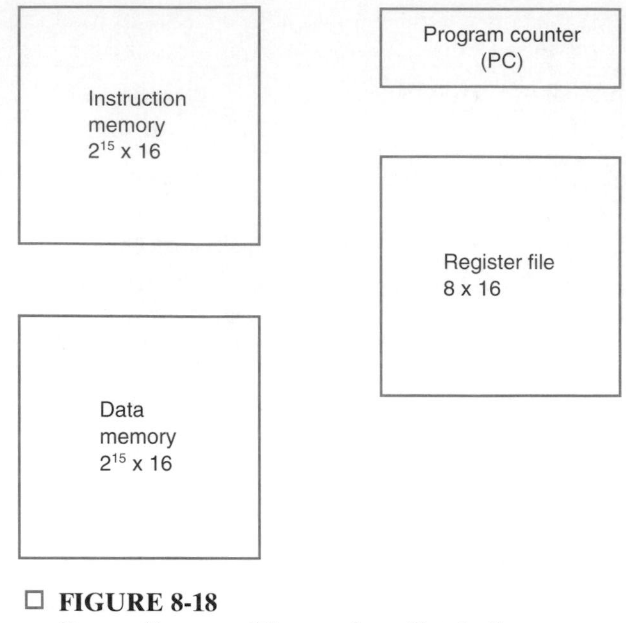

□ **FIGURE 8-18**
Storage Resource Diagram for a Simple Computer

stored in binary, in the computer's memory. The control unit in the computer uses the address or addresses provided by the program counter to retrieve the instruction from memory. It then decodes the opcode bits and other information in the instruction to perform the required microoperations for the execution of the instruction. So in a microprogrammed control, a microprogram is used to execute the instruction. In contrast, a microoperation is specified by the bits in a control word produced by hardware or read from control memory. The control word is decoded by the computer hardware to execute the microoperation.

Storage Resource Diagram

The storage resource diagram for a simple computer is shown in Figure 8-18. The diagram depicts the computer structure as viewed by a user programming it in a language that directly specifies the instructions to be executed. It gives the resources the user sees available for storing information. Note that the architecture includes two memories, one for storage of instructions and the other for storage of data. These may actually be different memories, or they may be the same memory, but viewed as different from the standpoint of the CPU. (See Chapter 12.)

Also visible to the programmer in the diagram is a register file with eight 16-bit registers and the 16-bit program counter. We use the diagram and the two formats presented in Figure 8-16 to define the simple computer. The hardware organization of the computer will take three forms, two of which will use the datapath and datapath control word from Figure 7-19 with minor modifications. The third form of organization will be based on the pipelined datapath from Figure 7-23.

Each form of organization will have a different control method: hardwired, microprogrammed, or pipelined. In addition to exploring these control methods, we will be dealing with some different approaches to the timing of execution of instructions. For simplicity, we focus only on implementing operations that use the datapath for each of the control methods, leaving the part of the control that has to do with general sequencing and decision making when executing programs to Chapter 10.

8-7 SINGLE-CYCLE HARDWIRED CONTROL

The block diagram for a computer that has a hardwired control unit and that fetches and executes an instruction in a single clock cycle is shown in Figure 8-19. We refer to this computer as the single-cycle computer. The datapath shown is the same as that in Figure 7-19. The data memory *M* is attached to the Address out, Data out, and Data in connections on the datapath. Although not usually thought of as part of the control unit, the instruction memory, together with its address inputs and instruction outputs, is shown for convenience with the control unit. In our discussions, we will only be reading words from the instruction memory, so one should view it as a combinational rather than a sequential component. The *PC* provides the instruction address to the instruction memory, and the instruction output from the instruction memory goes to the control logic, which, in this case, is the instruction decoder. The output from the instruction memory also goes to the zero fill, which provides input Constant in to the datapath.

All parts of the computer that are sequential are shown in blue. Note that there is no sequential logic in the control part other than the *PC*. Thus, aside from providing the address to the instruction memory, the control logic is combinational in this case. That fact, combined with the structure of the datapath and the use of separate instruction and data memories, allows the single-cycle computer to obtain and execute an instruction from the instruction memory all in a single clock cycle.

The *PC* holds the address of the current instruction. It increments its content by one at the beginning of the clock cycle in which an instruction is being executed. To fetch an instruction for execution, the content of the *PC* is applied to the address input of the instruction memory. The instruction then appears on the instruction output of the instruction memory and is applied to the instruction decoder and the zero fill.

Zero fill appends 13 zeros to the left of the operand (OP) field of the instruction to form a 16-bit unsigned operand for use in the datapath. For example, operand value 111 becomes 0000000000000111 or +7. Zero fill enters the datapath through the Constant in path.

Instruction Decoder

The instruction decoder is a combinational circuit that provides all of the control words for the datapath, based on the contents of the fields of the instruction. The instruction formats and opcodes were chosen very carefully by the computer

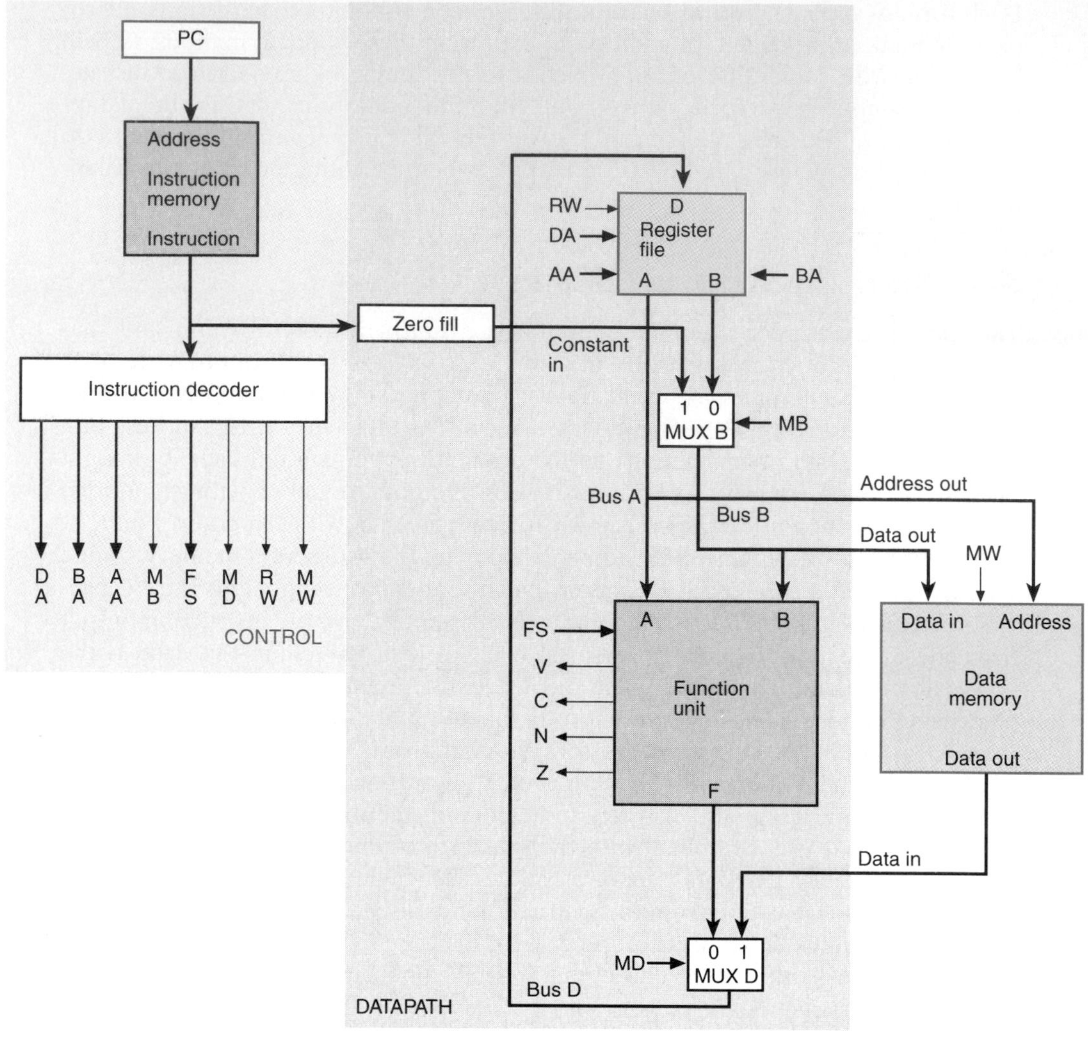

□ **FIGURE 8-19**
Block Diagram for a Single-Cycle Computer

designer to keep the control logic shown in Figure 8-20 very simple.The control logic maps the instructions into the values for the control word fields. Note that the DR, SA, and SB fields and the rightmost five bits of the opcode map directly through to the fields DA, AA, BA, and FS, respectively, of the control word. As a result of the careful choice of opcodes to match the FS field, any function unit operation can be specified just by the rightmost five bits of the opcode.

Bits 13 through 15, the leftmost three bits of the opcode, are fed into logic to generate the control lines for MB, MD, RW, and MW. These instruction bits select the source of the contents to be loaded into a register in the register file (MD) and

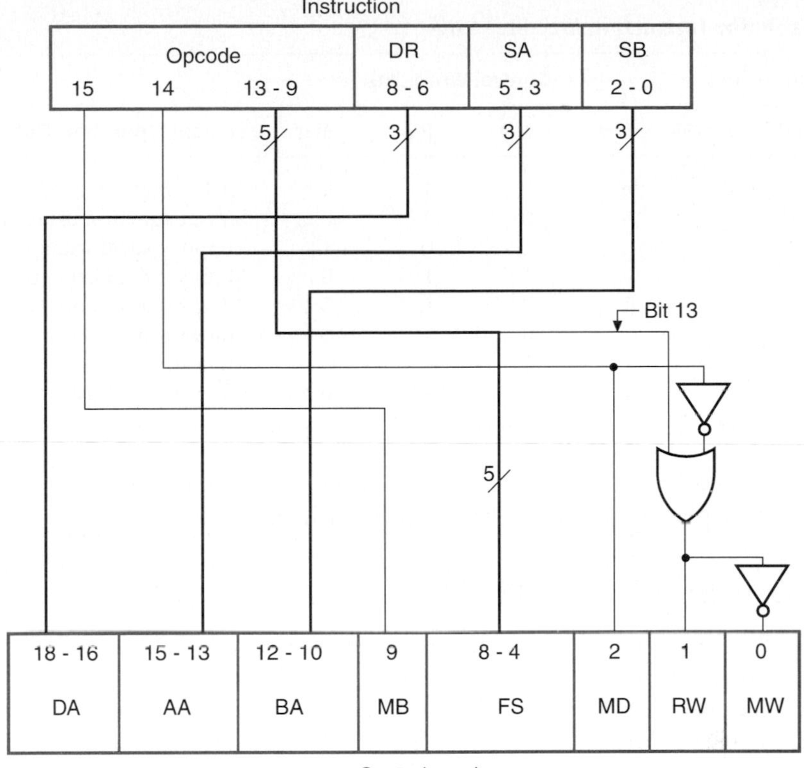

□ **FIGURE 8-20**
Diagram of Instruction Decoder

determine whether the register file is written (RW) to. They also determine whether the memory is written (MW), and whether a constant or the contents of a register appear on Bus *B* (MB). Clearly, these bits play a major role in controlling the datapath.

We will not detail the design process for the control logic here: however, we give the truth table used for the design as Table 8-7. If a given combination of values is needed on MB through MW, this table can be used to determine what values are needed in bits 13 through 15 of the instruction. To help understand the use for each of the combinations of values, the categories of typical instructions supported by a combination are listed at the right of the table. The four control word bits mapped back to bits 13 through 15 of the instruction determine a part of the definition of the opcode value. Note that bit 13 is used for controlling memory write, as well as being the first bit in the FS control word field. This overlap places some constraints on instructions that affect status bit values. Otherwise, virtually any single-cycle operation possible in the datapath can be defined by specifying the seven bits of the opcode. Note that "Shifter function using registers" occurs twice in the

□ **TABLE 8-7**
Truth Table for Instruction Decoder Logic

Instruction Bits			Control Word Bits				
Bit 15	**Bit 14**	**Bit 13**	**MB**	**MD**	**RW**	**MW**	**Typical Operation Category**
0	0	0	0	0	1	0	ALU function using registers
0	0	1	0	0	1	0	Shifter function using registers
0	1	0	0	1	0	1	Memory write using register data
0	1	1	0	1	1	0	Memory read using register data
1	0	0	1	0	1	0	ALU operation using a constant
1	0	1	1	0	1	0	Shifter function using registers
1	1	0	1	1	0	1	Memory write using constant data
1	1	1	1	1	1	0	Memory read using constant data

table. The repetition results from the use of don't-care conditions in the original table used for the design of instruction decoder logic.

Sample Instructions and Program

Six instructions for the single-cycle computer are listed in Table 8-8. The symbolic names associated with the instructions are useful for listing programs in symbolic form rather than in binary code. Because of the importance of instruction decoding, the rightmost four columns of the table show critical control signal values for each instruction, based on the values in Table 8-7.

Now suppose that the first instruction, "Add Immediate" (ADI), is present on the output of the instruction memory shown in Figure 8-19. Then, based on the first three bits of the opcode, from Table 8-8, the outputs of the instruction decoder have the values MB = 1, MD = 0, RW = 1, and MW = 0. The last three bits of the instruction, I(2:0), are extended to 16 bits by zero fill. We denote this in a register transfer statement by zf. Since MB is 1, this zero-filled value is placed on Bus *B*. With MD equal to 0, the function unit output is selected, and since the last five bits of the opcode, 00010, specify field FS, the operation is $A + B$. So the sign-extended value on Bus *B* is added to the contents of register SA, with the result presented on Bus *D*. The value on Bus *D* is written into register DR. Finally, with MW = 0, no write into memory occurs. This entire operation takes place in a single clock cycle. At the beginning of the next cycle, the destination register is written to and the *PC* is incremented to point to the next instruction.

The second instruction, LD, is a load from memory with opcode 0110000. The first three bits of this opcode give control signal values MD = 1, RW = 1, and MW = 0. These values, plus the register source field SA and register destination field DR, fully specify this instruction, which loads the contents of the memory address specified by register SA into register DR.

The third instruction, ST, stores the contents of a register in memory. The first three bits of the opcode give control signal values MB = 0, RW = 0, and MW = 1.

□ **TABLE 8-8**
Six Instructions for the Single-Cycle Computer

Operation code	Symbolic name	Format	Description	Function	MB	MD	RW	MW
1000010	ADI	Immediate	Add immediate operand	$R[\mathrm{DR}] \leftarrow R[\mathrm{SA}] + \mathrm{zf}\ I(2{:}0)$	1	0	1	0
0110000	LD	Register	Load memory content into register	$R[\mathrm{DR}] \leftarrow M[R[\mathrm{SA}]]$	0	1	1	0
0100000	ST	Register	Store register content in memory	$M[R[\mathrm{SA}]] \leftarrow R[\mathrm{SB}]$	0	1	0	1
0000001	INC	Register	Increment register	$R[\mathrm{DR}] \leftarrow R[\mathrm{SA}] + 1$	0	0	1	0
0001110	NOT	Register	Complement register	$R[\mathrm{DR}] \leftarrow \overline{R[\mathrm{SA}]}$	0	0	1	0
0000010	ADD	Register	Add registers	$R[\mathrm{DR}] \leftarrow R[\mathrm{SA}] + R[\mathrm{SB}]$	0	0	1	0

This causes a memory write operation, with the address and data from the register file. Also, it prevents the registers from being written to. The address for the memory write comes from the register selected by field SA, and the data for the memory write come from the register selected by SB. The DR field, although present, is not used, since no write occurs to any register.

Because this computer has load and store instructions and does not combine loading and storing of data operands with other operations, it is referred to as having a *load/store* architecture. The use of such an architecture simplifies the execution of instructions.

The final three instructions are ALU based without immediate operands. The last five bits of the opcode become the value for the FS field of the control word to specify the ALU operation. For the first two of these instructions, only one source register and a destination register are involved, so the SB field is not used. The final instruction uses all of the instruction fields, since three registers are involved.

To demonstrate how instructions such as these can be used in a simple program, consider the arithmetic expression $83 - (2 + 3)$. The following program performs this computation, assuming that register $R3$ contains 248, location 248 in data memory contains 2, location 249 contains 83, and the result is to be placed in location 250:

LD	R1, R3	Load $R1$ with contents of location 248 in memory ($R1 = 2$)
ADI	R1, R1, 3	Add 3 to $R1$ ($R1 = 5$)
NOT	R1, R1	Complement $R1$
INC	R1, R1	Increment $R1$ ($R1 = -5$)
INC	R3, R3	Increment the contents of $R3$ ($R3 = 249$)
LD	R2, R3	Load $R2$ with contents of location 249 in memory ($R2 = 83$)
ADD	R2, R2, R1	Add contents of $R1$ to contents of $R2$ ($R2 = 78$)
INC	R3, R3	Increment the contents of $R3$ ($R3 = 250$)
ST	R3, R2	Store $R2$ in memory location 250 ($M[250] = 78$)

The subtraction in this case is done by taking the 2's complement of $(2 + 3)$ and adding it to 83. If a register field is not used in executing an instruction, its symbolic value is omitted. The symbolic values for the register-type instructions, when the latter are present, are in the order DR, SA, and SB. For immediate types, the fields are in the order DR, SA, and OP. To store this program in the instruction memory, it is necessary to convert all of the symbolic names and decimal numbers used to their corresponding binary codes.

Although there may be instances in which single-cycle computer timing and control strategy is useful, it has a number of shortcomings. One shortcoming is in the area of performing complex operations. For example, suppose that an instruction is desired that executes unsigned binary multiplication using an add-and-shift algorithm. With the given datapath, this cannot be accomplished by a microoperation that can be executed in a single clock cycle. Thus, a control organization that provides multiple clock cycles for the execution of instructions is needed.

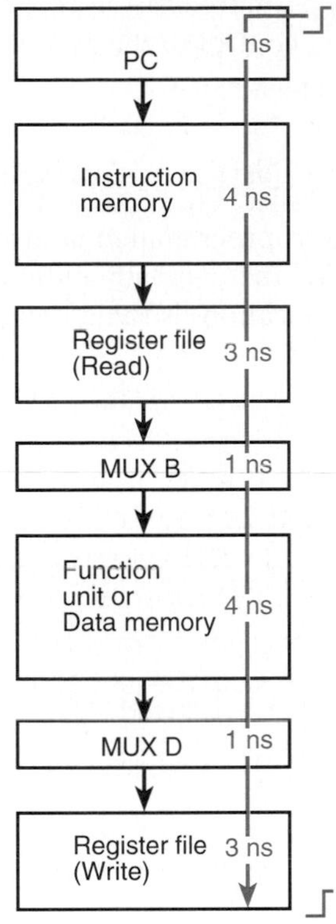

□ **FIGURE 8-21**
Worst Case Delay Path in Single-Cycle Computer

Also, the single-cycle computer has two distinct 16-bit memories, one for instructions and one for data. For a simple computer with instructions and data in the same 16-bit memory, two read accesses of memory are required to execute an instruction that loads a data word from memory into a register. The first access obtains the instruction, the second the data word. Since two different addresses must be applied to the memory address inputs, at least two clock cycles, one for each address, are required for obtaining and executing the instruction. This can also be accomplished easily with multiple-cycle control.

Finally, the single-cycle computer has a lower limit on the clock period based on a long worst case delay path. This path is shown in blue in the simplified diagram of Figure 8-21. The total delay along the path is 17 ns. This limits the clock frequency to 58.8MHz, which, although it may be adequate for some applications, is too slow for a modern computer CPU. In order to have a higher clock frequency, either the delays of the components on the path or the number of components in the path must be reduced. If the delays of the components cannot be reduced, reducing the number of components in the path is the only alternative. In Section 7-11, pipelining of the data path reduced the number of components in the longest combinational delay path and permitted the clock frequency to be increased. Thus, Section 8-9, we add a pipelined control to the pipelined datapath and demonstrate the resulting improved CPU performance.

8-8 MULTIPLE-CYCLE MICROPROGRAMMED CONTROL

To demonstrate the use of microprogrammed control, we use the architecture of the simple computer and modify the datapath and memory used in the single-cycle computer. The goal of the modifications is to demonstrate the use of a single mem-

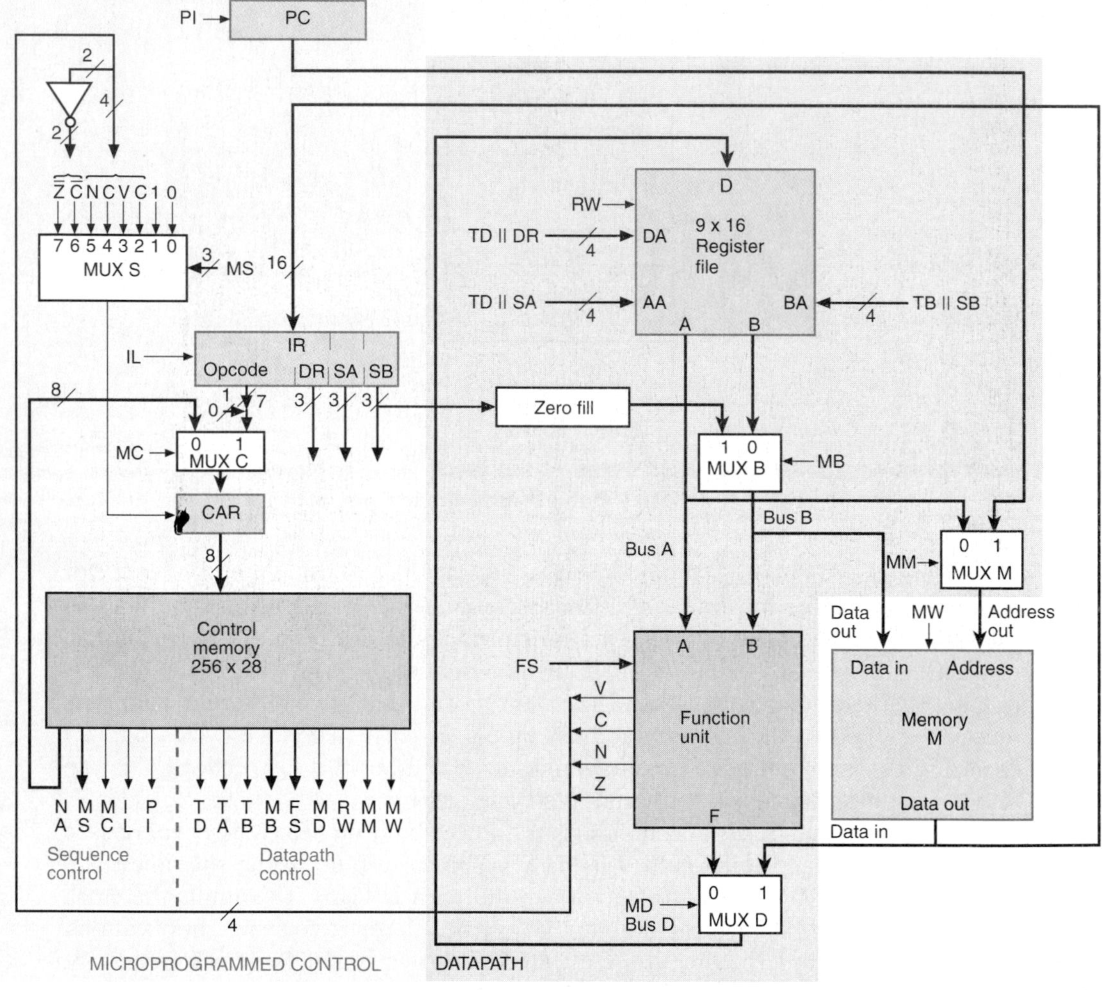

□ **FIGURE 8-22**
Multiple-Cycle Microprogrammed Computer

27 26 25 24 23 22 21 20 19 18 17 16 15 14 13 12 11 10 9 8 7 6 5 4 3 2 1 0

NA	MS	MC	IL	PI	TD	TA	TB	MB	FS	MD	RW	MM	MW

□ **FIGURE 8-23**
Format for Microinstruction

ory for both data and instructions and to demonstrate how more complex instructions can be implemented. The block diagram in Figure 8-22 shows the datapath and memory modifications and the microprogrammed control, and Figure 8-23 gives the instruction format for the microinstruction control words.

The separate instruction memory and data memory are replaced in Figure 8-22 by a single memory *M*. To fetch instructions, the *PC* is the address source for the memory, and to fetch data, Bus *A* is the address source. At the address input to memory, multiplexer MUX *M* selects between these two address sources. MUX *M* requires an additional control signal, MM, in the control word format. Since instructions from the memory are needed in the control unit, a path is added from memory back to the control unit.

In executing an instruction across multiple clock cycles, data generated during the current cycle is often needed in a later cycle. This data can be temporarily stored in a register from the time it is generated until the time it is used. Registers used for such temporary storage during the execution of the instruction are usually not visible to the user. To avoid adding significant complexity to this simple implementation, we add a single temporary register, *R*8, to the register file. To address *R*8, the register file in Figure 8-22 has a bit added on the left of each of its address inputs. Address inputs SA, SB, and DR are combined with new 1-bit control signals TA, TB, and TD, respectively, to form TA‖SA, TB‖SB, and TD‖DR. These combinations are symbolically represented on the register file address inputs in the figure. Using the 4-bit address AA as an example, we see that, for TA = 0, AA = 0 ‖ SA addresses one of the first eight registers in the register file, But for TA = 1, AA addresses the only register having an address beginning with 1—the temporary register, *R*8. Table 8-9 gives the specification for the new datapath control signals MM, TA, TB, and TD; it gives symbolic names and a description of the corresponding selection for each of the binary values of the new control signals.

Because of the multiple cycles of the modified computer, the instruction needs to be held in a register for use during its execution. The register used for this purpose is the *instruction register IR* in Figure 8-22. Since the *IR* loads only when an instruction is being read from memory, it has a load-enable signal IL that is added to the control word. Likewise, since the *PC* increments only when an instruction is fetched, the control word contains an increment-enable signal PI.

Note three of the microinstruction fields given in Section 7-10—DA, AA, and BA—are not in the microinstruction format in Figure 8-23, but go directly from the *IR* as DR, SA, and SB, respectively, to the register file. Obtaining these register addresses directly from the instruction in the *IR* reduces the length of the microinstruction control word and also reduces greatly the number of microinstructions required in the control memory.

□ **TABLE 8-9**
Control Word Information for Datapath

TD	TA	TB	MB		FS		MD	RW	MM	MW	
Select	Select	Select	Select	Code	Function	Code	Select	Function	Select	Function	Code
R[DR]	R[SA]	R[SB]	Register	0	$F = A$	00000	FnUt	No write(NW)	Address	No write(NW)	0
$R8$	$R8$	$R8$	Constant	1	$F = A + 1$	00001	Data In	Write(WR)	PC	Write(WR)	1
					$F = A + B$	00010					
					$F = A + B + 1$	00011					
					$F = A + \overline{B}$	00100					
					$F = A + \overline{B} + 1$	00101					
					$F = A - 1$	00110					
					$F = A$	00111					
					$F = A \wedge B$	01000					
					$F = A \vee B$	01010					
					$F = A \oplus B$	01100					
					$F = \overline{A}$	01110					
					$F = A$	10000					
					$F = \text{sl } A$	10010					
					$F = \text{sr } A$	10100					
					$F = 0$	10110					

The microprogram control in Figure 8-22 has a structure similar to the control for the binary multiplier in Figure 8-15. The microinstruction control words are stored in a control memory that is addressed by the *CAR*. Whether the *CAR* is incremented or loaded is determined by next-address logic, which is a bit more complex than that for the binary multiplier. The source of the *CAR* address is selected by multiplexer MUX *C*. One source is the 8-bit contents of the next-address field NA in the current microinstruction; the other is the 7-bit opcode field in the *IR* with one zero appended on the left. These sources are selected by control signal MC. The opcode loaded into the *CAR* points to the beginning of the microprogram in control memory used to execute the instruction. For example, if the opcode is 0000001, for load (LD), then the address at which the microprogram begins is 00000001.

The decision to increment or load the *CAR* is made using multiplexer MUX *S* in Figure 8-22. This multiplexer has three select inputs controlled by the 3-bit MS field in the microinstruction. Depending on the value of MS, one of eight decision variables or constants, shown on MUX *S*, is selected. If the resulting value from MUX *S* is 1, then the *CAR* is loaded; otherwise, the *CAR* is incremented. For example, if the value of MS is 010 (2), then the next control address will depend on the value of *C*, the carry status bit. If *C* is 1, then the output of MUX *S* is 1, and the next control address will be loaded from the output of MUX *C*. If *C* is 0, then the output of MUX *S* is 0, and the next control address will be *CAR* + 1. Table 8-10 gives the information for MUX *S* for use in microprogramming the control. The table shows the action and the symbolic notation for each of the binary values of MS, as well as the action and notation for MC, IL, and PI.

Microprogram Design

We now have all of the information necessary to begin writing microprograms to fetch and execute instructions. Rather than proceeding to this step immediately, we will first use an ASM chart to describe the register transfers and sequencing of microinstructions required for implementing fetches and the execution of a small set of instructions. We will then use this chart to produce the microcode. The ASM chart is given in Figure 8-24. Processing of each of the instructions consists of two steps: instruction fetch and instruction execution. The instruction fetch occurs in state IF in the upper right corner of the chart. The *PC* contains the address of the instruction in memory *M*. This address is applied to the memory, and the word read from memory is loaded into the *IR* on the clock pulse that ends state IF. The same clock pulse causes the program counter *PC* to be incremented in preparation for fetching the next instruction and also causes the new state to become EX0. Note that a binary code is associated with each of the states in the ASM chart. This code is the contents of *CAR*, which, along with the contents of the *IR*, represent the state of the control unit.

In EX0, a single 0 followed by the opcode—i.e., the contents of the leftmost seven bits of the *IR*—are loaded into the current address register *CAR*. This action causes the *CAR* to address 2^7, or 128, different locations, 0 through 127, depending on the opcode values in *IR*. These 128 different addresses represent the beginning

TABLE 8-10
Control Information for Sequence Control Fields

MS			MC		IL		PI		
Action	**Symbolic Notation**	**Code**	**Select**	**Symbolic Notation**	**Action**	**Symbolic Notation**	**Action**	**Symbolic Notation**	**Code**
Increment *CAR*	CNT	000	NA	NXA	No load	NLI	No load	NLP	0
Load *CAR*	NXT	001	Opcode	OPC	Load instr.	LDI	Increment PC	INP	1
If $C = 1$, load *CAR*; else increment *CAR*	BC	010							
If $V = 1$, load *CAR*; else increment CAR	BV	011							
If $Z = 1$, load *CAR*; else increment *CAR*	BZ	100							
If $N = 1$, load *CAR*; else increment *CAR*	BN	101							
If $C = 0$, load *CAR*; else increment *CAR*	BNC	110							
If $Z = 0$, load *CAR*, else increment *CAR*	BNZ	111							

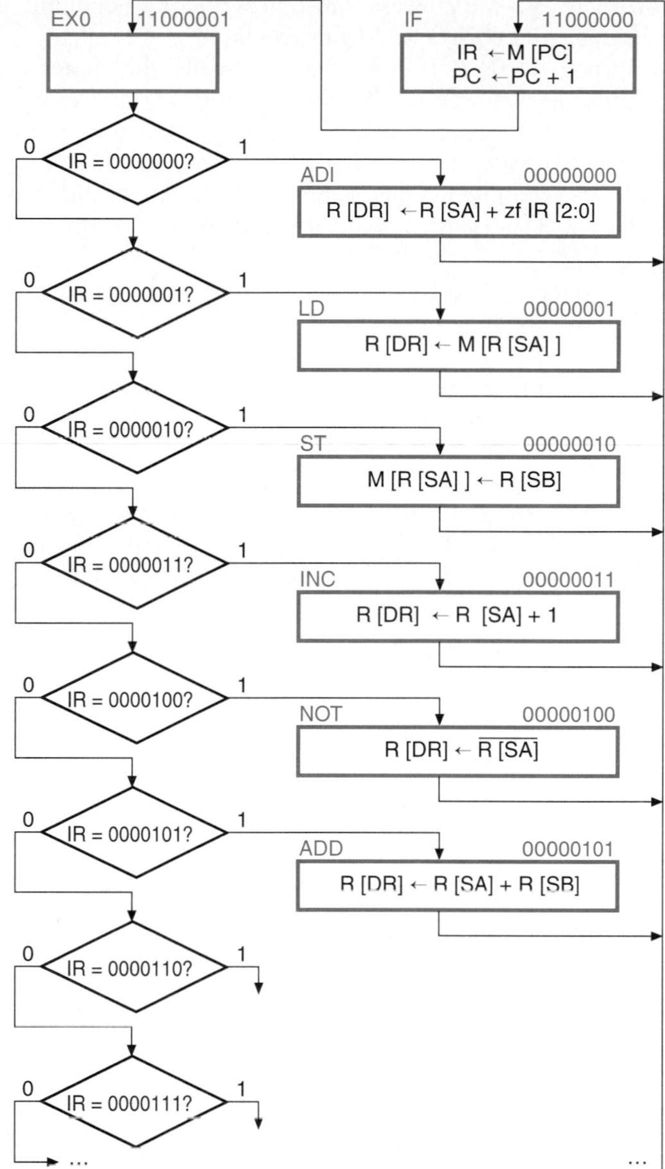

□ **FIGURE 8-24**
ASM Chart for Multiple-Cycle Microprogrammed Computer

of 128 potential microprograms, one for executing each of the instructions that can be specified by the opcode. The effect of this loading of the *CAR* is represented in Figure 8-24 by 127 decision boxes in a chain. Only the first 8 boxes, for opcodes 0000000 through 0000111, are shown. Note that, in contrast to the situation with hardwired control, the selection of opcodes is more flexible. Each of the 127 decision boxes corresponds to the address that represents the beginning of the microprogram that executes the corresponding instruction.

The first opcode, 0000000, is defined as an "Add Immediate" (ADI) instruction. This instruction, as well as the other five that follow, requires only one state in addition to IF and EX0 for its execution, In this case, the state is called ADI and has a *CAR* state code formed from a single 0 followed by the opcode. The execution of ADI adds the zero-filled value of the last three bits of the instruction to the contents of the register specified by field SA and places the result in the register specified by field DR.

The next two opcodes involve the use of memory and will cause two memory accesses to occur, one for fetching the instruction and one for loading or storing the result. Opcode 0000001 is "Load" (LD). It loads the word from memory addressed by the contents of the register SA into register DR. Opcode 0000010 is "Store" (ST). It stores the contents of register SB into the memory word addressed by register SA.

The next two opcodes, in sequence, are "Increment" (INC), and the logical complement (NOT). Both operations use instruction field SA to specify the register containing the source operand and instruction field DR to specify the destination register for the result. The final opcode is for addition (ADD), which performs 2's complement addition of the source operands specified by SA and SB, with the resulting sum loaded into the destination register DR.

The symbolic microprogram in Table 8-11 is based on the ASM chart. Here, we initially use the state names for addresses and next addresses. The symbolic names for the microinstruction control words come from Table 8-9 and Table 8-10. Many of the entries in Table 8-11 are blank, since the resource that is controlled is not used in the execution of a given microinstruction. It is useful for one to determine how each of the entries in Table 8-11 is obtained based on Figure 8-24, Table 8-9, and Table 8-10.

The symbolic microprogram of Table 8-11 is translated into a binary microprogram in Table 8-12. In this table, the *CAR* addresses given in binary in Figure 8-24 replace the state names in the first two columns. These addresses are in decimal for convenience. All blank entries from Table 8-11 are, by assumption, replaced with the appropriate number of 0's in Table 8-12.

It is interesting to briefly compare the timing of the execution of instructions in this organization with that for the single-cycle computer. Each instruction requires three clock cycles to fetch and execute, compared to one clock cycle for the single-cycle computer. But the long delay path from Figure 8-21 is broken up, so that the clock periods are considerably shorter. Nevertheless, due to setup time requirements for the added flip-flops in the *IR* and *CAR*, the overall time taken to execute an instruction will be slightly longer than in the single-cycle computer. So what is the benefit of this organization other than use of a single memory? The next two instructions give the answer.

The first instruction to be added is a "load register indirect" (LRI), with opcode 0000110. In this instruction, the contents of register SA address a word in memory. The word, which is known as an *indirect address*, is then used to address

☐ **TABLE 8-11**
Symbolic Microprogram for Fetch and Execution of Six Instructions

Address	NXT ADD	MS	MC	IL	PI	TD	TA	TB	MB	FS	MD	RW	MM	MW
IF	EX0	CNT	—	LDI	INP	—	—	—	—	—	—	NW	PC	NW
EXO	—	NXT	OPC	NLI	NLP	—	—	—	—	—	—	NW	—	NW
ADI	IF	NXT	NXA	NLI	NLP	DR	SA	—	Constant	$F = A + B$	FnUt	WR	—	NW
LD	IF	NXT	NXA	NLI	NLP	DR	SA	—	—	—	Data	WR	MA	NW
ST	IF	NXT	NXA	NLI	NLP	—	SA	SB	Register	—	—	NW	MA	WR
INC	IF	NXT	NXA	NLI	NLP	DR	SA	—	—	$F = A + 1$	FnUt	WR	—	NW
NOT	IF	NXT	NXA	NLI	NLP	DR	SA	—	—	$F = \overline{A}$	FnUt	WR	—	NW
ADD	IF	NXT	NXA	NLI	NLP	DR	SA	SB	Register	$F = A + B$	FnUt	WR	—	NW

☐ **TABLE 8-12**
Binary Microprogram for Fetch and Execution of Six Instructions

Address	NXT ADD	MS	MC	IL	PI	TD	TA	TB	MB	FS	MD	RW	MM	MW
192	193	000	0	1	1	0	0	0	0	00000	0	0	1	0
193	000	001	1	0	0	0	0	0	0	00000	0	0	0	0
000	192	001	0	0	0	0	0	0	1	00010	0	1	0	0
001	192	001	0	0	0	0	0	0	0	00000	1	1	0	0
002	192	001	0	0	0	0	0	0	0	00000	0	0	0	1
003	192	001	0	0	0	0	0	0	0	00001	0	1	0	0
004	192	001	0	0	0	0	0	0	0	01110	0	1	0	0
005	192	001	0	0	0	0	0	0	0	00010	0	1	0	0

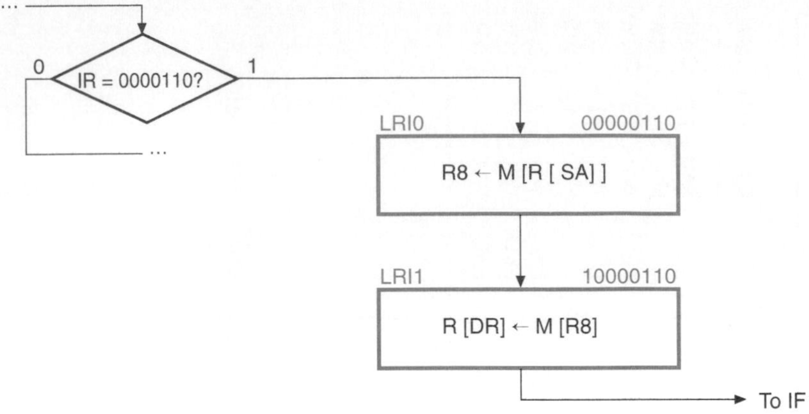

□ **FIGURE 8-25**
ASM Chart for Register Indirect Instruction

the word in memory that is loaded into register DR. This can all be represented symbolically as

$$R[DR] \leftarrow M[M[R[SA]]]$$

The ASM chart for the execution of this instruction is given in Figure 8-25. Following the instruction fetch and the transfer of the opcode 0000110 from the *IR* to the *CAR*, the state becomes LRI0. In this state, register SA addresses the memory to obtain the indirect address, which is then placed in temporary register *R*8. In the next state, the next memory access occurs with the address from *R8*. The operand obtained is placed in register DR to complete the operation. The ASM then returns to state IF to fetch the next instruction. Clearly, this instruction could not be executed with the given datapath in a single clock cycle. Also, to avoid disturbing the contents of registers *R*1 through *R*7 (except for the register being loaded), the use of register *R8* for temporary storage is essential. The LRI instruction requires four clock cycles for its execution. To accomplish the same operation without this instruction requires two LD operations and six clock cycles. So the LRI instruction gives an improvement in execution time.

The final instruction to be added is "shift right multiple" (SRM), with opcode 0000111. For this instruction to execute properly, SA must be equal to DR. The instruction specifies that the contents of register DR are to be shifted to the right by the number of positions given by the three bits of the OP field. The ASM chart for this operation is given in Figure 8-26. Register *R8* is used to store the number of bit positions remaining to be shifted. Initially, the shift amount, which ranges from 0 to 7 bit positions, is placed in *R*8. At the same time, the value being loaded into *R*8 is checked to see whether it is 0, meaning that no shift is required. If it is 0, the ASM flow returns to IF. If not, then a right-shift operation is performed on the contents of register SA, which are the same as those of register DR, with the result

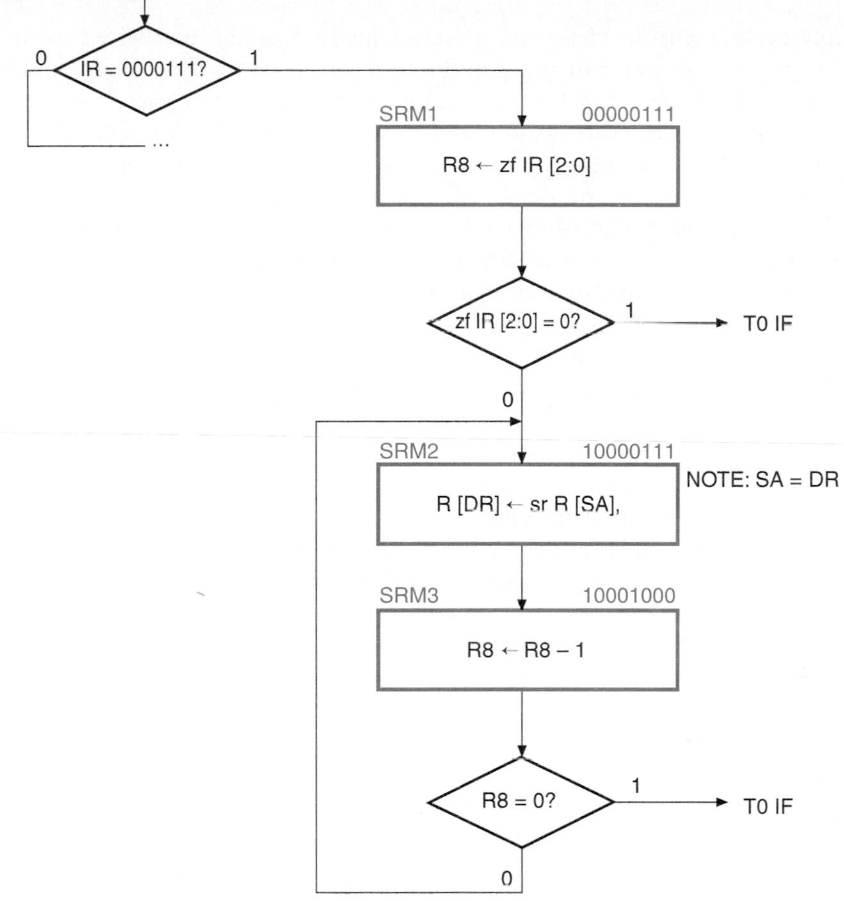

□ **FIGURE 8-26**
ASM Chart for Right-Shift Multiple Instruction

placed in register DR. $R8$ is decremented and tested to see whether it will be 0. If $R8 \neq 0$, then the shift and decrement are repeated. If $R8 = 0$, then register DR has been shifted right by the number of bit positions given by the OP field, so the ASM flow returns to state IF. This instruction requires $2s + 3$ clock cycles, where s is number of positions shifted. The range of clock cycles required is from 3 to 23. If the same operation were performed by using a right-shift instruction s times, then $3s$ instructions would be required. The improvement in the number of clock cycles required is $s - 3$, so for shifts up to three bit positions, there is no improvement. For shifts of 4 to 7 positions, 1 to 4 clock cycles are saved. Also, $s - 1$ fewer memory locations are required for storage for the SRM instruction, in contrast to s shift-right instructions.

The Hardwired Alternative

Multiple-cycle controls can also be implemented using hardwired control. A sequencer plus a decoder or one flip-flop per state can be used to implement the sequential part of the control, with added logic for the output part. It is possible to provide additional structure to the control unit of the sequencer-plus-decoder by using a counter or shift register as the sequence register. Depending on the complexity of the control to be designed, different features may be needed for the sequencer. For example, in some cases, a counter with only synchronous reset may suffice. In other cases, signals that disable counting or enable parallel loading may be needed. We next illustrate a design using a counter with reset for the first seven instructions of the multiple-cycle simple computer. Since these instructions do not use the outputs of the status bits, we will not implement that part of the control. Inclusion of the instruction SRM, which requires more states and involves a decision on status bits, is given as Problem 8-33 at the end of the chapter.

The ASM chart for this control is shown in Figure 8-27. Note that since we do not load the *CAR* from the *IR* and can use conditional output boxes, we can complete each of the first six instructions in just two clock cycles. The last instruction, LRI, requires three clock cycles. A block diagram of the control unit is shown in Figure 8-28. The counter is decoded to produce the three required states IF, EX0, and EX1. Since we are assuming only these seven instructions, with opcodes 0000000 through 0000110, only bits 9 through 11 of the *IR* have been decoded. The decoded state and instruction signals serve as inputs to combinational control logic that produces the control signals.

Let us now illustrate the derivation of some of the control signals from the ASM chart in Figure 8-27. First of all, the signal CR, the synchronous counter reset, determines the sequencing in the ASM chart. Note that the code for IF is 00; thus, any time we wish to return the state to IF, we simply reset the counter. For the first six instructions, IF is the state after EX0; for the last instruction, IF is the state after EX1. Based on this, the equation for CR, the counter reset is

$$CR = EX0 \cdot \overline{LRI} + EX1$$

As a second example, consider the signal RW. The register file is written in state EX0 for instructions ADI, LD, INC, NOT, ADD, and LRI. In EX1, the register file is also written. So the equation for RW is

$$RW = EX0 \cdot \overline{ST}$$

The remaining equations can be similarly derived from the ASM chart. If new instructions are added, it is important to review the previous equations to determine the changes required.

One advantage of microprogrammed control is that, if a versatile enough hardware structure has been established, there should be no need for further hardware or wiring changes outside of the microprogrammed control memory contents. If we want to establish different control sequences for the system, all we need to do is specify a different set of microinstructions for the control memory and replace

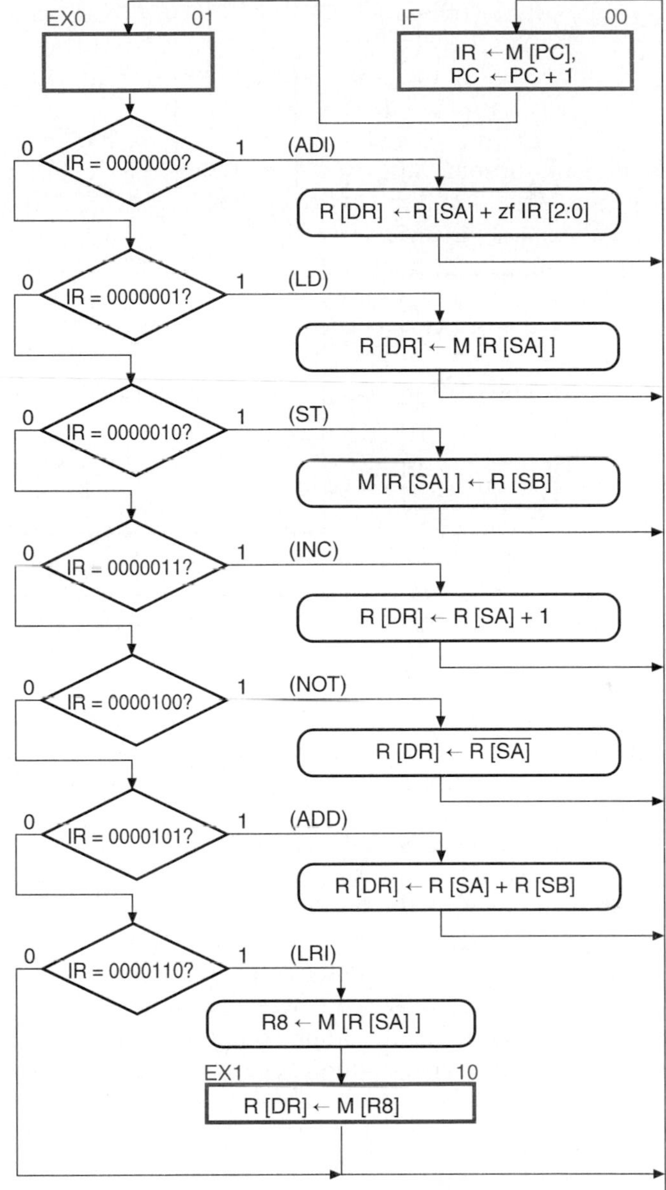

□ **FIGURE 8-27**
ASM Chart for Multiple-Cycle, Decoder-Based Computer

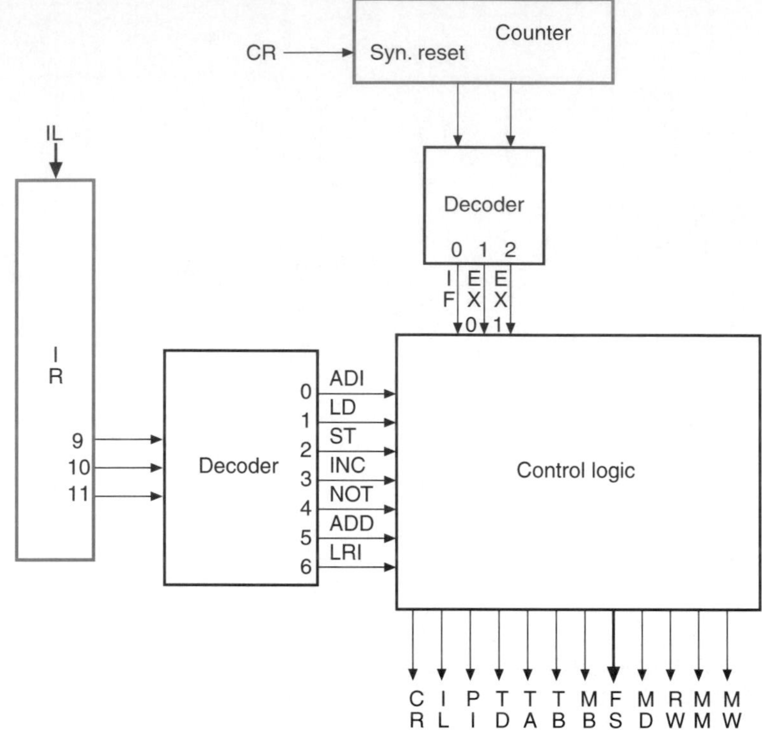

□ **FIGURE 8-28**
Block Diagram of Hardwired Counter and Decoder-Based, Multiple-Cycle Control Unit

the ROMs. A second advantage is that the ROM can be large enough to store many complex sequences of control words. If such sequences were to be constructed using hardwired control, the hardware cost would likely be very high, so microprogrammed control offers a cost-effective alternative. (One other alternative is to incorporate the complexity into the software for a programmable system.) On the other hand, if the control sequences are quite simple, the microprogrammed approach may be too costly, compared to the hardwired approach. This is due to the cost of including a next-address generator, a *CAR*, and a ROM in the design.

8-9 Pipelined Control

As indicated earlier, the clock period could be considerably shortened by using registers to break up long delay paths in the circuit. The microprogrammed computer design just completed uses temporary storage that is invisible to the programmer. This storage, however, does not shorten the clock period, and as a consequence, we did not get an improvement in performance for simple instruc-

□ **FIGURE 8-29**
Assembly Line Analogy to Computer Pipeline

tions. We can achieve such an improvement by using the pipelined datapath introduced in Section 7-11. Since the instruction must be fetched from a memory as well as executed, we add a stage to the analogous assembly line used for illustration in that section. Also, to model the data memory to be added to the pipelined datapath, we add a warehouse to the assembly line. The modified assembly line is shown in Figure 8-29.

Components and partially competed assemblies are stored in bins, as before. However, now all components come initially from the warehouse. If there are not enough bins for partially completed assemblies, they are taken to the warehouse and returned before they are needed for further steps. Also, after completion, the final products are stored in the warehouse. Warehousing operations take place at the third stage of the assembly line.

Analogous to the instruction fetch from the instruction memory, the operations on the assembly line are specified by order sheets produced by the assembly line programmer and placed in a file cabinet. At the first stage of the assembly line, an order is removed from the cabinet. The reader uses the order to tell the next person on the line which assembly and components to obtain from the bins, what operation is to be performed on the assembly and components (including taking them to or bringing them back from the warehouse), and which bin to place the assembly in after the operation. The person at the second stage of the assembly line passes on the information from the order needed by the third and fourth persons. Finally, the information from the order needed by the fourth person is passed on by the third person. This information accompanies the associated assembly as it moves down the line.

Note that one assembly or warehousing operation is completed at each advance of the belt. At any given time, four assembly or warehousing operations are in some stage of completion. Is this any faster than having only one stage, as in the case of the single-cycle computer? Suppose that with a single stage line with one person doing everything, it took one minute to do the operation and that 15 seconds was devoted to each of the four tasks. How many assemblies come from the one-stage line in one minute? Just one. How many assemblies come from the four-stage line in one minute? Four. So the throughput of the four-stage line is four times that of the one-stage line. Although both of these assembly lines take one minute to complete one assembly, the four-stage assembly line produces four-times as many assemblies in a given amount of time, giving a throughput that is four times that of the one-stage line.

For the analogy, the order-fetching stage on the assembly line corresponds to instruction fetch from instruction memory. The passing of the order information from the first person on the line to the second person corresponds to instruction decoding. Provision must also be made for the order information to be passed from stage to stage in the pipeline control, just as it is from person to person in the assembly line.

Figure 8-30 shows the block diagram of a pipelined computer based on the single-cycle computer. The datapath is that of Figure 7-23. The control has an added stage for instruction fetch that includes the *PC* and instruction memory. This becomes stage 1 of the combined pipeline. The instruction decoder and register file read are now in stage 2, the function unit and data memory read and write in stage 3, and the register file write is in stage 4. These stages are labeled at their boundaries with appropriate abbreviations. In order to provide the mechanism corresponding to the conveyor belt for our hypothetical assembly line, we place registers between the stages, as in Chapter 7. In the figure, we have added registers to the pipeline platforms between stages, as necessary to pass the decoded instruction information through the pipeline along with the data being processed. These additional registers serve to pass along the instruction information, just as order information was passed along verbally on the assembly line.

The added first stage is the instruction fetch stage, denoted by IF, which lies wholly in the control. In this stage, the instruction is fetched from the instruction memory, and the value in the *PC* is updated. Between the first-stage and the second stage is an inter-stage pipeline platform that plays the role of instruction register, so it has been labeled *IR*.

In the second-stage, DOF for decode and operand fetch, decoding of the *IR* into control signals takes place. Among the decoded signals, the register file addresses AA and BA and the multiplexer control signal MB are used in this stage for operand fetch. All other decoded control signals are passed on to the next pipeline platform, to be used later. Following the DOF stage is the second pipeline platform, whose registers store control signals to be used later. The third stage of the pipeline is the execution stage, denoted EX. In this stage, an ALU operation, a shift operation, or a memory operation is executed for most instructions. Thus, the control signals used in this stage are FS and MW. The read part of the data memory *M* is considered a part of the stage. For a memory read, the value of the word addressed is read to Data out from the data memory. All of the results produced from this stage, plus the control signals for the last stage, are captured by the third pipeline platform. The write part of data memory *M* is considered a part of this platform, so a memory write may occur here. The control information held in the final

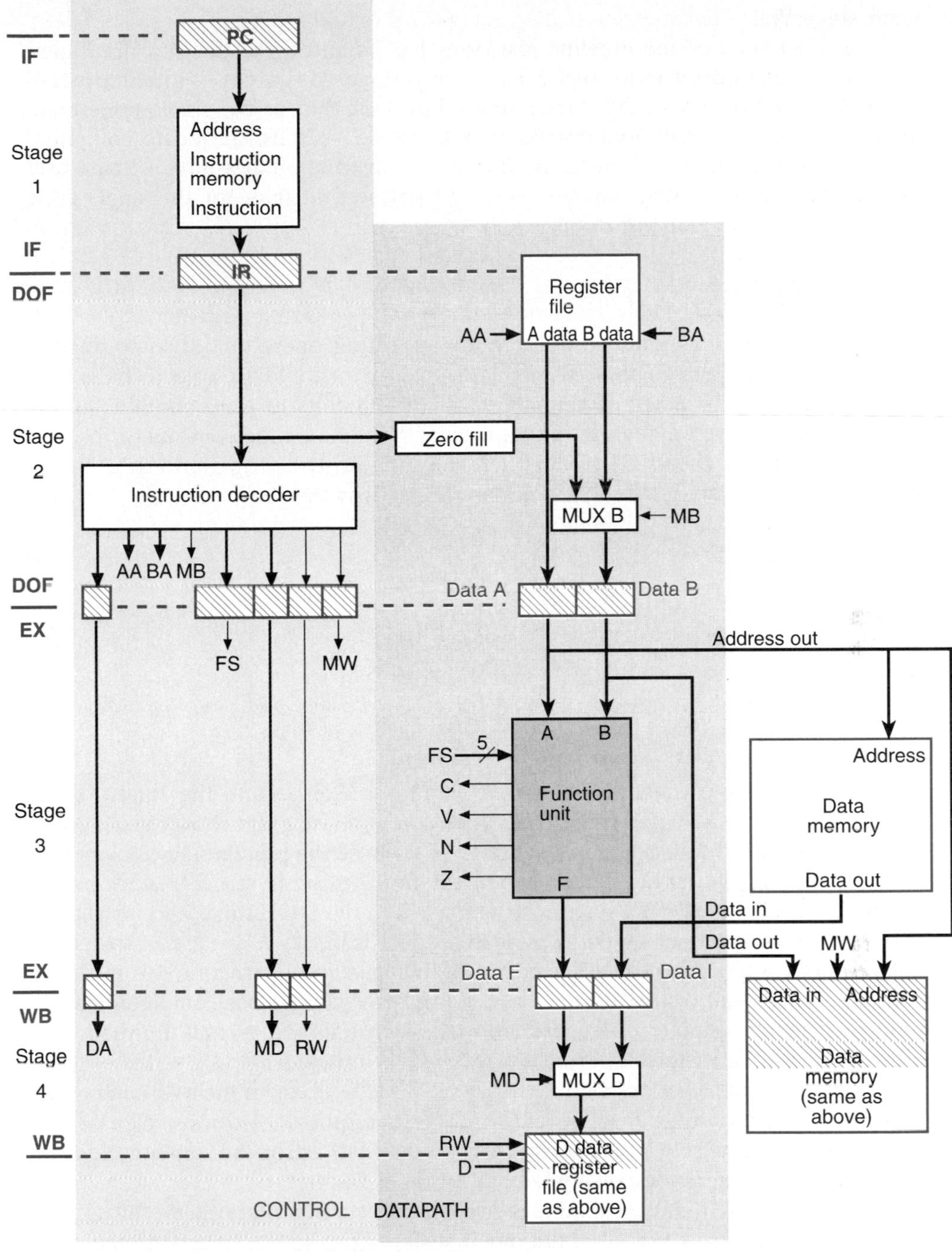

☐ **FIGURE 8-30**
Block Diagram of Pipelined Computer

pipeline platform consists of DA, MD, and RW, which are used in the final write-back stage, WB.

The location of the pipeline platforms has balanced the delays from Figure 8-21 which exceed no more that 5 ns for any stage. This gives a potential maximum clock frequency of 200 MHz, about 3.4 times that of the single-cycle computer. Note, however, that an instruction takes $4 \times 5 = 20$ ns to execute, compared to 17 ns for the simple computer. So if only one instruction at a time is being executed, even fewer instructions are executed per second than for the single-cycle computer.

Pipeline Programming and Performance

In our hypothetical assembly line, there are up to four operations at some stage of completion at any given time. By analogy, then, we should be able to have four instructions at some stage of completion in the pipeline of our computer at any given time. Suppose we consider a simple calculation: Load the constants 1 through 7 into the seven registers $R1$ through $R7$, respectively. The program to do this is as follows (the number on the left is a number to identify the instruction):

```
1    LDI R1, 1
2    LDI R2, 2
3    LDI R3, 3
4    LDI R4, 4
5    LDI R5, 5
6    LDI R6, 6
7    LDI R7, 7
```

Let us examine the execution of this program with respect to the stages of the pipeline in Figure 8-30. We use the pipeline execution diagram shown in Figure 8-31. In clock period 1, instruction 1 is in the IF stage of the pipeline. In clock period 2, instruction 1 is in the DOF stage and instruction 2 is the IF stage. In clock period 3, instruction 1 is in the EX stage, instruction 2 is in the DOF stage, and instruction 3 is in the IF stage. In clock period 4, instruction 1 is in the WB stage, instruction 2 is in the EX stage, instruction 3 is in the DOF stage, and instruction 4 is in the IF stage. So at the end of the fourth clock period, instruction 1 has completed execution, instruction 2 is three-fourths finished, instruction 3 is half finished, and instruction 4 is one-fourth finished. So we have completed $1 + 3/4 + 1/2 + 1/4 = 2.5$ instructions in four clock periods, or 20 ns. We can see from the overall diagram that the complete program of seven instructions requires 10 clock cycles to execute. Thus, the time required is 50 ns, compared to 119 ns for the single-cycle computer, and the program is executed about 2.4 times faster.

Now suppose that we examine the pipeline execution pattern carefully. In the first three clock cycles, not all of the pipeline stages are active, since the pipeline is *filling*. In the next four clock cycles, all stages of the pipeline are active, as indicated in blue, and the pipeline is fully utilized. In the last three clock cycles, not all stages of the pipeline are active, since the pipeline is *emptying*. If we want to find the max-

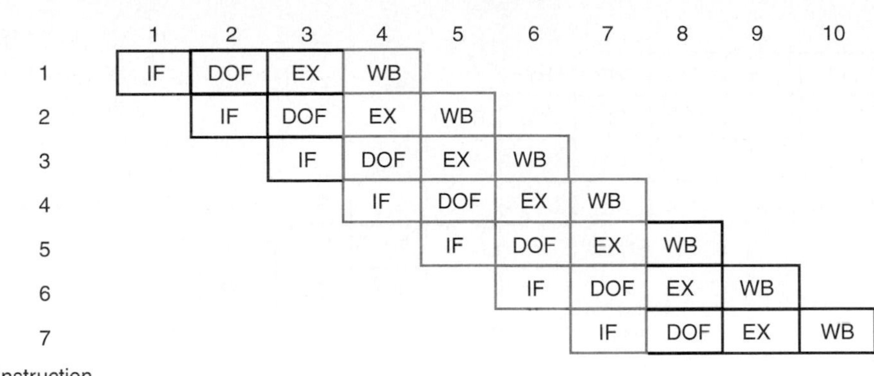

☐ **FIGURE 8-31**
Pipeline Execution Pattern of Register Number Program

imum possible improvement of the pipelined computer over the single-cycle computer, we compare the two in the situation where the pipeline is fully utilized. Over these four clock cycles, or 20 ns, the pipeline executes $(4 \times 4) \div 4 = 4.0$ instructions. In the same time, the single-cycle computer executes $20 \div 17 = 1.2$ instructions. So in the best case, the pipelined computer executes $4 \div 1.2 = 3.3$ times as many instructions in a given time as the single-cycle computer does. In this ideal situation, we say that the throughput of the pipelined computer is 3.3 times that of the single-cycle computer. Note that even though the pipeline has four stages, the pipelined computer is not four times as fast as the single-cycle computer, because the delays of the latter cannot be divided exactly into four equal pieces and the delays of the added pipeline platforms. Also, filling and emptying the pipeline reduces its speed enough that the speed of the pipelined computer is less than the ideal maximum speed of 3.3 times as fast as the single-cycle computer.

Before leaving our study of the pipelined computer, we will examine one more factor that causes its performance to be below the ideal. Suppose that we execute a program that sums the contents of the eight registers as follows:

```
1   ADD R1, R0, R1
2   ADD R3, R2, R3
3   ADD R5, R4, R5
4   ADD R7, R6, R7
5   ADD R3, R1, R3
6   NOP
7   ADD R7, R5, R7
8   NOP
9   NOP
10  ADD R7, R3, R7
```

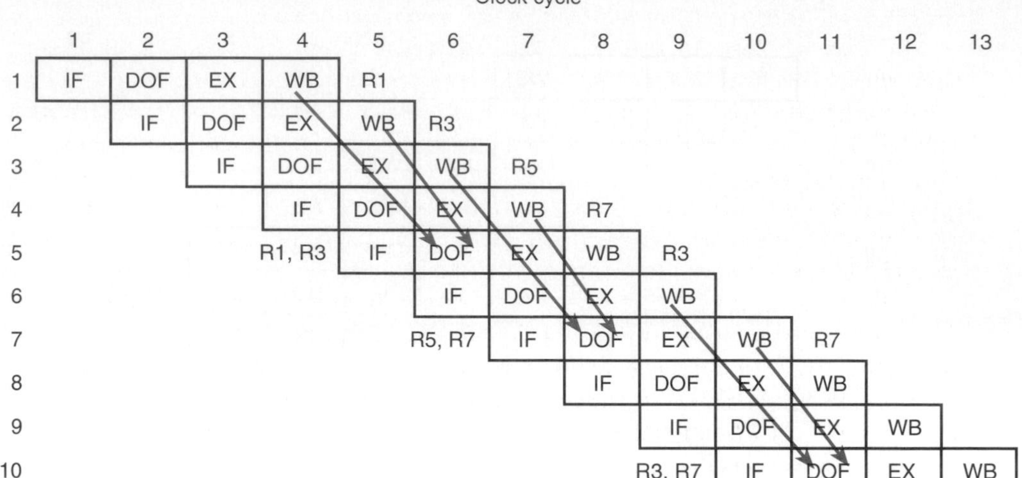

□ **FIGURE 8-32**
Pipeline Execution Pattern of Register Sum Program

In this program, the contents of four pairs of registers (R_{i-1}, R_i), i = 1, 3, 5, and 7, are added together, with the result going into the register R_i. Then two pairs of the result registers are added. Finally, these results are added to complete the sum in $R7$. In this program, there are three NOP instructions added—i. e., instructions that do nothing. Why are these instructions here? To answer this question, we examine the pipeline execution pattern for the program, shown in Figure 8-32. In addition to the usual components of the diagram, for some of the instructions the destination register is shown on the right, and for some of the instructions the pair of source registers is given on the left. Note that in clock cycle 6, instruction 5 fetches the contents of registers $R1$ and $R3$ for the addition in clock cycle 7. This works fine, since the write-back of $R1$ with the sum of $R0$ and $R1$ for instruction 1 occurs in cycle 4, and the write-back of $R3$ with the sum of $R2$ and $R3$ for instruction 2 occurs in cycle 5. So the data needed for instruction 5 from the previous instructions is written back before being used. But what if the next addition—that of $R5$ and $R7$—were instruction 6, instead of the first NOP? Then the fetch of $R5$ and $R7$ would occur in clock cycle 7. But instruction 4 would not yet have written its result back into $R7$. So instruction 6 would, in error, add the new contents of $R5$ to the old contents of $R7$. This is called a *data hazard*. To avoid such errors in this simple pipeline, the programmer inserted NOPs into the program. In general, for the program to be correct for the intended design, any pending write-back to a register or memory location from a previous instruction must be completed before an operand fetch from that register or memory location is executed in the current instruction. The proper relationship between write-backs and operation fetches is shown by the blue arrows in Figure 8-32. Careful examination of when the write-backs and operand fetches occur shows that the removal of any of the NOPs in this program will disturb this relationship and cause a data hazard.

Clearly, the insertion of a NOP, which does nothing, slows down useful execution. If we compare the execution of the sum-of-registers program using the pipelined computer with that using the single-cycle computer, the throughput is increased by a factor of only 1.8. Even if we ignore the losses due to filling and emptying the pipeline, the throughput improves by only a factor of 2.3, versus the ideal 3.3. So the reduction in performance caused by dealing with data hazards is significant. Nevertheless, on balance, use of a pipeline has been found to be a valuable technique for improving computer performance.

The study of the pipelined computer completes our examination of three computer control organizations to see different approaches and their effects. Both the datapaths and the controls we have studied are simplified and missing elements, such as program sequence control. In subsequent chapters, these missing elements will be added to provide a more complete picture of how a computer works and how it is designed.

8-10 Chapter Summary

This chapter has examined the interaction between datapaths and control units and the difference between programmed and nonprogrammed systems. The algorithmic state machine (ASM) is a means for representing and specifying control functions. Two approaches—sequencer plus decoder and one flip-flop per state—are added to sequential circuit design basics to implement ASM charts. The use of microprogrammed control to implement such charts is a more structured alternative for complex designs. A control unit for a datapath dedicated to integer multiplication illustrates these three design methods for nonprogrammed systems.

In the second part of the chapter, we presented control design for programmed systems, examining three different implementations of basic control units for a simple computer architecture. We introduced the concept of instructions and defined instruction formats for the simple computer. One implementation of this computer is capable of executing any instruction in a single clock cycle. Aside from having a program counter, the control unit of this computer, consists of a combinational decoder circuit.

Among the shortcomings of the single-cycle computer are limitations on the complexity of the instructions that can be executed on it, problems with the interface to a single memory, and the relatively low clock frequencies attained. To deal with the first two of these shortcomings, we examined a microprogrammed version of the simple computer, looking at microprogram segments for a number of instructions, the performance of microprogrammed control, and a hardwired alternative to microprogramming.

In the final section of the chapter, we presented pipelined control, which addresses the problem of the low clock frequency of the single-cycle computer.

By continuing the analogy to a manufacturing assembly line from Chapter 7, we expanded the pipeline control concept to include instruction fetch as a part of a pipelined control unit. We examined the basic performance of the pipelined computers, as well as the data hazard, one of a number of pitfalls that degrade the performance of pipelined computers.

REFERENCES

1. MANO, M. M. *Computer Engineering: Hardware Design.* Englewood Cliffs, NJ: Prentice Hall, 1988.
2. HAMACHER, V. C., VRANESIC, Z. G., AND ZAKY, S. G. *Computer Organization,* 3rd ed. New York, NY: McGraw-Hill, 1990.
3. HENNESSY, J. L., AND PATTERSON, D. A. *Computer Architecture: A Quantitative Approach,* 2nd ed. San Francisco, CA: Morgan Kaufmann, 1995.
4. MANO, M. M. *Computer System Architecture,* 3rd ed. Englewood Cliffs, NJ: Prentice Hall, 1993.
5. PATTERSON, D. A., AND HENNESSY, J. L. *Computer Organization and Design: The Hardware/Software Interface.* San Mateo, CA: Morgan Kaufmann, 1994.
6. WEISS, S., AND SMITH, J. E. *POWER and PowerPC.* San Mateo, CA: Morgan Kaufmann, 1994.

PROBLEMS

8–1. A state diagram of a sequential circuit is given in Figure 8-33. Find the corresponding ASM chart. Use a minimum number of decision boxes. The inputs to the circuit are X_1 and X_2, and the outputs are Z_1 and Z_2.

8–2. An ASM chart is given in Figure 8-34. Find the state table for the corresponding sequential circuit.

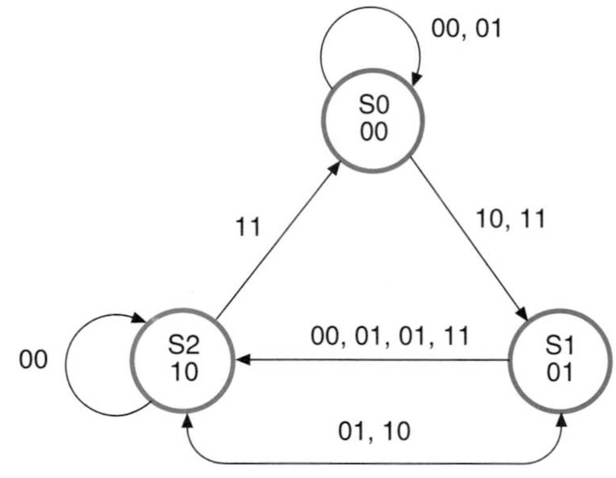

□ **FIGURE 8-33**
State Diagram for Problem 8-1

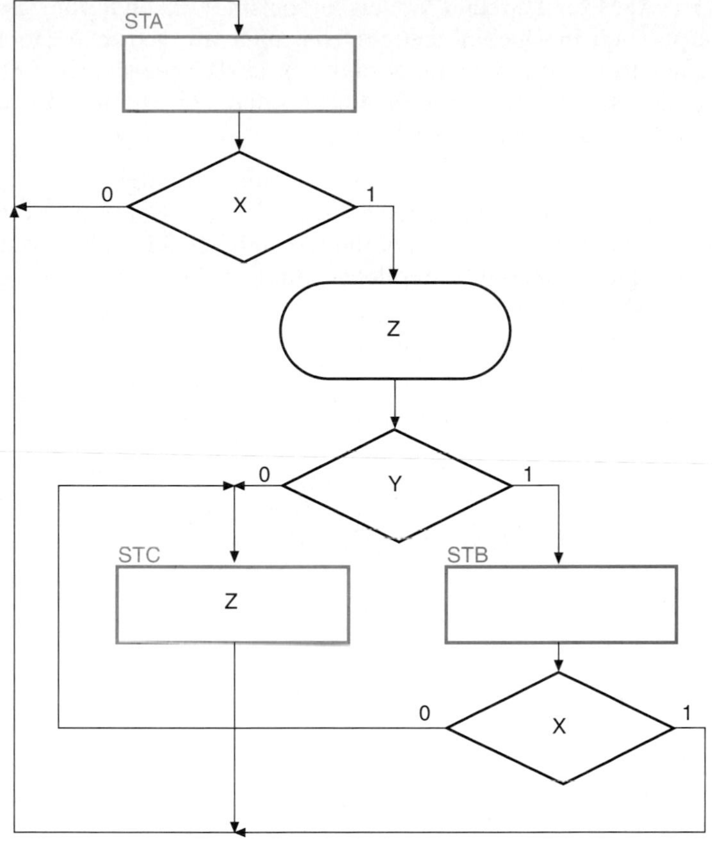

☐ **FIGURE 8-34**
ASM Chart for Problem 8-2 and Problem 8-3

8–3. Find the response for the ASM chart in Figure 8-34 to the following sequence of inputs (assume that the initial state is STA):

X: 0 1 0 0 1 0

Y: 1 1 1 1 0 1

State: STA

Z:

8–4. Find the ASM chart corresponding to the following description: There are two states, *A* and *B*. If in state *A* and input *X* is 0, then the next state is *A*. If in state *A* and input *X* is 1, then the next state is *B*. If in state *B* and input *Y* is 1, then the next state is *B*. If execution is in state *B* and input *Y* is 0, then the next state is *A*. Output *Z* is equal to 1 while the circuit is in state *B*.

8–5. Find the ASM for a circuit that detects transitions in an input signal X. In any clock cycle in which X changes from a 0 to a 1 or from a 1 to a 0, output Z is equal to 1 for the remainder of the cycle after the change. Otherwise, output Z is 0. Note that a 1 on Z indicates that a transition has occurred in X during the current clock cycle.

8–6. Find an ASM chart for a traffic light controller that works as follows: A timing signal T is the input to the controller. T defines the yellow light interval, as well as the changes of the red and green lights. The outputs to the signals are defined by the following table:

Output	Light Controlled
GN	Green Light, North/South Signal
YN	Yellow Light, North/South Signal
RN	Red Light, North/South Signal
GE	Green Light, East/West Signal
YE	Yellow Light, East/West Signal
RE	Red Light, East/West Signal

While $T = 0$, the green light is on for one signal and the red light for the other. With $T = 1$, the yellow light is on for the signal that was previously green, and the signal that was previously red remains red. When T becomes 0, the signal that was previously yellow becomes red, and the signal that was previously red becomes green. This pattern of alternating changes in color continues. Assume that the controller is synchronous with a clock that changes much more frequently than input T.

8–7. Implement the state table derived in Problem 8–2 from the ASM chart in Figure 8-34 by using the normal sequential circuit design method.

8–8. Implement the ASM chart in Figure 8-34 by using a sequence register and decoder.

8–9. Implement the ASM chart in Figure 8-34 by using one flip-flop per state.

8–10. Implement the ASM chart derived in Problem 8–6 by using one flip-flop per state.

8–11. Multiply the two unsigned binary numbers 100110 (multiplicand) and 110101 (multiplier) by using both the hand method and the hardware method.

8–12. Manually simulate the process of multiplying the two unsigned binary numbers 111111 (multiplicand) and 101011 (multiplier). List the contents of registers A, Q, P, and C and the control state, using the system in Figure 8-6 with n equal to 6 and with the hardwired control in Figure 8-11.

8–13. Determine the time it takes to process the multiplication operation in the digital system described in Figure 8-6 and Figure 8-9. Assume that the Q register has n bits and the interval for a clock cycle is f nanoseconds.

8–14. Prove that the multiplication of two n-bit numbers gives a product of no more than $2n$ bits. Show that this condition implies that no overflow in the final result can occur in the multiplier defined in Figure 8-6.

8–15. Consider the block diagram of the multiplier shown in Figure 8-6. Assume that the multiplier and multiplicand consist of 16 bits each.

(a) How many bits can be expected in the product, and where is it available?

(b) How many bits are in the P counter, and what is the binary number that must be loaded into it initially?

(c) Design the combinational circuit that checks for zero in the P counter.

8–16. Design a digital system with three 16-bit registers AR, BR, and CR to perform the following operations:

(a) Transfer two 16-bit signed numbers to AR and BR after a go signal G is enabled.

(b) If the number in AR is negative, multiply the contents of AR by two and transfer the result to register CR.

(c) If the number in AR is positive but nonzero, multiply the contents of BR by two and transfer the result to register CR.

(d) If the number in AR is zero, reset register CR to 0.

8–17. *Design a digital system that multiplies two unsigned binary numbers by the repeated addition method. For example, to multiply 5 by 4, the digital system adds the multiplicand four times: $5 + 5 + 5 + 5 = 20$. Let the multiplicand be in register BR, the multiplier in register AR, and the product in register PR. An adder circuit adds the contents of BR to PR. A zero-detection circuit Z checks when AR becomes zero after each time that it is decremented. Design the control by the hardwired method.

8–18. *Modify the multiplier design in Figure 8-6 and the ASM chart in Figure 8-7 to perform 2's complement signed-number multiplication using Booth's algorithm, which employs an adder-subtractor. The decision to add, to subtract, or to do nothing is made on the basis of the current least signigicant bit (LSB) in the Q register and on the previous LSB bit from the Q register before Q was shifted right. Thus, a flip-flop must be provided to store the previous LSB from the Q register. The initial value of the previous least significant bit is to be 0. The following table defines the decisions:

LSB of Q	Previous LSB of Q	Action
0	0	Leave partial product unchanged
0	1	Add multiplicand to partial product
1	0	Subtract multiplicand from partial product
1	1	Leave partial product unchanged

8–19. Define the following terms in your own words: **(a)** datapath; **(b)** control unit; **(c)** hardwired control; **(d)** microprogrammed control; **(e)** control memory; **(f)** control word; **(g)** microinstruction; **(h)** microprogram; **(i)** microoperation.

8–20. A microprogrammed control unit is similar to the one shown in Figure 8-15, except that multiplexer MUX2 has 15 input status bits instead of 3; and the control memory has 1,024 words. Formulate the microinstruction format, and specify the number of bits in each field. How many bits are required in each control memory word and in the entire control memory?

8–21. Repeat Problem 8–12, using the microprogrammed control in Figure 8-21 and the microprogram in Table 8-5 and Table 8-6. Give the *CAR* contents for each step, as well as the contents of the datapath registers.

8–22. A computer has a 32-bit instruction word broken into fields as follows: opcode, 8 bits; two register fields, 5 bits each; and one immediate operand/register field, 14 bits.

(a) What is the maximum number of operations that can be specified?
(b) How many registers can be addressed?
(c) What is the range of unsigned immediate operands that can be provided?
(d) What is the range of signed immediate operands that can be provided, assuming that bit 13 is the sign bit?

8–23. A digital computer has a memory unit with 24 bits per word. The instruction set consists of 150 different operations. There is only one type of instruction format, with an opcode part and an immediate operand part. Each instruction is stored in one word of memory.

(a) How many bits are needed for the opcode part of the instruction?
(b) How many bits are left for the immediate part of the instruction?
(c) If the immediate operand is used as an unsigned address to memory, what is the maximum number of words that can be addressed in memory?
(d) What are the largest and the smallest algebraic values of signed 2's complement binary numbers that can be accommodated as an immediate operand?

8–24. Manually simulate the single-cycle computer in Figure 8-19 for the following sequence of instructions, assuming that each register initially contains contents equal to its index (i.e., *R*0 contains 0, *R*1 contains 1, etc.):

```
ADD R0, R1, R2
ADD R3, R3, R4
ADD R5, R5, R6
ADD R0, R0, R3
ADD R0, R0, R5
ADD R0, R0, R7
STR  R7, R0
```

ADI R0, R0, 0
ADI R3, R3, 3
ADI R5, R5, 5

Give **(a)** the binary value of the instruction on the current line of the results and **(b)** the contents of any register changed by the instruction or the location and contents of any memory location changed by the instruction on the next line of the results. The results are positioned in this fashion because the new values do not appear in a register or memory, due to the execution of an instruction until after a positive clock edge has occurred.

8–25. The single-cycle computer in Figure 8-19 is to have five instructions added as described by the five register transfers. These are somewhat strange instructions in that they operate on specific registers.

(a) Complete the following table, giving the binary instruction decoder outputs for executing each of the instructions, as described by its register transfer. (See Figure 7-19 and Table 7-10 for datapath field codes. If a given field is not used, leave it blank.)

Instruction — Register Transfer	DA	AA	BA	MB	FS	MD	RW	MW
$R[0] = R[6] \oplus R[7]$								
$R[7] \leftarrow M[R[6]]$								
$R[7] \leftarrow R[7] + 2$								
$R[7] \leftarrow \text{sl } R[7]$								
$R[6] \leftarrow R[7]$								

(b) Complete the following table, giving the instruction for execution of each of the instructions, as described by its register transfer. (See Figure 8-20 and Table 8-7 for determining the opcode. If any field or bit is not used, give it the value 0.)

Instruction — Register Transfer	Opcode	DR	SA	SB or Operand
$R[0] = R[6] \oplus R[7]$				
$R[7] \leftarrow M[R[6]]$				
$R[7] \leftarrow R[7] + 2$				
$R[7] \leftarrow \text{sl } R[7]$				
$R[6] \leftarrow R[7]$				

8–26. A microinstruction stored in address 35 in the control memory of Figure 8-22 performs the operation

$$R1 \leftarrow R1 + R2, CAR \leftarrow CAR + 1$$

Give the microinstruction in symbolic form.

8–27. Give a symbolic microinstruction that resets register $R4$ to 0 and updates the Z and N status bits based on the value 0 transferred to $R4$. (*Hint*: Try the exclusive-OR. By examining the detailed ALU logic, determine what will be the values of the V and C status bits.?

8–28. List the symbolic and binary microinstructions similar to those given in Table 8-11 and Table 8-12 for the following register transfer statements:

(a) $R3 \leftarrow R1 - R2, CAR \leftarrow 17$

(b) $R5 \leftarrow \text{sr } R5, CAR \leftarrow CAR + 1$

(c) If $(Z = 0)$ then $(CAR \leftarrow 21)$ else $(CAR \leftarrow CAR + 1)$

(d) $R6 \leftarrow R6, C \leftarrow 0, CAR \leftarrow CAR + 1$ (update Z and S)

8–29. Write a microprogram for the CPU in Figure 8-22 to compute the average integer value of the four 8-bit unsigned binary numbers stored in registers $R1$, $R2$, $R3$, and $R4$. The average value is to be stored in register $R5$. Do not disturb the contents of the other two registers.

8–30. *A microprogram for the CPU in Figure 8-22 that compares two unsigned binary numbers stored in registers R[SA] and R[SB] is to be written. The register containing the smaller number is reset to 0. If the two numbers are equal, both registers are reset to 0. Is it possible to write this microprogram? Justify your answer.

8–31. Specify an instruction format for the microprogrammed computer that performs the operation

$$R[DR] \leftarrow R[SB] + M[SA]$$

Find the ASM chart for implementing the instruction you have found, assuming that 0000111 is the opcode. (See Figure 8-24.) List the sequence of microoperations for executing this instruction, including the handling of the *CAR* values.

8–32. Repeat Problem 8–31 for the instruction

$$M[SA] \leftarrow M[SA] + R[SB]$$

8–33. *Design a hardwired decoder-based controller that includes the shift-right multiple (SRM) instruction by doing the following:
(a) Modify the ASM chart in Figure 8-27 to include execution of SRM.
(b) Modify the decoder-based control diagram in Figure 8-28, based on the ASM chart.
(c) Write all control logic equations for the design.

8–34. A sequence of 25 instructions is to be executed in a four-stage pipelined computer with a clock period of 10 ns. How long will it take for the sequence of instructions to be fetched and executed?

8–35. The sequence of seven LDI instructions in the register number program with the pipeline execution pattern given in Figure 8-31 is fetched and executed.
Simulate the execution by giving the values in pipeline registers *PC*, *IR*, Data *A*, Data *B*, Data *F*, Data *I*, and in any register in the register file having its value changed for each clock cycle. Assume that all register file registers initially contain −1 (all 1's).

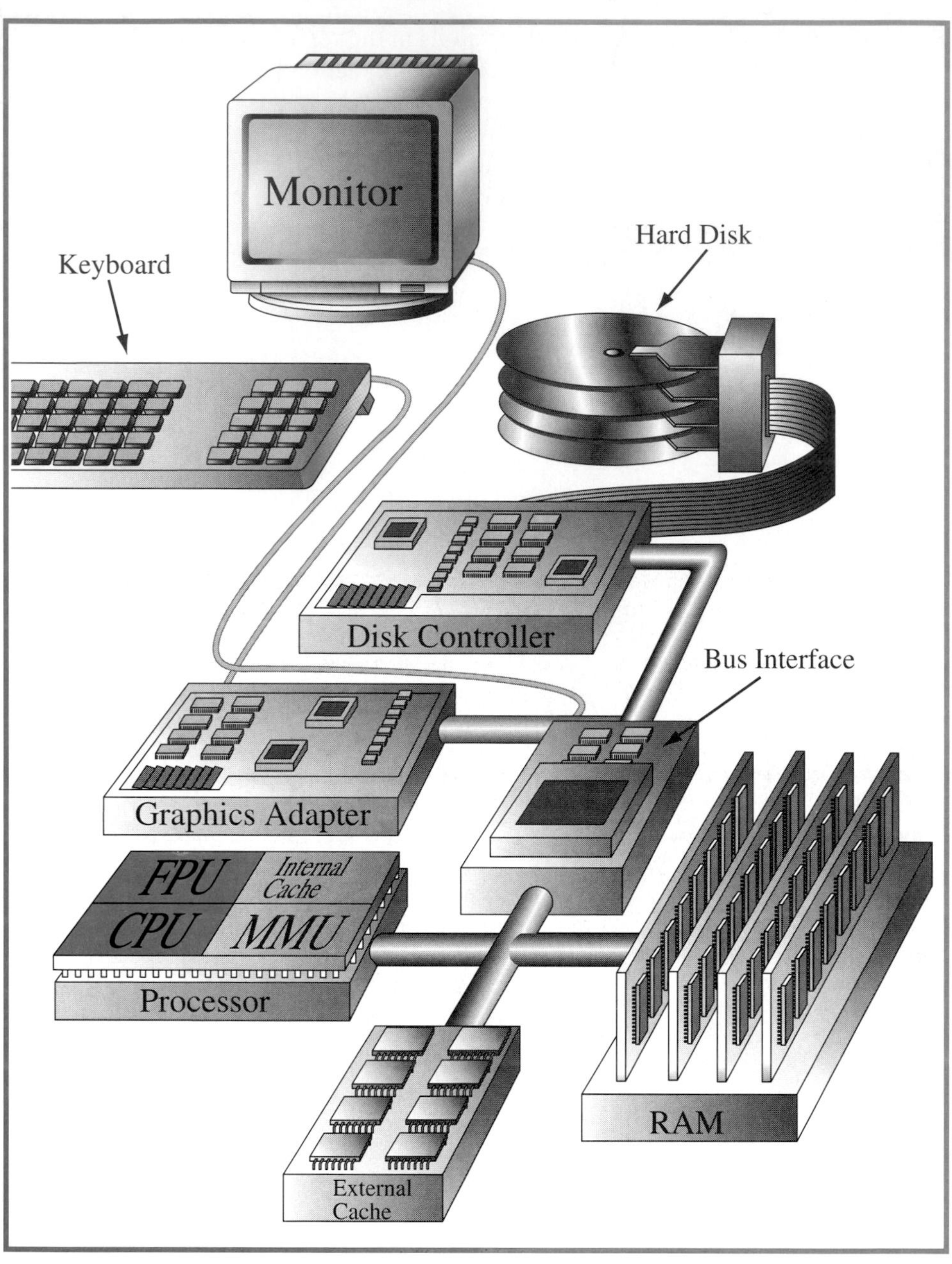
Monitor
Keyboard
Hard Disk
Disk Controller
Bus Interface
Graphics Adapter
FPU
Internal Cache
CPU
MMU
Processor
RAM
External Cache

CHAPTER

9

INSTRUCTION SET ARCHITECTURE

Up to this point, much of what we have studied has focused on digital system design, with computer components used as examples. In this chapter, the material studied becomes decidedly more specialized, dealing with instruction set architecture for general-purpose computers. We will examine the operations that the instructions perform and focus particularly on how the operands are obtained and where the results are stored. In our studies, we will contrast two distinct classes of architectures: reduced instruction set computers (RISCs) and complex instruction set computers (CISCs). We will classify elementary instructions into three categories: data transfer, data manipulation, and program control. In each of these categories, we will elaborate on typical elementary instructions.

In light of this change in focus, the general-purpose parts of the generic computer, including the central processing unit (CPU) and the accompanying floating-point unit (FPU), are heavily shaded. In addition, since a small general-purpose microprocessor may be present for controlling keyboard and monitor functions, we have lightly shaded these components. Aside from addressing used to access memory and I/O components, the concepts studied apply less to other areas of the computer. Increasingly, however, small CPUs are appearing more and more in the I/O components, giving a changing picture of the role of general-purpose instruction set architectures in the generic computer.

9-1 COMPUTER ARCHITECTURE CONCEPTS

The binary language in which instructions are defined and stored in memory is referred to as *machine language*. A symbolic language that replaces binary opcodes and addresses with symbolic names and that provides other features helpful to the programmer is referred to as *assembly language*. The logical structure of computers is normally described in assembly language reference manuals. Such manuals explain various internal elements of the computer that are of interest to the programmer, such as processor registers. The manuals list all hardware-implemented instructions, specify the symbolic names and binary code format of the instructions, and provide a precise definition of each instruction. In the past, this information represented the *architecture* of the computer. A computer was composed of its architecture, plus a specific *implementation* of that architecture. The implementation was separated into two parts: the organization and the hardware. The *organization* consists of structures such as datapaths, control units, memories, and the buses that interconnect them. *Hardware* refers to the logic, the electronic technologies employed, and the various physical design aspects of the computer. As computer designers pushed for higher and higher performance, and as increasingly more of the computer resided within a single IC, the relationships among architecture, organization, and hardware became so intertwined that a more integrated viewpoint became necessary. According to this new viewpoint, architecture as previously defined is more restrictively called *instruction set architecture*, and the term *architecture* is used to encompass the whole of the computer, including instruction set architecture, organization, and hardware. This unified view enables intelligent design trade-offs to be made that are apparent only in a tightly coupled design process. These trade-offs have the potential for producing better computer designs. In this chapter, we focus on instruction set architecture. In the next, we will look at two distinct instruction set architectures with a focus on implementation using two very different organizations.

A computer usually has a variety of instructions and multiple instruction formats. It is the function of the control unit to decode each instruction and provide the control signals needed to process it. A few simple examples of instructions and instruction formats were presented in Section 8-6. We will now expand this presentation by introducing typical instructions found in commercial general-purpose computers. We will also investigate the various instruction formats that may be encountered in a typical computer, with an emphasis on the addressing of operands. The format of an instruction is depicted in a rectangular box symbolizing the bits of the binary instruction. The bits are divided into groups called *fields*. Typical fields found in instruction formats are:

1. An *opcode field*, which specifies the operation to be performed.
2. An *address field*, which provides either a memory address or an address for selecting a processor register.
3. A *mode field*, which specifies the way the address field is to be interpreted.

Other special fields are sometimes employed under certain circumstances—for example, a field that gives the number of positions to shift in a shift-type instruction or an operand field in an immediate operand instruction.

Basic Computer Operation Cycle

In order to comprehend the various addressing concepts to be presented in the next two sections, we need to understand the basic operation cycle of the computer. The control unit of a computer is designed to execute each instruction of a program in the following sequence of steps:

1. Fetch the instruction from memory into a control register.
2. Decode the instruction.
3. Locate the operands used by the instruction.
4. Fetch operands from memory (if necessary).
5. Execute the operation in processor registers.
6. Store the results in the proper place.
7. Go back to step 1 to fetch the next instruction.

As explained in Section 8-6, there is one register in the computer called the program counter (PC) that keeps track of the instructions in the program stored in memory. The *PC* holds the address of the instruction to be executed next and is incremented by one each time a word is read from the program in memory. The decoding done in Step 2 determines the operation to be performed and the addressing mode of the instruction. The operands in Step 3 are located from the addressing mode and the address field of the instruction. The computer executes the instruction, storing the result, and returns to Step 1 to fetch the next instruction in sequence.

Register Set

The *register set* consists of all registers in the CPU that are accessible to the programmer. These registers are typically those mentioned in assembly language programming reference manuals. In the simple CPUs we have dealt with so far, the register set has consisted of the programmer-accessible portion of the register file and the *PC*. The CPUs can also contain other registers, such as the instruction register, registers in the register file that are accessible only to microprograms, and pipeline registers. These registers, however, are not directly accessible to the programmer and, as a consequence, are not a part of the register set, which represents the stored information in the CPU that instructions can access. Thus, the register set has a considerable influence on instruction set architecture.

The register set for a realistic CPU can become quite complex. For the discussion in this chapter, we will add two registers to the set we have used thus far: the *processor status register* (*PSR*) and the *stack pointer* (*SP*). The processor status register contains flip-flops that are selectively set by status values C, N, V, and Z from the ALU. These stored status bits will be used to make decisions that determine the program flow, based on ALU results or the contents of registers. The stored status bits in the processor status register are also referred to as the *condition codes* or the *flags*. Additional bits in the *PSR* will be discussed when we cover associated concepts in this chapter.

9-2 OPERAND ADDRESSING

Consider an instruction such as ADD, which specifies the addition of two operands to produce a result. Suppose that the result of the addition is treated as just another operand. Then the ADD instruction has three operands: the addend, the augend, and the result. An operand residing in memory is specified by its address. An operand residing in a processor register is specified by a register address, a binary code of n bits that specifies one of 2^n registers in the register file. Thus, a computer with 16 processor registers, say, $R0$ through $R15$, has in its instructions one or more register address fields of four bits. The binary code 0101, for example, designates register $R5$.

Some operands, however, are not explicitly addressed, because their location is specified either by the opcode of the instruction or by an address assigned to one of the other operands. In such a case, we say that the operand has an *implied address*. If the address is implied, then there is no need for a memory or register address field for the operand in the instruction. On the other hand, if an operand has an address in the instruction, then we say that the operand is explicitly addressed or has an *explicit address*.

The number of operands explicitly addressed for a data manipulation operation such as ADD is an important factor in defining the instruction set architecture for a computer. An additional factor is the number of such operands that can be explicitly addressed in memory by the instruction. These two factors are so important in defining the nature of instructions, that they act a means of distinguishing different instruction set architectures. They also govern the length of computer instructions.

We begin by illustrating simple programs with different numbers of explicitly addressed operands per instruction. Since each explicitly addressed operand has up to three memory or register addresses per instruction, we will label the instructions as having three, two, one, or zero addresses. Note that, of the three operands needed for an instruction such as ADD, the addresses of all operands not having an address in the instruction are implied.

To illustrate the influence of the number of operands on computer programs, we will evaluate the arithmetic statement

$$X = (A + B)(C + D)$$

using three, two, one, and zero address instructions. We will assume that the operands are in memory addresses symbolized by the letters A, B, C, and D and must not be changed by the program. The result is to be stored in memory at a location with address X. The arithmetic operations to be used in the instructions are addition, subtraction, and multiplication, denoted by ADD, SUB, and MUL, respectively. Further, three operations needed to transfer data during the evaluation are move, load, and store, denoted by MOVE, LD, and ST, respectively. LD moves an operand from memory to a register and ST from a register to memory. Depending on the addresses permitted, MOVE can transfer data between registers, between memory locations, or from memory to register or register to memory.

Three-address Instructions

A program that evaluates $X = (A + B)(C + D)$ using three-address instructions is the following (an equivalent register transfer statement is shown for each instruction):

ADD T1, A, B	$M[T1] \leftarrow M[A] + M[B]$
ADD T2, C, D	$M[T2] \leftarrow M[C] + M[D]$
MUL X, T1, T2	$M[X] \leftarrow M[T1] \times M[T2]$

The symbol $M[A]$ denotes the operand stored in memory at the address symbolized by A. The symbol $\times$ designates multiplication. T1 and T2 are temporary storage locations in memory.

This same program can use registers as the temporary storage locations:

ADD R1, A, B	$R1 \leftarrow M[A] + M[B]$
ADD R2, C, D	$R2 \leftarrow M[C] + M[D]$
MUL X, R1, R2	$M[X] \leftarrow R1 \times R2$

An advantage of the three-address format is that it results in short programs for evaluating expressions. A disadvantage is that the binary coded instructions require more bits to specify three addresses, particularly if they are memory addresses.

Two-address Instructions

For two-address instructions, each address field can again specify either a possible register or a memory address. The first operand address listed in the symbolic instruction is the implied address to which the result of the operation is transferred. The program is as follows:

MOVE T1, A	$M[T1] \leftarrow M[A]$
ADD T1, B	$M[T1] \leftarrow M[T1] + M[B]$
MOVE X, C	$M[X] \leftarrow M[C]$
ADD X, D	$M[X] \leftarrow M[X] + M[D]$
MUL X, T1	$M[X] \leftarrow M[X] \times M[T1]$

If a temporary storage register R1 is available, it can replace T1. Note that this program takes five instructions instead of the three used by the three-address instruction program.

One-address Instructions

To perform instructions such as ADD, a computer with one-address instructions uses an implied address—such as a register called an *accumulator ACC*—for obtaining one of the operands and as the location of the result. The program to evaluate the arithmetic statement is as follows:

LD	A	$ACC \leftarrow M[A]$
ADD	B	$ACC \leftarrow ACC + M[B]$
ST	X	$M[X] \leftarrow ACC$
LD	C	$ACC \leftarrow M[C]$
ADD	D	$ACC \leftarrow ACC + M[D]$
MUL	X	$ACC \leftarrow ACC \times M[X]$
ST	X	$M[X] \leftarrow ACC$

All operations are done between the ACC register and a memory operand. In this case, the number of instructions in the program has increased to seven.

Zero-address Instructions

To perform an ADD instruction with zero addresses, all three addresses in the instruction must be implied. A conventional way of achieving this goal is to use a structure referred to as a *stack*, which is a mechanism or structure that stores information such that the item stored last is the first retrieved. Because of its "last in, first out" nature, a stack is also called a *last in, first out* (*LIFO*) queue. The operation of a computer stack is analogous to that of a stack of trays or plates in which the last tray placed on top of the stack is the first to be taken off. Data manipulation operations such as ADD are performed on the stack. The word at the top of the stack is referred to as TOS. The word below it in the stack is referred to as TOS_{-1}. When one or more words are used as operands for an operation, they are removed from the stack. The word below them then becomes the new TOS. When a resulting word is produced, it is placed on the stack and becomes the new TOS. Thus, TOS and a few locations below it are the implied addresses for operands, and TOS is the implied address for the result. For example, the instruction that specifies an addition is simply

ADD

The resulting register transfer action is $TOS \leftarrow TOS + (TOS_{-1})$. Thus, there are no registers or register addresses used for data manipulation instructions in a stack architecture. Memory addressing, however, is used in such architectures for data transfers. For instance, the instruction

PUSH X

results in $TOS \leftarrow M[X]$, a transfer of the word in address X in memory to the top of the stack. A corresponding operation,

POP X

results in $M[X] \leftarrow TOS$, a transfer of the entry at the top of the stack to address X.

The program for evaluating the sample arithmetic statement for the zero-address situation is as follows:

PUSH	A	$TOS \leftarrow M[A]$
PUSH	B	$TOS \leftarrow M[B]$
ADD		$TOS \leftarrow TOS + TOS_{-1}$
PUSH	C	$TOS \leftarrow M[C]$
PUSH	D	$TOS \leftarrow M[D]$
ADD		$TOS \leftarrow TOS + TOS_{-1}$
MUL		$TOS \leftarrow TOS \times TOS_{-1}$
POP	X	$M[X] \leftarrow TOS$

This program requires eight instructions—one more than the number required by the previous one-address program. However, it uses no addressed memory locations or registers to execute data manipulation instructions.

Addressing Architectures

The programs just presented will change if the number of addresses to the memory in the instructions is restricted or if the memory addresses are restricted to specific instructions. These restrictions, combined with the number of operands addressed, define addressing architectures. We can illustrate such architectures with the evaluation of an arithmetic statement in a three-address architecture that has all of the accesses to memory. Such an addressing scheme is called a *memory-to-memory architecture*. This architecture has only control registers, such as the program counter in the CPU. All operands come directly from memory, and all results are sent directly to memory. The format of data transfer and manipulation instructions contains from one to three address fields, all of which are used for memory addresses. For the previous example, three instructions are required, but if an extra word must appear in the instruction for each memory address, then up to four memory reads are required to fetch each instruction. Including the fetching of operands and storing of results, the program to perform the addition would require 21 accesses to memory. If memory accesses take more than one clock cycle, the execution time would be in excess of 21 clock periods. Thus, even though the instruction count is low, the execution time is potentially high. Also, providing the capability for all operations to access memory increases the complexity of the control structures and may lengthen the clock cycle. Thus, this memory-to-memory architecture is typically not used in new designs.

In contrast, the three-address *register-to-register* or *load/store architecture*, which allows only one memory address and restricts its use to load and store types of instructions, is typical in modern processors. Such an architecture requires a sizeable register file, since all data manipulation instructions use register operands. With this architecture, the program to evaluate the sample arithmetic statement is as follows:

LD	R1, A	$R1 \leftarrow M[A]$
LD	R2, B	$R2 \leftarrow M[B]$

ADD	R3, R1, R2	$R3 \leftarrow R1 + R2$
LD	R1, C	$R1 \leftarrow M[C]$
LD	R2, D	$R2 \leftarrow M[D]$
ADD	R1, R1, R2	$R1 \leftarrow R1 + R2$
MUL	R1, R1, R3	$R1 \leftarrow R1 \times R3$
ST	X, R1	$M[X] \leftarrow R1$

Note that the instruction count increases to eight from three for the three-address, memory-to-memory case. Note also that the operations are the same as for the stack case except for the need for register addresses. By using registers, the number of accesses to memory for instructions, addresses, and operands is reduced from 21 to 18. If addresses can be obtained from registers instead of memory, as discussed in the next section, this number can be further reduced.

Variations of the previous two addressing architectures include three-address instructions and two-address instructions with one or two of the addresses to memory. The program lengths and number of memory accesses tend to be intermediate between the previous two architectures. An example of a two-address instruction with a single memory address allowed is

ADD R1, A $R1 \leftarrow R1 + M[A]$

This type of architecture is a *register-memory* architecture and remains prevalent among the current instruction set architectures, primarily to provide compatibility with older software using a specific architecture.

The program with one-address instructions illustrated previously gives the *single-accumulator architecture*. Since this architecture has no register file, its single address is for accessing memory. It requires 21 accesses to memory to evaluate the sample arithmetic statement. In more complex programs, significant additional memory accesses would be needed for temporary storage locations in memory. Because of its large number of memory accesses, this architecture is inefficient and, as a consequence, is restricted to use in CPUs for simple, low-cost applications that do not require high performance.

The zero-address instruction case using a stack supports the concept of a *stack architecture*. Data manipulation instructions such as ADD use no address, since they are performed on the top few elements of the stack. Single memory-address load and store operations, as shown in the program to evaluate the sample arithmetic statement, are used for data transfer. Since most of the stack is located in memory, one or more memory accesses may be required for each stack operation. If one assumes that the entire stack is in memory, the program requires 32 memory accesses. Practical stack architectures have placed a few of the locations at the top of the stack in registers, with the rest of the stack in memory, to reduce this problem somewhat. Nevertheless, the tight coupling of memory access of operands to data manipulation sequences on the stack tends to degrade the performance of the stack architecture. Thus, this type of architecture is used in new designs only if it is essential for compatibility with older software. The stack concept, however, has important uses when stacks are embedded in other addressing architectures, as we will demonstrate in subsequent sections.

9-3 Addressing Modes

The operation field of an instruction specifies the operation to be performed. This operation must be executed on data stored in computer registers or memory words. How the operands are selected during program execution is dependent on the addressing mode of the instruction. The addressing mode specifies a rule for interpreting or modifying the address field of the instruction before the operand is actually referenced. The address of the operand produced by the application of such a rule is called the *effective address.* Computers use addressing-mode techniques for the purpose of accommodating one or both of the following provisions:

1. To give programming flexibility to the user via such capabilities as pointers to memory, counters for loop control, indexing of data, and relocation of programs.
2. To reduce the number of bits in the address fields of the instruction.

The availability of various addressing modes gives the experienced programmer the ability to write programs that require fewer instructions. The effect, however, on throughput and execution time must be carefully weighed. For example, the presence of more complex addressing modes may actually result in lower throughput and longer execution time. Also, most machine-executable programs are produced by compilers that often do not use complex addressing modes effectively.

In some computers, the addressing mode of the instruction is specified by a distinct binary code. Other computers use a common binary code that designates both the operation and the addressing mode of the instruction. Instructions may be defined with a variety of addressing modes, and sometimes two or more addressing modes are combined in one instruction.

An example of an instruction format with a distinct addressing-mode field is shown in Figure 9-1. The opcode specifies the operation to be performed. The mode field is used to locate the operands needed for the operation. There may or may not be an address field in the instruction. If there is an address field, it may designate a memory address or a processor register. Moreover, as discussed in the previous section, the instruction may have more than one address field. In that case, each address field is associated with its own particular addressing mode.

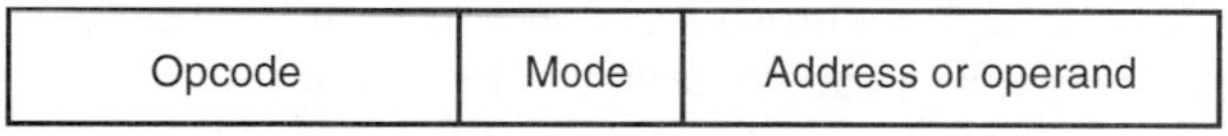

Opcode	Mode	Address or operand

□ **FIGURE 9-1**
Instruction Format with Mode Field

Implied Mode

Although most addressing modes modify the address field of the instruction, there is one mode that needs no address field at all: the implied mode. In this mode, the operand is specified implicitly in the definition of the opcode. It is the implied mode that provides the location for the two-operand-plus-result operations when fewer than three addresses are contained in the instruction. For example, the instruction "complement accumulator" is an implied-mode instruction because the operand in the accumulator register is implied in the definition of the instruction. In fact, any instruction that uses an accumulator without a second operand is an implied-mode instruction. For example, data manipulation instructions in a stack computer, such as ADD, are implied-mode instructions, since the operands are implied to be on top of stack.

Immediate Mode

In the immediate mode, the operand is specified in the instruction itself. In other words, an immediate-mode instruction has an operand field rather than an address field. The operand field contains the actual operand to be used in conjunction with the operation specified in the instruction. Immediate-mode instructions are useful, for example, for initializing registers to a constant value.

Register and Register-Indirect Modes

Earlier, we mentioned that the address field of the instruction may specify either a memory location or a processor register. When the address field specifies a processor register, the instruction is said to be in the register mode. In this mode, the operands are in registers that reside within the processor of the computer. The particular register is selected from a register address field in the instruction format.

In the register-indirect mode, the instruction specifies a register in the processor whose content gives the address of the operand in memory. In other words, the selected register contains the memory address of the operand, rather than the operand itself. Before using a register-indirect mode instruction, the programmer must ensure that the memory address is available in the processor register. A reference to the register is then equivalent to specifying a memory address. The advantage of register-indirect mode is that the address field of the instruction uses fewer bits to select a register than would have been required to specify a memory address directly.

An autoincrement or autodecrement mode is similar to the register-indirect mode, except that the register is incremented or decremented after (or before) its address value is used to access memory. When the address stored in the register refers to an array of data in memory, it is convenient to increment the register after each access to the array. This can be achieved by using a register-increment instruction. However, because it is such a common requirement, some computers incorporate an autoincrement mode that increments the content of the register containing the address after the memory data are accessed.

In the following instruction, an autoincrement mode is used to add the constant value 3 to the elements of an array addressed by register $R1$:

$$\text{ADD} \quad (\text{R1})+, 3 \qquad M[R1] \leftarrow M[R1] + 3, R1 \leftarrow R1 + 1$$

$R1$ is initialized to the address of the first element in the array. Then the ADD instruction is repeatedly executed until the addition of 3 to all elements of the array has occurred. The register transfer statement accompanying the instruction shows the addition of 3 to the memory location addressed by $R1$ and the incrementing of $R1$ in preparation for the next execution of the ADD on the next element in the array.

Direct Addressing Mode

In the direct addressing mode, the address field of the instruction gives the address of the operand in memory in a data transfer or data manipulation instruction. An example of a data transfer instruction is shown in Figure 9-2. The instruction in memory consists of two words. The first, at address 250, has the opcode for "load to *ACC*" and a mode field specifying a direct address. The second word of the instruction, at address 251, contains the address field, symbolized by ADRS, and is equal to 500. The *PC* holds the address of the instruction, which is brought from memory using two memory accesses. Simultaneously with or after the completion of the first access, the *PC* is incremented to 251. Then the second access for ADRS occurs and the *PC* is again incremented. The execution of the instruction results in the operation

$$ACC \leftarrow M[\text{ADRS}]$$

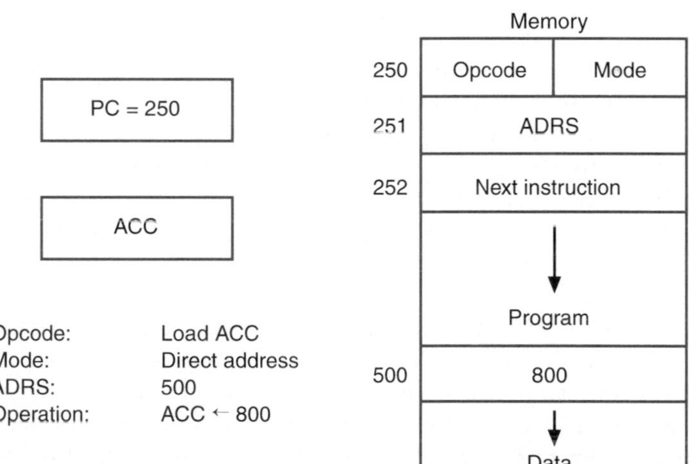

FIGURE 9-2
Example Demonstrating Direct Addressing for a Data Transfer Instruction

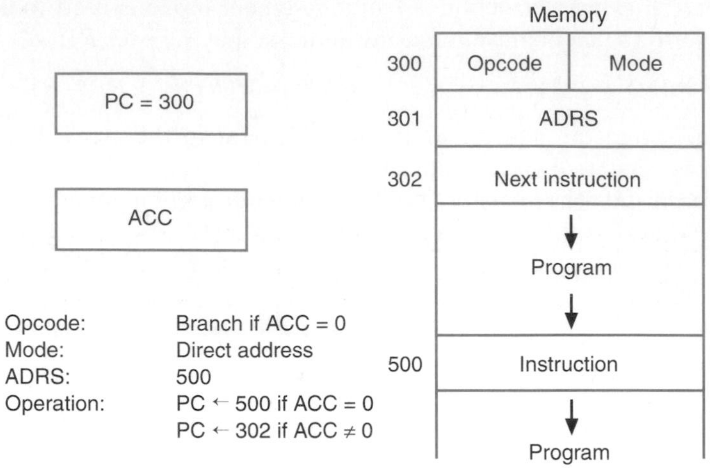

□ **FIGURE 9-3**
Example Demonstrating Direct Addressing in a Branch Instruction

Since ADRS = 500 and M[500] = 800, the *ACC* receives the number 800. After the instruction is executed, the *PC* holds the number 252, which is the address of the next instruction in the program.

Now consider a branch-type instruction, as shown in Figure 9-3. If the contents of *ACC* equal 0, control branches to ADRS; otherwise, the program continues with the next instruction in sequence. When *ACC* = 0, the branch to address 500 is accomplished by loading the value of the address field ADRS into the *PC*. Control then continues with the instruction at address 500. When *ACC* ≠ 0, no branch occurs, and the *PC*, which was incremented twice during the fetch of the instruction, holds the address 302, the address of the next instruction in sequence.

Sometimes the value given in the address field is the address of the operand, but sometimes it is just an address from which the address of the operand is calculated. To differentiate among the various addressing modes, it is useful to distinguish between the address part of the instruction, as given in the address field, and the address used by the control when executing the instruction. Recall that we refer to the latter as the effective address.

Indirect Addressing Mode

In the indirect addressing mode, the address field of the instruction gives the address at which the effective address is stored in memory. The control unit fetches the instruction from memory and uses the address part to access memory again in order to read the effective address. Consider the instruction "load to *ACC*" given in Figure 9-2. If the mode specifies an indirect address, the effective address is stored in M[ADRS]. Since ADRS = 500 and M[ADRS] = 800, the effective address is 800. This means that the operand loaded into the *ACC* is the one found in memory at address 800 (not shown in the figure).

Relative Addressing Mode

Some addressing modes require that the address field of the instruction be added to the content of a specified register in the CPU in order to evaluate the effective address. Often the register used is the *PC*. In the relative addressing mode, the effective address is calculated as follows:

$$\text{Effective address} = \text{Address part of the instruction} + \text{Contents of } PC$$

The address part of the instruction is considered to be a signed number that can be either positive or negative. When this number is added to the contents of the *PC*, the result produces an effective address whose position in memory is relative to the address of the next instruction in the program.

To clarify this with an example, let us assume that the *PC* contains the number 250 and the address part of the instruction contains the number 500, as in Figure 9-2, with the mode field specifying a relative address. The instruction at location 250 is read from memory during the fetch phase of the operation cycle, and the *PC* is incremented by 1 to 251. Since the instruction has a second word, the control unit reads the address field into a control register, and the *PC* is incremented to 252. The computation of the effective address for the relative addressing mode is 252 + 500 = 752. The result is that the operand associated with the instruction is 500 locations away, relative to the location of the next instruction.

Relative addressing is often used in branch-type instructions when the branch address is in a location close to the instruction word. Relative addressing produces more compact instructions, since the relative address can be specified with fewer bits than are required to designate the entire memory address.

Indexed Addressing Mode

In the indexed addressing mode, the content of an index register is added to the address part of the instruction to obtain the effective address. The index register may be a special CPU register or simply a register in a register file. We illustrate the use of indexed addressing by considering an array of data in memory. The address field of the instruction defines the beginning address of the array. Each operand in the array is stored in memory relative to the beginning address. The distance between the beginning address and the address of the operand is the index value stored in the register. Any operand in the array can be accessed with the same instruction, provided that the index register contains the correct index value. The index register can be incremented to facilitate access to consecutive operands.

Some computers dedicate one CPU register to function solely as an index register. This register is addressed implicitly when an index-mode instruction is used. In computers with many processor registers, any CPU register can be used as an index register. In such a case, the index register to be used must be specified with a register field within the instruction format.

A specialized variation of the index mode is the base-register mode. In this mode, the contents of a base register are added to the address part of the instruction to obtain the effective address. This is similar to indexed addressing, except

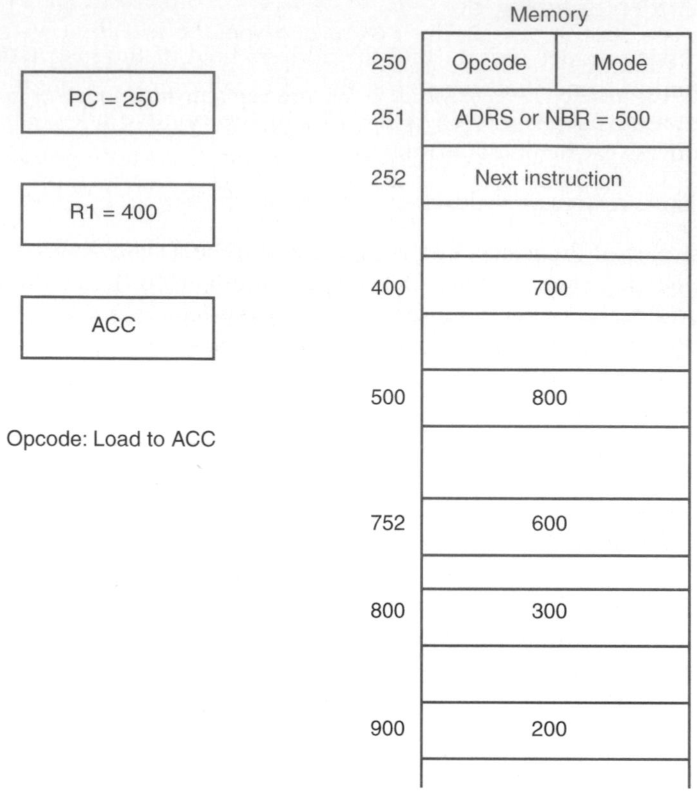

Memory

250	Opcode	Mode
251	ADRS or NBR = 500	
252	Next instruction	
400	700	
500	800	
752	600	
800	300	
900	200	

PC = 250

R1 = 400

ACC

Opcode: Load to ACC

FIGURE 9-4
Numerical Example for Addressing Modes

that the register is called a base register instead of an index register. The difference between the two modes is in the way they are used rather than in the way addresses are computed: an index register is assumed to hold an index number that is relative to the address field of the instruction; a base register is assumed to hold a base address, and the address field of the instruction gives a displacement relative to the base address.

Summary of Addressing Modes

In order to show the differences between the various modes, we will investigate the effect of the addressing mode on the instruction shown in Figure 9-4. The instruction in addresses 250 and 251 is "load to *ACC*," with the address field ADRS (or an operand NBR) equal to 500. The *PC* has the number 250 for fetching this instruction. The contents of a processor register *R*1 are 400, and the *ACC* receives the result after the instruction is executed. In the direct mode, the effective address is 500, and the operand to be loaded into the *ACC* is 800. In the immediate mode, the operand 500 is loaded into the *ACC*. In the indirect mode, the effective address is

800, and the operand is 300. In the relative mode, the effective address is 500 + 252 = 752, and the operand is 600. In the index mode, the effective address is 500 + 400 = 900, assuming that *R*1 is the index register. In the register mode, the operand is in *R*1, and 400 is loaded into the *ACC*. In the register-indirect mode, the effective address is the contents of *R*1, and the operand loaded into the *ACC* is 700.

Table 9-1 lists the value of the effective address and the operand loaded into the *ACC* for seven addressing modes. The table also shows the operation with a register transfer statement and a symbolic convention for each addressing mode. LDA is the symbol for the load-to-accumulator opcode. In the direct mode, we use the symbol ADRS for the address part of the instruction. The # symbol precedes the operand NBR in the immediate mode. The symbol ADRS enclosed in square brackets symbolizes an indirect address, which some compilers or assemblers designate with the symbol @. The symbol $ before the address makes the effective address relative to the *PC*. An index-mode instruction is recognized by the symbol of a register placed in parentheses after the address symbol. The register mode is indicated by giving the name of the processor register following LDA. In the register-indirect mode, the name of the register that holds the effective address is enclosed in parentheses.

□ **TABLE 9-1**
Symbolic Convention for Addressing Modes

			Refers to Figure 9-4	
Addressing mode	**Symbolic convention**	**Register transfer**	**Effective address**	**Contents of *ACC***
Direct	LDA ADRS	$ACC \leftarrow M[ADRS]$	500	800
Immediate	LDA #NBR	$ACC \leftarrow NBR$	251	500
Indirect	LDA [ADRS]	$ACC \leftarrow M[M[ADRS]]$	800	300
Relative	LDA $ADRS	$ACC \leftarrow M[ADRS + PC]$	752	600
Index	LDA ADRS (R1)	$ACC \leftarrow M[ADRS + R1]$	900	200
Register	LDA R1	$ACC \leftarrow R1$	—	400
Register indirect	LDA (R1)	$ACC \leftarrow M[R1]$	400	700

9-4 Instruction Set Architectures

Computers provide a set of instructions to permit computational tasks to be carried out. The instruction sets of different computers differ in several ways from each other. For example, the binary code assigned to the opcode field varies widely for different computers. Likewise, although a standard exists (see Reference 7), the symbolic name given to instructions varies for different computers. In comparison to these minor differences, however, there are two major types of instruction set architectures that differ markedly in the relationship of hardware to software: *Complex instruction set computers* (CISCs) provide hardware support for high-level language operations and have compact programs. *Reduced instruction set computers* (RISCs) emphasize simple instructions and flexibility that, when combined, provide higher throughput and faster execution. These two architectures can be distinguished by considering the properties that characterize their instruction sets.

A *RISC architecture* has the following properties:

1. Memory accesses are restricted to load and store instructions, and data manipulation instructions are register to register.
2. Addressing modes are limited in number.
3. Instruction formats are all of the same length.
4. Instructions perform elementary operations.

The goal of a RISC architecture is high throughput and fast execution. To achieve these goals, accesses to memory, which typically take longer than other elementary operations, are to be avoided, except for fetching instructions. A result of this view is the need for a relatively large register file. Because of the fixed instruction length, limited addressing modes, and elementary operations, the control unit of a RISC is comparatively simple and is typically hardwired. In addition, the underlying organization is universally a pipeline design.

A purely *CISC architecture* has the following properties:

1. Memory access is directly available to most types of instruction.
2. Addressing modes are substantial in number.
3. Instructions formats are of different lengths.
4. Instructions perform both elementary and complex operations.

The goal of the CISC architecture is to match more closely the operations used in programming languages and to provide instructions that facilitate compact programs and conserve memory. In addition, efficiencies in performance may result through a reduction in the number of instruction fetches from memory, compared to the number of elementary operations performed. Because of the high memory accessibility, the register files in a CISC are smaller than in a RISC. Also, because of the complexity of the instructions and the variability of the instruction formats, microprogrammed control is often used. In the quest for speed, the microprogrammed control in newer designs is likely to be controlling a pipelined datapath. CISC instructions are converted to a sequence of RISC-like operations that are processed by the RISC-like pipeline.

Actual instruction set architectures range between those which are purely RISC and those which are purely CISC. Nevertheless, there is a basic set of elementary operations that most computers include among their instructions. In this chapter, we will focus primarily on elementary instructions that are included in both CISC and RISC instruction sets. Most elementary computer instructions can be classified into three major categories:

1. Data transfer instructions.
2. Data manipulation instructions.
3. Program control instructions.

Data transfer instructions cause transfer of data from one location to another without changing the binary information content. Data manipulation instructions perform arithmetic, logic and shift operations. Program control instructions provide decision-making capabilities and change the path taken by the program when executed in the computer. In addition to the basic instruction set, a computer may have other instructions that provide special operations for particular applications.

9-5 DATA TRANSFER INSTRUCTIONS

Data transfer instructions move data from one place in the computer to another without changing the content of the data. Typical transfers are between memory and processor registers, between processor registers and input and output registers, and among the processor registers themselves.

Table 9-2 gives a list of eight typical data transfer instructions used in many computers. Accompanying each instruction is a mnemonic symbol, the assembly language abbreviation recommended by an IEEE standard (Reference 7). Different computers, however, may use different mnemonics for the same instruction name. The load instruction is used to designate a transfer from memory to a processor register. The store instruction designates a transfer from a processor register into a memory word. The move instruction is used in computers with multiple processor registers to designate a transfer from one register to another. It is also used for data transfer between registers and memory and between two memory words. The exchange instruction exchanges information between two registers, between a register and a memory word, or between two memory words. The push and pop instructions are for stack operations described next.

□ **TABLE 9-2**
Typical Data Transfer Instructions

Name	Mnemonic
Load	LD
Store	ST
Move	MOVE
Exchange	XCH
Push	PUSH
Pop	POP
Input	IN
Output	OUT

Stack Instructions

The stack architecture introduced earlier possesses features that facilitate a number of data-processing and control tasks. A stack is used in some electronic calculators and computers for the evaluation of arithmetic expressions. Unfortunately, because

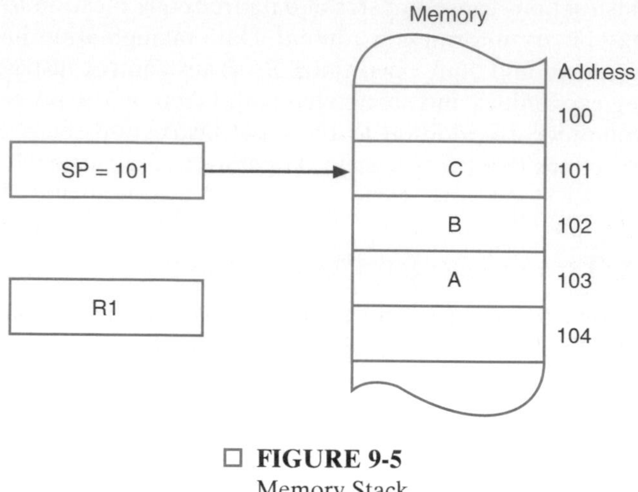

□ **FIGURE 9-5**
Memory Stack

of the negative effects on performance of having the stack reside primarily in memory, a stack in a computer typically handles only state information related to procedure calls and returns and interrupts, as explained in Section 9-8 and Section 9-9.

The stack instructions push and pop transfer data between a memory stack and a processor register or memory. The *push* operation places a new item onto the top of the stack. The *pop* operation removes one item from the stack so that the stack pops up. However, nothing is really physically pushed or popped in the stack. Rather, the memory stack is essentially a portion of a memory address space accessed by an address that is always incremented or decremented before or after the memory access. The register that holds the address for the stack is called a *stack pointer* (*SP*) because its value always points to the item at the top of the stack. Push and pop operations are implemented by decrementing or incrementing the stack pointer.

Figure 9-5 shows a portion of a memory organized as a stack that grows from higher to lower addresses. The stack pointer *SP* holds the binary address of the item that is currently on top of the stack. Three items are presently stored in the stack: A, B, and C, in consecutive addresses 103, 102, and 101, respectively. Item C is on top of the stack, so *SP* contains 101. To remove the top item, the stack is popped by reading the item at address 101 and incrementing *SP*. Item B is now on top of the stack, since *SP* contains address 102. To insert a new item, the stack is pushed by first decrementing *SP* and then writing the new item on top of the stack. Note that item C has been read out of the stack, but is not physically removed from it. This does not matter as far as the stack operation is concerned, because when the stack is pushed, a new item is written on top of it regardless of what was there before.

We assume that the items in the stack communicate with a data register *R*1 or a memory location X. A new item is placed on to the stack with the push operation as follows:

$$SP \leftarrow SP - 1$$

$$M[SP] \leftarrow R1$$

The stack pointer is decremented so that it points at the address of the next word. A memory write microoperation inserts the word from $R1$ onto the top of the stack. Note that SP holds the address of the top of the stack and that $M[SP]$ denotes the memory word specified by the address presently in SP. An item is deleted from the stack with a pop operation as follows:

$$R1 \leftarrow M[SP]$$

$$SP \leftarrow SP + 1$$

The top item is read from the stack into $R1$. The stack pointer is then incremented to point at the next item in the stack, which is the new top of the stack.

The two microoperations needed for either the push or the pop operation are an access to memory through SP and an update of SP. Which microoperation is done first, and whether SP is updated by incrementing or decrementing it, depends on the organization of the stack. In Figure 9-5, the stack grows by decreasing the memory address. By contrast, a stack may be constructed to grow by increasing the memory address. In such a case, SP is incremented for the push operation and decremented for the pop operation. A stack may also be constructed so that SP points to the next empty location above the top of the stack. In that case, the sequence of microoperations must be interchanged.

A stack pointer is loaded with an initial value, which must be the bottom address of an assigned stack in memory. From then on, SP is automatically decremented or incremented with every push or pop operation. The advantage of a memory stack is that the processor can refer to it without having to specify an address, since the address is always available and automatically updated in the stack pointer.

The final pair of data transfer instructions, input and output, depend on the type of input-output used as described next.

Independent versus Memory-Mapped I/O

Input and output (I/O) instructions transfer data between processor registers and input and output devices. These instructions are similar to load and store instructions, except that the transfers are to and from external registers instead of memory words. The computer is considered to have a certain number of input and output ports, with one or more ports dedicated to communication with a specific input or output device. A *port* is typically a register with input and/or output lines attached to the device. The particular port is chosen by an address, similarly to the way an address selects a word in memory. Input and output instructions include an address field in their format, for specifying the particular port selected for the transfer of data.

Port addresses are assigned in two ways. In the *independent I/O system*, the address ranges assigned to memory and I/O ports are independent from each other. The computer has distinct input and output instructions, as listed in Table 9-2 and containing a separate address field that is interpreted by the control and used to select a particular I/O port. Independent I/O addressing isolates memory and I/O selection, so that the memory address range is not affected by the port address assignment. For this reason, the method is also referred to as an *isolated I/O configuration*.

In contrast to independent I/O is *memory-mapped I/O*, which assigns a subrange of the memory addresses for addressing I/O ports. There are no separate addresses for handling input and output transfers, since I/O ports are treated as memory locations in one common address range. Each I/O port is regarded as a memory location, similar to a memory word. Computers that adopt the memory-mapped scheme have no distinct input or output instructions, because the same instructions are used for manipulating both memory and I/O data. For example, the load and store instructions used for memory transfer are also used for I/O transfer, provided that the address associated with the instruction is assigned to an I/O port and not to a memory word. The advantage of this scheme is the simplicity that results with the same set of instructions serving for both memory and I/O access.

9-6 Data Manipulation Instructions

Data manipulation instructions perform operations on data and provide the computational capabilities of the computer. In a typical computer, data manipulation instructions are usually divided into three basic types:

1. Arithmetic instructions.
2. Logical and bit manipulation instructions.
3. Shift instructions.

A list of elementary data manipulation instructions looks very much like the list of microoperations given in Chapter 7. However, an instruction is typically processed by executing a *sequence* of one or more microinstructions. A microoperation is an elementary operation executed by the hardware of the computer under the control of the control unit. In contrast, an instruction may involve several elementary operations that fetch the instruction, bring the operands from appropriate processor registers, and store the result in the specified location.

Arithmetic Instructions

The four basic arithmetic instructions are addition, subtraction, multiplication, and division. Most computers provide instructions for all four operations. Some small computers, however, have only addition and subtraction instructions; on such computers, multiplication and division must be carried out by means of programs. The four basic arithmetic operations are sufficient for formulating solutions to any numerical problem when they are used with numerical analysis methods.

A list of typical arithmetic instructions is given in Table 9-3. The increment instruction adds one to the value stored in a register or memory word. A common characteristic of the increment operation, when executed on a computer word, is that a binary number of all 1's produces a result of all 0's when incremented. The decrement instruction subtracts one from a value stored in a register or memory word. When decremented, a number of all 0's produces a number of all 1's.

☐ TABLE 9-3
Typical Arithmetic Instructions

Name	Mnemonic
Increment	INC
Decrement	DEC
Add	ADD
Subtract	SUB
Multiply	MUL
Divide	DIV
Add with carry	ADDC
Subtract with borrow	SUBB
Subtract reverse	SUBR
Negate	NEG

The add, subtract, multiply, and divide instructions may be available for different types of data. The data type assumed to be in processor registers during the execution of these arithmetic operations is included in the definition of the opcode. An arithmetic instruction may specify unsigned or signed integers, binary or decimal numbers, or floating-point data. The arithmetic operations with binary integers were presented in Chapter 1 and Chapter 3. The floating-point representation is used for scientific calculations and is presented in the next section.

The number of bits in any register is finite; therefore, the results of arithmetic operations are of finite precision. Most computers provide special instructions to facilitate double-precision arithmetic. A carry flip-flop is used to store the carry from an operation. The instruction "add with carry" performs the addition with two operands plus the value of the carry from the previous computation. Similarly, the "subtract with borrow" instruction subtracts two operands and a borrow that may have resulted from a previous operation. The subtract reverse instruction reverses the order of the operands, performing $B - A$ instead of $A - B$. The negate instruction performs the 2's complement of a signed number, which is equivalent to multiplying the number by -1.

Logical and Bit Manipulation Instructions

Logical instructions perform binary operations on words stored in registers or memory words. They are useful for manipulating individual bits or a group of bits that represent binary-coded information. Logical instructions consider each bit of the operand separately and treat it as a Boolean variable. By proper application of the logical instructions, it is possible to change bit values, to clear a group of bits, or to insert new bit values into operands stored in registers or memory.

Some typical logical and bit manipulation instructions are listed in Table 9-4. The clear instruction causes the specific operand to be replaced by 0's. The set instruction causes the operand to be replaced by 1's. The complement instruction inverts all the bits of the operand. The AND, OR, and XOR instructions produce the corresponding logical operations on individual bits of the operand. Although logical instructions perform Boolean operations, when used on words they often are viewed as performing bit manipulation operations. There are three bit manipulation operations possible: A selected bit can be cleared to 0, set to 1, or complemented. The three logical instructions are usually applied to do just that.

The AND instruction is used to clear a bit or a selected group of bits of an operand to 0. For any Boolean variable X, the relationship $X \cdot 0 = 0$ dictates that a binary variable ANDed with a 0 produces a 0; and similarly, the relationship $X \cdot 1 = X$ dictates that the variable does not change when ANDed with a 1. Therefore, the AND instruction is used to selectively clear bits of an operand by ANDing the operand with a word that has 0's in the bit positions that must be cleared and l's in the bit positions that must remain the same. The AND instruction is also called a *mask* because, by inserting 0's, it masks a selected portion of an operand. AND is also sometimes referred to as a *bit clear* instruction.

□ **TABLE 9-4**
Typical Logical and Bit Manipulation Instructions

Name	Mnemonic
Clear	CLR
Set	SET
Complement	NOT
AND	AND
OR	OR
Exclusive-OR	XOR
Clear carry	CLRC
Set carry	SETC
Complement carry	COMC

The OR instruction is used to set a bit or a selected group of bits of an operand to 1. For any Boolean variable X, the relationship $X + 1 = 1$ dictates that a binary variable ORed with a 1 produces a l; and similarly, the relationship $X + 0 = X$ dictates that the variable does not change when ORed with a 0. Therefore, the OR instruction can be used to selectively set bits of an operand by ORing the operand with a word with l's in the bit positions that must be set to 1. The OR instruction is sometimes called a *bit set* instruction.

The XOR instruction is used to selectively complement bits of an operand. This is because of the Boolean relationships $X \oplus 1 = \overline{X}$ and $X \oplus 0 = X$. A binary variable is complemented when XORed with a 1, but does not change value when XORed with a 0. The XOR instruction is sometimes called a *bit complement* instruction.

Other bit manipulation instructions included in Table 9-4 clear, set, or complement the carry bit. Additional instructions clear, set, or complement other status bits or flag bits in a similar manner.

Shift Instructions

Instructions to shift the content of an operand are provided in several varieties. Shifts are operations in which the bits of the operand are moved to the left or to the right. The incoming bit shifted in at the end of the word determines the type of shift. Instead of using just a 0, as for sl and sr in Chapter 7, here we add further possibilities. The shift instructions may specify either logical shifts, arithmetic shifts, or rotate-type operations.

□ **TABLE 9-5**
Typical Shift Instructions

Name	Mnemonic
Logical shift right	SHR
Logical shift left	SHL
Arithmetic shift right	SHRA
Arithmetic shift left	SHLA
Rotate right	ROR
Rotate left	ROL
Rotate right with carry	RORC
Rotate left with carry	ROLC

Table 9-5 lists four types of shift instructions. The logical shift inserts 0 into the incoming bit position after the shift. Arithmetic shifts conform to the rules for shifting two's complement signed numbers. The arithmetic shift right instruction preserves the sign bit in the leftmost position. The value of the sign bit is shifted to the right together with the rest of the number, but the sign bit itself remains unchanged. The arithmetic shift left instruction inserts 0 into the incoming bit in the rightmost position and is identical to the logical shift left instruction. The two instructions may differ, however, in that an arithmetic shift left may set the overflow status bit V, while a logical shift left does not affect V.

The rotate instructions produce a circular shift: the values shifted out of the outgoing bit of the word are not lost, as in a logical shift, but are rotated back into the incoming bit. The rotate-with-carry instructions treat the carry bit as an extension of the register whose word is being rotated. Thus, a rotate left with carry transfers the carry bit into the incoming bit in the rightmost bit position of the register, transfers the outgoing bit from the leftmost bit of the register into the carry, and shifts the entire register to the left. Some computers have a multiple-field format for the shift instruction. One field contains the opcode, and the others specify the type of shift and the number of positions that an operand is to be shifted. A shift instruction may include the following five fields:

OP REG TYPE RL COUNT

OP is the opcode field for specifying a shift, and REG is a register address that specifies the location of the operand. TYPE is a 2-bit field that specifies one of the four types of shifts (logical, arithmetic, rotate, and rotate with carry), while RL is a 1-bit field that specifies whether a shift is to the right or the left. COUNT is a k-bit field that specifies shifts of up to $2^k - 1$ positions. With such a format, it is possible to specify the type of shift, the direction of the shift, and the number of positions to be shifted, all in one instruction.

9-7 FLOATING-POINT COMPUTATIONS

In many scientific calculations, the range of numbers is very large. In a computer, the way to express such numbers is in floating-point notation. The floating-point number has two parts, one containing the sign of the number and a *fraction* (sometimes called a *mantissa*) and the other designating the position of the radix point in the number and called the *exponent*. For example, the decimal number +6132.789 is represented in floating-point notation as

Fraction	Exponent
+ .6132789	+ 04

The value of the exponent indicates that the actual position of the decimal point is four positions to the right of the indicated decimal point in the fraction. This representation is equivalent to the scientific notation $+.6132789 \times 10^{+4}$. Decimal floating-point numbers are interpreted as representing a number in the form

$$F \times 10^E$$

where F is the fraction and E the exponent. Only the fraction and the exponent are physically represented in computer registers; radix 10 and the decimal point of the fraction are assumed and are not shown explicitly. A floating-point binary number is represented in a similar manner, except that it uses radix 2 for the exponent. For example, the binary number + 1001.11 is represented with an 8-bit fraction and 6-bit exponent as

Fraction	Exponent
01001110	000100

The fraction has a 0 in the leftmost position to denote a plus. The binary point of the fraction follows the sign bit, but is not shown in the register. The exponent has the equivalent binary number +4. The floating-point number is equivalent to

$$F \times 2^E = +(0.1001110)_2 \times 2^{+4}$$

A floating-point number is said to be *normalized* if the most significant digit of the fraction is nonzero. For example, the decimal fraction 0.350 is normalized, but 0.0035 is not. Normalized numbers provide the maximum possible precision for the floating-point number. A zero cannot be normalized because it does not have a nonzero digit; it is usually represented in floating-point by all 0's in both the fraction and the exponent.

Floating-point representation increases the range of numbers that can be accommodated in a given register. Consider a computer with 48-bit registers. Since one bit must be reserved for the sign, the range of signed integers will be $\pm(2^{47} - 1)$, which is approximately $\pm 10^{14}$. The 48 bits can be used to represent a floating-point number, with one bit for the sign, 35 bits for the fraction, and 12 bits for the exponent. The largest positive or negative number that can be accommodated is thus

$$\pm(1 - 2^{-35}) \times 2^{+2047}$$

This number is derived from a fraction that contains 35 1's, and an exponent with a sign bit and 11 1's. The maximum exponent is $2^{11} - 1$, or 2047. The largest number that can be accommodated is approximately equivalent to decimal 10^{615}. Although a much larger range is represented, there are still only 48 bits in the representation. As a consequence, exactly the same number of numbers are represented. Hence, the range is traded for the precision of the numbers, which is reduced from 48 bits to 35 bits.

Arithmetic Operations

Arithmetic operations with floating-point numbers are more complicated than with integer numbers, and their execution takes longer and requires more complex hardware. Adding and subtracting two numbers requires that the radix points be aligned, since the exponent parts must be equal before adding or subtracting the fractions. The alignment is done by shifting one fraction and correspondingly adjusting its exponent until it is equal to the other exponent. Consider the sum of the following floating-point numbers:

$$\begin{array}{r} .5372400 \times 10^{2} \\ + .1580000 \times 10^{-1} \end{array}$$

It is necessary that the two exponents be equal before the fractions can be added. We can either shift the first number three positions to the left or shift the second number three positions to the right. When the fractions are stored in registers, shifting to the left causes a loss of the most significant digits. Shifting to the right causes a loss of the least significant digits. The second method is preferable because it only reduces the precision, whereas the first method may cause an error. The usual alignment procedure is to shift the fraction with the smaller exponent to the right by a number of places equal to the difference between the exponents. After this is done, the fractions can be added:

$$\begin{array}{r} .5372400 \times 10^2 \\ +\ .0001580 \times 10^2 \\ \hline .5373980 \times 10^2 \end{array}$$

When two normalized fractions are added, the sum may contain an overflow digit. An overflow can be corrected by shifting the sum once to the right and incrementing the exponent. When two numbers are subtracted, the result may contain most significant zeros in the fraction, as shown in the following example:

$$\begin{array}{r} .56780 \times 10^5 \\ -\ .56430 \times 10^5 \\ \hline .00350 \times 10^5 \end{array}$$

A floating-point number that has a 0 in the most significant position of the fraction is not normalized. To normalize the number, it is necessary to shift the fraction to the left and decrement the exponent until a nonzero digit appears in the first position. In the preceding example, it is necessary to shift left twice to obtain $.35000 \times 10^3$. In most computers, a normalization procedure is performed after each operation to ensure that all results are in normalized form.

Floating-point multiplication and division do not require an alignment of the fractions. Multiplication can be performed by multiplying the two fractions and adding the exponents. Division is accomplished by dividing the fractions and subtracting the exponents. In the examples shown, we used decimal numbers to demonstrate arithmetic operations on floating-point numbers. The same procedure applies to binary numbers, except that the base of the exponent is 2 instead of 10.

Biased Exponent

The sign and fraction part of a floating-point number is usually a signed-magnitude representation. The exponent representation employed in most computers is known as a *biased exponent*. The bias is an excess number added to the exponent so that, internally, all exponents become positive. As a consequence, the sign of the exponent is removed from being a separate entity.

Consider, for example, the range of decimal exponents from −99 to +99. This is represented by two digits and a sign. If we use an excess 99 bias, then the biased exponent e will be equal to $e = E + 99$, where E is the actual exponent. For $E = -99$, we have $e = -99 + 99 = 0$; and for $E = +99$, we have $e = 99 + 99 = 198$. In this way, the biased exponent is represented in a register as a positive number in the range from 000 to 198. Positive-biased exponents have a range of numbers from 099 to 198. Subtraction of the bias, 99, gives the positive values from 0 to +99. Negative-biased exponents have a range from 098 to 000. Subtraction of 99 gives the negative values from −1 to −99.

The advantage of biased exponents is that the resulting floating-point numbers contain only positive exponents. It is then simpler to compare the relative

1	8	23
s	e	f

□ **FIGURE 9-6**
IEEE Floating-Point Operand Format

magnitude between two numbers without being concerned with the signs of their exponents. Another advantage is that the most negative exponent converts to a biased exponent with all 0's. The floating-point representation of zero is then a zero fraction and a zero biased exponent, which is the smallest possible exponent.

Standard Operand Format

Arithmetic instructions that perform operations with floating-point data often use the suffix F. Thus, ADDF is an add instruction with floating-point numbers. There are two standard formats for representing a floating-point operand: the single-precision data type, consisting of 32 bits, and the double-precision data type, consisting of 64 bits. When both types of data are available, the single-precision instruction mnemonic uses an FS suffix, and the double precision uses FL (for "floating-point long").

The format of the IEEE standard (see Reference 8) single-precision floating-point operand is shown in Figure 9-6. It consists of 32 bits. The sign bit s designates the sign for the fraction. The biased exponent e contains 8 bits and uses an excess 127 number. The fraction f consists of 23 bits. The binary point is assumed to be immediately to the left of the most significant bit of the f field. In addition, an implied 1 bit is inserted to the left of the binary point, which, in effect, expands the number to 24 bits representing a value from 1.0_2 to $1.11...1_2$. The component of the binary floating-point number that consists of a leading bit to the left of the implied binary point, together with the fraction in the field, is called the *significand*. Following are some examples of field values and the corresponding significands:

f Field	Significand	Decimal Equivalent
100 . . . 0	1.100 . . . 0	1.50
010 . . . 0	1.010 . . . 0	1.25
000 . . . 0	1.000 . . . 0*	1.00*

*Assuming the exponent is not equal to 00 . . . 0.

Even though the f field by itself may not be normalized, the significand is always normalized because it has a nonzero bit in the most significant position. Since normalized numbers must have a nonzero most significant bit, this 1 bit is not included explicitly in the format, but must be inserted by the hardware during arithmetic computations. The exponent field uses an excess 127 bias value for normalized numbers. The range of valid exponents is from −126 (represented as

□ TABLE 9-6
Evaluating Biased Exponents

Exponent E in decimal	Biased exponent $e = E + 127$	
	Decimal	Binary
− 126	− 126 + 127 = 1	00000001
− 001	− 001 + 127 = 126	01111110
000	000 + 127 = 127	01111111
+ 001	001 + 127 = 128	10000000
+ 126	126 + 127 = 253	11111101
+ 127	127 + 127 = 254	11111110

00000001) through +127 (represented as 11111110). The maximum (11111111) and minimum (00000000) values that the e field can take are reserved to indicate exceptional conditions. Table 9-6 shows the biased and actual values of some exponents.

Normalized numbers are numbers that can be expressed as floating-point operands in which the e field is neither all 0's nor all 1's. The value of the number is derived from the three fields in the format of Figure 9-6 using the formula

$$(-1)^s 2^{e-127} \times (1.f)$$

The most positive normalized number that can be obtained has a 0 for the sign bit for a positive sign, a biased exponent equal to 254, and an f field with 23 1's. This gives an exponent $E = 254 - 127 = 127$. The significand is equal to $1 + 1 - 2^{-23} = 2 - 2^{-23}$. The maximum positive number that can be accommodated is

$$+2^{127} \times (2 - 2^{-23})$$

The smallest positive normalized number has a biased exponent equal to 00000001 and a fraction of all 0's. The exponent is $E = 1 - 127 = -126$, and the significand is equal to 1.0. The smallest positive number that can be accommodated is $+2^{-126}$. The corresponding negative numbers are the same, except that the sign bit is negative. As mentioned before, exponents with all 0's or all 1's (decimal 255) are reserved for special conditions:

1. When $e = 255$ and $f = 0$, the number represents plus or minus infinity. The sign is determined from the sign bit s.
2. When $e = 255$ and $f \neq 0$, the representation is considered to be *not a number*, or NaN, regardless of the sign value. NaNs are used to signify invalid operations, such as the multiplication of zero by infinity.
3. When $e = 0$ and $f = 0$, the number denotes plus or minus zero.
4. When $e = 0$, and $f \neq 0$, the number is said to be denormalized. This is the name given to numbers with a magnitude less than the minimum value that is represented in the normalized format.

9-8 Program Control Instructions

The instructions of a program are stored in successive memory locations. When processed by the control, the instructions are read from consecutive memory locations and executed one by one. Each time an instruction is fetched from memory, the *PC* is incremented so that it contains the address of the next instruction in sequence. In contrast, a program control instruction, when executed, may change the address value in the *PC* and cause the flow of control to be altered. In the control units in Chapter 8, we temporarily ignored this ability to alter the flow of the program, in order to keep the designs presented there simple. The change in the *PC* as a result of the execution of a program control instruction causes a break in the sequence of execution of instructions. This is an important feature of digital computers, since it provides control over the flow of program execution and a capability of branching to different program segments, depending on previous computations.

Some typical program control instructions are listed in Table 9-7. The branch and jump instructions are often used interchangeably to mean the same thing, although sometimes they are used to denote different addressing modes. For example, the jump may use direct or indirect addressing, whereas the branch uses relative addressing. The branch (or jump) is usually a one-address instruction. When executed, the branch instruction causes a transfer of the effective address into the *PC*. Since the *PC* contains the address of the instruction to be executed next, the next instruction will be fetched from the location specified by the effective address.

☐ TABLE 9-7
Typical Program Control Instructions

Name	Mnemonic
Branch	BR
Jump	JMP
Skip next instruction	SKP
Call procedure	CALL
Return from procedure	RET
Compare (by subtraction)	CMP
Test (by ANDing)	TEST

Branch and jump instructions may be conditional or unconditional. An unconditional branch instruction causes a branch to the specified effective address without any conditions. The conditional branch instruction specifies a condition that must be met in order for the branch to occur, such as the value in a specified register being negative. If the condition is met, the *PC* is loaded with the effective address, and the next instruction is taken from this address. If the condition is not met, the *PC* is not changed, and the next instruction is taken from the next location in sequence.

The skip instruction does not need an address field. A conditional skip instruction will skip the next instruction if the specified condition is met. This is accomplished by incrementing the *PC* during the execute phase of the instruction, in addition to incrementing it during the fetch phase. If the condition is not met, control proceeds to the next instruction in sequence, at which point the programmer may insert an unconditional branch instruction. Thus, a conditional skip instruction followed by an unconditional branch instruction causes a branch if the condition is not met. This contrasts with a single conditional branch instruction, which causes a branch if the condition *is* met. Since the skip involves the execution of two instructions, it is slower and uses more instruction memory.

The call and return instructions are used in conjunction with procedures. Their performance and implementation are discussed later in this section.

The compare instruction performs a comparison via a subtraction, with the difference not retained. Instead, the comparison causes a conditional branch, changes the contents of a register, or sets or resets stored status bits. Similarly, the test instruction performs the logical AND of two operands without retaining the result and executes one of the actions listed for the compare instruction.

Based on their three possible actions, compare and test instructions are viewed to be of three distinct types, depending upon the way in which conditional decisions are handled. The first type executes the entire decision as a single instruction. For example, the contents of two registers can be compared and a branch or jump taken if the contents are equal. Since there are two register addresses and a memory address involved, such an instruction requires three addresses. The second type of compare and test instruction also uses three addresses, all of which are register addresses. Considering the same example, if the contents of the first two registers are equal, a 1 is placed in the third register. If the contents are not equal, then a 0 is placed in the third register. These two types of instruction avoid the use of stored status bits. In the first case, no such bit is required, and in the second case, a register is used to simulate the presence of a status bit. The third type of compare and test, with the most complex structure, has compare and test operations that set or reset stored status bits. Branch or jump instructions are then used to conditionally change the program sequence. This third type of compare and test instruction is the focus of discussion in the next subsection.

Conditional Branch Instructions

A conditional branch instruction is a branch instruction that may or may not cause a transfer of control, depending on the value of stored bits in the *PSR*. Each conditional branch instruction tests a different combination of status bits for a condition. If the condition is true, control is transferred to the effective address. If the condition is false, the program continues with the next instruction.

Table 9-8 gives a list of conditional branch instructions that depend directly on the bits in the *PSR*. In most cases, the instruction mnemonic is constructed with the letter B (for "branch") and a letter for the name of the status bit. The letter N (for "not") is included if the status bit is tested for a 0 condition. Thus, BC is a branch if carry = 1, and BNC is branch if carry = 0.

□ **TABLE 9-8**
Conditional Branch Instructions Relating to Status Bits in the PSR

Branch condition	Mnemonic	Test condition
Branch if zero	BZ	$Z = 1$
Branch if not zero	BNZ	$Z = 0$
Branch if carry	BC	$C = 1$
Branch if no carry	BNC	$C = 0$
Branch if minus	BN	$N = 1$
Branch if plus	BNN	$N = 0$
Branch if overflow	BV	$V = 1$
Branch if no overflow	BNV	$V = 0$

The zero status bit Z is used to check whether the result of an ALU operation is equal to zero. The carry bit C is used to check the carry after the addition or the borrow after the subtraction of two operands in the ALU. It is also used in conjunction with shift instructions to check the value of the outgoing bit. The sign bit N reflects the state of the leftmost bit of the output from the ALU. $N = 0$ denotes a positive sign and $N = 1$ a negative sign. These instructions can be used to check the value of the leftmost bit, whether it represents a sign or not. The overflow bit V is used in conjunction with arithmetic operations with signed numbers.

As stated previously, the compare instruction performs a subtraction of two operands, say, $A - B$. The result of the operation is not transferred into a destination register, but the status bits are affected. The status bits provide information about the relative magnitude between A and B. Some computers provide special branch instructions that can be applied after the execution of a compare instruction. The specific conditions to be tested depend on whether the two numbers are considered to be unsigned or signed.

The relative magnitude between two unsigned binary numbers A and B can be determined by subtracting $A - B$ and checking the C and Z status bits. Most commercial computers consider the C status bit as a carry after addition and a borrow after subtraction. A borrow occurs when $A < B$ because the most significant position must borrow a bit to complete the subtraction. A borrow does not occur if $A \geq B$, because the difference $A - B$ is positive. The condition for borrowing is the inverse of the condition for carrying when the subtraction is done by taking the 2's complement of B. Computers that use the C status bit as a borrow after a subtraction complement the output carry after adding the 2's complement of the subtrahend and call this bit a borrow. The technique is typically applied to all instructions that use subtraction within the functional unit, not just the subtract instruction. For example, it applies to compare instructions.

The conditional branch instructions for unsigned numbers are listed in Table 9-9. It is assumed that a previous instruction has updated status bits C and Z after a subtraction $A-B$ or some other similar instruction. The words "higher," "lower," and "equal" are used to denote the relative magnitude between two unsigned num-

☐ TABLE 9-9
Conditional Branch Instructions for Unsigned Numbers

Branch condition	Mnemonic	Condition	Status bits*
Branch if higher	BH	$A > B$	$C + Z = 0$
Branch if higher or equal	BHE	$A \geq B$	$C = 0$
Branch if lower	BL	$A < B$	$C = 1$
Branch if lower or equal	BLE	$A \leq B$	$C + Z = 1$
Branch if equal	BE	$A = B$	$Z = 1$
Branch if not equal	BNE	$A \neq B$	$Z = 0$

*Note that C here is a borrow bit.

bers. The two numbers are equal if $A = B$. This is determined from the zero status bit Z, which is equal to 1 because $A - B = 0$. A is lower than B and the borrow $C = 1$ when $A < B$. For A to be lower than or equal to B $(A \leq B)$, we must have $C = 1$ or $Z = 1$. The relationship $A > B$, is the inverse of $A \leq B$ and is detected from the comp1emented condition of the status bits. Similarly, $A \geq B$ is the inverse of $A < B$ and $A \neq B$, is the inverse of $A = B$.

The conditional branch instructions for signed numbers are listed in Table 9-10. Again, it is assumed that a previous instruction has updated the status bits N, V, and Z after a subtraction $A - B$. The words "greater," "less," and "equal" are used to denote the relative magnitude between two signed numbers. If $N = 0$, the sign of the difference is positive, and A must be greater than or equal to B, provided that $V = 0$, indicating that no overflow occurred. An overflow causes a sign reversal, as discussed in Section 3-10. This means that if $N = 1$ and $V = 1$, there was a sign reversal, and the result should have been positive, which makes A greater than or equal to B. Therefore, the condition $A \geq B$ is true if both N and V are equal to 0 or both are equal to 1. This is the complement of the exclusive-OR operation.

☐ TABLE 9-10
Conditional Branch Instructions for Signed Numbers

Branch condition	Mnemonic	Condition	Status bits
Branch if greater	BG	$A > B$	$(N \oplus V) + Z = 0$
Branch if greater or equal	BGE	$A \geq B$	$N \oplus V = 0$
Branch if less	BL	$A < B$	$N \oplus V = 1$
Branch if less or equal	BLE	$A \leq B$	$(N \oplus V) + Z = 1$

For A to be greater than but not equal to B $(A > B)$, the result must be positive and nonzero. Since a zero result gives a positive sign, we must ensure that the Z bit is 0 to exclude the possibility that $A = B$. Note that the condition $(N \oplus V) + Z = 0$ means that both the exclusive-OR operation and the Z bit must be equal to 0. The other two conditions in the table can be derived in a similar manner. The conditions BE (branch on equal) and BNE (branch on not equal) given for unsigned numbers apply to signed numbers as well and can be determined from $Z = 1$ and $Z = 0$, respectively.

Procedure Call and Return Instructions

A *procedure* is a self-contained sequence of instructions that performs a given computational task. During the execution of a program, a procedure may be called to perform its function many times at various points in the program. Each time the procedure is called, a branch is made to the beginning of the procedure to start executing its set of instructions. After the procedure has been executed, a branch is made again to return to the main program. A procedure is also called a *subroutine*.

The instruction that transfers control to a procedure is known by different names, including a call procedure, call subroutine, jump to subroutine, branch to subroutine, and branch and link. We will refer to the routine containing the procedure call as the calling procedure. The call procedure instruction has a one address field and performs two operations. First, it stores the value of the *PC*, which is the address following the call procedure instruction, in a temporary location. This address is called the *return address*, and the corresponding instruction is the *continuation point* in the calling procedure. Second, the address specified in the call procedure instruction—the address of the first instruction in the procedure—is loaded into the *PC*. When the next instruction is fetched, it comes from the called procedure.

The final instruction in every procedure must be a return to the calling procedure. The return instruction takes the address that was stored by the call procedure instruction and places it in the *PC*. This results in a transfer of program execution back to the continuation point in the calling procedure.

Different computers use different temporary locations for storing the return address. Some computers store it in a fixed location in memory, some store it in a processor register, and some store it in a memory stack. The advantage of using a stack for the return address is that, when a succession of procedures are called, the sequential return address can be pushed onto the stack. The return instruction causes the stack to pop, and the contents of the top of the stack are then transferred to the *PC*. In this way, a return is always to the program that last called the procedure. A procedure call instruction using a stack is implemented with the following microoperations:

$SP \leftarrow SP - 1$	Decrement stack pointer
$M[SP] \leftarrow PC$	Store return address on stack
$PC \leftarrow$ Effective address	Transfer control to procedure

The return instruction is implemented by popping the stack and transferring the return address to the *PC*.

$PC \leftarrow M[SP]$	Transfer return address to PC
$SP \leftarrow SP + 1$	Increment stack pointer

By using a procedure stack, all return addresses are automatically stored by the hardware in the memory stack. Thus, the programmer does not have to be concerned about managing the return addresses for procedures called from within procedures.

9-9 Program Interrupt

A program interrupt is used to handle a variety of situations that require a departure from the normal program sequence. A program interrupt transfers control from a program that is currently running to another service program as a result of an externally- or internally-generated request. Control returns to the original program after the service program is executed. In principle, the interrupt procedure is similar to a call procedure, except in three respects:

1. The interrupt is usually initiated at an unpredicatble point in the program by an external or internal signal, rather than the execution of an instruction.
2. The address of the service program that processes the interrupt request is determined by a hardware procedure, rather than the address field of an instruction.
3. In response to an interrupt, it is necessary to store information that defines all or part of the contents of the register set, rather than storing only the program counter.

After the computer has been interrupted and the appropriate service program executed, the computer must return to exactly the same state that it was in before the interrupt occurred. Only if this happens will the interrupted program be able to resume exactly as if nothing happened. The state of the computer at the end of an execution of an instruction is determined from the contents of the register set. In addition to containing the condition codes, the *PSR* can specify what interrupts are allowed to occur and whether the computer is operating in user or system mode. Most computers have a resident operating system that controls and supervises all other programs. When the computer is executing a program that is part of the operating system, the computer is placed in system mode, in which certain instructions are privileged and can be executed in the system mode only. The computer is in user mode when it executes user programs, in which case it cannot execute the privileged instructions. The mode of the computer at any given time is determined from a special status bit or bits in the *PSR*.

Some computers store only the program counter when responding to an interrupt. In such computers, the program that performs the data processing for servicing the interrupt must include instructions to store the essential contents of the register set. Other computers store the entire register set automatically in response to an interrupt. Some computers have two sets of processor registers, so that when the program switches from user to system mode in response to an interrupt, it is not necessary to store the contents of processor registers because each computer mode employs its own set of registers.

The hardware procedure for processing interrupts is very similar to the execution of a procedure call instruction. The contents of the register set of the processor are temporarily stored in memory, typically by being pushed onto a memory stack, and the address of the first instruction of the interrupt service program is loaded into the *PC*. The address of the service program is chosen by the hardware. Some computers assign one memory location for the beginning address of the service program: the service program must then determine the source of the interrupt and proceed to service it. Other computers assign a separate memory location for each possible interrupt source. Sometimes, the interrupt source hardware itself supplies the address of the service routine. In any case, the computer must possess some form of hardware procedure for selecting a branch address for servicing the interrupt.

Most computers will not respond to an interrupt until the instruction that is in the process of being executed is completed. Then, just before going to fetch the next instruction, the control checks for any interrupt signals. If an interrupt has occurred, control goes to a hardware interrupt cycle. During this cycle, the contents of some part or all of the register set are pushed onto the stack. The branch address for the particular interrupt is then transferred to the *PC*, and the control goes to fetch the next instruction, which is the beginning of the interrupt service routine. The last instruction in the service routine is a return from the interrupt instruction. When this return is executed, the stack is popped to retrieve the return address, which is transferred to the *PC* as well as any stored contents of the rest of the register set, which are transferred back to the appropriate registers.

Types of Interrupts

The three major types of interrupts that cause a break in the normal execution of a program are as follows:

1. External interrupts.
2. Internal interrupts.
3. Software interrupts.

External interrupts come from input or output devices, from timing devices, from a circuit monitoring the power supply, or from any other external source. Conditions that cause external interrupts are an input or output device requesting a transfer of data, an external device completing a transfer of data, the time-out of an event, or an impending power failure. A time-out interrupt may result from a program that is in an endless loop and thus exceeds its time allocation. A power failure interrupt may have as its service program a few instructions that transfer the complete contents of the register set of the processor into a nondestructive memory such as a disk in the few milliseconds before power ceases.

Internal interrupts arise from the illegal or erroneous use of an instruction or data. Internal interrupts are also called *traps*. Examples of interrupts caused by internal conditions are an arithmetic overflow, an attempt to divide by zero, an invalid opcode, a memory stack overflow, and a protection violation. A *protection violation* is an attempt to address an area of memory that is not supposed to be accessed by the currently executing program. The service programs that process internal interrupts determine the corrective measure to be taken in each case.

External and internal interrupts are initiated by the hardware of the computer. By contrast, a *software interrupt* is initiated by executing an instruction. The software interrupt is a special call instruction that behaves like an interrupt rather than a procedure call. It can be used by the programmer to initiate an interrupt procedure at any desired point in the program. Typical use of the software interrupt is associated with a system call instruction. This instruction provides a means for switching from user mode to system mode. Certain operations in the computer may be performed by the operating system only in system mode. For example, a complex input or output procedure is done in system mode. In contrast, a program written by a user must run in user mode. When an input or output transfer is required, the user program causes a software interrupt, which stores the contents of the *PSR* (with the mode bit set to "user"), loads new *PSR* contents (with the mode bit set to "system"), and initiates the execution of a system program. The calling program must pass information to the operating system in order to specify the particular task that is being requested.

An alternative term for an interrupt is an *exception*, which may apply only to internal interrupts or to all interrupts, depending on the particular computer manufacturer. As an illustration of the use of the two terms, what one programmer calls interrupt-handling routines may be referred to as exception-handling routines by another programmer.

Processing External Interrupts

External interrupts may have single or multiple interrupt input lines. If there are more interrupt sources than there are interrupt inputs in the computer, two or more sources are ORed to form a common line. An interrupt signal may originate at any time during program execution. To ensure that no information is lost, the computer usually acknowledges the interrupt only after the execution of the current instruction is completed and only if the state of the processor warrants it.

Figure 9-7 shows a simplified external interrupt configuration. Four external interrupt sources are ORed to form a single interrupt input signal. Within the CPU is an enable-interrupt flip-flop (*EI*) that can be set or reset with two program instructions: enable interrupt (ENI) and disable interrupt (DSI). When *EI* is 0, the interrupt signal is neglected. When *EI* is 1 and the CPU is at the end of executing an instruction, the computer acknowledges the interrupt by enabling the interrupt acknowledge output *INTACK*. The interrupt source responds to *INTACK* by providing an interrupt vector address *IVAD* to the CPU. The program-controlled *EI* flip-flop allows the programmer to decide whether to use the interrupt facility. If a DSI instruction to reset *EI* has been inserted in the program, it means that the programmer does not want the program to be interrupted. The execution of an ENI

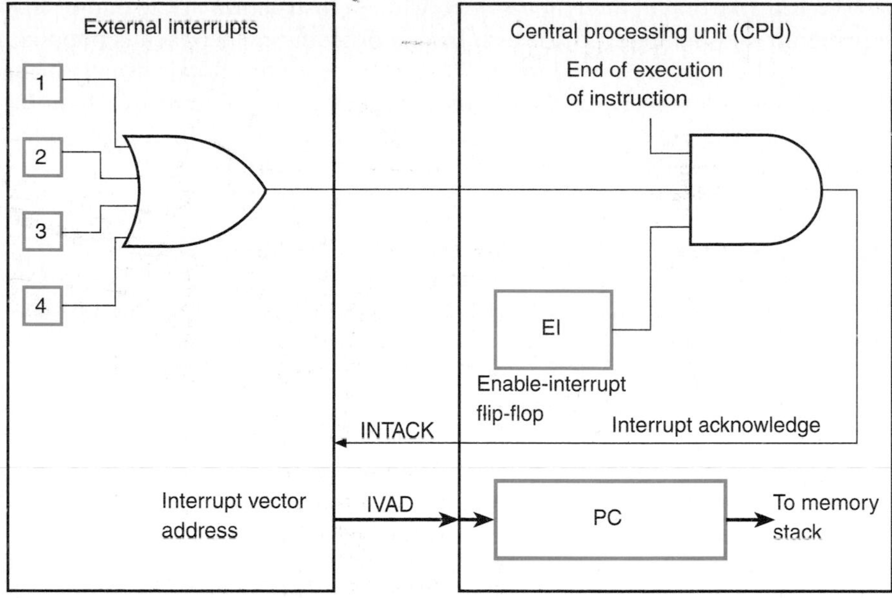

□ **FIGURE 9-7**
Example of External Interrupt Configuration

instruction to set *EI* indicates that the interrupt facility will be active while the program is running.

The computer responds to an interrupt request signal if $EI = 1$ and execution of the present instruction is completed. Typical microinstructions that implement the interrupt are as follows:

$SP \leftarrow SP - 1$	Decrement stack pointer
$M[SP] \leftarrow PC$	Store return address on stack
$SP \leftarrow SP - 1$	Decrement stack pointer
$M[SP] \leftarrow PSR$	Store processor status word on stack
$EI \leftarrow 0$	Reset enable-interrupt flip-flop
$INTACK \leftarrow 1$	Enable interrupt acknowledge
$PC \leftarrow IVAD$	Transfer interrupt vector address to PC
	Go to fetch phase.

The return address available in the *PC* is pushed onto the stack, and the *PSR* contents are pushed onto the stack. *EI* is reset to disable further interrupts. The program that services the interrupt can set *EI* with an instruction whenever it is appropriate to enable other interrupts. The CPU assumes that the external source will provide an *IVAD* in response to an *INTACK*. The *IVAD* is taken as the address of the first instruction of the program that services the interrupt. Obviously, a program must be written for that purpose and stored in memory.

The return from an interrupt is done with an instruction at the end of the service program that is similar to a return from a procedure. The stack is popped, and the return address is transferred to the *PC*. Since the *EI* flip-flop is usually included in the *PSR*, the value of *EI* for the original program is returned to *EI* when the old value of the *PSR* is returned. Thus, the interrupt system is enabled or disabled for the original program, as it was before the interrupt occurred.

9-10 Chapter Summary

In this chapter, we defined the concepts of instruction set architecture and the components of an instruction and explored the effects on programs of the maximum address count per instruction, using both memory addresses and register addresses. This led to the definitions of four types of addressing architecture: memory-to-memory, register to register, single-accumulator, and stack. Addressing modes specify how the information in an instruction is interpreted in determining the effective address of an operand.

Reduced instruction set computers (RISCs) and complex instruction set computers (CISCs) are two broad categories of instruction set architecture. A RISC has as its goals high throughput and fast execution of instructions. In contrast, a CISC attempts to closely match the operations used in programming languages and facilitates compact programs.

Three categories of elementary instructions are data transfer, data manipulation, and program control. In elaborating data transfer instructions, the concept of the memory stack appears. Transfers between the CPU and I/O are addressed by two different methods: independent I/O, with a separate address space, and memory-mapped I/O, which uses part of the memory address space. Data manipulation instructions fall into three classes: arithmetic, logical, and shift. Floating-point formats and operations handle broader ranges of operand values for arithmetic operations.

Program control instructions include basic unconditional and conditional transfers of control, the latter of which may or may not use condition codes. Procedure calls and returns permit programs to be broken up into procedures that perform useful tasks. Interruption of the normal sequence of program execution is based on three types of interrupts: external, internal, and software. Also referred to as exceptions, interrupts require special processing actions upon the initiation of routines to service them and upon returns to execution of the interrupted programs.

References

1. Mano, M. M. *Computer Engineering: Hardware Design.* Englewood Cliffs, NJ: Prentice Hall, 1988.
2. Goodman, J., and Miller, K. *A Programmer's View of Computer Architecture.* Fort Worth, TX: Saunders College Publishing, 1993.

3. Hamacher, V. C., Vranesic, Z. G., and Zaky, S. G. *Computer Organization,* 3rd ed. New York, NY: McGraw-Hill, 1990.
4. Hennessy, J. L., and Patterson, D. A. *Computer Architecture: A Quantitative Approach.* San Francisco, CA: Morgan Kaufmann, 1996.
5. Mano, M. M. *Computer System Architecture,* 3rd Ed. Englewood Cliffs, NJ: Prentice Hall, 1993.
6. Patterson, D. A., and Hennessy, J. L. *Computer Organization and Design: The Hardware/Software Interface.* San Mateo, CA: Morgan Kaufmann, 1994.
7. *IEEE Standard for Microprocessor Assembly Language.* (IEEE Std 694-1985.) New York, NY: The Institute of Electrical and Electronics Engineers.
8. *IEEE Standard for Binary Floating-Point Arithmetic.* (ANSI/IEEE Std 754-1985.) New York, NY: The Institute of Electrical and Electronics Engineers.
9. Kane, G., and Heinrich, J. *MIPS RISC Architecture.* Englewood Cliffs, NJ: Prentice Hall, 1992.
10. SPARC International. *The SPARC Architecture Manual, Version 8.* Englewood Cliffs, NJ: Prentice Hall, 1992.
11. Weiss, S., and Smith, J. E. *POWER and PowerPC.* San Mateo, CA: Morgan Kaufmann, 1994.

Problems

The asterisk (*) indicates a more advanced problem.

9–1. Write a program to evaluate the arithmetic statement

$$X = (A \times (B + C)) \div (D \times E - F)$$

Assume that all operands are initially in memory.

(a) Assume a register-to-register architecture with three-address instructions.

(b) Assume a memory-to-memory architecture with two-address instructions.

(c) Assume a single-accumulator computer with one-address instructions.

9–2. A two-word instruction is stored in memory at an address designated by the symbol W. The address field of the instruction (stored at $W + 1$) is designated by the symbol Y. The operand used during the execution of the instruction is stored at an address symbolized by Z. An index register contains the value X. State how Z is calculated from the other addresses if the addressing mode of the instruction is (a) direct; (b) indirect; (c) relative; (d) indexed.

9–3. A two-word relative mode branch-type instruction is stored in memory at location 310 and 311 (decimal). The branch is made to an address equivalent to decimal 225. Let the address field of the instruction (stored at address 311) be designated by X.

(a) Determine the value of X in decimal.

(b) Determine the value of X in binary, using 16 bits. (Note that the number is negative and must be in 2's complement notation. Why?)

9–4. How many times does the control unit refer to memory when it fetches and executes a two-word indirect addressing-mode instruction if the instruction is **(a)** a computational type requiring an operand from memory; **(b)** a branch type.

9–5. What must be the address field of an indexed addressing-mode instruction to make it the same as a register indirect-mode instruction?

9–6. An instruction is stored at location 200 with its address field at location 201. The address field has the value 300. A processor register $R1$ contains the number 150. Evaluate the effective address if the addressing mode of the instruction is **(a)** direct; **(b)** immediate; **(c)** relative; **(d)** register indirect; **(e)** indexed with $R1$ as the index register.

9–7. A computer has a 32-bit word length, and all instructions are one word in length. The register file of the computer has 32 registers.

(a) For a format with no mode fields and three register addresses, what is the maximum number of opcodes possible?

(b) For a format with two register address fields, one memory field, and a maximum of 256 opcodes, what is the maximum number of memory address bits available?

9–8. A computer with a register file, but without PUSH and POP instructions, is to be used to implement a stack. The computer does have the following register indirect modes:

Register indirect + increment:

LD R2 R1	$R2 \leftarrow M[R1]$
	$R1 \leftarrow R1 + 1$
ST R2 R1	$M[R1] \leftarrow R2$
	$R1 \leftarrow R1 + 1$

Decrement + register indirect:

LD R2 R1	$R1 \leftarrow R1 - 1$
	$R2 \leftarrow M[R1]$
ST R2 R1	$R1 \leftarrow R1 - 1$
	$M[R1] \leftarrow R2$

Show how these instructions can be used to provide the equivalent of PUSH and POP by using the instructions and register $R6$ as the stack pointer.

9–9. A first-in, first-out (FIFO) memory is organized to store information in such a manner that the item that stored first is the first item retrieved. A write counter WC holds the address for writing into memory. A read counter RC holds the address for reading from memory. An available storage counter

ASC indicates the number of words stored in the FIFO. *ASC* is incremented for every word stored and decremented for every item retrieved. Illustrate how a FIFO memory operates with these three counters. Assume that the FIFO is initially empty and that all counters contain zero. Then execute a number of reads and writes, showing the new contents of each counter after the read or write occurs.

9–10. A complex instruction, push registers (PSHR), pushes the contents of all of the registers onto the stack. There are eight registers, $R0$ through $R7$, in the CPU. A corresponding instruction, POPR, pops the saved contents of the registers back from the stack into the registers.
(a) Write a register transfer description for the execution of PSHR.
(b) Write a register transfer description for the execution of POPR.

9–11. Write a program to evaluate the arithmetic statement given in Problem 9–1 using computer stack instructions.

9–12. Give a register transfer description for the execution of the complex instruction, move string (MOVS). This instruction has three fields, two of which are memory addresses. Field LGH specifies the length of the string of words to be moved. Address field SRCA specifies the beginning address of the string, and address field DSTA specifies the beginning address of the new location of the string. The string is stored in consecutive locations in memory. For example, the string is initially in memory locations SRCA through SRCA + LGH − 1.

9–13. A computer with an independent I/O system has the input and output instructions

IN ADRS

OUT ADRS

where ADRS is the address of an I/O register port. Give the equivalent instructions for a computer with memory-mapped I/O.

9–14. Assuming an 8-bit computer, for the multiple-precision addition of two 32-bit unsigned numbers,

2A D3 67 8B + 12 68 5C 7E

(a) Write a program to execute the addition, using add and add with carry instructions, and
(b) Execute the program for the given operands. Each byte is expressed as a 2-digit hexadecimal number.

9–15. Perform the logic AND, OR, and XOR with the two bytes 01101110 and 10110110.

9–16. Given the 16-bit value 1010 1100 1110 0011, what operation must be performed in order to
(a) set the last 8 bits to 1?
(b) clear the first 8 bits to 0?
(c) complement the middle 8 bits?

9–17. An 8-bit register contains the value 00110110, and the carry bit is equal to 1. Perform the eight shift operations given by the instructions listed in Table 9-5 as a sequence of operations on this register.

9–18. Show how the following two floating-point numbers are to be added to get a normalized result:

$$(-.12467 \times 10^{+4}) + (+.71340 \times 10^{-1})$$

9–19. A 36-bit floating-point number consists of 24 bits plus sign for the fraction and 10 bits plus sign for the exponent. What are the largest and smallest positive non-zero quantities for normalized numbers?

9–20. A 32-bit register holds a floating-point decimal number in BCD. The fraction occupies 20 bits, for four decimal digits and a sign. The exponent occupies 12 bits, for two decimal digits and a sign. The plus sign is represented by 0000 and the − sign by 1001. What are the largest and smallest non-zero positive quantities for normalized decimal numbers?

9–21. A 4-bit exponent uses an excess 7 number for the bias. List all biased binary exponents from +8 through −7.

9–22. The IEEE standard double-precision floating-point operand format consists of 64 bits. The sign occupies 1 bit, the exponent has 11 bits, and the fraction occupies 52 bits. The exponent bias is 1,023. There is an implied bit to the left of the binary point in the fraction. Infinity is represented with a biased exponent equal to 2,047 and a fraction of 0.

(a) Give the formula for evaluating normalized numbers.

(b) List a few biased exponents in binary, as is done in Table 9-6.

(c) Calculate the largest and smallest positive normalized numbers that can be accommodated.

9–23. Prove that if the equality $2^x = 10^y$ holds, then $y = 0.3x$. Using this relationship, calculate the largest and smallest normalized floating-point numbers in decimal that can be accommodated in the single-precision IEEE format.

9–24. It is necessary to branch to ADRS if the bit in the least significant position of the operand in a 16-bit register is equal to 1. Show how this can be done with the TEST (Table 9-7) and BNZ (Table 9-8) instructions.

9–25. Consider the two 8-bit numbers $A = 00110001$and $B = 10011100$.

(a) Give the decimal equivalent of each number, assuming that (1) they are unsigned and (2) they are signed.

(b) Add the two binary numbers and interpret the sum, assuming that the numbers are (1) unsigned and (2) signed.

(c) Determine the values of the C (carry), Z (zero), N (sign), and V (overflow) status bits after the addition.

(d) List the conditional branch instructions from Table 9-8 that will have a true condition.

9–26. The program in a computer compares two unsigned numbers A and B by performing a subtraction $A - B$ and updating the status bits.
Let $A = 00110101$ and $B = 11000100$.
(a) Evaluate the difference and interpret the binary result.
(b) Determine the values of status bits C (borrow) and Z (zero).
(c) List the conditional branch instructions from Table 9-9 that will have a true condition.

9–27. The program in a computer compares two signed numbers A and B by performing subtraction $A - B$ and updating the status bits.
Let $A = 00110101$ and $B = 11000100$.
(a) Evaluate the difference and interpret the binary result.
(b) Determine the value of status bits N (sign), Z (zero), and V (overflow).
(c) List the conditional branch instructions from Table 9-10 that will have a true condition.

9–28. The top of a memory stack contains 6210. The stack pointer SP contains 3920. A two-word procedure call instruction is located in memory at address 1500, followed by the address field of 4389 at location 1501. What are the contents of PC, SP, and the top of the stack:
(a) before the call instruction is fetched from memory.
(b) after the call instruction is executed.
(c) after the return from the procedure.

9–29. A computer has no stack, but instead uses register $R6$ as a link register; i.e., the computer stores the return address in $R6$.
(a) Show the register transfers for a branch and link instruction.
(b) Assuming that another branch and link is present in the procedure called, what action must be taken by software before the branch and link occurs?

9–30. What are the basic differences between a branch, a procedure call, and a program interrupt?

9–31. Give five examples of external interrupts and five examples of internal interrupts. What is the difference between a software interrupt and a procedure call?

9–32. A computer responds to an interrupt request signal by pushing onto the stack the contents of the PC and the current PSR. The computer then reads new PSR contents from memory from the location given by the interrupt vector address ($IVAD$). The first address of the service program is taken from memory at location $IVAD + 1$.
(a) List the sequence of microoperations implementing the interrupt.
(b) List the sequence of microoperations implementing the return from interrupt.

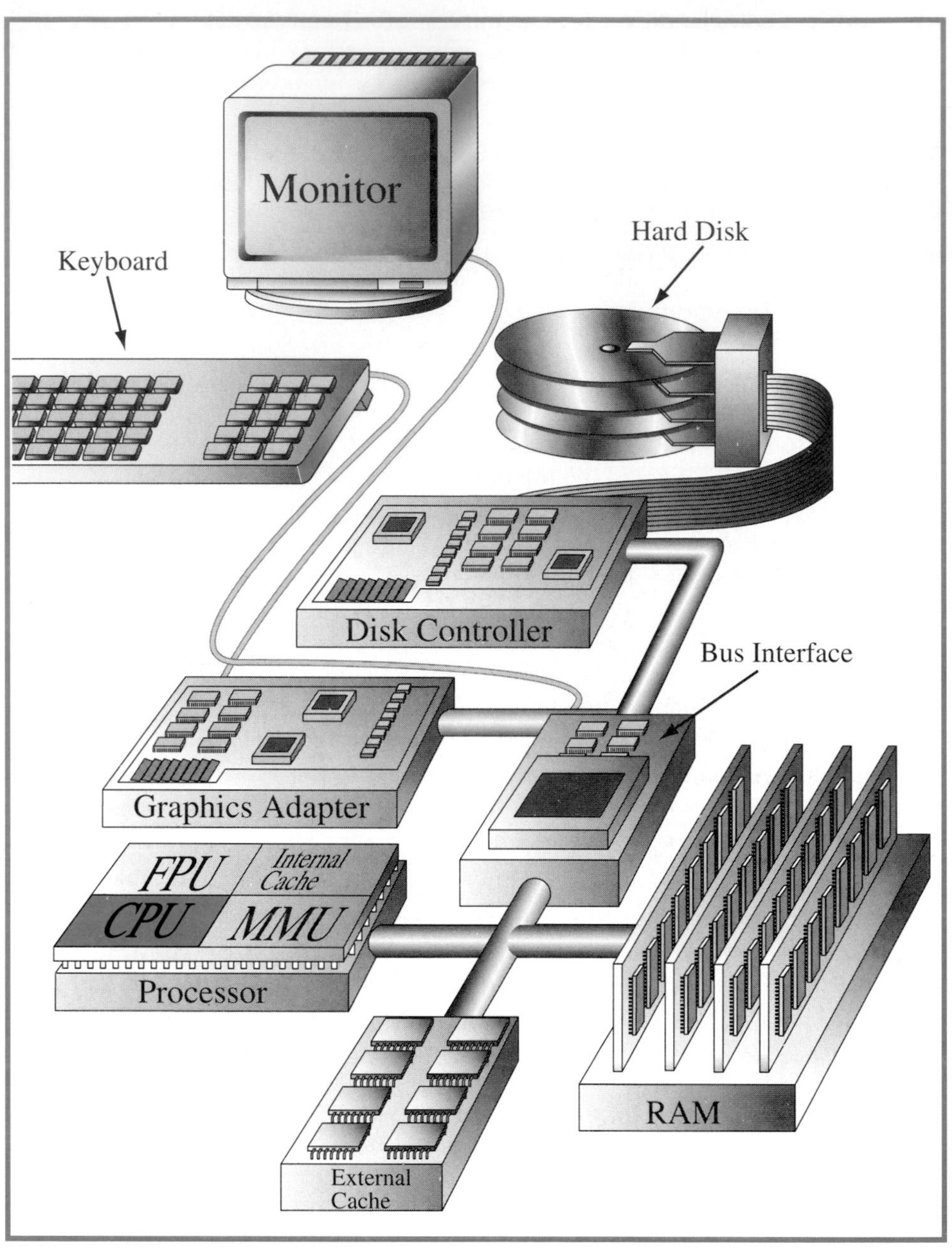

Monitor
Hard Disk
Keyboard
Disk Controller
Bus Interface
Graphics Adapter
FPU
Internal Cache
CPU
MMU
Processor
RAM
External Cache

CHAPTER

10

CENTRAL PROCESSING UNIT DESIGNS

The CPU is the key component of a digital computer. Its purpose is to decode instructions received from memory and perform transfer, arithmetic, logic, and control operations with data stored in internal registers, memory, or I/O interface units. Externally, the CPU provides one or more buses for transferring instructions, data, and control information to and from the components connected to it.

In the generic computer, the CPU is a part of the processor and is heavily shaded. CPUs, however, may also appear in computers. Small, relatively simple computers called microcontrollers are used in computers and in other digital systems to perform limited or specialized tasks. For example, a microcontroller is present in the keyboard and in the monitor in the generic computer; thus, these components are also shaded. In such microcontrollers, the CPU may be quite different from those discussed in this chapter. The word lengths may be short (say, four or eight bits), the number of registers small, and the instruction sets limited. Performance, relatively speaking, is poor, but adequate for the task. Most important, the cost of these microcontrollers is very low, making their use cost effective.

In the following pages, we consider two computer CPUs, one for a complex instruction set computer (CISC) and the other for a reduced instruction set computer (RISC). After a detailed examination of the designs, we compare the performance of the two CPUs and present a brief overview of some of the methods used to enhance that performance. Finally, we relate the design ideas discussed to general digital system design.

10-1 TWO CPU DESIGNS

As mentioned in previous chapters, a typical CPU is usually divided into two parts: the datapath and the control unit. The datapath consists of a function unit, registers, and internal buses that provide pathways for the transfer of information between the registers, the function unit, and other computer components. The datapath may or may not be pipelined. The control unit consists of a program counter, an instruction register, and control logic, and may be either hardwired or microprogrammed. If the datapath is pipelined, the control unit may also be a pipeline. The computer of which the CPU is a part is either a CISC or a RISC, each with its own instruction set architecture.

The purpose of this chapter is to present two CPU designs that illustrate combinations of architectural characteristics of the instruction set, the datapath, and the control unit. The designs will be top down, but with reuse of prior component designs, illustrating the influence of the instruction set architecture on the datapath and control units, and the influence of the datapath on the control unit. The material makes extensive use of tables and diagrams. Although we reuse and modify component designs from Chapters 7, 8, and 9, background information from these chapters is not repeated here. References, however, are given to earlier sections of the book, where detailed information can be found.

The two CPUs presented are for a CISC using a nonpipelined datapath with a microprogrammed control unit and a RISC using a pipelined datapath with a hardwired pipelined control unit. These represent two quite distinct combinations of instruction set architecture, datapath, and control unit.

10-2 The Complex Instruction Set Computer

The first design we present is for a complex instruction set computer with a non-pipelined datapath and microprogrammed control unit. We begin by describing the instruction set architecture, including the CPU register set, instruction formats, and addressing modes. The CISC nature of the instruction set architecture is demonstrated by its memory-to-memory access for data manipulation instructions, eight addressing modes, two instruction format lengths, and instructions that require significant sequences of operations for their execution.

We design a datapath for implementing the CISC architecture. This datapath is based on the one initially described in Section 7-9 and incorporated into a CPU in Section 8-7. Modifications are made to the register file, the function unit, and the buses to support the present instruction set architecture.

Once the datapath has been specified, a control unit is designed to complete the implementation of the instruction set architecture. The design of the control unit must involve a coordinated definition of both the hardware organization and the microprogram organization. In particular, dividing the microprogram into microroutines, while at the same time designing the sequencer with which they interact, is a key part of the design. Even the instruction fields and opcodes are tied to this coordinated effort. Following the definition of the hardware and microcode organizations, we detail essential parts of the microcode and the microroutines for representative operations.

Instruction Set Architecture

Figure 10-1 shows the CISC register set accessible to the programmer. All registers have 16 bits. The register file has eight registers, $R0$ through $R7$. $R0$ is a special reg-

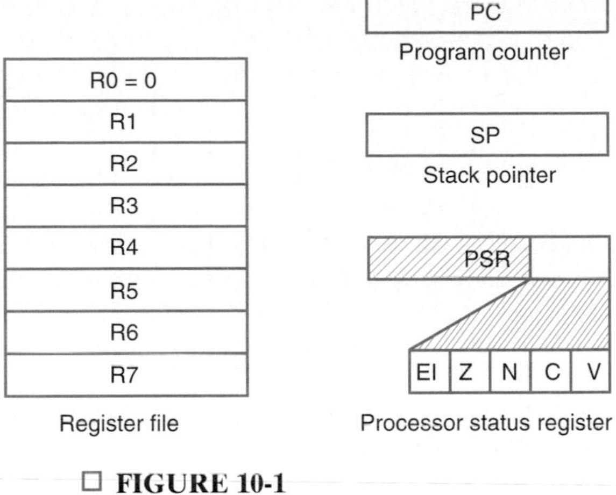

□ **FIGURE 10-1**
CPU Register Set Diagram for CISC

ister that always supplies the value zero when it is used as a source and discards the result when it is used as a destination.

In addition to the register file, there is a program counter *PC* and stack pointer *SP*. The presence of a stack pointer indicates that a memory stack is a part of the architecture. The final register is the processor status register *PSR*, which contains information only in its rightmost five bits; the remainder of the register is assumed to contain zero. The *PSR* contains the four stored status bit values *Z*, *N*, *C*, and *V* in positions 3 through 0, respectively. In addition, a stored interrupt enable bit *EI* appears in position 4.

Table 10-1 contains the 42 operations performed by the instructions. Each operation has a mnemonic and a carefully selected opcode. The operations are divided into four groups based on the number of explicit operands and whether the operation is a branch. In addition, the status bits affected by the operation are listed.

Figure 10-2 gives the instruction formats for the CPU. The generic instruction format has five fields. The first, OPCODE, specifies the operation. The next two, MODE and S, are used to determine the addresses of the operands. The last two fields, SRC and DST, are the 3-bit source register and destination register address fields, respectively. In addition, there is an optional second word W that appears with some instructions as an operand or an address, but not with others.

The first two bits of OPCODE, *IR*(15:14), determine the number of explicit operands and how the fields of the format are used. When these bits are 00, either no operand is required or the location of the operand is implied by OPCODE. Only the OPCODE field is needed, as shown in Figure 10-2(b). The four rightmost OPCODE bits can specify up to 16 operations without operands or with implied operand addresses.

If *IR*(15:14) is 01, the instruction has one operand and is a data transfer or data manipulation instruction. Since there is an operand, the MODE field specifies

TABLE 10-1
CISC Instruction Operations

Instruction	Mnemonic	Opcode	Status Effect
Zero-operand Instructions			
No operation	NOP	000000	None
Push registers	PSHR	000001	None
Pop registers	POPR	000010	None
Move string	MVS	000011	None
Return from procedure	RET	000100	None
Return from interrupt	RTI	000101	From stack
Invalid	000110 through 001111		
One-operand Instructions			
Push	PUSH	010000	None
Pop	POP	010001	None
Increment	INC	010010	ZCNV
Decrement	DEC	010011	ZCNV
Negate	NEG	010100	ZCNV
Complement	COM	010101	ZN
Logical shift right	SHR	010110	ZC
Logical shift left	SHL	010111	ZC
Arithmetic shift right	SHRA	011000	ZCNV
Arithmetic shift left	SHLA	011001	ZCNV
Rotate right	ROR	011010	ZC
Rotate left	ROL	011011	ZC
Rotate right with carry	RORC	011100	ZC
Rotate left with carry	ROLC	011101	ZC
Invalid	011110 through 011111		

Instruction	Mnemonic	Opcode	Status Effect
Two-operand Instructions			
Move	MOVE	100000	None
Exchange	XCH	100001	None
Add	ADD	100010	ZCNV
Add with carry	ADDC	101011	ZCNV
Subtract	SUB	100100	ZCNV
Subtract with borrow	SUBB	100101	ZCNV
Multiply	MUL	100110	ZCNV
Divide	DIV	100111	ZCNV
Compare	CMP	101000	ZCNV
AND	AND	101001	ZN
OR	OR	101010	ZN
Exclusive-OR	XOR	101011	ZN
Invalid	101100 through 101111		
Branch Instructions			
Jump	JMP	110000	None
Call procedure	CALL	110001	None
Branch if zero	BZ	111000	None
Branch if no zero	BNZ	111001	None
Branch if carry	BC	111010	None
Branch if no carry	BNC	111011	None
Branch if negative	BN	111100	None
Branch if no negative	BNN	111101	None
Branch if overflow	BV	111110	None
Branch if no overflow	BNV	111111	None
Invalid	110010 through 110111		

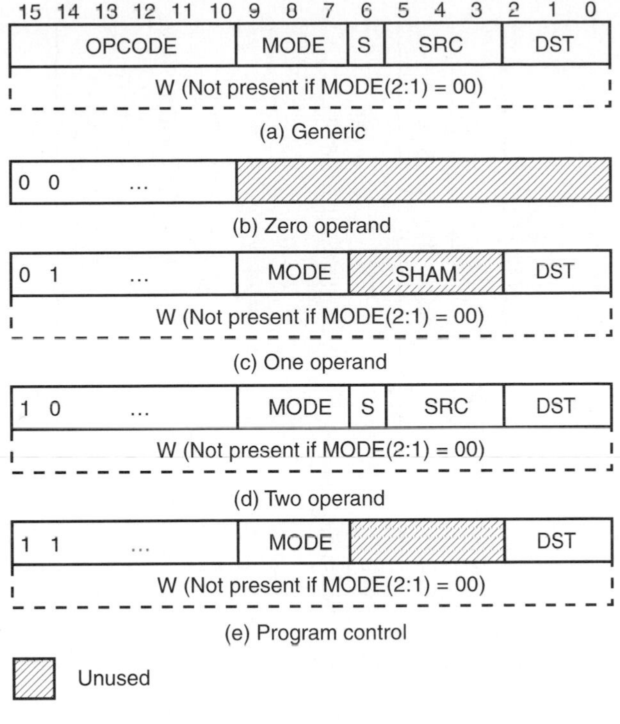

□ **FIGURE 10-2**
Instruction Formats

the addressing mode for obtaining it. The single address may involve the DST register address in its formation, so the DST field is also present. The S field and SRC field relate to the presence of two operands and so are not used. The only fields used appear in Figure 10-2(c). There are sufficient OPCODE bits for 16 instructions with a single operand. Table 10-2 gives the addressing modes specified by the MODE field. The first two bits of MODE specify four different types of addressing: register, immediate, indexed, and relative to the *PC*. The third bit of MODE specifies whether the address generated by these modes is used as an indirect address. The one exception to this is direct addressing, which is obtained by applying indirection to the immediate type. Otherwise, if the third bit equals 0, indirect addressing does not apply whereas, if it equals 1, indirect addressing does apply. For the register type of instruction, MODE(2:1) = 00 and the W word is not needed, since the operand or address comes from a register. For all other modes, the W word appears as the second word of the instruction. The third column of the table provides register transfer statements for each of the addressing modes for the one–operand instructions. Finally, in Figure 10-2, for one-operand shift instructions, normally unused bits contain a 4-bit *shift amount field* SHAM that specifies the number of bit positions to shift the operand.

TABLE 10-2
Addressing Modes

MODE	Address Mode	Register Transfer Description of Operands		
		IR(15:14) = 01 or 11	IR(15:14) = 10, S = 0	IR(15:14) = 10, S = 1
000	Register	$R[DST]$	$R[SRC], R[DST]$	$R[DST], R[SRC]$
001	Register Indirect	$M[R[DST]]$	$M[R[SRC]], R[DST]$	$M[R[DST]], R[SRC]$
010	Immediate	W	$W, R[DST]$	$W, R[SRC]$
011	Direct	$M[W]$	$M[W], R[DST]$	$M[W], R[SRC]$
100	Indexed	$M[R[DST] + W]$	$M[R[SRC] + W], R[DST]$	$M[R[DST] + W], R[SRC]$
101	Indexed Indirect	$M[M[R[DST] + W]]$	$M[M[\mathrm{R}[SRC] + W]], R[DST]$	$M[M[R[DST] + W]], R[SRC]$
110	Relative	$M[PC + W]$	$M[PC + W], R[DST]$	$M[PC + W], R[SRC]$
111	Relative Indirect	$M[M[PC + W]]$	$M[M[PC + W]], R[DST]$	$M[M[PC + W]], R[SRC]$

If *IR*(15:14) is equal to 10, then the instruction has two addresses. All fields of the generic instruction, including S and SRC, are used for this case. One of the addresses, either the source or the destination, uses the addressing modes. If S = 0, then the source uses the addressing mode specified by MODE, and the destination is a register. If S = 1, then the destination uses the addressing mode, and the source is a register. Register transfer descriptions of the resulting addresses are given in the fourth and fifth columns of Table 10-2. Again, depending on the contents of the MODE field, the second instruction word W, which is an address or an immediate operand, may or may not be present.

Instructions with *IR*(15:14) = 11 are branches. Aside from lack of the optional SHAM field for shifts, the format is the same as for *IR*(15:14) = 01. For all instructions of this type, the destination address (not the operand) becomes the new address placed in the program counter *PC*. As a consequence, the register mode is invalid for branch instructions.

Before proceeding to the next step, which defines the datapath to support the instruction set architecture, we will briefly note the characteristics of the architecture that define it as CISC or RISC. Most of the operations given in Chapter 9 are included in the instruction set. A number of operations that do not appear are redundant. The same actions can be achieved by using proper addressing modes with instructions that do appear. For example, LD, ST, IN, and OUT can all be achieved by using the MOVE instruction in a memory-mapped structure. By looking at the formats for the instructions, we find that most of the instructions can operate directly on operands from memory. There are eight addressing modes and two different lengths of instruction formats. In addition, some of the instructions perform complex operations which can be viewed as operations that are likely to take more than one clock cycle for the execution step. These characteristics clearly identify this as a CISC architecture.

Datapath Organization

Rather than beginning from scratch, we will reuse the nonpipelined datapath employed with the microprogrammed control in Section 8-8, with modifications. That datapath was shown in Figure 8-22, and the new, modified datapath based on it is given in Figure 10-6. We treat each modification in turn, beginning with the register file.

In Figure 8-22, register *R*8 was used as a temporary storage location. In the new microprogrammed architecture, there are complex instructions spanning many clock cycles and performing complicated operations. Thus, more temporary storage is needed for use by the microprograms. To meet this need, we expand the register file from 9 registers to 16. The first 8 registers, *R*0 through *R*7, are visible to the computer programmer. The second eight registers, *R*8 through *R*15, are used as temporary storage for microprogram operands and are hidden from the programmer. Figure 10-3 provides a map of the expanded register file with the temporary registers shaded. As indicated previously, register *R*0 supplies the constant 0, registers *R*1 through *R*7 are available to the programmer for use, and registers *R*8 through *R*15 provide general temporary storage for use by microprograms. The last

Address	Register
0	R0 = 0
1	R1
2	R2
3	R3
4	R4
5	R5
6	R6
7	R7
8	R8
9	R9
10	R10
11	R11
12	Source Address SA
13	Source Data SD
14	Destination Address DA
15	Destination Data DD

□ **FIGURE 10-3**
Register File Map

four registers, $R12$ through $R15$, have special uses: To keep the microcode simple, standard locations are essential for storing the operands and addresses used by execution microcode for most instructions. Thus, $R12$ is the location for the source address (SA), $R13$ for the source data (SD), $R14$ for the destination address (DA), and $R15$ for the destination data (DD).

We cannot access the eight temporary registers based on the 3-bit register addresses available in the instruction. To deal with this problem, we provide, first, 4-bit register addresses from the microinstruction, and second, a microinstruction bit to choose between these addresses and those from the instruction. In addition, the flexibility to allow the register addressed by DST to be a source and by SRC to be a destination is needed to permit results of operations to be placed directly in memory. To accomplish these goals, we modify the register file by adding the logic shown in Figure 10-4(a). The instruction set architecture uses two addresses, one for a source operand and the other for the other source as well as the destination. The register file uses the B address for a source, and the A and D addresses on the file are connected together, giving the same address for the other source and the destination. Although this reduction from three to two addresses is not essential at the microinstruction level, it decreases the number of bits needed for register addresses in the microinstruction and matches the use of the register fields in the instruction formats.

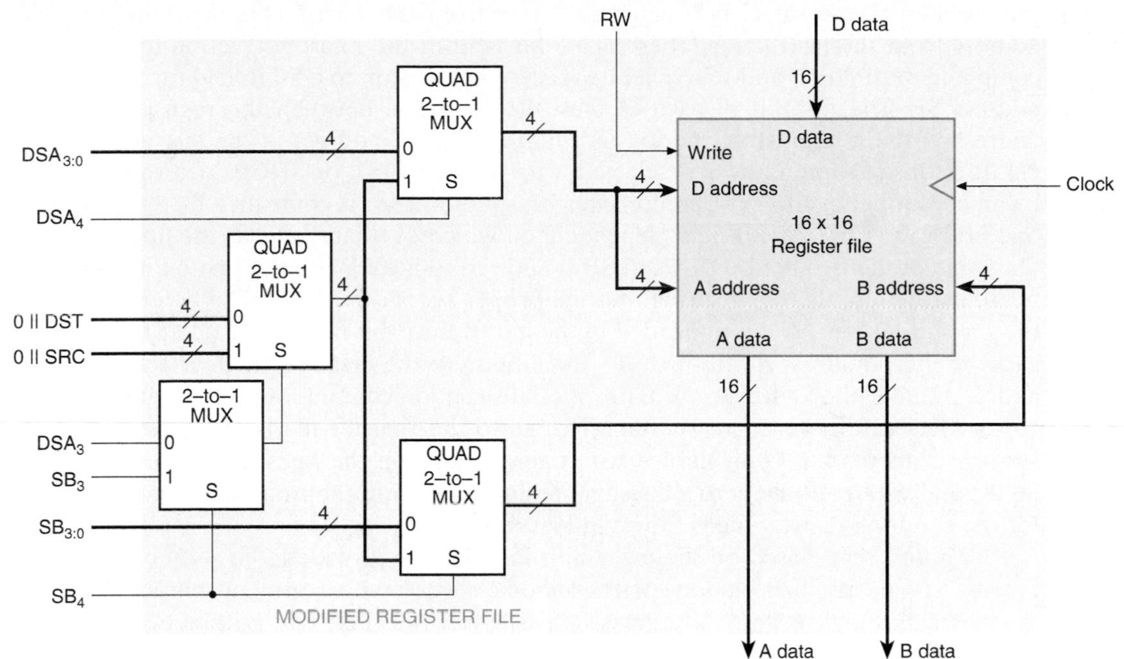

(a) Logic diagram

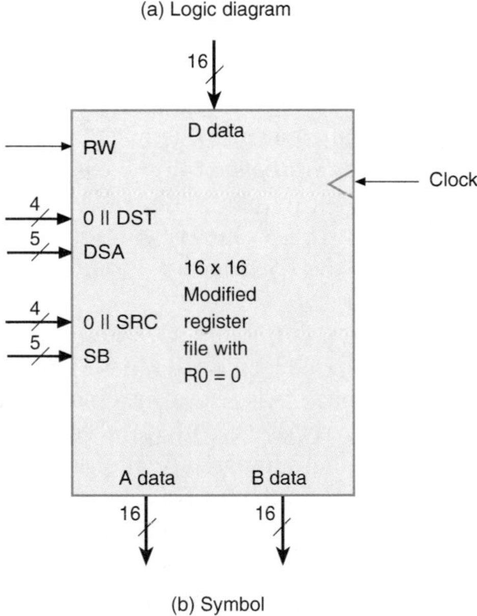

(b) Symbol

□ **FIGURE 10-4**
Modifications to Register File

A quad 2–to–1 multiplexer is attached to each of the two address inputs to the register file, to select between an address from the microinstruction and an address from the instruction. There is a 5-bit field in the microinstruction for the combined destination and source address DSA, in addition to a 5-bit field for the *B* address SB. The first bit of each of these fields selects between the register file address in the microinstruction (0) and the register file address in the instruction (1). If an instruction address is selected, whether it is DST or SRC is determined by an additional quad 2–to–1 multiplexer. This multiplexer is controlled by the second bit of the DSA or SB field, depending on which of them has 1 as the first bit. Only one of the two fields DSA and SB is allowed to have a 1 in the first bit in any microinstruction, therby ensuring that the proper second bit is used to determine the register address. A 0 is appended to the left of the 3-bit fields DST and SRC to cause them to address *R*0 through *R*7. In addition to the first bit, which selects the address source, the addresses from the microinstruction contain four bits so that all 16 registers can be reached. The final change to the register file is to replace the storage elements for *R*0 in the file with open circuits on the lines that were their inputs and with constant zero values on the lines that were their outputs. A symbol for the resulting register file is shown in Figure 10-4(b).

We find that, based on the eight shift instructions provided, the shifter from Figure 7-16 needs to be modified. The modifications involve the end bits of the shift logic. For logical shifts, a 0 is inserted, as before. For the right arithmetic shift, the sign bit is the incoming bit, and for the left arithmetic shift, 0 is the incoming bit. Rotates require that the bit from the opposite end of the shifter be fed around. Finally, rotates with carry require that the carry flip-flop output be provided as an input on both ends of the shifter.

The inputs are furnished by two 4–to–1 multiplexers, MUX *R* and MUX *L*, added to a basic 16-bit shifter, all shown in Figure 10-5(a). Also, the appropriate end bits from the input operand must be sent to the carry flip-flop. A 2–to–1 multiplexer MUX *SO* selects the end bit to pass to the carry flip-flop *C*. The symbol for the new shifter, which replaces the basic shifter from Figure 7-16, appears in Figure 10-5 (b). FS_2, FS_1, and FS_0 from the FS field drive the control inputs S_2, S_1 and S_0, respectively.

The instruction set requires the ability to store the contents of the *PC* in memory and load it back into the *PC*. Likewise, the contents of the *PSR* must be stored to memory and loaded back. Also, it is necessary to be able to store and load the *SP*. So, for these three registers, paths are established to and from the datapath buses. The paths into the datapath for the *PC* and *SP* are provided by inserting a 4–to–1 multiplexer MUX *A* into Bus *A*, as shown in Figure 10-6. The fourth input to MUX *A* comes from Data *B* of the register file. This input is needed to allow register-to-register transfers in the absence of a transfer *B* operation for the function unit. The path from *PSR* into the datapath is handled by changing MUX *B* to a 4–to–1 multiplexer. The *PSR* is placed on MUX *B*'s second input, the SHAM field from the *IR* is placed on its third input, and the zero-filled 4-bit constant from the microprogrammed control is placed on its fourth input. Paths are also added from Bus *D* to the *PSR*, *PC*, and *SP*.

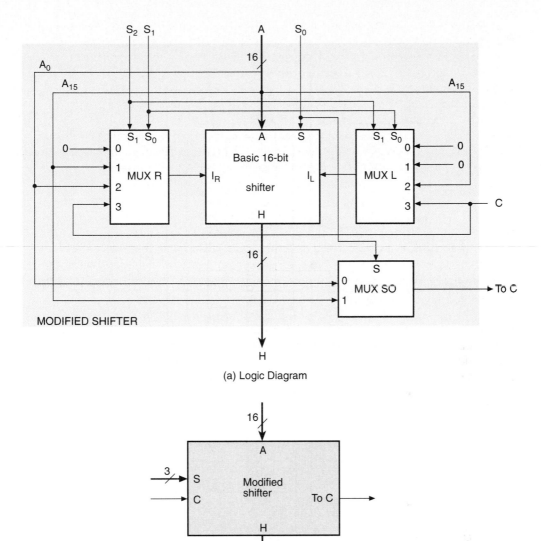

(a) Logic Diagram

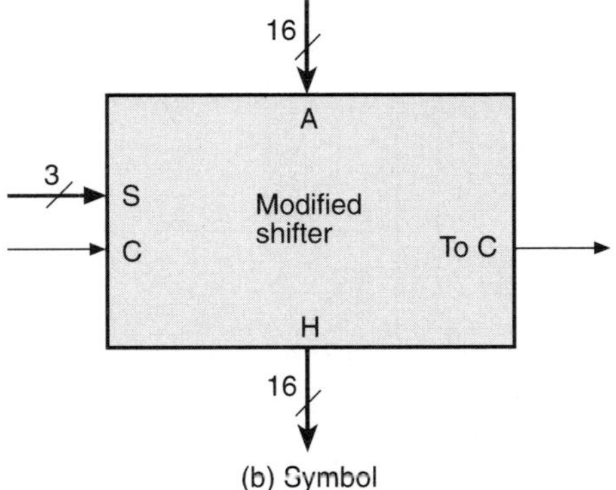

(b) Symbol

□ **FIGURE 10-5**
16-Bit Shifter and Symbol

The final area for modification of the datapath is the status bit hardware that lies at the boundary between the datapath and the control unit. Added there are five flip-flops storing program status bits *EI*, *Z*, *N*, *C*, and *V*, which did not appear in prior datapath designs and which make up the *PSR*. The values of these status bits are used to make decisions at the program level, as illustrated in Section 9-8. To distinguish the status values coming from the function unit from these stored val-

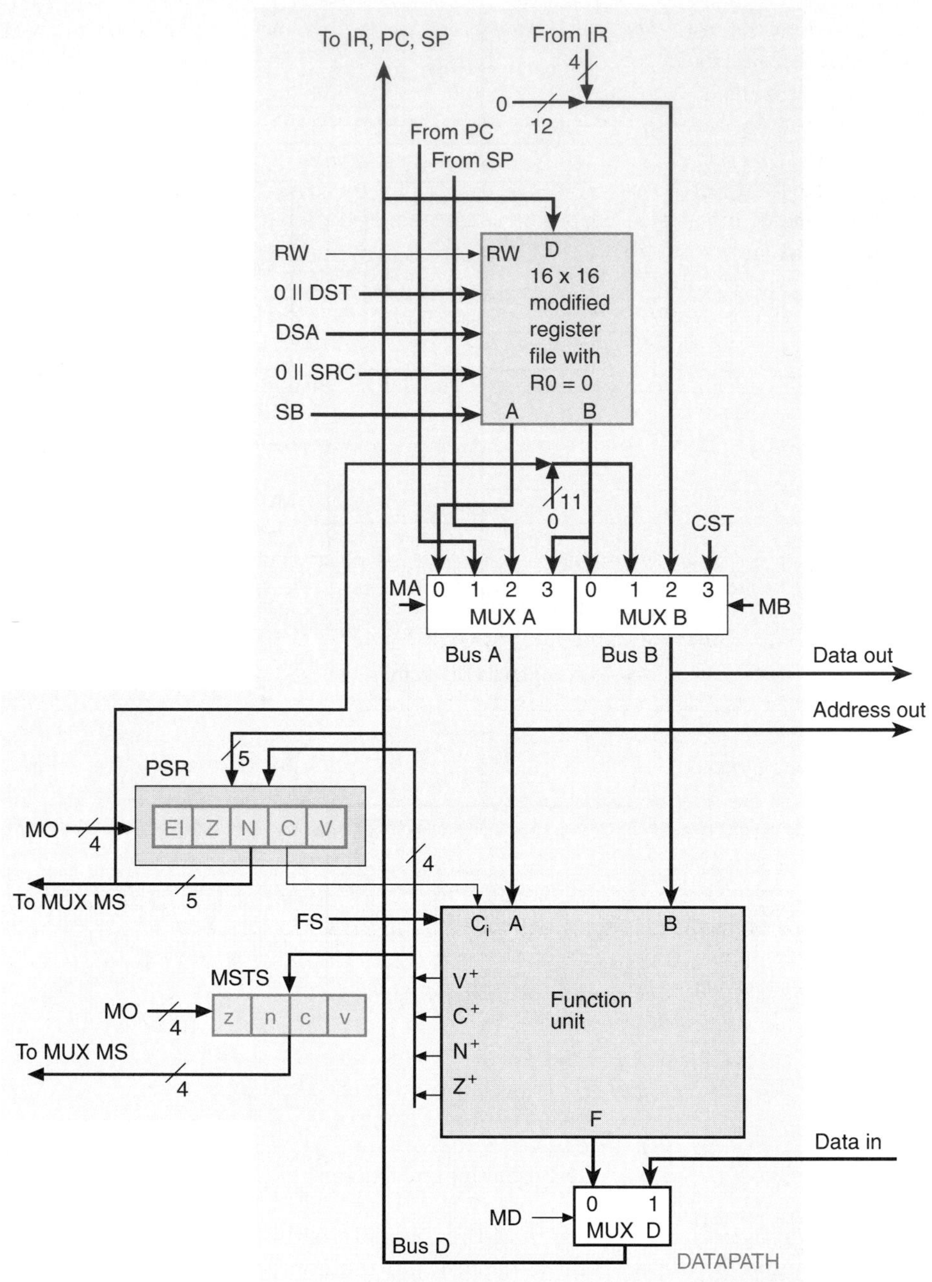

□ **FIGURE 10-6**
CISC Datapath

ues, the function unit outputs are labeled with a superscript plus sign (+). The logic to selectively control the loading of the status bits in accordance with Table 10-1 is represented by the box surrounding the bits. Also, we find it necessary to make stored status values available for use by the microprogram routines without disturbing the stored *PSR* values for the program level. Thus, a second flip-flop register is provided for temporary storage of the values Z^+, N^+, C^+, and V^+. The bits of this register are labeled z, n, c, and v, and they constitute the *microstatus register MSTS*. These additions, plus attachments to the buses, are shown in Figure 10-6. The four control bits labeled MO (miscellaneous operations) are decoded to control the sets of status bits to be enabled for loading.

All of the modifications to the original datapath are represented in Figure 10-6. As a part of the design process, the new datapath needs to be checked to make sure that it has all of the capabilities necessary for implementing the instruction set and addressing modes. Certainly, some decisions have been made that have not been discussed. For example, there is no dedicated multiplication or division hardware, so these operations must be implemented by microprograms controlling the datapath.

Microprogrammed Control Organization

The microprogrammed control unit accompanies the datapath of Figure 10-6 in Figure 10-7. The control consists of four principal parts. One is the control ROM, which has 256 words of 31 bits each. There are three control unit registers: the instruction register *IR*, the program counter *PC*, and the stack pointer *SP*. In some designs the *PC* and *SP* are logically included in the register file and thus are a part of the datapath. Here, since they are separate from the register file and are used primarily for program control, we have included them with the control. Sequencing within the control unit is provided by the microsequencer, which contains two registers: *the control address register CAR* and the *subroutine branch register SBR*. The program counter for the microprogram, the *CAR* simply counts up to the next address in sequence or loads in parallel. With a parallel load, the address can be set to any value and the next-address comes from three sources including the next-address field in the current microinstruction.

Microroutines have subroutines, just as programs do. To distinguish them, we call subroutines for microprograms *microsubroutines*. The *SBR* is used to store the next address for the *CAR* at the time a microsubroutine is entered. This return address is then retrieved at the end of the microsubroutine in order to return microprogram execution to the next microinstruction in the calling microroutine. The final part of the control unit is the instruction decoder, which consists of combinational logic and is also a next address source for the CAR.

The microinstructions stored in the control ROM use the two different formats shown in Figure 10-7 and detailed in Figure 10-8. The first microinstruction field MC selects the format used. If MC is 00, 01, or 10, format A applies. The microinstructions in this format perform data transfers and manipulation and decode instructions. They also handle returns from a microsubroutine. If MC is 11,

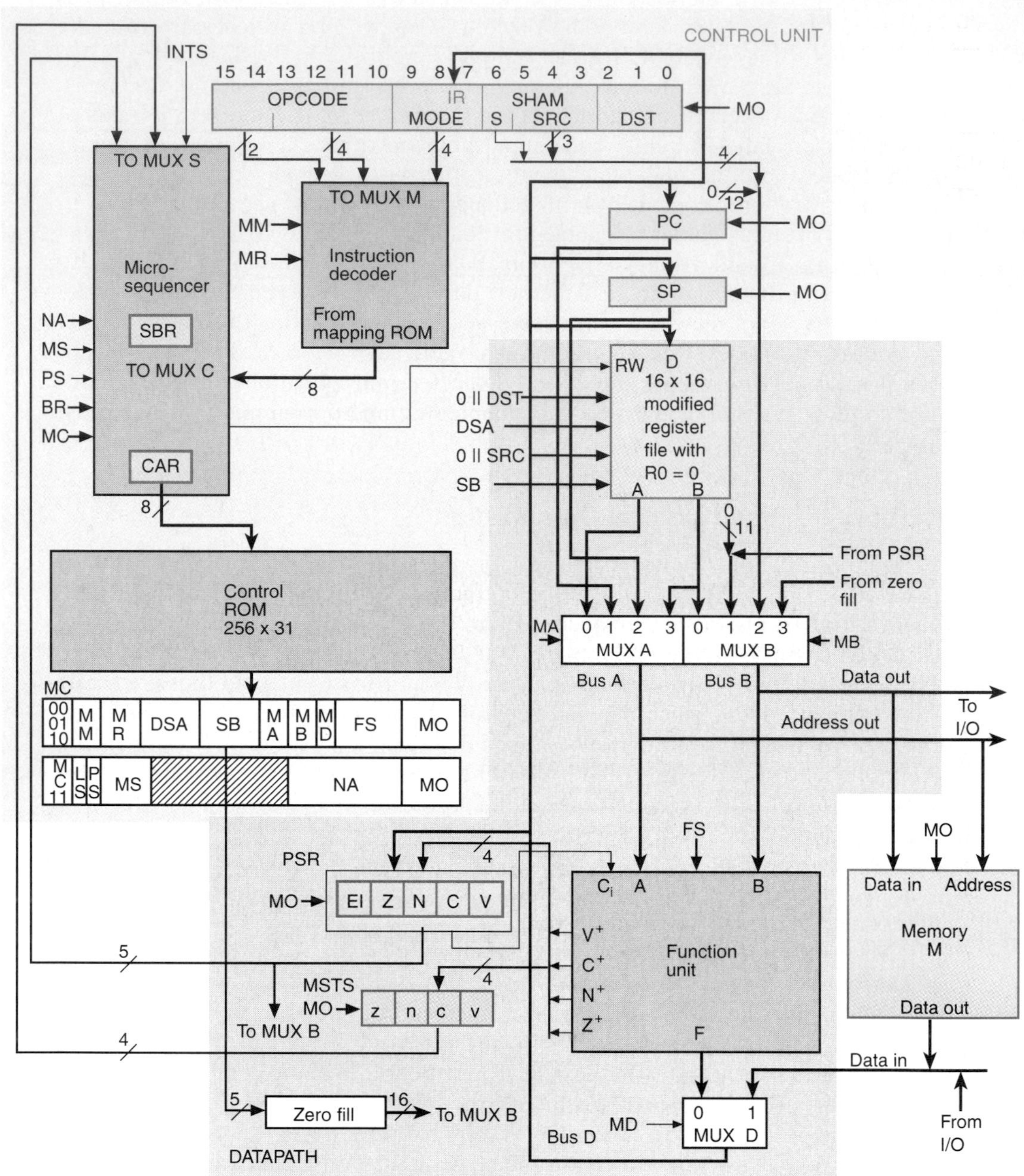

□ **FIGURE 10-7**
CISC CPU

30 29	28 27	26 25 24	23 22 21 20 19	18 17 16 15 14	13 12	11 10	9	8 7 6 5 4	3 2 1 0
MC	MM	MR	DSA	SB	MA	MB	MD	FS	MO

(a) Format A (MC = 00, 01, 11)

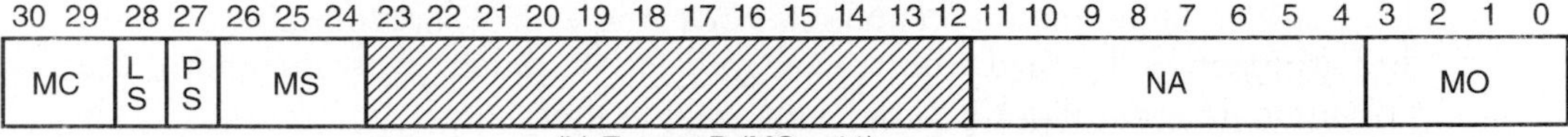

(b) Format B (MC = 11)

FIGURE 10-8
Microinstruction Formats

format B applies. Microinstructions in this format change the flow of the microprogram by implementing branches and a microsubroutine call.

In format A, the fields in bits 23 through 4, DSA through FS, control the datapath. The codes for these fields appear in Table 10-3. The actions of the DSA and SB fields have already been described in conjunction with the modified register file. The MA and MB fields control MUX *A* and MUX *B*, respectively, whose

TABLE 10-3
Control Word Encoding for Microinstruction Format A: Datapath Part

DSA, SB		MA	MB		MD		FS	
$R0 = 0$	00000	Register A	Register B	00	Function	0	$F = A$	00000
$R1$	00001	PC	PSR	01	Data in	1	$F = A + 1$	00001
$R2$	00010	SP	ICST	10			$F = A + B$	00010
$R3$	00011	Register B	MCST	11			$F = A + B + 1$	00011
$R4$	00100						$F = A + \overline{B}$	00100
$R5$	00101						$F = A + \overline{B} + 1$	00101
$R6$	00110						$F = A - 1$	00110
$R7$	00111						$F = A$	00111
$R8$	01000						$F = A \wedge B$	01000
$R9$	01001						$F = A \vee B$	01010
$R10$	01010						$F = A \oplus B$	01100
$R11$	01011						$F = \overline{A}$	01110
$R12$ (SA)	01100						$F = \text{lsl } A$	10000
$R13$ (SD)	01101						$F = \text{lsr } A$	10001
$R14$ (DA)	01110						$F = \text{asl } A$	10010
$R15$ (DD)	01111						$F = \text{asr } A$	10011
$R[DST]$	10XXX*						$F = \text{rol } A$	10100
$R[SRC]$	11XXX*						$F = \text{ror } A$	10101
							$F = \text{rolc } A$	10110
							$F = \text{rorc } A$	10111

*Only one of *DSA* and *SB* may contain either of these patterns in any microinstruction. The other must contain a pattern beginning with a 0.

selections were described in the discussion of the datapath. Here notation is added: MCST for a zero-filled constant coming from the SB field of a microinstruction and ICST for a zero-filled constant coming from the SHAM field of instruction. The MD field controls MUX *D*, as in previous designs. The two codes for shifts originally in the FS field have been replaced by the eight codes now required for the modified shifter. Shift operation notations beginning with "l" are logical shifts, with "a" are arithmetic shifts, and with "r" are rotates. An added "c" at the end of the shift notation indicates that the carry is included in the rotate.

All of the remaining fields have some relationship to the operation of the control unit. In order to discuss these fields, we need to examine the parts of the control unit. The heart of the control in Figure 10-7 is the microsequencer, which includes the *SBR,* the *CAR*, and the address determination logic. The microsequencer control fields are given in Table 10-4 and Table 10-5. Initially, we will discuss the microsequencer in terms of the operations performed: we then briefly look at the detailed logic.

Table 10-4 gives key information on the microsequencer operation, since MC specifies the source of the next address to appear in the *CAR*. For code 00, the contents of the *CAR* are incremented to point to the next microinstruction in sequence. For code 01, the next address comes from *SBR*, which holds the return address for a microsubroutine. For code 10, the next address is determined by the instruction decoder. The decoder is capable of executing an unconditional branch or a 4-way, 8-way, or 16-way conditional branch based on the values of the OPCODE or MODE bits. These multiple-way branches, to be detailed later, are faster than using a sequence of 2-way branches. For the first three MC codes, format A is used. Thus, these branches for decoding instructions can be performed simultaneously with data transfer and manipulation.

□ **TABLE 10-4**
Control Word Information for MC and LS

Action	Format	Symbolic Notation	Codes	
			MC	LS
Increment *CAR*	A	NXT	00	—
Return from subroutine	A	RET	01	—
Map instruction into *CAR*	A	MAP	10	—
Jump to *NA* if *ST* bit is satisfied; else increment *CAR*	B	BR	11	0
Call subroutine at *NA* if *ST* bit is satisfied; else increment *CAR*	B	CALL	11	1

For MC equal to 11, format B is used, and an unconditional or conditional branch is specified. If the branch is unconditional or if the condition is satisfied, then a jump to the address specified in the 8-bit next-address field NA occurs. If the condition is not satisfied, then the *CAR* is simply incremented. In addition, if the jump occurs and LS equals 1, the incremented version of the *CAR* is stored in *SBR*, the microsubroutine return register. Since there is only a single such register and no way to save its contents elsewhere, only a single level of microsubroutine calls is permitted. Also, for any format B microinstruction, the register file is not to be written.

The remaining fields for the microsequencer appear in Table 10-5. MS is used to select either an unconditional branch or the conditions upon which to branch, just as is done in Chapter 8. Here, however, there are many more conditions on which to branch. For MS equal to 0000, the branch or subroutine call is unconditional. The next five values of MS use the five bits of the *PSR—EI*, *Z*, *N*, *C*, and *V*—as conditions. The field PS determines whether the jump occurs when the value of the condition is 1 or 0. If PS is 0, then the jump occurs when the value of the condition is 1; if PS is 1, then the jump occurs when the value of the condition is 0. The next four values of MS cause branches on the microstatus bits *z*, *n*, *c*, and *v* of *MSTS*. The final value of MS branches on *INTS*, the interrupt status bit.

☐ TABLE 10-5
Control Word Information for Polarity Bit and Multiplexer S in Format B Microinstructions

PS			MS		
Action	**Symbolic Notation**	**Code**	**Condition Status Signal**	**Symbolic Notation**	**Code**
Pass status bit unchanged	TS	0	Constant 1 for unconditional transfer	BU	0000
Complement status bit	CS	1	Zero *PSR* bit	BZ	0001
			Negative *PSR* bit	BN	0010
			Carry *PSR* bit	BC	0011
			Overflow *PSR* bit	BV	0100
			Enable-interrupt *PSR* bit	EI	0101
			Zero *MSTS* bit	Bz	0110
			Negative *MSTS* bit	Bn	0111
			Carry *MSTS* bit	Bc	1000
			Overflow *MSTS* bit	Bv	1001
			Interrupt signal *INTS*	BI	1010

We now complete our discussion of the microsequencer by examining briefly the implementing hardware in Figure 10-9. The selection of the next microaddress to place in the *CAR* is performed by MUX *C*, which is controlled by MC. The inputs to MC correspond to the desired sources of the next addresses specified in Table 10-4. The selection of branch conditions given in Table 10-5 is accomplished by MUX *S*. Its output enters an exclusive-OR gate that complements the value of the condition based on the value of PS. The resulting exclusive-OR output signal *ST* drives the select input of MUX *R*, which selects between next address NA and the incremented *CAR* value, thereby implementing the conditional branch. To stop

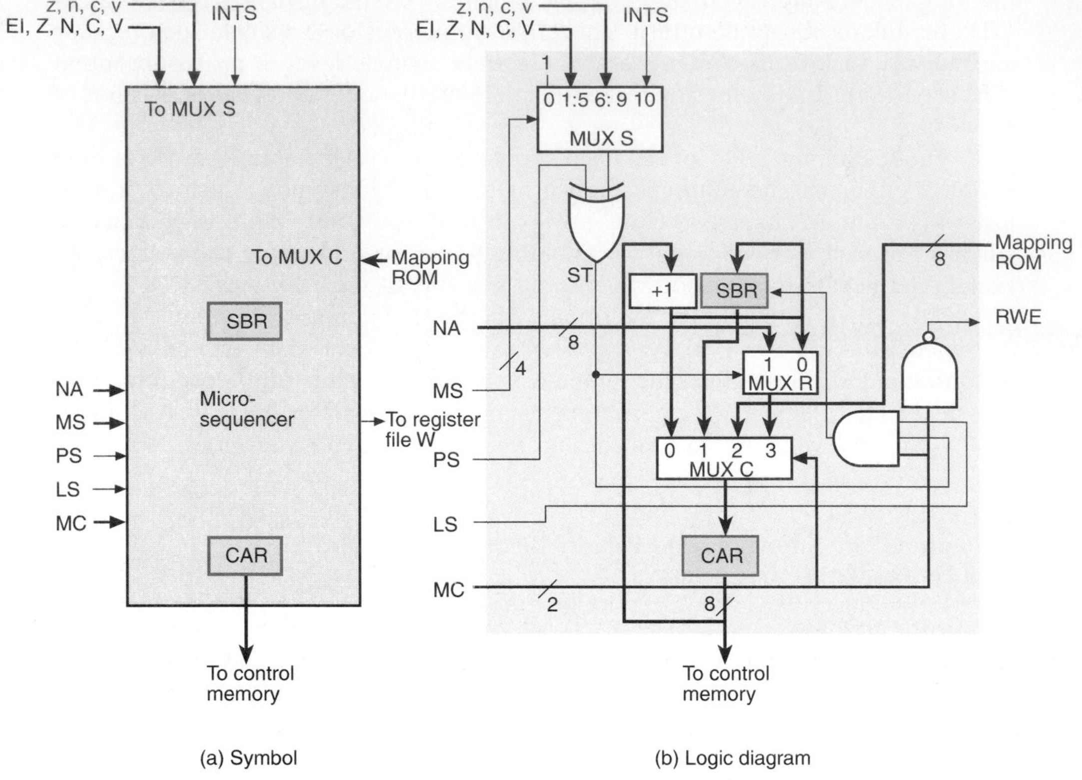

□ **FIGURE 10-9**
The Microsequencer

the register file from being written for format B (MC = 11) microinstructions, a two-input NAND gate produces *RWE* (register write enable) = 0 for MC = 11. By ANDing *RWE* with the signal otherwise driving *RW* on the register file, we cause MC = 11 to prevent the writing of the file. The loading of *SBR* for subroutine calls is accomplished by logic consisting of one additional AND gate. The inputs to the AND consist of the two MC bits, LS and ST. The AND output is 1 for MC = 11, LS = 1, and ST = 1. This causes *SBR* to be loaded with the incremented *CAR* value for a microsubroutine call instruction with the branch taken.

The instruction decoder produces control ROM addresses based on its control fields and fields of the instruction in the *IR*. By using control fields, different control ROM addresses can be produced for the same values of the instruction fields. This allows distinct microroutines to be executed in distinct parts of the execution of a given instruction. The control fields for the instruction decoder are given in Table 10-6. MM defines which field of the instruction is involved in determining the address provided. If MM equals 00, then the left two bits of OPCODE are used to determine the address; if MM equals 01, the right four bits of

□ **TABLE 10-6**
Control Word Encoding for Instruction Decoder

MM		MR	
Select	**Code**	**Region**	**Code**
OPCODE(5:4)	00	0	000
OPCODE(3:0)	01	1	001
MODE ∥ S	10	2	010
MODE ∥ S	11	3	011
		4	100
		5	101
		6	110
		7	111

OPCODE are used to determine the address. Since there are two bits of OPCODE that can take on arbitrary values for MM equal to 00, four different addresses can result. Likewise, for MM equal to 01, there are four bits of OPCODE that can assume arbitrary values, so 16 different addresses can result. Thus, using just five microinstructions, with execution of only two of them in sequence, it is possible to produce a 64-way branch giving a different address for each of the 64 possible OPCODE values. In contrast, if 2-way branches were used, 63 instructions with six microinstructions in sequence would be required to perform the same decoding to unique addresses.

To handle the MODE and S fields, MM equal to 10 or 11 uses the 4-bit field pair MODE ∥ S to provide up to 16 different addresses. The microprogram region field MR provides distinct sets of addresses for the same *IR* fields. For example, it may be necessary to have several distinct sets of addresses, with each set resulting from values given in the rightmost four bits of OPCODE. The MR field provides a means for the microinstruction to select these various sets. Since this field has three bits, up to eight distinct sets of addresses can be selected for each of the four binary values of MM.

The internal implementation of the instruction decoder, whose symbol is given in Figure 10-10(a) is made up of a quad 4–to–1 multiplexer MUX *M* and a ROM, as shown in Figure 10-10(b). The ROM has nine inputs and eight outputs. MUX *M* is controlled by MM, and the ROM has MM, MR, and the MUX *M* outputs as its inputs. The ROM maps its input values to outputs that are the addresses to which the microsequencer is to jump. Thus, the ROM is referred to as a *mapping ROM*. Because the addresses can be assigned arbitrarily, the location of the various microprogram routines can be determined by the designer and then implemented by the mapping ROM. The content of the ROM is tightly dependent, however, on the relationship between *IR* bit values and the microroutines, so its specification will be deferred until we consider the latter. Although in this case we have chosen a ROM for the decoding process, gate logic or a PLA could also be used.

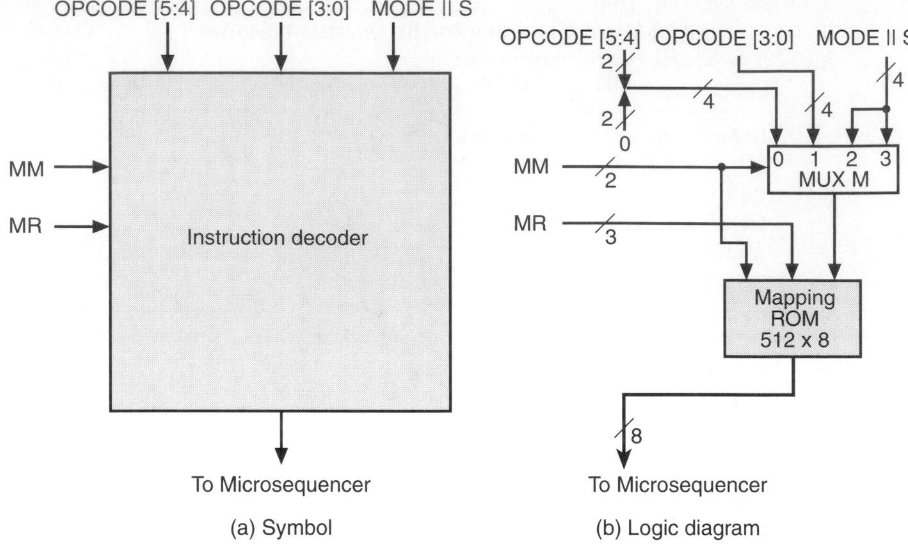

□ **FIGURE 10-10**
Instruction Decoder

The final field, MO, for miscellaneous operations is present in both formats A and B. Table 10-7 gives the functions performed for the codes in this field. These codes are carefully defined to provide the control necessary for implementing the instruction set architecture. The codes control the loading of memory, *PC*, *IR*, *SP*, *PSR*, and *MSTS*. In addition, code 1011 replaces the C_{in} value from the FS field with the value stored in the *C* status bit and enables update of the *PSR*. The C_{in} replacement is needed for the ADDC and SUBB instructions. So that information intended for memory or other registers is not written into the register file, writing to the file is blocked for memory write and loads into the *PC*, *IR PSR*, and *SP*. An RW field in the microinstruction has greater flexibility, but requires an extra bit. Note that only one MO code can be used at a time, so there must be a code for each combination of the various operations required.

Microprogram Structure

We approach the microprogram design top down. The top level consists of an ASM- like chart giving a flow of microroutines. These routines have labels similar to the stages in the pipelined CPU in Section 8-9. In this case, however, rather than being performed in a single clock cycle with combinational logic, the routines require the use of the same hardware over multiple cycles. The flow between and, to some extent, within the routines is intimately tied to the instructions and their decoding. Since the mapping ROM can be used for branching simultaneously with a format A data transfer or manipulation operation, it is convenient to control the

☐ **TABLE 10-7**
Control Information for Miscellaneous Operations for Format A and B Microinstructions

MO		
Operations	**Symbolic Notation**	**Code**
No operation	—	0000
$INACK = 1$	INCK	0001
Memory write*	WRITE	0010
Load *PC**	LPC	0011
Load *IR* and increment *PC**	DPC	0100
Increment *PC*	IPC	0101
Load *PSR**	LST	0110
Load *SP**	LSP	0111
Decrement *SP*	DSP	1000
Decrement *SP* and memory write*	DSM	1001
Increment *SP*	ISP	1010
Select C as C_{in} for arithmetic	CIN	1011
Enable update of status bits Z, N, C, and V	EST	1100
Enable update of status bits Z and C	EZC	1101
Enable update of status bits N and Z	ENZ	1110
Enable update of microstatus bits z, n, c, and v	EMS	1111

* Prevents write to register file

flow between microroutines entirely by using the mapping ROM. This flow is shown in Figure 10-11; the chart is not strictly an ASM chart, since each rectangular box corresponds to microroutines representing multiple states rather than a single state and to multiple clock cycles rather than a single one.

The execution of each instruction begins with the Instruction Fetch microroutine. The *PC* provides the address and is updated to the next address. The instruction fetched is placed in the *IR*. Then the instruction-decoding process begins, using MUX *M* and the mapping ROM. For MM equal to 00, only the first two bits of OPCODE are used, with the remaining two bits set to 0, so that a four-way branch results. This branch is represented by the three binary decision boxes in the figure. Since the bits of OPCODE that are used denote the number of operands for the instruction being decoded, the destinations of the branches are, in two cases, microroutines to fetch the operands. In another case, program branch addresses are fetched. In the final case, the branch in the chart goes directly to execution. Note that the first two bits of OPCODE also help to define the operation, so there are four separate paths to the four execution blocks. These paths preserve the information from the decoding of the two bits of OPCODE in the Instruction Fetch. Thus, there is no need to examine these bits again later in the microprogram in order to determine the operation that is to be executed.

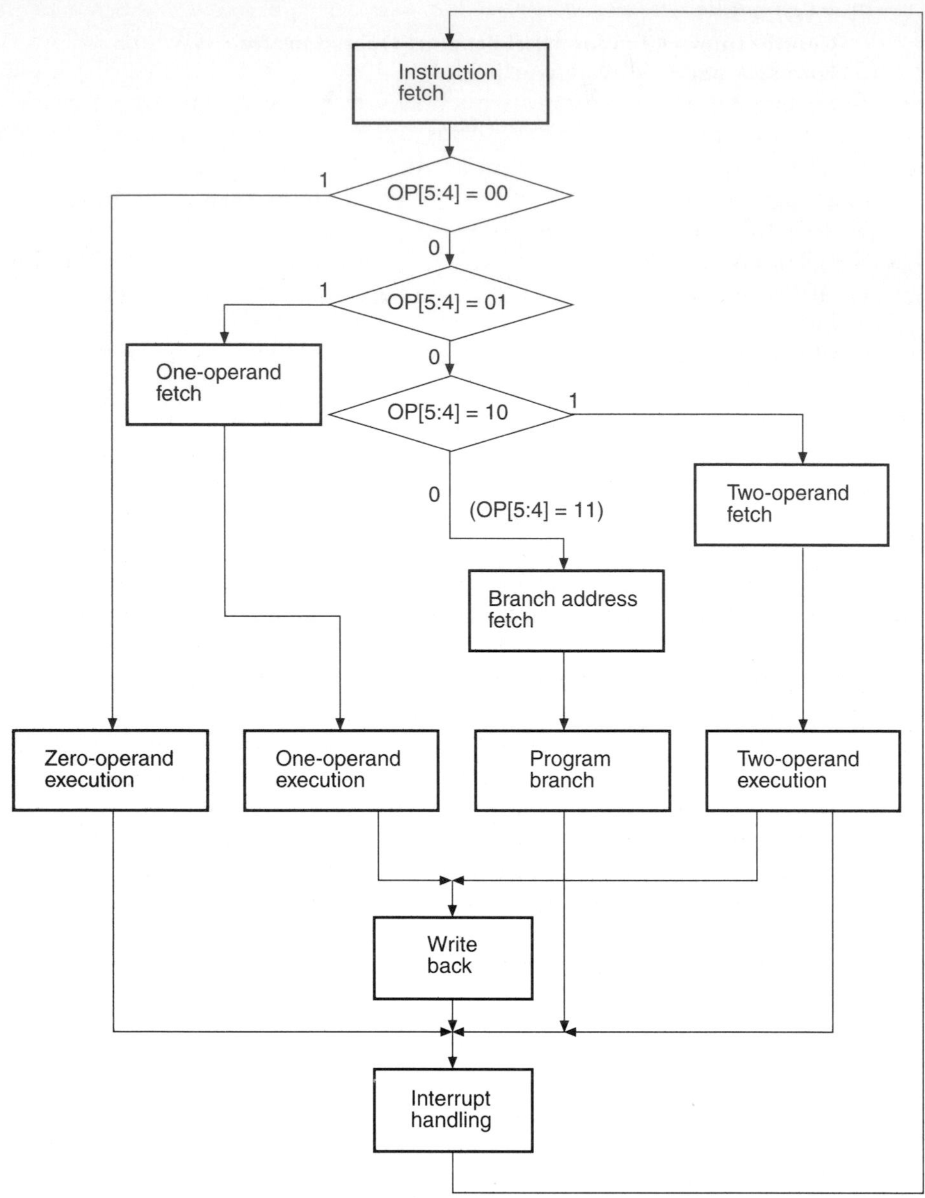

□ **FIGURE 10-11**
ASM-like Chart of Microroutines

In three of the four decisions, an operand fetch routine is performed. Depending upon the first two bits of OPCODE, either a single operand, two operands, or a branch address is fetched. The operand and address values are placed in locations reserved for them in registers *R*12 through *R*15 (*SA*, *SD*, *DA*, and *DD*) . The four execution routines find the operands and addresses in these standard register locations and, in most cases, use them to produce a result that is left in standard location *DD*. The Write Back routine also uses the standard register locations to find the result and its address.

Following their execution, it is necessary for most operations to place the result in its destination. This is accomplished by the Write-Back microroutine. Some of the operations, however, do not have a result to be written in that routine. The existence of these operations is apparent from the paths leading directly from Zero-operand Execution, Program Branch and Two-operand Execution to Interrupt Handling. After each execution microroutine, the program enters the Interrupt Handling microroutine to check for an interrupt before fetching the next instruction.

The flow just described demonstrates the use of the mapping ROM in the Instruction Fetch microroutine. All mappings performed in Instruction Fetch and other microroutines are represented in Table 10-8, the programming table for the mapping ROM. This ROM is used in the execution of particular microinstructions that have the value MAP (10) for MC. The symbolic addresses of these microinstructions are designated in the leftmost column of the table. If the mapping ROM is used in executing a microinstruction, the microinstruction specifics the values of the MM and MR fields in order to define the pattern to be matched to the MUX *M* input to the ROM. In addition to the patterns, the origin in the *IR* of the bits is given for clarity. Where *X*'s appear in bits of the *IR* matching patterns, any possible combination of 1's and 0's on these bits gives the single corresponding symbolic address as the ROM output. If there are four X_s', then 16 ROM rows are represented by a single table row. Where *O*'s or *M*'s with subscripts appear, or where *S* appears, the mapping ROM is being used for an unconditional multiway branch, so these rows really correspond to 8 and 16 different rows giving 8 or 16 symbolic output addresses, respectively. Each of these addresses causes the *CAR* to execute the next microinstruction from a different location. In the case where *O*'s are used, the OPCODE is being decoded into *CAR* addresses for the various operations. Where *M*'s and *S* appear, the MODE and S values are respectively generating the addresses for the microroutines that find the effective addresses specified in the instruction being executed.

Microroutines

We are now prepared to look at the microroutines that implement the CISC CPU. We will use only symbolic addresses for the microinstructions. Assigning binary addresses is quite straightforward because of the addressing flexibility provided by the mapping ROM. Register transfer descriptions are given for each microinstruction. In fact, the microprogram is initially written in terms of these descriptions, and the binary code is added afterward. We use the same position in the tables for fields in both the A and B formats. The field name in the A format is followed by a slash (/), with the name in the B format after the slash. The B format names apply to entries in the tables only for MC equal to 3. Binary values in the fields are given base 16.

☐ **TABLE 10-8**
Programming Table for Mapping ROM

Location of Microinstructions Performing a Mapping	ROM Inputs			ROM Outputs—Location of Next Microinstruction
Symbolic Microaddress	**MM**	**MR**	**IR Bits and Match Field**	**Symbolic Microaddress**
IF1	00	000	$OPCODE(5:4) \parallel 00_2 = 0000_2$	0EX
IF1	00	000	$OPCODE(5:4) \parallel 00_2 = 0100_2$	1OF
IF1	00	000	$OPCODE(5:4) \parallel 00_2 = 1000_2$	2OF
IF1	00	000	$OPCODE(5:4) \parallel 00_2 = 1100_2$	BAF
In 1OF Microroutine	00	001	$OPCODE(5:4) \parallel 00_2 = \mathrm{XXXX}$	1EX
In 2OF Microroutine	00	010	$OPCODE(5:4) \parallel 00_2 = \mathrm{XXXX}$	2EX
In BAF Microroutine	00	011	$OPCODE(5:4) \parallel 00_2 = \mathrm{XXXX}$	BEX
0EX	01	000	$OPCODE(3:0) = O_3O_2O_1O_0$	16 0-Op EX Microinstruction Addresses
1EX	01	001	$OPCODE(3:0) = O_3O_2O_1O_0$	16 1-Op EX Microinstruction Addresses
2EX	01	010	$OPCODE(3:0) = O_3O_2O_1O_0$	16 2-Op Ex Microinstruction Addresses
BEX	01	011	$OPCODE(3:0) = O_3O_2O_1O_0$	16 Br-Op EX Microinstruction Addresses
In EX Microroutines	01	100	$OPCODE(3:0) = \mathrm{XXXX}$	WB0
In EX Microroutines	01	101	$OPCODE(3:0) = \mathrm{XXXX}$	INT0
1OF	10	001	$MODE \parallel S = M_2M_1M_0\mathrm{X}$	8 1-Op OF Microinstruction Addresses
2OF	10	010	$MODE \parallel S = M_2M_1M_0S$	16 2-Op OF Microinstruction Addresses
BAF	10	011	$MODE \parallel S = M_2M_1M_0\mathrm{X}$	8 Br-Op Microinstruction Addresses
XCH2	10	100	$MODE \parallel S = \mathrm{XXX1}$	XCH3
XCH2	10	100	$MODE \parallel S = 0000$	XCH3
XCH2	10	100	$MODE \parallel S = \mathrm{1XX0}$	XCH4
XCH2	10	100	$MODE \parallel S = \mathrm{X1X0}$	XCH4
XCH2	10	100	$MODE \parallel S = \mathrm{XX10}$	XCH4
WB0	11	000	$MODE \parallel S = \mathrm{XXX0}$	WB1
WB0	11	000	$MODE \parallel S = 0001$	WB1
WB0	11	000	$MODE \parallel S = \mathrm{1XX1}$	WB2
WB0	11	000	$MODE \parallel S = \mathrm{X1X1}$	WB2
WB0	11	000	$MODE \parallel S = \mathrm{XX11}$	WB2
In WB microroutine	11	001	$MODE \parallel S = \mathrm{XXXX}$	INT0
In INT microroutine	11	010	$MODE \parallel S = \mathrm{XXXX}$	IF0

The Instruction Fetch microroutine is shown in Table 10-9. The instruction fetch occurs in microaddress IF0, where the instruction is fetched from memory and placed in the *IR*. The *PC* is also simultaneously updated to point to the next

□ **TABLE 10-9**
Instruction Fetch Microroutine

Sym Add	Register Transfer Description	MC	MM /LS	MR /PS	DSA /MS	SB	MA	MB	MD	FS /NA	MO
IF0	$IR \leftarrow M[PC], PC \leftarrow PC + 1$	0	0	0	00	00	1	0	1	00	4
IF1	$CAR \leftarrow ROM[00000_2 \parallel OPCODE(5{:}4) \parallel 00_2]$	2	0	0	00	00	0	0	0	00	0

instruction. In IF1, instruction decoding begins, with the first two bits of the OPCODE used by the mapping ROM to determine the number of operands in the instruction. According to rows labeled IF1 in Table 10-8, the MM and MR fields must contain 00 and 000. The next microinstruction to be executed, based on the first two bits of OPCODE, is the first microinstruction in one of the following microroutines in Figure 10-11: Zero-operand Execution, One-operand Fetch, Two-operand Fetch, and Branch Address Fetch.

Suppose that OPCODE is 010000 and that MODE is 011. Then, from the second row of Table 10-8, the next microinstruction is in microroutine One-operand Fetch, which begins with microinstruction 1OF. This microroutine is given in Table 10-10. The first instruction involves the use of the mapping ROM to decode the combined MODE and S fields of the instruction in order to determine the addressing mode. Since there is only a single operand, the S field has no effect on the microroutines. Consequently, we will find the same microroutines used for both S equal to 0 and S equal to 1. This means that there are only 8 addresses to which the MODE values are mapped, instead of 16. Since MODE has the value 011, the mode is direct addressing. The microcode for this mode begins in 1DR0, the address the mapping ROM will provide from 1OF. In 1DR0, the *PC* points to the word W of the instruction. This word is fetched from memory *M* and placed in the destination address register *DA*. Simultaneously, the *PC* is updated by 1. In 1DR1, the address in *DA* is then used as a direct address to fetch the operand for register *DD* from memory *M*. Simultaneously, the ROM maps its inputs to address 1EX, to begin the executing instruction. This mapping uses MM equal to 00 and MR equal to 001, as shown in Table 10-8. Since the value of the OPCODE(5:4) $\parallel 00_2$ field to be matched in the ROM is all X's, the contents of OPCODE(5:4) have no effect on the mapping, making it an unconditional jump. In Table 10-10, rather than include this detail, we simply show the next address for the *CAR* as 1EX and use "(ROM)" to indicate that this address was produced by the mapping ROM. The values of MM and MR needed for the mapping ROM appear in the microcode portion of the table in 1DR1.

Otherwise, the Table 10-10 contains eight routines accessed from the 8-way branch in address 1OF, one for each of the eight addressing modes. The first two letters in the symbolic address denote the addressing mode: RG for register, IM for immediate, ID for indexed, and RL for relative. The presence of an I as the third letter denotes the use of indirect addressing in that mode. The one exception to this notation is the use of DR instead of IMI (immediate indirect) for direct addressing.

□ **TABLE 10-10**
One-operand Fetch Microroutine

Sym Add	Register Transfer Description	MC	MM /LS	MR /PS	DSA /MS	SB	MA	MB	MD	FS /NA	MO
1OF	$CAR \leftarrow PLA[10001_2 \parallel MODE \parallel S]$	2	2	1	00	00	0	0	0	00	0
1RG	$DD \leftarrow R[DST], CAR \leftarrow$ 1EX(ROM)	2	0	1	0F	10	0	0	0	00	0
1RGI0	$DA \leftarrow R[DST]$	0	0	0	0E	10	0	0	0	00	0
1RGI1	$DD \leftarrow M[DA], CAR \leftarrow$ 1EX(ROM)	2	0	1	0F	0E	3	0	1	00	0
1IM	$DD \leftarrow M[PC], PC \leftarrow PC + 1,$ $CAR \leftarrow$ EX1(PLA)	2	0	1	0F	00	1	0	1	00	5
1DR0	$DA \leftarrow M[PC], PC \leftarrow PC + 1$	0	0	0	0E	00	1	0	1	00	5
1DR1	$DD \leftarrow M[DA], CAR \leftarrow$ 1EX(ROM)	2	0	1	0F	0E	3	0	1	00	0
1ID0	$DA \leftarrow M[PC], PC \leftarrow PC + 1$	0	0	0	0E	00	1	0	1	00	5
1ID1	$DA \leftarrow DA + R[DST]$	0	0	0	0E	10	0	0	0	02	0
1ID2	$DD \leftarrow M[DA], CAR \leftarrow$ 1EX(ROM)	2	0	1	0F	0E	3	0	1	00	0
1IDI0	$DA \leftarrow M[PC], PC \leftarrow PC + 1$	0	0	0	0E	00	1	0	1	00	5
1IDI1	$DA \leftarrow DA + R[DST]$	0	0	0	0E	10	0	0	0	02	0
1IDI2	$DA \leftarrow M[DA]$	0	0	0	0E	00	0	0	1	00	0
1IDI3	$DD \leftarrow M[DA], CAR \leftarrow$ 1EX(ROM)	2	0	1	0F	0E	3	0	1	00	0
1RL0	$DA \leftarrow M[PC], PC \leftarrow PC + 1$	0	0	0	0E	00	1	0	1	00	5
1RL1	$DA \leftarrow DA + PC$	0	0	0	0E	0E	1	0	0	02	0
1RL2	$DD \leftarrow M[DA], CAR \leftarrow$ 1EX(ROM)	2	0	1	0F	0E	3	0	1	00	0
1RLI0	$DA \leftarrow M[PC], PC \leftarrow PC + 1$	0	0	0	0E	00	1	0	1	00	5
1RLI1	$DA \leftarrow DA + PC$	0	0	0	0E	0E	1	0	0	02	0
1RLI2	$DA \leftarrow M[DA]$	0	0	0	0E	00	0	0	1	00	0
1RLI3	$DD \leftarrow M[DA], CAR \leftarrow$ 1EX(ROM)	2	0	1	0F	0E	3	0	1	00	0

The microroutine Two-operand Fetch is similar to One-operand Fetch, with two exceptions: A second operand that is always located in a register must be fetched for execution. The S bit determines to which of the two operands the addressing modes are applied and which lies in a register. If S is 0, the addressing mode is applied to the source operand, and the destination operand and result are in a register. If S is 1, the addressing mode is applied to the destination operand and result, and the source is a register. Two-operand Fetch is illustrated for the direct addressing mode in Table 10-11. The mapping ROM is used in 2OF to perform a 16-way branch. In the table, however, the microcode is shown only for the case of direct addressing, MODE = 011. Microinstructions in 2DR0 through 2DR2 handle direct addressing for S = 0 and in addresses 2DRS0 through 2DRS2 handle direct addressing for S = 1. For the case S = 0, the contents of the W word in *M*[*PC*] are transferred to source address register *SA*, and the *PC* is updated. Then, *SA* is used as the direct address to fetch the source operand in *M*[*SA*]. Finally, the contents of the destination register *R*[*DST*] are transferred to destination data register *DD* for execution. If there is a result, it will be written into *R*[*DST*] in the Write-Back routine. For the case S = 1, direct addressing using word W is applied to obtain destination address *DA* and destination data *DD*. The contents of the

□ **TABLE 10-11**
Two-operand Fetch Microroutine Examples

Sym Add	Register Transfer Description	MC	MM /LS	MR /PS	DSA /MS	SB	MA	MB	MD	FS /NA	MO
2OF	$CAR \leftarrow ROM[10010_2 \parallel MODE \parallel S]$	2	2	2	00	00	0	0	0	00	0
2DR0	$SA \leftarrow M[PC], PC \leftarrow PC+1$	0	0	0	0C	00	1	0	1	00	5
2DR1	$SD \leftarrow M[SA]$	0	0	0	0D	0C	3	0	1	00	0
2DR2	$DD \leftarrow R[DST], CAR \leftarrow$ 2EX(ROM)	2	0	2	0F	10	3	0	0	00	0
2DRS0	$DA \leftarrow M[PC], PC \leftarrow PC+1$	0	0	0	0E	00	1	0	1	00	0
2DRS1	$DD \leftarrow M[DA]$	0	0	0	0F	0E	3	0	1	00	0
2DRS2	$SD \leftarrow R[SRC], CAR \leftarrow$ 2EX(ROM)	2	0	2	0D	11	3	0	0	00	0

source register $R[SRC]$ are transferred as the source data to register SD for execution. In this case, in the Write-Back microroutine, a result will be written into memory M at the location addressed by DA.

Branch address fetch is illustrated for direct addressing in Table 10-12. The microinstruction in location BAF performs an 8-way branch based on the MODE field, so that the instruction-specified mode is used to obtain the branch address. In general, the microcode resembles that for the routine One-operand Fetch. However, since we are interested in the destination address rather than the destination operand, the routine jumps to branch execution beginning at location BEX as soon as the destination address has been placed in DA. In the case of direct addressing that is illustrated, the word W is transferred into DA from the address given by the PC, and the PC is updated. The contents of DA will then be transferred into the PC in the branch execution routine. Note that since DA is used, register mode 000 is invalid for branch instructions, because in this case the address is not a memory address, but a CPU register. If the contents of a register are to be used as the branch address, then register indirect, 001, is the proper mode to use.

□ **TABLE 10-12**
Example of Branch Address Fetch

Sym Add	Register Transfer Description	MC	MM/ LS	MR/ PS	DSA/ MS	SB	MA	MB	MD	FS /NA	MO
BAF	$CAR \leftarrow ROM[10011_2 \parallel MODE \parallel$ S]	2	2	3	00	00	0	0	0	00	0
BDR0	$DA \leftarrow M[PC], PC \leftarrow PC+1,$ $CAR \leftarrow$ BEX(ROM)	2	0	3	0E	00	0	1	1	00	5

The next set of routines perform the actual execution of instructions. In Table 10-13, three instructions not having explicit operands are detailed. In location 0EX, a 16-way branch based on the last four bits of OPCODE jumps to the microroutine for the instruction to be executed. Note that address 0EX is entered, from the Instruction Fetch microroutine which implies that the first two bits of

□ TABLE 10-13
Zero-operand Execution Microroutine Examples

Sym Add	Register Transfer Description	MC	MM /LS	MR /PS	DSA /MS	SB	MA	MB	MD	FS /NA	MO
0EX	$CAR \leftarrow ROM[01000_2 \parallel OPCODE[3:0]]$	2	1	0	00	00	0	0	0	00	0
PSHR0	$SP \leftarrow SP - 1$	0	0	0	00	00	0	0	0	00	8
PSHR1	$M[SP] \leftarrow R1, SP \leftarrow SP - 1$	0	0	0	00	01	2	0	0	00	9
PSHR2	$M[SP] \leftarrow R2, SP \leftarrow SP - 1$	0	0	0	00	02	2	0	0	00	9
PSHR3	$M[SP] \leftarrow R3, SP \leftarrow SP - 1$	0	0	0	00	03	2	0	0	00	9
PSHR4	$M[SP] \leftarrow R4, SP \leftarrow SP - 1$	0	0	0	00	04	2	0	0	00	9
PSHR5	$M[SP] \leftarrow R5, SP \leftarrow SP - 1$	0	0	0	00	05	2	0	0	00	9
PSHR6	$M[SP] \leftarrow R6, SP \leftarrow SP - 1$	0	0	0	00	06	2	0	0	00	9
PSHR7	$M[SP] \leftarrow R7, CAR \leftarrow$ INT0(ROM)	2	1	5	00	07	2	0	0	00	2
RET0	$PC \leftarrow M[SP]$	0	0	0	00	00	2	0	1	00	3
RET1	$SP \leftarrow SP + 1, CAR \leftarrow$ INT0(ROM)	2	1	5	00	00	0	0	0	00	A
RTI0	$PSR \leftarrow M[SP]$	0	0	0	00	00	2	0	1	00	6
RTI1	$SP \leftarrow SP + 1$	0	0	0	00	00	0	0	0	00	A
RTI3	$PC \leftarrow M[SP]$,	0	0	0	00	00	2	0	1	00	3
RTI4	$SP \leftarrow SP + 1, CAR \leftarrow$ INT0(ROM)	2	1	5	00	00	0	0	0	00	A

OPCODE are 00. Thus, the microcode executing the instruction is determined on the basis of all six bits of OPCODE.

The first instruction implemented in the table is push registers, PSHR. This instruction pushes the seven programmer-accessible registers onto the stack, with *R*1 first. Recall that the stack grows from higher addresses toward lower addresses; thus, the decrement is used when pushing items onto the stack. First, the stack pointer *SP* is decremented to point to a vacant location. Then, each in a series of six microinstructions saves one register in turn on the stack and decrements the stack pointer *SP* to provide a location on the stack for the contents of the next register. The final microinstruction saves *R*7 and includes a jump to the interrupt microroutine to check for interrupts. Because of the many writes required and the fact that writes are to the stack, the Write Back microroutine is bypassed, and the writes are performed instead in the execution microroutine.

The second instruction implemented in the table is return from procedure, RET. In this instruction, the stored value of the *PC*, which points to the instruction after a call procedure, is returned to the *PC* from the stack. The top item is transferred from the stack to the *PC*, in RET0. In RET1, the stack pointer is incremented and control is transferred to INT0.

The final instruction in the table is return from interrupt, RTI. This instruction is similar to RET, except that when the interrupt occurred, two words—the contents of the *PC* and *PSR*—were placed on the stack. Thus, the value of the *PSR* must be retrieved from the stack before that of the *PC* is retrieved. Note that when the value of the *PSR* is retrieved, the enable interrupt *EI* is restored to the value it had before the interrupt occurred. If *EI* is 1, the interrupt is enabled. As for all zero-operand instructions, the Write-Back microroutine is bypassed, and the

□ **TABLE 10-14**
One-operand Execution Microroutine Examples

Sym Add	Register Transfer Description	MC	MM /LS	MR /PS	DSA /MS	SB	MA	MB	MD	FS /NA	MO
1EX	$CAR \leftarrow ROM[01001_2 \parallel OPCODE(3:0)]$	2	1	1	00	00	0	0	0	00	0
INC	$DD \leftarrow DD + 1, CAR \leftarrow$ WB0(ROM)	2	1	4	0F	00	0	0	0	01	C
DEC	$DD \leftarrow DD - 1, CAR \leftarrow$ WB0(ROM)	2	1	4	0F	00	0	0	0	06	C
NEG0	$DD \leftarrow \overline{DD}$	0	0	0	0F	00	0	0	0	0E	0
NEG	$DD \leftarrow \underline{DD} + 1, CAR \leftarrow$ WB0(ROM)	2	1	4	0F	00	0	0	0	01	C
COM	$DD \leftarrow \overline{DD}, CAR \leftarrow$ WB0(ROM)	2	1	4	0F	00	0	0	0	0E	E
SHR0	$R9 \leftarrow 0$	0	0	0	09	00	3	0	0	00	0
SHR1	$R9 \leftarrow 0 \parallel SHAM$	0	0	0	09	00	0	2	0	02	F
SHR2	$z: CAR \leftarrow$ INT0	3	0	0	06	00	0	0	0	INT0	0
SHR3	$[DD \leftarrow 0 \parallel DD(15:1)]$	0	0	0	0F	00	0	0	0	11	0
SHR4	$R9 \leftarrow R9 - 1$	0	0	0	09	00	0	0	0	06	F
SHR5	$\bar{z}: CAR \leftarrow$ SHR3	3	0	1	06	00	0	0	0	SHR3	0
SHR6	$DD \leftarrow DD, (C)AR \leftarrow$ WB0(ROM)	2	1	4	0F	00	0	0	0	00	D

microprogram flow goes to INT0, the beginning of the Interrupt Handling microroutine.

Next, we examine a sample of microroutines for the execution of one-operand instructions in Table 10-14. The microroutines begin in 1EX with a 16-way branch based on the last four bits of OPCODE. This completes instruction decoding and selects one of 16 microcode segments that executes the instruction. Note that in the case of the one-operand instructions, the operand is placed in register *DD*, and the result is to be left in that register for the Write Back microroutine. Also, as *DD* is loaded, the codes in MO cause status bits to be set. The three instructions increment (INC), decrement (DEC), and complement (COM) are each executed by a single microinstruction using register *DD*. The last instruction, logical shift right (SHR), in contrast, requires six microinstructions and from 2 to 49 clock cycles for its execution. In SHR0 and SHR1, the four bits in the shift amount field SHAM in the instruction are transferred to *R*9, which will be used in a loop to count the number of shifts remaining to be performed. *MSTS* load is enabled during this transfer, and if SHAM is 0, there is no shifting to be done. In this case, in SHR2, the microroutine jumps to the interrupt microroutine with *DD* unchanged. In SHR3, *DD* is shifted to the right by one bit position, giving the incoming bit the value 0. The composite register (||) notation is used here because it easily describes the incoming bit value. Next, *R*9 is decremented to indicate that one less shift remains to be done. *MSTS* load is enabled during this operation, and in SHR5, the z bit is examined. If it is 0, the number of shifts remaining is nonzero, resulting in a jump to SHR3 to perform another shift. If z is 1, then there are no remaining shifts. In SHR6, the status values Z and C are determined and a jump to WB0 occurs, so that the resulting value in *DD* can be stored in the destination location.

□ **TABLE 10-15**
Two-operand Execution Microroutine Examples

Sym Add	Register Transfer Description	MC	MM /LS	MR /PS	DSA /MS	SB	MA	MB	MD	FS /NA	MO
2EX	$CAR \leftarrow ROM[01010_2 \,\|\, OPCODE(0{:}3)]$	2	1	2	00	00	0	0	0	00	0
MOVE	$DD \leftarrow SD, CAR \leftarrow \text{WB0(ROM)}$	2	1	4	0F	0D	3	0	0	00	0
XCH0	$R9 \leftarrow SD$	0	0	0	09	0F	3	0	0	00	0
XCH1	$SD \leftarrow DD$	0	0	0	0E	0F	3	0	0	00	0
XCH2	$DD \leftarrow R9, CAR \leftarrow ROM[10100_2 \,\|\, MODE \,\|\, S]$	2	2	4	0F	09	3	0	0	00	0
XCH3	$R[SRC] \leftarrow SD, CAR \leftarrow \text{WB0(ROM)}$	2	1	4	11	0D	3	0	0	00	0
XCH4	$M[SA] \leftarrow SD, CAR \leftarrow \text{WB0(ROM)}$	2	1	4	0C	0D	0	0	0	00	2
ADD	$DD \leftarrow DD + SD, CAR \leftarrow \text{WB0(ROM)}$	2	1	4	0F	0D	0	0	0	02	0
ADDC	$DD \leftarrow DD + SD + C, CAR \leftarrow \text{WB0(ROM)}$	2	1	4	0F	0D	0	0	0	03	B
CMP	$DD \leftarrow DD - SD, CAR \leftarrow \text{INT0(ROM)}$	2	1	5	0F	0D	0	0	0	05	C

The implementation of the execution of two-operand instructions is illustrated in Table 10-15. As with other execution microroutines, 2EX contains a 16-way branch based on the last four bits of OPCODE. Two-operand instructions expect the source address to be in register *SA*, the source operand in *SD*, the destination address in *DA*, and the destination operand in *DD*. Thus, the first instruction illustrated, MOVE, is executed by simply transferring the contents of *SD* to *DD*. The Two-operand Fetch microroutine has done the hard job of obtaining the source operand and of producing the destination address. The Write-Back routine will do the job of placing the new destination data in the destination address. This same notion applies to the add (ADD) and add with carry (ADDC) microcode appearing in the table. The compare (CMP) instruction sets status bits in *PSR*, but is not to change the destination data. Thus, the microcode bypasses Write Back, going directly to Interrupt Handling.

The most interesting of the two-operand instructions is exchange (XCH), which exchanges the source and destination data. This microcode segment is a bit unusual, since it has to write results to two locations, yet Write Back can handle only one. As a consequence, the writing of the result into the source is done in the execution microroutine, with the writing of the destination left, as usual, to Write Back. The exchange of the data occurs in XCH0 through XCH2. In addition, in XCH2, the mapping ROM is used to determine whether the addressing modes apply to the source or the destination by examining the value of S in the *IR*. If S is 0, it must be determined whether MODE is 000. If S is 1 or MODE is 000, then the contents of *SD* are returned to register *R*[*SRC*]. Otherwise, for all other modes with S = 0, the contents of *SD* are transferred to *M*[*SA*]. This reasoning corresponds to the five rows labeled with XCH2 (MM = 10 and MR = 100) in the mapping ROM contents in Table 10-8. For the first row, if S = 1, the addressing modes do not apply to the source, so microinstruction XCH3 is executed, to transfer the contents of *SD* to *R*[*SRC*]. For the second row, if S = 0 and MODE = 000, the transfer from *SD* to *R*[*SRC*] in XCH3 is executed, since this is register mode. Otherwise, if S = 0 with at least one of the bits in MODE equal to 1, the contents of *SD* are placed in memory location *SA* using XCH4. This situation corresponds to the last three rows for XCH2 in Table 10-8.

The microinstructions for three branch instructions are presented in Table 10-16. In BEX, there is again a 16-way branch based on OPCODE(3:0). The first instruction, an unconditional jump (JMP), simply takes the effective address from *DA* and places it in the *PC* as the next address. Since, with all branches, there is no Write Back, a jump occurs in the microcode to the Interrupt Handling microroutine. The second instruction, call procedure (CALL), first saves the updated contents of the *PC* onto the stack. It then transfers the destination address to the *PC* to execute the jump. The final two instructions illustrate conditional branches in which the contents of the *PC* are changed only if the condition is satisfied. The first jump occurs for *Z* equal to 1, the second for *Z* equal to 0.

TABLE 10-16
Program Branch Microroutine Examples

Sym Add	Register Transfer Description	MC	MM /LS	MR /PS	DSA /MS	SB	MA	MB	MD	FS /NA	MO
BEX	$CAR \leftarrow ROM[01011_2 \parallel OPCODE(3:0)]$	2	1	3	00	00	0	0	0	00	0
JMP	$PC \leftarrow DA, CAR \leftarrow$ INT0(ROM)	2	1	5	0E	00	0	0	0	00	3
CALL0	$R8 \leftarrow PC$	0	0	0	08	00	1	0	0	00	0
CALL1	$SP \leftarrow SP - 1$	0	0	0	0	00	0	0	0	00	8
CALL2	$M[SP] \leftarrow R8$	0	0	0	0	08	2	0	0	00	2
CALL3	$PC \leftarrow DA, CAR \leftarrow$ INT0(ROM)	2	1	5	0E	00	0	0	0	00	3
BZ0	Z: $CAR \leftarrow$ BRA	3	0	0	1	—	—	—	—	BRA	0
BZ1	$CAR \leftarrow$ INT0(ROM)	2	1	5	00	00	0	0	0	00	0
BRA	$PC \leftarrow DA, CAR \leftarrow$ INT0(ROM)	2	1	5	0E	00	0	0	0	00	3
BNZ0	$\overline{Z}$: $CAR \leftarrow$ BRA	3	0	1	1	—	—	—	—	BRA	0
BNZ1	$CAR \leftarrow$ INT0(ROM)	2	1	5	0E	00	0	0	0	00	0

For those instructions with results to be stored, the Write-Back microroutine in Table 10-17 is executed. To determine where to put the contents of *DD*, it is necessary to examine the value of S and MODE. If S = 0, then the destination address is register *R*[*DST*]. If S = 1, then the destination is register *R*[*DST*] if MODE =

TABLE 10-17
Write Back Microroutine

Sym Add	Register Transfer Description	MC	MM /LS	MR /PS	DSA /MS	SB	MA	MB	MD	FS /NA	MO
WB0	$CAR \leftarrow ROM[11000_2 \parallel MODE \parallel S]$	2	3	0	00	00	0	0	0	00	0
WB1	$R[DST] \leftarrow DD, CAR \leftarrow$ INT0(ROM)	2	1	5	10	0F	3	0	0	00	0
WB2	$M[DA] \leftarrow DD, CAR \leftarrow$ INT0(ROM)	2	1	5	0E	0F	0	0	0	00	2

000. Otherwise, the destination of the result is memory location *SA*. This is the same as the situation in the exchange instruction XCH, except that the specified value of S is opposite. The mapping information is given in those rows of Table 10-8 labeled with WB0. Regardless of the Write-Back operation performed, the Interrupt Handling microroutine is executed next.

The final microroutine, for Interrupt Handling, is given in Table 10-18. In INT0, if *INTS* is 0, indicating that no interrupt is pending, a jump occurs to IF0, since the processing of instructions is complete. If *INTS* is 1, then the next seven microinstructions are executed to save the values of the *PC* and *PSR* on the stack, disable the interrupts, send an interrupt acknowledge, and load the interrupt vector that results into the *PC*, as described in Chapter 9. This last action starts processing at the beginning of the routine that will service the interrupt. Note that the Interrupt-Handling microroutine is based on the assumption that the interrupt is from an external source: if internal sources are to be considered, they require additional microcode.

□ **TABLE 10-18**
Interrupt-Handling Microroutine

Sym Add	Register Transfer Description	MC	MM /LS	MR /PS	DSA /MS	SB	MA	MB	MD	FS /NA	MO
INT0	$\overline{INTS}: CAR \leftarrow \text{IF0}$	3	0	1	A	00	00	00	00	IF1	0
INT1	$R8 \leftarrow PC$	0	0	0	08	00	1	0	0	00	0
INT2	$SP \leftarrow SP - 1$	0	0	0	00	00	0	0	0	00	8
INT3	$M[SP] \leftarrow R8, SP \leftarrow SP - 1$	0	0	0	00	08	2	0	0	00	9
INT4	$M[SP] \leftarrow PSR$	0	0	0	00	00	2	1	0	00	9
INT5	$PSR \leftarrow 0$	0	0	0	00	00	0	0	0	00	7
INT6	$INACK \leftarrow 1$	0	0	0	00	00	0	0	0	00	1
INT7	$PC \leftarrow IVAD, CAR \leftarrow \text{IF0(ROM)}$	2	3	2	00	00	0	0	1	00	3

10-3 The Reduced Instruction Set Computer

The second design we examine is for a reduced instruction set computer with a pipelined datapath and control unit. We begin by describing the instruction set architecture for this particular RISC design, which is characterized by load/store memory access, four addressing modes, a single instruction format length, and instructions that require only elementary operations. We then design a datapath and control to implement the architecture. The datapath is based on the pipelined datapath initially described in Figure 7-23. In order to implement the instruction set architecture, modifications are made to the register file and the function unit in that figure. The control unit is based on the pipelined control unit added to the datapath in Section 8-9. Due to data and control hazards associated with pipelined designs, not only modifications to the control unit, but further modifications to the datapath, are required.

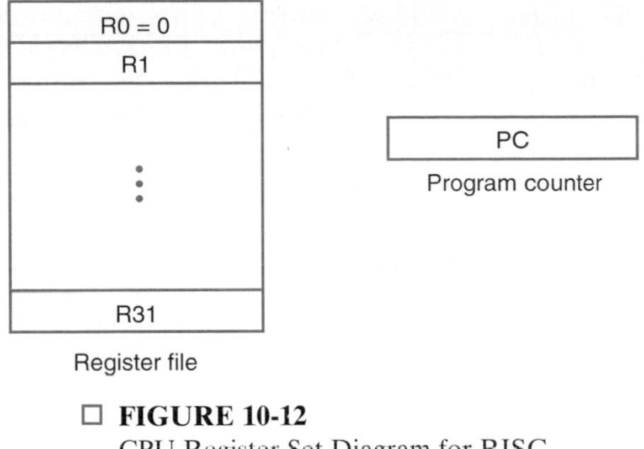

□ **FIGURE 10-12**
CPU Register Set Diagram for RISC

Instruction Set Architecture

Figure 10-12 shows the CPU registers accessible to the programmer in this RISC. All registers are 32 bits. The register file has 32 registers, *R*0 through *R*31. *R*0 is a special register that always supplies the value zero when it is used as a source and discards the result when it is used as a destination. The size of the programmer-accessible register file is larger in the RISC than in the CISC because of the load/store instruction set architecture. Since the data manipulation operations can use only register operands, many currently active operands should be in the register file. Otherwise, numerous stores and loads would be needed to temporarily save operands in the data memory between data manipulation operations. In addition, in many real pipelines these stores and loads require more than one clock cycle for their execution. To prevent these factors from degrading the performance of the RISC, at least 32 registers are required in the register file.

In addition to the register file, only a program counter *PC* is provided. If stack pointer-based or processor status register-based operations are required, they are implemented by sequences of instructions using the register file.

Figure 10-13 gives the three instruction formats for the RISC CPU. The formats use a single word of 32 bits. This longer word length is essential in order to provide realistic address values, since the additional address word used in the CISC CPU is not available in the RISC CPU. The first format specifies three registers. The two registers addressed by the 5-bit source register fields SA and SB contain two operands. The third register, addressed by a 5-bit destination register field DR, specifies the register location for the result. A 7-bit OPCODE provides for a maximum of 128 operations.

The remaining two formats replace the second register with a 15-bit constant. In the two-register format, the constant acts as an immediate operand; in the branch format, the constant is a *target offset*. *Target address* is another name for the effective address, particularly if the address is used in a branch instruction. The target address is formed by the addition of the target offset to the contents of the *PC*. Thus, branching uses relative addressing based on the updated value of the *PC*.

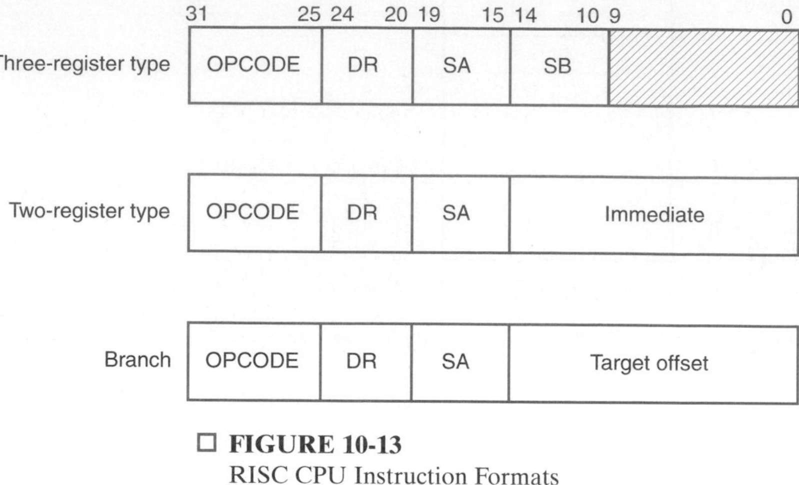

□ **FIGURE 10-13**
RISC CPU Instruction Formats

The branch instructions specify source register SA. Whether the branch or jump is taken is based on whether the contents of the source register are zero. The DR field is used to specify the register in which to store the return address for the procedure call.

Table 10-19 contains the 27 operations to be performed by the instructions. A mnemonic, an opcode, and a register transfer description are given for each operation. All of the operations are elementary and can be described by a single register transfer statement. The only operations that can access memory are Load and Store. A significant number of immediate instructions help to reduce data memory accesses and speed up execution when constants are employed. Since the immediate field of the instruction is only 15 bits, the leftmost 17 bits must be filled to form a 32-bit operand. In addition to using zero fill, a second method is used called *sign extension*. The most significant bit of the immediate operand, bit 14 of the instruction, is viewed as a sign bit. To form a 32-bit 2's-complement operand, this bit is copied into the 17 bits. In Table 10-19, the sign extension of the immediate field is denoted by se *IM*. The same notation, se *IM*, also represents the sign extension of the target offset field.

The absence of stored versions of status bits is handled by the use of three instructions: Branch if Zero (BZ), Branch if Nonzero (BNZ), and Set if Less Than (SLT). BZ and BNZ are single instructions that determine whether a register operand is zero or nonzero and branch accordingly. SLT stores a value in register $R[DR]$ that acts like a negative status bit. If $R[SA]$ is less than $R[SB]$, a 1 is placed in register $R[DR]$; if $R[SA]$ is greater than or equal to $R[SB]$, a 0 is placed in $R[DR]$. The register $R[DR]$ can then be examined by a subsequent instruction to see whether it is zero (0) or nonzero (1). Thus, using two instructions, the relative values of two operands or the sign of one operand (by letting $R[SB]$ equal to $R0$) can be determined.

□ **TABLE 10-19**
RISC Instruction Operations

Operation	Symbolic Notation	Op Code	Action
No Operation	NOP	0000000	None
Add	ADD	0000010	$R[DR] \leftarrow R[SA] + R[SB]$
Subtract	SUB	0000101	$R[DR] \leftarrow R[SA] + R[SB] + 1$
Set if Less Than	SLT	1100101	If $R[SA] < R[SB]$ then $R[DR] = 1$
AND	AND	0001000	$R[DR] \leftarrow R[SA] \wedge R[SB]$
OR	OR	0001010	$R[DR] \leftarrow R[SA] \vee R[SB]$
Exclusive-OR	XOR	0001100	$R[DR] \leftarrow R[SA] \oplus R[SB]$
Store	ST	0000001	$M[R[SA]] \leftarrow R[SB]$
Load	LD	0100001	$R[DR] \leftarrow M[R[SA]]$
Add Immediate	ADI	0100010	$R[DR] \leftarrow R[SA] + \text{se } IM$
Subtract Immediate	SBI	0100101	$R[DR] \leftarrow R[SA] + \overline{(\text{se } IM)} + 1$
Complement	NOT	0101110	$R[DR] \leftarrow \overline{R[SA]}$
AND Immediate	ANI	0101000	$R[DR] \leftarrow R[SA] \wedge (0 \Vert IM)$
OR Immediate	ORI	0101010	$R[DR] \leftarrow R[SA] \vee (0 \Vert IM)$
Exclusive-OR Immediate	XRI	0101100	$R[DR] \leftarrow R[SA] \oplus (0 \Vert IM)$
Add Immediate Unsigned	AIU	1100010	$R[DR] \leftarrow R[SA] + (0 \Vert IM)$
Subtract Immediate Unsigned	SIU	1100101	$R[DR] \leftarrow R[SA] + \overline{(0 \Vert IM)} + 1$
Move	MOV	1000010	$R[DR] \leftarrow R[SA]$
Logical Left Shift by SH Bits	LSL	0110000	$R[DR] \leftarrow \text{lsl } R[SA] \text{ by } SH$
Logical Right Shift by SH Bits	LSR	0110001	$R[DR] \leftarrow \text{lsr } R[SA] \text{ by } SH$
Jump Register	JMR	1100001	$PC \leftarrow R[SA]$
Branch on Zero	BZ	0100000	If $R[SA] = 0$, then $PC \leftarrow PC + \text{se } IM$
Branch on Nonzero	BNZ	1100000	If $R[SA] \neq 0$, then $PC \leftarrow PC + \text{se } IM$
Jump	JMP	1000100	$PC \leftarrow PC + \text{se } IM$
Jump and Link	JML	1000000	$PC \leftarrow PC + \text{se } IM$, $R[DR] \leftarrow PC + 1$

The Jump and Link (JML) instruction provides a mechanism for implementing procedures. The value in the *PC* after updating is stored in register $R[DR]$, and then the sum of the *PC* and the sign-extended target offset from the instruction is placed in the *PC*. The return from a called procedure can use the Jump Register instruction with SA equal to DR for the calling procedure. If a procedure is to be called from within a called procedure, then each successive procedure that is called will need its own register for storing the return value. A software stack that moves return addresses from $R[DR]$ to memory at the beginning of a called procedure and restores them to $R[SA]$ before the return can also be used.

Addressing Modes

The four addressing modes in the RISC are register, register indirect, immediate, and relative. The mode is specified by the operation code, rather than by a separate mode field. As a consequence, the mode for a given operation is fixed and cannot be varied. The three-operand data manipulation instructions use register mode addressing. Register indirect, however, applies only to the load and store instructions, the only instructions that access data memory. Instructions using the two-register format have an immediate value that replaces register address SB. Relative addressing applies exclusively to branch and jump instructions and so produces addresses only for the instruction memory.

When programmers want to use an addressing mode, such as indexed addressing, not provided by the instruction set architecture, they must use a sequence of RISC instructions. For example, for an indexed address for a load operation, the desired transfer is

$$R15 \leftarrow M[R5 + 0 \,||\, I]$$

This transfer can be accomplished by executing two instructions:

AIU R9, R5, I
LD R15, R9

The first instruction, Add Immediate Unsigned, forms the address by appending 17 0's to the left of I and adding the result to $R5$. The resulting effective address is then temporarily stored in $R9$. Next, the Load instruction uses the contents of $R9$ as the address at which to fetch the operand and places the operand in the destination register $R15$. Since, for indexed addressing, I is regarded as a positive offset in memory, the use of unsigned addition is appropriate. This sequence of two operations for indexed addressing is one justification for having unsigned immediate addition available.

Datapath Organization

We use the pipelined datapath in Section 7-11 and Section 8-9 as the basis for the datapath here and deal only with modifications. These modifications affect the register file, the function unit, and the bus structure. The reader should also refer to the datapath in Figure 8-30 and the new datapath shown later in Figure 10-15 in order to understand fully the discussion that follows. We treat each modification in turn, beginning with the register file.

In Figure 8-30, there are 16 16-bit registers, and all registers are identical in function. In the new datapath, there are 32 32-bit registers. Also, the contents of $R0$ are to be read as the constant zero. If a write is attempted into $R0$, the data will be lost. These changes are implemented in the new register file in Figure 10-15. All data inputs and the data output are 32 bits. To correspond to the 32 registers, the address inputs are all 5 bits. The value zero when reading $R0$ is implemented by replacing the storage elements for $R0$ in the register file with open circuits on the lines that were their inputs and with constant zero values on the lines that were their outputs.

A second major modification to the datapath is the replacement of the single-bit position shifter with a barrel shifter to speed up the execution of multiple-position shifting. This barrel shifter can perform a logical right or logical left shift of from 0 to 31 positions. A block diagram for the barrel shifter appears in Figure

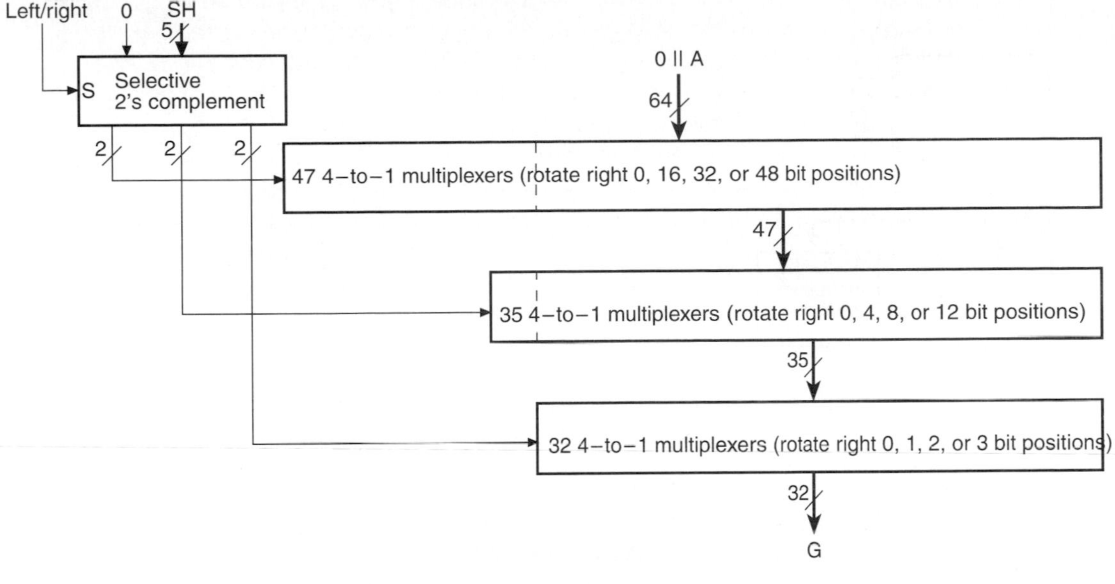

□ **FIGURE 10-14**
32-bit Barrel Shifter

10-14. The data input is 32-bit operand A, and the output is 32-bit result G. Left/$\overline{\text{right}}$, a control signal decoded from OPCODE, selects a left or right shift. The shift amount field $SH = IR(4{:}0)$ specifies the number of bit positions to shift the data input and takes on values from 0 through 31. A logical shift of p bit positions involves inserting p zeros into the result. In order to provide these zeros and simplify the design of the shifter, we will perform both the left and right shift by using a right rotate. The input to this rotate will be the input data A with 32 zeros concatenated to its left. A right shift is performed by rotating the input p positions to the right; a left shift is performed by rotating $64 - p$ positions to the right. This number of positions can be obtained by taking the 2's complement of the 6-bit value of $0 \,||\, SH$.

The 63 different rotates can be obtained by using three levels of 4–to–1 multiplexers, as shown in Figure 10-14. The first level shifts by 0, 16, 32, or 48 positions, the second level by 0, 4, 8, or 12 positions, and the third level by 0, 1, 2, or 3 positions. The number of positions for A to be shifted, 0 through 63, can be implemented by representing A as a three-digit base-4 integer. From left to right, the digits have weights $4^2 = 16$, $4^1 = 4$, and $4^0 = 1$. The digit values in each of the positions are 0, 1, 2, and 3. Each digit controls a level of the 4–to–1 multiplexers, the most significant digit controlling the first level, the least significant the third level. Due to the presence of 32 zeros in the 64-bit input, fewer than 64 multiplexers can be used in each level. A level requires the number of multiplexers to be 32 plus the total number of positions its output can be shifted by subsequent levels. The output of the first level can be shifted at most $12 + 3 = 15$ positions to the right. Thus, this level requires $32 + 15 = 47$ multiplexers. The output of the second level can be

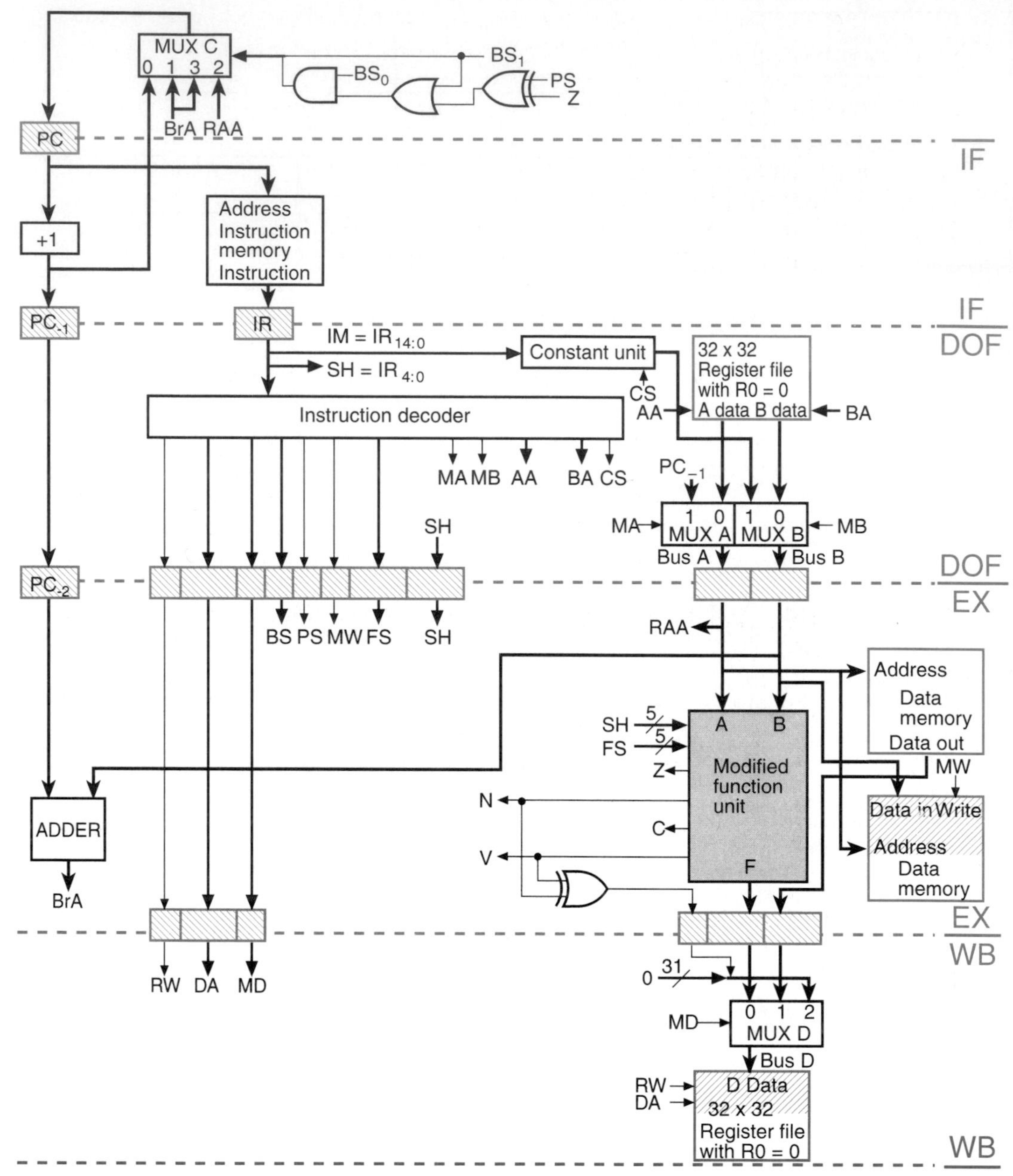

FIGURE 10-15
Pipelined RISC CPU

shifted at most 3 positions, giving 32 + 3 = 35 multiplexers. The final level cannot be shifted further and so needs just 32 multiplexers.

In the function unit, the ALU is expanded to 32 bits, and the barrel shifter replaces the single position shifter. The resulting modified function unit uses the same function codes as in Chapters 7 and 8, except that the two codes for shifts are now labeled as logical shifts. The shift amount SH is a new 5-bit input to the modified function unit in Figure 10-15.

The remaining datapath changes are shown in Figure 10-15. Beginning at the top of the datapath, zero fill has been replaced by the constant unit. The constant unit performs zero fill for $CS = 0$ and sign extension for $CS = 1$. MUX A is added to provide a path for the updated PC, PC_{-1}, to the register file for implementation of the Jump and Link (JML) instruction.

One other change in the figure helps implement the Set if Less Than (SLT) instruction. This logic provides a 1 to be loaded into $R[DA]$ if $R[AA] - R[BA] < 0$ and a 0 to be loaded into $R[DA]$ if $R[AA] - R[BA] \geq 0$. It is implemented by adding an additional input to MUX D. The leftmost 31 bits of the input are 0; the rightmost bit is 1 if N is 1 and V is 0, i.e., if the result of the subtraction is negative and there is no overflow. It is also 1 if N is 0 and V is 1, i.e., if the result of the subtraction is positive and there is an overflow. These represent all cases in which $R[AA]$ is greater than $R[BA]$ and which can be implemented by an exclusive-OR of N and V.

A final difference in the datapath is that the register file is no longer edge triggered and no longer a part of a pipeline platform at the end of the write back (WB) stage. Instead, the register file uses latches and is written much earlier than the positive clock edge. Special timing signals are provided that permit the register file to be written in the first half and to be read in the last half of the cycle. In particular, in the second half of the cycle, it is possible to read data written into the register file during the first half of the same clock cycle. This is called a *read-after-write* register file, and it both avoids added complexity in the logic used for handling hazards and reduces the cost of the register file.

Control Organization

The control organization in the RISC is modified from that in Chapter 8. The modified instruction decoder is essential to deal with the new instruction set. In Figure 10-15, CS and SH are added from the IR, one bit is added to MD, and there is a new pipeline platform for SH and 2 bit platforms for MD.

The remaining control signals are included to handle the new control logic related to the PC. All of this logic relates to loading addresses into the PC in order to implement branches and jumps. MUX C selects from three different sources for the next value of PC. The updated PC is used to move sequentially through a program. A branch target address BrA that is formed from the sum of the updated PC value for the branch instruction and the sign-extended target offset is used for branches and jumps. The value in $R[AA]$ is used for the register jump. The selection of these values is controlled by the field BS. If $BS0 = 0$, then the updated PC is selected by $BS1 = 0$, and $R[AA]$ is selected by $BS1 = 1$. If $BS0 = 1$ and $BS1 = 1$, then BrA is selected unconditionally. If $BS0 = 1$ and $BS1 = 0$, then, for $PS = 0$, a branch to BrA occurs for $Z = 1$, and for $PS = 1$, a branch to BrA occurs for $Z = 0$. This implements the two conditional branch instructions BZ and BNZ.

In order to have the value of the updated PC for the branch and jump instructions when they reach the execution stage, two pipeline registers, PC_{-1} and PC_{-2}, are added. PC_{-2} and the value from the constant unit are inputs to the dedicated adder that forms BrA in the execution stage. Note that MUX C and the attached control logic are in the EX stage, although shown above the PC. The related clock cycle difference causes problems with instructions following branches that we will deal with in later subsections.

The heart of the control unit is the instruction decoder. This is combinational circuitry that converts the operation code in the IR into the control signals necessary for the datapath and control unit. In Table 10-20, each instruction is identified by its mnemonic. A register transfer statement and the opcode are given for the instruction. The opcodes are selected such that the least significant five of the seven

☐ TABLE 10-20
Control Words for Instructions

Symbolic Notation	Action	Op Code	Control Word Values								
			RW	MD	BS	PS	MW	FS	MB	MA	CS
NOP	None	0000000	0	—	00	—	0	—	—	—	—
ADD	$R[DR] \leftarrow R[SA] + R[SB]$	0000010	1	00	00	—	0	00010	0	0	—
SUB	$R[DR] \leftarrow R[SA] + \overline{R[SB]} + 1$	0000101	1	00	00	—	0	00101	0	0	—
SLT	If $R[SA] < R[SB]$ then $R[DR] = 1$	1100101	1	10	00	—	0	00101	0	0	—
AND	$R[DR] \leftarrow R[SA] \wedge R[SB]$	0001000	1	00	00	—	0	01000	0	0	—
OR	$R[DR] \leftarrow \mathrm{R}[SA] \vee \mathrm{R}[SB]$	0001010	1	00	00	—	0	01010	0	0	—
XOR	$R[DR] \leftarrow \mathrm{R}[SA] \oplus \mathrm{R}[SB]$	0001100	1	00	00	—	0	01100	0	0	—
ST	$M[R[SA]] \leftarrow R[SB]$	0000001	0	00	00	—	1	—	0	0	—
LD	$R[DR] \leftarrow M[R[SA]]$	0100001	1	01	00	—	0	—	—	0	—
ADI	$R[DR] \leftarrow R[SA] + \text{se}\, IM$	0100010	1	00	00	—	0	00010	1	0	1
SBI	$R[DR] \leftarrow R[SA] + \overline{(\text{se}\, IM)} + 1$	0100101	1	00	00	—	0	00101	1	0	1
NOT	$R[DR] \leftarrow \overline{R[SA]}$	0101110	1	00	00	—	0	01110	—	0	—
ANI	$R[DR] \leftarrow R[SA] \wedge (0 \Vert IM)$	0101000	1	00	00	—	0	01000	1	0	0
ORI	$R[DR] \leftarrow R[SA] \vee (0 \Vert IM)$	0101010	1	00	00	—	0	01010	1	0	0
XRI	$R[DR] \leftarrow R[SA] \oplus (0 \Vert IM)$	0101100	1	00	00	—	0	01100	1	0	0
AIU	$R[DR] \leftarrow R[SA] + (0 \Vert IM)$	1100010	1	00	00	—	0	00010	1	0	0
SIU	$R[DR] \leftarrow R[SA] + \overline{(0 \Vert IM)} + 1$	1100101	1	00	00	—	0	00101	1	0	0
MOV	$R[DR] \leftarrow R[SA]$	1000000	1	00	00	—	0	00000	—	0	—
LSL	$R[DR] \leftarrow$ lsl $R[SA]$ by SH	0110000	1	00	00	—	0	10100	—	0	—
LSR	$R[DR] \leftarrow$ lsr $R[SA]$ by SH	0110001	1	00	00	—	0	11000	—	0	—
JMR	$PC \leftarrow R[SA]$	1100001	0	—	10	—	0	00000	—	—	—
BZ	If $R[SA] = 0$, then $PC \leftarrow PC + 1 + \text{se}\, IM$	0100000	0	—	01	0	0	00000	1	0	1
BNZ	If $R[SA] \neq 0$, then $PC \leftarrow PC + 1 + \text{se}\, IM$	1100000	0	—	01	1	0	00000	1	0	1
JMP	$PC \leftarrow PC + 1 + \text{se}\, IM$	1000100	0	—	11	—	0	—	1	—	1
JML	$PC \leftarrow PC + 1 + \text{se}\, IM, R[DR] \leftarrow PC + 1$	0000111	1	00	11	—	0	00111	1	1	1

bits match the bits in the control field FS. This leads to simpler decoding. The register file addresses AA, BA, and DA come directly from SA, SB, and DR, respectively, in the *IR*.

Otherwise, to determine the control codes, the CPU is viewed much as is the single-cycle CPU in Figure 8-19. The pipeline platforms can be ignored in this determination; however, it is important to examine the timing carefully to be sure that various parts of the register transfer statement for the operation are taking place in the right stage of the pipeline. For example, note that the adder for the *PC* is in stage EX. This adder is connected to MUX *C* and its attached control logic and to the incrementer +1 for the *PC*. Thus, all of this logic is in the EX stage, and the loading of the *PC* that begins the IF stage is controlled from the EX stage. Likewise, the input *R*[*AA*] is in the same combinational block of logic and comes not from the *A* Data output of the register file, but from Bus *A* in the EX stage, as shown.

Table 10-20 can serve as the basis for the design of the instruction decoder. It contains the values for all control signals, except the register addresses from *IR*. In contrast to the decoder in Section 8-7, the logic is complex and is most easily designed by using a computer-based logic synthesis program.

Data Hazards

In Section 8-9, we examined a pipeline execution diagram and found that filling and flushing of the pipeline reduced the throughput below the maximum level achievable. Unfortunately, there are other problems with pipeline operation that reduce throughput. In this and the next subsection, we will examine two such problems: data hazards and control hazards. Hazards are timing problems that arise because the execution of an operation in a pipeline is delayed by one or more clock cycles from the time at which the instruction containing the operation was fetched. If a subsequent instruction tries to use the result of the operation as an operand before the result is available, it uses the old or stale value, which is very likely to give a wrong result. To deal with the two types of hazards, we will illustrate at least two solutions, one that uses software and another that uses hardware.

Two data hazards are illustrated by examining the execution of the following program:

```
1  MOV  R1, R5
2  ADD  R2, R1, R6
3  ADD  R3, R1, R2
```

The execution diagram of this program appears in Figure 10-16(a). The MOV instruction places the contents of *R*5 into *R*1 in the first half of WB in cycle 4. But, as shown by the blue arrow, the first ADD instruction reads *R*1 in the last half of DOF in cycle 3, one cycle before it is written. Thus, the ADD instruction uses the stale value in *R*1. The result of this operation is placed in *R*2 in the first half of WB in cycle 5. The second ADD instruction, however, reads both *R*1 and *R*2 in the second half of DOF in cycle 4. In the case of *R*1, the value read was written in the first half of WB in cycle 4. So the value read in the second half of cycle 4 is the new

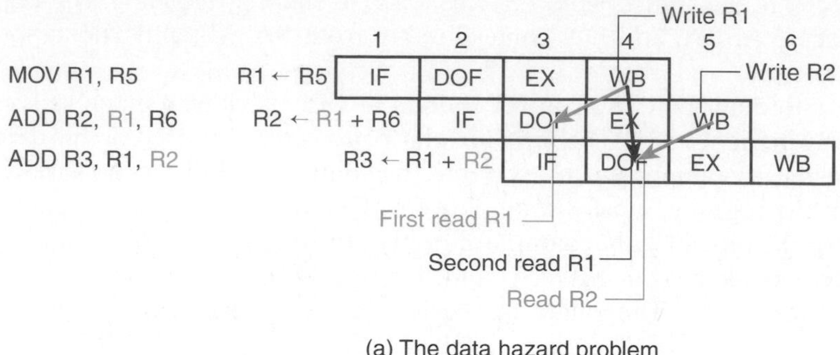

(a) The data hazard problem

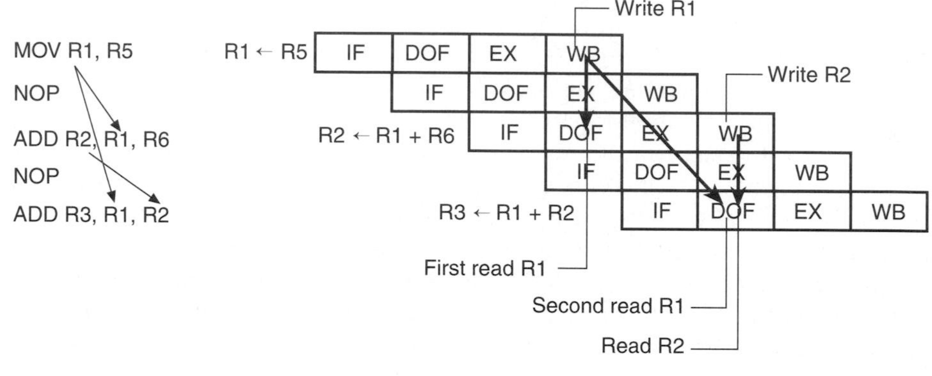

(b) A program-based solution

□ **FIGURE 10-16**
Example of Data Hazard

value. The write-back of *R*2, however, occurs in the first half of cycle 5, after it is read by the next instruction during cycle 4. So *R*2 has not been updated to the new value at the time it is read. This gives two data hazards, as indicated by the large blue arrows in the figure. The registers that are not properly updated to new values are highlighted in blue in the program and in the register transfer statements in the figure. In each of these cases, the read of the involved register occurs one clock cycle too soon with respect to the write of that register.

One possible remedy for data hazards is to have the compiler or programmer generate the machine code to delay instructions so that new values are available. The program is written so that any pending write to a register occurs in the same or an earlier clock cycle than a subsequent read from the register. To accomplish this, the programmer or compiler needs to have detailed information on how the pipeline operates. Figure 10-16(b) illustrates a modification of the simple three-line program that solves the problem. No-operation (NOP) instructions are inserted between the first and second instructions, and between the second and third instructions to delay the respective reads relative to the writes by one clock cycle. The execution diagram shows that, at worst, this approach has writes and subse-

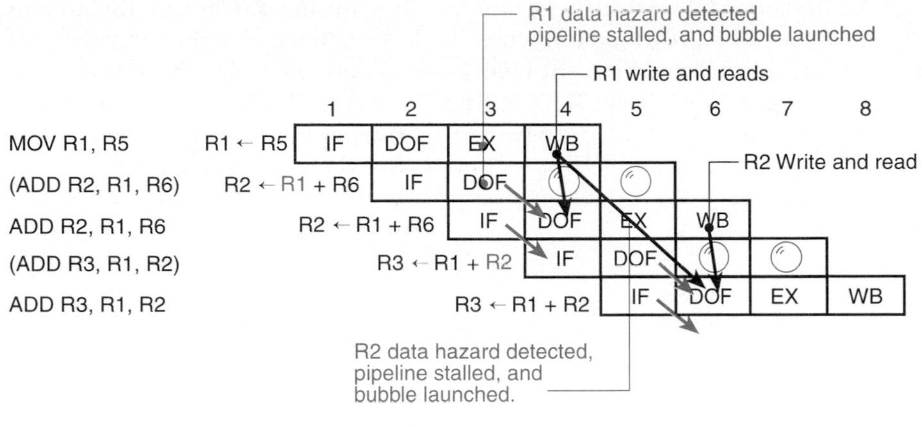

□ **FIGURE 10-17**
Example of Data Hazard Stall

quent reads in the same clock cycle. This is indicated by the pairs consisting of a register write and a subsequent register read connected by a black arrow in the diagram. Because of the read-after-write assumption for the register file, the timing shown permits the program to be executed on correct operands.

This approach solves the problem, but what is the cost? First of all, the program is obviously longer, although it may be possible to place other, unrelated instructions in the NOP positions instead of just wasting them. Also, the program takes two clock cycles longer and reduces the throughput from 0.5 instruction per cycle to 0.375 instruction per cycle with the NOPs in place.

Figure 10-17 illustrates an alternative solution involving added hardware. Instead of the programmer or compiler putting NOPs in the program, the hardware inserts the NOPs automatically. When an operand is found at the DOF stage that has not been written back yet, the associated execution and write-back are prevented, and the pipeline flow in IF and DOF is stalled for one clock cycle. Then the flow resumes with completion of the instruction, and a new instruction is fetched as usual. The delay of one cycle is enough to permit a result to be written before it is read as an operand.

When the actions associated with an instruction flowing through the pipe are prevented from happening at a given point, the pipeline is said to contain a *bubble* in subsequent clock cycles and stages for that instruction. In Figure 10-17, when the flow for the first ADD instruction is prevented beyond the DOF stage, in the next two clock cycles a bubble passes through the EX and the WB stages, respectively. The holding of the pipeline flow in the IF and DOF stages delays the microoperations taking place in these stages for one clock cycle. In the figure, this delay is represented by two diagonal blue arrows from the initial location in which the completion of the microoperation is prevented to the location one clock cycle later in which the microoperation is performed. When the pipeline flow is held in IF and DOF for an extra clock cycle , the pipeline is said to be *stalled*, and if the cause of the stall is a data hazard, then the stall is referred to as a *data hazard stall.*

An implementation of data hazard handling for the pipelined RISC that uses data hazard stalls is presented in Figure 10-18. The added or modified hardware is shown in the areas shaded in light blue. For this particular pipeline stage arrangement, a data hazard will occur for a register file read if there is a destination register at the execution stage that is to be written back in the next clock cycle and that is to be read at the current DOF stage as either of the two operands. So we have to determine whether such a register exists. This is done by evaluating the Boolean equations

$$HA = \overline{MA_{\text{DOF}}} \cdot (DA_{\text{EX}} = AA_{\text{DOF}}) \cdot RW_{\text{EX}} \cdot \sum_{i=0}^{4} (DA_{\text{EX}})_i$$

$$HB = \overline{MB_{\text{DOF}}} \cdot (DA_{\text{EX}} = BA_{\text{DOF}}) \cdot RW_{\text{EX}} \cdot \sum_{i=0}^{4} (DA_{\text{EX}})_i$$

and

$$DHS = HA + HB$$

The following events must all occur for HA, which represents a hazard for the A data, to equal 1:

1. MA in the DOF stage must be 0, meaning that the A operand is coming from the register file.
2. AA in the DOF stage equals DA in the EX stage, meaning that there is potentially a register being read in the DOF stage that is to be written in the next clock cycle.
3. RW in the EX stage is 1, meaning that register DA in the EX stage will definitely be written in WB during the next clock cycle.
4. The OR (Σ) of all bits of DA is 1, meaning that the register to be written is not $R0$ and so is a register that must be written before being read. ($R0$ has the same value 0 regardless of any writes to it.)

If all these conditions hold, there is a write pending for the next clock cycle to a register that is the same as one being read and used on Bus A. Thus, a data hazard exists for the A operand from the register file. HB represents the same combination of events for the B data. If either of the HA or HB terms equals 1, there is a data hazard and DHS is 1, meaning that a data hazard stall is required.

The logic implementing the above equations is shown in the shaded area in the center of Figure 10-18. The blocks marked "Comp" are equality comparators that have output 1 if and only if the two 5-bit inputs are equal. The OR gate with DA entering it ORs together the five bits of DA and has output 1 as long as DA is not 00000 ($R0$).

DHS is inverted and the inverted signal is used to initiate a bubble in the pipeline for the instruction currently in the IR as well as to stop the PC and IR from changing. The bubble, which prevents actions from occurring as the instruction passes through the EX and WB stages, is produced by using AND gates to force RW

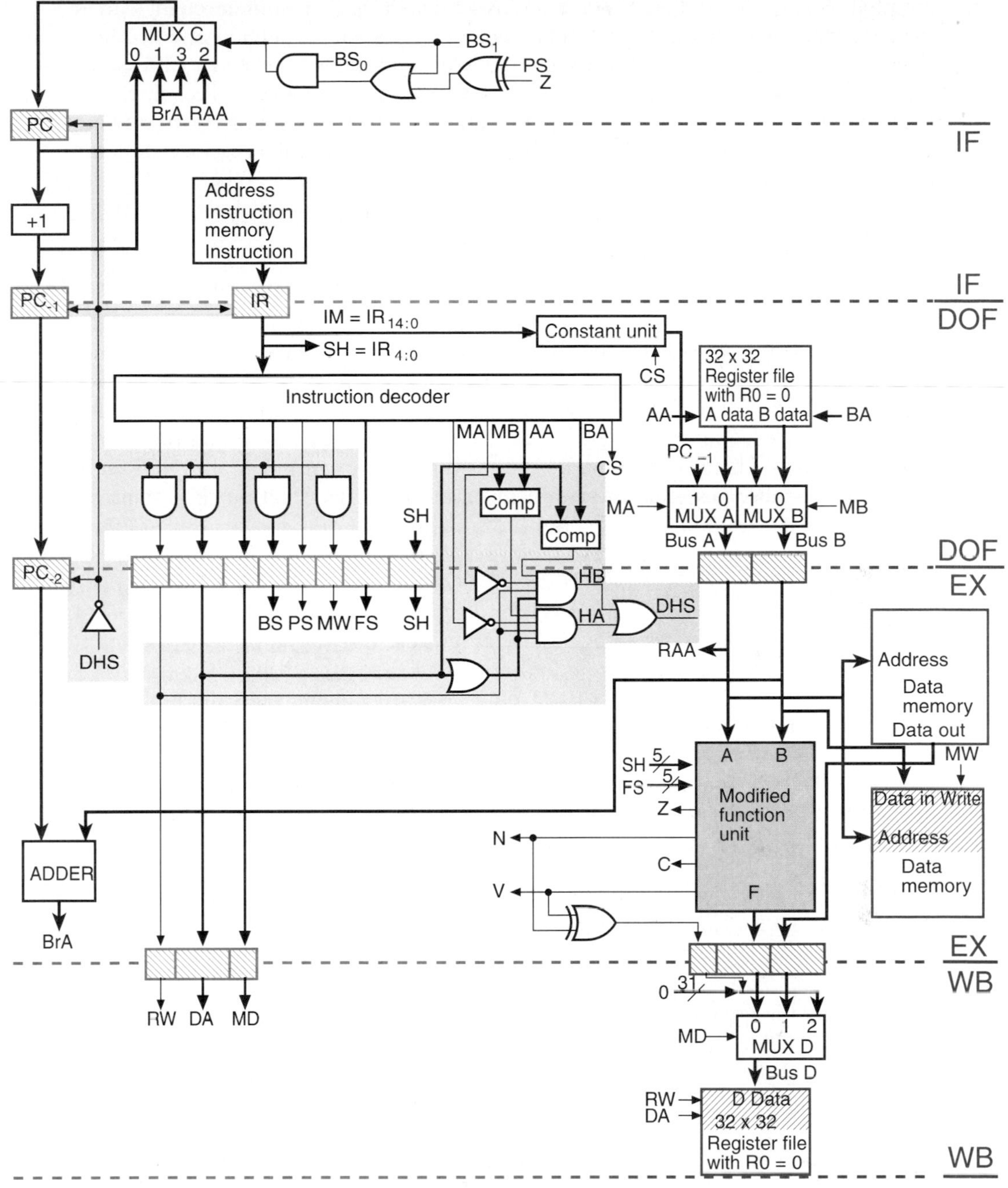

□ **FIGURE 10-18**
Pipelined RISC: Data Hazard Stall

and MW to 0. These 0s prevent the instruction from writing the register file and the memory. AND gates also force BS to 0 causing the PC to be incremented instead of loaded during the EX stage for a jump register or branch instruction affected by a data hazard. Finally, to prevent the data stall from continuing for the next and subsequent clock cycles, AND gates force DA to 0 so that it appears that $R0$ is being written, giving a condition which does not cause a stall. The registers to remain unchanged in the stall are the PC, the PC_{-1}, PC_{-2}, and the IR. These registers are replaced with registers with load control signals driven by $\overline{DHS}$. When $\overline{DHS}$ goes to 0, requesting a stall, the load signals become 0 and these pipeline platform registers hold their contents unchanged for the next clock cycle.

Returning to Figure 10-17, we see that in cycle 3 the data hazard for $R1$ is detected, so that $\overline{DHS}$ goes to 0 before the next clock edge. RW, MW, BS, and DA are set to 0, and at the clock edge, a bubble is launched into the EX stage for the ADD. At the same clock edge, the IF and DOF stages are stalled, so the information in them now is associated with clock cycle 4 instead of 3. In clock cycle 4, since DA_{EX} is 0, there is no stall, so the execution of the stalled ADD instruction proceeds. The same sequence of events occurs for the next ADD. Note that the execution diagram is identical to that in Figure 10-16(b), except that the NOPs are replaced by stalled instructions, shown in parentheses. Thus, although it removes the need for programming NOPs into the software, the data hazard stall solution has the same throughput penalty as the program with the NOPs.

A second hardware solution, *data forwarding*, does not have this penalty. Data forwarding is based on the answer to the following question: When a data hazard is detected, is the result available somewhere else in the pipeline, so that it can be used immediately in the operation having the data hazard? The answer is "almost." The result will be on Bus D, but it is not available until the next clock cycle. The result is to be written into the destination register during that clock cycle. The information needed to form the result, however, is available on the inputs to the pipeline platform that provides the inputs to MUX D. All that is needed to form the result during the current clock cycle is a multiplexer to select from the three values, just as MUX D does. MUX D' is accordingly added to produce the result on Bus D'. In Figure 10-19, instead of reading the operand from the register file, we use data forwarding to replace the operand with the value on Bus D$'$. This replacement is implemented with an additional input to MUX A and to MUX B from Bus D' as shown. Essentially the same logic as before is used to detect the data hazard, except that the separate detection signals HA and HB are used directly for A data and B data, respectively, so that the replacement occurs for the operand that has the data hazard.

The data-forwarding execution diagram for the three-instruction example appears in Figure 10-20. The data hazard for $R1$ is detected in cycle 3. This causes the value to go into $R1$ in the next cycle, to be forwarded from the EX stage of the first instruction in cycle 3. The correct value of $R1$ enters the DOF/EX platform at the next clock edge so that execution of the first ADD can proceed normally. The data hazard for $R2$ is detected in cycle 4, and the correct value is forwarded from the EX stage of the second instruction in that cycle. This gives the correct value in the DOF/EX platform needed for the second ADD to proceed normally. In con-

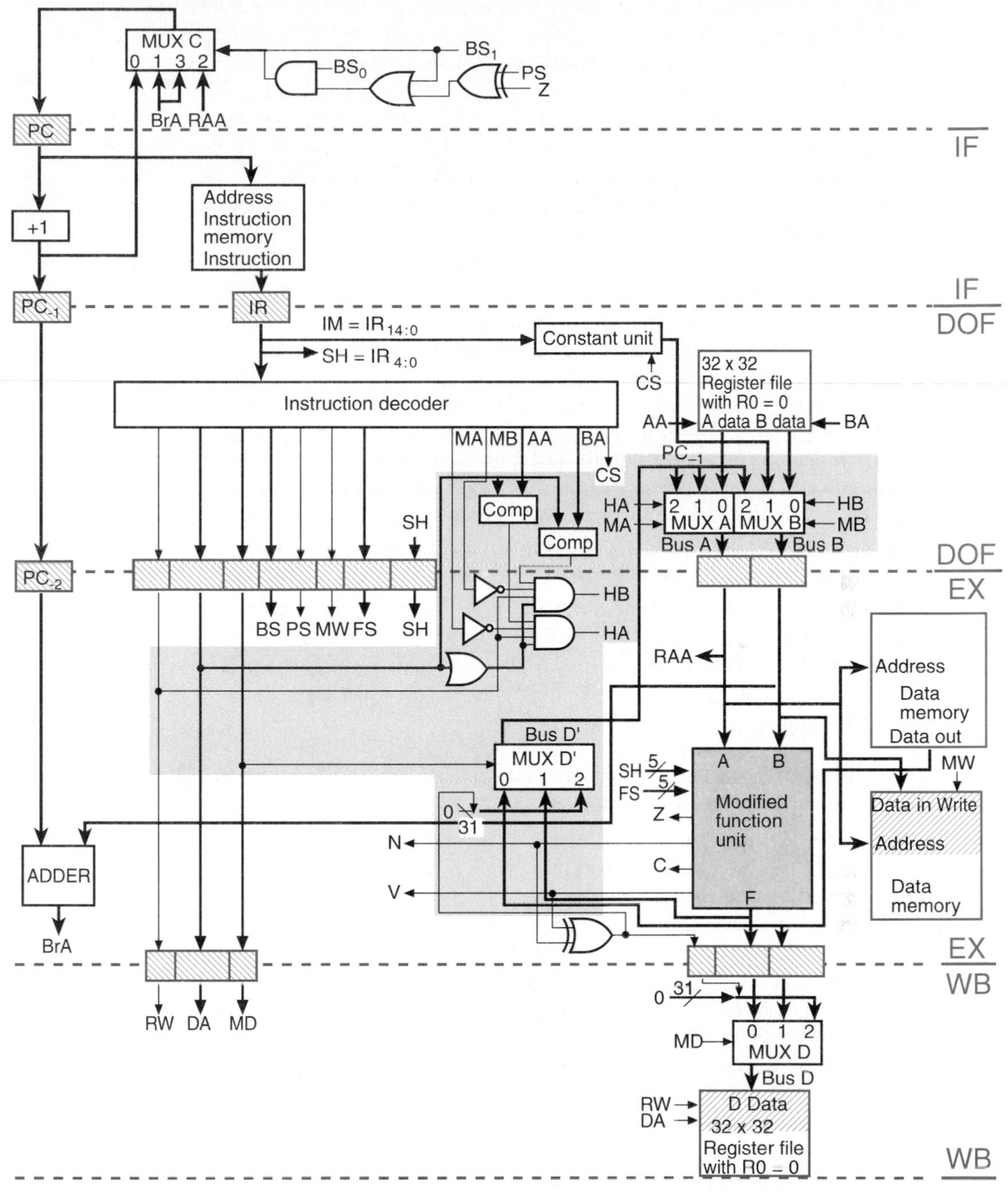

□ **FIGURE 10-19**
Pipeline RISC: Data Forwarding

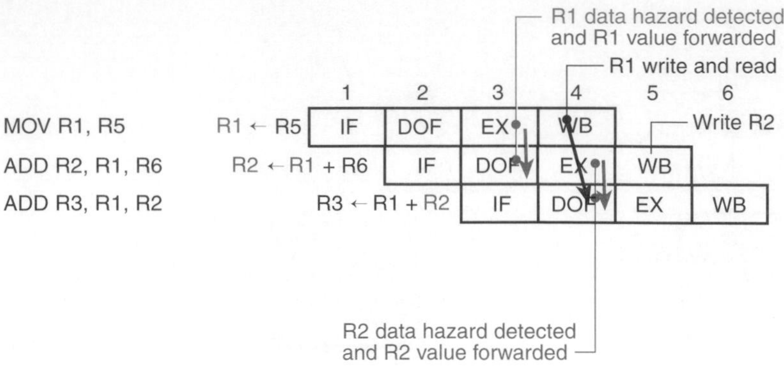

□ **FIGURE 10-20**
Example of Data Forwarding

trast to the data hazard stall method, data forwarding does not increase the number of clock cycles required to execute the program and hence does not affect the throughput in terms of the number of clock cycles required. It may, however, add combinational delay, causing the clock period to be somewhat longer.

Data hazards can also occur with memory access, as well as with register access. For the ST and LD instructions, it is not likely to be able to do a data memory read after a write in a single clock cycle. Further, some memory reads may take more than one clock cycle, in contrast to what we have assumed here. Thus, the reduction in throughput for a data hazard may be increased due to a longer delay before the data is available. (See Chapter 12.)

Control Hazards

Control hazards are associated with branches in the control flow of the program. The following program containing a conditional branch illustrates a control hazard:

1	BZ	R1, 18
2	MOV	R2, R3
3	MOV	R1, R2
4	MOV	R4, R2
20	MOV	R5, R6

The execution diagram for this program is given in Figure 10-21(a). If $R1$ is zero, then a branch to the instruction in location 20 (recall that addressing is PC relative) is to occur, skipping the instructions in locations 2 and 3. If $R1$ is non-zero, then the instructions in locations 2 and 3 are to be executed in sequence. Assume that the branch is taken to location 20 because $R1$ is equal to zero. The fact that $R1$ equals 0 is not detected until EX in cycle 3 of the first instruction in Figure 10-21(a). So the *PC* is set to 20 on the clock edge at the end of cycle 3. But the MOV instructions in locations 2 and 3 are into the EX and DOF stages, respectively, after the clock edge. Thus, unless corrective action is taken, these instructions will complete exe-

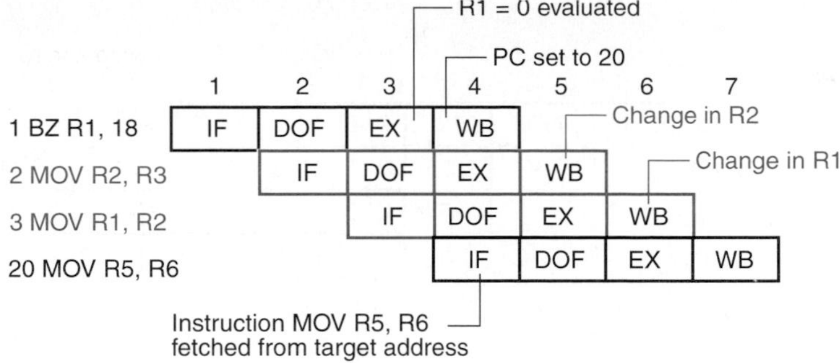

(a) Branch Hazard Problem

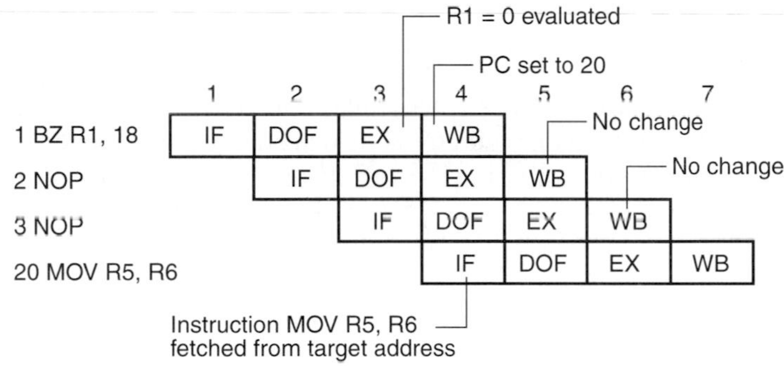

(b) Program-based Solution

☐ **FIGURE 10-21**
Example of Control Hazard

cution, even though the programmer's intention was for them to be skipped. This situation is one form of a *control hazard*.

NOP instructions can be used to deal with control hazards just as they were used with data hazards. The insertion of NOPs is performed by the programmer or compiler generating the machine language program. The program must be written so that only operations intended to be performed, regardless of whether the branch is taken, are introduced into the pipeline before the branch execution actually occurs. Figure 10-21(b) illustrates a modification of the simple three-line program that satisfies this condition. Two NOPs are inserted after the branch instruction BZ. These two NOPs can be performed regardless of whether the branch is taken in the EX stage of BZ in cycle 3 with no adverse effects on the correctness of the program. When control hazards in the CPU are handled in this manner by programming, the branch hazard dealt with by the NOPs is referred to as a *delayed branch*. Branch execution is delayed by two clock cycles in this CPU.

The NOP solution in Figure 10-21(b) increases the time required to process the simple program by two clock cycles, regardless of whether the branch is taken. Note,

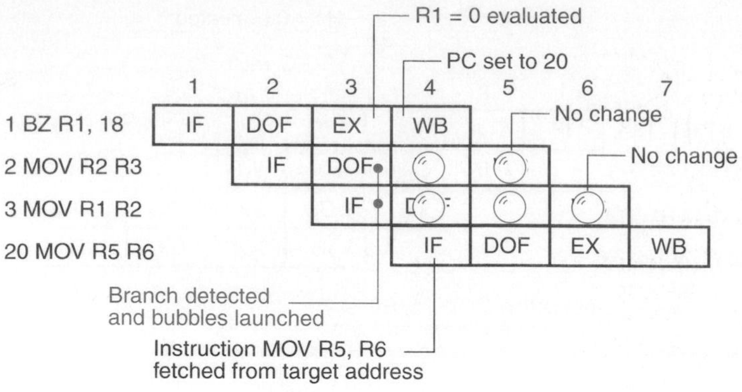

□ **FIGURE 10-22**
Example of Branch Prediction with Branch Taken

however, that these wasted cycles can sometimes be avoided by rearranging the order of instructions. Suppose that those instructions which are to be executed regardless of whether the branch is taken can be placed in the two locations following the branch instruction. In this situation, the lost throughput is completely recovered.

Just as in the case of the data hazard, a stall can be used to deal with the control hazard. But, also as in the case of the data hazard, the reduction in throughput will be the same as with the insertion of NOPs. This solution is referred to as a *branch hazard stall* and will not be presented here.

A second hardware solution is to use *branch prediction*. In its simplest form, this method predicts that branches will never be taken. Thus, instructions will be fetched and decoded and operands fetched on the basis of the addition of 1 to the value of the *PC*. These actions occur until it is known during the execution cycle whether the branch in question will be taken. If the branch is not taken, the instructions already in the pipeline due to the prediction will be allowed to proceed. If the branch is taken, the instructions following the branch instruction need to be cancelled. Usually, the cancellation is done by inserting bubbles into the execution and write-back stages for these instructions. This is illustrated for the four-instruction program in Figure 10-22. Based on the prediction that the branch will not be taken, the two MOV instructions after BZ are fetched, and the first one is decoded and its operands fetched. These actions take place in cycles 2 and 3. In cycle 3, the condition upon which the branch is based has been evaluated, and it is found that $R1 = 0$. Thus, the branch is to be taken. At the end of cycle 3, the *PC* is set to 20, and the instruction fetch in cycle 4 is performed using the new value of the *PC*. In cycle 3, the fact that the branch is taken has been detected, and bubbles are inserted into the pipeline for instructions 2 and 3. Proceeding through the pipeline, these bubbles have the same effect as two NOP instructions. However, because the NOPs are not present in the program, there is no delay or performance penalty when the branch is not taken.

The branch prediction hardware is shown in Figure 10-23. Whether a branch is taken is determined by looking at the selection values on the inputs to MUX *C*.

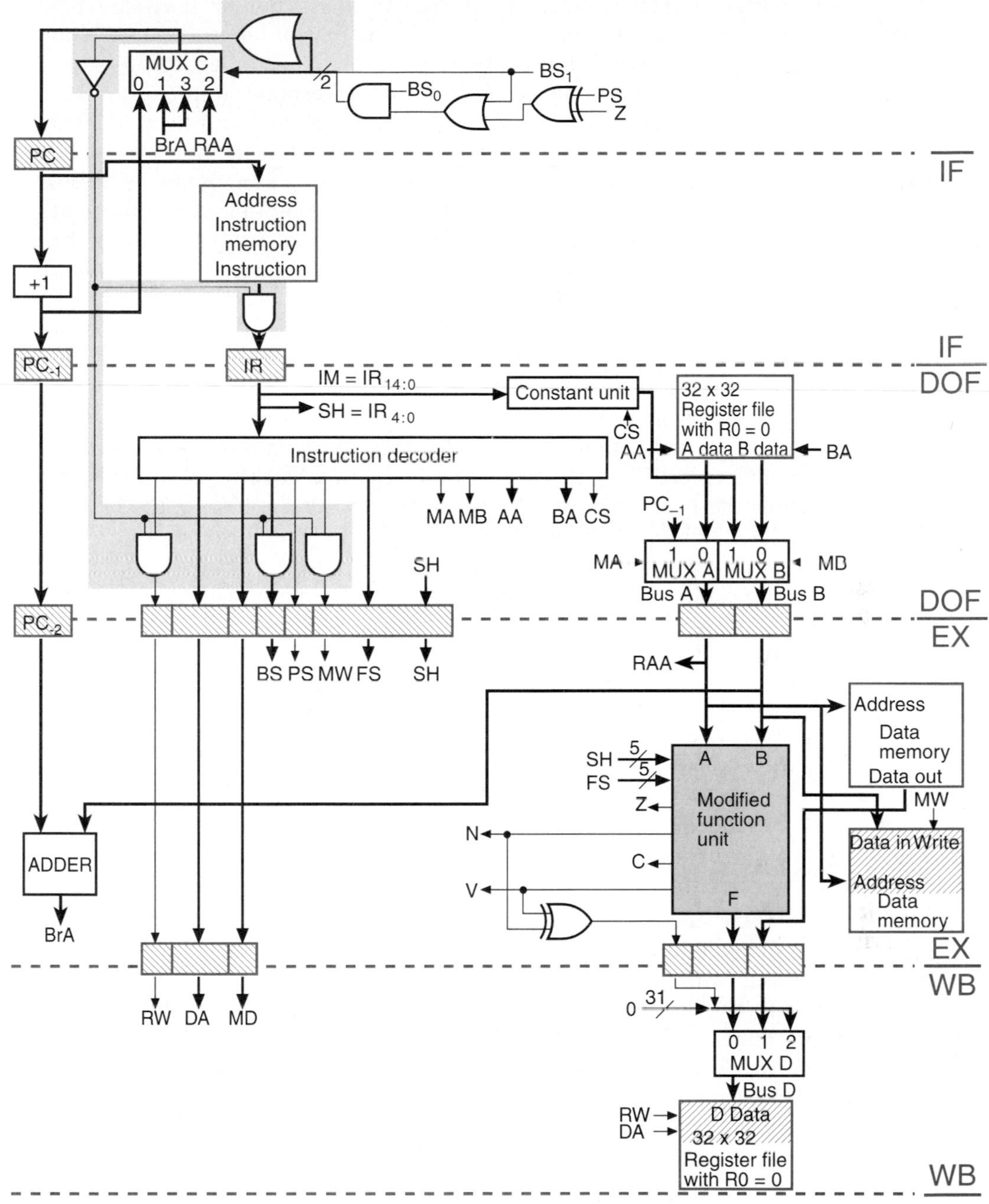

□ **FIGURE 10-23**
Pipelined RISC: Branch Prediction

If the pair of inputs is 01, then a conditional branch is being taken. If the pair is 10, then an unconditional JMR is occurring. If the pair is 11, then an unconditional JMP or JML is taking place. On the other hand, if the pair of inputs is 00, then no branch is occurring. Thus, a branch occurs for all combinations other than 00—i.e., for at least one 1—on the pair of lines. Logically, this corresponds to the OR of the lines, as shown in the figure. The output of the OR is inverted and then ANDed with the *RW* and *MW* fields so that the register file and the data memory cannot be written for the instruction following the branch instruction if the branch is taken. The inverted output is also ANDed with the *BS* field so that a branch in the next instruction is not executed. In order to cancel the second instruction following the branch, the inverted OR output is ANDed with the *IR* output. This gives an instruction of all 0's, for which the OPCODE field is defined as NOP. If the branch is not taken, however, the inverted OR output is 1, and the *IR* and the three control fields remain unchanged, given normal execution of the two instructions following the branch.

Branch prediction can also be done on the assumption that the branch is taken. In this case, the instructions must be fetched and operands fetched down the path of the branch target. Thus, the branch target address must be computed and used for fetching the instruction in the branch target location. In case the branch does not take place, however, the updated value of the *PC* must also be saved. As a consequence, this solution will require additional hardware to compute and store the branch target address. Nevertheless, if branches are more likely to be taken than not, the "branch taken" prediction may yield a more favorable cost-performance trade-off than the "branch not taken" prediction.

For simplicity of presentation, we have treated the hardware solutions for dealing with hazards one at a time. In an actual CPU, these solutions need to be combined. In addition, other hazards, such as those associated with writing and reading memory locations, need to be handled.

10-4 More on Design

The two designs considered in this chapter represent two different instruction set architectures and two different supporting CPU organizations. The CISC architecture matches well with the microprogrammed control organization, and the RISC architecture matches well with the pipelined control organization. In this section, we will deal with some of the issues surrounding the two architectures and the two organizations. We begin by doing a very simple comparison of the performance of the two architectures and organizations. Then we cover a few advanced concepts that build upon the foundations established to achieve very high performance. Finally, we relate the two organizations to more general digital systems design.

CISC-RISC Comparison

We compare the CISC and RISC CPUs on the basis of the execution of a simple series of instructions, all of which perform simple register-to-register operations on a pair of registers. Our primary focus will be throughput, i.e., the number of instructions executed per interval of time. Initially, we perform calculations based on clock cycles: then we add a crude estimate of the length of a clock cycle to introduce time into the calculations.

In the CISC processor, a simple register-to-register operation requires two cycles for instruction fetch, three cycles for operand fetch, two cycles for execution, two cycles for write-back, and one cycle for interrupt handing, for a total of 10 clock cycles. Since there is only one instruction being processed at a time, the throughput is 1/10 = 0.1 instruction per clock cycle. Based on the component delay values used in Section 8-7 plus added control and multiplexer delays the clock cycle length is 20 ns. This gives an instruction execution time of 200 ns and a throughput of 5.0 million instructions per second (MIPS).

In the RISC processor, assuming no data hazard or control hazard stalls, a simple register-to-register operation takes four clock cycles. But four instructions are being processed at once, so the throughput is 4/4 = 1 instruction per clock cycle. Based on the values used in Section 8-7, the clock cycle is 5 ns. Thus, resulting throughput is 200 MIPS.

On the basis of this analysis, it would appear that the RISC implementation is 40 times faster than the CISC implementation. But a number of factors have been neglected in the analysis. For one thing, a typical CISC instruction is able to do considerably more than a RISC instruction. For example, try adding the contents of a register to an operand from memory accessed by using indirect indexed addressing with the result returned to memory with the two architectures. The CISC architecture does this with a single instruction: the RISC architecture requires at least six instructions. For these types of instructions, the MIPS-based throughput ratio is cut to from 40 to about 9.

Also, our microprogrammed implementation was not designed specifically with performance in mind. For instance, it is possible to move the multiway branches currently located at the first address of the some microroutines to the address of the unconditional ROM-based jump at the end of some microroutines that precede the execution of those branches. By eliminating the ROM-based jumps, the number of cycles required for the simple register-to-register operation is reduced from 10 to 7. It is also possible to make the microprogrammed datapath and control into a 2-stage pipelined structure. This is done routinely in microprogrammed control designs and would reduce the clock cycle to 14 ns, reducing the original throughput ratio from 40 to 20. If six RISC instructions are required to perform the equivalent of a CISC instruction as previously illustrated, the throughput ratio is further reduced to about 4.7, ignoring the effects of branching that occur in instruction decoding.

Still, although the performance of the two designs now appears to be closer together, we must realize that the simplistic methods we just used for comparison are, in the final analysis, invalid. The most optimistic CISC performance is based on an instruction that makes full use of the power of the CISC architecture. Recall, however, that in doing a simple register-to-register instruction, the performance of the CISC was inferior to that of the RISC. Also, hazards interfere with the performance of the RISC. Truly, the performance of both architectures depends on the sequence of instructions executed in typical programs. Thus, we need to know how the two architectures compare in executing significant, typical, real programs. Such programs are referred to as *benchmarks*. The use of benchmarks, while far from perfect as an evaluation method, gives a better picture of the comparative performance of the two architectures. Even then, note that the comparisons depend on the methods used to assemble or compile the programs, since, unless the instruction set architectures are identical, the machine language code will be different.

In terms of benchmarking for general computation, pipeline-based implementations of RISC architectures often outperform comparable microprogram-controlled CISC implementations. Nevertheless, CISC architectures are very viable commercially because they avoid the cost of rewriting massive amounts of software developed for such architectures in the past. The inferiority in the performance of the CISC is partially remedied by using RISC-like pipelines within contemporary CISC implementations, as we illustrate later in this section.

Advanced CPU Concepts

Among the various methods used to design high-speed CPUs are multiple units organized as a pipeline-parallel structure, microprogramming with pipelines, super-pipelines, and superscalar architectures.

Consider the case in which an operation takes multiple clock cycles to execute, but the instruction fetch and write-back operations can be handled in a single cycle. Then it is possible to initiate an instruction every clock cycle, but not possible to complete the execution of an instruction every cycle. In such a situation, the performance of the CPU can be substantially improved by having multiple execution units in parallel. A high-level block diagram for this kind of system is shown in Figure 10-24. The instruction fetch, decoding, and operand fetch are carried out in the I-unit pipeline. In addition, the I-unit handles branches. When decoding of a non-branch instruction has been completed, the instruction and operands are *issued* to the appropriate E-unit. When execution of the instruction is completed by the E-unit, the write-back to the register file occurs. If a memory access is required, then the D-unit is used to execute the memory write. If the operation is a store, it goes immediately to the D-unit. Note that the actual execution units may be microprogrammed and may also involve internal pipelines.

Suppose that a sequence of three instructions—say, a multiplication, a 16-bit shift, and an addition—has no data hazards. Suppose further that there is a single pipelined E-unit that performs all of these operations, which take 17, 8, and 2 clock cycles, respectively, and that both the multiplication and the shift require multiple passes through portions of the E-unit pipeline. This situation allows only one clock cycle of overlap between pairs of the three instructions. Thus, the fastest that the sequence of operations executes in the E-unit is $17 + 8 + 2 - 2 = 25$ clock cycles. But with an E-unit for each operation, these operations can be executed in max $(17, 1 + 8, 2 + 2)$, which equals 17 clock cycles. The additional 1 and 2 are due to the issuing of one instruction per clock cycle to the E-unit set. The resulting execution throughput is improved by a factor of $25/17 = 1.5$.

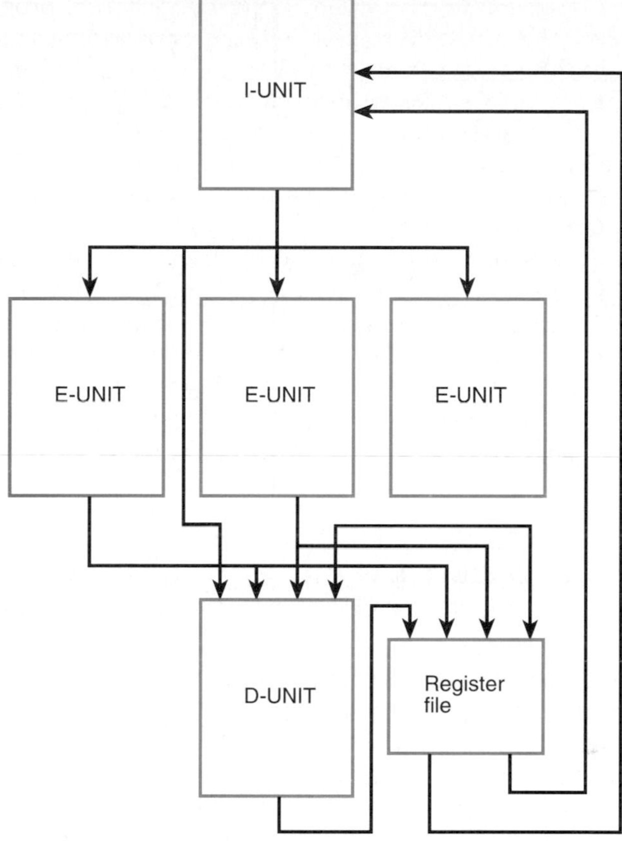

□ **FIGURE 10-24**
Multiple Execution Unit Organization

Suppose that we must implement a CISC architecture, but we are interested in initiating and completing close to one instruction per short RISC clock cycle for simple, frequently used instructions. We will use a pipelined data path and a combination of pipelined and microprogrammed control, as shown in Figure 10-25. An instruction is fetched and enters the Decode 1 stage. If it is a simple RISC-like instruction that executes completely in the normal Execute stage of the pipeline, it is partially decoded and passed on to the Decode 2 and operand fetch stage. There, decoding is completed, and the instruction is sent on down the pipe. On the other hand, the instruction requires multiple operations, multiple memory accesses, or sequences thereof, the Decode 1 stage produces a microcode address for the ROM. Also, it causes the Decode 2 stage to take microinstructions from the ROM, rather than from the partially decoded instruction from the Decode 1 stage. Execution of microinstructions from the ROM continues until execution of the instruction is completed.

Recall that to execute microinstructions, it is often necessary to have temporary registers in which to store information. An organization of this type will fre-

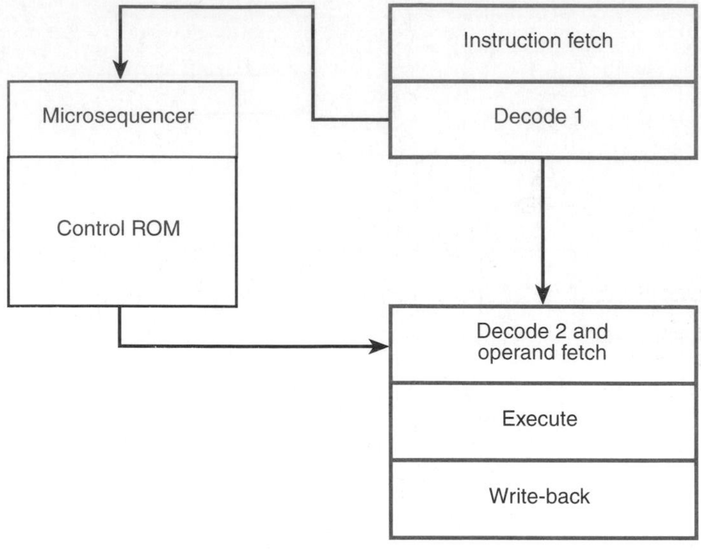

□ **FIGURE 10-25**
Combined CISC-RISC Oganization

quently supply temporary registers with a convenient mechanism for switching between temporary registers and the usual programmer-accessible register file.

The preceding organization supports an architecture that has combined CISC-RISC properties. It illustrate that pipelines and microprograms can be compatible and need not be viewed as mutually exclusive. The combined architecture allows new software to take advantage of a RISC architecture while preserving compatibility with older software dependent upon a CISC architecture.

In all of the methods considered thus far, the peak throughput possible is one instruction per clock cycle. With this limitation, it is desirable to maximize the clock rate by minimizing the maximum pipeline stage delay. If, as a consequence, a large number of pipeline stages is used, the CPU is said to be *superpipelined*. A superpipelined CPU will generally have a very high clock frequency, in the range of a few hundred or more MHz. In such an organization, however, handling hazards effectively is critical, since any stalling or reinitialization of the pipeline will degrade the performance of the CPU significantly. Also, as more pipeline platforms are added, further dividing up the combinational logic, the setup and propagation delay times of the flip-flops begin to dominate the platform-to-platform delay and the speed of the clock. The improvement achieved is less, and when hazards are taken into account, the performance may actually become worse rather than better.

For fast execution, an alternative to superpipelining is the use of a *superscalar* organization. The goal of this kind of organization is to have a peak rate of initiating instructions in excess of one instruction per clock cycle. A superscalar CPU that fetches a pair of instructions simultaneously by using a double-word wide path from instruction memory is illustrated in Figure 10-26. The processor checks for hazards among the instructions, as well as available execution units in the instruction issue

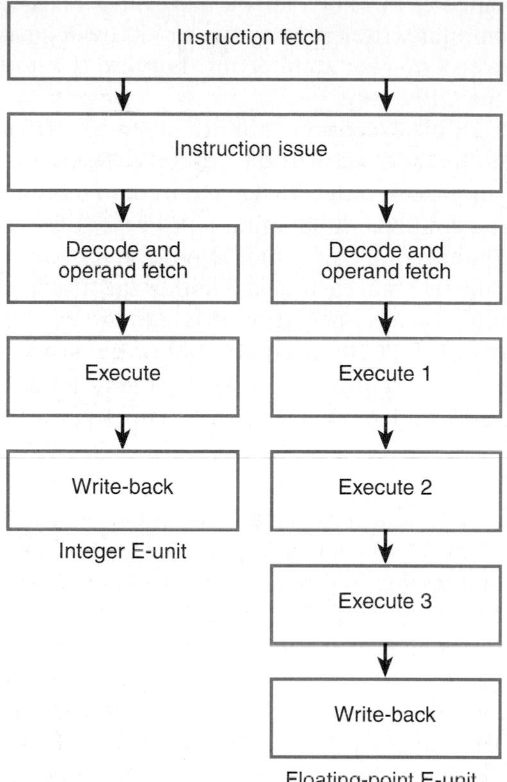

□ **FIGURE 10-26**
Superscalar Organization

stage of the pipeline. If there are hazards or busy execution units corresponding to the first instruction, then both instructions are held for later issuing. If the first instruction has no hazard and its E-unit is available, but there is a hazard or no available E-unit for the second instruction, then only the first instruction is issued. Otherwise, both instructions are issued in parallel. If a given superscalar architecture has the ability to issue up to four instructions simultaneously, then its peak execution rate is four instructions per clock cycle. If the clock cycle is 5 ns, then such a CPU has a peak execution rate of 800 MIPS. Note that the checks made for hazards between instructions in the execution stages and those in the issue stage become very complex as the maximum number of instructions issued simultaneously is increased. The resulting hardware complexity has the potential for increasing the clock cycle length, so the trade-offs in such a design need to be examined very carefully.

We close this section with two observations. First, as the quest for better performance causes us to design increasingly complex organizations, hazards cause the order of the instructions to play a more important role in the throughput that is achievable. Also, improved performance can be achieved by reducing the number of hazard-producing instructions, such as branches. As a consequence, to fully exploit the performance capabilities of the hardware, the assembly language programmer and the compiler writer need to be very knowledgeable about the behavior of not only the instruction set architecture, but also the underlying organization of the hardware of the CPU.

Finally, when multiple execution units are involved, very often the CPU design we have been considering here actually becomes the design for the entire processor, as is shown for the generic computer. This is apparent in the superscalar organization in Figure 10-26, which contains the floating-point unit (FPU). The FPU, the MMU, and the portion of the internal cache that handles data are effectively types of E-units. The portion of the internal cache that handles the instructions can be viewed as a part of the I-Unit that fetches instructions. Thus, in the quest for higher and higher throughput, the realm of the CPU becomes that of the processor, as in the generic computer. With the ever-growing complexity of integrated circuits, the processor is encompassing more and more of the electronic components of a computer.

Digital Systems

The two sizable digital system designs we have examined in this chapter are general-purpose CPUs. How does their design relate to that of other digital systems? First of all, each digital system has an architecture. Although that architecture may not in any way deal with instructions to be executed, it is likely that it still can be described by using register transfer descriptions and, possibly, one or more algorithmic state machines. On the other hand, it might have instructions, but they may be quite different from those for a CPU. The system may have no datapath at all or may have several datapaths. There is likely to be some form of control unit, and there may be multiple control units that interact. The system may or may not include memories. Thus, the total spectrum of digital systems has a very wide range of architectural possibilities.

So what is the connection of the general digital system to the content of this chapter? Simply stated, the connection is design techniques. To illustrate, consider that we have shown in detail how a system with instructions can be implemented using a datapath and a control unit. From here, it is relatively easy to implement a simpler, instructionless system. We have shown how an instruction can be easily decoded using multiplexers and a ROM. This technique can be applied to other digital systems having instructions. We have also shown how microprogramming has been used to implement controls for complex functions. If a system has one or more very complex functions, whether programmable or not, then a microprogrammed control is a possibility. Finally, we have shown how high speeds can be achieved by using pipelines or parallel execution units. Thus, if the goal of a system is high speed, then pipelining or parallel units are techniques to consider.

10-5 Chapter Summary

In this chapter, we examined two CPU designs: a CISC with a conventional datapath and microprogrammed control unit and a RISC with a pipelined datapath and control unit. The instruction set of the CISC uses many operations, with memory access supported by eight addressing modes. The CISC also has many operations that are complex in the sense that they require many clock cycles for their execution. The CISC uses multiple instruction lengths. The RISC, in contrast, has its memory access restricted to load and store operations and has only four addressing modes, most of which are restricted to specific operations. RISC operations are all simple in the sense that only *one* clock cycle is required for their execution. The RISC employs instructions of a single length.

The nonpipelined datapath for the CISC CPU employs a versatile shifter and status bits and uses modified register file addressing to provide temporary storage registers. In addition, it requires an added multiplexer capability in datapath buses. The RISC datapath is widened to 32 bits and has an enlarged register file. A barrel shifter provides multiple position shifts in a single clock cycle, and miscellaneous added features supply instruction set–specific capabilities.

The CISC control unit includes a stack pointer in addition to the program counter. Control microprograms reside in ROM, and a combination of a multiplexer and a ROM provides fast instruction decoding. The control unit also has extensive jump and conditional branching capabilities, including one level of microsubroutines. The microprogram for the control is modularized to permit many microsubroutines to be shared in implementing the microprogram for the instructions.

The RISC control unit is pipelined and has special hardware added to deal with branches. Pipelined CPUs have both data and control hazard problems. We examined one of each type of hazard, as well as software and hardware solutions for each.

After discussing CISC and RISC performance, we touched on some advanced concepts, including parallel execution units, a combination of microprogrammed control with a pipeline, superpipelined CPUs, and superscalar CPUs. Finally, we related the design techniques in this chapter to more general digital system design.

REFERENCES

1. DIETMEYER, D. L., *Logic Design of Digital Systems,* 3rd ed. Boston, MA: Allyn-Bacon, 1988.
2. MANO, M. M. *Computer Engineering: Hardware Design.* Englewood Cliffs, NJ: Prentice Hall, 1988.
3. HAMACHER, V. C., VRANESIC, Z. G., AND ZAKY, S. G. *Computer Organization,* 3rd ed. New York, NY: McGraw-Hill, 1990.
4. HENNESSY, J. L., AND PATTERSON, D. A. *Computer Architecture: A Quantitative Approach,* 2nd ed. San Francisco, CA: Morgan Kaufmann, 1996.
5. KANE, G., AND HEINRICH, J. *MIPS RISC Architecture.* Englewood Cliffs, NJ: Prentice Hall, 1992.
6. SPARC INTERNATIONAL, INC. *The SPARC Architecture Manual: Version 8.* Englewood Cliffs, NJ: Prentice Hall, 1992.

7. MANO, M. M. *Computer System Architecture,* 3rd ed. Englewood Cliffs, NJ: Prentice Hall, 1993.

8. PATTERSON, D. A., AND HENNESSY, J. L. *Computer Organization and Design: The Hardware/Software Interface.* San Mateo, CA: Morgan Kaufmann, 1994.

9. WEISS, S., AND SMITH, J. E. *POWER and PowerPC.* San Mateo, CA: Morgan Kaufmann, 1994.

10. WYANT, G., AND HAMMERSTROM, T. *How Microprocessors Work.* Emeryville, CA: Ziff-Davis Press, 1994.

PROBLEMS

The asterisk (*) indicates a more advanced problem.

10–1. Write a register transfer statement that describes the operation performed by each of the following instructions for the CISC design by using information from Table 10-1. In those cases in which a second word W is part of the instruction, it is given. Use hexadecimal integers for representing addresses and operands.

(a) 0001 0000 0000 0000
(b) 0100 1001 0000 0110   0000 1111 0000 1111
(c) 1000 0110 0101 0011   0000 0000 1111 1111
(d) 1100 0111 1000 0000   0000 0000 1111 0000

10–2. In the CISC formats in Figure 10-2, the format with $IR(15{:}14) = 11$ has bits $IR(3)$ through $IR(6)$ unused. Currently, this format is able to handle 16 operations. How many operations could be supported if the unused bits were added to the OPCODE? What modifications would be required to the instruction decoder in the CISC design to support this change?

10–3. *Suppose that the CISC instruction format in Figure 10-2 is changed to the following:

15 14 13 12	11 10 9	8 7 6	5 4 3	2 1 0
OPCODE	SMODE	DMODE	SRC	DST
W (Present if MOD ≠ 00)				

SMODE is a mode field for the source operand, and DMODE is a mode field for the destination. Due to the presence of these fields, the S field is no longer necessary.

(a) Based on this new format, how many operations can be specified using OPCODE alone?

(b) How many operations can be specified by: (1) using 000X as the OPCODE specifier for zero-address instruction, with vacant bits $IR(11)$ through $IR(0)$ used as additional OPCODE bits, (2) using 01XX as the OPCODE specifier for one-address instructions, with vacant bits $IR(11)$ through $IR(9)$ as additional OPCODE bits, (3) using 1XXX as the OPCODE specifier for two-address operations, and (4) using 001X as the OPCODE specifier for branch operations, with vacant bits $IR(11)$ through $IR(9)$ as additional OPCODE bits?

10–4. The following instruction in location 2001_{10} is executed for each of the eight addressing modes in Table 10-2.

0 1 0 0 1 1	MODE	0 0 0 0	0 1 1
W (Present if MOD ≠ 00X) = 1000_{10}			

Assuming R_i contains i_{10} and M_i contains i_{10}, give the eight effective addresses that result.

10–5. Using the CISC instruction set, write an efficient assembly language program to add the contents of:
(a) registers $R1$ through $R6$, with the result placed in $R7$.
(b) memory locations 100_{10} through 120_{10}, with the result placed in 125_{10}.
In both cases, all register contents are to remain unchanged, except for those of $R7$.

10–6. For each of the CISC shift operations given, shifter input A equals $F0C6_{16}$, and C equals 1. Using the shifter in Figure 10-5, find shifter output H in hexadecimal for each of the following shift operations
(a) Logical shift left (lsl)
(b) Rotate right with carry (rorc)
(c) Arithmetic shift right (asr)
(d) Rotate left (rol)

10–7. *Design the *PSR* and *MSTS* hardware for the CISC design based on Figure 10-7, Table 10-1, and Table 10-7. Use D flip-flops, gates, and a ROM or PLA. Note carefully all sources that may modify the carry bit.

10–8. Draw an ASM chart for each of the following:
(a) the logical shift right microcode given in Table 10-14.
(b) the microcode for a rotate left with carry operation.

10–9. Write microcode for each of the following two-operand addressing modes with S = 0. Give both a register transfer description and a hexadecimal representation for binary code for each microinstruction as in Table 10-14.
(a) Register indirect
(b) Immediate
(c) Indexed indirect
(d) Relative

10–10. Write microcode for the execution part of each of the following instructions. Give both a register transfer description and a hexadecimal representation for binary code for each microinstruction as in Table 10-14.
(a) Arithmetic shift right
(b) Compare
(c) Add with carry
(d) Branch if plus

10–11. Write microcode for the execution part of each of the following instructions. Give both a register transfer description and the hexadecimal representation for the binary code for each microinstruction as in Table 10-14.
(a) Rotate left with carry

(b) Exclusive-OR
(c) Branch if no carry
(d) Subtract with borrow

10–12. *Write a microprogram for the execution part of the multiply operation that uses the add and shift right algorithm. Assume that all operands are unsigned integers and that you are to provide both register transfer statements and hexadecimal microcode as in Table 10-14. The multiplicand is in the destination location, and the multiplier is in the source location. After the multiplication is complete, the least significant half of the result is in the source location, and the most significant half is in the destination location. (*Hint:* You will find the rotate right with carry microoperation particularly useful.)

10–13. *Write a microprogram for the divide operation. Assume that all operands are unsigned integers and that you are to provide both register transfer statements and hexadecimal microcode as in Table 10-14. The 16-bit dividend is in the destination location, and the divisor is in the source location. After the division is complete, the quotient is in the source location, and the remainder is in the destination location.

10–14. Write a microprogram for the move string operation. This operation is to copy a string of words of length L in memory locations A through $A + L - 1$ into memory locations B through $B + L - 1$. The three integers A, B, and L are stored on the top of the stack, with A the topmost element. Provide register transfer statements and hexadecimal microcode as in Table 10-14.

10–15. For each of the RISC operations in Table 10-20, list the addressing mode or modes used.

10–16. Simulate the operation of the barrel shifter in Figure 10-14 for each of the following shifts and $A = 0FC84210_{16}$. List the hexadecimal values on the 47 lines, 35 lines, and 32 lines out of the three levels of the shifter.
(a) Left, $SH = 10$
(b) Right, $SH = 13$
(c) Left, $SH = 27$

10–17. For the RISC CPU in Figure 10-15, simulate, in hexadecimal, the processing of the instruction ADI R1 R17 2F00 located in $PC = 10F$. Assume that $R17$ contains 0000000F. Show the contents of each of the pipeline platforms and of the register file (the latter only when a change occurs) for each of the clock cycles.

10–18. Repeat Problem 10–17 for the instruction SLT R31 R17 R16 with $R16$, containing 0000100F.

10–19. Repeat Problem 10–17 for the instruction LSL R1 R17 000F.

10–20. *Use a computer-based logic minimization program to design the instruction decoder for a RISC from Table 10-20. The field FS need not be done, since it can be wired directly from OPCODE.

10–21. For the RISC design, draw the execution diagram for the following RISC program, and indicate any data hazards that are present:

```
1 MOV    R7, R6
2 SUB    R8, R8, R6
3 AND    R8, R8, R7
```

10–22. For the RISC design , draw the execution diagram for the following RISC program (with the contents of *R*7 nonzero after the subtraction), and indicate any data or control hazards that are present:

```
1 SUB    R7, R7, R6
2 BNZ    R7, 000F
3 AND    R8, R7
4 OR     R5, R7
```

10–23. Rewrite the RISC programs in Problem 10–21 and Problem 10–22 using NOPs to avoid all data and control hazards.

10–24. Draw the execution diagrams for the program in Problem 10–21, assuming
(a) the RISC CPU with data stall given in Figure 10-18.
(b) the RISC CPU with data forwarding in Figure 10-19.

10–25. Simulate the processing of the program in Problem 10–21 using the RISC CPU with data hazard stall in Figure 10-18. Give the contents of each pipeline platform and the register file (the latter only whenever a change occurs) for each clock cycle. Initially, *R*6 contains 00000010_{16}, *R*7 contains 00000020_{16}, *R*8 contains 00000030_{16}, and the *PC* contains 00000001_{16}. Is the data hazard avoided?

10–26. Repeat Problem 10–25 using the RISC CPU with data forwarding in Figure 10-19.

10–27. Draw the execution diagram for the program in Problem 10–22, assuming the combination of the RISC CPU with branch prediction in Figure 10-23 and the RISC CPU with data forwarding in Figure 10-19.

10–28. Assuming that the CISC CPU clock cycle is four times that of the RISC CPU clock cycle,
(a) Write a RISC program for the RISC CPU with data forwarding that will execute a CISC INC operation with the destination addressed using index register-indirect mode.
(b) Compare the processing time for the RISC program with the processing time for the CISC instruction. Include instruction fetch in both calculations.

10–29. *Assuming that the CISC CPU clock cycle is four times that of the RISC CPU clock cycle,
(a) Write the microcode for a two-operand fetch for the relative indirect mode.
(b) Write a RISC program for the RISC CPU with data forwarding that will execute a CISC ADD operation with the destination operand in a register and the source operand addressed using relative indirect mode.
(c) Compare the processing time for the RISC program with the processing time for the CISC ADD instruction. Include instruction fetch in both calculations.

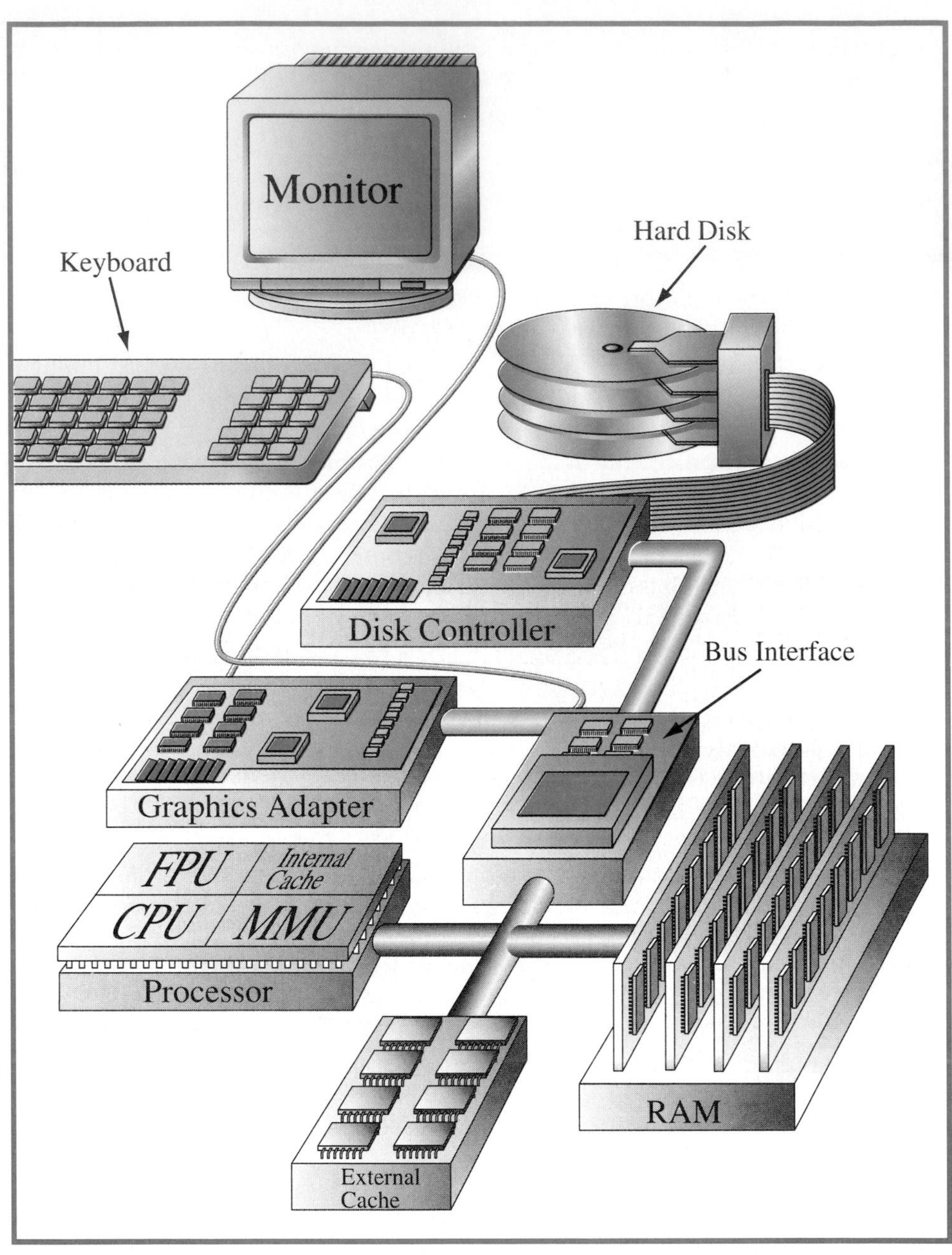
Monitor
Hard Disk
Keyboard
Disk Controller
Bus Interface
Graphics Adapter
FPU
Internal Cache
CPU
MMU
Processor
RAM
External Cache

CHAPTER 11

INPUT-OUTPUT AND COMMUNICATION

In this chapter, we give an overview of selected aspects of computer input-output (I/O) and communication between the CPU and I/O devices, I/O interfaces, and I/O processors. Because of the wide variety of different I/O devices and the quest for faster handling of programs and data, I/O is one of the most complex areas of computer design. As a consequence, we are able to present only selected pieces of the I/O puzzle. We illustrate just three devices in detail: a keyboard, a hard disk, and a graphics display. We then introduce the I/O bus and the I/O interfaces that connect to these devices. We consider serial communication and use the I/O structure for the keyboard as an illustration. We discuss four modes for performing data transfers: program-controlled transfer, interrupt-initiated transfer, direct memory access, and the use of an I/O processor.

In terms of the generic computer, it is apparent from the blue shading that I/O involves a very large part of the computer. Only the processor, external cache, and RAM are not as highly involved, although they, too, are used extensively in directing and performing I/O transfers. Even the generic computer, which has fewer I/O devices than most PC systems, has a diverse set of such devices requiring significant digital electronic hardware for support.

11-1 COMPUTER I/O

The input and output subsystem of a computer provides an efficient mode of communication between the CPU and the outside environment. Programs and data must be entered into the memory for processing, and results obtained from computations must be recorded or displayed. Among the input and output devices that are commonly found in computer systems are keyboards, monitors, printers, magnetic disks, and compact disk read-only memory (CD-ROM) drives. Other input and output devices frequently encountered are modems or other communication interfaces, magnetic tape drives, scanners, and sound cards with speakers and microphones. Significant numbers of computers, such as those used in automobiles, have analog-to-digital converters, digital-to-analog converters, and other data acquisition and control components.

The I/O facility of a computer is a function of its intended application. This results in a wide diversity of attached devices and corresponding differences in the needs for interacting with them. Since each device behaves differently, it would be time consuming to dwell on the detailed interconnections needed between the computer and each peripheral. We will, therefore, examine just three peripherals that appear in most computers: the keyboard, the hard disk, and the graphics display. These represent typical points in the range of data transfer rates required for peripherals. In addition, we present some of the common characteristics found in the I/O subsystem of computers, as well as the various techniques available for transferring data either in parallel, using many conducting paths, or serially, through communication lines.

11-2 Sample Peripherals

Devices that the CPU controls directly are said to be connected *on-line*. These devices communicate directly with the CPU or transfer binary information into or out of the memory upon command from the CPU. Input or output devices attached to the computer on-line are called *peripherals*. In this section, we examine three peripheral devices: a keyboard, a hard disk, and a graphics display. We also use the keyboard as an example to illustrate I/O concepts in a later section. We introduce the hard disk both to motivate the need for direct memory access and to provide background for the role of the device in Chapter 12 as a component in a memory hierarchy. We include the graphics display to illustrate the very high potential transfer rate requirements of contemporary applications.

Keyboard

The keyboard is among the simplest of the electromechanical devices attached to the typical computer. Since it is manually controlled, it has one of the slowest data rates of any peripheral.

The keyboard consists of a collection of keys that can be depressed by the user. It is necessary to detect which of the keys have been depressed. To do this, a *scan matrix* that lies beneath the keys is used, as shown in Figure 11-1. This two-dimensional matrix is conceptually similar to matrix used in RAM. The matrix shown in the figure is 8×16, giving 128 intersections, so it can handle up to 128 keys. A decoder drives the X lines of the matrix, which are analogous to the word lines of a RAM. A multiplexer is attached to the Y lines of the matrix, which are analogous to the bit lines of a RAM. The decoder and the multiplexer are con-

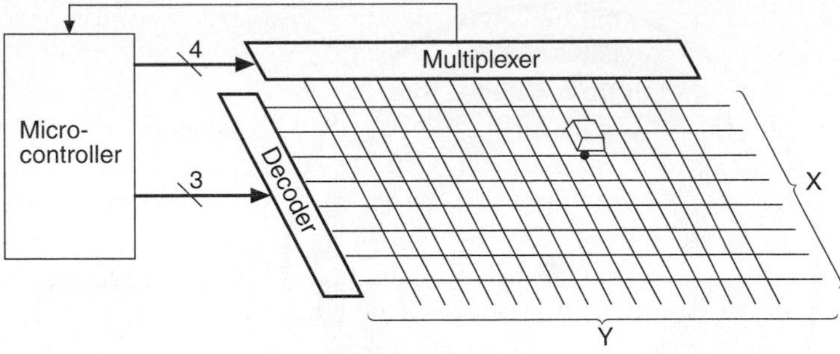

☐ **FIGURE 11-1**
Keyboard Scan Matrix

trolled by a microcontroller, a tiny computer that contains RAM, ROM, a timer, and simple I/O interfaces.

The microcontroller is programmed to periodically scan all intersections in the matrix by manipulating the control inputs of the decoder and multiplexer. If the key is depressed at an intersection, a signal path is closed from an output of the *X* decoder to an input of the *Y* multiplexer. The existence of this path is sensed at an input to the microcontroller. The 7-bit control code applied to the decoder and multiplexer at the time identifies the key. To allow for "rollover" in typing, in which multiple keys are depressed before any of them is released, the microcontroller actually identifies the depressing and release of the keys. Whether a key is depressed or released, the control code at the time of the event is sensed and is translated by the microcontroller into a *K-scan code*. When a key is depressed, a *make code* is produced; when a key is released, a *break code* is produced. Thus, there are two codes for each key, one for when the key is depressed and one for when the key is released. Note that the scanning of the entire keyboard occurs hundreds of times per second, so there is no danger of missing any depressions or releases of keys.

After presenting a number of I/O interface concepts, we will revisit the keyboard to see what happens to the K-scan codes before they are finally translated to ASCII characters.

Hard Disk

The hard disk is the primary intermediate-speed, nonvolatile, writable storage medium for most computers. The typical hard drive stores information serially on a nonremovable disk with a few many platters, as shown in the upper right of the generic computer. Each platter is magnetizable on one or both surfaces. There are one or more read/write *heads* per recording surface; for the remainder of our discussion, we will assume a single head per surface. Each platter is divided into concentric *tracks*, as illustrated in Figure 11-2. The set of tracks that are at the same distance from the center of the disk on all platter surfaces is referred to as a *cylinder*. Each track is divided into *sectors* containing a fixed number of bytes. The num-

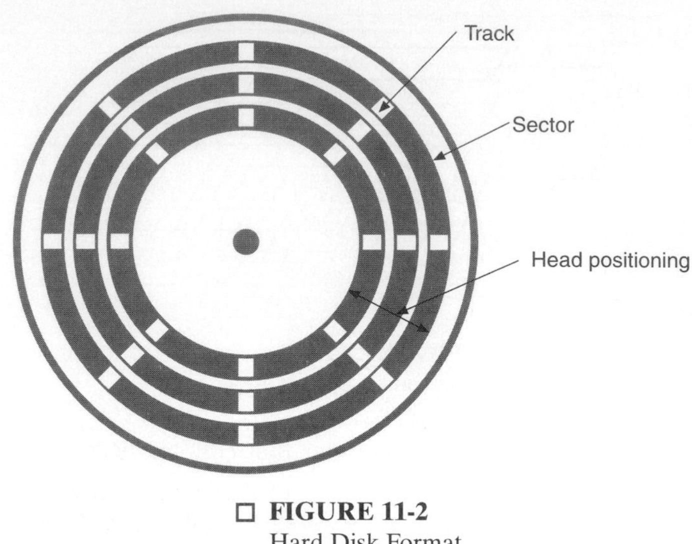

□ **FIGURE 11-2**
Hard Disk Format

ber of bytes per sector typically ranges from 256 to 4K. The typical byte address includes the cylinder number, head number, sector number, and word offset within the sector. The addressing assumes that the number of sectors per track is fixed. In modern, high-capacity disks, more sectors are included in the longer outer tracks than in the shorter inner tracks. In addition, a number of spare sectors are reserved to take the place of defective sectors. As a consequence of these design choices, the actual physical address of a sector on the disk is likely to be different from the address of the sector sent to the disk controller. The mapping from this address to the physical address is typically accomplished in the disk controller or drive electronics.

To enable information to be accessed, the set of heads is mounted on an actuator that can move the heads radially over the disk, as shown in the generic computer. The time required to move the heads from the current cylinder to the desired cylinder is called the *seek time.* The time required to rotate the disk from its current position to that having the desired sector under the heads is called the *rotational delay.* In addition, a certain amount of time is required by the disk controller to access and output information. This time is the *controller time.* The time required to locate a word on the disk is the *disk access time*, which is the sum of the controller time, the seek time, and the rotational delay. Average values over all possibilities are used for these four parameters. Words may be transferred singly, but as we will see in Chapter 12, they are often accessed in blocks. The transfer rate for a block of words, once the block has been located, is the *disk transfer rate,* typically specified in megabytes/second (MB/s). The transfer rate required by the CPU-memory bus to transfer a sector from disk is the number of bytes in the sector divided by the length of time taken to read a sector from the disk. The length of time required to read a sector is equal to the proportion of the cylinder occupied by the sector divided by the rotational speed of the disk. For example, with 63 sec-

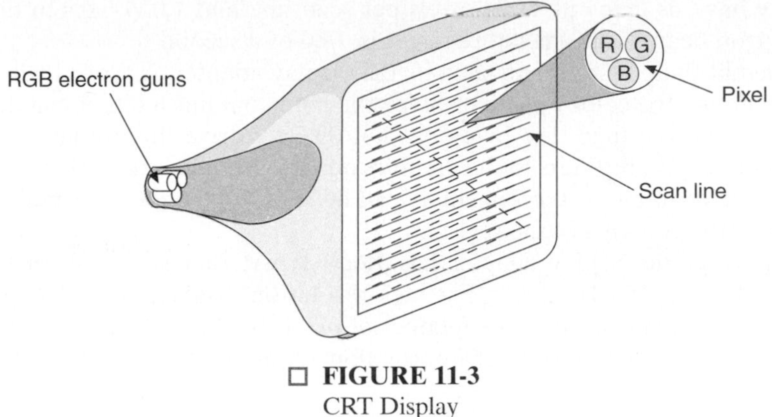

□ **FIGURE 11-3**
CRT Display

tors, 512 B per sector a rotational speed of 5,400 rpm, and allowance for the gap between sectors, this time is about 0.15 ms, giving a transfer rate of 512/0.15 ms = 3.4 MB/s. The controller will store the information read from the sector in its memory. The sum of the disk access time and the disk transfer rate times the number of bytes per sector gives an estimate of the time required to transfer the information in a sector to or from the hard disk. Typical values in the mid-1990s are a seek time of 10 ms, a rotational delay of 6 ms, a sector transfer time of 0.15 ms, and a negligible controller time, giving an access time for an isolated sector of 16.15 ms.

Graphics Display

The graphics display is the primary output device for the interactive use of computers. Displays use a number of different technologies, the most prevalent of which is currently the cathode-ray tube (CRT), illustrated in Figure 11-3. The most modern versions of the CRT display are based on analog signals, which are generated on the display adapter board. The display is defined in terms of picture elements called *pixels*. The color display has three locations associated with each pixel on the screen. These locations correspond to the primary colors red, green, and blue (RGB). At each location, there is the corresponding colored phosphor. A phosphor emits light of its color when excited by a beam of electrons. In order to excite the three phosphors simultaneously, three electron guns are used, one for red, one for green, and one for blue—hence the RGB electron guns shown in the figure. The color that results for a given pixel is determined by the intensity of the electron beams striking the phosphors within the pixel.

The electron beams are scanned across the screen to form a set of horizontal lines called *scan lines*. This set of lines is referred to as a *raster*. The lines are scanned from top to bottom, beginning at the upper left and ending at the lower right. The electron guns remain at zero intensity as they scan from right to left in preparation for drawing the next scan line. The resolution of the information displayed is given in terms of the number of pixels per scan line and the number of scan lines in the raster. A high-resolution super video graphics array (SVGA) display may have as many as 1,280 pixels per scan line and 1,024 lines in the raster. The electron beams scan the entire raster in 1/60 of a second.

Each of the pixels is controlled by the display adapter. A typical adapter uses a byte to define the color of a pixel. Since the byte contains 8 bits, it can define 256 colors at any given time. The byte does not directly drive the display, but instead selects 1 out of 256 registers in the graphics adapter to define the color. Each register is 20 bits or more, so the 256 colors can be selected from over 1 million colors by defining the contents of the registers.

Typically, the display adapter has video RAM that stores all of the bytes which control the display pixels. For high-resolution display with 1,280 pixels per scan line and 1,024 scan lines, the number of pixels is $1{,}280 \times 1{,}024 = 1{,}310{,}720$. So for 256 colors, a single screen of information requires at least 1.25 MB of video RAM.

I/O Transfer Rates

An indicated earlier, the three peripheral devices discussed in this section give a sense of the range of peak I/O transfer rates. The keyboard data transfer rate is less than 10 bytes/s. For the hard disk, while the disk controller is capturing the data arriving rapidly from the disk in the sector buffer, the transfer of data from the buffer to main memory is impossible. Thus, in the case in which the next sector is to be read immediately, all of the data from the sector buffer needs to be stored in main memory during the time the gap on the disk between the sectors passes under the disk head. For 63 sectors and a rotational speed of 5,400 rpm, this time is about 25 μs. Thus, the peak transfer rate required is 512B/25 ms = 20 MB/s. For a display with 256 colors, if a screen is to be changed entirely every 1/60 of a second, 1.25 MB of data must be delivered to the video RAM from the CPU in that amount of time. This requires a data rate of $1.25 \text{ MB} \times 60 = 75 \text{ MB/s}$.

Based on the preceding, we can conclude that the peak data rates required by the particular peripherals we have considered have a wide range. The rates for the hard disk and the display are high enough compared to the maximum rate of transfer on the computer buses to provide a challenge to designers. Attempts to meet this challenge use techniques in the disk controller and the graphics adapter to reduce the peak transfer rates required and use fast bus designs between the peripheral interfaces and memory.

11-3 I/O Interfaces

Peripherals connected to a computer need special communication links to interface them with the CPU. The purpose of these links is to resolve the differences in the properties of the CPU and memory and the properties of each peripheral. The major differences are as follows:

1. Peripherals are often electromechanical devices whose manner of operation is different from that of the CPU and memory, which are electronic devices. Therefore, a conversion of signal values may be required.

2. The data transfer rate of peripherals is usually different from the clock rate of the CPU. Consequently, a synchronization mechanism may be needed.
3. Data codes and formats in peripherals differ from the word format in the CPU and memory.
4. The operating modes of peripherals differ from each other, and each must be controlled in a way that does not disturb the operation of other peripherals connected to the CPU.

To resolve these differences, computer systems include special hardware components between the CPU and the peripherals to supervise and synchronize all input and output transfers. These components are called *interface units* because they interface between the bus from the CPU and the peripheral device. In addition, each device has its own controller to supervise the operations of the particular mechanism of that peripheral. For example, the controller in a printer attached to a computer controls the motion of the paper, the timing of the printing, and the selection of the characters to be printed.

I/O Bus and Interface Unit

A typical communication structure between the CPU and several peripherals is shown in Figure 11-4. Each peripheral has an interface unit associated with it. The common bus from the CPU is attached to all peripheral interfaces. To communicate with a particular device, the CPU places a device address on the address bus. Each interface attached to the common bus contains an address decoder that monitors the address lines. When the interface detects its own address, it activates the path between the bus lines and the device that it controls. All peripherals with addresses that do not correspond to the address on the bus are disabled by their interface. At the same time that the address is made available on the address bus, the CPU provides a function code on the control lines. The selected interface responds to the function code and proceeds to execute it. If data must be transferred, the interface communicates with both the device and the CPU data bus to synchronize the transfer.

In addition to communicating with the I/O devices, the CPU of a computer must communicate with the memory unit through an address and data bus. There are three ways that external computer buses communicate with memory and I/O. One method uses common data, address, and control buses for both memory and I/O. We have referred to this configuration as *memory-mapped I/O*. The common address space is shared between the interface units and memory words, each having distinct addresses. Computers that adopt the memory-mapped scheme read and write from interface units as if they were assigned memory addresses by using the same instructions that read from and write to memory.

The second alternative is to share a common address bus and data bus, but use different control lines for memory and I/O. Such computers have separate read and write lines for memory and I/O. To read or write from memory, the CPU activates the memory read or memory write control. To perform input to or output from an interface, the CPU activates the read I/O or write I/O control, using spe-

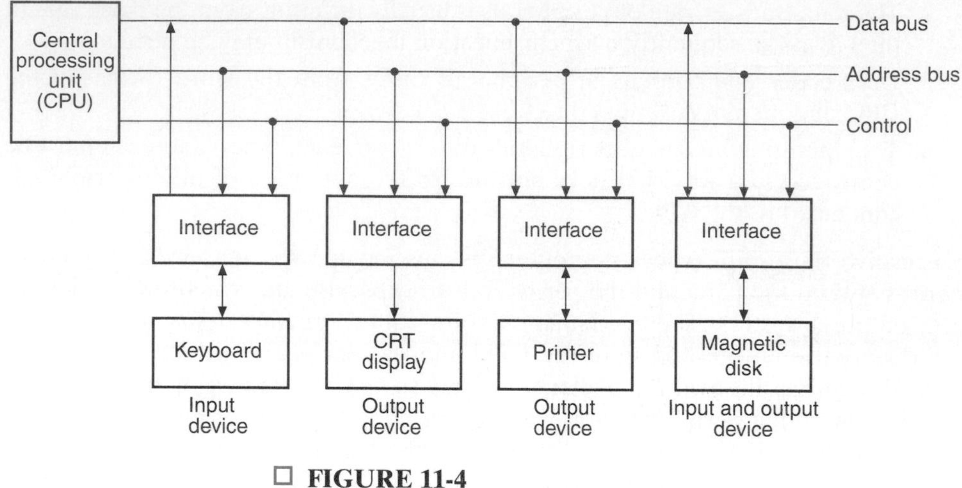

□ **FIGURE 11-4**
Connection of I/O Devices to CPU

cial instructions. In this way, the addresses assigned to memory and I/O interface units are independent from each other and are distinguished by separate control lines. This method is referred to as the *isolated I/O configuration.*

The third alternative is to have two independent sets of data, address, and control buses. This is possible in computers that include an *I/O processor* in the system in addition to the CPU. The memory communicates with both the CPU and I/O processor through a common memory bus. The I/O processor communicates with the input and output devices through separate address, data, and control lines. The purpose of the I/O processor is to provide an independent pathway for the transfer of information between external devices and internal memory. The I/O processor is sometimes called a *data channel.*

Example of I/O Interface

A typical I/O interface unit is shown in block diagram form in Figure 11-5. It consists of two data registers called *ports*, a control register, a status register, a bidirectional data bus, and timing and control circuits. The function of the interface is to translate the signals between the CPU buses and the I/O device and to provide the needed hardware to satisfy the two sets of timing constraints.

The I/O data from the device can be transferred into either port A or port B. The interface may operate with an output device, with an input device, or with a device that requires both input and output. If the interface is connected to a printer, it will only output data; if it services a scanner, it will only input data. A hard disk transfers data in both directions, but not at the same time; so the interface needs only one set of I/O bidirectional data lines.

The *control register* receives control information from the CPU. By loading appropriate bits into this register, the interface and the device can be placed in a variety of operating modes. For example, port A may be defined as an input port

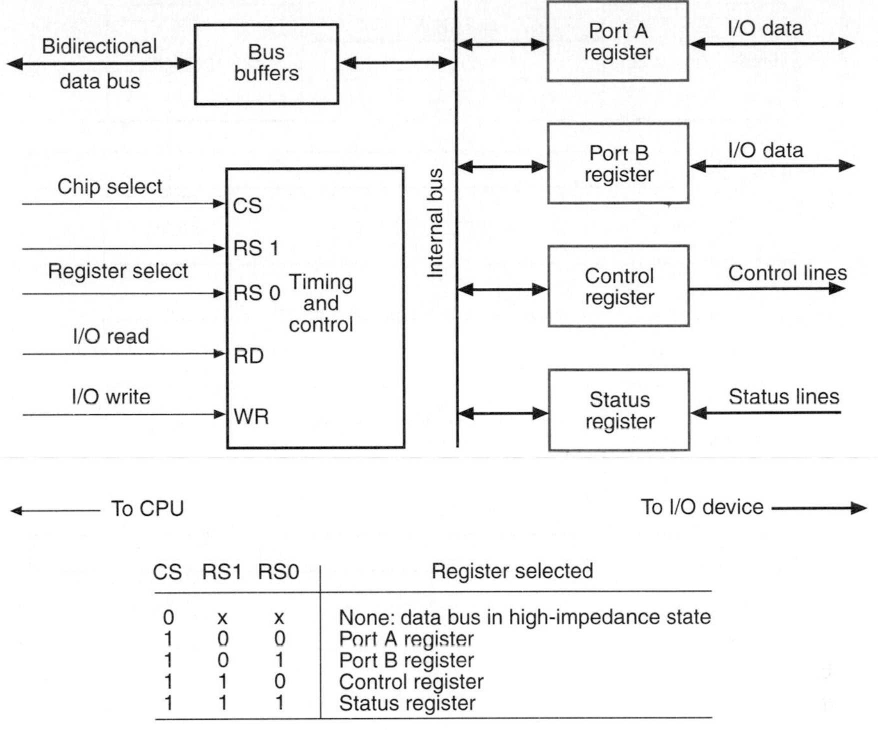

CS	RS1	RS0	Register selected
0	x	x	None: data bus in high-impedance state
1	0	0	Port A register
1	0	1	Port B register
1	1	0	Control register
1	1	1	Status register

□ **FIGURE 11-5**
Example of I/O Interface Unit

only. A magnetic tape unit may be instructed to rewind the tape or to start the tape moving in the forward direction. The bits in the status register are used for status conditions and for recording errors that may occur during data transfer. For example, a status bit may indicate that port A has received a new data item from the device, while another bit in the status register may indicate that a parity error has occurred during the transfer.

The interface registers communicate with the CPU through the bidirectional data bus. The address bus selects the interface unit through the chip select input and the two register select inputs. A circuit (usually a decoder or a gate) detects the address assigned to the interface registers. This circuit enables the chip select (CS) input when the interface is selected by the address bus. The two *register select inputs* $RS1$ and $RS0$ are usually connected to the two least significant lines of the address bus. These two inputs select one of the four registers in the interface, as specified in the table accompanying the diagram in Figure 11-5. The contents of the selected register are transferred into the CPU via the data bus when the I/O read signal is enabled. The CPU transfers binary information into the selected register via the data bus when the I/O write input is enabled.

The CPU, interface, and I/O device are likely to have different clocks that are not synchronized with each other. Thus, these units are said to be *asynchronous*

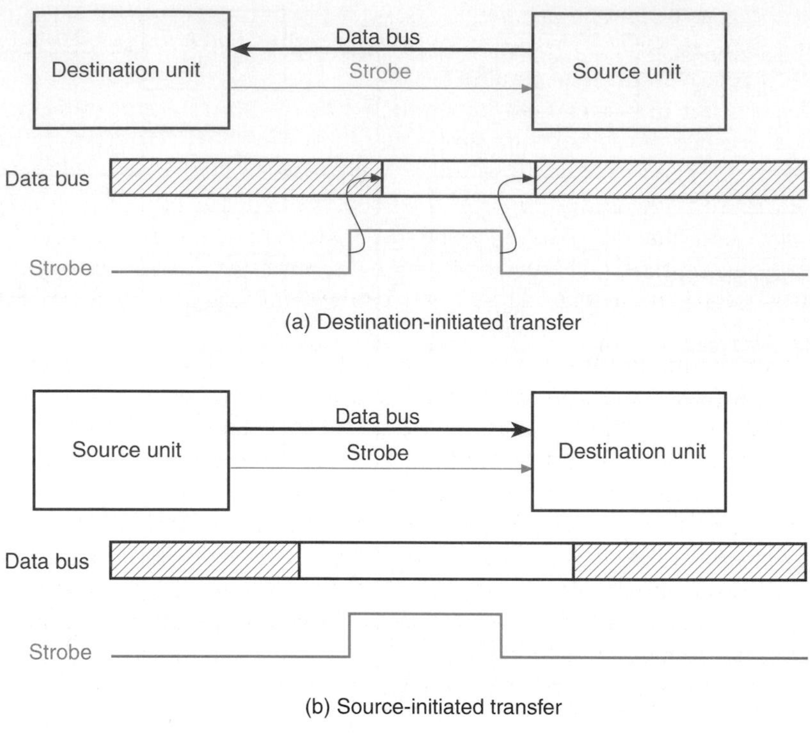

□ **FIGURE 11-6**
Asynchronous Transfer Using Strobing

with respect to each other. Asynchronous data transfer between two independent units requires that control signals be transmitted between the units to indicate the time at which data is being transmitted. In the case of CPU-to-interface communication, control signals must also indicate the time at which the address is valid. We will look at two methods for performing this timing, strobing, as it is called, and, handshaking. Initially, we will consider generic cases in which no addresses are involved; subsequently, we will add addressing. The communicating units for the generic case will be referred to as the source unit and destination unit.

Strobing

Data transfers using *strobing* are shown in Figure 11-6. The data bus between the two units is assumed to be made bidirectional by the use of three-state buffers.

The transfer in Figure 11-6(a) is initiated by the destination unit. In the shaded area of the data signal, the data is invalid. Also, a change in Strobe at the tail of each arrow causes a change on the data bus at the head of the arrow. The destination unit changes the Strobe from 0 to 1. When the value 1 on Strobe reaches the source unit, the unit responds by placing the data on the data bus. The destination unit expects the data to be available, at worst, a fixed amount of time after Strobe goes to 1. At that time, the destination unit captures the data in a register and changes Strobe from 1 to 0. In response to the 0 value on Strobe, the source unit removes the data from the bus.

The transfer in Figure 11-6(b) is initiated by the source unit. In this case, the source unit places the data on the data bus. After the short time required for the data to settle on the bus, the source unit changes Strobe from 0 to 1. In response to Strobe equal to 1, the destination unit sets up the transfer to one of its registers. The source then changes Strobe from 1 to 0, which triggers the transfer into the register at the destination. Finally, after a short time required to ensure that the register transfer is done, the source removes the data from the data bus, completing the transfer.

Although simple, the strobe method of transferring data has several disadvantages. First, when the source unit initiates the transfer, there is no indication to it that the data was ever captured by the destination unit. It is possible, due to a hardware failure, that the source unit did not receive the change in Strobe. Second, when the destination unit performs the transfer, there is no indication to it that the source has actually placed the data on the bus. Thus, the destination unit could be reading arbitrary values from the bus rather than actual data. Finally, the speeds at which the various units respond may vary. If there are multiple units, the unit initiating a transfer must wait for the delay of the slowest of the attached communicating units before changing Strobe to 0. Thus, the time taken for every transfer is determined by the slowest unit with which a given unit initiates transfers.

Handshaking

The *handshaking* method uses two control signals to deal with the timing of transfers. In addition to the signal from the unit initiating the transfer, there is a second control signal from the other unit involved in the transfer.

The basic principle of a two-signal handshaking procedure for data transfer is as follows. One control line from the initiating unit is used to request a response from the other unit. The second control line from the other unit is used to reply to the initiating unit that the response is occurring. In this way, each unit informs the other of its status, and the result is an orderly transfer through the bus.

Figure 11-7 shows data transfer procedures using handshaking. In Figure 11-7(a), the transfer is initiated by the destination unit. The two handshaking lines are called Request and Reply. The initial state is when both Request and Reply are disabled and in the 00 state. The subsequent states are 10, 11, and 01. The destination unit initiates the transfer by enabling Request. The source unit responds by placing the data on the bus. After a short time for settling of the data on the bus, the source unit activates Reply to signal the presence of the data. In response to Reply, the destination unit captures the data in a register and disables Request. The source unit then disables Reply and the system goes to the initial state. The destination unit may not make another request until the source unit has shown its readiness to provide new data by disabling Reply. Figure 11-7(b) represents handshaking for the source-initiated transfer. In this case, the source

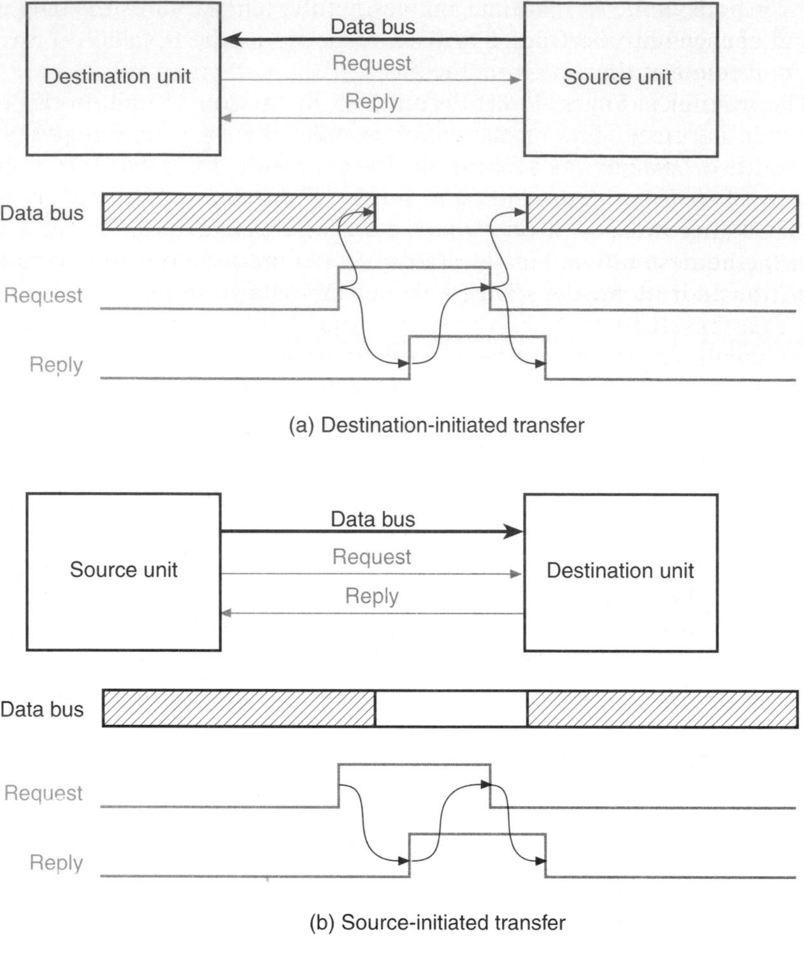

FIGURE 11-7
Asynchronous Transfer Using Handshaking

controls the interval between when the data is applied and when Request changes to 1 and between when Request changes to 0 and when the data is removed.

The handshaking scheme provides a high degree of flexibility and reliability because the successful completion of a data transfer relies on active participation by both units. If one unit is faulty, the data transfer will not be completed. Such an error can be detected by means of a time-out mechanism, which produces an alarm if the data transfer is not completed within a predetermined time interval. The time-out is implemented by means of an internal clock that starts counting time when the unit enables one of its handshaking control signals. If the return handshake does not occur within a given period, the unit assumes that an error occurred. The time-out signal can be used to interrupt the CPU and execute a service routine that takes appropriate error recovery action. Also, the timing is controlled by both units, not just the initiating unit. Within the time-out limits, the response of each unit to a change in the control signal of the other unit can take an arbitrary amount of time, and the transfer will still be successful.

The examples of transfers in Figure 11-6 and Figure 11-7 represent transfers between an interface and an I/O device and between a CPU and an interface. In the latter case, however, an address will be necessary to select the interface with which the CPU wishes to communicate and a register within the interface. In order to ensure that the CPU addresses the correct interface, the address must have settled on the address bus before the Strobe or Request signal changes from 0 to 1. Further, the address must remain stable until the change in the strobe or request from 1 to 0 has settled to 0 at the interface logic. If either of these conditions is violated, another interface may be falsely activated, causing an incorrect data transfer.

11-4 Serial Communication

The transfer of data between two units may be performed in parallel or serial. In parallel data transfer, each bit of the message has its own path, and the entire message is transmitted at one time. This means that an n-bit message is transmitted in parallel through n separate conductor paths. In serial data transmission, each bit in the message is sent in sequence, one at a time. This method requires the use of one or two signal lines. Parallel transmission is faster, but requires many wires. It is used for short distances and when speed is important. Serial transmission is slower, but less expensive, since it requires only one conductor.

One way that computers and terminals that are remote from each other are connected is via telephone lines. Since telephone lines were originally designed for voice communication, but computers communicate in terms of digital signals, some form of conversion is needed. The devices that do the conversion are called *data sets* or *modems* (modulator-demodulators). A modem converts digital signals into audio tones to be transmitted over telephone lines and also converts audio tones from the line to digital signals for use by a computer. There are various modulation schemes, as well as several different grades of communication media and transmission speeds. Serial data can be transmitted between two points in three different modes: simplex, half duplex, or full duplex. A *simplex* line carries information in one direction only. This mode is seldom used in data communication, because the receiver cannot communicate with the transmitter to indicate whether errors have occurred. Examples of simplex transmission are radio and television broadcasting.

A *half-duplex* transmission system is a system that is capable of transmitting in both directions, but only one direction at a time. A pair of wires is needed for this mode. A common situation is for one modem to act as the transmitter and the other as the receiver. When transmission in one direction is completed, the roles of the modems are reversed to enable transmission in the opposite direction. The time required to switch a half-duplex line from one direction to the other is called the *turnaround time.*

A *full-duplex* transmission can send and receive data in both directions simultaneously. This can be achieved by means of a two-wire plus ground link, with dif-

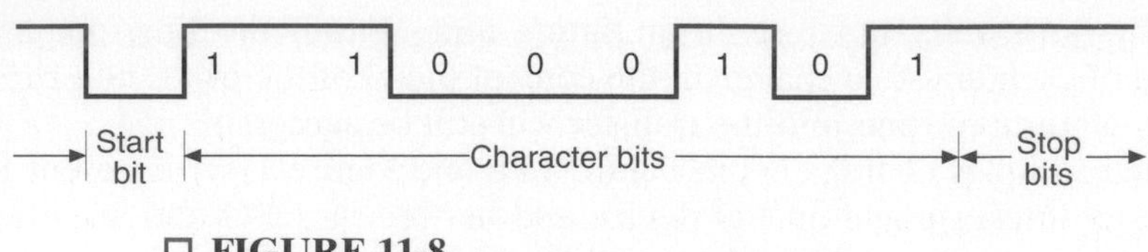

□ **FIGURE 11-8**
Format of Asynchronous Serial Transfer of Data

ferent wire dedicated to each direction of transmission. Alternatively, a single-wire circuit can support full-duplex communication if the frequency spectrum is subdivided into two nonoverlapping frequency bands to create separate receiving and transmitting channels in the same physical pair of wires.

The serial transmission of data can be synchronous or asynchronous. In *synchronous transmission*, the two units share a common clock frequency, and bits are transmitted continuously at that frequency. In long-distance serial transmission, the transmitter and receiver units are each driven by separate clocks of the same frequency. Synchronization signals are transmitted periodically between the two units to keep their clock frequencies in step with each other. In *asynchronous* transmission, binary information is sent only when it is available, and the line remains idle when there is no information to be transmitted. This is in contrast to synchronous transmission, in which bits must be transmitted continuously to keep the clock frequencies in both units synchronized.

Asynchronous Transmission

One of the most common applications of serial transmission is the communication of one computer with another via modems connected through the telephone system. Each character consists of an alphanumeric code of eight bits, with additional bits inserted at both ends of the code. In asynchronous serial transmission, each character consists of three parts: the start bit, the character bits, and the stop bits. The convention is for the transmitter to rest at the 1 state when no characters are transmitted. The first bit, called the start bit, is always 0 and is used to indicate the beginning of a character. An example of this format is shown in Figure 11-8.

A transmitted character can be detected by the receiver by applying the transmission rules. When a character is not being sent, the line is kept in the 1 state. The initiation of transmission is detected from the start bit, which is always 0. The character bits always follow the start bit. After the last bit of the character is transmitted, a stop bit is detected when the line returns to the 1 state for at least the time taken to transmit one bit. By means of these rules, the receiver can detect the start bit when the line goes from 1 to 0. By using a clock, the receiver examines the line at appropriate times to determine the bit values. The receiver knows the transfer rate of the bits and the number of character bits to accept.

After the character bits are transmitted, one or two stop bits are sent. The stop bits are always in the 1 state and frame the end of character to signify the idle or wait state. These bits allow both the transmitter and the receiver to resynchronize. The length of time that the line stays in the 1 state depends on the amount of time required for the equipment to resynchronize. Some older electromechanical terminals use two stop bits, but newer equipment often uses just one. The line remains in the 1 state until another character is transmitted. The stop time ensures that a new character will not follow for the time taken to transmit one or two bits.

As an illustration, consider serial transmission with a transfer rate of 10 characters per second. Suppose that each transmitted character consists of a start bit, 8 character bits, and 2 stop bits, for a total of 11 bits. If the bits are transmitted at a rate of 10 bits per second, then each bit takes 0.1 second for transfer. Since there are 11 bits to be transmitted, it follows that the *bit time* is 9.09 msec. The *baud rate* is defined as the maximum number of changes per second in the signal being transmitted. This is often, but not always, equivalent to the rate of data transfer in bits per second. Ten characters per second with an 11-bit format has a transfer rate of 110 baud.

Synchronous Transmission

Synchronous transmission does not use start or stop bits to frame characters. The modems employed in synchronous transmission have internal clocks that are set to the frequency at which bits are being transmitted. For proper operation, it is required that the clocks of the transmitter and receiver modems remain synchronized at all times. The communication line, however, carries only the data bits, from which information on the clock frequency must be extracted. Frequency synchronization is achieved by the receiving modem from the signal transitions that occur in the data that is received. Any frequency shift that may occur between the transmitter and receiver clocks is continuously adjusted by maintaining the receiver clock at the frequency of the incoming bit stream. In this way, the same rate is maintained in both the transmitter and the receiver.

Contrary to asynchronous transmission, in which each character can be sent separately with its own start and stop bits, synchronous transmission must send a continuous message in order to maintain synchronism. The message consists of a group of bits that form a block of data. The entire block is transmitted with special control bits at the beginning and the end, in order to frame the block into one unit of information.

The Keyboard Revisited

To this point, we have covered the basic nature of the I/O interface and serial transmission. With these two concepts available, we are now ready to continue with the example of the keyboard and its interface, as shown in Figure 11-9. The K-scan code produced by the keyboard microcontroller is to be transferred serially from the keyboard through the keyboard cable to the keyboard controller in the computer. The serial transfer on the Keyboard serial data line uses a format just like that shown for asynchronous transfer in Figure 11-8. In this case, however, a signal Keyboard clock is also sent through the cable. Thus, the transmission is synchronous with a transmitted clock signal, rather than asynchronous. These same signals are used to transmit control commands to the keyboard. In the keyboard control-

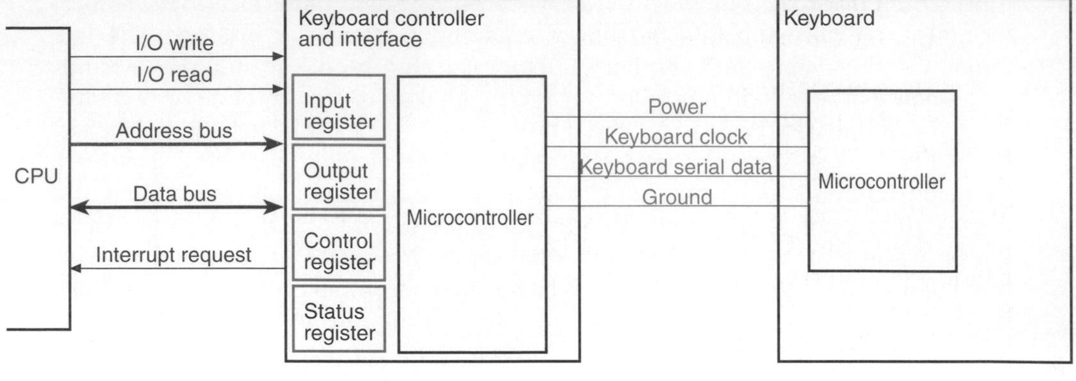

□ **FIGURE 11-9**
Keyboard Controller and Interface

ler, the microcontroller converts the K-scan code to a more standard *scan code*, which it then places in the Input register, at the same time sending an interrupt signal to the CPU indicating that a key has been pressed and a code is available. The interrupt-handling routine reads the scan code from the input register into a special area in memory. This area is manipulated by software stored in the BIOS that can translate the scan code into an ASCII character code for use by applications.

The Output register in the interface receives data from the CPU. The data can be passed on to control the keyboard—for example, setting the repetition rate when a key is held down. The Control register is used for commands to the keyboard controller. Finally, the Status register reports specific information on the status of the keyboard and the keyboard controller.

Perhaps one of the most interesting aspects of keyboard I/O is its high complexity. It involves two microcontrollers executing different programs, plus the main processor executing BIOS software—i.e., three different computers executing three distinct programs.

11-5 Modes of Transfer

Binary information received from an external device is usually stored in memory for later processing. Information transferred from the central computer into an external device originates in the memory. The CPU merely executes the I/O instructions and may accept the data temporarily, but the ultimate source or destination is the memory. Data transfer between the central computer and I/O devices may be handled in a variety of modes, some of which use the CPU as an intermediate path, while others transfer the data directly to and from the memory. Data transfer to and from peripherals may be handled in one of four possible modes:

1. Data transfer under program control.
2. Interrupt-initiated data transfer.

3. Direct memory access transfer.
4. Transfer through an I/O processor.

Program-controlled operations are the result of I/O instructions written in the computer program. Each transfer of data is initiated by an instruction in the program. Usually, the transfer is to and from a CPU register and peripheral. Other instructions are needed to transfer the data to and from the CPU and memory. Transferring data under program control requires constant monitoring of the peripheral by the CPU. Once a data transfer is initiated, the CPU is required to monitor the interface to see when a transfer can again be made. It is up to the programmed instructions executed in the CPU to keep close tabs on everything that is taking place in the interface unit and the external device.

In the program-controlled transfer, the CPU stays in a program loop called a *busy-wait loop* until the I/O unit indicates that it is ready for data transfer. This is a time-consuming process, since it keeps the processor busy needlessly. The loop can be avoided by using the interrupt facility and special commands to inform the interface to issue an interrupt request signal when the data is available from the device. This allows the CPU to proceed to execute another program. The interface, meanwhile, keeps monitoring the device. When the interface determines that the device is ready for data transfer, it generates an interrupt request to the computer. Upon detecting the external interrupt signal, the CPU momentarily stops the task it is performing, branches to a service program to process the data transfer, and then returns to the original task. This interrupt-initiated transfer is the type used for the keyboard controller shown in Figure 11-9.

Transferring data under program control is through the I/O bus and between the CPU and a peripheral. In *direct memory access* (DMA), the interface transfers data into and out of the memory unit through the memory bus. The CPU initiates the transfer by supplying the interface with the starting address and the number of words needing to be transferred and then proceeds to execute other tasks. When the transfer is made, the interface requests memory cycles through the memory bus. When the request is granted by the memory controller, the interface transfers the data directly into memory. The CPU merely delays memory operations to allow the direct memory I/O transfer. Since the speed of a peripheral is usually slower than that of a processor, I/O memory transfers are infrequent compared to processor access to memory. DMA transfer is discussed in more detail in Section 11-7.

Many computers combine the interface logic with the requirements for DMA into one unit called an *I/O processor* (IOP). The IOP can handle many peripherals through a DMA-and-interrupt facility. In such a system, the computer is divided into three separate modules: the memory unit, the CPU, and the IOP. I/O processors are presented in Section 11-8.

Example of Program-Controlled Transfer

A simple example of data transfer from an I/O device through an interface into the CPU is shown in Figure 11-10. The device transfers bytes of data one at a time as

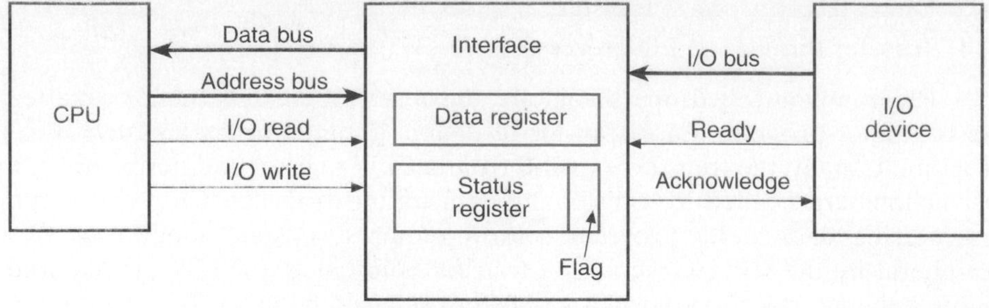

☐ **FIGURE 11-10**
Data Transfer from I/O Device to CPU

they are available. When a byte is available, the device places it on the I/O bus and enables Ready. The interface accepts the byte into its data register and enables Acknowledge. The interface sets a bit in the status register, which we will refer to as a *flag*. The device can now disable Ready, but it will not transfer another byte until Acknowledge is disabled by the interface, according to the handshaking procedure established in Section 11-3.

Under program control, the CPU must check the flag to determine whether there is a new byte in the interface data register. This is done by reading the contents of the status register into a CPU register and checking the value of the flag. If the flag is equal to 1, the CPU reads the data from the data register. The flag is then cleared to 0 either by the CPU or the interface, depending on how the interface circuits are designed. Once the flag is cleared, the interface disables Acknowledge, and the device can transfer the next data byte.

A flowchart of the program that must be written for the preceding transfer is shown in Figure 11-11. The flowchart assumes that the device is sending a sequence of bytes that must be stored in memory. The program continually examines the status of the interface until the flag is set to 1. Each byte is brought into the CPU and transferred to memory until all of the data have been transferred.

The program-controlled data transfer is used only in systems that are dedicated to monitor a device continuously. The difference in information transfer rate between the CPU and the I/O device makes this type of transfer inefficient. To see why, consider a typical computer that can execute the instructions to read the status register and check the flag in 100 ns. Assume that the input device transfers its data at an average rate of 100 bytes/s. This is equivalent to one byte every 10,000 μs, meaning that the CPU will check the flag 100,000 times between each transfer. Thus, the CPU is wasting time checking the flag instead of doing a useful processing task.

Interrupt-Initiated Transfer

An alternative to the CPU constantly monitoring the flag is to let the interface inform the computer when it is ready to transfer data. This mode of transfer uses

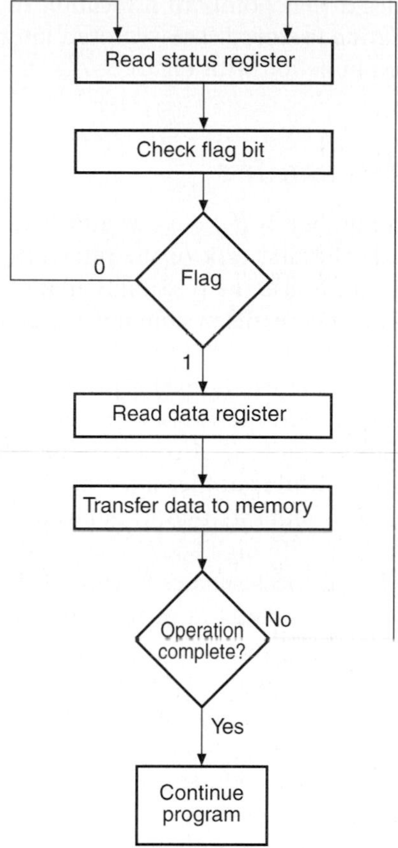

□ **FIGURE 11-11**
Flowchart for CPU Program to Input Data

the interrupt facility. While the CPU is running a program, it does not check the flag. However, when the flag is set, the computer is momentarily interrupted from proceeding with the current program and is informed of the fact that the flag has been set. The CPU drops what it is doing to take care of the input or output transfer. After the transfer is completed, the computer returns to the previous program to continue what it was doing before the interrupt. The CPU responds to the interrupt signal by storing the return address from the program counter into a memory stack or register, and then control branches to a service routine that processes the required I/O transfer. The way that the processor chooses the branch address of the service routine varies from one unit to another. In principle, there are two methods for accomplishing this: *vectored interrupt* and *nonvectored interrupt*. In a nonvectored interrupt, the branch address is assigned to a fixed location in memory. In a vectored interrupt, the source that interrupts supplies the branch address to the computer. This information is called the *vector address*. In some computers, the vector address is the first address of the service routine; in other computers, the vector address is an address that points to a location in memory where the first address of the service routine is stored. The vectored interrupt procedure was presented in Section 9-9 in conjunction with Figure 9-7.

11-6 Priority Interrupt

A typical computer has a number of I/O devices attached to it that are able to originate an interrupt request. The first task of the interrupt system is to identify the source of the interrupt. There is also the possibility that several sources will request service simultaneously. In this case, the system must decide which device to service first.

A priority interrupt system establishes a priority over the various interrupt sources to determine which interrupt request to service first when two or more arrive simultaneously. The system may also determine which requests are permitted to interrupt the computer while another interrupt is being serviced. Higher levels of priority are assigned to requests that, if delayed or interrupted, could have serious consequences. Devices with high-speed transfers such as magnetic disks are given high priority, and slow devices such as keyboards receive the lowest priority. When two devices interrupt the computer at the same time, the computer services the device with the higher priority first.

Establishing the priority of simultaneous interrupts can be done by software or hardware. Software uses a polling procedure to identify the interrupt source of highest priority. In this method, there is one common branch address for all interrupts. The program at the branch address takes care of interrupts by polling the interrupt sources in sequence. The priority of each interrupt source determines the order in which it is polled. The source with the highest priority is tested first, and if its interrupt signal is on, control branches to a routine which services that source. Otherwise, the source with the next lower priority is tested, and so on. Thus, the initial service routine for all interrupts consists of a program that tests the interrupt sources in sequence and branches to one of many other possible service routines. The particular service routine that is reached belongs to the highest priority device among all devices that interrupted the computer. The disadvantage of the software method is that if there are many interrupts, the time required to poll all the sources can exceed the time available to service the I/O device. In this situation, a hardware priority interrupt unit can be used to speed up the operation of the system.

A hardware priority interrupt unit functions as an overall manager in an interrupt system environment. The unit accepts interrupt requests from many sources, determines which of the incoming requests has the highest priority, and issues an interrupt request to the computer based on this determination. To speed up the operation, each interrupt source has its own interrupt vector address to access its own service routine directly. Thus, no polling is required, because all the decisions are made by the hardware priority interrupt unit. The hardware priority function can be established either by a serial or parallel connection of interrupt lines. The serial connection is also known as the daisy chain method.

Daisy Chain Priority

The *daisy chain* method of establishing priority consists of a serial connection of all devices that request an interrupt. The device with the highest priority is placed in the first position, followed by devices with lower priority in descending order, down to the device with the lowest priority, which is placed last in the chain. This method of connection between three devices and the CPU is shown in Figure 11-12. Interrupt request lines from all devices are ORed to form the interrupt line to the CPU. If any device has its Interrupt request at 1, the interrupt line goes to 1 and enables the interrupt input of the CPU. When no interrupts are pending, the interrupt line stays at 0, and no interrupts are recognized by the CPU. The CPU responds to an interrupt request by enabling Interrupt acknowledge. The signal that is produced is received by device 0 at its *PI* (priority in) input. The signal then passes on to the next device through the *PO* (priority out) output only if device 0 is not requesting an interrupt. If device 0 has a pending interrupt, it blocks the acknowledge signal from the next device by placing a 0 on the *PO* output and proceeds to insert its own interrupt vector address (*VAD*) onto the data bus for the CPU to use during the interrupt cycle.

A device with a 0 on its *PI* input generates a 0 on its *PO* output to inform the device with next lower priority that the acknowledge signal has been blocked. A device that is requesting an interrupt and has a 1 on its *PI* input will intercept the acknowledge signal by placing a 0 on its *PO* output. If the device does not have pending interrupts, it transmits the acknowledge signal to the next device by placing a 1 on its *PO* output. Thus, the device with $PI = 1$ and $PO = 0$ is the one with the highest priority that is requesting an interrupt, and this device places its *VAD* on the data bus. The daisy chain arrangement gives the highest priority to the device that receives the Interrupt acknowledge signal from the CPU. The farther the device is from the first position, the lower is its priority.

Figure 11-13 shows the internal logic that must be included within each device connected in the daisy chain scheme. The device sets its *RF* latch when it is about to interrupt the CPU. The output of the latch functionally enters the OR which drives the interrupt line. If $PI = 0$, both *PO* and the enable line to *VAD* are

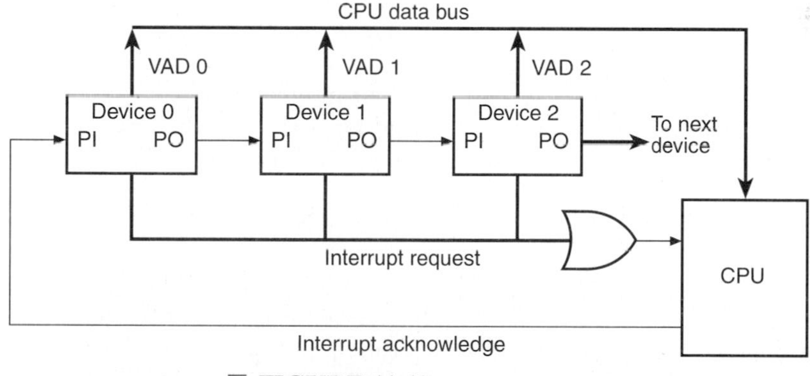

□ **FIGURE 11-12**
Daisy Chain Priority Interrupt

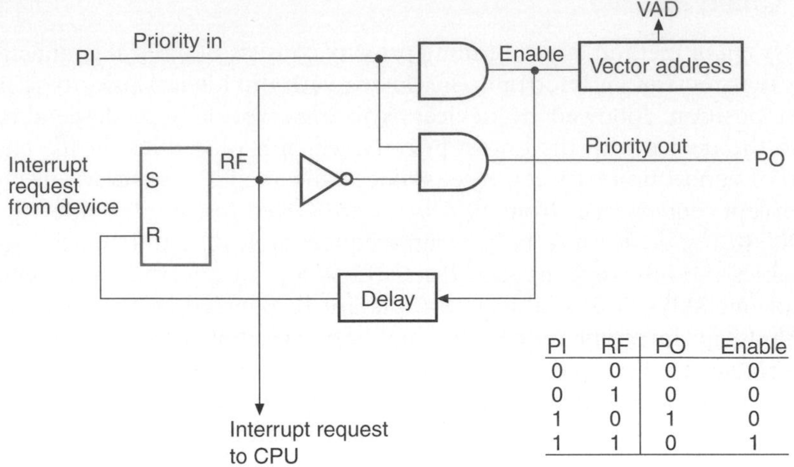

PI	RF	PO	Enable
0	0	0	0
0	1	0	0
1	0	1	0
1	1	0	1

□ **FIGURE 11-13**
One Stage of the Daisy Chain Priority Arrangement

equal to 0, irrespective of the value of *RF*. If $PI = 1$ and $RF = 0$, then $PO = 1$, the vector address is disabled, and the acknowledge signal passes to the next device through *PO*. The device is active when $PI = 1$ and $RF = 1$, which places a 0 on PO and enables the vector address onto the data bus. It is assumed that each device has its own distinct vector address. The *RF* latch is reset after a sufficient delay to ensure that the CPU has received the vector address.

Parallel Priority Hardware

The parallel priority interrupt method uses a register with bits set separately by the interrupt signal from each device. Priority is established according to the position of the bits in the register. In addition to the interrupt register, the circuit may include a mask register to control the status of each interrupt request. The mask register can be programmed to disable lower priority interrupts while a higher priority device is being serviced. It can also allow a high-priority device to interrupt the CPU while a lower priority device is being serviced.

The priority logic for a system with four interrupt sources is shown in Figure 11-14. The logic consists of an interrupt register with individual bits set by external conditions and cleared by program instructions. Interrupt input 3 has the highest priority, input 0 the lowest. The mask register has the same number of bits as the interrupt register. By means of program instructions, it is possible to set or reset any bit in the mask register. Each interrupt bit and its corresponding mask bit are applied to an AND gate to produce the four inputs to a priority encoder. In this way, an interrupt is recognized only if its corresponding mask bit is set to 1 by the program. The priority encoder generates two bits of the vector address, which is transferred to the CPU via the data bus. Output *V* of the encoder is set to 1 if an interrupt request that is not masked has occurred. This provides the interrupt signal for the CPU.

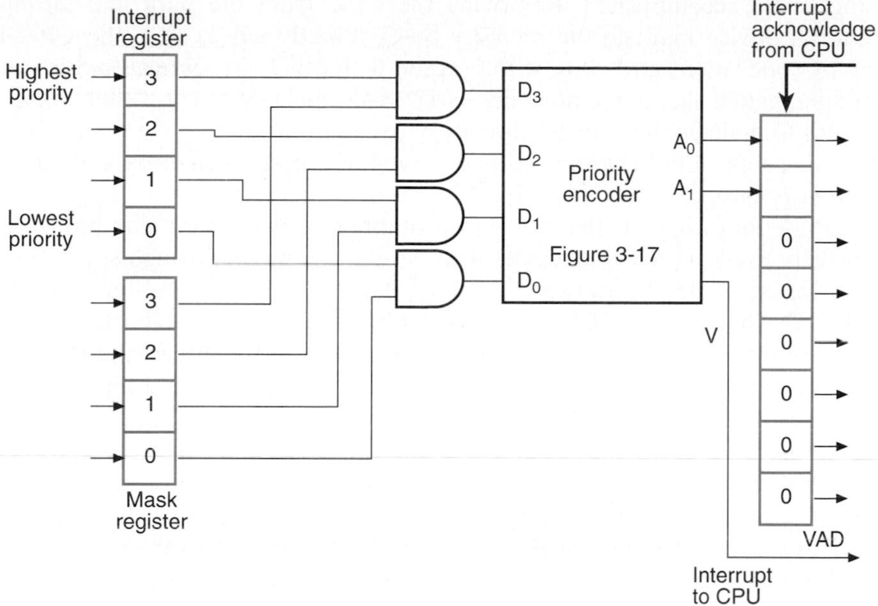

□ **FIGURE 11-14**
Parallel Priority Interrupt Hardware

The priority encoder is a circuit that implements the priority function. The logic of the priority encoder is such that, if two or more inputs occur at the same time, the input having the highest priority takes precedence. The circuit of a four-input priority encoder can be found in Section 3-6, and its truth table is listed in Table 3-6. Input D_3 has the highest priority so, regardless of the values of other inputs, when this input is 1, the output is $A_1 A_0 = 11$. D_2 has the next lower priority. The output is 10 if $D_2 = 1$, provided that $D_3 = 0$, regardless of the values of the other two lower priority inputs. The output is 01 when $D_1 = 1$, provided that the two higher priority inputs are equal to 0, and so on down the priority levels. The interrupt output labeled V is equal to 1 when one or more inputs are equal to 1. If all inputs are 0, V is 0, and the other two outputs of the encoder are not used. This is because the vector address is not transferred to the CPU when $V = 0$.

The output of the priority encoder is used to form part of the vector address of the interrupt source. The other bits of the vector address can be assigned any values. For example, the vector address can be found by appending six zeros to the outputs of the encoder. With this choice, the interrupt vectors for the four I/O devices are assigned the 8-bit binary numbers equivalent to decimal 0, 1, 2, and 3.

11-7 Direct Memory Access

The transfer of blocks of information between a fast storage device such as magnetic disk and the CPU can preoccupy the CPU and permit little, if any, other processing to be accomplished. Removing the CPU from the path and letting the peripheral device manage the memory buses directly will relieve the CPU from many I/O operations and allow it to proceed with other processing. In this transfer technique, called direct memory access (DMA), the DMA controller takes over the buses to manage the transfer directly between the I/O device and memory. As a consequence, the CPU is temporarily deprived of access to memory and control of the memory buses.

DMA may capture the buses in a number of ways. One common method extensively used in microprocessors is to disable the buses through special control signals. Figure 11-15 shows two control signals in a CPU that facilitate the DMA transfer. The bus request (*BR*) input is used by the DMA controller to request the CPU to relinquish control of the buses. When *BR* input is active, the CPU places the address bus, the data bus, and the read and write lines into a high-impedance state. After this is done, the CPU activates the bus granted (*BG*) output to inform the external DMA that it can take control of the buses. As long as the *BG* line is active, the CPU is unable to proceed with any operations requiring access to the buses. When the bus request input is disabled by the DMA, the CPU returns to its normal operation, disables the *BG* output, and takes control of the buses.

When the *BG* line is enabled, the external DMA controller takes control of the bus system in order to communicate directly with memory. The transfer can be made for an entire block of memory words, suspending operation of the CPU until the entire block is transferred, a process referred to as *burst transfer*. Or the transfer can be made one word at a time between executions of CPU instructions, a process called *single-cycle transfer* or *cycle stealing*. The CPU merely delays its bus operations for one memory cycle to allow the direct memory-I/O transfer to steal one memory cycle.

DMA Controller

The DMA controller needs the usual circuits of an interface to communicate with the CPU and the I/O device. In addition, it needs an address register, a word-count register, and a set of address lines. The address register and address lines are used for direct communication with memory. The word-count register specifies the number of words that must be transferred. The data transfer may be done directly between the device and memory under control of the DMA.

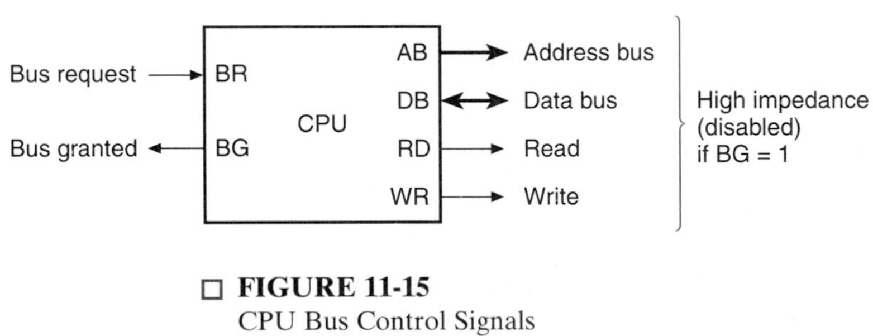

□ **FIGURE 11-15**
CPU Bus Control Signals

Figure 11-16 shows the block diagram of a typical DMA controller. The unit communicates with the CPU via the data bus and control lines. The registers in the DMA are selected by the CPU through the address bus by enabling the *DS* (DMA select) and *RS* (register select) inputs. The *RD* (read) and *WR* (write) inputs are bidirectional. When the *BG* (bus granted) input is 0, the CPU can communicate with the DMA registers through the data bus to read from or write to those registers. When $BG = 1$, the CPU has relinquished the buses, and the DMA can communicate directly with memory by specifying an address on the address bus and activating the *RD* or *WR* control. The DMA communicates with the external peripheral through the DMA request and DMA acknowledge lines by a prescribed handshaking procedure.

The DMA controller has three registers: an address register, a word-count register, and a control register. The address register contains an address to specify the desired location of a word in memory. The address bits go through bus buffers onto the address bus. The address register is incremented after each word is transferred to memory. The word-count register holds the number of words to be transferred. This register is decremented by one after each word transfer and internally tested for zero. The control register specifies the mode of transfer. All registers in the DMA appear to the CPU as I/O interface registers. Thus, the CPU can read from or write to the DMA registers under program control via the data bus.

After initialization by the CPU, the DMA starts and continues to transfer data between memory and the peripheral unit until an entire block is transferred. The initialization process is essentially a program consisting of I/O instructions that

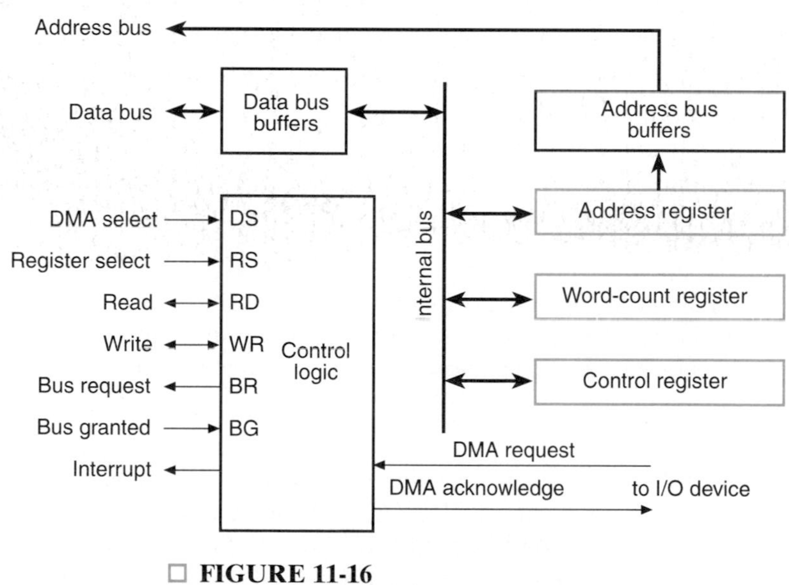

□ **FIGURE 11-16**
Block Diagram of a DMA Controller

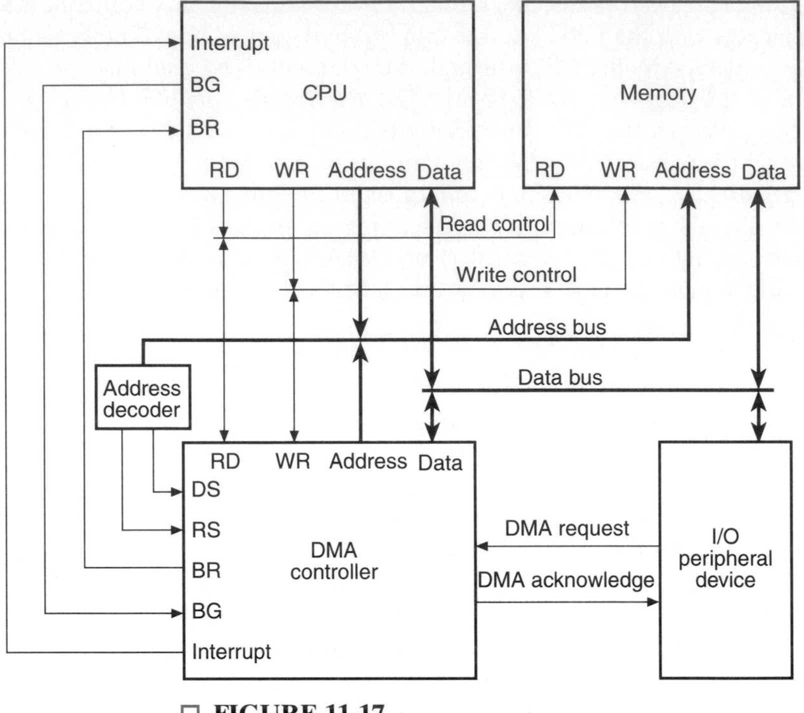

□ **FIGURE 11-17**
DMA Transfer in a Computer System

include the address for selecting particular DMA registers. The CPU initializes the DMA by sending the following information through the data bus:

1. The starting address of the memory block in which data is available (for reading) or data is to be stored (for writing).
2. The word count, which is the number of words in the memory block.
3. A control bit to specify the mode of transfer, such as read or write.
4. A control bit to start the DMA transfer.

The starting address is stored in the address register, the word count in the word-count register, and the control information in the control register. Once the DMA is initialized, the CPU stops communicating with it unless the CPU receives an interrupt signal or needs to check how many words have been transferred.

DMA Transfer

The position of the DMA controller among the other components in a computer system is illustrated in Figure 11-17. The CPU communicates with the DMA through the address and data buses, as with any interface unit. The DMA has its own address, which activates the *DS* and *RS* lines. The CPU initializes the DMA through the data bus. Once the DMA receives the start control bit, it can begin transferring data between the peripheral device and memory. When the peripheral device sends a DMA request, the DMA controller activates the *BR* line, informing the CPU that it is to relinquish the buses. The CPU responds with its *BG* line, informing the DMA that the buses are disabled. The DMA then puts the current value of its address register onto the address bus, initiates the *RD* or *WR* signal, and sends a DMA acknowledge to the peripheral device.

When the peripheral device receives a DMA acknowledge, it puts a word on the data bus (for writing) or receives a word from the data bus (for reading). Thus, the DMA controls the read or write operation and supplies the address for memory. The peripheral unit can then communicate with memory through the data bus for a direct transfer of data between the two units while the CPU access to the data bus is momentarily disabled.

For each word that is transferred, the DMA increments its address register and decrements its word-count register. If the word count has not reached zero, the DMA checks the request line coming from the peripheral. In a high-speed device, the line will be activated as soon as the previous transfer is completed. A second transfer is then initiated, and the process continues until the entire block is transferred. If the speed of the peripheral is slower, the DMA request line may be activated somewhat later. In this case, the DMA disables the bus request line so that the CPU can continue to execute its program. When the peripheral requests a transfer, the DMA requests the buses again.

If the word count reaches zero, the DMA stops any further transfer and removes its bus request. It also informs the CPU of the termination of the transfer by means of an interrupt. When the CPU responds to the interrupt, it reads the contents of the word-count register. A value of zero indicates that all the words were successfully transferred. The CPU can read the word-count register at any time, as well, to check the number of words already transferred.

A DMA controller may have more than one channel. In this case, each channel has a request and acknowledge pair of control signals that are connected to separate peripheral devices. Each channel also has its own address register and word-count register so that channels with high priority are serviced before channels with lower priority.

DMA transfer is very useful in many applications, including the fast transfer of information between magnetic disks and memory and between graphic displays and memory.

11-8 I/O Processors

Instead of having each interface communicate with the CPU, a computer may incorporate one or more external processors and assign them the task of communicating directly with all I/O devices. An input-output processor (IOP) may be classified as a processor with direct memory access capability that communicates with I/O devices. In this configuration, the computer system can be divided into a memory unit and a number of processors composed of the CPU and one or more IOPs. Each IOP takes care of input and output tasks, relieving the CPU of the "housekeeping" chores involved in I/O transfers. A processor that communicates with remote units over telephone and other communication media in a serial fashion is called a *data communication processor* (DCP). The benefit derived from using I/O processors is improved system performance, achieved through relieving the CPU of detailed tasks relating to I/O and assigning them to the appropriate I/O processors.

An IOP is similar to a CPU, except that it is designed to handle the details of I/O processing. Unlike the DMA controller, which must be set up entirely by the CPU, the IOP can fetch and execute its own instructions. IOP instructions are specifically designed to facilitate I/O transfers. In addition, the IOP can perform other processing tasks, such as arithmetic, logic, branching, and translation of code.

The block diagram of a computer with two processors is shown in Figure 11-18. The memory occupies a central position and can communicate with each processor by means of DMA. The CPU is responsible for processing data needed in the solution of computational tasks. The IOP provides a path for the transfer of data between various peripheral devices and the memory. The CPU is usually assigned the task of initiating the I/O program. From then on, the IOP operates independently of the CPU and continues to transfer data between external devices and memory. The data formats of peripheral devices often differ from those of memory and the CPU. The IOP must structure data words from many different sources. For example, it may be necessary to take four bytes from an input device and pack them into one 32-bit word before the transfer to memory. Data are gathered in the IOP at the device bit rate and bit capacity while the CPU is executing its own program. After assembly into a memory word, the data is transferred from the IOP directly into memory by stealing one memory cycle from the CPU. Similarly, an output word transferred from memory to the IOP is directed from the IOP to the output device at the device bit rate and bit capacity.

The communication between the IOP and the devices attached to it is similar to the program-controlled method of transfer. Communication with memory is similar to the DMA method. The way the CPU and IOP communicate with each other depends on the level of sophistication of the system. In very large-scale computers, each processor is independent of all the others, and any one processor can initiate an operation. In most computer systems, the CPU is the master, while the

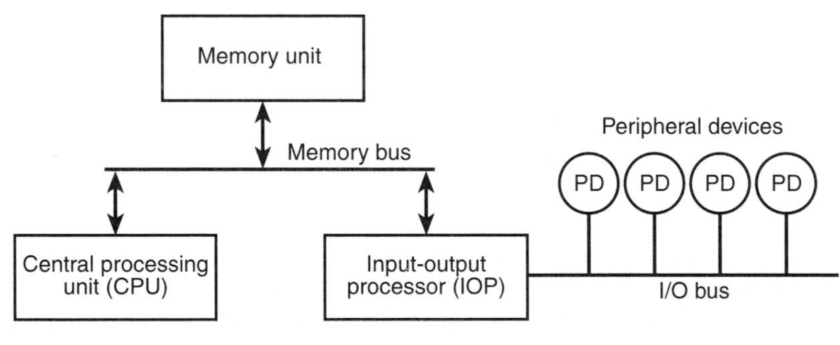

□ **FIGURE 11-18**
Block Diagram of a Computer with I/O Processor

IOP is a slave processor. The CPU is assigned the task of initiating all operations, but I/O instructions are executed in the IOP. CPU instructions provide operations to start an I/O transfer and also to test I/O status conditions needed for making decisions on various I/O activities. The IOP, in turn, typically asks for attention from the CPU by means of an interrupt. It also responds to CPU requests by placing a status word in a prescribed location in memory, to be examined later by a CPU program. When an I/O operation is desired, the CPU informs the IOP where to find the I/O program and then leaves the details of the transfer to the IOP.

Instructions that are read from memory by an IOP are sometimes called *commands*, to distinguish them from instructions that are read by the CPU. An instruction and a command have similar functions. Commands are prepared by programmers and are stored in memory. The command words constitute the program for the IOP. The CPU informs the IOP where to find commands in memory when it is time to execute the I/O program.

Communication between the CPU and the IOP may take different forms, depending on the particular computer used. In most cases, the memory acts as a message center where each processor leaves information for the other. To appreciate the operation of a typical IOP, we will illustrate the method by which the CPU and IOP communicate with each other. This simplified example omits many operating details in order to provide an overview of basic concepts.

The sequence of operations may be carried out as shown in the flowchart of Figure 11-19. The CPU sends an instruction to test the IOP path. The IOP responds by inserting a status word in memory for the CPU to check. The bits of the status word indicate the condition of the IOP and I/O device, such as "IOP overload condition," "device busy with another transfer," or "device ready for I/O transfer." The CPU refers to the status word in memory to decide what to do next. If all is in order, the CPU sends the instruction to start the I/O transfer. The memory address received with this instruction tells the IOP where to find its program.

The CPU can now continue with another program while the IOP is busy with the I/O program. Both programs refer to memory by means of DMA transfer. When the IOP terminates the execution of its program, it sends an interrupt request to the CPU. The CPU responds by issuing an instruction to read the status from the IOP. The IOP then responds by placing the contents of its status report into a specified memory location. The status word indicates whether the transfer has been completed or whether any errors occurred during the transfer. By inspecting the bits in the status word, the CPU determines whether the I/O operation was completed satisfactorily, without errors.

The IOP takes care of all data transfers between several I/O units and memory while the CPU is processing another program. The IOP and CPU compete for the use of memory, so the number of devices that can be in operation is limited by the access time of the memory. It is not possible for I/O devices to saturate the memory in most systems, as the speed of most devices is much slower than that of the CPU. However, multiple fast units, such as magnetic disks or graphics displays, can use an appreciable number of the available memory cycles. In that case, the speed of the CPU may deteriorate because the CPU will often have to wait for the IOP to conduct memory transfers.

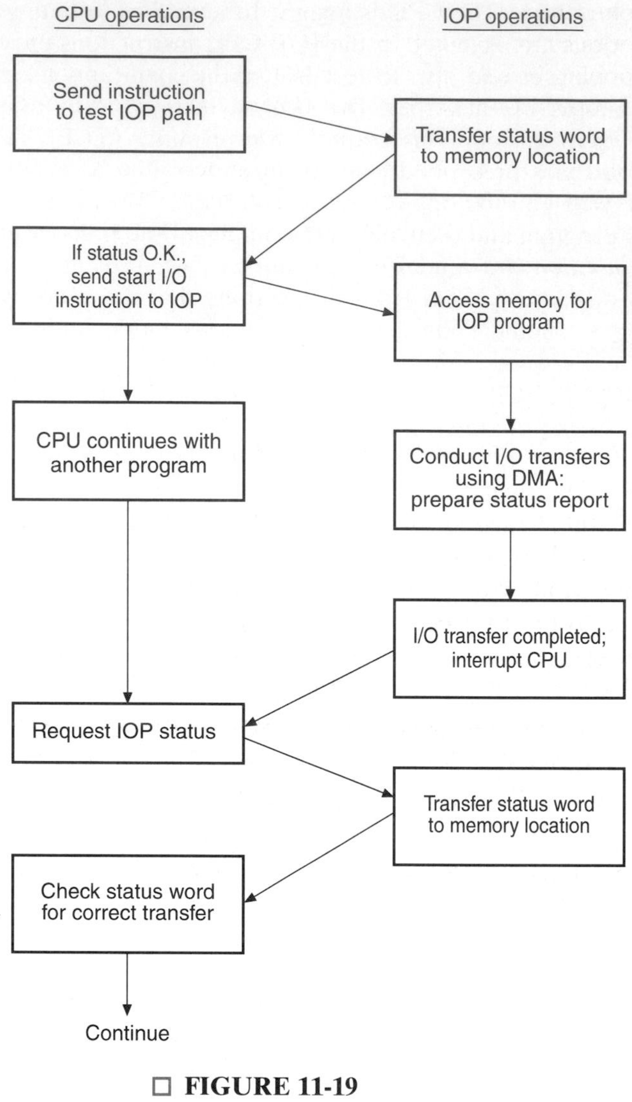
CPU operations
IOP operations
Send instruction to test IOP path
Transfer status word to memory location
If status O.K., send start I/O instruction to IOP
Access memory for IOP program
CPU continues with another program
Conduct I/O transfers using DMA: prepare status report
I/O transfer completed; interrupt CPU
Request IOP status
Transfer status word to memory location
Check status word for correct transfer
Continue

□ **FIGURE 11-19**
CPU-IOP Communication

11-9 CHAPTER SUMMARY

In this chapter, we introduced I/O devices, typically called peripherals, and their associated digital support structures, including I/O buses, interfaces, and controllers. We studied the structure of a keyboard, a hard disk, and a graphics display. We looked at an example of a generic I/O interface and examined the interface and I/O controller for the keyboard. We considered timing problems between systems with different clocks and the parallel and serial transmission of information.

We also looked at modes of transferring information and saw how the more complex modes came about, principally to relieve the CPU from extensive, performance-robbing handling of I/O transfers. Interrupt-initiated transfers with multiple I/O interfaces lead to means of establishing priority between interrupt sources. Priority can be handled by software, serial daisy chain logic, or parallel interrupt-priority logic. Direct memory access accomplishes the transfer of data directly between an I/O interface and memory, with little CPU involvement. Finally, the I/O processor provides even greater independence of the CPU in handling I/O.

References

1. Hwang, K., and Briggs, F. A. *Computer Architecture and Parallel Processing.* New York, NY: McGraw-Hill, 1984.
2. Lippiatt, A. G., and Wright, G. L. *The Architecture of Small Computer Systems.* 2nd ed. Englewood Cliffs, NJ: Prentice Hall, 1985.
3. Liu, Y. C., and Gibson, G. *Microcomputer Systems: The 8086/8088 Family.* 2nd ed. Englewood Cliffs, NJ: Prentice Hall, 1986.
4. Hamacher, V. C., Vranesic, Z. G., and Zaky, S. G. *Computer Organization.* 3rd ed. New York, NY: McGraw-Hill, 1990.
5. Mano, M. M. *Computer System Architecture.* 3rd ed. Englewood Cliffs, NJ: Prentice Hall, 1993.
6. Patterson, D. A., and Hennesy, J. L. *Computer Organization and Design: The Hardware/Software Interface.* San Mateo, CA: Morgan Kaufmann, 1994.
7. Van Gilluwe, F. *The Undocumented PC.* Reading, MA: Addison-Wesley, 1994.
8. Messmer, H-P. *The Indispensable PC Hardware Book.* 2nd ed. Reading, MA: Addison-Wesley, 1995.

Problems

The asterisk (*) indicates a more advanced problem.

11–1. Find the formatted capacity of the hard disks described in the following table:

Disk	Heads	Cylinders	Sectors/ Track	Bytes/ Sector
A	5	733	17	512
B	15	900	17	512
C	7	1,023	64	512

11–2. Estimate the time required to transfer 4 KB from disk to memory given the following disk parameters: seek time, 10 ms; rotational delay, 1ms; controller time, 2 μs; transfer rate, 4 MB/s.

11–3. The addresses assigned to the four registers of the I/O interface of Figure 11-5 are equal to the binary equivalent of 60, 61, 62, and 63. Show the external circuit that must be connected between an 8-bit I/O address from the CPU and the *CS*, *RS*0, and *RS*1 inputs of the interface.

11–4. Six interface units of the type shown in Figure 11-5 are connected to a CPU that uses an I/O address of eight bits. Each one of the six chip select (*CS*) inputs is connected to a different address line. Thus, the high-order address line is connected to the *CS* input of the first interface unit, and the sixth address line is connected to the *CS* input of the sixth interface unit. The two low-order address lines are connected to the *RS*1 and *RS*0 inputs of all six interface units. Determine the 8-bit address of each register in each interface (a total of 24 addresses).

11–5. A different type of I/O interface does not have the *RS*1 and *RS*0 inputs. Up to two registers can be addressed by using a separate I/O read signal and I/O write signal for each address available. Assume that 25% of the registers at the interface with the CPU are read only, 25% of the registers are write only, and 50% of the registers are both read and write (bidirectional). How many registers can be addressed if the address contains eight bits?

11–6. A commercial interface unit uses different names for the handshake lines associated with the transfer of data from the I/O device to the interface unit. The interface input handshake line is labeled *STB* (strobe), and the interface output handshake line is labeled *IBF* (input buffer full). A low-level signal on *STB* loads data from the I/O bus into the interface data register. A high-level signal on *IBF* indicates that the data has been accepted by the interface. *IBF* goes low after an I/O read signal from the CPU when it reads the contents of the data register.

(a) Draw a block diagram showing the CPU, the interface, and the I/O device, along with the pertinent interconnections between the three units.

(b) Draw a timing diagram for the handshaking transfer.

11–7. Assume that the transfers with strobing shown in Figure 11-6 is between a CPU on the left and an I/O interface on the right. There is an address coming from the CPU for each of the transfers, both of which are initiated by the CPU.

(a) Draw block diagrams showing the interconnections for the transfers.

(b) Draw the timing diagrams for the two transfers, assuming that the address must be applied some time before the strobe becomes 1 and removed some time after the strobe becomes 0.

11–8. Assume that the transfers with handshaking shown in Figure 11-7 is between a CPU on the left and an I/O interface on the right. There is an address coming from the CPU for each of the transfers, both of which are initiated by the CPU.

(a) Draw block diagrams, showing that interconnections for the transfers.

(b) Draw the timing diagrams, assuming that the address must be applied some time before the request becomes 1 and removed some time after the request becomes 0.

11–9. How many characters per second can be transmitted over a 9,600-baud line in each of the following modes? (Assume a character code of eight bits.)

(a) Asynchronous serial transmission with two stop bits.

(b) Asynchronous serial transmission with one stop bit.

(c) Repeat (a) for a 28,800-baud line.

(d) Repeat (b) for a 115,200-baud line.

11–10. Sketch the timing diagram of the 11 bits (similar to Figure 11-8) that are transmitted over an asynchronous serial communication line when the ASCII letter E is transmitted with even parity. Assume that the ASCII character code is transmitted least significant bit first, with the parity bit following the character code.

11–11. What is the difference between the synchronous and the asynchronous serial transfer of information?

11–12. What is the basic advantage of using interrupt-initiated data transfer over transfer under program control without an interrupt?

11–13. What happens in the daisy chain priority interrupt shown in Figure 11-12 when device 0 requests an interrupt after device 1 has sent an interrupt request to the CPU, but before the CPU responds with the interrupt acknowledge?

11–14. Consider a computer without priority interrupt hardware. Any one of many sources can interrupt the computer, and any interrupt request results in storing the return address and branching to a common interrupt routine. Explain how a priority can be established in the interrupt service program.

11–15. What should be done in Figure 11-14 to make the four *VAD* values equal to the binary equivalent of 76, 77, 78, and 79?

11–16. Design parallel priority interrupt hardware for a system with eight interrupt sources.

11–17. A priority structure is to be designed that provides vector addresses.

(a) Obtain the truth table of an 8×3 priority encoder.

(b) The three outputs x, y, z from the priority encoder are used to provide an 8-bit vector address in the form $101xyz00$. List the eight addresses, starting from the one with the highest priority.

11–18. Why are the read and write control lines in a DMA controller bidirectional? Under what condition and for what purpose are they used as inputs? Under what condition and for what purpose are they used as outputs?

11–19. It is necessary to transfer 256 words from a magnetic disk to a section of memory starting from address 1230. The transfer is by means of DMA as shown in Figure 11-17.

(a) Give the initial values that the CPU must transfer to the DMA controller.

(b) Give the step-by-step account of the actions taken during the input of the first two words.

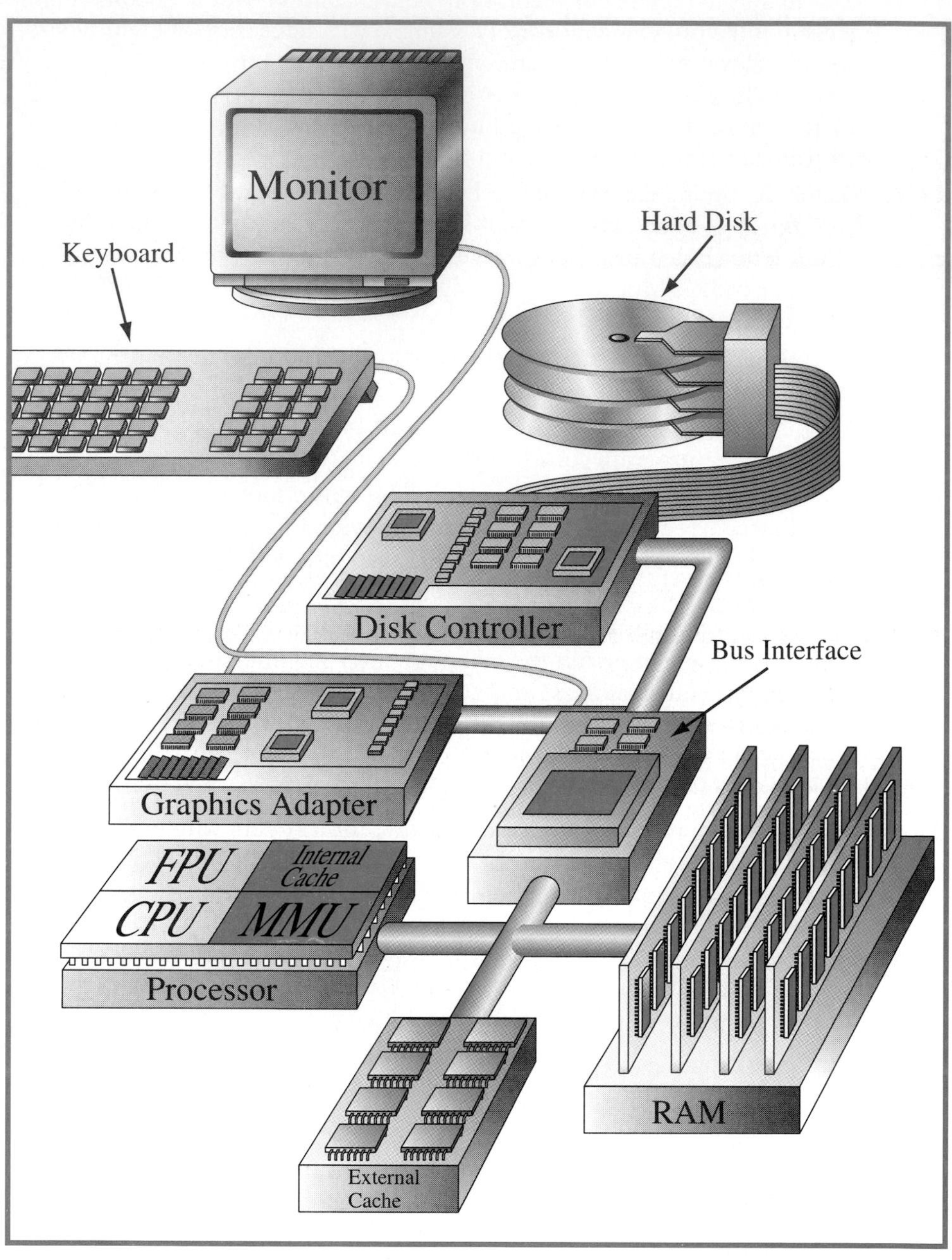

Monitor
Hard Disk
Keyboard
Disk Controller
Bus Interface
Graphics Adapter
FPU
Internal Cache
CPU
MMU
Processor
RAM
External Cache

CHAPTER

12

MEMORY SYSTEMS

In Chapter 6, we discussed basic RAM technology for implementing memory systems, including SRAMs and DRAMS. In the current chapter, we probe more deeply into what really constitutes a computer memory system. We begin with the premise that a fast, large memory is desirable and demonstrate that a straightforward implementation of such a memory for the typical computer is too costly and too slow. As a consequence, we study a more elegant solution in which most accesses to memory are fast (but some are slow) and the memory appears to be large. This solution employs two concepts: cache memory and virtual memory. A cache memory is a small, fast memory with special control hardware that permits it to handle a significant proportion of all accesses required by the CPU with an access time close to the CPU clock period. Virtual memory, implemented in software and hardware, using an intermediate-sized main memory (typically, DRAM), gives the appearance of a large main memory with access time similar to the main memory for the vast majority of accesses. The actual storage medium for most of the code and data in the virtual memory is a hard disk. Because there is a progression of components in the memory system having larger and larger storage capability, but slower and slower access (cache, main memory, and hard disk), the term *memory hierarchy* is applied.

As represented by the areas of the generic computer shaded in blue, a number of components are heavily involved in the memory hierarchy. Within the processor, there is the memory management unit (MMU), which is hardware provided to support virtual memory. Also in the processor, the internal cache appears. Since this cache is too small to fully support the cache function, there is also an external cache attached to the CPU bus. Of course, the RAM is involved, and due to the presence of virtual memory, the hard disk, the bus interface, and the disk controller all have a role as a part of the memory system.

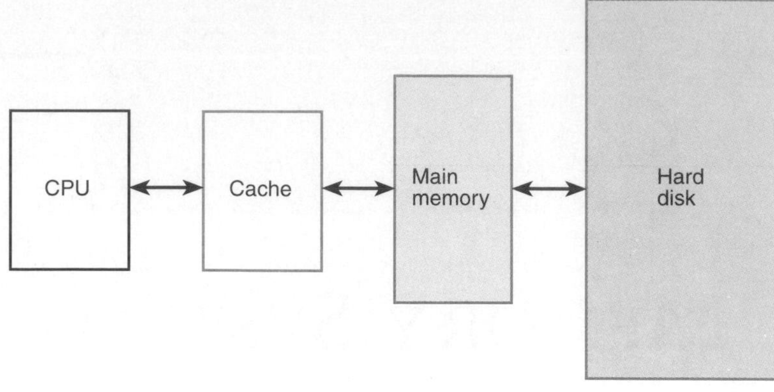

□ **FIGURE 12-1**
Memory Hierarchy

12-1 Memory Hierarchy

Figure 12-1 shows a generic block diagram for a memory hierarchy. The lowest level of the hierarchy is a small, fast memory called a *cache.* For the hierarchy to function well, a very large proportion of the CPU instruction and operand fetches are expected to be from the cache. At the next level upward in the hierarchy is the *main memory.* The main memory serves directly most of the CPU instruction and operand fetches not satisfied by the cache. In addition, the cache fetches all of its data, some portion of which is passed on to the CPU, from the main memory. At the top level of the hierarchy is the *hard disk*, which is accessed only in the very infrequent cases in which a CPU instruction or operand fetch is not found in main memory.

With this memory hierarchy, since the CPU fetches most of the instructions and operands from the cache, it "sees" a fast memory, most of the time. Occasionally, when a word must come from main memory, a fetch takes somewhat longer. Very infrequently, when a word must be fetched from the hard disk, the fetch takes a very long time. In this last case, the CPU is likely to experience an interrupt that passes execution to a program which brings in a block of words from the hard disk. On balance, the situation is usually satisfactory, providing an average fetch time close to that of the cache. Moreover, the CPU sees a memory address space considerably larger than that of main memory.

With this general notion of a memory hierarchy kept in mind, we will proceed to consider an example that illustrates the potential power of such a hierarchy. However, there is one issue to be clarified first. In most instruction set architectures, the smallest of the objects that are addressed is a byte rather than a word. For a given load or store operation, whether a byte or word is affected is typically determined by the opcode. Addressing to bytes brings with it some assumptions and hardware details that are important, but, if used up to this point in the text, would have unnecessarily complicated much of the material covered. Consequently, for simplicity, we have assumed up to now that an addressed location con-

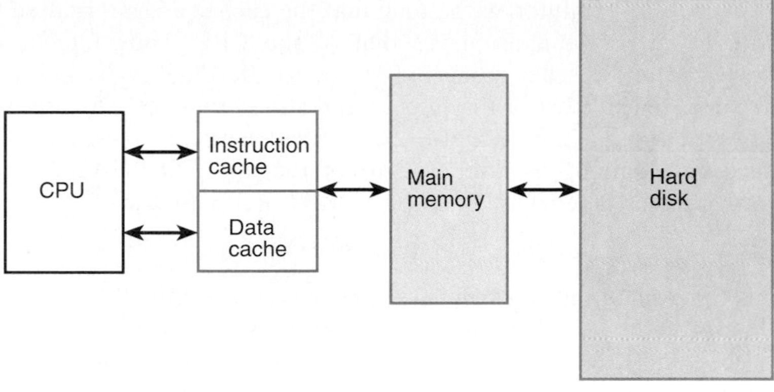

□ **FIGURE 12-2**
Example of Memory Hierarchy

tains a word. By contrast, in this chapter we will assume that addresses are defined for bytes, to match current practice. Nevertheless, we will still assume that data is moved around as words or sets of words, to avoid messy explanations relating to the manipulation of bytes. This assumption simply hides some hardware details that would distract from the main focus of our discussion, but nevertheless must be handled by the hardware designer. To accomplish the simplification, if there are 2^b bytes per word, we will ignore the last b bits of the address. Since these bits are not needed to address a word, we will show their values as 0's. For the examples we will present, b is always equal to 2, so two 0's are shown.

In Section 10-3, the pipelined CPU had a memory address with 32 bits and was able to access an instruction and data, if necessary, in each of the 5-ns clock cycles. Also, we assumed that the instruction and the data were, in effect, fetched from two different memories. To support this assumption in this chapter, we will suppose initially that the memory is divided in half—one-half for instructions and one-half for data. Each half of the memory must have an access time of 5 ns. In addition, if we utilize all the bits in the 32-bit address, then the memory can contain up to 2^{32} bytes, or 4 gigabytes (GB), of information. So the goal is to have two 2-GB memories, each with an access time of 5 ns.

Is such a memory realistic in terms of current (1996) computer technology? The typical memory is constructed of DRAM modules ranging in size from 8 to 64 Mbytes and costing about \$20 per Mbyte. The typical access time is about 80 ns. Thus, our two 2-GB memories would cost \$81,920 for DRAM alone and would have an access time of somewhat more than 80 ns per word. This kind of memory both is too costly and operates at only one-sixteenth the desired speed. So our goal must be achieved another way, leading us to explore a memory hierarchy.

We begin by assuming a hierarchy with two caches, one for instructions and one for data, as shown in Figure 12-2. The use of these two caches permits one instruction and one operand to be fetched, or one instruction to be fetched and one result to be stored, in a single clock cycle if the caches are fast enough. In terms of the generic computer, we assume that the caches are internal, so that they can operate at speeds comparable to that of the CPU. Thus, fetches from the instruction cache and fetches from and stores to the data cache can be accomplished in 5 ns. Hence, most of the fetches and stores for the CPU are from or to these caches. Suppose, then, that we are satisfied with most—say, 95%—of the memory accesses taking 5 ns. Suppose further that most of the remaining 5% of the memory accesses take 80 ns. Then the average access time is

$$0.95 \times 5 + 0.05 \times 80 = 8.75 \text{ ns}$$

This means that, on 19 out of every 20 memory accesses, the CPU operates at full speed, while the CPU will have to wait for 16 clock cycles for 1 out of every 20 memory accesses. This wait can be accomplished by stalling the CPU pipeline. Thus, we have accomplished our goal of "most" memory accesses taking 5 ns. But there is still the problem of the cost of the large memory.

Now suppose that, in addition to infrequently accepting a wait for a word from main memory that will take 80 ns, we are also willing to accept a very infrequent wait for a hard disk access taking 16 ms = 1.6×10^7 ns. Suppose that we have data indicating that about 95% of the fetches will be from a cache and about 4.999995% of the fetches will be from main memory. With this information, we can estimate the average access time as

$$0.95 \times 5 + 0.04999995 \times 80 + 5 \times 10^{-8} \times 1.6 \times 10^{7} = 9.55 \text{ ns}$$

Thus, the average access time is about twice the 5-ns clock period, but is about one-eighth of the 80-ns access time for main memory, again with 19 out of 20 of the accesses taking place in 5 ns. So we have achieved an average access time of about 9.6 ns for a memory structure with a capacity of 2^{32} bytes, quite close to the original goal.

Now what about the cost? We assume 5-ns SRAM plus other logic within the processor chip for the two caches. We also assume 80-ns DRAM for main memory and a hard drive with an access time of 16 ms. If we suppose that the two caches increase the cost of a processor chip by $100, that there are 32MB of DRAM at $20 per MB, and that the 4-GB hard disk is $800, then the total cost of the memory hierarchy is about $1,540, around one-fiftieth that of the large main memory approach.

It therefore appears that the original goal of the appearance of a fast, large memory has been achieved by the memory hierarchy at a reasonable cost. But along the way, we made some assumptions, namely, that 95% of the time the word desired would come from what we are now calling the cache and that 99.999995% of the time the words would come from either cache or main memory, with the remainder from hard disk. In the rest of this chapter, we will explore why assumptions similar to these usually hold and will examine the hardware and associated software components needed to achieve the goals of the memory hierarchy.

12-2 Locality of Reference

In the previous section, we indicated that the success of the memory hierarchy is based on assumptions that are critical to achieving the appearance of a large, fast memory. We will now deal with the foundation for making these assumptions, which is called *locality of reference.* Here "reference" means reference to memory for accessing instructions and for reading or writing operands. The term "locality" refers to the the relative times at which instructions and operands are accessed (*temporal locality*) and the relative locations at which they reside in main memory (*spatial locality*).

Let us consider first the nature of the typical program. A program frequently contains many loops. In a loop, a sequence of instructions is executed many times before the program exits the loop and moves on to another loop or straight-line code not in a loop. In addition, loops are often nested in a hierarchy in which loops are contained in loops, etc. Suppose we have a loop of eight instructions that is to be executed 100 times. Then for 800 executions, all instruction fetches will occur from just eight addresses in memory. Thus, each of the eight addresses is visited 100 times during the time the loop is executed. This is an example of temporal locality in the sense that an address which is accessed is likely to be accessed many times in the near future. Also, it is likely that the addresses of the instructions will be in sequential order. Thus, if an address is accessed for an instruction, nearby addresses are going to be addressed during the execution of the loop. This is an example of spatial locality.

In terms of accessing operands, similar temporal and spatial localities also occur. For example, in a computation on an array of numbers, there are multiple visits to the locations of many of the operands, giving temporal locality. Also, as the computation proceeds, when a particular address is accessed for a number, sequential addresses near to it are likely to be accessed for other numbers in the array, giving spatial locality.

From the prior discussion, we can conjecture that there is significant locality of reference in computer programs. To verify this decisively, it is necessary to study the patterns of execution of real programs. Such studies have demonstrated the presence of significant temporal and spatial locality of reference and play an important role in the design of caches and virtual memory systems.

The next question to answer is: What is the relation of locality of reference to the memory hierarchy? To examine this issue, we consider again the instruction fetch within a loop and look at the relationship between the cache and main memory. Initially, we assume that instructions are present only in main memory and that the cache is empty. When the CPU fetches the first instruction in a loop, it obtains the instruction from main memory. But the instruction and a portion of its address called the *address tag* are also placed in the cache. What then happens for the next 99 executions of this instruction? The answer is that the instruction can be fetched from the cache, which provides a much faster access. This is temporal locality at work: the instruction that was fetched once will tend to be used again and is now present in the cache for fast access.

Additionally, when the CPU fetches the instruction from main memory, the cache fetches nearby instructions into its SRAM. Now suppose that the nearby instructions include the entire loop of eight instructions presented in our example. Then all of the instructions are in the cache. By bringing in such a block of instructions, the cache is able to exploit spatial locality: it takes advantage of the fact that the execution of the first instruction implies the execution of instructions with nearby addresses by making the latter instructions available for fast access.

In our example, each of the instructions is fetched from main memory exactly once for the 100 executions of the loop. All other instruction fetches come from the cache. Thus, in this particular example, at least 99% of the instructions being executed are fetched from the cache, so that the rate of execution of instructions is governed almost completely by the cache access time and CPU speed, and very little by the main memory access time. Without temporal locality, many more accesses to main memory would occur, slowing down the system.

A relationship similar to that between cache and the main memory can exist between main memory and the hard disk. Again, both temporal and spatial locality of reference are of interest, except this time on a much larger scale. Programs and data are fetched from the hard disk, and data is written to the hard disk in blocks that range from kilowords to megawords. Ideally, once the code and initial data for a program reside in main memory, the hard disk need not be accessed except for storing final results of the program. But this can happen only if all of the code and data, including intermediate data used by the program, reside fully in main memory. If not, then it will be necessary to bring in code from the hard disk and to read and write data from and to the hard disk during program execution. Words are read from and written to the disk in blocks referred to as *pages*. If the movement of pages between main memory and hard disk is transparent to the programmer, then it will appear as if main memory is large enough to hold the entire program and all of the data. Hence, this automated arrangement is referred to as *virtual memory*. During the execution of the program, if an instruction to be executed is not in main memory, the CPU program flow is diverted to bring the page containing the instruction into main memory. Then the instruction can be read from main memory and executed. The details of this operation and the hardware and software actions required for it will be covered in Section 12-4.

In sum, locality of reference is absolutely key to the success of the concepts of cache memory and virtual memory. In the case of most programs, locality of reference is present to a fairly high degree. But occasionally, one does encounter a program that, for example, requires frequent access to a large body of data which cannot be accommodated in main memory. In such a case, the computer spends almost all of its time moving information between main memory and the hard disk and does little other computation. Continuous sounds emanating from the hard disk as the heads move from track to track is a telltale sign of this phenomenon, which is referred to as *thrashing*.

12-3 Cache Memory

To illustrate the concept of cache memory, we assume a very small cache of eight 32-bit words and a small main memory with 1 KB (256 words), as shown in Figure 12-3. Both of these are too small to be realistic, but their size makes illustration of the concepts easier. The cache address contains 3 bits, the memory address 10. Out of the 256 words in main memory, only 8 at a time may lie in the cache. In order

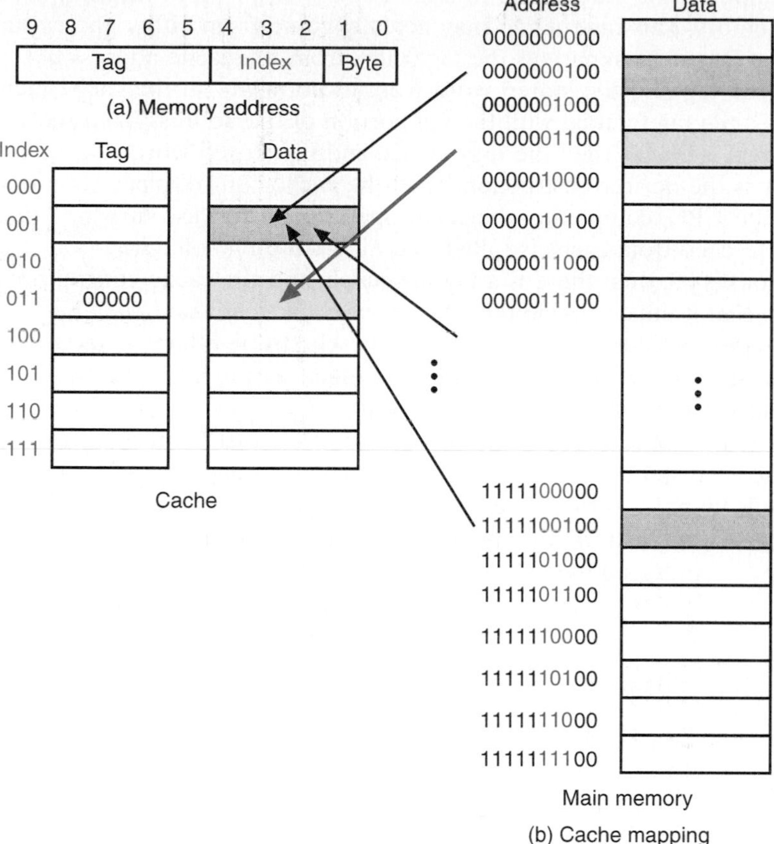

☐ **FIGURE 12-3**
Direct Mapped Cache

for the CPU to address a word in the cache, there must be information in the cache to identify the address of the word in main memory. If we consider the example of the loop in the last section, clearly, we find it desirable to contain the entire loop within the cache, so that all of the instructions can be fetched from the cache while the program is executing most of the passes through the loop. The instructions in the loop lie in consecutive word addresses. Thus, it is desirable for the cache to have words from consecutive addresses in main memory present simultaneously. A simple way to facilitate this feature is to make bits 2 through 4 of the main memory address be the cache address. We refer to these bits as the *index*, as shown in Figure 12-3. Note that the data from address 0000001100 in main memory must be stored in cache address 011. The upper 5 bits of the main memory address, called the *tag*, are stored in the cache along with the data. Continuing the example, we find that for main memory address 0000001100, the tag is 00000. The tag combined with the index (or cache address) and byte field identify an address in main memory.

Suppose that the CPU is to fetch an instruction from location 000001100 in main memory. This instruction may actually come from either the cache or main memory. The cache separates the tag 00000 from the cache address 011, internally fetches the tag and the stored word from location 011 in the cache memory, and compares the tag fetched with the tag portion of the address from the CPU. If the tag fetched is 00000, then the tags match, and the stored word fetched from cache memory is the desired instruction. Thus, the cache control places this word on the bus to the CPU, completing the fetch operation. This case in which the memory word is fetched from cache is called a *cache hit*. If the tag fetched from cache memory is not 00000, then there is a tag mismatch, and the cache control notifies main memory that it must provide the memory word, which is not available in the cache. This situation is called a *cache miss*. For a cache to be effective, the slower fetches from main memory must be avoided as much as possible, making considerably more cache hits than cache misses necessary.

When a cache miss occurs on a fetch, the word from main memory is not placed just on the bus for the CPU. The cache also captures the word and its tag and stores them for future access. In our example, the tag 00000 and the word from memory will be written in cache location 011 in anticipation of future accesses to the same memory address. The handling of writes to memory will be dealt with later in the chapter.

Cache Mappings

The example we just considered uses a particular association or mapping between the main memory address and the cache address; namely, the last three bits of the main memory word address are the cache address. Additionally, there is only one location in the cache for the 2^5 locations in main memory that have their last three bits in common. This mapping in Figure 12-3 in which only one location in the cache can contain the word from a particular main memory location is called *direct mapping*.

Direct mapping for a cache, however, does not always produce the most desirable situation. In our loop instruction fetch example, suppose that instructions and data are in the same cache and that data from location 1111101100 is frequently used. Then when the instruction in 0000001100 is fetched, location 011 in the cache is likely to contain the data from 1111101100 and tag 11111. A cache miss occurs and causes tag 11111 to be replaced in the cache with tag 00000 and the data to be replaced with the instruction. But the next time the data is needed, another cache miss occurs, since the location in the cache is now occupied by the instruction. Throughout the execution of the loop, both instruction fetch and data fetch cause many cache misses, significantly slowing CPU processing. To solve this problem, we explore alternative cache mappings.

In direct mapping, 2^5 addresses in main memory map to the single address in the cache that matches their last three bits. These locations are highlighted in gray in Figure 12-3 for index 001. As is illustrated, only one of the 2^5 addresses can have its word in cache address 001 at any time. In contrast, suppose that we let locations in main memory map into an arbitrary location in the cache. Then any location in

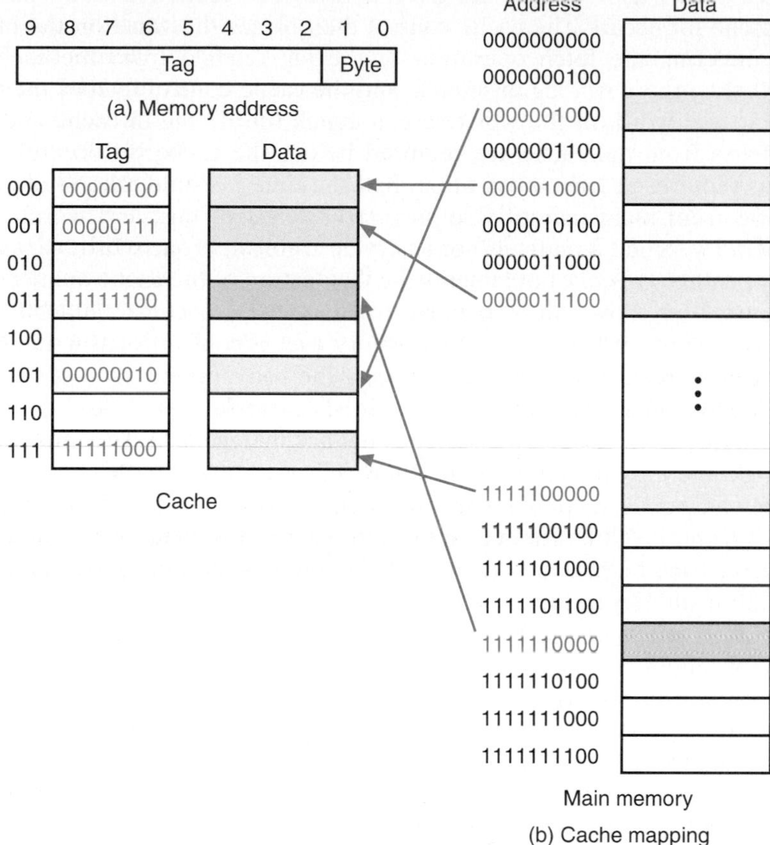

□ **FIGURE 12-4**
Fully Associative Cache

memory can be mapped to any one of the eight addresses in the cache. This means that the tag will now be the full main memory word address. We examine the operation of such a cache having a *fully associative* mapping in Figure 12-4. Note that in this case there are two main memory addresses, 0000010000 and 1111110000, with bits 2 through 4 equal to 100 among the cache tags. These two addresses cannot be present simultaneously in the direct-mapped cache, as they would both occupy the cache address 100. Thus, a succession of cache misses due to alternate fetching of an instruction and data with the same index is avoided here, since both can be in the cache.

Now suppose that the CPU is to fetch an instruction from location 0000010000 in main memory. This instruction may actually be returned from either the cache or main memory. Since the instruction might lie in the cache, the cache must compare 00000100 to each of its eight tags. One way to do this is to successively read each tag and the associated word from the cache memory and compare the tag to 00000100.

If a match occurs, as it will for the given address and cache location 000 in Figure 12-4, a cache hit occurs. The cache control then places the word on the bus to the CPU, completing the fetch operation. If the tag fetched from the cache is not 00000100, then there is a tag mismatch, and the cache control fetches the next successive tag and word. In the worst case, a match on the tag in cache address 111, eight fetches from the cache are required before the cache hit occurs. At 5 ns a fetch, this requires at least 40 ns, about half the time it would take to obtained the instruction from main memory. So successive reads of tags and words from the cache memory to find a match is not a very desirable approach. Instead, a structure called *associative memory* implements the tag portion of the cache memory.

Figure 12-5 shows an associative memory for a cache with 4-bit tags. The mechanism for writing tags into the memory uses a conventional write. Likewise, the tags can be read from the memory using the conventional memory read. Thus, the associative memory can use the bit slice model for RAM presented in Chapter 6. In addition, each tag storage row has match logic. The implementation of this logic and its connection to the RAM cells are shown in the figure. The match logic does an equality comparison or match between the tag T and the applied address A from the CPU. The match logic for each tag is composed of an exclusive-OR gate for each bit and a NOR gate that combines the outputs of the exclusive-ORs. If all of the bits of the tag and the address match, then the outputs of all the exclusive-ORs are 0 and the NOR output is a 1, indicating a match. If there is a mismatch between any of the bits in the tag and the address, then at least one exclusive-OR has a 1 output, which causes the output of the NOR gate to be 0, indicating a mismatch.

Since all tags are unique, only two situations can arise in the associative memory: there will be a match, with a 1 on the output of the match logic for one matching tag and a 0 on the remaining match logic outputs; or there will be no match, and all of the match logic outputs will be 0. With an associative memory holding the cache tags, the outputs of the match logic drive the word lines for the data memory words to be read. A signal must indicate whether a hit or a miss has occurred. If this signal is 1 for a hit and 0 for a miss, then it can be generated by using the OR of the match outputs. In the case of a hit, a 1 on Hit/$\overline{\text{miss}}$ places the word on the memory bus to the CPU; in the case of a miss, a 0 on Hit/$\overline{\text{miss}}$ tells the main memory that it is to provide the word addressed.

As in the case of the direct-mapped cache discussed earlier, the fully associative cache must capture the data word and its address tag and store them for future accesses. But now a new problem arises: Where in the cache are the tag and data to be placed? In addition to selecting a cache mapping, the cache designer must select a replacement approach that determines the location in the cache to be used for the incoming tag and data. One possibility is to select a *random replacement* location. The 3-bit address can be read from a simple hardware structure that generates a number which satisfies certain properties of random numbers. A somewhat more thoughtful approach is to use a *FIFO* location. In this case, the location selected for replacement is the one that has occupied the cache for the longest time, based on the notion that the use of this oldest entry is

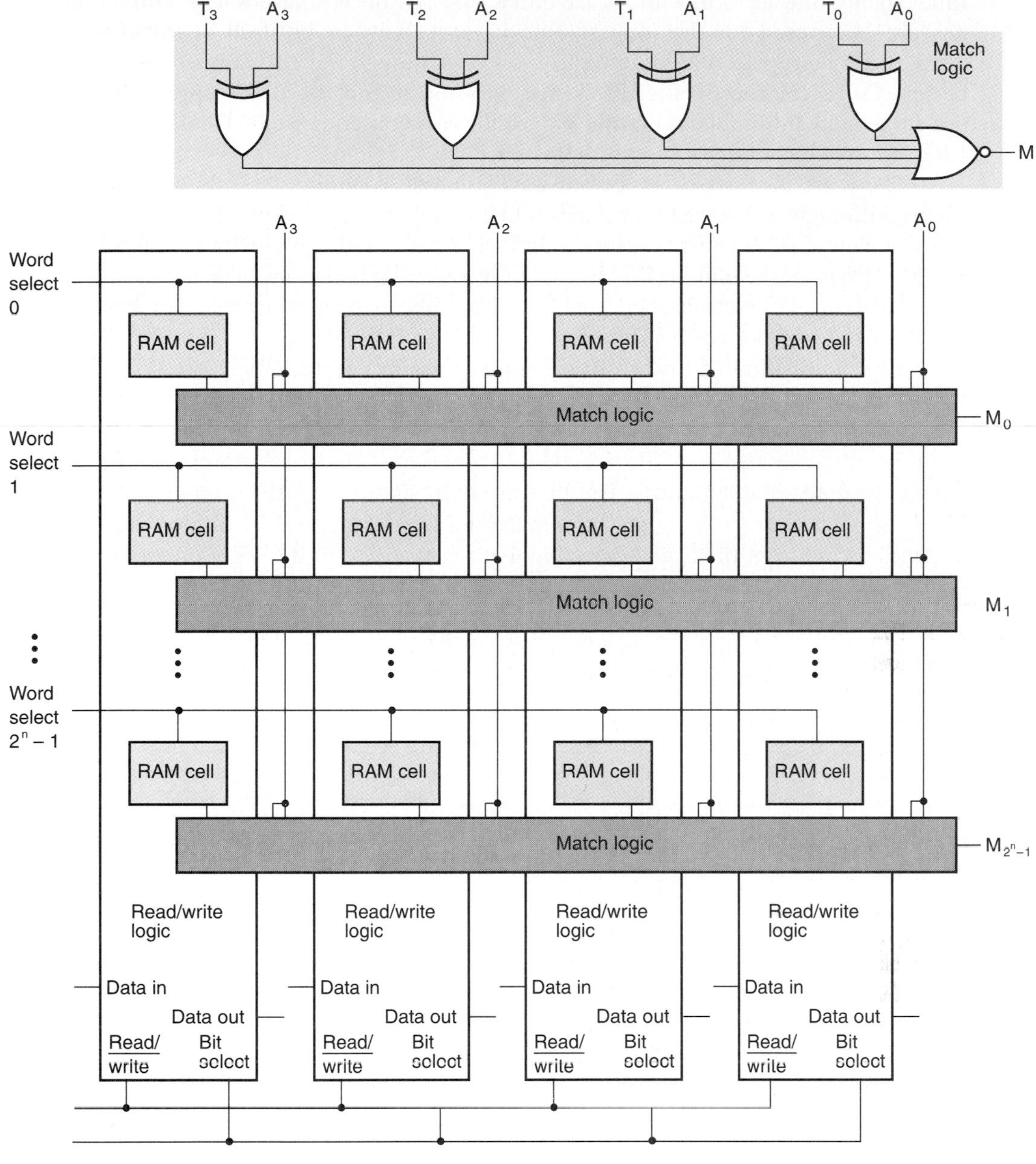

□ **FIGURE 12-5**
Associative Memory for 4-bit Tags

likely to be finished. An approach that appears to attack the replacement problem even more directly is the *least recently used* (LRU) location approach. The goal of this approach is to replace the entry that has been unused for the longest time—hence the least recently used entry. The reason is that a cache entry that has not been used for the longest time is least likely to be used in the future. Thus, it can be replaced by a new cache entry. Although the LRU approach yields better results for caches, the difference between it and the other approaches is not large, and full implementation is costly. As a consequence, if used at all, the LRU approach is often only approximated.

There are also performance and cost issues surrounding the fully associative cache. Although such a cache provides maximum flexibility and good performance, it is not clear that the cost is justified. In fact, an alternative mapping that has better performance and eliminates the cost of most of the matching logic is a compromise between a direct-mapped cache and a fully associative cache. For such a mapping, lower order address bits act much as they do in direct mapping; however, for each combination of lower order address bits, instead of having one location, there is a *set* of *s* locations. As with direct mapping, the tags and words are read from the cache memory locations addressed by the lower order address bits. For example, if the *set size s* equals two, then two tags and the two accompanying data words are read simultaneously. The tags are then simultaneously compared to the CPU-supplied address using just two matching logic structures. If one of the tags matches the address, then the associated word is returned to the CPU on the memory bus. If neither tag matches the address, then the two 0 matching values are used to send a miss signal to the CPU and main memory. Since there are sets of locations and associativity is used on sets, this technique is called *set-associative mapping*. Such a mapping with a set size *s* is called an *s-way* set-associative mapping.

Figure 12-6 shows a two-way set-associative cache. There are eight cache locations arranged in four rows of two locations each. The rows are addressed by a 2-bit index and contain tags made up of the remaining six bits of the main memory address. The cache entry for a main memory address must lie in a specific row of the cache, but can be in either of the two columns. In the figure, the addresses are the same as are in the fully associative cache in Figure 12-4. Note that no mapping is shown for main memory address 1111100000, since the two cache cells in set 00 are already occupied by addresses 0000010000 and 1111110000. In order to accommodate 1111100000, the set size would need to be at least three. This example illustrates a case in which the reduced flexibility of a set-associative cache, compared to a fully associative cache, has an impact. The impact declines as the set size increases.

Figure 12-7 is a section of a hardware block diagram for the set-associative cache of Figure 12-6. The index is used to address each row of the cache memory. The two tags read from the tag memories are compared to the tag part of the address on the address bus from the CPU. If a match occurs, then the three-state buffer on the corresponding data memory output is activated, placing the data onto the data bus to the CPU. In addition, the match signal causes the output of the $\text{Hit}/\overline{\text{miss}}$ OR gate to become 1, indicating a hit. If a match does not occur, then $\text{Hit}/\overline{\text{miss}}$ is 0, informing the main memory that it must supply the word to the CPU and informing the CPU that the word will be delayed.

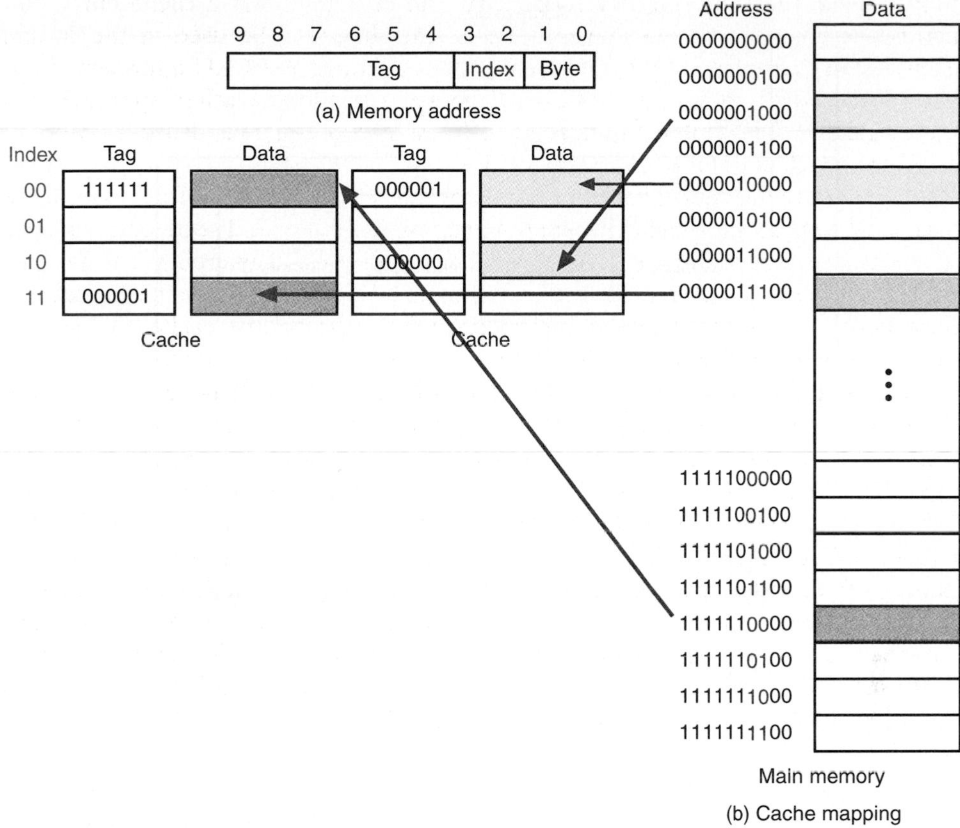

☐ **FIGURE 12-6**
Two-way Set-associative Cache

Line Size

To this point, we have assumed that each cache entry consists of a tag and a single memory word. In real caches, spatial locality is to be exploited, so additional words close to the one addressed are included in the cache entry. Then, rather than a single word being fetched from main memory when a cache miss occurs, a block of *l* words called a *line* is fetched. The number of words in a line is a power of two, and the words are aligned on address boundaries. For example, if four words are included in a line, then the addresses of the words in the line differ only in bits 2 and 3. The use of a block of words changes the makeup of the fields into which the cache divides the address. The new field structure is shown in Figure 12-8(a). Bits 2 and 3, the Word field, are used to address the word within the line. In this case, two bits are used, so there are four words per line. The next field, Index, identifies the set. Here there are two bits used, so there are four sets of tags and lines. The remainder of the address word is the Tag field, which contains the remaining four bits of the 10-bit memory address.

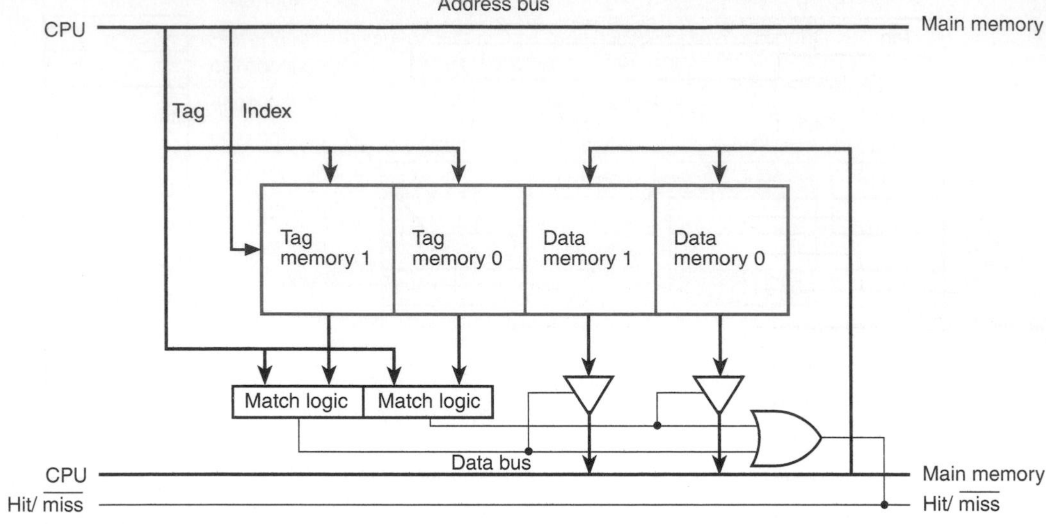

□ **FIGURE 12-7**
Partial Hardware Block Diagram for Set-associative Cache

The resulting cache structure is shown in Figure 12-8(b). The tag memory has eight entries, two in each of the four sets. Corresponding to each of the tag entries is a line of four data words. To ensure fast operation, Index is applied to the tag memory to read two tags, one for each of the set entries, simultaneously. At the same time, Index and the Word address are applied to read out two words from the cache data memory that correspond to the two tags. Matching logic provided for each of the two set elements compares each tag to the CPU-supplied address. If a match occurs, then the associated cache data word already read is placed on the memory bus to the CPU. Otherwise, a cache miss is signaled, and the word addressed is returned from main memory to the CPU. The line containing the word and its tag are also loaded into the cache. To facilitate loading the entire line of words, the width of the memory bus between main memory and the cache, as well as the cache load path, is made more than one word wide. Ideally, for our example the path is $4 \times 32 = 128$ bits wide. This allows the entire line to be placed in the cache in a single main memory read cycle. If the path is narrower, then a sequence of several reads from main memory is required.

An additional decision that the cache designer has to make is to determine the line size. A wide path to memory can affect both cost and performance, and a narrower path can slow transfer of the line to the cache. These features encourage a smaller cache line size, while spatial locality of reference encourages a larger line.

Cache Loading

Before any words and tags have been loaded into the cache, all locations contain invalid information. If a hit occurs on the cache at this time, then the word fetched

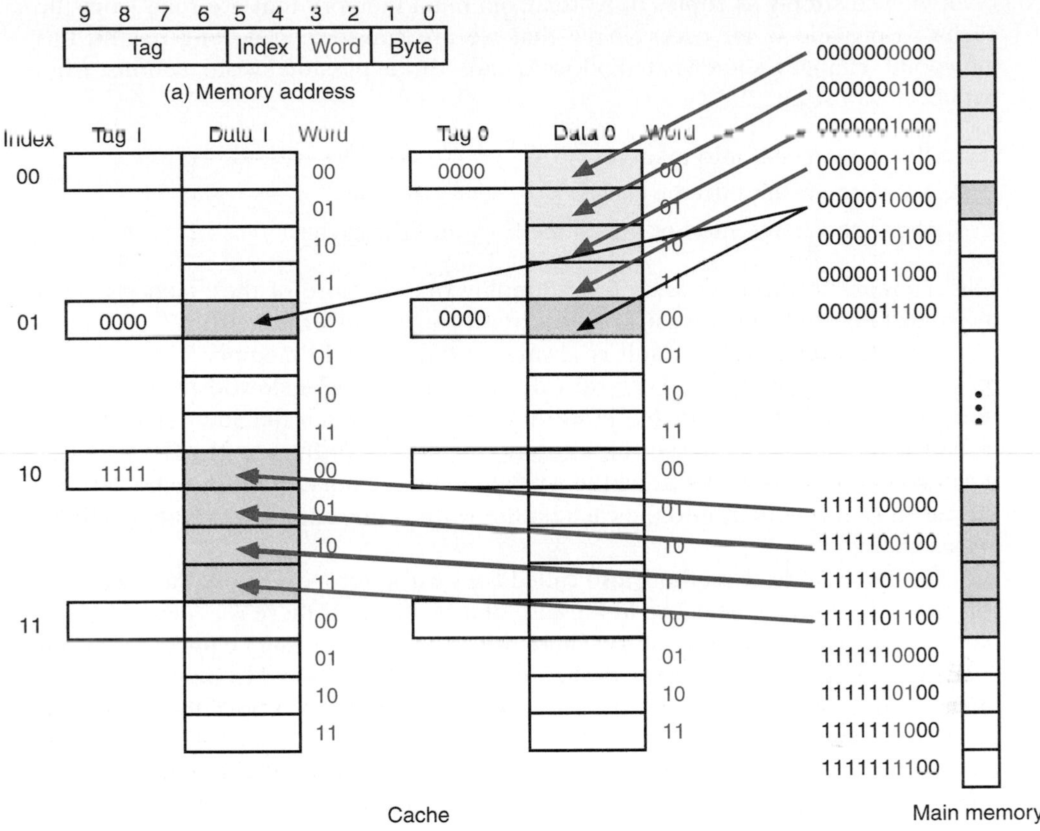

☐ **FIGURE 12-8**
Set-associative Cache with 4-word Lines

and sent to the CPU cannot have come from main memory and is invalid. As lines are fetched from main memory into the cache, cache entries become valid, but there is no way to distinguish valid from invalid entries. To deal with this problem, in addition to the tag, a bit is added to each cache entry. This *valid bit*, indicates that the associated cache line is valid (1) or invalid (0). It is read out of the cache along with the tag. If the valid bit is 0, then a cache miss occurs even if the tag matches the address from the CPU, requiring the addressed word to be taken from main memory.

Write Methods

We have focused so far on reading instructions and operands from the cache. What happens when a write occurs? Recall that up to now the words in a cache have been viewed simply as copies of words from main memory that are read from the cache to provide faster access. Now that we are considering writing results, this viewpoint changes somewhat. Following are three possible write actions from which we can select:

1. Write the result into main memory.
2. Write the result into the cache.
3. Write the result into both main memory and the cache.

Various realistic cache write methods employ one or more of these actions. Such methods fall into two main categories: write-through and write-back.

In *write-through*, the result is always written to main memory. This uses the main memory write time and can slow down processing. The slowdown can be partially avoided by using *write buffering*, a technique in which the address and word to be written are stored in special registers called write buffers by the CPU so that it can continue processing during the write to main memory. In most cache designs, the result is also written into the cache if the word is present there—that is, if there is a cache hit.

In the *write-back* method, also called *copy-back*, for a cache hit, the CPU performs a write only to the cache in the case of a cache bit. If there is a miss, the CPU performs a write to main memory. There are two possible design choices for when a cache miss occurs. One is to read the line containing the word to be written from main memory into the cache, with the new word written into both the cache and main memory. This is referred to as *write-allocate.* It is done with the hope that there will be additional writes to the same block which will result in write hits and thus avoid writes to main memory. The other choice on a write miss is simply to write to main memory. In what follows, we will assume that write-allocate is used.

The goal of a write-back cache is to be able to write at the writing speed of the cache whenever there is a cache hit. This avoids having all writes performed at the slower writing speed of main memory. In addition, it reduces the number of accesses to main memory, making it more accessible to DMA, an I/O processor, or another CPU in the system. A disadvantage of write-back is that main memory entries corresponding to words in the cache that have been written are invalid. Unfortunately, this can cause a problem with respect to I/O processors or another CPU in the system accessing the same main memory, due to "stale" data in the memory.

The implementation of the write-back concept requires a write-back operation from the cache location to be used to store a new line being brought from main memory on a read miss. If the location in the cache contains a word that has been written into, then the entire line from the cache must be written back into main memory in order to release the location for the new line. This write-back requires additional time whenever a read miss occurs. To avoid a write-back on every read miss, an additional bit is added to each cache entry. This bit, called the *dirty bit*, is a 1 if the line in the cache has been written and a 0 if it has not been written. Write-back must be performed only if the dirty bit is a 1. With write-allo-

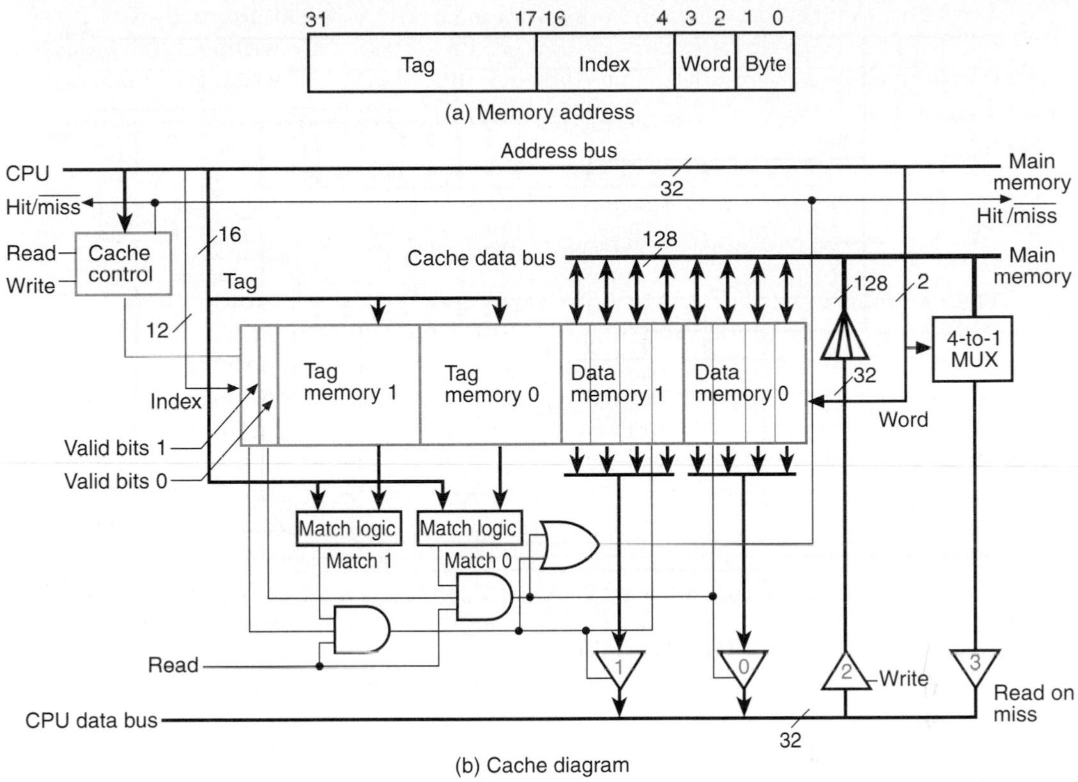

☐ **FIGURE 12-9**
Detailed Block Diagram for 256K Cache

cate used in a write-back cache, a write-back operation may also be required on a write miss.

Many other issues affect the choice of cache design parameters, particularly in the case of caches in a system in which the main memory may be read or written by a device other than the CPU for which the cache is provided.

Integration of Concepts

We now put together the basic concepts we have examined to determine the block diagram for a 256KB, two-way set-associative cache with write-through. The memory address shown in Figure 12-9(a) contains 32 bits using byte addressing with line size l = 16 bytes. The index contains 13 bits. Since 4 bits are used for addressing words and bytes, and 13 bits are used for the index, the tag contains the remaining 15 bits of the 32-bit address. The cache contains 16,384 entries consisting of 2^{13} = 8,192 sets. Each cache entry contains 16 bytes of data, a 15-bit tag, and a valid bit. The replacement strategy is random replacement.

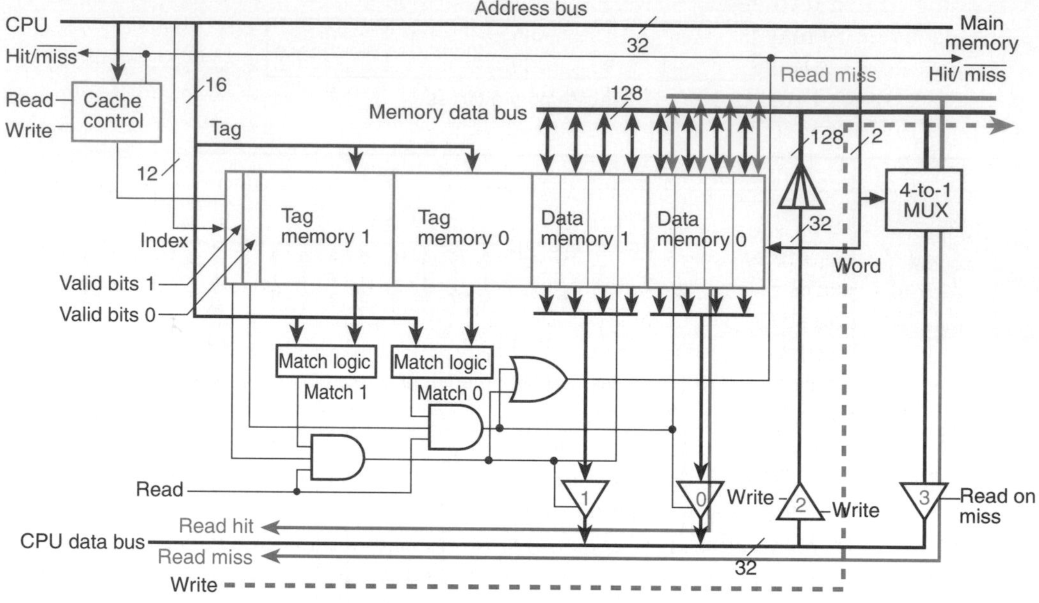

□ **FIGURE 12-10**
256K Cache: Read and Write Operations

Figure 12-9(b) gives the block diagram for the cache. There are two data memories and two tag memories, since the cache is two-way set associative. Each of these memories contains 2^{13} = 8,192 entries. Each entry in the data memory consists of 16 bytes. Since 32-bit words are assumed, there are four words in each data memory entry. Thus, each of the data memories consists of four 8,192 × 32 memories in parallel with the index as their common address. In order to read a single word from these four memories on a cache hit, a 4–to–1 selector using 3-state memory outputs selects the word, based on the two bits in the Word field of the address. The two tag memories are 8,192 × 15; in addition to them, a valid bit is associated with each cache entry. These bits are stored in an 8,192 × 2 memory and read out during a cache access with the data and tags. Note that the path between the cache and main memory is 128 bits wide. This allows us to assume that an entire cache line can be read from main memory in a single main memory cycle, an assumption that does not necessarily hold in practice. To understand the elements of the cache and how they work together, we will look at three possible cases of reading and writing. For each of these cases, we assume that the address from the CPU is $0F3F4024_{16}$. This gives Tag = 000011110011111_2 = $079F_{16}$, Index = 1010000000010_2 = 1402_{16}, and Word = 01_2.

First we assume a read hit—a read operation in which the data word lies in a cache entry, as in Figure 12-10. The cache uses the Index field to read out two tag entries from location 1402_{16} in Tag memory 1 and Tag memory 0. The match logic compares the tags of the entries, and in this case we assume that Tag 0 matches, causing Match 0 to be 1. This does not necessarily mean that we have a hit, since the cache entry may be invalid. Thus, the Valid 0 from location 1402_{16} bit is ANDed with Match 0. Also, the data can be placed on the CPU data bus only if the operation is a read. Thus, Read is ANDed with the Match 0 bit and the Valid 0 bit to form the control signal for three-state buffer 0. In this case, the control signal for the buffer 0 is 1. The data memories have used the Index field to read out eight words from location 1402_{16} at the same times the tags were read. The Word field selects the two of the eight words with word = 01_2 to place on the data buses going into the three-state buffers 1 and 0. Finally, with three-state buffer 0 turned on, the word addressed is placed on the CPU data bus. Also, the Hit/$\overline{\text{miss}}$ signal sends a 1 to the CPU and the main memory, notifying them of the hit.

In the second case, also shown in Figure 12-10, we assume a read miss—a read operation in which the data word is not in a cache entry. As before, the Index field address reads out the tag and valid entries, two tag comparisons are made, and two valid bits are checked. For both entries, a miss has occurred and is signaled by Hit/$\overline{\text{miss}}$ at 0. This means that the word must be fetched from main memory. Accordingly, the cache control selects the cache entry to be replaced, and four words read from main memory are applied simultaneously by the memory data bus to the cache inputs and are written into the cache entry. At the same time, the 4-to-1 multiplexer selects the word addressed by the Word field and places it on the CPU data bus using the three-state buffer 3.

In the third case in Figure 12-10, we assume a write operation. The word from the CPU is fanned out to appear in all four of the word positions of the 128-bit memory data bus. The address to which the word is to be written is provided by the address bus to main memory for the write operation into the addressed word only. If the address causes a hit on the cache, the word addressed is also written into the cache.

Instruction and Data Caches

In most of the designs in previous chapters, we assumed that it was possible to fetch an instruction and to read an operand or write a result in the same clock cycle. To do this, however, we need a cache that can provide access to two distinct addresses in a single clock cycle. In response to this need, we discussed in a prior subsection an *instruction cache* and a *data cache*. In addition to easily providing multiple accesses per clock, the use of two caches permits caches that have different design parameters. The design parameters for each cache can be selected to fit the different characteristics of access for fetching instructions or reading and writing data. Because the demands on each of these caches are typically less than those on a single cache, a simpler design can be used. For example, a single cache may require a four-way set-association structure, whereas an instruction cache needs only direct mapping, and a data cache may need only a two-way set-associative structure.

In other instances, a single cache for both instructions and data may be used. Such a *unified cache* is typically as large as the instruction and data caches combined. The unified cache allows cache entries to be shared by instructions and data freely. Thus, at one time more entries can be occupied by instructions, and at another time more entries can be occupied by data. This flexibility has the potential for increasing the number of cache hits. This higher hit rate may be misleading, however, since the unified cache supports only one access at a time, and separate caches support two simultaneous accesses as long as one is for instructions and one is for data.

Multiple-Level Caches

It is possible to extend the depth of the memory hierarchy by adding additional levels of cache. Two levels of cache, often referred to as L1 and L2, with L1 closest to the CPU, are often used. In order to satisfy the demand of the CPU for instruction and operands, a very fast L1 cache is needed. To achieve the necessary speed, the delay that occurs when crossing IC boundaries is intolerable. Thus, the L1 cache is placed in the processor IC together with the CPU and is referred to as the *internal cache*, as in the generic computer processor. But the area in the IC is limited, so the L1 cache is typically small and inadequate if it is the only cache. Thus, a larger L2 cache is added outside of the processor IC.

The design of a two-level cache is more complex than that of a single-level cache. Two sets of parameters are specified. The L1 cache can be designed to specific CPU access needs including the possibility of separate instruction and data caches. Also, the constraint of external pins between the CPU and L1 cache is removed. In addition to permitting faster reads, the path between the CPU and the L1 cache can be quite wide, allowing, for example, multiple instructions to be fetched simultaneously. On the other hand, the L2 cache occupies the typical external cache environment. It differs, however, from the typical external cache in that, rather than providing instructions and operands to a CPU, it primarily provides instructions and operands to the first-level cache L1. Since the L2 cache is accessed only on L1 misses, the access pattern is considerably different than that for a CPU, and the design parameters are accordingly different.

12-4 Virtual Memory

In our quest for a large, fast memory, we have achieved the appearance of a fast, medium-sized memory through the use of a cache. In order to have the appearance of a large memory, we now explore the relationship between main memory and hard disk. Because of the complexity of managing transfers between these two media, the control of such transfers involves the use of data structures and programs. Initially, we will discuss the most basic data structure used and the necessary hardware and software actions. Then we will deal with special hardware used to implement time-critical hardware actions.

With respect to large memory, not only do we want the entire virtual address space to appear to be main memory, but in most cases we would also like this complete space to appear to be available to each program that is executing. Thus, each program will "see" a memory the size of the virtual address space. Equally important to the programmer is the fact that real address space in main memory and real disk addresses are replaced by a single address space that has no restrictions on its use. With this arrangement, virtual memory can be used not only to provide the appearance of large main memory, but also to free up the programmer from having to consider the actual locations of the program and data in main memory and on the hard disk. The job of the software and hardware that implement virtual memory is to map each *virtual address* for each program into a *physical address* in the main memory. In addition, with a virtual address space for each program, it is possible for a virtual address from one program and a virtual address from another program to map to the same physical address. This allows code and data to be shared by multiple programs, thereby reducing the size of the main memory space and disk space required.

To permit the software to map virtual addresses to physical addresses, and to facilitate the transfer of information between main memory and hard disk, the virtual address space is divided into blocks of addresses, typically of a fixed size. These blocks, called *pages*, are larger than, but analogous to, lines in a cache. The physical address space in memory is divided into blocks called *page frames* that are the same size as the pages. When a page is present in the physical address space, it occupies a page frame. For purposes of illustration, we assume that a page consists of 4K bytes (1K words of 32 bits). Further, we assume that there are 32 address bits in the virtual address space. There are 2^{20} pages, maximum, in the virtual address space, and assuming a main memory of 16M bytes, there are 2^{12} page frames in main memory. Figure 12-11 shows the fields of virtual and physical addresses. The portion of the virtual address used to address words or bytes within a page is the *page offset*, which is the only part of the address that the virtual and physical addresses share. Note that words are assumed to be aligned in terms of their location with respect to their byte addresses such that each word address ends in binary 00. Likewise, pages are assumed to be aligned with respect to the byte addresses such that the page offset of the first byte in the page is 000_{16} and the page offset of the last byte in the page is FFF_{16}. The 20-bit portion of the virtual address used to select pages from the virtual address space is the *virtual page number*. The 12-bit portion of the physical address used to select pages in main memory is the *page frame number*. The figure shows a hypothetical mapping from the virtual address space into the physical address space. The virtual and physical page numbers are given in hexadecimal. A virtual page can be mapped to any physical page frame. Six mappings of pages from virtual memory to physical memory are shown. These pages constitute a total of 24K bytes. Note that there are no virtual pages mapped to physical page frames FFC and FFE. Thus, any data present in these pages is invalid.

Page Tables

In general, there may be a very large number of virtual pages, each of which must be mapped to either main memory or hard disk. The mappings are stored in a data structure called a *page table*. There are many ways to structure page tables and access them; we will suppose that page tables themselves are also kept in pages. Assuming that the representation of each mapping requires one word, 2^{10}, or 1K,

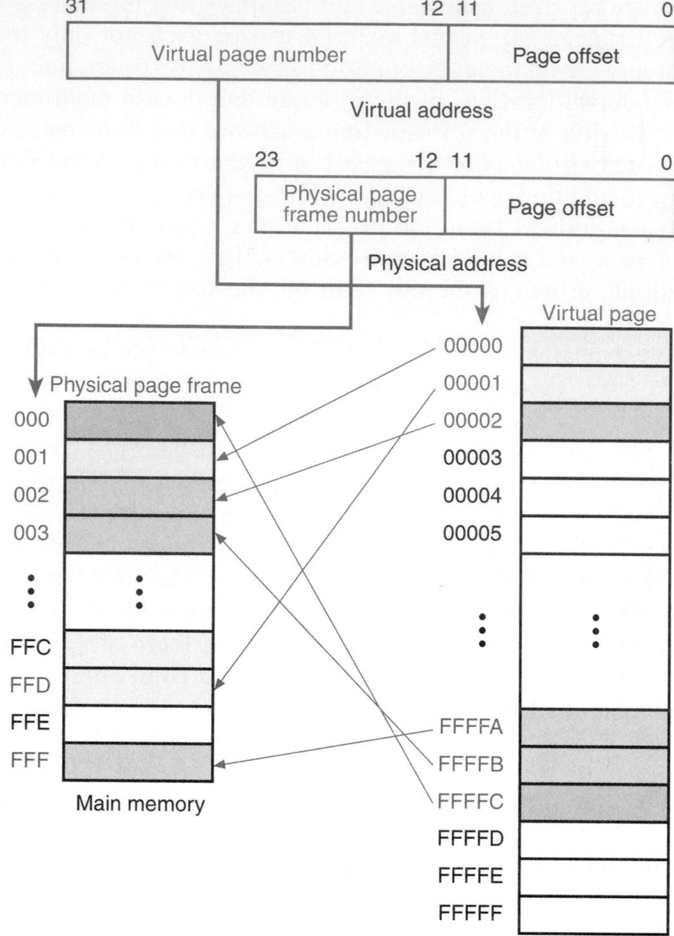

□ **FIGURE 12-11**
Virtual and Physical Address Fields and Mapping

mappings can be contained in a 4 KB page. Thus, the mappings for the entire address space for a program of 2^{22} bytes (4 MB) can be contained in one 4 KB page. A special table for each program called a *directory page* provides the mappings used to locate the 4 KB program page tables.

A sample format for a page table entry is given in Figure 12-12. Twelve bits are used for the page frame number in which the page is located in main memory. In addition, there are three single bit fields: Valid, Dirty, and Used. If Valid is 1, then the page frame in memory is valid; if Valid is 0, the page frame in memory is invalid, meaning that it does not correspond to correct code or data. If Dirty is 1, then there has been a write to at least one byte in the page since it was placed in main memory. If Dirty is 0, there have been no writes to the page since it entered main memory. Note that the Valid and Dirty bits correspond exactly to those in

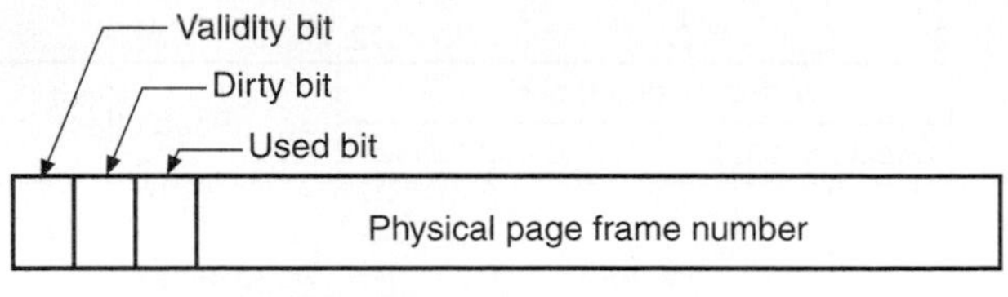

□ **FIGURE 12-12**
Format for Page Table Entries

a cache which uses write-back. When it is necessary for a page to be removed from main memory and the Dirty bit is 1, then the page is copied back to the hard disk. If the Dirty bit is 0, indicating that the page in main memory has not been written into, then the page coming into the same page frame is simply written over the present page. This can be done because the disk version of the present page is still correct. In order to use this feature, the software keeps a record of the location of the page on the disk elsewhere when it places the page in main memory. The Used bit is a simple mechanism for implementing a crude approximation to an LRU replacement scheme. Some additional bit positions in a page entry may be reserved for flags used by the computer operating system. For example, a few flags might represent the read and write protection status of a page and whether the page can be accessed in user mode or supervisor mode.

The page table structure we have just described is shown in Figure 12-13. The *directory page pointer* is a register that points to the location of the directory page in main memory. The directory page contains the locations of up to 1K page tables associated with the program that is executing. These page tables may be in main memory or on the hard disk. The page table to be accessed is derived from the most significant 10 bits of the virtual page number, which we call the *directory offset*. Assuming that the page table selected is in main memory, it can be accessed by the *page table page number*. The least significant 10 bits of the virtual page number, which we call the *page table offset*, can be used to access the entry for the page to be accessed. If the page is in main memory, the page offset is used to locate the physical location of the byte or word to be accessed. If either the page table or the desired page is not in main memory, it must first be fetched by software from the hard disk to main memory before the word within it is accessed. Note that combining the offsets with register or table entries is done by simply setting the offset to the right of the page frame number, rather than adding the two together. This approach requires no delay, whereas addition would cause significant delay.

Translation Lookaside Buffer

From the preceding discussion, we note that virtual memory has a considerable performance penalty even in the best case, when the directory, the page table, and the page to be accessed are in main memory. For our assumed page table approach, three successive accesses to main memory occur in order to fetch a single operand or instruction:

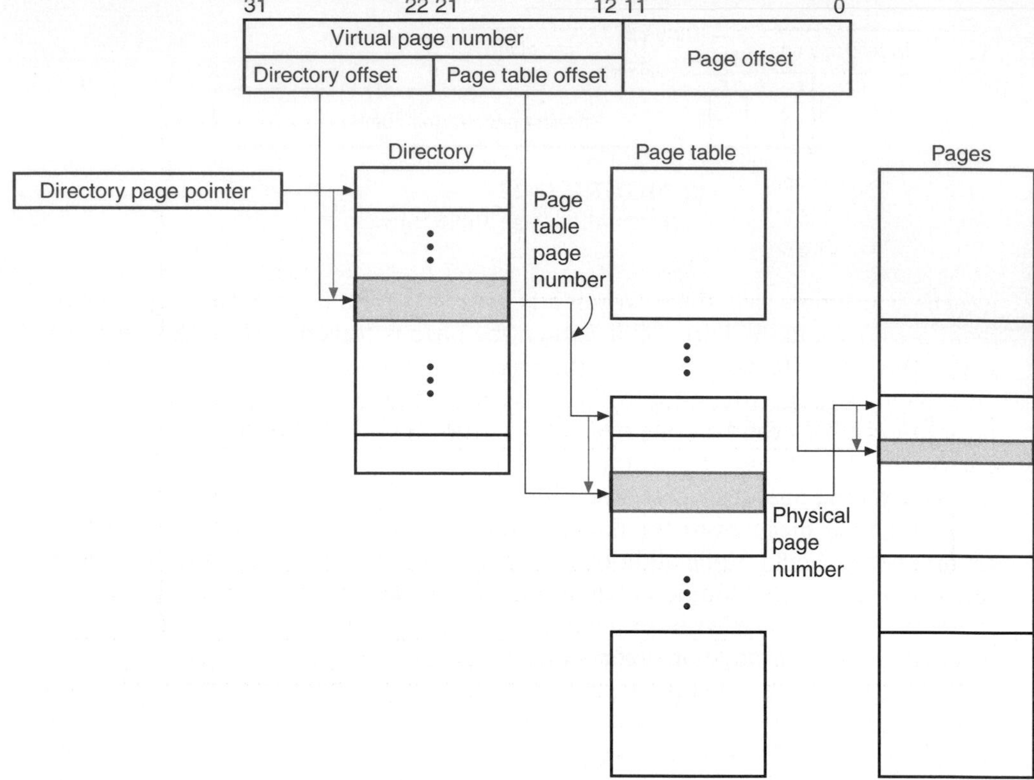

□ **FIGURE 12-13**
Example of Page Table Structure

1. Access for the directory entry.
2. Access for the page table entry.
3. Access for the operand or instruction.

Note that these accesses are performed automatically by hardware that is part of the MMU in the generic computer. Thus, to make virtual memory feasible, we need to drastically reduce accesses to main memory. If we have a cache, and if all of the entries are in the cache, then the time for each access is reduced. Nevertheless, three accesses are needed to the cache. To reduce the number of accesses, we will employ yet another cache for the purpose of translating the virtual address directly into a physical address. This new cache is called a *translation lookaside buffer* (TLB). It holds the locations of recently addressed pages to speed access to cache or main memory. Figure 12-14 gives an example of a TLB, which is typically fully associative or set associative, since it is necessary to compare the virtual page number from the CPU with a number of virtual page number tags. In addition to the latter, a cache entry includes the physical page number for those pages in main memory and a Valid bit. If the page is in main memory, the Dirty bit also appears.

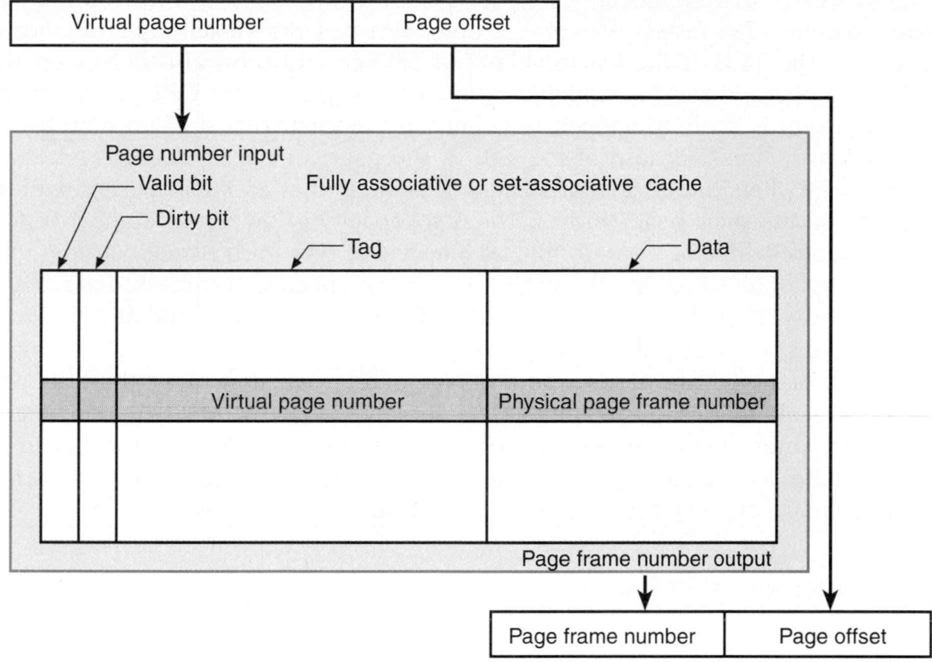

□ **FIGURE 12-14**
Example of Translation Lookaside Buffer

The Dirty bit serves the same function for a page in main memory as discussed previously for a line in a cache.

We now briefly look at a memory access using the TLB in Figure 12-14. The virtual page number is applied to the page number input to the cache. Within the cache, this page number is compared simultaneously with all of the virtual page number tags. If a match occurs and the Valid bit is a 1, then a TLB hit has occurred, and the physical page frame number appears on the page number output of the cache. This operation can be performed very quickly and produces the physical address required to access memory or a cache. On the other hand, if there is a TLB miss, then it is necessary to access main memory for the directory table entry and the page table entry. If there is a physical page in main memory, then the page table entry is brought into the TLB cache and replaces one of the entries there. Overall, three memory accesses are required, including the one for the operand. If the physical page does not exist in main memory, then a *page fault* occurs. In this case, a software-implemented action fetches the page from its hard disk location to main memory. During the time required to complete this action, the CPU may execute a different program rather than waiting until the page has been placed in main memory.

Noting the prior hierarchy of actions based on the presentation of a virtual address, we see that the effectiveness of virtual memory depends on temporal and spatial locality. The fastest response is possible when the virtual page number is present in the TLB. If the hardware is fast enough and a hit also occurs on the cache, the operand can be available in as little as one or two CPU clock cycles. Such an event is likely to happen frequently if the same virtual pages tend to get accessed over time. Because of the size of the pages, if one operand is accessed from a page, then, due to spatial locality, it is likely that another operand will be accessed on the same page. With the limited capacity of the TLB, the next fastest action requires three accesses to main memory and slows processing considerably. In the worst of all situations, the page table and the page to be accessed are not in main memory. Then, lengthy transfers of two pages—the page table and the page from hard disk—are required.

Note that the basic hardware for implementing virtual memory, the TLB, and other optional features for memory access are included in the MMU in the generic computer. Among the other features is hardware support for an additional layer of virtual addressing called segmentation and for protection mechanisms to permit appropriate isolation and sharing of programs and data.

Virtual Memory and Cache

Although we have considered the cache and virtual memory separately, in an actual system they are both very likely to be present. In that case, the virtual address is converted to the physical address, and then the physical address is applied to the cache. Assuming that the TLB takes one clock cycle and the cache takes one clock cycle, in the best of cases fetching an instruction or operand requires two CPU clock cycles. As a consequence, in many pipelined CPU designs, two or more clock cycles are allowed for an operand fetch. Since instruction fetch addresses are more predictable, it is possible to modify the CPU pipeline and consider the TLB and cache to be a two-stage pipeline segment, so that an instruction fetch appears to require only one clock cycle.

12-5 Chapter Summary

In this chapter, we examined the components of a memory hierarchy. Two concepts fundamental to the hierarchy are cache memory and virtual memory.

Based on the concept of locality of reference, a cache is a small, fast memory that, holds the operands and instructions most likely to be used by the CPU. Typically, a cache gives the appearance of a memory the size of main memory with a speed close to that of the cache. A cache operates by matching the tag portion of the CPU address with the tag portions of the addresses of the data in the cache. If a match occurs and other specific conditions are satisfied, a cache hit occurs, and the data can be obtained from the cache. If a cache miss occurs, the data must be obtained from the slower main memory. The cache designer must determine the values of a number of parameters, including the mapping of main memory addresses to cache addresses, the selection of the line of the cache to be replaced when a new line is added, the size of the cache, the size of the cache line, and the method for performing memory writes. There may be more than one cache in a memory hierarchy, and instructions and data may have separate caches.

Virtual memory is used to give the appearance of a large memory—much larger than the main memory—at a speed that is, on average, close to that of the main memory. Most of the virtual address space is actually on hard disk. To facilitate the movement of information between the memory and the hard disk, both are divided up in fixed size address blocks called page frames and pages, respectively. When a page is placed in main memory, its virtual address must be translated to a physical address. The translation is done using one or more page tables. In order to perform the translation on each memory access without a severe performance penalty, special hardware is employed. This hardware, called a translation lookaside buffer (TLB), is a special cache that is a part of the memory management unit (MMU) of the computer.

Together with main memory, the cache and the TLB give the illusion of a large, fast memory that is in fact a hierarchy of memories of different capacities, speeds, and technologies, with hardware and software performing automatic transfers between levels.

REFERENCES

1. MANO, M. M. *Computer Engineering: Hardware Design.* Englewood Cliffs, NJ: Prentice Hall, 1988.
2. HAMACHER, V. C., VRANESIC, Z. G., AND ZAKY, S. G. *Computer Organization,* 3rd ed. New York: McGraw-Hill, 1990.
3. HENNESSY, J. L., AND PATTERSON, D. A. *Computer Architecture: A Quantitative Approach.* San Mateo, CA: Morgan Kaufmann, 1990.
4. BARON, R. J., AND HIGBIE, L. *Computer Architecture.* Reading, MA: Addison-Wesley, 1992.
5. HANDY, J. *Cache Memory Book.* San Diego: Academic Press, 1993.
6. MANO, M. M. *Computer System Architecture,* 3rd Ed. Englewood Cliffs, NJ: Prentice Hall, 1993.
7. PATTERSON, D. A., AND HENNESSY, J. L. *Computer Organization and Design: The Hardware/Software Interface.* San Mateo, CA: Morgan Kaufmann, 1994.
8. WYANT, G., AND HAMMERSTROM, T. *How Microprocessors Work.* Emeryville, CA: Ziff-Davis Press, 1994.
9. MESSMER, H.-P., *The Indispensable PC Hardware Book,* 2nd ed. Wokingham, U.K.: Addison-Wesley, 1995.

PROBLEMS

The asterisk (*) indicates a more advanced problem.

12–1. A CPU produces the following sequence of read addresses in hexadecimal:

54, 58, 104, 5C, 108, 60, F0, 64, 54, 58, 10C, 5C, 110, 60, F0, 64

Supposing that the cache is empty to begin with, and assuming an LRU replacement, determine whether each address produces a hit or a miss for each of the following caches:
(a) direct mapped in Figure 12-3,
(b) fully associative in Figure 12-4, and
(c) two-way set associative in Figure 12-6.

12–2. Repeat Problem 12-1 for the following sequence of read addresses:

0, 4, 8, 10, 14, 18, 1C, 24, 28, 2C, 30, 34, 38, 3C, 40, 44, 48, 4C, 50, 54, 58, 5C

12–3. A computer has a 32-bit address and a direct-mapped cache. Addressing is to the byte level. The cache has a capacity of 256K bytes and uses lines that are 32 bytes. It uses write-through and so does not require a dirty bit.
(a) How many bits are in the index for the cache?
(b) How many bits are in the tag for the cache?
(c) What is the total number of bits of storage in the cache, including the valid bits, the tags, and the cache lines?

12–4. A two-way set-associative cache in a system with 32-bit addresses has one 4-byte word per line and a capacity of 128K bytes. Addressing is to the byte level.
(a) How many bits are there in the index and the tag?
(b) Indicate the value of the index in hexadecimal for cache entries from the following main memory addresses in hexadecimal: 0284A482, 01148C89, 0038CF00 and 0038CF01.
(c) Can all of the cache entries from part (b) be in the cache simultaneously?

12–5. Discuss the advantages and disadvantages of:
(a) separate instruction and data caches versus a unified cache for both.
(b) a write-back cache versus a write-through cache.

12–6. Explain why write-allocate is typically not used in a write-through cache.

12–7. A high-speed workstation has 64-bit words and 64-bit addresses with address resolution to the byte level.
(a) How many bytes can be in the address space of the workstation?
(b) Assuming a direct-mapped cache with 8,192 64-byte lines, how many bits are in each of the following address fields for the cache: (1) Byte, (2) Index, and (3) Tag?

12–8. A cache memory has an access time from the CPU of 8 ns, and the main memory has an access time from the CPU of 85 ns. What is the effective access time for the cache–main memory hierarchy if the hit ratio is:

(a) 0.87?
(b) 0.90?
(c) 0.95?

12–9. Redesign the cache in Figure 12-7 so that it is the same size, but is four-way set associative rather than two-way set associative.

12–10. *The cache in Figure 12-9 is to be redesigned to use write-back with write-allocate rather than write-through. Respond to the following requests, making sure to deal with all of the address and data issues involved in the write-back operation.
(a) Draw the new block diagram.
(b) Explain the sequence of actions you propose for a write miss and for a read miss.

12–11. A virtual memory system uses 4K byte pages, 32-bit words, and a 32-bit virtual address. A particular program and its data require 3,657 pages.
(a) What is the minimum number of page tables required?
(b) What is the minimum number of entries required in the directory page?
(c) Based on your answers to (a) and (b), how many entries are there in the last page table?

12–12. A computer can accommodate a maximum of 64M bytes of main memory. It has a 32-bit word and a 32-bit virtual address and uses 4K byte pages. The TLB only contains entries that include the Valid, Dirty, and Used bits, the virtual page number, and the physical page number. Assuming that the TLB is fully associative and has 32 entries:
(a) How many bits of associative memory are required for the TLB?
(b) How many bits of SRAM are required for the TLB?

12–13. Five programs are concurrently executing in a multitasking computer with virtual memory pages having 4K bytes. Each page table entry is 32 bits. What is the minimum numbers of bytes of main memory occupied by the directory pages and page tables for the five programs if the numbers of pages per program, in decimal, are as follows: 4598, 8792, 6345, 12742, 142688.

12–14. In caches, we use both write-through and write-back as potential writing approaches. But for virtual memory, only an approach that resembles write-back is used. Give a sound explanation of why this is so.

12–15. Explain clearly why both the cache memory concept and the virtual memory concept would be ineffective if locality of reference of memory-addressing patterns did not hold.

INDEX